W9-AUY-660

84 Business Park Drive, Armonk, NY 504   Phone (914) 273-2233
Fax (914) 273-2227   Toll Free (888) 698-TIM (8463)   www.its-about-time.com

**It's About Time
President**
Tom Laster

**Director of Product
Development**
Barbara Zahm, Ph.D.

Creative/Art Director
John Nordland

Design/Production
Kathleen Bowen
Burmar Technical
Corporation
Nancy Delmerico
Kadi Sarv
Jon Voss

Illustrations
Tomas Bunk
Dennis Falcon

Project Editor
Ruta Demery
*EarthComm, Active Physics,
Active Chemistry, Active Biology,
Coordinated Science
for the 21st Century*

Project Managers
Ruta Demery
EarthComm, Active Physics

Barbara Zahm
*Active Physics
Active Chemistry, Active Biology,
Coordinated Science
for the 21st Century*

Project Coordinators
Loretta Steeves
*Coordinated Science
for the 21st Century*

Emily Crum
Matthew Smith
EarthComm

Technical Art
Stuart Armstrong
EarthComm

Burmar Technical Corporation
Kadi Sarv
*Active Physics, Active Chemistry
Active Biology*

Photo Research
Caitlin Callahan
Kathleen Bowen
Jon Voss
Kadi Sarv
Jennifer Von Holstein

Safety Reviewers
Ed Robeck, Ph.D.
EarthComm, Active Biology

Gregory Puskar
Active Physics

Jack Breazale
Active Chemistry

This project was supported, in part, by the
National Science Foundation
Opinions expressed are those of the authors and not necessarily those of the National Science
Foundation or the donors of the American Geological Institute Foundation.

Coordinated Science
for the 21st Century™

An Integrated, Project-Based Approach

Arthur Eisenkraft, Ph.D.

Ruta Demery

Gary Freebury

Robert Ritter, Ph.D.

Michael Smith, Ph.D.

John B. Southard, Ph.D.

IT's ABOUT TIME®

HERFF JONES EDUCATION DIVISION

Coordinated Science
for the 21st
Century™

An Integrated, Project-Based Approach

Coordinated Science for the 21st Century is an innovative core curricula assembled from four proven inquiry-based programs. It is supported by the National Science Foundation and was developed by leading educators and scientists. Unit 1, *Active Physics*, was developed by the American Association of Physics Teachers and the American Institute of Physics.

Both *Active Physics* and *Active Chemistry*, Units 1 and 2, are projects directed by Arthur Eisenkraft, Ph.D., past president of the NSTA.

Active Biology, Unit 3, was developed to follow the same Active-Learning Instructional Model as *Active Physics* and *Active Chemistry*.

EarthComm, Unit 4, was developed by the American Geological Institute, under the guidance of Michael Smith, Ph.D., former Director of Education and Outreach, and John Southard, Ph.D., of MIT.

Each unit of this course has been designed and built on the National Science Education Standards. Each utilizes the same instructional model and the same inquiry-based approach.

Project Director, Active Physics and Active Chemistry

Arthur Eisenkraft has taught high school physics for over 28 years and is currently the Distinguished Professor of Science Education and a Senior Research Fellow at the University of Massachusetts, Boston. Dr. Eisenkraft is the author of numerous science and educational publications. He holds U.S. Patent #4447141 for a Laser Vision Testing System (which tests visual acuity for spatial frequency).

Dr. Eisenkraft has been recognized with numerous awards including: Presidential Award for Excellence in Science Teaching, 1986 from President Reagan; American Association of Physics Teachers (AAPT) Excellence in Pre-College Teaching Award, 1999; AAPT Distinguished Service Citation for "excellent contributions to the teaching of physics", 1989; Science Teacher of the Year, Disney American Teacher Awards in their American Teacher Awards program, 1991; Honorary Doctor of Science degree from Rensselaer Polytechnic Institute, 1993; Tandy Technology Scholar Award 2000.

In 1999 Dr. Eisenkraft was elected to a 3-year cycle as the President-Elect, President and Retiring President of the National Science Teachers Association (NSTA), the largest science teacher organization in the world. In 2003, he was elected a fellow of the American Association for the Advancement of Science (AAAS).

Dr. Eisenkraft has been involved with a number of projects and chaired many competition programs, including: the Toshiba/NSTA ExploraVisions Awards (1991 to the present); the Toyota TAPESTRY Grants (1990 to the present); the Duracell/NSTA Scholarship Competitions (1984 to 2000). He was a columnist and on the Advisory Board of *Quantum* (a science and math student magazine that was published by NSTA as a joint venture between the United States and Russia; 1989 to 2001). In 1993, he served as Executive Director for the XXIV International Physics Olympiad after being Academic Director for the United States Team for six years. He has served on a number of committees of the National Academy of Sciences including the content committee that helped write the National Science Education Standards.

Dr. Eisenkraft has appeared on *The Today Show*, *National Public Radio*, *Public Television*, *The Disney Channel* and numerous radio shows. He serves as an advisor to the ESPN Sports Figures Video Productions.

He is a frequent presenter and keynote speaker at National Conventions. He has published over 100 articles and presented over 200 papers and workshops. He has been featured in articles in *The New York Times*, *Education Week*, *Physics Today*, *Scientific American*, *The American Journal of Physics* and *The Physics Teacher*.

Content Specialist, Active Chemistry

Gary Freebury, a noted chemistry teacher, educator, and writer worked as the Project Manager and Editor for the 3 Prototype chapters of *Active Chemistry* which are currently being field tested but are also in print, and he will continue to serve as the Project Manager and Editor to the project. He will be responsible for the writing of any introductory materials, producing the table of contents, indices, glossary, and reference materials. He will have a critical role in maintaining the integrity of safety standards across all units. He will also coordinate any modifications, changes, and additions to the materials based upon pilot and field-testing results.

Mr. Freebury has been teaching chemistry for more than 35 years. He has been the Safety Advisor for Montana Schools, past director of the Chemistry Olympiad, past chairman of the Montana Section of the American Chemical Society (ACS), member of the Executive Committee of the Montana Section of the ACS, and a past member of the Montana Science Advisory Council. Mr. Freebury has been the regional director and author of Scope, Sequence and Coordination (SS&C) – Integrated Science Curriculum and Co-director of the National Science Foundation supported Chemistry Concepts four-year program. He earned a B.S. degree at Eastern Montana College in mathematics and physical science, and an M.S. degree in chemistry at the University of Northern Iowa.

Acknowledgements

Principal Investigator, EarthComm

Michael Smith, Ph.D., is a former Director of Education at the American Geological Institute in Alexandria, Virginia. Dr. Smith worked as an exploration geologist and hydrogeologist. He began his Earth Science teaching career with Shady Side Academy in Pittsburgh, PA in 1988 and most recently taught Earth Science at the Charter School of Wilmington, DE. He earned a doctorate from the University of Pittsburgh's Cognitive Studies in Education Program and joined the faculty of the University of Delaware School of Education in 1995. Dr. Smith received the Outstanding Earth Science Teacher Award for Pennsylvania from the National Association of Geoscience Teachers in 1991, served as Secretary of the National Earth Science Teachers Association, and is a reviewer for Science Education and The Journal of Research in Science Teaching. He worked on the Delaware Teacher Standards, Delaware Science Assessment, National Board of Teacher Certification, and AAAS Project 2061 Curriculum Evaluation programs.

Senior Writer, EarthComm

John B. Southard, Ph.D., received his undergraduate degree from the Massachusetts Institute of Technology in 1960 and his doctorate in geology from Harvard University in 1966. After a National Science Foundation postdoctoral fellowship at the California Institute of Technology, he joined the faculty at the Massachusetts Institute of Technology, where he is currently Professor of Geology Emeritus. He was awarded the MIT School of Science teaching prize in 1989 and was one of the first cohorts of the MacVicar Fellows at MIT, in recognition of excellence in undergraduate teaching. He has taught numerous undergraduate courses in introductory geology, sedimentary geology, field geology, and environmental Earth Science both at MIT and in Harvard's adult education program. He was editor of the Journal of Sedimentary Petrology from 1992 to 1996, and he continues to do technical editing of scientific books and papers for SEPM, a professional society for sedimentary geology. Dr. Southard received the 2001 Neil Miner Award from the National Association of Geoscience Teachers.

Primary Author, Active Biology
Project Editor,
EarthComm, Active Physics,
Active Chemistry and Active Biology

Ruta Demery has helped bring to publication several National Science Foundation (NSF) projects. She was the project editor for *EarthComm*, *Active Physics*, *Active Chemistry*, and *Active Biology*. She was also a contributing writer for *Active Physics* and *Active Biology*, both students' and teachers' editions. Besides participating in the development and publishing of numerous innovative mathematics and science books for over 30 years, she has worked as a classroom science and mathematics teacher in both middle school and high school. She brings to her work a strong background in curriculum development and a keen interest in student assessment. When time permits, she also leads workshops to familiarize teachers with inquiry-based methods.

Contributing Author,
Active Biology and Active Physics

Bob Ritter is presently the principal of Holy Trinity High School in Edmonton, Alberta. Dr. Ritter began his teaching career in 1973, and since then he has had a variety of teaching assignments. He has worked as a classroom teacher, Science Consultant, and Department Head. He has also taught Biological Science to student teachers at the University of Alberta. He is presently involved with steering committees for "At Risk High School Students" and "High School Science." Dr. Ritter is frequently a presenter and speaker at national and regional conventions across Canada and the United States. He has initiated many creative projects, including establishing a science-mentor program in which students would have an opportunity to work with professional biologists. In 1993 Dr. Ritter received the Prime Minister's Award for Science and Technology Teaching. He has also been honored as Teacher of the Year and with an Award of Merit for contribution to science education.

Coordinated Science for the 21st Century was developed by teams of leading science educators, university educators and classroom teachers with financial support from the National Science Foundation.

NSF Program Officer

Gerhard Salinger
Instructional Materials Development (IMD)

UNIT 1: ACTIVE PHYSICS

Principal Investigators

Bernard V. Khoury
American Association
of Physics Teachers

Dwight Edward
Neuenschwander
American Institute
of Physics

Project Director

Arthur Eisenkraft
University of
Massachusetts

Primary and Contributing Authors

Richard Berg
University of Maryland
College Park, MD

Howard Brody
University of
Pennsylvania
Philadelphia, PA

Chris Chiaverina
New Trier Township
High School
Crystal Lake, IL

Ron DeFronzo
Eastbay Ed.
Collaborative
Attleboro, MA

Ruta Demery
Blue Ink Editing
Stayner, ON

Carl Duzen
Lower Merion
High School
Havertown, PA

Jon L. Harkness
Active Physics
Regional Coordinator
Wausau, WI

Ruth Howes
Ball State University
Muncie, IN

Douglas A. Johnson
Madison West
High School
Madison, WI

Ernest Kuehl
Lawrence High School
Cedarhurst, NY

Robert L. Lehrman
Bayside, NY

Salvatore Levy
Roslyn High School
Roslyn, NY

Tom Liao
SUNY Stony Brook
Stony Brook, NY

Charles Payne
Ball State University
Muncie, IN

Mary Quinlan
Radnor High School
Radnor, PA

Harry Rheam
Eastern Senior
High School
Atco, NJ

Bob Ritter
University of Alberta
Edmonton, AB

John Roeder
The Calhoun School
New York, NY

John J. Rusch
University of Wisconsin
Superior
Superior, WI

Patty Rourke
Potomac School
McLean, VA

Ceanne Tzimopoulos
Omega Publishing
Medford, MA

Larry Weathers
The Bromfield School
Harvard, MA

David Wright
Tidewater Comm.
College
Virginia Beach, VA

Consultants

Peter Brancazio
Brooklyn College
of CUNY
Brooklyn, NY

Robert Capen
Canyon del Oro
High School
Tucson, AZ

Carole Escobar

Earl Graf
SUNY Stony Brook
Stony Brook, NY

Jack Hehn
American Association
of Physics Teachers
College Park, MD

Donald F. Kirwan
Louisiana State
University
Baton Rouge, LA

Gayle Kirwan
Louisiana State
University
Baton Rouge, LA

James La Porte
Virginia Tech
Blacksburg, VA

Charles Misner
University of Maryland
College Park, MD

Robert F. Neff
Suffern, NY

Ingrid Novodvorsky
Mountain View
High School
Tucson, AZ

John Robson
University of Arizona
Tucson, AZ

Mark Sanders
Virginia Tech
Blacksburg, VA

Brian Schwartz
Brooklyn College
of CUNY
New York, NY

Bruce Seiger
Wellesley High School
Newburyport, MA

Clifford Swartz
SUNY Stony Brook
Setauket, NY

Barbara Tinker
The Concord
Consortium
Concord, MA

Robert E. Tinker
The Concord
Consortium
Concord, MA

Joyce Weiskopf
Herndon, VA

Donna Willis
American Association
of Physics Teachers
College Park, MD

Safety Reviewer

Gregory Puskar
University of
West Virginia
Morgantown, WV

Equity Reviewer

Leo Edwards
Fayetteville State
University
Fayetteville, NC

Physics at Work

Alex Straus, writer
New York, NY
Mekea Hurwitz
photographer

Physics InfoMall

Brian Adrian
Bethany College
Lindsborg, KS

**First Printing
Reviewer**

John L. Hubisz
North Carolina State
University
Raleigh, NC

Unit Reviewers

Robert Adams
Polytech High School
Woodside, DE

George A. Amann
F.D. Roosevelt
High School
Rhinebeck, NY

Patrick Callahan
Catasaugua High School
Center Valley, PA

Beverly Cannon
Science and Engineering
Magnet High School
Dallas, TX

Barbara Chauvin

Elizabeth Chesick
The Baldwin School
Haverford, PA

Chris Chiaverina
New Trier Township
High School
Crystal Lake, IL

Andria Erzberger
Palo Alto Senior
High School
Los Altos Hills, CA

Elizabeth Farrell
Ramseyer
Niles West High School
Skokie, IL

Mary Gromko
President of Council of
State Science
Supervisors
Denver, CO

Thomas Guetzloff

Jon L. Harkness
Active Physics
Regional Coordinator
Wausau, WI

Dawn Harman
Moon Valley
High School
Phoenix, AZ

James Hill
Piner High School
Sonoma, CA

Bob Kearney

Claudia Khourey-
Bowers
McKinley Senior
High School

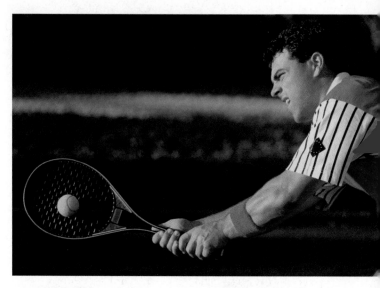

Steve Kliewer
Bullard High School
Fresno, CA

Ernest Kuehl
Roslyn High School
Cedarhurst, NY

Jane Nelson
University High School
Orlando, FL

Mary Quinlan
Radnor High School
Radnor, PA

John Roeder
The Calhoun School
New York, NY

Patty Rourke
Potomac School
McLean, VA

Gerhard Salinger
Fairfax, VA

Irene Slater
La Pietra School
for Girls

Pilot Test Teachers

John Agosta

Donald Campbell
Portage Central
High School
Portage, MI

John Carlson
Norwalk Community
Technical College
Norwalk, CT

Veanna Crawford
Alamo Heights
High School
New Braunfels, TX

Janie Edmonds
West Milford
High School
Randolph, NJ

Eddie Edwards
Amarillo Area Center
for Advanced Learning
Amarillo, TX

Arthur Eisenkraft
University of
Massachusetts

Tom Ford

Bill Franklin

Roger Goerke
St. Paul, MN

Tom Gordon
Greenwich
High School
Greenwich, CT

Ariel Hepp

John Herrman
College of Steubenville
Steubenville, OH

Linda Hodges

Ernest Kuehl
Lawrence High School
Cedarhurst, NY

Fran Leary
Troy High School
Schenectady, NY

Harold Lefcourt

Cherie Lehman
West Lafayette
High School
West Lafayette, IN

Kathy Malone
Shady Side Academy
Pittsburgh, PA

Bill Metzler
Westlake High School
Thornwood, NY

Elizabeth Farrell
Ramseyer
Niles West High School
Skokie, IL

Daniel Repogle
Central Noble High
School, Albion, IN

Evelyn Restivo
Maypearl High School
Maypearl, TX

Doug Rich
Fox Lane High School
Bedford, NY

John Roeder
The Calhoun School
New York, NY

Tom Senior
New Trier Township
High School
Highland Park, IL

John Thayer
District of Columbia
Public Schools
Silver Spring, MD

Carol-Ann Tripp
Providence Country Day
East Providence, RI

Yvette Van Hise
High Tech High School
Freehold, NJ

Jan Waarvick

Sandra Walton
Dubuque Senior High
School, Dubuque, IA

Larry Wood
Fox Lane High School
Bedford, NY

Field Test Coordinator

Marilyn Decker
Northeastern University
Acton, MA

Field Test Workshop Staff

John Carlson

Marilyn Decker

Arthur Eisenkraft

Douglas Johnson

John Koser

Ernest Kuehl

Mary Quinlan

Elizabeth Farrell
Ramseyer

John Roeder

Field Test Evaluators

Susan Baker-Cohen

Susan Cloutier

George Hein

Judith Kelley

all from Lesley College,
Cambridge, MA

Field Test Teachers and Schools

Rob Adams
Polytech High School
Woodside, DE

Benjamin Allen
Falls Church High
School, Falls Church,
VA

Robert Applebaum
New Trier High School
Winnetka, IL

Joe Arnett
Plano Sr. High School
Plano, TX

Bix Baker
GFW High School
Winthrop, MN

Debra Beightol
Fremont High School
Fremont, NE

Patrick Callahan
Catasaugua High School
Catasaugua, PA

George Coker
Bowling Green
High School
Bowling Green, KY

Janice Costabile
South Brunswick
High School
Monmouth Junction, NJ

Stanley Crum
Homestead High School
Fort Wayne, IN

Russel Davison
Brandon High School
Brandon, FL

Christine K. Deyo
Rochester Adams
High School
Rochester Hills, MI

Jim Doller
Fox Lane High School
Bedford, NY

Jessica Downing
Esparto High School
Esparto, CA

Douglas Fackelman
Brighton High School
Brighton, CO

Rick Forrest
Rochester High School
Rochester Hills, MI

Mark Freeman
Blacksburg High School
Blacksburg, VA

Jonathan Gillis
Enloe High School
Raleigh, NC

Karen Gruner
Holton Arms School
Bethesda, MD

Larry Harrison
DuPont Manual High
School, Louisville, KY

Alan Haught
Weaver High School
Hartford, CT

Steven Iona
Horizon High School
Thornton, CO

Phil Jowell
Oak Ridge High School
Conroe, TX

Deborah Knight
Windsor Forest High
School, Savannah, GA

Thomas Kobilarcik
Marist High School
Chicago, IL

Sheila Kolb
Plano Senior High
School, Plano, TX

Todd Lindsay
Park Hill High School
Kansas City, MO

Malinda Mann
South Putnam
High School
Greencastle, IN

Steve Martin
Maricopa High School
Maricopa, AZ

Nancy McGrory
North Quincy High
School, N. Quincy, MA

David Morton
Mountain Valley High
School, Rumford, ME

Charles Muller
Highland Park
High School
Highland Park, NJ

Fred Muller
Mercy High School
Burlingame, CA

Vivian O'Brien
Plymouth Regional
High School
Plymouth, NH

Robin Parkinson
Northridge High School
Layton, UT

Donald Perry
Newport High School
Bellevue, WA

Francis Poodry
Lincoln High School
Philadelphia, PA

John Potts
Custer County District
High School
Miles City, MT

Doug Rich
Fox Lane High School
Bedford, NY

John Roeder
The Calhoun School
New York, NY

Consuelo Rogers
Maryknoll Schools
Honolulu, HI

Lee Rossmaessler
Mott Middle College
High School
Flint, MI

John Rowe
Hughes Alternative
Center, Cincinnati, OH

Rebecca
Bonner Sanders
South Brunswick
High School
Monmouth Junction, NJ

David Schlipp
Narbonne High School
Harbor City, CA

Eric Shackelford
Notre Dame
High School
Sherman Oaks, CA

Robert Sorensen
Springville-Griffith
Institute
and Central School
Springville, NY

Teresa Stalions
Crittenden County High
School, Marion, KY

Roberta Tanner
Loveland High School
Loveland, CO

Anthony Umelo
Anacostia
Sr. High School
Washington, D.C.

Judy Vondruska
Mitchell High School
Mitchell, SD

Deborah Waldron
Yorktown High School
Arlington, VA

Ken Wester
The Mississippi
School for
Mathematics
and Science
Columbus, MS

Susan Willis
Conroe High School
Conroe, TX

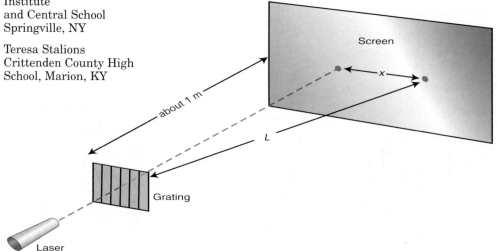

UNIT 2: ACTIVE CHEMISTRY

**Primary and
Contributing
Authors**

Gary Freebury
Kalispell, MT

Mary Gromko
Colorado Springs, CO

Carl Heltzel
Transylvania University
Lexington, KY

John Roeder
The Calhoun School
New York, NY

Hannah Sevian
University of
Massachusetts – Boston
Campus, Boston, MA

Sandra Smith
Colorado Springs, CO

Michael Tinnesand
American Chemical
Society
Washington, D.C.

Reviewers

James Davis
Chemistry Professor
Emeritus Harvard
University

George Miller
University of California
at Irvine
Irvine, CA

Carlo Parravano
Merck Institute for
Science Education
Rahway, NJ

Maren Reeder
Merck Institute for
Science Education
Rahway, NJ

**Active Chemistry
Pilot Testers**

Rob Adams
Wyoming, DE

Ina Ahern
Plymouth Regional
High School
Plymouth, NH

John Bibb
Georgetown, DE

Amy Biddle
Pinkerton Academy
Derry, NH

Robert Dayton
Rush-Henrietta
High School
Henrietta, NY

Barbara Duch
Education Resource Ctr.
Newark, DE

Gabriel Duque
North Miami
Senior High School
Miami, FL

Liz Garcia
Carson High School
Carson, CA

Laura Hajdukiewicz
The Bromfield School
Harvard, MA

Jonathon Haraty
SAGE School
Springfield, MA

Carl Heltzel
Transylvania University
Lexington, KY

Natalie Hiller
Philadelphia
Public Schools
Philadelphia, PA

Penny Hood
Stanwood High School
Stanwood, WA

Tamilyn Ingram
Menifee County
High School
Frenchburg, KY

Barbara Jeffries
Casey Co. High School
Liberty, KY

Diane Johnson
Lewis County
High School
Vanceburg, KY

Gerry LaFontaine
Toll Gate High School
Warwick, RI

Jo Larmore
Laurel, DE

Jeffrey Little
Pikeville
High School
Pikeville, KY

Kathy Lucas
Casey Co.
High School
Liberty, KY

Barbara Malkas
Taconic High School
Pittsfield, MA

Barbara Martin
Reading High School
Reading, MA

Robert Mayton
Allen Central
High School
Eastern, KY

Vicki Mockbee
Caesar Rodney
High School
Camden, DE

Brenda Mullins
Knott County
Central High School
Hindman, KY

Jim Nash
Stanwood High School
Stanwood, WA

Barry North
Rivendell Interstate
Regional High School
Orford, NH

Angela Pence
Wolfe County
High School
Campton, KY

Roy Penix
Prestonburg
High School
Prestonburg, KY

Robin Ringland
Stanwood High School
Stanwood, WA

George Robertson
Stanwood High School
Stanwood, WA

Lance Rudiger
Potsdam High School
Potsdam, NY

John Scali
Newark, DE

Noreen Scarpitto
Reading High School
Reading, MA

Fred Schiess
Stanwood High School
Stanwood, WA

Hannah Sevian
University of
Massachusetts

Angela Skaggs
Magoffin County
High School
Salyersville, KY

Mary Stoukides
Toll Gate High School
Warwick, RI

Jim Swanson
Saugus High School
Saugus, MA

Kathy Swingle
Rehoboth, DE

Jeffrey Scott Townsend
Powell Co. High School
Stanton, KY

Josh Underwood
Deming High School
Mount Olivet, KY

Elaine Weil
Berlin, MD

Jen Wilson
Wilmington, DE

UNIT 3: ACTIVE BIOLOGY

Primary and Contributing Authors

Ruta Demery

Bob Ritter, Ph.D.

Reviewers

Arthur Eisenkraft, Ph.D.
Project Director of
Active Physics and

Active Chemistry
Past President of
the National Science
Teachers Association
(NSTA)

William H. Leonard Ph.D.
Clemson University
Professor of
Education and Biology
Co-Author *Biology:
A Community Context*

Philip Estrada
Biology Teacher
Hollywood
High School, LAUSD

Marissa Hipol
Biology Teacher
Hollywood
High School, LAUSD

Laura Hajdukiewicz
Biology Teacher
Andover, MA

UNIT 4: EARTHCOMM

Primary and Contributing Authors

Julie Bartley
University of
West Georgia

Chuck Bell
Deer Valley High School
Glendale, AZ

Daniel J. Bisaccio
Souhegan High School
Amherst, NH

Lori Borroni-Engle
Taft High School
San Antonio, TX

Richard M. Busch
West Chester University
West Chester, PA

Steve Carlson
Middle School, OR

Kathleen Cochrane
Our Lady of Ransom
School, Niles, IL

Cathey Donald
Auburn High School

Warren Fish
Paul Revere School
Los Angeles, CA

Miriam Fuhrman
Carlsbad, CA

Robert A. Gastaldo
Colby College

Jay Hackett
Colorado Springs, CO

John Kemeny
University of Arizona

John Kounas
Westwood High School
Sloan, IA

William Leonard
Clemson University

Tim Lutz
West Chester University

Laurie Martin-Vermilyea
American Geological
Institute

Steve Mattox
Grand Valley State
University

Keith McKain
Milford Senior
High School
Milford, DE

Mary McMillan
Niwot High School
Niwot, CO

Carolyn Collins Petersen
C. Collins Petersen
Productions
Groton, MA

Mary Poulton
University of Arizona

Bill Romey
Orleans, MA

David Shah
Deer Valley High School
Glendale, AZ

Janine Shigihara Shelley
Junior High School
Shelley, ID

Michael Smith
American Geological
Institute

Tom Vandewater
Colton, NY

Reviewers

Gary Beck
BP Exploration

Phil Bennett
University of Texas
Austin, TX

Steve Bergman
Southern Methodist
University

Samuel Berkheiser
Pennsylvania Geologic
Survey

Arthur Bloom
Cornell University

Craig Bohren
Penn State University

Bruce Bolt
University of California,
Berkeley, CA

John Callahan
Appalachian State
University

Sandip Chattopadhyay
R.S. Kerr Environmental
Research Center

Beth Ellen Clark
Cornell University

Jimmy Diehl
Michigan Technological
University

Sue Beske-Diehl
Michigan Technological
University

Neil M. Dubrovsky
United States
Geological Survey

Frank Ethridge
Colorado State
University

Catherine Finley
University of
Northern Colorado

Ronald Greeley
Arizona State
University

Michelle Hall-Wallace
University of Arizona

Judy Hannah
Colorado State
University

Blaine Hanson
Dept. of Land, Air, and
Water Resources

James W. Head III
Brown University

Patricia Heiser
Ohio University

John R. Hill
Indiana Geological
Survey

Travis Hudson American
Geological Institute

Jackie Huntoon
Michigan Tech.
University

Teresa Jordan
Cornell University

Allan Juhas
Lakewood, Colorado

Robert Kay
Cornell University

Chris Keane
American Geological
Institute

Bill Kirby
United States
Geological Survey

Mark Kirschbaum
United States
Geological Survey

Dave Kirtland
United States
Geological Survey

Jessica Elzea Kogel
Thiele Kaolin Company

Melinda Laituri
Colorado State
University

Martha Leake
Valdosta State
University

Donald Lewis
Happy Valley, CA

Steven Losh
Cornell University

Jerry McManus
Woods Hole
Oceanographic
Institution

Marcus Milling
American Geological
Institute

Alexandra Moore
Cornell University

Jack Oliver
Cornell University

Don Pair
University of Dayton

Mauri Pelto
Nicolas College

Bruce Pivetz
ManTech Environmental
Research Services Corp.

Stephen Pompea
Pompea & Associates

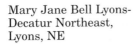

Peter Ray
Florida State University

William Rose
Michigan Technological
University

Lou Solebello
Macon, Georgia

Robert Stewart
Texas A&M University

Ellen Stofan
NASA

Barbara Sullivan
University of Rhode
Island

Carol Tang
Arizona State University

Bob Tilling
United States Geological
Survey

Stanley Totten
Hanover College

Scott Tyler
University of Nevada
Reno, NV

Michael Velbel
Michigan State
University

Ellen Wohl
Colorado State University

David Wunsch
State Geologist of New
Hampshire

Pilot Test Evaluator

Larry Enochs
Oregon State University

Pilot Test Teachers

Rhonda Artho
Dumas High School
Dumas, TX

Mary Jane Bell Lyons-
Decatur Northeast,
Lyons, NE

Rebecca Brewster
Plant City High School,
Plant City, FL

Terry Clifton
Jackson High School
Jackson, MI

Virginia Cooter
North Greene High
School, Greeneville, TN

Monica Davis
North Little Rock
High School
North Little Rock, AR

Joseph Drahuschak
Troxell Jr. High School,
Allentown, PA

Ron Fabick
Brunswick High School
Brunswick, OH

Virginia Jones
Bonneville High School
Idaho Falls, ID

Troy Lilly
Snyder High School
Snyder, TX

Sherman Lundy
Burlington High School
Burlington, IA

Norma Martof Fairmont
Heights High School
Capitol Heights, MD

Keith McKain
Milford Senior High
School, Milford, DE

Mary McMillan
Niwot High School
Niwot, CO

Kristin Michalski
Mukwonago High School,
Mukwonago, WI

Dianne Mollica
Bishop Denis J. O'Connell
High School
Arlington, VA

Arden Rauch
Schenectady High School,
Schenectady, NY

Laura Reysz Lawrence
Central High School
Indianapolis, IN

Floyd Rogers
Palatine High School
Palatine, IL

Ed Ruszczyk
New Canaan High School,
New Canaan, CT

Jane Skinner
Farragut High School
Knoxville, TN

Shelley Snyder
Mount Abraham High
School, Bristol, VT

Joy Tanigawa
El Rancho High School
Pico Rivera, CA

Dennis Wilcox
Milwaukee School of
Languages
Milwaukee, WI

Kim Willoughby
SE Raleigh High School
Raleigh, NC

Field Test Workshop Staff

Don W. Byerly
University of Tennessee

Derek Geise
University of Nebraska

Michael A. Gibson
University of Tennessee

David C. Gosselin
University of Nebraska

Robert Hartshorn
University of Tennessee

William Kean
University of Wisconsin

Ellen Metzger
San Jose State University

Tracy Posnanski
University of Wisconsin

J. Preston Prather
University of Tennessee

Ed Robeck
Salisbury University

Richard Sedlock
San Jose State University

Bridget Wyatt
San Jose State University

Field Test Evaluators

Bob Bernoff
Dresher, PA

Do Yong Park
University of Iowa

Field Test Teachers

Kerry Adams Alamosa
High School Alamosa, CO

Jason Ahlberg
Lincoln High School
Lincoln, NE

Gregory Bailey
Fulton High School
Knoxville, TN

Mary Jane Bell Lyons-
Decatur Northeast
Lyons, NE

Rod Benson
Helena High School
Helena, MT

Sandra Bethel
Greenfield High School
Greenfield, TN

John Cary
Malibu High School
Malibu, CA

Elke Christoffersen
Poland Regional High
School, Poland, ME

Tom Clark
Benicia High School
Benicia, CA

Julie Cook
Jefferson City
High School
Jefferson City, MO

Virginia Cooter
North Greene High
School, Greeneville, TN

Mary Cummane
Perspectives Charter
Chicago, IL

Sharon D'Agosta
Creighton Preparatory
Omaha, NE

Mark Daniels
Kettle Morraine High
School, Milwaukee, WI

Beth Droughton
Bloomfield High School
Bloomfield, NJ

Steve Ferris
Lincoln High School
Lincoln, NE

Bob Feurer
North Bend Central
Public, North Bend, NE

Sue Frack
Lincoln Northeast High
School, Lincoln, NE

Rebecca Fredrickson
Greendale High School
Greendale, WI

Sally Ghilarducci
Hamilton High School
Milwaukee, WI

Kerin Goedert
Lincoln High School
Ypsilanti, MI

Martin Goldsmith
Menominee Falls High
Menominee Falls, WI

Randall Hall
Arlington High School
St. Paul, MN

Theresa Harrison
Wichita West High
School, Wichita, KS

Gilbert Highlander
Red Bank High School
Chattanooga, TN

Jim Hunt
Chattanooga School
of Arts & Sciences
Chattanooga, TN

Patricia Jarzynski
Watertown High School
Watertown, WI

Pam Kasprowicz Bartlett
High School Bartlett, IL

Caren Kershner Moffat
Consolidated Moffat, CO

Mary Jane Kirkham
Fulton High School

Ted Koehn
Lincoln East High School
Lincoln, NE

Philip Lacey
East Liverpool
High School
East Liverpool, OH

Joan Lahm
Scotus Central Catholic
Columbus, NE

Erica Larson
Tipton Community

Michael Laura
Banning High School
Wilmington, CA

Fawn LeMay
Plattsmouth High
Plattsmouth, NE

Christine Lightner
Smethport Area High
School, Smethport, PA

Nick Mason
Normandy High School
St. Louis, MO

James Matson
Wichita West High
Wichita, KS

Jeffrey Messer
Western High School
Parma, MI

Dave Miller
Parkview High School
Springfield, MO

Rick Nettesheim
Waukesha South
Waukesha, WI

John Niemoth Niobrara
Public Niobrara, NE

Margaret Olsen
Woodward Academy
College Park, GA

Ronald Ozuna
Roosevelt High
Los Angeles, CA

Paul Parra
Omaha North High
Omaha, NE

D. Keith Patton
West High School
Denver, CO

Phyllis Peck
Fairfield High School
Fairfield, CA

Randy Pelton
Jackson High School
Massillon, OH

Reggie Pettitt
Holderness High School
Holderness, NH

June Rasmussen
Brighton High School
South Brighton, TN

Russ Reese
Kalama High School
Kalama, WA

Janet Ricker
South Greene
High School
Greeneville, TN

Wendy Saber
Washington Park High
Racine, WI

Garry Sampson
Wauwatosa West High
Tosa, WI

Daniel Sauls Chuckey-
Doak High School
Afton, TN

Todd Shattuck
L.A. Center for Enriched
Studies, Los Angeles, CA

Heather Shedd
Tennyson High School
Hayward, CA

Lynn Sironen
North Kingstown High
North Kingstown, RI

Jane Skinner
Farragut High School
Knoxville, TN

Sarah Smith
Garringer High
Charlotte, NC

Aaron Spurr
Malcolm Price Laboratory
Cedar Falls, IA

Karen Tiffany
Watertown High School
Watertown, WI

Tom Tyler
Bishop O'Dowd High
School, Oakland, CA

Valerie Walter
Freedom High School
Bethlehem, PA

Christopher J. Akin
Williams Milford
Mill Academy
Baltimore, MD

Roseanne Williby
Skutt Catholic High
School, Omaha, NE

Carmen Woodhall
Canton South High
Canton, OH

**Field Test
Coordinator**

William Houston
American Geological
Institute

Advisory Board

Jane Crowder
Bellevue, WA

Arthur Eisenkraft
University of
Massachusetts

Tom Ervin
LeClaire, IA

Mary Kay Hemenway
University of Texas at
Austin

Bill Leonard
Clemson University

Don Lewis
Lafayette, CA

Wendell Mohling National
Science Teachers
Association

Harold Pratt
Littleton, CO

Barb Tewksbury
Hamilton College

Laure Wallace
USGS

AGI Foundation

Jan van Sant
Executive Director

Acknowledgements

The American Geological Institute and EarthComm

Imagine more than 500,000 Earth scientists worldwide sharing a common voice, and you've just imagined the mission of the American Geological Institute. Our mission is to raise public awareness of the Earth sciences and the role that they play in mankind's use of natural resources, mitigation of natural hazards, and stewardship of the environment. For more than 50 years, AGI has served the scientists and teachers of its Member Societies and hundreds of associated colleges, universities, and corporations by producing Earth science educational materials, *Geotimes*–a geoscience news magazine, GeoRef–a reference database, and government affairs and public awareness programs.

So many important decisions made every day that affect our lives depend upon an understanding of how our Earth works. That's why AGI created *EarthComm*. In your *EarthComm* classroom, you'll discover the wonder and importance of Earth science by studying it where it counts—in your community. As you use the rock record to investigate climate change, do field work in nearby beaches, parks, or streams, explore the evolution and extinction of life, understand where your energy resources come from, or find out how to forecast severe weather, you'll gain a better understanding of how to use your knowledge of Earth science to make wise personal decisions.

We would like to thank the AGI Foundation Members that have been supportive in bringing Earth science to students. These AGI Foundation Members include: American Association of Petroleum Geologists Foundation, Anadarko Petroleum Corp., The Anschutz Foundation, Apache Canada Ltd., Baker Hughes Foundation, Barrett Resources Corp., Elizabeth and Stephen Bechtel, Jr. Foundation, BP Foundation, Burlington Resources Foundation, CGG Americas, Inc., ChevronTexaco Corp., Conoco Inc., Consolidated Natural Gas Foundation, Devon Energy Corp., Diamond Offshore Co., Dominion Exploration & Production, Inc., EEX Corp., Equitable Production Co., ExxonMobil Foundation, Five States Energy Co., Geological Society of America Foundation, Global Marine Drilling Co., Halliburton Foundation, Inc., Kerr McGee Foundation, Maxus Energy Corp., Noble Drilling Corp., Occidental Petroleum Charitable Foundation, Ocean Energy, Optimistic Petroleum Co., Parker Drilling Co., Phillips Petroleum Co., Santa Fe Snyder Corp., Schlumberger Foundation, Shell Oil Company Foundation, Southwestern Energy Co., Texas Crude Energy, Inc., Unocal Corp. USX Foundation (Marathon Oil Co.).

We at AGI wish you success in your exploration of the Earth System and your Community.

Michael J. Smith
Director of Education, AGI

Marcus E. Milling
Executive Director, AGI

You can do science. Here are the reasons why...

The following features make it that easier to understand the science principles you will be studying. Whether it's physics, chemistry, biology, or Earth science, the way you go about learning each of them is the same. Every step along the way prepares you for the next step. Using all these features together will help you actually learn about science and see how it works for you every day, everywhere.

Look for these features in each chapter of *Coordinated Science for the 21st Century*.

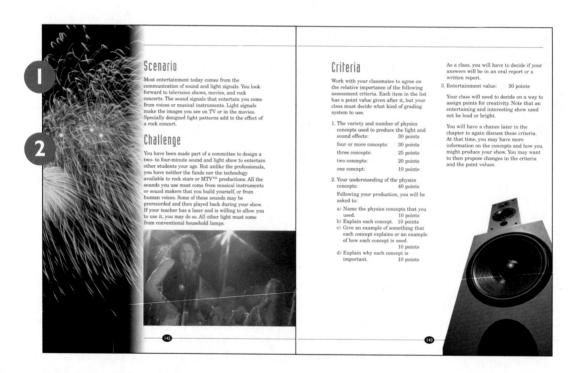

1 Scenario

Each chapter begins with a realistic event or situation. You might actually have experienced the event, or you can imagine yourself participating in a similar situation at home, in school, or in your community. Chances are you probably never thought about the science involved in each case.

2 Challenge

This feature presents you with a challenge 0you can expect to complete by the end of the chapter. This challenge gives you the opportunity to learn science as you complete a realistic, science-based project. As you progress through the chapter, you will accumulate all the scientific knowledge you need to successfully complete the challenge.

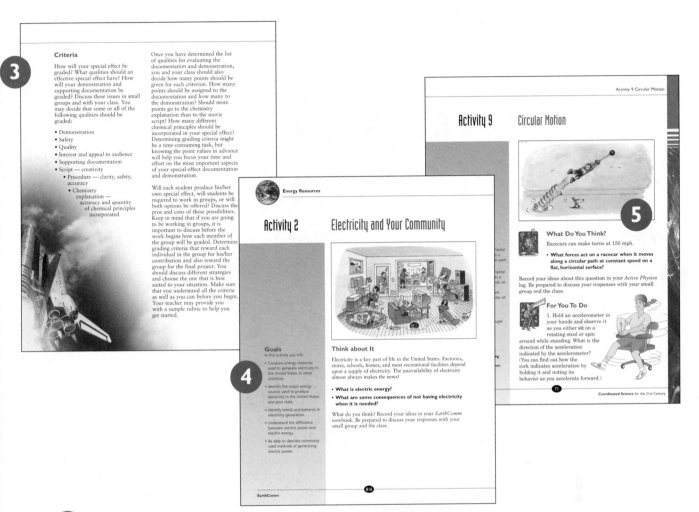

3 Criteria

Before you begin the chapter and the challenge, you and your classmates, along with your teacher, will explore exactly how you will be graded. You will review the criteria and expectations for solving the challenge, and make decisions about how your work should be evaluated.

4 Goals

At the beginning of each activity you are provided with a list of goals that you should be able to achieve by completing your science inquiry.

5 What Do You Already Know?

Before you start each activity you will be asked one or two questions to consider. You will have a chance to discuss your ideas with your group and your class. You are not expected to come up with the "right" answer, but to share your current understanding and reasoning.

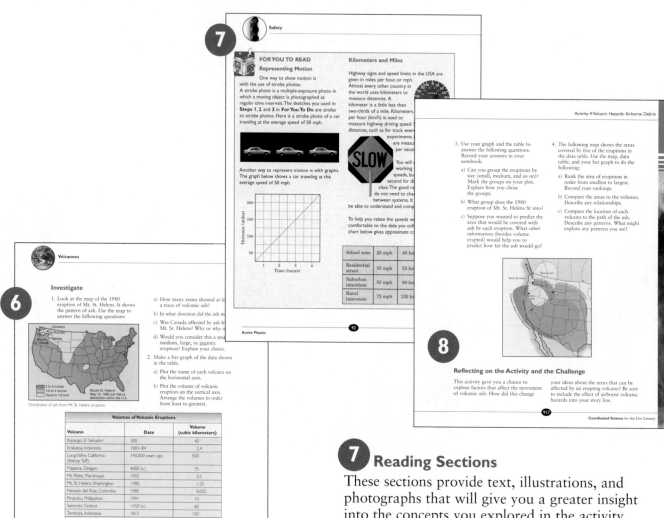

7 Reading Sections

These sections provide text, illustrations, and photographs that will give you a greater insight into the concepts you explored in the activity. Equations and formulas are provided with easy-to-understand explanations. Science Words that may be new or unfamiliar to you are defined and explained. In some chapters Checking Up questions are included to guide you in your reading.

6 Investigate

In *Coordinated Science for the 21st Century* you learn by doing science. In your small groups, or as a class, you will take part in scientific inquiry by doing hands-on experiments, participating in fieldwork, or searching for answers using the Internet and reference materials.

8 Reflecting on the Activity and the Challenge

Each activity helps to prepare you to be successful in the challenge. This feature gives you a brief summary of the activity. It will help you relate the activity that you just completed to the challenge. It's another piece of the chapter jigsaw puzzle.

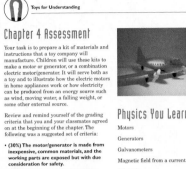

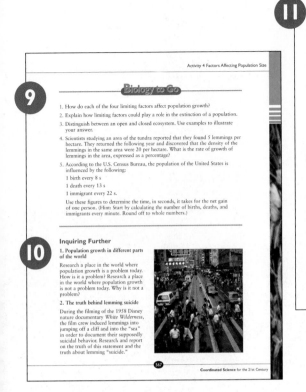

9 Science to Go

Questions in this feature ask you to use the key principles and concepts introduced in the activity. You may also be presented with new situations in which you will be asked to apply what you have learned. They are excellent as a study guide, helping you review and understand what is most important from the activity. You will also be provided with suggestions for ways in which you can organize your work and get ready for the challenge.

10 Inquiring Further

This feature stretches your thinking. It provides lots of suggestions for deepening your understanding of the concepts and skills developed in the activity. Also, if you're looking for more challenging or in-depth problems, questions, and exercises, you'll find them right here.

11 Chapter Assessment

How do you measure up? Here is your opportunity to share what you have actually learned. Using the activities as a guide, you can now complete the challenge you were presented at the beginning of the chapter.

12 Science at Work

Science is an integral part of many fascinating careers. This feature introduces some people working in different fields that involve the principles of science.

Unit 1

ACTIVE PHYSICS

Table of Contents

Screen

Lens

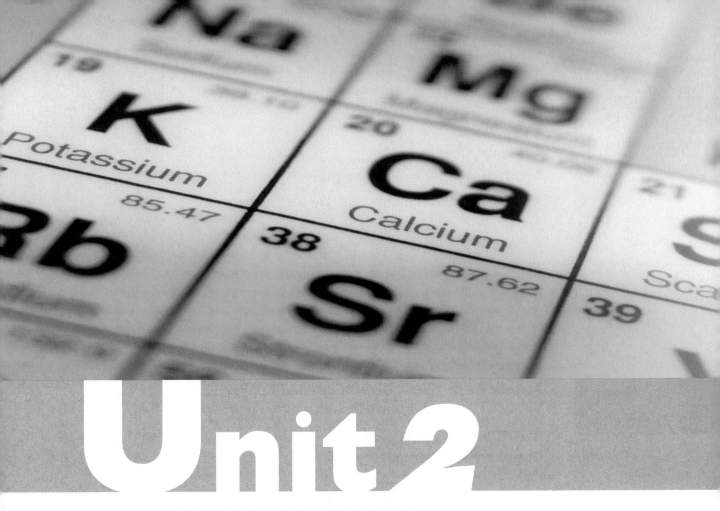

Unit 2

ACTIVE CHEMISTRY

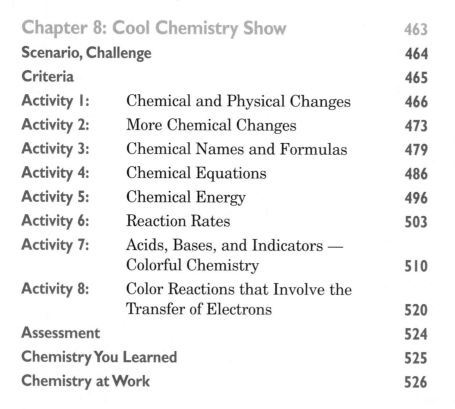

Unit 3

ACTIVE BIOLOGY

Unit 4

EARTHCOMM

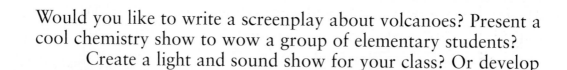

Would you like to write a screenplay about volcanoes? Present a cool chemistry show to wow a group of elementary students? Create a light and sound show for your class? Or develop a booklet to help people to decide how to vote? This year science will be filled with many fascinating challenges found in this book, *Coordinated Science for the 21st Century*. The course comes from other programs based on learning by doing: *Active Physics*, *Active Chemistry*, *Active Biology*, and *EarthComm*. It's a science course that introduces you to the fundamental laws and theories of four large disciplines of science and asks that you use your creativity, interest, and expertise to complete challenges in the same way that professionals do.

When scientists, artists, musicians, writers, and engineers are hired to solve a problem, their first responsibility is to understand the nature of the assignment. Often, they must do research to better acquaint themselves with the topic. The person assigning the job has a sense of what is needed, but does not know exactly what the final product will be. For instance, a movie

producer may decide that a film about volcanoes would be thrilling, yet informative. The producer would then hire a team of screenwriters to provide the script. This group of talented people will need a wide range of expertise. The team must image an exciting, yet realistic plot. They must also rely on someone who has knowledge about volcanoes, where they can occur, how they

can be monitored, and the hazards that they can produce. In one of the chapters, you will be required to write a screenplay about volcanoes. You will be helped along the way with activities that will give you the knowledge and expertise to be successful.

As you complete chapter challenges, you will begin to combine your science knowledge with your other interests and create something that will represent the best efforts of you and your team. You will be learning about the disciplines of physics, chemistry, biology, and Earth science. You will investigate how waves produce music. You will discover how the behavior of atoms and electrons can make some foods delicious and others poisonous. You will understand that the extinction of one organism can affect an entire ecosystem. And you will learn about volcanoes and earthquakes, and how they affect all the Earth systems.

Hundreds of teachers, professors, artists, editors, and science education professionals who have worked on these materials have taken into account the best information on how students can achieve and how students can be engaged in their own learning. We have taken the philosophy and recommendations of the National Science Education Standards and created a course that will engage you in asking the types of questions that scientists ask; will have you explore the world in the way that scientists explore the world; and will require you to show us that you really understand the meaning and implications of the scientific principles.

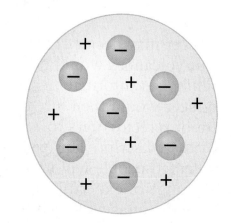

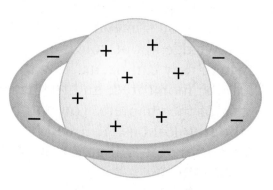

The development of the chapters in this book has been funded by the National Science Foundation. They have gone through a process that included feedback from students much like you in classrooms across the country. Based on that feedback, the materials have been improved so that you can learn science, enjoy what you are learning, and be successful in the course.

This year will be an introduction to four major sciences. Scientists in each

discipline have ways of learning that are unique to their disciplines and yet the sciences are also similar. The energy released in an earthquake travels in waves, so does sound, as does the light that is given off when electrons move from one orbit to another. Measuring the waves involved in each of these examples requires different tools and different apparatus. You will be introduced to these sciences and will become acquainted with the similarities and differences among the disciplines.

You will begin to think like a scientist.

Safety in the Science Classroom

Chemistry is a laboratory science. During this course you will be doing many activities in which safety is a factor. To ensure the safety of you and all students, the following safety rules will be followed. You will be responsible for abiding by these rules at all times. After reading the rules, you and a parent or guardian must sign a Safety Contract acknowledging that you have read and understood the rules and will follow them at all times.

General Rules

1. There will be no running, jumping, pushing, or other behavior considered inappropriate in the science laboratory. You must behave in an orderly and responsible way at all times.

2. Eating, drinking, chewing gum, or applying cosmetics is strictly prohibited.

3. All spills and accidents must be reported to your teacher immediately.

4. You must follow all directions carefully and use only materials and equipment provided by your teacher. Only activities approved by your teacher may be carried out in the chemistry laboratory.

5. No loose, hanging clothing is allowed in the laboratory; long sleeves must be rolled up; bulky jackets, as well as jewelry, must be removed.

6. Never work in the lab unless your teacher or an approved substitute is present.

7. Identify and know the location of a fire extinguisher, fire blanket, emergency shower, eyewash, gas and water shut-offs, and telephone.

Equipment Rules

1. All equipment must be checked out and returned properly.

2. Do not touch any equipment until you are instructed to do so.

3. Do not use glassware that is broken or cracked. Alert your teacher to any glassware that is broken or cracked.

Working with Chemicals

1. Never touch or smell chemical unless specifically instructed to do so by your teacher. Never taste chemicals.

2. Safety goggles must be worn at all times.

3. Carefully read all labels to make sure you are using the correct chemicals and use only the amount of chemicals instructed by your teacher.

4. Keep your hands away from your face and thoroughly wash with soap and water before exiting the classroom.

5. Contact lenses can absorb certain chemicals. Advise your teacher if you wear contact lenses.

6. Never add water to an acid and always add acid slowly to water.

7. Follow your teachers' instructions for the correct disposal of chemicals. Do not dispose of any chemical waste, including paper towels used for chemical spills, in the trash basket or down a sink drain.

Flame Safety

1. Use extreme caution when using any type of flame. Keep your hands, hair, and clothing away from flames.

2. Long hair must be tied back at all times.

3. Keep all flammable materials away from open flames. Some winter jackets are extremely flammable and should be removed before entering the laboratory.

4. Always point the mouth of a test tube away from yourself or any other person when heating a substance.

5. Extinguish the flame as soon as you are finished.

6. Always use heat-resistant gloves when working with an open flame.

Work Area

1. When working in the laboratory all materials should be removed from the workstation except for instructions and data tables. Materials should not be removed from the desktop to the floor as this is a hazard for someone walking with glassware or chemicals.

2. The work area should be kept clean at all times. After completing an activity, wipe down the area.

3. Notify your teacher of any spills immediately so they can be properly taken care of.

Unit 1

Active Physics®

Making Connections

You are about to begin a very active year continuing your exploration of science. This year, you will get chances to explore the world around you in a very meaningful way. *Coordinated Science for the 21st Century* includes chapters from *Active Physics*, *Active Chemistry*, *Active Biology*, and *EarthComm*.

To organize the understanding and investigation of the world, scientists have made up categories like physics, chemistry, biology, and Earth science. All of these fields are related and information from one field is often used to develop new ideas in a different field. There are now fields like biochemistry, physical chemistry, and environmental science, which require an understanding of more than one branch of science.

What Do You Think?

• What is science?

• How are physics, chemistry, biology and Earth science similar? How are they different?

Record your ideas about this in your log. Be prepared to discuss your responses with your small group and the class.

For You To Try

• Bertrand Russell (1872-1970), an English philosopher and mathematician, said that, "Science is what you know. Philosophy is what you don't know." Research to find quotes from other famous people about what they think that science is. Which one do you think comes closest to what you think that science is?

• Rarely does a week go by without a new scientific study or discovery being reported in the media. Review the news for the last few months to find what new studies or discoveries have been made recently. Research and report on one of these.

Chapter 1

PHYSICS IN ACTION

Scenario

Have you ever dreamed of being a sports analyst for Monday Night Football with millions of people listening to every word you say? What about the sports commentator for the Summer Olympics? Imagine interviewing the Most Valuable Player (MVP) after an NBA championship game or interviewing the Olympic Gold Medalist in women's figure skating. What type of credentials are needed to have such a glamorous career in sportscasting? Should you major in journalism in college or be a retired professional athlete if you desire to land such a lucrative and exciting job? Could the study of physics be a key to becoming a sports analyst? Could a student with a knowledge of physics bring to the TV viewer a different perspective that might provide a new outlook on sporting events?

Challenge

PBS has decided that it wants to televise certain sporting events and that they would like these programs to have some educational as well as entertainment value. As a test of this idea, you are to provide the voice-over on a sports video and to explain the physics of the action appearing on the screen. Here is your chance to audition for a job in sportscasting. Each student (or group of students) will do a "science commentary" on a short (2–3 minute) sports video.

To assess how well you understand this material, you (or your group) are to do one of these:

- **submit a written script**
- **narrate live**
- **dub onto the video soundtrack**
- **record on an audiocassette**

Your task is not to give a play-by-play description of the sporting event or give the rules of the game but rather to go a step beyond and educate the audience by describing to them the rules of nature that govern the event. This approach will give the viewer (and you) a different perspective on both sports and physics. The laws of physics cover not only obscure phenomena in the lab, but everyday events in the real world as well.

Criteria

What criteria should be used to evaluate a voice-over dialogue or script of a sporting event? Since the intention is to provide an analysis of and interest in the physics of sports, the voice-over should include the use of physics terms and physics principles. All of these terms and principles should be used correctly. How many of these terms and principles would constitute an excellent job? Would it be enough to use one physics term correctly and explain how one physics principle is illustrated in the sport? Should use of one physics term and one physics principle be a minimum standard to get minimal credit for this assessment? Discuss in your small groups and your class and decide on reasonable expectations for the physics criteria for the assessment.

Since the assessment requires a product that will be a part of television, another aspect of the criteria for success would be the entertainment quality of the voice-over. Does a commentator who adds humor or drama receive a higher rating than someone who has similar physics content but has added no excitement or interest to the broadcast? How does one weigh the value of the entertainment quality and the value of relevant physics? What are reasonable expectations for the entertainment aspect of the voice-over? Discuss and decide as a class.

Although many people may be in the broadcast booth, a voice-over becomes the product of one person—the commentator or the scriptwriter. Although you will be working in cooperative groups during the chapter, each person will be responsible for a voice-over or script for a sporting event. As a team of two or three, you may wish to work together and share different aspects of the job, but the output of work per person should be the same. That is why one voice-over will be required of each person irrespective of whether individuals prefer to work independently or in groups.

Activity 1

A Running Start and Frames of Reference

What Do You Think?

Many things that happen in athletics are affected by the amount of "running start" speed an athlete can produce.

- **What determines the amount of horizontal distance a basketball player travels while "hanging" to do a "slam dunk" during a fast break?**

- **How do figure skaters keep moving across the ice at high speeds for long times while seldom "pumping" their skates?**

Record your ideas about these questions in your *Active Physics* log. Be prepared to discuss your responses with your small group and the class.

GOALS

In this activity you will:

- Understand and apply Galileo's Principle of Inertia.

- Understand and apply Newton's First Law of Motion.

- Recognize inertial mass as a physical property of matter.

For You To Do

1. Use a salad bowl and a ball to explore the question, "When a ball is released to roll down the inside surface of a salad bowl, is the motion of the ball up the far side of the bowl the 'mirror image' of the ball's downward motion?" Use a nonpermanent pen to mark a starting position for the ball near the top edge of the bowl. Use a

flexible ruler to measure, in centimeters, the distance along the bowl's curved surface from the bottom-center of the bowl to the mark.

a) Make a table similar to the one below in your log.

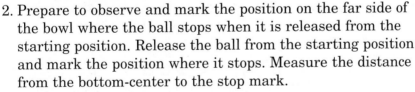

Start	Trial	Starting Distance (cm)	Recovered Distance (cm)	Recovered Distance / Starting Distance
High	1			
	2			
	3			
Medium	1			
	2			
	3			
Low	1			
	2			
	3			

b) Record the measured distance in your table as the High Starting Distance.

Do not use a glass bowl, if possible.

2. Prepare to observe and mark the position on the far side of the bowl where the ball stops when it is released from the starting position. Release the ball from the starting position and mark the position where it stops. Measure the distance from the bottom-center to the stop mark.

a) Record the distance in your table as the High Recovered Distance.

3. Repeat **Step 2** above two more times to see if the results are consistent.

a) Record the data for all three trials.

4. Mark two more starting positions on the surface of the bowl, one for Medium Starting Distance and another for Low Starting Distance.

a) Measure and record each of the new distances.

b) Observe, mark, measure and record the recovered distances for three trials at each of the medium and low starting positions.

c) Complete the table by calculating and recording the value of the ratio of the Recovered Distance to the Starting Distance for each trial. (The ratio is the Recovered Distance divided by the Starting Distance.)

> Example:
>   If the Recovered Distance is 6.0 cm for a Starting Distance of 10.0 cm, the value of the ratio is $\frac{6.0 \text{ cm}}{10.0 \text{ cm}} = 0.6$.

d) For each of the three starting distances, to what extent is the motion of the ball up the far side of the bowl the "mirror image" of the downward motion? Use data as evidence for your answer.

e) Does the fraction of the starting distance "recovered" when going up the far side of the bowl depend on the amount of starting distance? Describe any pattern of data that supports your answer.

5. Repeat the activity but roll the ball along varying slopes during its upward motion. Make a track that has the same slope on both sides, as shown below. Your teacher will suggest how high the ends of the track sections should be elevated. This time, concentrate on comparing the vertical height of the ball's release position to the vertical height of the position where the ball stops.

a) Measure and record the vertical height (not the distance along the track) from which the ball will be released at the top end of the left-hand section of track.

b) Prepare to observe and mark the position on the right-hand section of track where the ball stops when it is released from the starting position. Release the ball from the top end of the left-hand section of track and mark the position where it stops. Measure and record the vertical height of the position where the ball stops.

✎ c) Calculate the ratio of the recovered height to the starting height. How is this case, and the result, similar to what you did when using the salad bowl? How is it different?

6. Leave the left-hand starting section of track unchanged, but change the right-hand section of track so that it has less slope and is at least long enough to allow the ball to recover the starting height. The track should be arranged approximately as shown below.

✎ a) Predict the position where the ball will stop on the right-hand track if it is released from the same height as before on the left-hand track. Mark the position of your guess on the right-hand track and explain the basis for your prediction in your log.

7. Release the ball from the same height on the left-hand section of track as before and mark the position where the ball stops on the right-hand section of track.

✎ a) How well did you guess the position? Why do you think your guess was "on" or "off"?

✎ b) Measure the vertical height of the position where the ball stopped and again calculate the ratio of the recovered height to the starting height. Did the ratio change? Why, do you think, did the ratio change or not change?

8. Imagine what would happen if you again did not change the left-hand starting section of track, but changed the right-hand section of track so that it would be horizontal, as shown below.

✎ a) How far along the horizontal track would the ball need to roll to recover its starting height (or most of it)? How far do you think the ball would roll?

✎ b) When rolling on the horizontal track, what would "keep the ball going"?

FOR YOU TO READ

Inertia

Italian philosopher Galileo Galilei (1564–1642), who can be said to have introduced science to the world, noticed that a ball rolled down one ramp seems to seek the same height when it rolls up another ramp. He also did a "thought experiment" in which he imagined a ball made of extremely hard material set into motion on a horizontal, smooth surface, similar to the final track in **For You To Do**. He concluded that the ball would continue its motion on the horizontal surface with constant speed along a straight line "to the horizon" (forever). From this, and from his observation that an object at rest remains at rest unless something causes it to move, Galileo formed the Principle of **Inertia**:

Inertia is the natural tendency of an object to remain at rest or to remain moving with constant speed in a straight line.

Isaac Newton, born in England on Christmas day in 1642 (within a year of Galileo's death), used Galileo's Principle of Inertia as the basis for developing his First Law of Motion, presented in **Physics Talk**. Crediting Galileo and others for their contributions to his thinking, Newton said, "If I have seen farther than others, it is because I have stood on the shoulders of giants."

Running Starts

Running starts take place in many sporting activities. Since there seems to be this prior motion in many sports, there must be some advantage to it.

In sports where the objective is to maximize the speed of an object or the distance traveled in air, the prior motion may be essential. When a javelin is thrown, at the instant of release it has the same speed as the hand that is propelling it.

- The hand has a forward speed relative to the elbow, the elbow has a forward speed relative to the shoulder (because the arm is rotating around the elbow and shoulder joints), and the shoulder has a forward speed relative to the ground because the body is rotating and the body is also moving forward.

- The javelin speed then is the sum of each of the above speeds. If the thrower is not running forward, that speed does not add into the equation.

You can write a **velocity** equation to show the speeds involved.

$$v_{javelin} = v_{hand} + v_{elbow} + v_{shoulder} + v_{ground}$$

Motion captures everyone's attention in sports. Starting, stopping, and changing direction (**accelerations**) are part of the motion story, and they are exciting components of many sports. Ordinary, straight-line motion is just as important but is easily overlooked.

PHYSICS TALK

Newton's First Law of Motion

Isaac Newton included Galileo's Principle of Inertia as part of his **First Law of Motion**:

In the absence of an unbalanced force, an object at rest remains at rest, and an object already in motion remains in motion with constant speed in a straight-line path.

Newton also explained that an object's mass is a measure of its inertia, or tendency to resist a change in motion.

Here is an example of how Newton's First Law of Motion works:

Inertia is expressed in kilograms of mass. If an empty grocery cart has a mass of 10 kg and a cart full of groceries has a mass of 100 kg, which cart would be more difficult to move (have a greater tendency to remain at rest)? If both carts already were moving at equal speeds, which cart would be more difficult to stop (would have a greater tendency to keep moving)? Obviously in both cases, the answer is the cart with more mass.

Physics Words

inertia: the natural tendency of an object to remain at rest or to remain moving with constant speed in a straight line.

acceleration: the change in velocity per unit time.

frame of reference: a vantage point with respect to which position and motion may be described.

FOR YOU TO READ

Frames of Reference

In this activity, you investigated Newton's First Law. In the absence of external forces, an object at rest remains at rest and an object in motion remains in motion. If you were challenged to throw a ball as far as possible, you would probably now be sure to ask if you could have a running start. If you run with the ball prior to throwing it, the ball gets your speed before you even try to release it. If you can run at 5 m/s, then the ball will get the additional speed of 5 m/s when you throw it. When you do throw the ball, the ball's speed is the sum of your speed before releasing the ball, 5 m/s, and the speed of the release.

It may be easier to understand this if you think of a toy cannon that could be placed on a skateboard. The toy cannon always shoots a small ball forward at 7 m/s. This can be checked with multiple trials. The toy cannon is then attached to the skateboard. A release mechanism is set up so that the cannon continues to shoot the ball forward at 7 m/s when the skateboard is at rest. When the skateboard is given an initial push, the skateboard is able to travel at 3 m/s. If the cannon releases the ball while the skateboard is moving, the ball's speed is now measured to be 10 m/s. From where did the additional speed come? The ball's speed is the sum of the ball's speed from the cannon plus the speed of the skateboard. 7 m/s + 3 m/s = 10 m/s.

You may be wondering if the ball is moving at 7 m/s or 10 m/s. Both values are correct — it depends on your **frame of reference**. The ball is moving at 7 m/s relative *to* the skateboard. The ball is moving at 10 m/s relative *to* the Earth.

Imagine that you are on a train that is stopped at the platform. You begin to walk toward the front of the train at 3 m/s. Everybody in the train will agree that you are moving at 3 m/s toward the front of the train. This is your speed *relative to* the train. Everybody looking into the train from the platform will also agree that you are moving at 3 m/s toward the front of the train. This is your speed *relative to* the platform.

Imagine that you are on the same train, but now the train is moving past the platform at 9 m/s. You begin to walk toward the front of the train at 3 m/s. Everybody in the train will agree that you are moving at 3 m/s toward the

front of the train. This is your speed *relative to* the train. Everybody looking into the train from the platform will say that you are moving at 12 m/s (3 m/s + 9 m/s) toward the front of the train. This is your speed relative *to* the platform.

Whenever you describe speed, you must always ask, "*Relative to what?*" Often, when the speed is relative to the Earth, this is assumed in the problem. If your frame of reference is the Earth, then it all seems quite obvious. If your frame of reference is the moving train, then different speeds are observed.

In sports where you want to provide the greatest speed to a baseball, a javelin, a football, or a tennis ball, that speed could be increased if you were able to get on a moving platform. That being against the rules and inappropriate for many reasons, an athlete will try to get the body moving with a running start, if allowed. If the running start is not permitted, the athlete tries to move every part of his or her body to get the greatest speed.

Sample Problem I

A sailboat has a constant velocity of 22 m/s east. Someone on the boat prepares to toss a rock into the water.

a) Before being tossed, what is the speed of the rock with respect to the boat?

b) Before being tossed, what is the speed of the rock with respect to the shore?

c) If the rock is tossed with a velocity of 16 m/s east, what is the rock's velocity with respect to shore?

d) If the rock is tossed with a velocity of 16 m/s west, what is the rock's velocity with respect to shore?

Strategy: Before determining a velocity, it is important to check the frame of reference. The rock's velocity with respect to the boat is different from the velocity with respect to the shore. The direction of the rock also impacts the final answer.

Givens:

$$v_b = 22 \text{ m/s east}$$
$$v_r = 16 \text{ m/s (direction varies)}$$

Solution:

a) With respect to the boat, the rock's velocity is 0 m/s.
   The rock is moving at the same speed as the boat, but you wouldn't notice this velocity if you were in the boat's frame of reference.

b) With respect to shore, the rock's velocity is 22 m/s east.
   The rock is on the boat, which is traveling at 22 m/s east. Relative to the shore, the boat and everything on it act as a system traveling at the same velocity.

c) With respect to the shore, the rock's velocity is now 38 m/s east.
   It is the sum of the velocity values. Since each is directed east, the relative velocity is the sum of the two.

$$v = v_b + v_r$$
$$= 22 \text{ m/s east} + 16 \text{ m/s east}$$
$$= 38 \text{ m/s east}$$

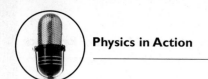

d) With respect to shore, the rock's velocity is now 6 m/s east.

Since the directions are opposite, the relative velocity is the difference between the two.

$$v = v_b - v_r$$

$$= 22 \text{ m/s east} - 16 \text{ m/s west}$$

$$= 6 \text{ m/s east}$$

Sample Problem 2

A quarterback on a football team is getting ready to throw a pass. If he is moving backward at 1.5 m/s and he throws the ball forward at 10.0 m/s, what is the velocity of the ball relative to the ground?

Strategy: Use a negative sign to indicate the backward direction. Add the two velocities to find the velocity relative to the ground.

Givens:

$$\xleftarrow{\hspace{3cm}} \xrightarrow{\hspace{6cm}}$$
−1.5 m/s   10.0 m/s

Solution:

Add the velocities.

$$10.0 \text{ m/s} + (-1.5 \text{ m/s}) = 8.5 \text{ m/s}$$

The ball is moving forward at 8.5 m/s relative to the ground.

Reflecting on the Activity and the Challenge

Running starts can be observed in many sports. Many observers may not realize the important role that inertia plays in preserving the speed already established when an athlete engages in activities such as jumping, throwing, or skating from a running start. "Immovable objects," such as football linemen, illustrate the tendency of highly massive objects to remain at rest and can be observed in many sports. You should have no problem finding a great variety of video segments that illustrate Newton's First Law.

Physics To Go

1. Provide three illustrations of Newton's First Law in sporting events. Describe the sporting event and which object when at rest stays at rest, or when in motion stays in motion. Describe these same three illustrations in the manner of an entertaining sportscaster.

2. Find out about a sport called curling (it is an Olympic competition that involves some of the oldest Olympians) and how this sport could be used to illustrate Newton's First Law of Motion.

3. When a skater glides across the ice on only one skate, what kind of motion does the skater have? Use principles of physics as evidence for your answer.

4. Use what you have learned in **Activity 1** to describe the motion of a hockey puck between the instant the puck leaves a player's stick and the instant it hits something. (No "slap shot" allowed; the puck must remain in contact with the ice.)

5. Why do baseball players often slide into second base and third base, but almost never slide into first base after hitting the ball? (The answer depends on both the rules of baseball and the laws of physics.)

6. Do you think it is possible to arrange conditions in the "real world" to have an object move, unassisted, in a straight line at constant speed forever? Explain why or why not.

7. You are pulling your little brother in his red wagon. He has a ball, and he throws it straight up into the air while you are pulling him forward at a constant speed.

   a) What will the path of the ball look like to your little brother in the wagon?
   b) What will the path of the ball look like to a little girl who is standing on the sidewalk watching you?
   c) If your brother throws the ball forward at a velocity of 2.5 m/s while you are pulling the wagon at a velocity of 4.5 m/s, at what speed does the girl see the ball go by?

8. A track and field athlete is running forward with a javelin at a velocity of 4.2 m/s. If he throws the javelin at a velocity relative to him of 10.3 m/s, what is the velocity of the javelin relative to the ground?

9. You are riding the train to school. Since the train car is almost empty, you and your friend are throwing a ball back and forth. The train is moving at a velocity of 5.6 m/s. Suppose you throw the ball to each other at the same speed, 2.4 m/s.

   a) What is the velocity of the ball relative to the tracks when the ball is moving toward the front of the car?
   b) What is the velocity of the ball relative to the tracks when it is moving toward the back of the car?
   c) What if you and your friend throw the ball perpendicular to the aisle of the train? What is the ball's velocity then?

10. Two athletes are running toward the pole vault. One is running at 3.8 m/s and the other is running at 4.3 m/s.

    a) What is their velocity relative to each other?
    b) If they leave the ground at their respective velocities, which one has the energy to go higher in the vault? Explain.

11. While riding a horse, a competitor shoots an arrow toward a target. The speed of the arrow as it reaches the target is 85 m/s. If the horse was traveling at 18 m/s, at what speed did the arrow leave the bow? (Assume the horse and arrow are traveling in the same direction.)

Activity 2   Push or Pull—Adding Vectors

What Do You Think?

Moving a football one yard to score a touchdown requires strategy, timing, and many forces.

- **What is a force?**
- **Can the same force move a bowling ball and a ping-pong ball?**

Record your ideas about these questions in your *Active Physics* log. Be prepared to discuss your responses with your small group and the class.

For You To Do

1. Use a flexible ruler as a "force meter". Use washers to make a scale of measurement for (that is, to calibrate) the meter inpennyweights. The force you are using to calibrate the meter is gravity, the force with which Earth pulls downward

GOALS

In this activity you will:

- Recognize that a force is a push or a pull.
- Identify the forces acting on an object.
- Determine when the forces on an object are either balanced or unbalanced.
- Calibrate a force meter in arbitrary units.
- Use a force meter to apply measured amounts of force to objects.
- Compare amounts of acceleration semiquantitatively.
- Understand and apply Newton's Second Law of Motion.
- Understand and apply the definition of the newton as a unit of force.
- Understand weight as a special application of Newton's Second Law.

on every object near its surface. *Carefully* clamp the plastic strip into position as shown in the diagram on the previous page.

2. Draw a line on a piece of paper. Hold the paper next to the plastic strip so that the line is even with the edge of the strip. Mark the position of the end of the strip on the reference line and label the position as the "zero" mark.

3. Place one washer on the top surface of the strip near the strip's outside end. Notice that the strip bends downward and then stops. Hold the paper in the original position and mark the new position of the end of the strip. Label the mark as "1 pennyweight."

4. Repeat **Step 3** for two, three, and four coins placed on the strip. In each case mark and label the new position of the end of the strip.

   a) Copy the reference line and the calibration marks from the piece of paper into your log.

5. Practice holding one end of the "force meter" (plastic strip) in your hand and pushing the free end against an object until you can bend the strip by forces of 1, 2, 3, and 4-pennyweight amounts. To become good at this, you will need to check the amount of bend in the strip against your calibration marks as you practice.

6. Use the force meter to push an object such as a tennis ball with a continuous 1-pennyweight force. You will need to keep up with the object as it moves and to keep the proper bend in the force meter. You may need to practice a few times to be able to do this.

   a) In your log, record the amount of force used, a description of the object, and the kind of motion the object seemed to have.

7. Repeat **Step 6** three more times, pushing on the same object with steady (constant) 2, 3, and 4-pennyweight amounts of force.

   a) Record the results in your log for each amount of force.

8. Based on your observations, complete the statement: "The greater the constant, unbalanced force pushing on an object,..."

✍ a) Write the completed statement in your log.

9. Select an object that has a small mass. Use the force meter to push on the object with a rather large, steady force such as 3- or 4-pennyweight amounts.

✍ a) Record the amount of force used, a description of the object pushed (especially including its mass, compared to the other objects to be pushed) and the kind of motion the object seemed to have.

10. Repeat **Step 9** using the same amount of force to push objects of greater and greater mass.

✍ a) Record the results in your log for each object.

11. Based on your observations, complete the statement: "When equal amounts of constant, unbalanced force are used to push objects having different masses, the more massive object..."

✍ a) Write the completed statement in your log.

Coordinated Science for the 21st Century

Physics Words

Newton's Second Law of Motion: if a body is acted upon by an external force, it will accelerate in the direction of the unbalanced force with an acceleration proportional to the force and inversely proportional to the mass.

weight: the vertical, downward force exerted on a mass as a result of gravity.

PHYSICS TALK

Newton's Second Law of Motion

Based on observations from experiments similar to yours, Isaac Newton wrote his **Second Law of Motion**:

The acceleration of an object is directly proportional to the unbalanced force acting on it and is inversely proportional to the object's mass. The direction of the acceleration is the same as the direction of the unbalanced force.

If 1 N (newton) is defined as the amount of unbalanced force that will cause a 1-kg mass to accelerate at 1 m/s^2 (meter per second every second), the law can be written as an equation:

$$F = ma$$

where F is expressed in newtons (symbol N), mass is expressed in kilograms (kg), and acceleration is expressed in meters per second every second (m/s^2).

By definition, the unit "newton" can be written in its equivalent form: $(\text{kg})\text{m/s}^2$.

Newton's Second Law can be arranged in three possible forms:

$$F = ma \qquad a = \frac{F}{m} \qquad m = \frac{F}{a}$$

FOR YOU TO READ

Weight and Newton's Second Law

Newton's Second Law explains what "weight" means, and how to measure it. If an object having a mass of 1 kg is dropped, its free fall acceleration is roughly 10 m/s^2.

Using Newton's Second Law,

$$F = ma$$

the force acting on the falling mass can be calculated as

$$F = ma$$
$$= 1 \text{ kg} \times 10 \text{ m/s}^2 \text{ or } 10 \text{ N}$$

The 10-N force causing the acceleration is known to be the gravitational pull of Earth on the 1-kg object. This gravitational force is given the special name **weight**. Therefore, it is correct to say, "The weight of a 1-kg mass is ten newtons."

What is the weight of a 2-kg mass? If dropped, a 2-kg mass also would accelerate due to gravity (as do all objects in free fall) at about 10 m/s^2. Therefore, according to Newton's Second Law, the weight of a 2-kg mass is equal to

$$2 \text{ kg} \times 10 \text{ m/s}^2 \text{ or } 20 \text{ N}$$

In general, to calculate the numerical value of an object's weight in newtons, it is necessary only to multiply the numerical value of its mass by the numerical value of the g (acceleration due to gravity), which is about 10m/s^2.

$$\text{Weight} = mg$$

The preceding equation is the "special case" of Newton's Second Law that must be applied to any situation in which the force causing an object to accelerate is Earth's gravitational pull.

Where There's Acceleration, There Must Be an Unbalanced Force

There are lots of different everyday forces. You just read about the force of gravity. There is also the force of a spring, the force of a rubber band, the force of a magnet, the force of your hand, the force of a bat hitting a ball, the force of friction, the buoyant force of water, and many more. Newton's Second Law tells you that accelerations are caused by unbalanced external forces. It doesn't matter what kind of force it is or how it originated. If you observe an acceleration (a change in velocity), then there must be an unbalanced force causing it.

When you apply a force, if the object has a small mass, the acceleration may be quite large for a given force. If the object has a large mass, the acceleration will be smaller for the same applied force. Occasionally, the mass is so large that you are not even able to measure the acceleration because it is so small.

If you push on a go-cart with the largest force possible, the cart will accelerate a great deal. If you push on a car with that same force, you

21

will measure a much smaller acceleration. If you were to push on the Earth, the acceleration would be too small to measure. Can you convince someone that a push on the Earth moves the Earth? Why should you believe something that you can't measure? If you were to assume that the Earth does not accelerate when you push on it, then you would have to believe that Newton's Second Law stops working when the mass gets too big. If that were so, you would want to determine how big is "too big." When you conduct such experiments, you find that the acceleration gets less and less as the mass gets larger and larger. Eventually, the acceleration gets so small that it is difficult to measure. Your inability to measure it doesn't mean that it is zero. It just means that it is smaller than your best measurement. In this way, you can assume that Newton's Second Law is always valid.

All of these statements are summarized in Newton's Second Law as you read in **Physics Talk**:

$$F = ma$$

or in forms that emphasize the acceleration and the mass

$$a = \frac{F}{m} \text{ and } m = \frac{F}{a}$$

Sample Problem 1

A tennis racket hits a ball with a force of 150 N. While the 275-g ball is in contact with the racket, what is its acceleration?

Strategy: Newton's Second Law relates the force acting on an object, the mass of the object, and the acceleration given to it by the force. Use the form of the equation that emphasizes acceleration to find the acceleration. The force unit, the newton, is defined as the amount of force needed to give a mass of 1.0 kg an acceleration of 1.0 m/s^2. Therefore, you will need to change the grams to kilograms.

Givens:
$$F = 150.0 \text{ N}$$
$$m = 275 \text{ g}$$

Solution:
$$275 \text{ g} = 0.275 \text{ kg}$$
$$a = \frac{F}{m}$$
$$= \frac{150 \text{ N}}{0.275 \text{ kg}}$$
$$= 545 \text{ m/s}^2$$

Sample Problem 2

As the result of a serve, a tennis ball ($m_t = 58$ g) accelerates at 43 m/s^2.

a) What force is responsible for this acceleration?

b) Could an identical force accelerate a 5.0-kg bowling ball at the same rate?

Strategy: Newton's Second Law states that the acceleration of an object is directly proportional to the applied force and indirectly proportional to the mass ($F = ma$).

Givens:
$$a = 43 \text{ m/s}^2$$
$$m_t = 58 \text{ g} = 0.058 \text{ kg}$$
$$m_b = 5.0 \text{ kg}$$

Solution:
a)
$$F = m_t a$$
$$= 0.058 \text{ kg} \times 43 \text{ m/s}^2$$
$$= 2.494 \text{ N or } 2.5 \text{ N}$$

b) Since the mass of the bowling ball is much greater than that of the tennis ball, an identical force will result in a smaller acceleration.
(You can calculate the acceleration.)

$$a = \frac{F}{m_b}$$

$$= \frac{2.5\ N}{5.0\ kg}$$

$$= 0.50\ m/s^2$$

Adding Vectors

A vector is a quantity that has both magnitude and direction. Velocity is a vector. In the previous activity you found that the direction in which an object was traveling and the speed at which it was moving are equally important.

Force is also a vector because you can measure how big it is (its magnitude) and its direction. Acceleration is also a vector. The equation for acceleration reminds you that the force and the acceleration must be in the same direction.

Often, more than one force acts on an object. If the two forces are in the same direction, the sum of the forces is simply the addition of the two forces. A 30-N force by one person and a force of 40 N by a second person (pushing in the same direction) on the same desk provides a 70-N force on the desk. If the two forces are in opposite directions, then you give one of the forces a negative value and add them again. If one student pushes on a desk to the right with a force of 30 N and a second student pushes on the same desk to the left with a force of 40 N, the net force on the desk will be 10 N to the left. Mathematically, you would state

that 30 N + (−40 N) = − 10 N where the negative sign denotes "to the left."

30 N
40 N
30 N + 40 N = 70 N

30 N   −40 N
30 N + (−40 N) = −10 N

Occasionally, the two forces acting on an object are at right angles. For instance, one student may be kicking a soccer ball with a force of 30 N ahead toward the goal, while the second student kicks the same soccer ball with a force of 40 N toward the sideline. To find the net force on the ball and the direction the ball will travel, you must use vector addition. You can do this by using a vector diagram or the Pythagorean Theorem.

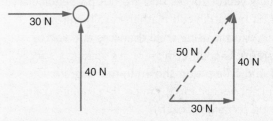

In the vector diagram shown above, the two force vectors are shown as arrows acting on the soccer ball. The magnitudes of the vectors are drawn to scale. The 30-N force may be drawn as 3.0 cm and the 40-N force may be drawn as 4.0 cm, if the scale is 10 N = 1 cm. To add the vectors, slide them so that the tip of the 30-N vector can be placed next to the tail of the 40-N vector (tip to tail method). The sum of the two vectors is then drawn from the tail of the 30-N vector to the tip of the 40-N vector. This *resultant* vector is measured and is found to be 5.0 cm, which is equivalent to 50 N. The angle is measured with a protractor and is found to be 53°.

Coordinated Science for the 21st Century

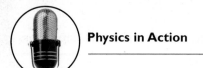
A second method of finding the resultant vector is to recognize that the 30-N and 40-N force vectors form a right triangle. The resultant is the hypotenuse of this triangle. Its length can be found by the Pythagorean Theorem.

$$a^2 + b^2 = c^2$$
$$30 \text{ N}^2 + 40 \text{ N}^2 = c^2$$
$$900 \text{ N}^2 + 1600 \text{ N}^2 = c^2$$
$$c = \sqrt{2500} \text{ N}^2$$
$$c = 50 \text{ N}$$

The angle can be found by using the tangent function.

$$\tan \theta = \frac{\text{opposite}}{\text{adjacent}} = \frac{40 \text{ N}}{30 \text{ N}} = 1.33$$
$$\theta = 53°$$

Adding vector forces that are not perpendicular is a bit more difficult mathematically, but no more difficult using scale drawings and vector diagrams. Two other players are kicking a soccer ball in the direction shown in the diagram. The resultant vector force can be determined using the tip to tail approach.

The two arrows in the left diagram correspond to the two players kicking the ball at different angles. The diagram at the right shows the two vectors being added "tip to tail." The resultant vector (shown as a dotted line) represents the net force and is the direction of the acceleration of the soccer ball.

Sample Problem 3

One player applies a force of 125 N in a north direction. Another player pushes with a force of 125 N west. What is the magnitude and direction of the resultant force?

Strategy: Since the forces are acting at right angles, you can use the Pythagorean Theorem to find the resultant force. The direction of the force can be found using the tangent function.

Givens:
$$F_1 = 125 \text{ N}$$
$$F_2 = 125 \text{ N}$$

Solution:

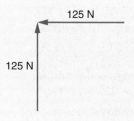

125 N

125 N

$$F_R^2 = F_1^2 + F_2^2$$
$$F_R = \sqrt{125 \text{ N}^2 + 125 \text{ N}^2}$$
$$= \sqrt{31{,}250 \text{ N}^2}$$
$$= 177 \text{ N}$$
$$\tan \theta = \frac{\text{opposite}}{\text{adjacent}} = \frac{125 \text{ N}}{125 \text{ N}} = 1$$
$$\theta = 45°$$

The resultant force is 177 N, 45° west of north.

Reflecting on the Activity and the Challenge

What you learned in this activity really increases the possibilities for interpreting sports events in terms of physics. Now you can explain why accelerations occur in terms of the masses and forces involved. You know that forces produce accelerations. Therefore, if you see an acceleration occur, you know to look for the forces involved. You can apply this to the sport that you will describe.

Also, you can explain, in terms of mass and weight, why gravity has no "favorite" athletes; in every case of free fall in sports, g has the same value, about 10 m/s^2.

Physics To Go

1. Copy and complete the following table using Newton's Second Law of Motion. Be sure to include the unit of measurement for each missing item.

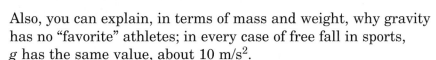

Newton's Second Law:	F	=	m	×	a
Sprinter beginning 100-meter dash	?		70 kg		5 m/s^2
Long jumper in flight	800 N		?		10 m/s^2
Shot put ball in flight	70 N		7 kg		?
Ski jumper going down hill before jumping	400 N		?		5 m/s^2
Hockey player "shaving ice" while stopping	−1500 N		100 kg		?
Running back being tackled	?		100 kg		−30 m/s^2

2. The following items refer to the table in **Question 1**:

   a) In which cases in the table does the acceleration match "g," the acceleration due to gravity 10 m/s^2?
   Are the matches to g coincidences or not? Explain.

   b) The force on the hockey player stopping is given in the table as a negative value. Should the player's acceleration also be negative? What do you think it means for a force or an acceleration to be negative?

   c) The acceleration of the running back being tackled also is given as negative. Should the unbalanced force acting on him also be negative? Explain.

d) In your mind, "play" an imagined video clip that illustrates the event represented by each horizontal row of the preceding table. Write a brief voice-over script for each video clip that explains how Newton's Second Law of Motion is operating in the event. Use appropriate physics terms, equations, numbers, and units of measurement in the scripts.

3. What is the acceleration of a 0.30-kg volleyball when a player uses a force of 42 N to spike the ball?

4. What force would be needed to accelerate a 0.040-kg golf ball at 20.0 m/s^2?

5. Most people can throw a baseball farther than a bowling ball, and most people would find it less painful to catch a flying baseball than a bowling ball flying at the same speed as the baseball. Explain these two apparent facts in terms of:
a) Newton's First Law of Motion.
b) Newton's Second Law of Motion.

6. Calculate the weight of a new fast-food sandwich that has a mass of 0.1 kg. Think of a clever name for the sandwich that would incorporate its weight.

7. In the United States, people measure body weight in pounds. Write down the weight, in pounds, of a person who is known to you. (This could be your weight or someone else's.)
a) Convert the person's weight in pounds to the international unit of force, newtons. To do so, use the following conversion equation:
Weight in newtons = Weight in pounds × 4.38 newtons/pound
b) Use the person's body weight, in newtons, and the equation
$$\text{Weight} = mg$$
to calculate the person's body mass, in kilograms.

8. Imagine a sled (such as a bobsled or luge used in Olympic competitions) sliding down a 45° slope of extremely slippery ice. Assume there is no friction or air resistance (not really possible). Even under such ideal conditions, it is a fact that gravity could cause the sled to accelerate at a maximum of only 7.1 m/s^2. Why would the "ideal" acceleration of the sled not be g, 10 m/s^2? Your answer is expected only to suggest reasons why, on a 45° hill, the ideal free fall acceleration is "diluted" from 10 m/s^2 to about 7 m/s^2; you are not expected to give a complete explanation of why the "dilution" occurs.

26

9. If you were doing the voice-over for a tug-of-war, how would you explain what was happening? Write a few sentences as if you were the science narrator of that athletic event.

10. You throw a ball. When the ball is many meters away from you, is the force of your hand still acting on the ball?

11. Carlo and Sara push on a desk in the same direction. Carlo pushes with a force of 50 N, and Sara pushes with a force of 40 N. What is the total resultant force acting on the desk?

12. A car is stuck in the mud. Four adults each push on the back of the car with a force of 200 N. What is the total force on the car?

13. During a football game, two players try to tackle another player. One player applies a force of 50.0 N to the east. A second player applies a force of 120.0 N to the north. What is the total applied force? (Since force is a vector, you must give both the magnitude and direction of the force.)

14. In auto racing, a crash occurs. A red car hits a blue car from the front with a force of 4000 N. A yellow car also hits the blue car from the side with a force of 5000 N. What is the total force on the blue car? (Since force is a vector, you must give both the magnitude and direction of the force.)

15. A baseball player throws a ball. While the 700.0-g ball is in the pitcher's hand, there is a force of 125 N on it. What is the acceleration of the ball?

16. If the acceleration due to gravity at the surface of the Earth is approximately 9.8 m/s^2, what force does the gravitational attraction of the Earth exert on a 12.8-kg object?

17. A force of 30.0 N acts on an object. At right angles to this force, another force of 40.0 N acts on the same object.

a) What is the net force on the object?
b) What acceleration would this give a 5.6-kg wagon?

18. Bob exerts a 30.0 N force to the left on a box ($m = 100.0$ kg). Carol exerts a 20.0 N force on the same box, perpendicular to Bob's.

a) What is the net force on the box?
b) Determine the acceleration of the box.
c) At what rate would the box accelerate if both forces were to the left?

Activity 3   Center of Mass

GOALS

In this activity you will:

- Locate the center of mass of oddly shaped two-dimensional objects.

- infer the location of the center of mass of symmetrical three-dimensional objects.

- Measure the approximate location of the center of mass of body.

- Understand that the entire mass of an object may be thought of as being located at the object's center of mass.

What Do You Think?

The center of mass of a high jumper using the "Fosbury Flop" (arched back) technique passes below the bar as the jumper's body successfully passes over the bar.

- **What is "center of mass"? What does it mean?**
- **Where is your body's center of mass?**

Record your ideas about these questions in your *Active Physics* log. Be prepared to discuss your responses with your small group and the class.

For You To Do

1. You will be provided four objects made from thin sheets of material in the shapes shown below.

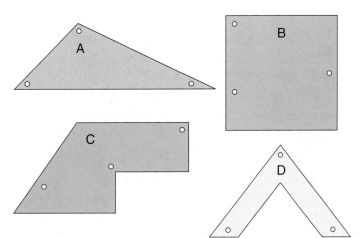

✎ a) Make a sketch in your log to show each shape in a reduced scale of size.

2. Use your intuition and trial and error to locate a point on the flat surfaces of objects *A*, *B*, and *C* where the object will balance on your fingertip. Mark the "balance point" of each object using a nonpermanent method.

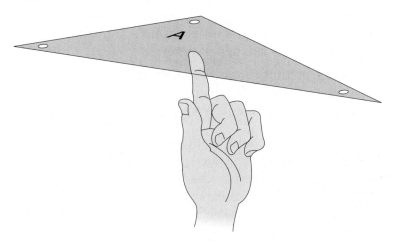

✎ a) Mark the balance points on the sketches in your log.

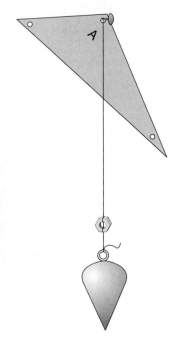

3. To check on the balance points found by the above method for objects A, B and C, use one of the small holes in object A to hang it from a pin as shown, and, also as shown, hang a "plumb bob" (a weight on a string) from the same pin.

a) Does the string pass over the balance point you marked for object A when you used your finger to balance the object? Should this happen? Write why you think it should or should not happen in your log.

b) Use a different hole in object A to suspend it from the pin and again hang the plumb bob from the pin. Does the string pass over the balance point marked before? Should it? Write your responses in your log.

4. The intersection of the two lines made where the string passed over the surface of object A could have been used to predict the balance point without first trying to balance the object on your finger. Use the suspension and plumb bob method to check the correspondence of the two methods of finding the balance point for objects B and C.

a) Record your findings about how well the two methods agree in your log.

5. Locate an "imaginary" balance point for object D. Tape a lightweight piece of paper between the "open arms" of object D and suspend the object and the plumb bob from the pin. Trace the path of the string across the piece of paper. Suspend the object from a different hole and trace the path of the string across the paper again.

a) Do you agree that the intersection of the two lines on the paper mark the balance point of object D? What is special about this balance point? Write your answers in your log.

6. The above "balance points" that you found for two-dimensional, or "flat," objects A, B, C, and D were, in each case, the location of the object's "center of mass." Do a "thought experiment" (an experiment in your mind) to determine the location of the center of mass of each of the following objects:

Shot put ball (solid steel)   Basketball
Banana   Planet Earth
Baseball bat   Hockey stick

🖎 a) For each object, describe in your log how you decided upon the location of the center of mass.

Physics Words

center of mass: the point at which all the mass of an object is considered to be concentrated for calculations concerning motion of the object.

7. The technique that was used to find the center of mass (C of M) relied on the fact that the C of M always lies beneath the point of support when an object is hanging. Similarly, when an object is balanced, the C of M is always above the point of support. To find your C of M, carefully balance on one foot and then the other. Try to keep your arms and legs in roughly identical positions as you shift your weight. Your C of M is located where a vertical meter stick from one foot and the other intersect. Locate this point. The actual C of M is inside your body, since nobody has zero thickness.

🖎 a) Record the location of your C of M.

8. Your teacher will balance a hammer on a finger to locate the hammer's C of M and make an obvious mark on the hammer at the C of M. As your teacher drops the hammer into a catch box on the floor, and it twists and turns, notice the movement of the C of M.

🖎 a) How does the movement of the C of M compare to the motion of the entire hammer?

Reflecting on the Activity and the Challenge

The **center of mass** is an important concept in any sports activity. The motion of the center of mass of a diver or gymnast is much easier to observe than the movements of the entire body. The sure-fire way of having a football player fall is to move his center of mass away from his support.

Think about the possibilities for using a transparent plastic cover on a TV monitor and using a pen to trace the motion of the center of mass of an athlete executing a free fall jump or dive. This could be used to simulate the light-pen technique used by TV commentators when they comment on football replays. This would seem a good way to add an interesting feature to your TV sports commentary.

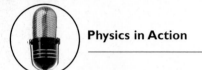

Physics To Go

1. When applying a force to make an object move, why is it most effective to have the applied force "aimed" directly at the object's center of mass?

2. "Center of gravity" means essentially the same thing as "center of mass." Why is it often said to be desirable for football players to have a low center of gravity?

3. Stand next to a wall facing parallel to the wall. With your right arm at your side pushing against the wall and with the right edge of your right foot against the wall at floor level, try to remain standing as you lift your left foot. Why is this impossible to do?

4. Think of positions for the human body for which the center of mass might be located outside the body. Describe each position and where you think the center of mass would be located relative to the body for each position.

5. An object tends to rotate (spin) if it is pushed on by a force that is not aimed at the center of mass. How do athletes use this fact to initiate spins before they fly through the air, as in gymnastics, skating, and diving events?

6. Could the suspension technique for finding the center of mass used in **For You To Do** be adapted to locate the center of mass of a three-dimensional object? If you had a crane that you could use to suspend an automobile from various points of attachment, how could you locate the auto's center of mass?

7. Find the center of mass of a baseball bat using the technique that you learned in class.

8. Carefully balance a light object (not too massive) over a table or catch box. Notice that the C of M is directly over the point of support. Move the support a little bit. Explain how this technique can be adapted to tackling in football.

9. Cut out a piece of cardboard in the shape of your state or a country. Find the geographic center of mass of your shape.

Activity 4   Defy Gravity

GOALS

In this activity you will:

- Measure changes in height of the body's center of mass during a vertical jump.

- Calculate changes in the gravitational potential energy of the body's center of mass during a vertical jump.

- Understand and apply the definition of work.

- Recognize that work is equivalent to energy.

- Understand and apply the joule as a unit of work and energy using equivalent forms of the joule.

- Apply conservation of work and energy to the analysis of a vertical jump, including weight, force, height, and time of flight.

What Do You Think?

No athlete can escape the pull of gravity.

- **Does the "hang time" of some athletes defy the above fact?**

- **Does a world-class skater defy gravity to remain in the air long enough to do a triple axel?**

Record your ideas about these questions in your *Active Physics* log. Be prepared to discuss your responses with your small group and the class.

For You To Do

1. Your teacher will show you a slow-motion video of a world-class figure skater doing a triple axel jump. The image of the skater will appear to "jerk," because a video camera completes one "frame," or one complete picture, every $\frac{1}{30}$ s. When the video is played at normal speed, you perceive the action as continuous; played at slow motion, the individual frames can be detected and counted. The duration of each frame is $\frac{1}{30}$ s.

33

a) Count and record in your log the number of frames during which the skater is in the air.

b) Calculate the skater's "hang time." (Show your calculation in your log.)

$$\text{Time in air (s)} = \text{Number of frames} \times \frac{1}{30}\,\text{s}$$

c) Did the skater "hang" in the air during any part of the jump, appearing to "defy gravity"? If necessary, view the slow-motion sequence again to make the observations necessary to answer this question in your log. If your observations indicate that hanging did occur, be sure to indicate the exact frames during which it happened.

2. Your teacher will show you a similar slow-motion video of a basketball player whose hang time is believed by many fans to clearly defy gravity.

a) Using the same method as above for the skater, show in your log the data and calculations used to determine the player's hang time during the "slam dunk."

b) Did the player hang? Cite evidence from the video in your answer.

3. How much force and energy does a person use to do a vertical jump? A person uses body muscles to "launch" the body into the air, and, primarily, it is leg muscles that provide the force. First, analyze only the part of jumping that happens before the feet leave the ground. Find your body mass, in kilograms, and your body weight, in newtons, for later calculations. If you wish not to use data for your own body, you may use the data for another person who is willing to share the information with you. (See **Activity 2**, **Physics To Go**, **Question 7**, for how to convert your body weight in pounds to weight in newtons and mass in kilograms.)

a) Record your weight, in newtons, and mass, in kilograms, in your log.

4. Recall the location of your body's center of mass from **Activity 3**. Place a patch of tape on either the right or left side of your clothing (above one hip) at the same level as your body's center of mass. Crouch as if you are ready to make a vertical jump. While crouched, have an assistant measure the vertical distance, in meters, from the floor to the level of your body's center of mass (C of M).

Ready position

✎ a) In your log, record the distance, in meters, from the floor to your C of M in the ready position.

5. Straighten your body and rise to your tiptoes as if you are ready to leave the floor to launch your body into a vertical jump, but don't jump. Hold this launch position while an assistant measures the vertical distance from the floor to the level of your center of mass.

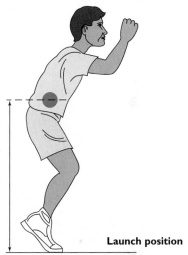

✎ a) In your log, record the distance, in meters, from the floor to your C of M in the launch position.

✎ b) By subtraction, calculate and record the vertical height through which you used your leg muscles to provide the force to lift your center of mass from the "ready" position to the "launch" position. Record this in your log as legwork height.

Launch position

Legwork height = Launch position − Ready position

6. Now it's time to jump! Have an assistant ready to observe and measure the vertical height from the floor to the level of your center of mass at the peak of your jump. When your assistant is ready to observe, jump straight up as high as you can. (Can you hang at the peak of your jump for a while to make it easier for your assistant to observe the position of your center of mass? Try it, and see if your assistant thinks you are successful.)

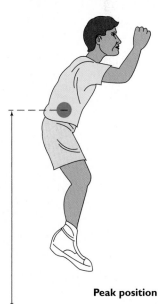

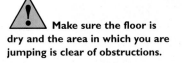 **Make sure the floor is dry and the area in which you are jumping is clear of obstructions.**

Peak position

Coordinated Science for the 21st Century

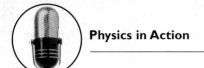
Physics Words

work: the product of the displacement and the force in the direction of the displacement; work is a scalar quantity.

potential energy: energy that is dependent on the position of the object.

kinetic energy: the energy an object possesses because of its motion.

a) In your log, record the distance from the floor to C of M at peak position.

b) By subtraction, calculate and record the vertical height through which your center of mass moved during the jump.

Jump height = Peak position – Launch position

7. The information needed to analyze the muscular force and energy used to accomplish your jump—and an example of how to use sample data from a student's jump to perform the analysis—is presented in **Physics Talk** and the Example Analysis.

a) Use the information presented in the **Physics Talk** and Example Analysis sections and the data collected during above **Steps 4** through **6** to calculate the hang time and the total force provided by *your* leg muscles during your vertical jump. Show as much detail in your log as is shown in the Example Analysis.

8. An ultrasonic ranging device coupled to a computer or graphing calculator, which can be used to monitor position, speed, acceleration, and time for moving objects, may be available at your school. If so, it could be used to monitor a person doing a vertical jump. This would provide interesting information to compare to the data and analysis that you already have for the vertical jump. Check with your teacher to see if this would be possible.

FOR YOU TO READ

Conservation of Energy

In this activity you jumped and measured your vertical leap. You went through a chain of energy conversions where the total energy remained the same, in the absence of air resistance. You began by lifting your body from the crouched "ready" position to the "launch" position. The **work** that you did was equal to the product of the applied force and the distance. The work done must have lifted you from the ready position to the launch position (an increase in **potential energy**) and also provided you with the speed to continue moving up (the **kinetic energy**). After you left

the ground, your body's potential energy continued to increase, and the kinetic energy decreased. Finally, you reached the peak of your jump, where all of the energy became potential energy. On the way down, that potential energy began to decrease and the kinetic energy began to increase.

When you are in the ready position, you have elastic potential energy. As you move toward the launch position, you have exchanged your elastic potential energy for an increase in gravitational potential energy and an increase in kinetic energy. As you rise in the air, you lose the kinetic energy and gain more gravitational potential energy. You can show this in a table.

Energy → Position ↓	Elastic potential energy	Gravitational potential energy = mgh	Kinetic energy = $\frac{1}{2}mv^2$
ready position	maximum	0	0
launch position	0	some	maximum
peak position	0	maximum	0

The energy of the three positions must be equal. In this first table, the sum of the energies in each row must be equal. The launch position has both gravitational potential energy and kinetic energy. Using the values in the activity, the total energy at each position is 410 J.

Energy → Position ↓	Elastic potential energy	Gravitational potential energy = mgh	Kinetic energy = $\frac{1}{2}mv^2$
ready position	410 J	0	0
launch position	0	150 J	260 J
peak position	0	410 J	0

In the ready position, all 410 J is elastic potential energy. In the peak position, all 410 J is gravitational potential energy. In the launch position, the total energy is still 410 J but 150 J is gravitational potential energy and 260 J is kinetic energy.

Consider someone the same size, who can jump much higher. Since that person can jump much higher, the peak position is greater, and therefore the gravitational potential energy of the jumper is greater. In the example shown below, the gravitational potential energy is 600 J. Notice that this means the elastic potential energy of the jumper's legs must be 600 J. And when the jumper is in the launch position, the total energy (potential plus kinetic) is also 600 J.

Energy → Position ↓	Elastic potential energy	Gravitational potential energy = mgh	Kinetic energy = $\frac{1}{2}mv^2$
ready position	600 J	0	0
launch position	0	150 J	450 J
peak position	0	600 J	0

A third person of the same size is not able to jump as high. What numbers should be placed in blank areas to preserve the principle of conservation of energy?

Total energy must be conserved. Therefore, in the launch position the kinetic energy of the jumper must be 50 J. In the peak position, all the energy is in potential energy and must be 200 J.

The conservation of energy is a unifying principle in all science. It is worthwhile to practice solving problems that will help you to see the variety of ways in which energy conservation appears.

Energy → Position ↓	Elastic potential energy	Gravitational potential energy = mgh	Kinetic energy = $\frac{1}{2}mv^2$
ready position	200 J	0	0
launch position	0	150 J	50 J
peak position	0	200 J	0

A similar example to jumping from a hard floor into the air is jumping on a trampoline (or your bed, when you were younger). If you were to jump on the trampoline, the potential energy from the height you are jumping would provide kinetic energy when you landed on the trampoline. As you continued down, you would continue to gain speed because you would still be losing gravitational potential energy. The trampoline bends and/or the springs holding the trampoline stretch. Either way, the trampoline or springs gain elastic potential energy at the expense of the kinetic energy and the changes in potential energy.

Energy → Position ↓	Elastic potential energy	Gravitational potential energy = mgh	Kinetic energy $= \frac{1}{2}mv^2$
High in the air position	0	2300 J	0
Landing on the trampoline position	0	500 J	1800 J
Lowest point on the trampoline position	2300 J	0	0

A pole-vaulter runs with the pole. The pole bends. The pole straightens and pushes the vaulter into the air. The vaulter gets to his highest point, goes over the bar, and then falls back to the ground, where he lands on a soft mattress. You can analyze the pole-vaulter's motion in terms of energy conservation. (Ignore air resistance.)

A pole-vaulter runs with the pole. (*The vaulter has kinetic energy.*) The pole bends. (*The vaulter loses kinetic energy, and the pole gains elastic potential energy as it bends.*) The pole unbends and pushes the vaulter into the air. (*The pole loses the elastic potential energy, and the vaulter gains kinetic energy and gravitational potential energy.*) The vaulter gets to his highest point (*the vaulter has almost all gravitational potential energy*) goes over the bar, and then falls back to the ground (*the gravitational potential energy becomes kinetic energy*), where he lands on a soft mattress (*the kinetic energy becomes the elastic potential energy of the mattress, which then turns to heat energy*). The height the pole-vaulter can reach is dependent on the total energy that he starts with. The faster he runs, the higher he can go.

The conservation of energy is one of the great discoveries of science. You can describe the energies in words (elastic potential energy, gravitational potential energy, kinetic energy, and heat energy). There is also sound energy, →

light energy, chemical energy, electrical energy, and nuclear energy. The words do not give the complete picture. Each type of energy can be measured and calculated. In a closed system, the total of all the energies at any one time must equal the total of all the energies at any other time. That is what is meant by the conservation of energy.

If you choose to look at one object in the system, that one object can gain energy. For example, in the collision between a player's foot and a soccer ball, the soccer ball can gain kinetic energy and move faster. Whatever energy the ball gained, you can be sure that the foot lost an equal amount of energy. The ball gained energy, the foot lost energy, and the "ball and foot" total energy remained the same. The ball gained energy because work (force x distance) was done on it. The foot lost energy because work (force x distance) was done on it. The total system of "ball and foot" neither gained nor lost energy.

Physics provides you with the means to calculate energies. You may wish to practice some of these calculations now. Never lose sight of the fact that you can calculate the energies because the sum of all of the energies remains the same.

The equations for work, gravitational potential energy, and kinetic energy are given below.

The equation for work is:

$$W = F \cdot d$$

Work is done only when the force and displacement are (at least partially) in the same (or opposite) directions.

The equation for gravitational potential energy is:

$$PE_{gravitational} = mgh = wh$$

The w represents the weight of the object in newtons, where $w = mg$. On Earth's surface, when dealing with g in this course, consider it to be equal to 9.8 m/s^2.(Sometimes we use 10 m/s^2 for ease of calculations.)

The equation for kinetic energy is:

$$KE = \frac{1}{2}mv^2$$

Sample Problem

A trainer lifts a 5.0-kg equipment bag from the floor to the shelf of a locker. The locker is 1.6 m off the floor.

a) How much force will be required to lift the bag off the floor?

b) How much work will be done in lifting the bag to the shelf?

c) How much potential energy does the bag have as it sits on the shelf?

d) If the bag falls off the shelf, how fast will it be going when it hits the floor?

Strategy: This problem has several parts. It may look complicated, but if you follow it step by step, it should not be difficult to solve.

Part (a): Why does it take a force to lift the bag? It takes a force because the trainer must act against the pull of the gravitational field of the Earth. This force is called weight, and you can solve for it using Newton's Second Law.

Part (b): The information you need to find the work done on an object is the force exerted on it and the distance it travels. The distance was given and you calculated the force needed. Use the equation for work.

Part (c): The amount of potential energy depends on the mass of the object, the acceleration due to gravity, and the height of the object above what is designated as zero height (in this case, the floor). You have all the needed pieces of information, so you can apply the equation for potential energy.

Part (d): The bag has some potential energy. When it falls off the shelf, the potential energy becomes kinetic energy as it falls. When it strikes the ground in its fall, it has zero potential energy and all kinetic energy. You calculated the potential energy. Conservation of energy tells you that the kinetic energy will be equal to the potential energy. You know the mass of the bag so you can calculate the velocity with the kinetic energy formula.

Givens:
$m = 5.0$ kg
$h = 1.6$ m
$a = 9.8$ m/s^2

Solution:

a)
$$F = ma$$
$$= (5.0 \text{ kg})(9.8 \text{ m/s}^2)$$
$$= 49 \text{ kg} \cdot \text{m/s}^2 \text{ or } 49 \text{ N}$$

b)
$$W = F \cdot d$$
$$= (49 \text{ N})(1.6 \text{ m})$$
$$= 78.4 \text{ Nm or } 78 \text{ J (Nm = J)}$$

c)
$$PE_{gravitational} = mgh$$
$$= (5.0 \text{ kg})(9.8 \text{ m/s}^2)(1.6 \text{ m})$$
$$= 78 \text{ J}$$

Should you be surprised that this is the same answer as **Part (b)**? No, because you are familiar with energy conservation. You know that the work is what gave the bag the potential energy it has. So, in the absence of work that may be converted to heat because of friction, which you did not have in this case, the work equals the potential energy.

d)
$$KE = \frac{1}{2}mv^2$$
$$v^2 = \frac{KE}{\frac{1}{2}m}$$
$$= \frac{78 \text{ J}}{\frac{1}{2}(5.0 \text{ kg})}$$
$$= 31 \text{ m}^2/\text{s}^2$$
$$v = 5.6 \text{ m/s}$$

PHYSICS TALK

Work

When you lifted your body from the ready (crouched) position to the launch (standing on tiptoes) position before takeoff during the vertical jump activity, you performed what physicists call work. In the context of physics, the word *work* is defined as:

The work done when a constant force is applied to move an object is equal to the amount of applied force multiplied by the distance through which the object moves in the direction of the force.

You used symbols to write the definition of work as:

$$W = F \cdot d$$

where F is the applied force in newtons, d is the distance the object moves in meters, and work is expressed in joules (symbol, J). At any time it is desired, the unit "joule" can be written in its equivalent form as force times distance, "(N)(m)."

The unit "newton" can be written in the equivalent form "(kg)m/s^2." Therefore, the unit joule also can be written in the equivalent form (kg)m^2/s^2. In summary, the units for expressing work are:

$$1 \text{ J} = 1 \text{ (N)(m)} = 1 \text{ (kg)m}^2/\text{s}^2$$

As you read, it is very common in sports that work is transformed into kinetic energy, and then, in turn, the kinetic energy is transformed into gravitational potential energy. This chain of transformations can be written as:

$$\text{Work} = KE = PE$$

$$Fd = \frac{1}{2}mv^2 = mgh$$

These transformations are used in the analysis of data for a vertical jump.

42

Example:
Calculation of Hang Time and Force During Vertical Jump

DATA: Body Weight = 100 pounds = 440 N
       Body Mass = 44 kg
       Legwork Height = 0.35 m
       Jump Height = 0.60 m

Analysis:
Work done to lift the center of mass from ready position to launch position without jumping ($W_{R\,to\,L}$):

$$W_{R\,to\,L} = Fd = (\text{Body Weight}) \times (\text{Legwork Height})$$
$$= 440 \text{ N} \times 0.35 \text{ m} = 150 \text{ J}$$

Gravitational Potential Energy gained from jumping from launch position to peak position (PE_J):

$$PE_J = mgh = (\text{Body Mass}) \times (g) \times (\text{Jump Height})$$
$$= 44 \text{ kg} \times 10 \text{ m/s}^2 \times 0.60 \text{ m}$$
$$= 260 \text{ (kg)m}^2/\text{s}^2 = 260 \text{ (N)(m)} = 260 \text{ J}$$

The jumper's kinetic energy at takeoff was transformed to increase the potential energy of the jumper's center of mass by 260 J from launch position to peak position. Conservation of energy demands that the kinetic energy at launch be 260 J:

$$KE = \frac{1}{2}mv^2 = 260 \text{ J}$$

This allows calculation of the jumper's launch speed:

$$v = \sqrt{2(KE)/m} = \sqrt{2(260 \text{ J})/(44 \text{ kg})} = 3.4 \text{ m/s}$$

From the definition of acceleration, $a = \Delta v/\Delta t$, the jumper's time of flight "one way" during the jump was:

$$\Delta t = \Delta v/a = (3.4 \text{ m/s}) / (10 \text{ m/s}^2) = 0.34 \text{ s}$$
Therefore, the total time in the air (hang time) was 2×0.34 s = 0.68 s.

→

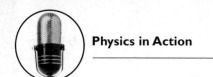
The total work done by the jumper's leg muscles before launch, W_T, was the work done to lift the center of mass from ready position to launch position without jumping, $W_{R\ to\ L}$ = 150 J, plus the amount of work done to provide the center of mass with 260 J of kinetic energy at launch, a total of 150 J + 260 J = 410 J. Rearranging the equation $W = F \cdot d$ into the form $F = W/d$, the total force provided by the jumper's leg muscles, F_T was:

$$F_T = \frac{W_T}{(\text{Legwork Height})}$$

$$= 410 \text{ J} / 0.35 \text{ m}$$

$$= 1200 \text{ N}$$

Approximately one-third of the total force exerted by the jumper's leg muscles was used to lift the jumper's center of mass to the launch position, and approximately two-thirds of the force was used to accelerate the jumper's center of mass to the launch speed.

Reflecting on the Activity and the Challenge

Work, the force applied by an athlete to cause an object to move (including the athlete's own body as the object in some cases), multiplied by the distance the object moves while the athlete is applying the force explains many things in sports. For example, the vertical speed of any jumper's takeoff (which determines height and "hang time") is determined by the amount of work done against gravity by the jumper's muscles before takeoff. You will be able to find many other examples of work in action in sports videos, and now you will be able to explain them.

Physics To Go

1. How much work does a male figure skater do when lifting a 50-kg female skating partner's body a vertical distance of 1 m in a pairs competition?

2. Describe the energy transformations during a bobsled run, beginning with team members pushing to start the sled and ending when the brake is applied to stop the sled after crossing the finish line. Include work as one form of energy in your answer.

3. Suppose that a person who saw the video of the basketball player used in **For You To Do** said, "He really can hang in the air. I've seen him do it. Maybe he was just having a 'bad hang day' when the video was taken, or maybe the speed of the camera or VCR was 'off.' How do I know that the player in the video wasn't a 'look-alike' who can't hang?" Do you think these are legitimate statements and questions? Why or why not?

4. If someone claims that a law of physics can be defied or violated, should the person making the claim need to provide observable evidence that the claim is true, or should someone else need to prove that the claim is not true? Who do you think should have the burden of proof? Discuss this issue within your group and write your own personal opinion in your log.

5. Identify and discuss two ways in which an athlete can increase his or her maximum vertical jump height.

6. Calculate the amount of work, in joules, done when:
   a) a 1.0-N weight is lifted a vertical distance of 1.0 m.
   b) a 1.0-N weight is lifted a vertical distance of 10 m.
   c) a 10-N weight is lifted a vertical distance of 1.0 m.
   d) a 0.10-N weight is lifted a vertical distance of 100 m.
   e) a 100-N weight is lifted a distance of 0.10 m.

7. List how much gravitational potential energy, in joules, each of the weights in **Question 6** above would have when lifted to the height listed for it.

45

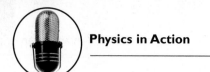

8. List how much kinetic energy, in joules, each of the weights in **Questions 6** and **7** would have at the instant before striking the ground if each weight were dropped from the height listed for it.

9. How much work is done on a go-cart if you push it with a force of 50.0 N and move it a distance of 43 m?

10. What is the kinetic energy of a 62-kg cyclist if she is moving on her bicycle at 8.2 m/s?

11. A net force of 30.00 N acts on a 5.00-kg wagon that is initially at rest.

    a) What is the acceleration of the wagon?
    b) If the wagon travels 18.75 m, what is the work done on the wagon?

12. Assume you do 40,000 J of work by applying a force of 3200 N to a 1200-kg car.

    a) How far will the car move?
    b) What is the acceleration of the car?

13. A baseball ($m = 150.0$ g) is traveling at 40.0 m/s. How much work must be done to stop the ball?

14. A boat exerts a force of 417 N pulling a water-skier ($m = 64.0$ kg) from rest. The skier's speed is now 15.0 m/s. Over what distance was this force exerted?

Activity 5   Run and Jump

What Do You Think?

The men's high jump record is over 8 feet.

• Pretend that you have just met somebody who has never jumped before. What instructions could you provide to get the person to jump up (that is, which way do you apply the force)?

Record your ideas about this question in your *Active Physics* log. Be prepared to discuss your responses with your small group and the class.

For You To Do

1. Carefully stand on a skateboard or sit on a wheeled chair near a wall. By touching only the wall, not the floor, cause yourself to move away from the wall to "coast" across the floor. Use words and diagrams to record answers to the following questions in your log:

a) When is your motion accelerated? For what distance does the accelerated motion last? In what direction do you accelerate?

GOALS

In this activity you will:

- Understand the definition of acceleration.
- Understand meters per second per second as the unit of acceleration.
- Use an accelerometer to detect acceleration.
- Use an accelerometer to make semiquantitative comparisons of accelerations.
- Distinguish between acceleration and deceleration.

Physics Words

Newton's Third Law of Motion: forces come in pairs; the force of object A on object B is equal and opposite to the force of object B on object A.

b) When is your motion at constant speed? Neglecting the effects of friction, how far should you travel? (Remember Galileo's Principle of Inertia when answering this question.)

c) Newton's Second Law, $F = ma$, says that a force must be active when acceleration occurs. What is the source of the force, the push or pull, that causes you to accelerate in this case? Identify the object that does the pushing on your mass (body plus skateboard) to cause the acceleration. Also identify the direction of the push that causes you to accelerate.

d) Obviously, you do some pushing, too. On what object do you push? In what direction?

e) How do you think, on the basis of both amount and direction, the following two forces compare?
   • The force exerted by you on the wall
   • The force exerted by the wall on you

2. Do a "thought experiment" about the forces involved when you are running or walking on a horizontal surface. Use words and sketches to answer the following questions in your log:

a) With each step, you push the bottom surface of your shoe, the sole, horizontally backward. The force acts parallel to the surface of the ground, trying to scrape the ground in the direction opposite your motion. Usually, friction is enough to prevent your shoe from sliding across the ground surface.

b) Since you move forward, not rearward, there must be a force in the forward direction that causes you to accelerate. Identify where the forward force comes from, and compare its amount and direction to the rearward force exerted by your shoe with each step.

c) Would it be possible to walk or run on an extremely slippery skating rink when wearing ordinary shoes? Discuss why or why not in terms of forces.

3. Think about the vertical forces acting on you while you are standing on the floor.

a) Copy the diagram of a person at left in your log.

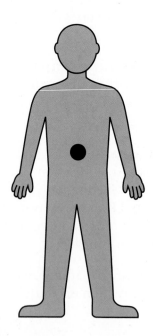

b) Identify all the vertical forces. Use an arrow to designate the size and direction of the force. Draw the forces from the dot.

c) How can you be sure that the force with which you push on the floor and the floor pushes on you are equal?

4. Set up a meter stick with a few books for support as shown.

5. Place a washer in the center of the meter stick.

a) In your log, record what happens.

6. Remove the washer and replace it with 100 g (weight of 100 g = 1.0 N). Continue to place 1.0 N weights on the center of the meter stick. Note what happens as you place each weight on the stick.

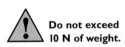
Do not exceed 10 N of weight.

a) Measure the deflection of the meter stick for each 1.0 N of weight and record the values for these deflections.

b) How does the deflection of the meter stick compare to the weight it is supporting? In your log, sketch a graph to show this relationship.

c) Write a concluding statement concerning the washer and the deflection of the meter stick.

PHYSICS TALK

Newton's Third Law of Motion

Newton's **Third Law of Motion** can be stated as:

For every applied force, there is an equal and opposite force.

If you push or pull on something, that something pushes or pulls back on you with an equal amount of force in the opposite direction. This is an inescapable fact; it happens every time.

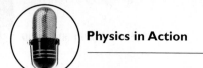

Reflecting on the Activity and the Challenge

According to Newton's Third Law, each time an athlete acts to exert a force on something, an equal and opposite force happens in return. Countless examples of this exist as possibilities to include in your video production. When you kick a soccer ball, the soccer ball exerts a force on your foot. When you push backward on the ground, the ground pushes forward on you (and you move). When a boxer's fist exerts a force on another boxer's body, the body exerts an equal force on the fist. Indeed, it should be rather easy to find a video sequence of a sport that illustrates all three of Newton's Laws of Motion.

Physics To Go

1. When an athlete is preparing to throw a shot put ball, does the ball exert a force on the athlete's hand equal and opposite to the force the hand exerts on the ball?

2. When you sit on a chair, the seat of the chair pushes up on your body with a force equal and opposite to your weight. How does the chair know exactly how hard to push up on you—are chairs intelligent?

3. For a hit in baseball, compare the force exerted by the bat on the ball to the force exerted by the ball on the bat. Why do bats sometimes break?

4. Compare the amount of force experienced by each football player when a big linebacker tackles a small running back.

5. Identify the forces active when a hockey player "hits the boards" at the side of the rink at high speed.

6. Newton's Second Law, $F = ma$, suggests that when catching a baseball in your hand, a great amount of force is required to stop a high-speed baseball in a very short time interval. The great amount of force is needed to provide the great amount of deceleration required. Use Newton's Third Law to explain why baseball players prefer to wear gloves for catching high-speed baseballs. Use a pair of forces in your explanation.

7. Write a sentence or two explaining the physics of an imaginary sports clip using Newton's Third Law. How can you make this description more exciting so that it can be used as part of your sports voice-over?

8. Write a sentence or two explaining the concept that a deflection of the ground can produce a force. How can you make this description more exciting so that it can be used as part of your sports voice-over?

Stretching Exercises

Ask the manager of a building that has an elevator for permission to use the elevator for a physics experiment. Your teacher may be able to help you make the necessary arrangements.

1. Stand on a bathroom scale in the elevator and record the force indicated by the scale while the elevator is:

   a) At rest.
   b) Beginning to move (accelerating) upward.
   c) Seeming to move upward at constant speed.
   d) Beginning to stop (decelerating) while moving upward.
   e) Beginning to move (accelerating) downward.
   f) Seeming to move downward at constant speed.
   g) Beginning to stop (decelerating) while moving downward.

2. For each of the above conditions of the elevator's motion, the Earth's downward force of gravity is the same. If you are accelerating up, the floor must be pushing up with a force larger than the acceleration due to gravity.

   a) Make a sketch that shows the vertical forces acting on your body.
   b) Use Newton's Laws of Motion to explain how the forces acting on your body are responsible for the kind of motion—at rest, constant speed, acceleration, or deceleration—that your body has.

Activity 6   The Mu of the Shoe

GOALS

In this activity you will:

- Understand and apply the definition of the coefficient of sliding friction, μ.

- Measure the coefficient of sliding friction between the soles of athletic shoes and a variety of floor surface materials.

- Calculate the effects of frictional forces on the motion of objects.

What Do You Think?

A shoe store may sell as many as 100 different kinds of sport shoes.

- **Why do some sports require special shoes?**

Record your ideas about this question in your *Active Physics* log. Be prepared to discuss your responses with your small group and the class.

For You To Do

1. Take an athletic shoe. Use a spring scale to measure the weight of the shoe, in newtons.

   a) Record a description of the shoe (such as its brand) and the shoe's weight, in your log.

2. Place the shoe on one of two horizontal surfaces (either rough or smooth) designated by your teacher to be used for testing. Attach the spring scale to the shoe as shown below so that the spring scale can be used to slide the shoe across the surface while, at the same time, the amount of force indicated by the scale can be read.

a) Record in your log a description of the surface on which the shoe is to slide.

b) Measure and record the amount of force, in newtons, needed to cause the shoe to slide on the surface at constant speed. Do not measure the force needed to start, or "tear the shoe loose," from rest. Measure the force needed, after the shoe has started moving, to keep it sliding at low, constant speed. Also, be careful to pull horizontally so that the applied force neither tends to lift the shoe nor pull downward on the shoe.

c) Use the data you have gathered to calculate μ, the coefficient of sliding friction for this particular kind of shoe on the particular kind of surface used. Show your calculations in your log.

The coefficient of sliding friction, symbolized by μ, is calculated using the following equation:

$$\mu = \frac{\text{force required to slide object on surface at constant speed}}{\text{perpendicular force exerted by the surface on the object}}$$

Example:

Brand X athletic shoe has a weight of 5 N. If 1.5 N of applied horizontal force is required to cause the shoe to slide with constant speed on a smooth concrete floor, what is the coefficient of sliding friction?

$$\mu_{x \text{ on concrete}} = \frac{1.5 \text{ N}}{5.0 \text{ N}} = 0.30$$

---

3. Add "filler" to the shoe to approximately double its weight and repeat the above procedure for measuring the μ of the shoe.

   a) Calculate μ for this surface, showing your work in your log.

   b) Taking into account possible errors of measurement, does the weight of the shoe seem to affect μ? Use data to answer the question in your log.

   c) How do you think the weight of an athlete wearing the shoe would affect μ? Why?

4. Place the shoe on the second surface designated by your teacher and repeat the procedure.

   a) Make another sketch to show the forces acting on the shoe.

   b) Calculate μ.

   c) How does the value of μ for this surface compare to μ for the first surface used? Try to explain any difference in μ.

   d) Would it make any difference if you used the empty shoe or the shoe with the filler to calculate μ in this activity? Explain your answer.

Reflecting on the Activity and the Challenge

Many athletes seem more concerned about their shoes than most other items of equipment, and for good reason. Small differences in the shoes (or skates or skis) athletes wear can affect performance. As everyone knows, athletic shoes have become a major industry because people in all "walks" of life have discovered that athletic shoes are great to wear, not only on a track but, as well, just about anywhere. Now that you have studied friction, a major aspect of what makes shoes function well when need exists to be "sure-footed," you are prepared to do "physics commentary" on athletic footgear and other effects of friction in sports. Your sports commentary may discuss the μ of the shoe, the change in friction when a playing field gets wet, and the need for friction when running.

PHYSICS TALK

Coefficient of Sliding Friction, μ

There are not enough letters in the English alphabet to provide the number of symbols needed in physics, so letters from another alphabet, the Greek alphabet, also are used as symbols. The letter μ, pronounced "mu," traditionally is used in physics as the symbol for the "coefficient of sliding friction."

The coefficient of sliding friction, symbolized by μ, is defined as the ratio of two forces:

$$\mu = \frac{\text{force required to slide object on surface at constant speed}}{\text{perpendicular force exerted by the surface on the object}}$$

Facts about the coefficient of sliding friction:

- **μ does not have any units because it is a force divided by a force; it has no unit of measurement.**

- **μ usually is expressed in decimal form, such as 0.85 for rubber on dry concrete (0.60 on wet concrete).**

- **μ is valid only for the pair of surfaces in contact when the value is measured; any significant change in either of the surfaces (such as the kind of material, surface texture, moisture, or lubrication on a surface, etc.) may cause the value of μ to change.**

- **Only when sliding occurs on a horizontal surface, and the pulling force is horizontal, is the perpendicular force that the sliding object exerts on the surface equal to the weight of the object.**

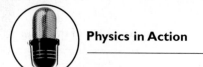

Physics To Go

1. Identify a sport and changing weather conditions that probably would cause an athlete to want to increase friction to have better footing. Name the sport, describe the change in conditions, and explain what the athlete might do to increase friction between the shoes and ground surface.

2. Identify a sport in which athletes desire to have frictional forces as small as possible and describe what the athletes do to reduce friction.

3. If a basketball player's shoes provide an amount of friction that is "just right" when she plays on her home court, can she be sure the same shoes will provide the same amount of friction when playing on another court? Explain why or why not.

4. A cross-country skier who weighs 600 N has chosen ski wax that provides $\mu = 0.03$. What is the minimum amount of horizontal force that would keep the skier moving at constant speed across level snow?

5. A racecar having a mass of 1000 kg was traveling at high speed on a wet concrete road under foggy conditions. The tires on the vehicle later were measured to have $\mu = 0.55$ on that road surface. Before colliding with the guardrail, the driver locked the brakes and skidded 100 m, leaving visible marks on the road. The driver claimed not to have been exceeding 65 miles per hour (29 m/s). Use the equation:

$$\text{Work} = \text{Kinetic Energy}$$

to estimate the driver's speed upon hitting the brakes. (Hint: In this case, the force that did the work to stop the car was the frictional force; calculate the frictional force using the weight of the vehicle, in newtons, and use the frictional force as the force for calculating work.)

6. Identify at least three examples of sports in which air or water have limiting effects on motion similar to sliding friction. Do you think forces of "air resistance" and "water resistance" remain constant or do they change as the speeds of objects (such as athletes, bobsleds, or rowing sculls) moving through them change? Use examples from your own experience with these forms of resistance as a basis for your answer.

7. If there is a maximum frictional force between your shoe and the track, does that set a limit on how fast you can start (accelerate) in a sprint? Does that mean you cannot have more than a certain acceleration even if you have incredibly strong leg muscles? What is done to solve this problem?

8. How might an athletic shoe company use the results of your experiment to "sell" a shoe? Write copy for such an advertisement.

9. Explain why friction is important to running. Why are cleats used in football, soccer, and other sports?

10. Choose a sport and describe an event in which friction with the ground or the air plays a significant part. Create a voice-over or script that uses physics to explain the action.

Activity 7   Concentrating on Collisions

GOALS

In this activity you will:

- Understand and apply the definition of momentum.

- Conduct semiquantitative analyses of the momentum of pairs of objects involved in one-dimensional collisions.

- Infer the relative masses of two objects by staging and observing collisions between the objects.

⚠ **Moving bowling balls can cause injury to people and property. Be careful!**

What Do You Think?

In contact sports, very large forces happen during short time intervals.

- **A football player runs toward the goal line, and a defensive player tries to stop him with a head-on collision. What factors determine whether the offensive player scores?**

Record your ideas about this question in your *Active Physics* log. Be prepared to discuss your responses with your small group and the class.

For You To Do

1. You will stage a head-on collision between two matched bocci balls. Set up a launch ramp for one ball, and find a level area clear of obstructions nearby where the other bocci ball can be at rest.

2. Temporarily remove the "target" bocci ball. Find a point of release within the first one-fourth of the ramp's total length that gives the ball a slow, steady speed across the floor. Mark the point of release on the ramp with a piece of tape.

3. Replace the target ball. Adjust the aim until a good approximation of a head-on collision is obtained. Stage the collision.

   ✍ a) Record the results in your log. Use a diagram and words to describe what happened to each ball.

4. Repeat the above type of collision, but this time move the release point up the ramp to at least double the ramp distance.

   ✍ a) Describe the results in your log.

   ✍ b) How did the results of the collision change from the first time?

   ✍ c) Identify a real-life situation that this collision could represent.

5. Arrange another head-on collision between the balls, but this time have both balls moving at equal speeds before the collision. Using a second, identical ramp, aim the second ramp so that the second ball's path is aligned with the first ball's path. Mark a release point on the second ramp at a height equal to the mark already made on the first ramp. This should ensure that the balls will have low, approximately equal speeds. On a signal, two persons should release the balls simultaneously from equal ramp heights.

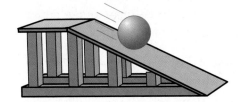 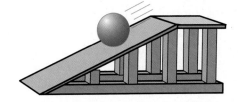

   ✍ a) Describe the results in your log.

   ✍ b) Identify a real-life situation that this collision could represent.

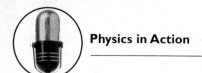
Physics Words

momentum: the product of the mass and the velocity of an object; momentum is a vector quantity.

6. Repeat **Steps 1, 2, and 3**, but replace the stationary bocci ball with a nerf ball.

a) Be sure to write all responses, including identification of a similar situation in real life, in your log.

7. Repeat **Steps 1, 2,** and **3**, but in this case have the nerf ball roll down the ramp to strike a stationary bocci ball.

a) Be sure to write all responses, including identification of a similar situation in real life, in your log.

8. Using your observations, determine the relative mass of a golf ball compared to a wiffle ball by staging collisions between them.

a) Which ball has the greater mass? How many times more massive is it than the other ball? Describe what you did to decide upon your answer.

b) Use a scale or balance to check your result. Comment on how well observing collisions between the balls worked as a method of comparing their masses.

FOR YOU TO READ

Momentum

Taken alone, neither the masses nor the velocities of the objects were important in determining the collisions you observed in this activity. The crucial quantity is **momentum** (mass × velocity). A soccer ball has less mass than a bocci ball, but a soccer ball can have the same momentum as a bocci ball if the soccer ball is moving fast. A soccer ball moving very fast can affect a stationary bocci ball more than a soccer ball moving very slowly. This is similar to the damage small pieces of sand moving at very high speeds can cause (such as when a sand blaster is used to clean various surfaces).

Sportscasters often use the term *momentum* in a different way. When a team is doing well, or "on a roll," that team has momentum.

A team can gain or lose momentum, depending on how things are going. This momentum clearly does not refer to the mass of the entire team multiplied by the team's velocity.

Other times, sportscasters use the term momentum to mean exactly how it is defined in the activity (mass × velocity), when they say things such as, "Her momentum carried her out of bounds."

Reflecting on the Activity and the Challenge

You already have identified several real-life situations that involve collisions, and many such situations happen in sports. Some involve athletes colliding with one another as in hockey and football. Others cases include athletes colliding with objects, such as when kicking a ball. Still others include collisions between objects such as a golf club, bat, or racquet with a ball. Some spectacular collisions in sports provide fun opportunities for demonstrating your knowledge about collisions during voice-over commentaries. Use the concept of momentum when describing collisions in your sports video.

Physics To Go

1. Sports commentators often say that a team has momentum when things are going well for the team. Explain the difference between that meaning of the word momentum and its specific meaning in physics.

2. Suppose a running back collides with a defending linebacker who has just come to a stop. If both players have the same mass, what do you expect to see happen in the resulting collision?

3. Describe the collision of a running back and a linebacker of equal mass running toward each other at equal speeds.

Coordinated Science for the 21st Century

4. Suppose that you have two baseball bats, a heavy (38-ounce) bat and a light (30-ounce) bat.

   a) If you were able to swing both bats at the same speed, which bat would allow you to hit the ball the farthest distance? Explain your answer.

   b) How fast would you need to swing the light bat to produce the same hitting effect as the heavy bat? Explain your answer.

5. Why do football teams prefer offensive and defensive linemen who weigh about 300 pounds?

6. What determines who will get knocked backward when a big hockey player checks a small player in a head-on collision?

7. A 100.0-kg athlete is running at 10.0 m/s. At what speed would a 0.10-kg ball need to travel in the same direction so that the momentum of the athlete and the momentum of the ball would be equal?

8. Use the words *mass*, *velocity*, and *momentum* to write a paragraph that gives a detailed "before and after" description of what happens when a moving shuffleboard puck hits a stationary puck of equal mass in a head-on collision.

9. Describe a collision in some sport by using the term *momentum*. Adapt this description to a 15-s dialogue that could be used as part of the voice-over for a video.

Activity 8

Conservation of Momentum

GOALS

In this activity you will:

- Understand and apply the Law of Conservation of Momentum.

- Measure the momentum before and after a moving mass strikes a stationary mass in a head-on, inelastic collision.

What Do You Think?

The outcome of a collision between two objects is predictable.

- **What determines the momentum of an object?**
- **What does it mean to "conserve" something?**

Record your ideas about these questions in your *Active Physics* log. Be prepared to discuss your responses with your small group and the class.

For You To Do

1. From the objects provided arrange to have a head-on collision between two objects of equal mass. Before the collision, have one object moving and the other object at rest. Arrange for the objects to stick together to move as a single object after the collision. Stage a head-on, sticky collision between equal masses. Measure the velocity, in meters per second, of the moving mass before the collision and the velocity of the combined masses after the collision.

63

a) Prepare a data table in your log similar to the one shown below. Provide enough horizontal rows in the table to enter data for at least four collisions.

Sticky Head-on Collisions: One Object Moving before Collision					
Mass of Object 1 (kg)	Mass of Object 2 (kg)	Velocity of Object 1 before Collision (m/s)	Velocity of Object 2 before Collision (m/s)	Mass of Combined Objects after Collision (kg)	Velocity of Combined Objects after Collision (m/s)
1.0	1.0		0.0	2.0	
2.0	1.0		0.0	3.0	
1.0	2.0		0.0	3.0	
			0.0		

b) Record the measured values of the velocities in the first row of the data table.

2. Stage other sticky head-on collisions using the masses listed in the second and third rows of the data table. Then stage one or more additional collisions using other masses. Measure the velocities before and after each collision.

a) Enter the measured values in the data table.

3. Organize a table for recording the momentum of each object before and after each of the above collisions.

a) Prepare a table similar to the following example in your log:

Momentum of Object before and after Collisions Momentum = Mass × Velocity		
Before the Collision		After the Collision
Momentum of Object 1 kg (m/s)	Momentum of Object 2 kg (m/s)	Momentum of Combined Objects 1 and 2 kg (m/s)

b) Calculate the momentum of each object before and after each of the above collisions and enter each momentum value in the table.

c) Calculate and compare the total momentum before each collision to the total momentum after each collision.

d) Allowing for minor variations due to errors of measurement, write in your log a general conclusion about how the momentum before a collision compares to the momentum afterward.

FOR YOU TO READ

The Law of Conservation of Momentum

In this activity, you investigated another conservation principle that is a hallmark of physics—the conservation of momentum. If you sum all of the momenta before a collision or explosion, you know that the sum of all the momenta after the collision will be the same.

If the momentum before a collision is 500 kg·m/s, then the momentum after the collision is 500 kg·m/s. A football player stops to catch a pass. The player is not moving and therefore has momentum equal to zero. If an opponent that has a momentum of 500 kg·m/s then hits the player, both players move off with (a combined) 500 kg·m/s of momentum. Any time you see a collision in sports, you can explain that collision using the conservation of momentum.

Conservation of momentum is an experimental fact. Physicists have compared the momenta before and after a collision between pairs of objects ranging from railroad cars slamming together to subatomic particles impacting one another at near the speed of light. Never have any exceptions been found to the statement, "The total momentum before a collision is equal to the total momentum after the collision if no external forces act on the system." This statement is known as the Law of Conservation of Momentum. In all collisions between cars and trucks, between protons and protons, between planets and meteors, the momentum before the collision equals the momentum after.

→

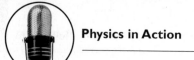

A single cue ball hits a rack of 15 billiard balls and they all scatter. It would seem like everything has changed. Physicists have discovered that in this collision, as in all collisions and explosions, nature does keep at least one thing from changing—the total momentum. The sum of the momenta of all of the billiard balls immediately after the collision is equal to the momentum of the original cue ball. Nature loves momentum. Irrespective of the changes you visually note, the total momentum undergoes no change whatsoever. The objects may move in new directions and with new speeds, but the momentum stays the same. There aren't many of these conservation laws that are known.

Conservation of momentum can be shown to emerge from Newton's Laws. Newton's Third Law states that if object A and object B collide, the force of object A on B must be equal and opposite to the force of object B on A.

$$F_{A \text{ on } B} = -F_{B \text{ on } A}$$

The negative sign shows mathematically that the equally sized forces are in opposite directions.

Newton's Second Law states that $F = ma$. Also, acceleration, a, equals the change in velocity divided by the change in time ($a = \Delta v / \Delta t$):

$$m_B a_B = -m_A \, a_A$$

$$m_B \frac{\Delta v_B}{\Delta t} = -m_A \frac{\Delta v_A}{\Delta t}$$

$$m_B \frac{(v_f - v_i)_B}{\Delta t} = -m_A \frac{(v_f - v_i)_A}{\Delta t}$$

Since the change in time must be the same for both objects (A acts on B for as long as B acts on A), then Δt can be eliminated from both sides of the equation.

Combining the initial velocities (v_i) on one side of the equation and the final velocities (v_f) on the other side of the equation:

$$m_A v_{iA} + m_B v_{iB} = m_A v_{fA} + m_B v_{fB}$$

$$(m_A v_A)_{before} + (m_B v_B)_{before} =$$
$$(m_A v_A)_{after} + (m_B v_B)_{after}$$

Newton's Laws have yielded the conservation of momentum. The momentum of object A *before* the collision plus the momentum of object B *before* the collision equals the momentum of object A *after* the collision plus the momentum of object B *after* the collision

This equation not only works in one-dimensional collisions, but works equally well in the extraordinarily complex two-dimensional collisions of billiard balls and three-dimensional collisions of bowling.

Solving conservation of momentum problems is easy. Calculate each object's momentum before the collision. Calculate each object's momentum after the collision. The totals before the collision must equal the total after the collision.

There are a variety of collisions involving two objects. In each collision, momentum is conserved and the same equation is used. The equation gets simpler when one of the objects is at rest and has zero momentum. You may wish to draw two sketches for each collision—one showing each object before the collision and one showing each object after the collision. By writing the momenta you know directly on the sketch, the calculations become easier.

Collision Type 1: One moving object hits a stationary object and both stick together and move off at the same speed:

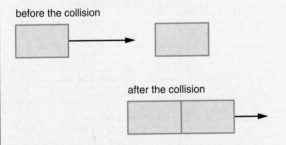

before the collision

after the collision

Collision Type 2: Two stationary objects explode and move off in opposite directions.

Collision Type 3: One moving object hits a stationary object. The first object stops, and the second object moves off.

Collision Type 4: One moving object hits a stationary object, and both move off at different speeds.

Collision Type 5: Two moving objects collide, and both objects move at different speeds after the collision.

Collision Type 6: Two moving objects collide, and both objects stick together and move off at the same speed.

Sample Problem 1

A 75.00-kg ice skater is moving to the east at 3.00 m/s toward his 50.00-kg partner, who is moving toward him (west) at 1.80 m/s. If he catches her up and they move away together, what is their final velocity?

Strategy: This is a problem involving the Law of Conservation of Momentum. The momentum of an isolated system before an interaction is equal to the momentum of the system after the interaction. As you are working through this problem, remember that the *v* in this expression is velocity and that it has direction as well as magnitude. Make east the positive direction, and then west will be negative.

Givens:
$$m_b = 75.00 \text{ kg}$$
$$m_g = 50.00 \text{ kg}$$
$$v_b = 3.00 \text{ m/s}$$
$$v_g = -1.80 \text{ m/s}$$

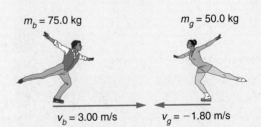

$m_b = 75.0$ kg $\qquad m_g = 50.0$ kg

$v_b = 3.00$ m/s $\qquad v_g = -1.80$ m/s

Solution:

$$(m_b v_b)_{before} + (m_g v_g)_{after} = [(m_b + m_g)v_{bg}]_{after}$$

(75.00 kg)(3.00 m/s) + (50.00 kg)(−1.80 m/s) =
(75.00 kg + 50.00 kg)v_{bg}

$$v_{bg} = \frac{225 \text{ kg·m/s} - 90.0 \text{ kg·m/s}}{125.00 \text{ kg}}$$

$$= 1.08 \text{ m/s}$$

Sample Problem 2

A steel ball with a mass of 2 kg is traveling at 3 m/s west. It collides with a stationary ball that has a mass of 1 kg. Upon collision, the smaller ball moves away to the west at 4 m/s. What is the velocity of the larger ball?

Strategy: Again, you will use the Law of Conservation of Momentum. Before the collision, only the larger ball has momentum. After the collision, the two balls move away at different velocities.

Givens:

before the collision

$m_1 = 2$ kg $m_2 = 1$ kg

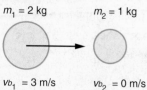

$v_{b_1} = 3$ m/s $v_{b_2} = 0$ m/s

after the collision

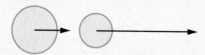

v_{a_1} after = ? m/s v_{a_2} after = 4 m/s

$m_1 = 2$ kg $v_{b_2} = 0$ m/s
$m_2 = 1$ kg $v_{a_2} = 4$ m/s
$v_{b_1} = 3$ m/s

Solution:

$$(m_1 v_1)_b + (m_2 v_2)_b = (m_1 v_1)_a + (m_2 v_2)_a$$

(2 kg)(3 m/s) + (1 kg)(0 m/s) =
      (2 kg)v_{a_1} + (1 kg)(4 m/s)

6 kg·m/s = (2v_{a_1}) kg + 4 kg·m/s

$v_{a_1} = 1$ m/s

Reflecting on the Activity and the Challenge

The Law of Conservation of Momentum is a very powerful tool for explaining collisions in sports and other areas. The law works even when, as often happens in sports, one of the objects involved in a collision "bounces back," reversing the direction of its velocity and, therefore, its momentum, as a result of a collision. When describing a collision between a bat and ball, or a collision between two people, you can describe how the total momentum is conserved.

Physics To Go

1. A railroad car of 2000 kg coasting at 3.0 m/s overtakes and locks together with an identical car coasting on the same track in the same direction at 2.0 m/s. What is the speed of the cars after they lock together?

2. In a hockey game, an 80.0-kg player skating at 10.0 m/s overtakes and bumps from behind a 100.0-kg player who is moving in the same direction at 8.00 m/s. As a result of being bumped from behind, the 100.0-kg player's speed increases to 9.78 m/s. What is the 80.0-kg player's velocity (speed and direction) after the bump?

3. A 3-kg hard steel ball collides head-on with a 1-kg hard steel ball. The balls are moving at 2 m/s in opposite directions before they collide. Upon colliding, the 3-kg ball stops. What is the velocity of the 1-kg object after the collision? (Hint: Assign velocities in one direction as positive; then any velocities in the opposite direction are negative.)

4. A 45-kg female figure skater and her 75-kg male skating partner begin their ice dancing performance standing at rest in face-to-face position with the palms of their hands touching. Cued by the start of their dance music, both skaters "push off" with their hands to move backward. If the female skater moves at 2.0 m/s relative to the ice, what is the velocity of the male skater? (Hint: The momentum before the skaters push off is zero.)

5. A 0.35-kg tennis racquet moving to the right at 20.0 m/s hits a 0.060-kg tennis ball that is moving to the left at 30.0 m/s. The racquet continues moving to the right after the collision, but at a reduced speed of 10.0 m/s. What is the velocity (speed and direction) of the tennis ball after it is hit by the racquet?

6. A stationary 3-kg hard steel ball is hit head-on by a 1-kg hard steel ball moving to the right at 4 m/s. After the collision, the 3-kg ball moves to the right at 2 m/s. What is the velocity (speed and direction) of the 1-kg ball after the collision? (Hint: Direction is important.)

7. A 90.00-kg hockey goalie, at rest in front of the goal, stops a puck ($m = 0.16$ kg) that is traveling at 30.00 m/s. At what speed do the goalie and puck travel after the save?

8. A 45.00-kg girl jumps from the side of a pool into a raft ($m = 0.08$ kg) floating on the surface of the water. She leaves the side at a speed of 1.10 m/s and lands on the raft. At what speed will the girl-raft system begin to travel across the pool?

9. Two cars collide head on. Initially, car A ($m = 1700.0$ kg) is traveling at 10.00 m/s north and car B is traveling at 25.00 m/s south. After the collision, car A reverses its direction and travels at 5.00 m/s while car B continues in its initial direction at a speed of 3.75 m/s. What is the mass of car B?

10. A proton ($m = 1.67 \times 10^{-27}$ kg) traveling at 2.50×10^5 m/s collides with an unknown particle initially at rest. After the collision the proton reverses direction and travels at 1.10×10^5 m/s. Determine the change in momentum of the unknown particle.

11. You shoot a 0.04-kg bullet moving at 200.0 m/s into a 20.00-kg block initially at rest on an icy pond.

   a) What is the velocity of the bullet-block combination?
   b) The coefficient of friction between the block and the ice is 0.15. How far would the block slide before coming to rest?

12. Write a 15- to 30-s voice-over that highlights the conservation of momentum in a sport of your choosing.

Activity 9   Circular Motion

GOALS

In this activity you will:

- Understand that a centripetal force is required to keep a mass moving in a circular path at constant speed.

- Understand that a centripetal acceleration accompanies a centripetal force, and that, at any instant, both the acceleration and force are directed toward the center of the circular path.

- Apply the equation for circular motion.

- Understand that centrifugal force is the reaction to centripetal force.

⚠ **To avoid becoming too dizzy, limit your spins while standing to about four.**

What Do You Think?

Racecars can make turns at 150 mph.

- **What forces act on a racecar when it moves along a circular path at constant speed on a flat, horizontal surface?**

Record your ideas about this question in your *Active Physics* log. Be prepared to discuss your responses with your small group and the class.

For You To Do

1. Hold an accelerometer in your hands and observe it as you either sit on a rotating stool or spin around while standing. What is the direction of the acceleration indicated by the accelerometer? (You can find out how the cork indicates acceleration by holding it and noting its behavior as you accelerate forward.)

71

a) Make a sketch in your log to simulate a snapshot photo taken from above as the accelerometer was moving along a circular path. Show the circular path, the accelerometer "frozen" at one instant, the cork "frozen" in leaning position, and an arrow to represent the velocity of the accelerometer at the instant represented by your sketch.

2. Review in your textbook and your log how you used a force meter to apply a constant force to objects to cause the objects to accelerate in **Activity 2**.

a) Based on the results of **Activity 2**, write a brief statement in your log that summarizes how the amount and direction of acceleration of an object depends on amount and direction of the force acting on the object.

3. Start a ball rolling across the floor. While it is rolling, catch up with the ball and use the force meter to push exactly sideways, or perpendicular, to the motion of the ball with a fixed amount of force. Carefully follow alongside the ball and, as will be necessary, keep adjusting the direction of push so that it is always perpendicular to the motion of the ball.

a) Make a top view sketch in your log that shows:

- a line to represent the straight-line path of the ball before you began pushing sideways on the ball

- a dashed line to represent the straight-line path on which the ball would have continued moving if you had not pushed sideways on it

- a line of appropriate shape to show the path taken by the ball as you pushed perpendicular to the direction of the ball's motion with a constant amount of force.

b) When you pushed on the ball exactly sideways to its motion, did you cause the ball to move either faster or slower? Explain your answer.

c) Assuming that friction could be eliminated to allow the ball to continue moving at constant speed, describe what you would need to do to make the ball keep moving on a circular path.

d) If you stop pushing on the ball, how does the ball move? Try it, and use a sketch and words in your log to describe what happens.

4. Review each of the items listed below. Copy each item into your log and write a statement to discuss how each item is related to an object moving along a circular path. If an item does not apply to circular motion, explain why.

Physics Words

centripetal acceleration: the inward radial acceleration of an object moving at a constant speed in a circle.

✎ a) Galileo's Principle of Inertia

✎ b) Newton's First Law of Motion

✎ c) Newton's Second Law of Motion

FOR YOU TO READ

The Unbalanced Force Required for Circular Motion

During the above activities you saw two things that are related by Newton's Second Law of Motion, $F = ma$. First, the accelerometer showed that when an object moves in a circular path there is an acceleration that at any instant is toward the center of the circle. This acceleration has a special name, **centripetal acceleration**. The word centripetal means "toward-the-center"; therefore, centripetal acceleration refers to acceleration toward the center of the circle when an object moves in a circular path.

You also saw that a centripetal force, a toward-the-center force, causes circular motion. When a centripetal force is applied to a moving object, the object's path curves; without the centripetal force, the object follows the tendency to move in a straight line. Therefore, a centripetal force, when applied, is an unbalanced force, meaning that it is not "balanced off" by another force.

Newton's Second Law seems to apply to circular motion just as well as it applies to accelerated motion along a straight line, but with a strange "twist." It is a clearly correct application of $F = ma$ to say that a centripetal force, F, causes a mass, m, to experience an acceleration, a. However, the strange part is that when an object moves along a circular path at constant speed, acceleration is happening with no change in the object's speed. The force changes the direction of the velocity.

Velocity describes both the amount of speed and the direction of motion of an object. Thinking about the velocity of an object moving with constant speed on a circular path, it is true that the velocity is changing from one instant to the next not in the amount of the velocity, but with respect to the direction of the velocity. The diagram shows an object moving at constant speed on a circular path. Arrows are used to represent the velocity of the object at several instants during one trip around the circle.

→

Physicists have shown that a special form of Newton's Second Law governs circular motion:

$$F_c = ma_c = \frac{mv^2}{r}$$

where F_c is the centripetal force in newtons, m is the mass of the object moving on the circular path in kilograms, a_c is the centripetal acceleration in m/s², v is the velocity in m/s, and r is the radius of the circular path in meters.

Sample Problem

Find the centripetal force required to cause a 1000.0-kg automobile travelling at 27.0 m/s (60 miles/hour) to turn on an unbanked curve having a radius of 100.0 m.

Strategy: This problem requires you to find centripetal force. You can use the equation that uses Newton's Second Law to calculate F_c.

Givens:

$m = 1000.0$ kg

$y = 27.0$ m/s

$r = 100.0$ m/s

Solution:

$$F_c = \frac{mv^2}{r}$$

$$= \frac{(1000.0 \text{ kg } (27.0 \text{ m/s})^2}{100.0 \text{ m}}$$

$$= \frac{(1000.0 \text{ kg} \times 730 \text{ m}^2/\text{s}^2)}{100.0 \text{ m}}$$

$$= 7300 \text{ N}$$

If the force of friction is less than the above amount, the car will not follow the curve and will skid in the direction in which it is travelling at the instant the tires "break loose."

Reflecting on the Activity and the Challenge

Both circular motion and motion along curved paths that are not parts of perfect circles are involved in many sports. For example, both the discus and hammer throw events in track and field involve rapid circular motion before launching a projectile. Track, speed skating, and automobile races are done on curved paths. Whenever an object or athlete is observed to move along a curved path, you can be sure that a force is acting to cause the change in direction. Now you are prepared to provide voice-over explanations of examples of motion along curved paths in sports, and in many cases you perhaps can estimate the amount of force involved.

Physics To Go

1. For the car used as the example in the **For You To Read**, what is the minimum value of the coefficient of sliding friction between the car tires and the road surface that will allow the car to go around the curve without skidding? (Hint: First calculate the weight of the car, in newtons.)

2. If you twirl an object on the end of a string, you, of course, must maintain an inward, centripetal force to keep the object moving in a circular path. You feel a force that seems to be pulling outward along the string toward the object. But the outward force that you detect, called the "centrifugal force," is only the reaction to the centripetal force that you are applying to the string. Contrary to what many people believe, there is no outward force acting on an object moving in a circular path. Explain why this must be true in terms of what happens if the string breaks while you are twirling an object.

3. A 50.0-kg jet pilot in level flight at a constant speed of 270.0 m/s (600 miles per hour) feels the seat of the airplane pushing up on her with a force equal to her normal weight, $50.0 \text{ kg} \times 10 \text{ m/s}^2 = 500 \text{ N}$. If she rolls the airplane on its side and executes a tight circular turn that has a radius of 1000.0 m, with how much force will the seat of the airplane push on her? How many "g's" (how many times her normal weight) will she experience?

4. Imagine a video segment of an athlete or an item of sporting equipment moving on a circular path in a sporting event. Estimate the mass, speed, and radius of the circle. Use the estimated values to calculate centripetal force and identify the source of the force.

5. Below are alternate explanations of the same event given by a person who was not wearing a seat belt when a car went around a sharp curve:

   a) "I was sitting near the middle of the front seat when the car turned sharply to the left. The centrifugal force made my body slide across the seat toward the right, outward from the center of the curve, and then my right shoulder slammed against the door on the passenger side of the car."

   b) "I was sitting near the middle of the front seat when the car turned sharply to the left. My body kept going in a straight line while, at the same time due to insufficient friction, the seat slid to the left beneath me, until the door on the passenger side of the car had moved far enough to the left to exert a centripetal force against my right shoulder."

   Are both explanations correct, or is one right and one wrong? Explain your answer in terms of both explanations.

6. People seem to be fascinated with having their bodies put in a state of circular motion. Describe an amusement park ride based on circular motion that you think is fun, and describe what happens to your body during the ride.

7. a) Explain why football players fall on a wet field while changing directions during a play.

   b) Include the concepts centripetal force and Newton's Laws in a revised explanation.

   c) In a new revision, make the explanation exciting enough to include in your sports video voice-over.

PHYSICS AT WORK

Dean Bell

TELEVISION PRODUCER USES SPORTS TO TEACH MATH AND PHYSICS

Dean Bell is an award-winning filmmaker and television writer, director, and producer. His show *Sports Figures* is a highly acclaimed ESPN educational television series designed to teach the principles of physics and mathematics through sports. His approach has been to tell a story, pose a problem, and then follow through with its mathematical and scientific explanation. "But always," he says, "you must make it fun. It has to be both educational and entertaining."

Dean began his career as a filmmaker after college. He landed the apprentice film editor's position on a Woody Allen film. From there, he worked his way up in the field, from assistant editor, to editor and finally writer, director, and producer.

"I've always been a fan of educational TV," he states, "although I never thought that was where my career would take me. It's one of life's little ironies that I've ended up producing this type of show. You see, my father worked in scientific optics and was very science oriented. He was always delighted in finding out how things worked, and was even on the Mr. Wizard TV show a few times."

Dean writes the script for each segment, working together with top educational science consultants. "We spend a day coming up with ideas and then researching each subject thoroughly. Our job is to illustrate the relationship between a sports situation and the related mathematical or physics principles."

"At the end of the day," says Dean, "it really is nice to be working on a show that means something and that is so worthwhile. I'm still getting ahead in my career as a film and TV producer, but now I'm also an educator."

Chapter 1 Assessment

Your big day has arrived. You will be meeting with the local television station to audition for a job as a "physics of sports" commentator. Whether you will get the job will be decided on the quality of your voice-over.

With what you learned in this chapter, you are ready to do your science commentary on a short sports video. Choose a videotape from a sports event, either a school event or a professional event. Each of you will be responsible for producing your own commentary, whether or not you worked in cooperative groups during the activities. You are not expected to give a play-by-play description, but rather describe the rules of nature that govern the event. Your viewers should come away with a different perspective of both sports and physics. You may produce one of the following:

- **a written script**
- **a live narrative**
- **a video soundtrack**
- **an audiocassette**

Review the criteria by which your voice-over dialogue or script will be evaluated. Your voice-over should:

- **use physics principles and terms correctly**
- **have entertainment value**

After reviewing the criteria, decide as a class the point value you will give to each of these criteria:

- **How important is the physics content? How many physics terms and principles should be illustrated to get the minimum credit? The maximum credit?**
- **What value would you place on the entertainment aspect? How do you fairly assess the excitement and interest of the broadcast?**

Physics You Learned

Galileo's Principle of Inertia

Newton's First Law of Motion

Newton's Second Law of Motion

Newton's Third Law of Motion

Weight

Center of mass; center of gravity

Friction between different surfaces

Momentum

Law of Conservation of Momentum

Centripetal acceleration

Centripetal force

Chapter 2

SAFETY

Scenario

Probably the most dangerous thing you will do today is travel to your destination. Transportation is necessary, but the need to get there in a hurry, and the large number of people and vehicles, have made transportation very risky. There is a greater chance of being killed or injured traveling than in any other common activity. Realizing this, people and governments have begun to take action to alter the statistics. New safety systems have been designed and put into use in automobiles and airplanes. New laws and a new awareness are working together with these systems to reduce the danger in traveling.

What are these new safety systems? You are probably familiar with many of them. In this chapter, you will become more familiar with most of these designs. Could you design or even build a better safety device for a car or a plane? Many students around the country have been doing just that, and with great success!

Challenge

Your design team will develop a safety system for protecting automobile, airplane, bicycle, motorcycle, or train passengers. As you study existing safety systems, you and your design team should be listing ideas for improving an existing system or designing a new system for preventing accidents. You may also consider a system that will minimize the harm caused by accidents.

Your final product will be a working model or prototype of a safety system. On the day that you bring the final product to class, the teams will display them around the room while class members informally view them and discuss them with members of the design team. During this time, class members will ask questions about each others products. The questions will be placed in envelopes provided to each team by the teacher. The teacher will use some of these questions during the oral presentations on the next day.

The product will be judged according to the following three parts:

1. The quality of your safety feature enhancement and the working model or prototype.

2. The quality of a five-minute oral report that should include:

   - **the need for the system**
   - **the method used to develop the working model**
   - **the demonstration of the working model**
   - **the discussion of the physics concepts involved**
   - **the description of the next-generation version of the system**
   - **the answers to questions posed by the class**

3. The quality of a written and/or multimedia report including:

   - **the information from the oral report**
   - **the documentation of the sources of expert information**
   - **the discussion of consumer acceptance and market potential**
   - **the discussion of the physics concepts applied in the design of the safety system**

Criteria

You and your classmates will work with your teacher to define the criteria for determining grades. You will also be asked to evaluate your own work. Discuss as a class the performance task and the points that should be allocated for each part. A starting point for your discussions may be:

- **Part 1 = 40 points**
- **Part 2 = 30 points**
- **Part 3 = 30 points**

Since group work is made up of individual work, your teacher will assign some points to each individual's contribution to the project. If individual points total 30 points, then parts 1, 2 and 3 must be changed so that the total remains at 100.

Activity 1 Response Time

GOALS

In this activity you will:

- Identify the parts of the process of stopping a car.
- Measure reaction time.
- Wire a series circuit.

What Do You Think?

Many deaths that occur on the highway are drivers and passengers in vehicles that did not cause the accident. The driver was not able to respond in time to avoid becoming a statistic.

• How long would it take you to respond to an emergency?

Record your ideas about this question in your *Active Physics* log. Be prepared to discuss your responses with your small group and the class.

For You To Do

1. To stop a car, you must move your foot from the gas pedal to the brake pedal. Try moving your right foot between imaginary pedals.

a Estimate how long it takes to move your foot between the imaginary pedals. Record your estimate.

2. The first step in stopping a car happens even before you move your foot to the brake. It takes time to see or hear something that tells you to move your foot. Test this by having a friend stand behind you and clap. When you hear the sound, move your foot between imaginary pedals.

a) Estimate how long it took you to respond to the loud noise. Record your estimate.

3. Create a simple electric circuit to test your response time. Your group will need a battery in a clip, two switches, a flashlight bulb in a socket, and connecting wires. Connect the wires from one terminal of the battery to the first switch, then to the second switch, to the light bulb, and back to the battery.

⚠ Have your teacher approve your circuit before proceeding to Step 4.

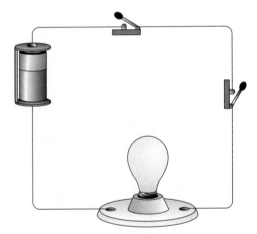

4. Close one switch while the other is open. Close the other switch. Take turns turning the light off and on with each person operating only one switch.

a) Record what happens in each case.

5. Try to keep the light on for exactly one second, then five seconds. You can estimate one second by saying "one thousand one."

a) How quickly do you think you can turn the light off after your partner turns it on? The time the bulb is lit is your response time. Record an estimate of your response time in your log.

83

6. Find your response time using the electric circuit.

a) How could you improve the accuracy of the measurement?

b) How would repeating the investigation improve the accuracy?

7. Test your response time with the other equipment set up in your classroom. Use a standard reaction time meter, such as one used in driver education. You will need to follow the directions for the model available in your class.

a) Record your response time.

8. Use two stopwatches. One person starts both stopwatches at the same time, and hands one to her lab partner. When the first person stops her watch, the lab partner stops his. The difference in the two times is the response time.

a) Record your response time.

9. Use a centimeter ruler. Hold the centimeter ruler at the top, between thumb and forefinger, with zero at the bottom. Your partner places thumb and forefinger at the lower end, but does not touch the ruler. Drop the ruler. Your partner must stop the ruler from falling by closing thumb and forefinger.

a) The position of your partner's fingers marks the distance the ruler fell while her nervous system was responding. Record the distance in your log.

b) The graph at the top of the next page shows the relationship between the distance the ruler fell and the time it took to stop it. Use the graph to find and record your response time.

Givens:

$$\Delta d = 400 \text{ mi.}$$

$$v_{av} = \frac{\Delta d}{\Delta t}$$

$$= \frac{400 \text{ mi.}}{8 \text{ h}}$$

$$= 50 \text{ mph (miles per hour)}$$

Your average speed is 50 mph. This does not tell you the fastest or slowest speed that you traveled. This also does not tell you how fast you were going at any particular moment.

Sample Problem 2

Elisha would like to ride her bike to the beach. From car trips with her parents, she knows that the distance is 30 mi. She thinks she can keep up an average speed of about 15 mph. How long will it take her to ride to the beach?

Strategy: You can use the equation for average speed.

$$v_{av} = \frac{\Delta d}{\Delta t}$$

However, you will first need to rearrange the terms to solve for elapsed time.

$$\Delta t = \frac{\Delta d}{v_{av}}$$

Solution:

$$\Delta t = \frac{\Delta d}{v_{av}}$$

$$= \frac{30 \text{ mi.}}{15 \text{ mph}}$$

$$= 2 \text{ h}$$

FOR YOU TO READ

Representing Motion

One way to show motion is with the use of strobe photos.

A strobe photo is a multiple-exposure photo in which a moving object is photographed at regular time intervals. The sketches you used in **Steps 1, 2**, and **3** in **For You To Do** are similar to strobe photos. Here is a strobe photo of a car traveling at the average speed of 50 mph.

Another way to represent motion is with graphs. The graph below shows a car traveling at the average speed of 50 mph.

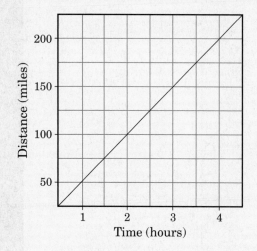

Kilometers and Miles

Highway signs and speed limits in the USA are given in miles per hour, or mph. Almost every other country in the world uses kilometers to measure distances. A kilometer is a little less than two-thirds of a mile. Kilometers per hour (km/h) is used to measure highway driving speed. Shorter distances, such as for track events and experiments in a science class, are measured in meters per second, m/s.

You will use mph when working with driving speeds, but meters per second for data you collect in class. The good news is that you do not need to change measures between systems. It is important to be able to understand and compare measures.

To help you relate the speeds with which you are comfortable to the data you collect in class, the chart below gives *approximate* comparisons.

School zone	25 mph	40 km/h	11 m/s
Residential street	35 mph	55 km/h	16 m/s
Suburban interstate	55 mph	90 km/h	25 m/s
Rural interstate	75 mph	120 km/h	34 m/s

Reflecting on the Activity and the Challenge

You now know how reaction time and speed affect the distance required to stop. You should be able to make a good argument about tailgating as part of the **Chapter Challenge**. If your car can be designed to limit tailgating or to alert drivers to the dangers of tailgating, it will add to improved safety.

Physics To Go

1. Describe the motion of each car moving to the right. The strobe pictures were taken every 3 s (seconds).

a)

b)

2. Sketch strobe pictures of the following:

a) A car starting at rest and reaching a final constant speed.
b) A car traveling at a constant speed then coming to a stop.

3. For each graph below, describe the motion of the car:

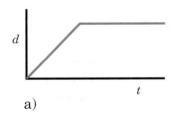

a)

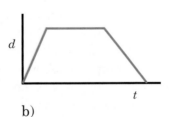

b)

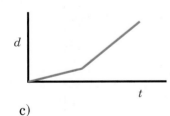

c)

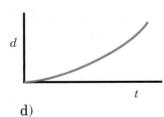

d)

4. A racecar driver travels at 110 m/s (that's almost 250 mph) for 20 s. How far has the driver traveled?

5. A salesperson drove the 215 miles from New York City to Washington, DC, in 412 hours.

   a) What was her average speed?
   b) How fast was she going when she passed through Baltimore?

6. If you planned to walk to a park that was 5 miles away, what average speed would you have to keep up to arrive in 2 hours?

7. Use your average response time from **Activity 1** to answer the following:

   a) How far does your car travel in meters during your response time if you are moving at 55 mph (25 m/s)?
   b) How far does your car travel during your response time if you are moving at 35 mph (16 m/s)? How does the distance compare with the distance at 55 mph?
   c) Suppose you are very tired and your response time is doubled. How far would you travel at 55 mph during your response time?

8. According to traffic experts, the proper following distance you should leave between your car and the vehicle in front of you is two seconds. As the vehicle in front of you passes a fixed point, say to yourself "one thousand one, one thousand two." Your car should reach the point as you complete the phrase. How can the experts be sure? Isn't two seconds a measure of time? Will two seconds be safe on the interstate highway?

9. You calculated the distance your car would move during your response time. Use that information to determine a safe following distance at:

   a) 25 mph
   b) 55 mph
   c) 75 mph

10. Apply what you learned in this activity to write a convincing argument that describes why following a car too closely (tailgating) is dangerous. Include the factors you would use to decide how close counts as "tailgating."

Stretching Exercises

Measure a distance of about 100 m. You can use a football field or get a long tape or trundle wheel to measure a similar distance. You also need a watch capable of measuring seconds. Determine your average speed traveling that distance for each of the following:

   a) a slow walk
   b) a fast walk
   c) running
   d) on a bicycle
   e) another method of your choice

Activity 3   Accidents

GOALS

In this activity you will:

- Evaluate your own understandings of safety.

- Evaluate the safety features on selected vehicles.

- Compare and contrast the safety features on selected vehicles.

- Identify safety features in selected vehicles.

- Identify safety features required for other modes of transportation (in-line skates, skateboards, cycling, etc.).

What Do You Think?

Chances are you will not be able to avoid being in an accident at some time in the future.

- **How can you protect yourself from serious injury, or even death, should an accident occur?**

- **What do you think is the greatest danger to you or the people in an accident?**

Record your ideas about these questions in your *Active Physics* log. Be prepared to discuss your responses with your small group and the class.

For You To Do

1. Many people think that they know the risks involved with day-to-day transportation. The "test" below will check your knowledge of automobile accidents. The statements are organized in a true and false format. Record a T in your log for each statement you believe is true and an F if you believe the statement is false. Your teacher will supply the correct answers for discussion at the end of the activity.

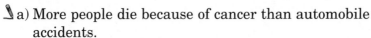

a) More people die because of cancer than automobile accidents.

b) Your chances of surviving a collision improve if you are thrown from the car.

c) The fatality rate of motorcycle accidents is less than that of cars.

d) A large number of people who are belted into their cars are killed in a burning or submerged car.

e) If you don't have a child restraint seat, you should place the child in your seat belt with you.

f) You can react fast enough during an accident to brace yourself in the car seat.

g) Most people die in traffic accidents during long trips.

h) A person not wearing a seat belt in your car poses a hazard to you.

i) Traffic accidents occur most often on Monday mornings.

j) Male drivers between the ages of 16 and 19 are most likely to be involved in traffic accidents.

k) Casualty collisions are most frequent during the winter months due to snow and ice.

l) More pedestrians than drivers are killed by cars.

m) The greatest number of roadway fatalities can be attributed to poor driving conditions.

n) The greatest number of females involved in traffic accidents are between the ages of 16 and 20.

o) Unrestrained occupant casualties are more likely to be young adults between the ages of 16 and 19.

2. Calculate your score. Give yourself two points for a correct answer, and subtract one point for an incorrect answer. You might want to match your score against the descriptors given below.

21 to 30 points: Expert Analyst

14 to 20 points: Assistant Analyst

9 to 13 points: Novice Analyst

8 points and below: Myth Believer

a) Record your score in your log. Were you surprised about the extent of your knowledge? Some of the reasons behind these facts will be better understood as you continue to travel through this chapter.

⚠ **Obtain permission from the cars' owners before proceeding.**

3. Survey at least three different cars for safety features. The list on the next page will allow you to evaluate the safety features of each of the cars. Place a check mark in the appropriate square.

Number 1 indicates very poor or nonexistent, 2 is minimum standard, 3 is average, 4 is good, and 5 is very good.

For example, when rating air bags: a car with no air bags could be given a 1 rating, a car with only a driver-side air bag a 2, a car with driver and passenger side air bags a 3, a car with slow release driver and passenger-side air bags a 4, and a car which includes side-door air bags to the previous list a 5. You may add additional safety features not identified in the chart. Many additional features can be added!

a) Copy and complete the table in your log.

b) Which car would you evaluate as being safest?

Car Tested: Make and Model _____	Year _____				
Safety Feature	Rating				
Padded front seats	1	2	3	4	5
Padded roof frame	1	2	3	4	5
Head rests	1	2	3	4	5
Knee protection	1	2	3	4	5
Anti-daze rear-view mirror that brakes on impact	1	2	3	4	5
Child proof safety locks on rear doors	1	2	3	4	5
Padded console	1	2	3	4	5
Padded sun visor	1	2	3	4	5
Padded doors and arm rests	1	2	3	4	5
Steering wheel with padded rim and hub	1	2	3	4	5
Padded gear lever					
Padded door pillars					
Air bags					

Reflecting on the Activity and the Challenge

Serious injuries in an automobile accident have many causes. If there are no restraints or safety devices in a vehicle, or if the vehicle is not constructed to absorb any of the energy of the collision, even a minor collision can cause serious injury. Until the early 1960s, automobile design and construction did not even consider passenger safety. The general belief was that a heavy car was a safe car. While there is some truth to that statement, today's lighter cars are far safer than the "tanks" of the past.

The safety survey may have provided ideas for constructing a prototype of a safety system used for transportation. If it has, write down ideas in your log that have been generated from this activity.

Physics To Go

1. Review and list all the safety features found in today's new cars. As you compile your list, write next to each safety feature one or more of the following designations:

   F: effective in a front-end collision.

   R: effective in a rear-end collision.

   S: effective in a collision where the car is struck on the side.

   T: effective when the car rolls over or turns over onto its roof.

2. Make a list of safety features that could be used for cycling.

3. Make a list of safety features that could be used for in-line skating.

4. Make a list of safety features that could be used for skateboarding.

5. Ask family members or friends if you may evaluate their car. Discuss and explain your evaluation to the car owners. Record your evaluation and their response in your log.

Stretching Exercises

1. Read a consumer report on car safety. Are any cars on the road particularly unsafe?

2. Collect brochures from various automobile dealers. What new safety features are presented in the brochures? How much of the advertising is devoted to safety?

Activity 4   Life (and Death) before Seat Belts

GOALS

In this activity you will:

• Understand Newton's First Law of Motion.

• Understand the role of safety belts.

• Identify the three collisions in every accident.

 **Perform the activity outside of traffic areas. Do not obstruct paths to exits. Do not leave carts lying on the floor.**

 What Do You Think?

Throughout most of the country, the law requires automobile passengers to wear seat belts.

• **Should wearing a seat belt be a personal choice?**

• **What are two reasons why there should be seat belt laws and two reasons why there should not?**

Record your ideas about these questions in your *Active Physics* log. Be prepared to discuss your responses with your small group and the class.

For You To Do

1. In this activity, you will investigate car crashes where the driver or passenger does not wear seat belts. Your model car is a laboratory cart. Your model passenger is molded from a lump of soft clay. With the "passenger" in place, send the "car" at a low speed into a wall.

a) Describe, in your log, what happens to the "passenger."

2. Repeat the collision at a high speed. Compare and contrast this collision with the previous one.

a) Compare and contrast requires you to find and record at least one similarity and one difference. A better response includes more similarities and differences.

3. You can conduct a more analytical experiment by having the cart hit the wall at varying speeds. Set up a ramp on which the car can travel. Release the car on the ramp and observe as it crashes into the wall. Repeat the collision for at least two ramp heights.

a) Record the heights of the ramp and describe the results of the collision. Describe the collision by noting the damage to the "passenger."

Physics Words

Newton's First Law of Motion: an object at rest stays at rest and an object in motion stays in motion unless acted upon by an unbalanced, external force.

inertia: the natural tendency of an object to remain at rest or to remain moving with constant speed in a straight line.

PHYSICS TALK

Newton's First Law of Motion

Newton's First Law of Motion (also called the Law of **Inertia**) is one of the foundations of physics. It states:

An object at rest stays at rest, and an object in motion stays in motion unless acted upon by a net external force.

There are three distinct parts to Newton's First Law.

Part 1 says that objects at rest stay at rest. This hardly seems surprising.

Part 2 says that objects in motion stay in motion. This may seem strange indeed. After looking at the collisions of this activity, this should seem clearer.

Part 3 says that Parts 1 and 2 are only true when no force is present.

FOR YOU TO READ

Three Collisions in One Accident!

Arthur C. Damask analyzes automobile accidents and deaths for insurance companies and police reports. This is how Professor Damask describes an accident:

Consider the occupants of a conveyance moving at some speed. If the conveyance strikes an object, it will rapidly decelerate to some lower speed or stop entirely; this is called the first collision. But the occupants have been moving at the same speed, and will continue to do so until they are stopped by striking the interior parts of the car (if not ejected); this is the second collision. The brain and body organs have also been moving at the same speed and will continue to do so until they are stopped by colliding with the shell of the body, i.e., the interior of the skull, the thoracic cavity, and the abdominal wall. This is called the third collision.

Newton's First Law of Motion explains the three collisions:

- First collision: the car strikes the pole; the pole exerts the force that brings the car to rest.
- Second collision: when the car stops, the body keeps moving; the structure of the car exerts the force that brings the body to rest.
- Third collision: the body stops, but the heart and brain keep moving; the body wall exerts the force that brings the heart and brain to rest.

Even with all the safety features in our automobiles, some deaths cannot be prevented. In one accident, only a single car was involved, with only the driver inside. The car failed to follow the road around a turn, and it struck a telephone pole. The seat belt and the air bag prevented any serious injuries apart from a few bruises, but the driver died. An autopsy showed that the driver's aorta had burst, at the point where it leaves the heart.

Reflecting on the Activity and the Challenge

In this activity you discovered that an object in motion continues in motion until a force stops it. A car will stop when it hits a pole but the passenger will keep on moving. If the car and passenger have a large speed, then the passenger will continue moving with this large speed. The passenger at the large speed will experience more damage from the fast-moving cart.

Have you ever heard someone say that they can prevent an injury by bracing themselves against the crash? They can't! Restraining devices help provide support. Without a restraining system, the force of impact is either absorbed by the rib, skull, or brain.

Use Newton's First Law of Motion to describe your design. How will your safety system protect passengers from low speed and higher speed collisions?

Physics To Go

1. Describe how Newton's First Law applies to the following situations:

   • You step on the brakes to stop your car.

   (Sample answer: You and the car are moving forward. The brakes apply a force to the tires and stop them from rotating. Newton's law states that an object in motion will remain in motion unless a force acts upon it. In this case, the force is friction between the ground and the tires. You remain in motion since the force that stopped the car did not stop you.)

   • You step on the accelerator to get going.

   • You turn the wheel to go around a curve. (Hint: You keep moving in a straight line.)

   • You step on the brakes, and an object in the back of the car comes flying forward.

2. Give two more examples of how Newton's First Law applies to vehicles or people in motion.

3. According to Newton's First Law, objects in motion will continue in motion unless acted upon by a force. Using Newton's First Law, explain why a cart that rolls down a ramp eventually comes to rest.

4. The skateboard, shown in the picture to the right, strikes the curb. Draw a diagram indicating the direction in which the person moves. Use Newton's First Law to explain the direction of movement.

5. Explain, in your own words, the three collisions during a single crash as described by Professor Damask in **For You To Read**.

6. Use the diagrams below to compare the second and third collisions described by Professor Damask with the impact of a punch during a boxing match.

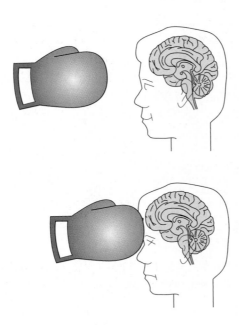

7. When was the law instituted requiring drivers to wear seat belts?

Stretching Exercises

1. Determine what opinions people in your community hold about the wearing of seat belts. Compare the opinions of the 60+ years old and 25 to 59 year old groups with that of the 15 to 24 year old group. Survey at least five people in each age group: Group A = 15 to 24 years, Group B = 25 to 59 years, and Group C = 60 years and older. (Survey the same number of individuals in each age group.) Ask each individual to fill out a survey card.

A sample questionnaire is provided below. You may wish to eliminate any question that you feel is not relevant. You are encouraged to develop questions of your own that help you understand what attitudes people in your community hold about wearing seat belts. The answers have been divided into three categories: 1 = agree; 2 = will accept, but do not hold a strong opinion; and 3 = disagree. Try to keep your survey to between five and ten questions.

Age group:	Date of Survey:		
Statement	Agree	No strong opinion	Disagree
1. I believe people should be fined for not wearing seat belts.	1	2	3
2. I wouldn't wear a seat belt if I didn't have to.	1	2	3
3. People who don't wear seat belts pose a threat to me when they ride in my car.	1	2	3
4. I believe that seat belts save lives.	1	2	3
5. Seat belts wrinkle my clothes and fit poorly so I don't wear them.	1	2	3

2. Make a list of reasons why people refuse to wear seat belts. Can you challenge these opinions using what you have learned about Newton's First Law of Motion?

Activity 5

Life (and Fewer Deaths) after Seat Belts

GOALS

In this activity you will:

- Understand the role of safety belts.
- Compare the effectiveness of various wide and narrow belts.
- Relate pressure, force and area.

What Do You Think?

In a collision, you cannot brace yourself and prevent injuries. Your arms and legs are not strong enough to overcome the inertia during even a minor collision. Instead of thinking about stopping yourself when the car is going 30 mph, think about catching 10 bowling balls hurtling towards you at 30 mph. The two situations are equivalent.

- **Suppose you had to design seat belts for a racecar that can go 200 mph. How would they be different from the ones available on passenger cars?**

Record your ideas about this question in your *Active Physics* log. Be prepared to discuss your responses with your small group and the class.

Coordinated Science for the 21st Century

For You To Do

1. In this activity you will test different materials for their suitability for use as seat belts. Your model car is, once again, a laboratory cart; your model passenger is molded from a lump of soft clay. Give your passenger a seat belt by stretching a thin piece of wire across the front of the passenger, and attaching it on the sides or rear of the car.

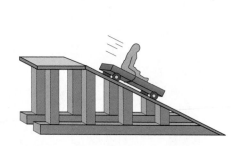

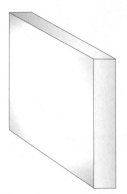

 **Perform the activity outside of traffic areas. Do not obstruct paths to exits. Do not leave carts lying on the floor.**

2. Make a collision by sending the car down a ramp. Start with small angles of incline and increase the height of the ramp until you see significant injury to the clay passenger.

a) In your log, note the height of the ramp at which significant injury occurs.

3. Use at least two other kinds of seat belts (ribbons, cloth, etc.). Begin by using the same incline of ramp and release height as in **Step 2**.

a) In your log, record the ramp height at which significant injury occurs to the "passenger" using the other kinds of seat belt material.

4. Crash dummies cost $50,000! Watch the video presentation of a car in a collision, with a dummy in the driver's seat. You may have to observe it more than once to answer the following questions:

a) In the collision, the car stops abruptly. What happens to the driver?

b) What parts of the driver's body are in the greatest danger? Explain what you saw in terms of the law of inertia (Newton's First Law of Motion).

Physics Words

force: a push or a pull that is able to accelerate an object; force is measured in newtons; force is a vector quantity.

pressure: force per surface area where the force is normal (perpendicular) to the surface; measured in pascals.

FOR YOU TO READ

Force and Pressure

When you repeated this experiment accurately each time, the **force** that each belt exerted on the clay was the same each time that the car was started at the same ramp height. Yet different materials have different effects; for example, a wire cuts far more deeply into the clay than a broader material does.

The force that each of the belts exerts on the clay is the same. When a thin wire is used, all the force is concentrated onto a small area. By replacing the wire with a broader material, you spread the force out over a much larger area of contact.

The force per unit area, which is called **pressure**, is much smaller with a ribbon, for example, than with a wire. It is the pressure, not the force, that determines how much damage the seat belt does to the body. A force applied to a single rib might be enough to break a rib. If the same force is spread out over many ribs, the force on each rib can be made too small to do any damage. While the total force does not change, the pressure becomes much smaller.

PHYSICS TALK

Pressure is the force per unit area:

$$P = \frac{F}{A}$$

where F is force in newtons (N)

A is area in meters squared (m^2)

and P is pressure in newtons per meter squared (N/m^2).

Force can be measured using a spring scale.

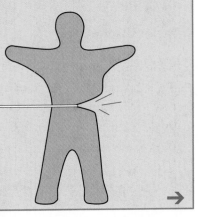

Coordinated Science for the 21st Century

Sample Problem

Two brothers have the same mass and apply a constant force of 450 N while standing in the snow. Brother A is wearing snow shoes that have a base area of 2.0 m². Brother B, without snowshoes, has a base area of 0.1 m². Why does the brother without snowshoes sink into the snow?

Strategy: This problem involves the pressure that is exerted on the snow surface by each brother. You can use the equation for pressure to compare the pressure exerted by each brother.

Givens:

$$F = 450 \text{ N}$$
$$A_1 = 2.0 \text{ m}^2$$
$$A_2 = 0.1 \text{ m}^2$$

Solution:

Brother A	Brother B
$P = \dfrac{F}{A}$	$P = \dfrac{F}{A}$
$= \dfrac{450 \text{ N}}{2.0 \text{ m}^2}$	$= \dfrac{450 \text{ N}}{0.1 \text{ m}^2}$
$= 225 \text{ N/m}^2 \text{ or } 230 \text{ N/m}^2$	$= 4500 \text{ N/m}^2$

The pressure that Brother B exerts on the snow is much greater.

Reflecting on the Activity and the Challenge

In this activity you gathered data to provide evidence on the effectiveness of seat belts as restraint systems. The material used for the seat belt and the width of the restraint affected the distortion of the clay figure. By applying the force over a greater area, the pressure exerted by the seat belt during the collision can be reduced.

It is important to note that not every safety restraint system will be a seat belt or harness, but that all restraints attempt to reduce the pressure exerted on an object by increasing the area over which a force is applied.

How will your design team account for decreasing pressure by increasing the area of impact? Think about ways that you could test your design prototype for the pressure created during impact. Your presentation of the design will be much more convincing if you have quantitative data to support your claims. Simply stating that a safety system works well is not as convincing as being able to show how it reduces pressure during a collision.

Physics To Go

1. Use Newton's First Law to describe a collision with the passenger wearing a seat belt during a collision.

2. What is the pressure exerted when a force of 10 N is applied to an object that has an area of:

   a) 1.0 m^2?
   b) 0.2 m^2?
   c) 15 m^2?
   d) 400 cm^2?

3. A person who weighs approximately 155 lb. exerts 700 N of force on the ground while standing. If his shoes cover an area of 400 cm^2 (0.0400 m^2), calculate:

   a) the average pressure his shoes exert on the ground
   b) the pressure he would exert by standing on one foot

4. For comparison purposes, calculate the pressure you exert in the situations described below. Divide your weight in newtons, by the area of your shoes. (To find your weight in newtons multiply your weight in pounds by 4.5 N/lb. You can approximate the area of your shoes by tracing the soles on a sheet of centimeter squared paper.)

a) How much pressure would you exert if you were standing in high heels?

b) How much pressure would you exert while standing on your hands?

c) If a bed of nails contains 5000 nails per square meter, how much force would one nail support if you were to lie on the bed? With this calculation you can now explain how people are able to lie on a bed of nails. It's just physic!

5. Describe why a wire seat belt would not be effective even though the force exerted on you by the wire seat belt is identical to that of a cloth seat belt.

6. Do you think there ought to be seat belt laws? How does not using seat belts affect the society as a whole?

7. Conduct a survey of 10 people. Ask each person what percentage of the time they wear a seat belt while in a car. Be prepared to share your data with the class.

Stretching Exercises

The pressure exerted on your clay model by a thin wire can be estimated quite easily. Loop the wire around the "passenger," and connect the wire to a force meter.

a) Pull the force meter hard enough to make the wire sink into the model just about as far as it did in the collision.

b) Record the force as shown on the force meter (in newtons).

c) Estimate the frontal area of the wire—its diameter times the length of the wire that contacts the passenger. Record this value in centimeters squared (cm^2).

d) Divide the force by the area. This is the pressure in newtons per centimeter squared (N/cm^2).

Activity 6   Why Air Bags?

GOALS

In this activity you will:

- Model an automobile air bag.
- Relate pressure to force and area.
- Demonstrate that the force of an impact can be reduced by spreading it out over a longer time.

What Do You Think?

Air bags do not take the place of seat belts. Air bags are an additional protection. They are intended to be used with seat belts to increase safety.

- **Why are air bags effective?**

- **How does the air bag protect you?**

Record your ideas about these questions in your *Active Physics* log. Be prepared to discuss your responses with your small group and the class.

For You To Do

1. You will use a large plastic bag or a partially inflated beach ball as a model for an air bag. Impact is provided by a heavy steel ball, or just a good-sized rock, dropped from a height of a couple of meters.

   Gather the equipment you will need for this activity. Your problem is to find out how long it takes the object to come to rest. What is the total time duration from when the object first touches the air bag until it bounces back?

2. With a camcorder, videotape the object striking the air bag from a given height, such as 1.5 m.

   ✎ a) Record the exact height, from which you dropped the object.

3. Play the sequence back, one frame at a time. Count the number of frames during the time the object is moving into the air bag—from the moment it first touches the bag until it comes to rest, before bouncing. Each frame stands for $\frac{1}{30}$s. (Check your manual.)

   ✎ a) In your log, record the number of frames and calculate how long it takes for the object to come to rest.

   If a camcorder is not available, the experiment may be performed, although less effectively, by attaching a ticker-tape timer to the falling object.

   After the object is dropped, with the object still attached, stretch the tape from the release position to the air bag. Mark the dot on the tape that was made just as the object touched the air bag.

   Now push the object into the air bag, about as far as it went just before it bounced. Mark the tape at the dot that was made as the object came to rest. The dots should be close together for a short interval at this point.

   Now count the time that passed between the two marks you made. (You must know how rapidly dots are produced by your timer.)

⚠ **Set up the activity in an area clear of obstruction. Arrange for containment of the dropped object.**

4. Repeat **Steps 2** and **3**, but this time drop the ball against a hard surface, such as the floor. Keep the height from which the object is dropped constant.

a) Record how long it takes for the object to come to rest on a floor.

5. Choose two other surfaces and repeat **Steps 2** and **3**.

a) Record how long it takes the object to come to rest each time.

b) In your log, list all the surfaces you tested in the order in which you expect the most damage to be done to a falling object, to the least damage.

c) Is there a relationship between the time it takes for the object to come to rest and the potential damage to the object landing on the surface? Explain this relationship in your log.

PHYSICS TALK

Force and Impulse

Newton's First Law states that an object in motion will remain in motion unless acted upon by a net external force. In this activity you were able to stop an object with a force. In all cases the object was traveling at the same speed before impact. Stopping the object was done quickly or gradually. The amount of damage is related to the time during which the force stopped the object. The air bag was able to stop the object with little damage by taking a long time. The hard surface stopped the object with more damage by taking a short time.

Physicists have a useful way to describe these observations. An **impulse** is needed to stop an object. That impulse is defined as the product (multiplication) of the force applied and the time that the force is applied.

→

Physics Words

impulse: the product of force and the interval of time during which the force acts; impulse results in a change in momentum.

115

Impulse = $F\Delta t$

where F is force in newtons (N)

Δt is the time interval during which the force is applied in seconds (s).

Impulse is calculated in newton seconds (Ns).

An object of a specific mass and a specific speed will need a definite impulse to stop. Any forces acting for enough time can provide that impulse.

If the impulse required to stop is 60 Ns, a force of 60 N acting for 1 s has the required impulse. A force of 10 N acting for 6 s also has the required impulse.

Force F	Time Interval Δt	Impulse $F\Delta t$
60 N	1 s	60 Ns
10 N	6 s	60 Ns
6000 N	0.01 s	60 Ns

The greater the force and the smaller the time interval, the greater the damage that is done.

Sample Problem

A person requires an impulse of 1500 Ns to stop. What force must be applied to the person to stop in 0.05 s?

Strategy: You can use the equation for impulse and rearrange the terms to solve for the force required.

$$\text{Impulse} = F\Delta t$$

$$F = \frac{\text{Impulse}}{\Delta t}$$

Givens:

Impulse = 1500 Ns

$\Delta t = 0.05$ s

Solution:

$$F = \frac{\text{Impulse}}{\Delta t}$$

$$= \frac{1500 \text{ Ns}}{0.05 \text{ s}}$$

$$= 30{,}000 \text{ N}$$

Reflecting on the Activity and the Challenge

People once believed that the heavier the automobile, the greater the protection it offered passengers. Although a heavy, rigid car may not bend as easily as an automobile with a lighter frame, it doesn't always offer more protection.

In this activity, you found that air bags are able to protect you by extending the time it takes to stop you. Without the air bags, you will hit something and stop in a brief time. This will require a large force, large enough to injure you. With the air bag, the time to stop is longer and the force required is therefore smaller.

Force and impulse must be considered in designing your safety system. Stopping an object gradually reduces damage. The harder a surface, the shorter the stopping distance and the greater the damage. In part this provides a clue to the use of padded dashboards and sun visors in newer cars. Understanding impulse allows designers to reduce damage both to cars and passengers.

Physics To Go

1. If an impulse of 60 Ns is required to stop an object, list in your log three force and time combinations (other than those given in the **Physics Talk**) that can stop an object.

2. A person weighing 130 lb. (60 kg) traveling at 40 mph (18 m/s) requires an impulse of approximately 1000 Ns to stop. Calculate the force on the person if the time to stop is:

   a) 0.01 s
   b) 0.10 s
   c) 1.00 s

3. Explain in your log why an air bag is effective. Use the terms force, impulse, and time in your response.

4. Explain in your log why a car hitting a brick wall will suffer more damage than a car hitting a snow bank.

5. There are several other safety designs that employ the concept of spreading out the time interval of a force. Describe in your log how the ones listed below perform this task:

   a) the bumper
   b) a collapsible steering wheel
   c) frontal "crush" zones
   d) padding on the dashboard

6. There are many other situations in which the force of an impulse is reduced by spreading it out over a longer time. Explain in your log how each of the actions below effectively reduces the force by increasing the time. Use the terms force, impulse, and time in your response.

   a) catching a hard ball
   b) jumping to the ground from a height
   c) bungee jumping
   d) a fireman's net

7. The speed of airplanes is considerably higher than the speed of automobiles. How might the design of a seat belt for an airplane reflect the fact that a greater impulse is exerted on a plane when it stops?

8. Airplanes have seat belts. Should they also have air bags?

Activity 7

Automatic Triggering Devices

What Do You Think?

An air bag must inflate in a sudden crash, but must not inflate under normal stopping conditions.

• How does the air bag "know" whether to inflate?

Record your ideas about this question in your *Active Physics* log. Be prepared to discuss your responses with your small group and the class.

GOALS

In this activity you will:

• Design a device that is capable of transmitting a digital electrical signal when it is accelerated in a collision.

For You To Do

Inquiry Investigation

1. Form engineering teams of three to five students. Meet with your engineering team to design an automatic air bag triggering device using a knife switch, rubber bands, string, wires, and a flashlight bulb. Other materials may also be supplied by you or your teacher.

119

Be sure to receive your teacher's approval before using any material.

2. The design parameters are as follows:

• The device must turn a flashlight bulb on, or turn it off. This will be interpreted as the trigger signal.
• The device must not trigger if the car is brought to a sudden stop from a slow speed.
• The device must trigger if the sudden stop is from a high speed.
• The car containing the device must be released down a ramp. The car will then strike a wall at the bottom of the ramp.
• The battery and bulb must be attached to the car along with the triggering device. The bulb does not have to remain in the final on or off state, but it must at least flash to show triggering.

3. Follow your teacher's guidelines for using time, space, and materials as you design your triggering device.

4. Demonstrate your design team's trigger for the class.

FOR YOU TO READ

Impulse and Changes in Momentum

It takes an unbalanced, opposing force to stop a moving car. **Newton's Second Law** lets you find out how much force is required to stop any car of any mass and any speed.

The overall idea can be shown using a concept map.

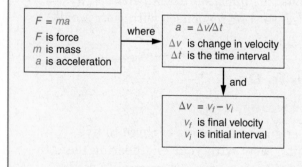

If you know the mass and can determine the acceleration, you can calculate the force using Newton's Second Law:

$$F = ma$$

A moving car has a forward **velocity** of 15 m/s. Stopping the car gives it a final velocity of 0 m/s.

The change in velocity $= v_{final} - v_{initial}$
$$= 0 - 15 \text{ m/s}$$
$$= -15 \text{ m/s}.$$

Any change in velocity is defined as **acceleration**.

In this case the change in velocity is −15 m/s. If the change in velocity occurs in 3 s, the acceleration is −15 m/s in 3 s, or −5 m/s every second, or −5 m/s². You can look at this as an equation:

$$a = \frac{\Delta v}{\Delta t} = \frac{v_f - v_o}{\Delta t} = \frac{0 \text{ m/s} - 15 \text{ m/s}}{3s} = -5 \text{ m/s}^2$$

If the car had stopped in 0.5 s, the change in speed is identical, but the acceleration is now:

$$a = \frac{\Delta v}{\Delta t} = \frac{v_f - v_o}{\Delta t} = \frac{0 \text{ m/s} - 15 \text{ m/s}}{0.5 \text{ s}} = -30 \text{ m/s}^2$$

Newton's Second Law informs you that unbalanced outside forces cause all accelerations. The force stopping the car may have been the frictional force of the brakes and tires on the road, or the force of a tree, or the force of another car. Once you know the acceleration, you can calculate the force using Newton's Second Law. If the car has a mass of 1000 kg, the unbalanced force for the acceleration of −5 m/s *every* second would be −5000 N. The force for the larger acceleration of −30 m/s *every* second would be −30,000 N. The negative sign tells you that the unbalanced force was opposite in direction to the velocity.

The change in velocity, acceleration, and force give a complete picture.

There is another, equivalent picture that describes the same collision in terms of **momentum** and impulse.

Any moving car has momentum. Momentum is defined as the mass of the car multiplied by its velocity $p = mv$.

The impulse/momentum equation tells you that the momentum of the car can be changed by applying a force for a given amount of time.

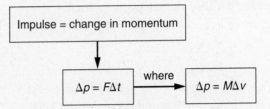

Impulse = change in momentum

$\Delta p = F\Delta t$ where $\Delta p = M\Delta v$

Using the impulse/momentum approach explains something different about a collision. Consider this question, "why do you prefer to land on soft grass rather than on hard concrete?" Soft grass is preferred because the force on your legs is less when you land on soft grass. Let's find out why using Newton's Second Law and then by using the impulse/momentum relation.

Newton's Second Law explanation:

Whether you land on concrete or soft grass, your change in velocity will be identical. Your velocity may decrease from 3 m/s to 0 m/s. On concrete, this change occurs very fast, while on soft grass this change occurs in a longer period of time. Your acceleration on soft grass is smaller because the change in velocity occurred in a longer period of time.

$$a = \frac{\Delta v}{\Delta t}$$

When the change in the period of time gets larger, the denominator of the fraction gets larger and the value of the acceleration gets smaller.

When landing on grass, Newton's Second Law then tells you that the force must be smaller because the acceleration is smaller for an identical mass. $F = ma$. Smaller acceleration on grass requires a smaller force. Smaller forces are easier on your legs and you prefer to land on soft grass.

Momentum/impulse explanation:
Whether you land on concrete or soft grass, your change in momentum will be identical. Your velocity will decrease from 3 m/s to 0 m/s on either concrete or grass.

$$F\Delta t = \Delta p$$

You can get this change in momentum with a large force over a short time or a small force over a longer time.

If your mass is 50 kg, the amount of your change in momentum may be 150 kg m/s, when you decrease your velocity from 3 m/s to 0 m/s. There are many forces and associated times that can give this change in the value of the momentum.

If you could land on a surface that requires 3 s to stop, it will only require 50 N. A more realistic time of 1 s to stop will require a larger force of 150 N. A hard surface that brings you to a stop in 0.01 s requires a much larger force of 15,000 N.

On concrete, this change in the value of the momentum occurs very fast (a short time) and requires a large force. It hurts. On soft grass this change in the value of the momentum occurs in a longer time and requires a small force: it is less painful and is preferred.

Change in value of momentum	Force	Change in Time Δt	$F\Delta t$
150 kg m/s	50 N	3 s	150 kg m/s
150 kg m/s	75 N	2 s	150 kg m/s
150 kg m/s	150 N	1 s	150 kg m/s
150 kg m/s	1500 N	0.1 s	150 kg m/s
150 kg m/s	15,000 N	0.01 s	150 kg m/s

Physics To Go

1. How do impulse and Newton's First Law (the Law of Inertia) play a role in your air bag trigger design?

2. Imagine a device where a weight is hung from a string within a metal can. If the weight hits the side of the can, a circuit is completed. How do impulse and the Law of Inertia work in this device?

3. In cars built before 1970, the dashboard was made of hard metal. After 1970, the cars were installed with padded dashboards like you find in cars today. In designing a safe car, it is better to have a passenger hit a cushioned dashboard than a hard metal dashboard.

   a) Explain why the padded dashboard is better using Newton's Second Law.
   b) Explain why the padded dashboard is better using impulse and momentum.

4. Why would you prefer to hit an air bag during a collision than the steering wheel?

   a) Explain why the air bag is better using Newton's Second Law.
   b) Explain why the air bag is better using impulse and momentum.

5. Explain why you bend your knees when you jump to the ground.

6. Catching a fast ball stings your hand. Why does wearing a padded glove help?

7. When a soccer ball hits your chest, you can stiffen your body and the ball will bounce away from you. In contrast, you can "soften" your body and the ball drops to your feet. Explain how the force on the ball is different during each play.

Stretching Exercises

1. How does a seat belt "know" to hold you firmly in a crash, but allow you to lean forward or adjust it without locking? Write your response in your log.

2. Go to a local auto repair shop, junk yard, or parts supply store. Ask if they can show you a seat belt locking mechanism. How does it work? Construct a poster to describe what you have learned.

Physics Words

Newton's Second Law of Motion: if a body is acted upon by an external force, it will accelerate in the direction of the unbalanced force with an acceleration proportional to the force and inversely proportational to the mass.

velocity: speed in a given direction; displacement divided by the time interval; velocity is a vector quantity, it has magnitude and direction.

acceleration: the change in velocity per unit time.

momentum: the product of the mass and the velocity of an object; momentum is a vector quantity.

Activity 8   The Rear-End Collision

GOALS

In this activity you will:

• Evaluate from simulated collisions, the effect of rear-end collisions on the neck muscles.

• Understand the causes of whiplash injuries.

• Understand Newton's Second Law of Motion.

• Understand the role of safety devices in preventing whiplash injury.

What Do You Think?

Whiplash is a serious injury that is caused by a rear-end collision. It is the focus of many lawsuits, loss of ability to work, and discomfort.

• **What is whiplash?**

• **Why is it more prominent in rear-end collisions?**

Record your ideas about these questions in your *Active Physics* log. Be prepared to discuss your responses with your small group and the class.

For You To Do

1. You will use two pieces of wood to represent the torso (the trunk of the body) and the head of a passenger. Attach a small piece of wood (about 1" x 2" x 2") to a larger piece of wood (about 1" x 3" x 10") with some duct tape acting like a hinge between the two pieces.

   a) Make a sketch to show your passenger. Label what each part of the model passenger represents.

2. Set up a ramp against a stack of books about 40 cm high, as shown in the diagram below. Place the wooden model passenger at the front of a collision cart positioned about 50 cm from the end of the ramp. Release a second cart from a few centimeters up the ramp.

   a) In your log record what happens to the head and torso of the wooden model.

3. With the first cart still positioned about 50 cm from the end of the ramp, release the second cart from the top of the ramp.

   a) Describe what happens to the head of the model passenger in this collision.

   b) Use Newton's First Law of Motion to explain your observations.

⚠ **Perform the activity outside of traffic areas. Do not obstruct paths to exits. Do not leave carts lying on the floor.**

 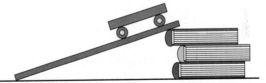

4. The duct tape represents the neck muscles and bones of the vertebral column. How large a force do the neck muscles exert to keep the head from flying off the body, and to return the head to the upright position? To answer this question, begin by estimating the mass of an average head.

   a) Estimate and record in your log the mass of an average human head. The mass would be close to the mass of a filled water container of the same size.

5. Mark off a distance about 30-cm long on the lab table or the floor. Obtain a piece of wood and attach it to a spring scale. Pull the wooden mass with the spring scale over the distance you marked.

   a) In your log record the force required to pull the mass and the time it took to cover the distance.

   b) Repeat the step, but vary the time required to pull the mass over the distance. Record the forces and the times in your log.

   c) Use your observations to complete the following statement:

   The shorter the time (that is, the greater the acceleration) the ▭ the force required.

6. The ratio of the mass of the wood to the estimated mass of the head is the same as the ratio of the forces required to pull them.

   a) Use the following ratio to calculate how large a force the neck muscles exert to keep the head from flying off the body, and to return the head to the upright position under different accelerations.

   $$\frac{\text{mass of head}}{\text{mass of wood}} = \frac{\text{force to move head}}{\text{force to move wood}}$$

7. Whiplash is a serious injury that can be caused by a rear-end collision. The back of the car seat pushes forward on the torso of the driver and the passengers and their bodies lunge forward. The heads remain still for a very short time. The body moving forward and the head remaining still causes the head to snap backwards. The neck muscles and bones of the vertebral column become damaged. The same muscles must then snap the head back to its place atop the shoulders.

   a) What type of safety devices can reduce the delay between body and head movement to help prevent injury?

   b) What additional devices have been placed in cars to help reduce the impact of rear-end collisions?

FOR YOU TO READ

Newton's Second Law of Motion

Newton's First Law of Motion is limited since it only tells you what happens to objects if net force acts upon them. Knowing that objects at rest have a tendency to remain at rest and that objects in motion will continue in motion does not provide enough information to analyze collisions. Newton's Second Law allows you to make predictions about what happens when an external force is applied to an object. If you were to place a collision cart on a level surface, it would not move. However, if you begin to push the cart, it will begin to move.

Newton's Second Law states:

If a body is acted on by a force, it will accelerate in the direction of the unbalanced force. The acceleration will be larger for smaller masses. The acceleration can be an increase in speed, a decrease in speed, or a change in direction.

Newton's Second Law of Motion indicates that the change in motion is determined by the force acting on the object, and the mass of the object itself.

Analyzing the Rear-End Collision

This activity demonstrated the effects of a rear-end collision. Newton's First Law and Newton's Second Law can help explain the "whiplash" injury that passengers suffer during this kind of collision.

Imagine looking at the rear-end collision in slow motion. Think about all that happens.

1. A car is stopped at a red light. This is the car in which the driver is going to be injured with whiplash. The driver is at rest within the car.

2. The stopped car gets hit from the rear.

3. The car begins to move. The back of the seat pushes the driver forward and his torso moves with the car. The driver's head is not supported and stays back where it is.

4. The neck muscles hold the head to the torso as the body moves forward. The muscles then "whip" the head forward. The head keeps moving until it gets ahead of the torso. The head is stopped by the neck muscles. The muscles pull the head back to its usual position. Ouch!

Let's repeat the description of the collision and insert all of the places where Newton's First Law applies. Newton's First Law states that *an object at rest stays at rest and an object in motion stays in motion unless acted upon by an unbalanced, outside force.*

1. A car is stopped at a red light. This is the car in which the driver is going to be injured with whiplash. The driver is at rest within the car. *Newton's First Law: an object (the driver) at rest stays at rest.*

2. The stopped car gets hit from the rear.

3. The car begins to move. The back of the seat pushes the driver forward and his torso moves with the car. *Newton's First Law: an object (the driver's torso) at rest stays at rest unless acted upon by an unbalanced, outside force.* The driver's head is not supported and stays back where it is. *Newton's First Law: an object (the driver's head) at rest stays at rest.*

4. The neck muscles hold the head to the body as the body moves forward. The muscles then "whip" the head forward. *Newton's First Law: an object (the head) at rest stays at rest unless acted upon by an unbalanced, outside force.* The head keeps moving until it gets ahead of the torso. *Newton's First Law: an object (the head) in motion stays in motion.* The head is stopped by the neck muscles. *Newton's First Law: an object (the head) in motion stays in motion unless acted upon by an unbalanced, outside force.* The muscles pull the head back to its usual position. *Newton's First Law: an object at rest stays at rest unless acted upon by an unbalanced, outside force.* Ouch!

Let's repeat the description of the collision and insert all of the places where Newton's Second Law applies. Newton's Second Law states that *all accelerations are caused by unbalanced, outside forces, F = ma.* An acceleration is any change in speed.

1. A car is stopped at a red light. This is the car in which the driver is going to be injured with whiplash. The driver is at rest within the car.

2. The stopped car gets hit from the rear.

3. The car begins to move. *Newton's Second Law: the car accelerates because of the unbalanced, outside force from the rear; F = ma.* The back of the seat pushes the driver forward and his torso moves with the car. *Newton's Second Law: the torso accelerates because of the unbalanced, outside force from the back of the seat; F = ma.* The driver's head is not supported and stays back where it is.

4. The neck muscles hold the head to the torso as the body moves forward. The muscles then "whip" the head forward. *Newton's Second Law: the head accelerates because of the unbalanced force of the muscles; F = ma.* The head keeps moving until it gets ahead of the torso. The head is stopped by the neck muscles. *Newton's Second Law: the head accelerates (slows down) because of the unbalanced force from the neck muscles; F = ma.* The muscles pull the head back to its usual position. *Newton's Second Law: the head accelerates because of the unbalanced force from the rear; F = ma.* Ouch!

Newton's Second Law informs you that all accelerations are caused by *unbalanced, outside* forces. It does not say that all forces cause accelerations. An object at rest may have many forces acting upon it. When you hold a book in your hand, the book is at rest. There is a force of gravity pulling the book down. There is a force of your hand pushing the book up. These forces are equal and opposite. The "net" force on the book is zero because the two forces balance each other. There is no acceleration because there is no "net" force.

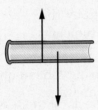

Both forces act through the center of the book. They are shifted a bit in the diagram to emphasize that the upward force of the hand acts on the bottom of the book and the downward force of gravity acts on the middle of the book.

As a car moves down the highway at a constant speed, there are forces acting on the car but there is no acceleration. This indicates that the net force must be zero. The force of the engine on the tires and road moving the car forward must be equal and opposite to the force of the air pushing the car backward. These forces balance each other in this case, where the speed is not changing. There is no net force and there is no acceleration. The car stays in motion at a constant speed. A similar situation occurs when you push a book across a table at constant speed. The push is to the right and the friction is to the left, opposing motion. If the forces are equal in size, there is no net force on the book and the book does not accelerate—it moves with a constant speed.

Reflecting on the Activity and the Challenge

The vertebral column becomes thinner and the bones become smaller as the column attaches to the skull. The attachment bones are supported by the least amount of muscle. Unfortunately, the smaller bones, with less muscle support, make this area particularly susceptible to injury. One of the greatest dangers following whiplash is the damage to the brainstem. The brainstem is particularly vital to life support because it regulates blood pressure and breathing movements. Consider how your safety device will help prevent whiplash following a collision. What part of the restraining device prevents the movement of the head?

Physics To Go

1. Why are neck injuries common after rear-end collisions?

2. Explain why the packages in the back move forward if a truck comes to a quick stop.

3. As a bus accelerates, the passengers on the bus are jolted toward the back of the bus. Indicate what causes the passengers to be pushed backward.

4. Why would the rear-end collision demonstrated by the laboratory experiment be most dangerous for someone driving a motorcycle?

5. Would headrests serve the greatest benefit during a head-on collision or a rear-end collision? Explain your answer.

6. A cork is attached to a string and placed in a jar of water as shown by the diagram to the right. Later, the jar is inverted.

   a) If the glass jar is pushed along the surface of a table, predict the direction in which the cork will move.

   b) If you place your left hand about 50 cm in front of the jar and push it with your right hand until it strikes your left hand, predict the direction in which the cork will move.

⚠ Be sure the outside of the jar is dry so it does not slip out of your hands.

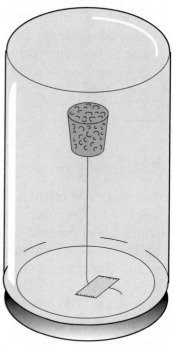

Activity 9

Cushioning Collisions (Computer Analysis)

GOALS

In this activity you will:

- Apply the concept of impulse in the analysis of automobile collisions.

- Use a computer's motion probe (sonic ranger) to determine the velocity of moving vehicles.

- Use a computer's force probe to determine the force exerted during a collision.

- Compare the momentum of a model vehicle before the collision with the impulse applied during the collision.

- Explore ways of using cushions to increase the time that a force acts during a primary collision.

What Do You Think?

The use of sand canisters around bridge supports and crush zones in cars are examples of technological systems that are designed to minimize the impact of collisions between a car and a stationary object or another car.

- **How do these technological systems reduce the impact of the primary collision?**

Record your ideas about this question in your *Active Physics* log. Be prepared to discuss your responses with your small group and the class.

For You To Do

1. In this investigation you will be using a force probe that is attached to a computer to determine the effectiveness of different types of cushions for a toy vehicle. Release a toy car

131

at the top of a ramp and measure the force of impact as it strikes a barrier at the bottom. A sonic ranger can be mounted on the ramp to measure the speed of the toy car prior to the collision. Open the appropriate computer files to prepare the sonic ranger to graph velocity vs. time and the force probe to graph force vs. time.

2. Mount the sonic ranger at the bottom of a ramp and place the force probe against a barrier about 10 cm from the bottom of the ramp, as shown in the diagram. Attach an index card to the back of the car, to obtain better reflection of the sound wave and improve the readings of the sonic ranger.

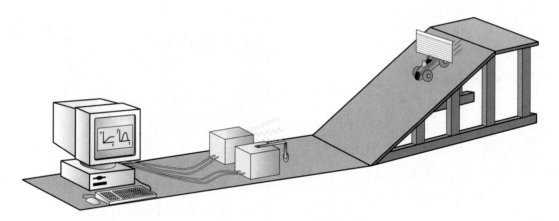

3. Conduct a few runs of the car against the force probe to ensure that the data collection equipment is working properly.

4. Attach a cushioning material to the front of the car. Conduct a number of runs with the same type of cushioning. Make sure that the car is coasting down the same slope from the same position each time.

✎ a) Make copies of the velocity vs. time and force vs. time graphs that are displayed on the computer.

5. Repeat **Step 4** using other types of cushioning materials.

✎ a) Record your observations in your log.

6. Use the graphical information you obtained in this activity to answer the following:

a) Compare the force vs. time graphs for the cushioned cars with those for the cars without cushioning.

b) Compare the areas under the force vs. time graphs for all of the experimental trials.

c) Compute the momentum of the car (the product of the mass and the velocity) prior to the collision and compare it with the area under the force vs. time graphs.

d) Summarize your comparisons in a chart.

e) How can impulse be used to explain the effectiveness of cushioning systems?

f) Describe the relationship between impulse ($F\Delta t$) and the change in momentum ($m\Delta v$).

PHYSICS TALK

Change in Momentum and Impulse

Momentum is the product of the mass and the velocity of an object.

$$p = mv$$

where p is the momentum,

m is the mass,

and v is the velocity.

Change in momentum is the product of mass and the change in velocity.

$$\Delta p = m\Delta v$$

Impulse = change in momentum

$$F\Delta t = m\Delta v$$

133

Sample Problem

A vehicle has a mass of 1500 kg. It is traveling at 15.0 m/s. Calculate the change in momentum required to slow the vehicle down to 5.0 m/s.

Strategy: You can use the equation for calculating the change in momentum.

$$\Delta p = m \Delta v$$

Recall that the Δ symbol means "the change in." If you know the final and initial velocities you can write this equation as:

$$\Delta p = m(v_f - v_i)$$

where v_f is the final velocity and

v_i is the initial velocity.

Givens:

$m = 1500$ kg

$v_f = 5.0$ m/s

$v_i = 15.0$ m/s

Solution:

$$\Delta p = m \ (v_f - v_i)$$

$$= 1500 \text{ kg } (5.0 \text{ m/s} - 15.0 \text{ m/s})$$

$$= 1500 \text{ kg } (-10.0 \text{ m/s})$$

$$= -15{,}000 \text{ kg} \cdot \text{m/s}$$

Reflecting on the Activity and the Challenge

What you learned in this activity better prepares you to defend the design of your safety system. The principles of momentum and impulse must be used to justify your design. Previously, you discovered objects with greater mass are more difficult to stop than smaller ones. You determined that increasing the velocity of objects also makes them more difficult to stop. Objects that have a greater mass or greater velocity have greater momentum.

Linking the two ideas together allows you to begin examining the relationship between momentum and impulse. For a large momentum change in a short time, a large force is required. A crushed rib cage or broken leg bones often result. The change in the momentum can be defined by the impulse on the object.

What device will you use to increase the stopping time for the **Chapter Challenge** activity? Make sure that you include impulse and change in momentum in your report. Your design features must be supported by the principles of physics.

Physics To Go

1. Helmets are designed to protect cyclists. How would the designer of helmets make use of the concept of impulse to improve their effectiveness?

2. The Congress of the United States periodically reviews federal legislation that relates to the design of safer cars. For many years, one regulation was that car bumpers must be able to withstand a 5 mph collision. What was the intent of this regulation? The speed was later lowered to 3 mph. Why? Should it be changed again?

3. If a car has a mass of 1200 kg and an initial velocity of 10 m/s (about 20 mph) calculate the change in momentum required to:

   a) bring it to rest
   b) slow it to 5 m/s (approximately 10 mph)

4. If the braking force for a car is 10,000 N, calculate the impulse if the brake is applied for 1.2 s. If the car has a mass of 1200 kg, what is the change in velocity of the car over this 1.2 s time interval?

5. A 1500-kg car, traveling at 5.0 m/s after braking, strikes a power pole and comes to a full stop in 0.1 s. Calculate the force exerted by the power pole and brakes required to stop the car.

6. For the car described in **Question 5**, explain why a breakaway pole that brings the car to rest after 2.8 s is safer than the conventional power pole.

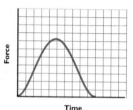

7. Write a short essay relating your explanation for the operation of the cushioning systems to the explanation of the operation of the air bags.

8. Explain why a collapsible steering wheel is able to help prevent injuries during a car crash.

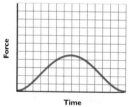

9. Compare and contrast the two force vs. time graphs shown.

Stretching Exercises

Package an egg in a small container so that the egg will not break upon impact. Your teacher will provide the limitations in the construction of your package. You may be limited to two pieces of paper and some tape. You may be limited to a certain size package or a package of a certain weight. Bring your package to class so that it can be compared in a crash test with the other packages.

(Hint: Place each egg in a plastic bag before packaging to help avoid a messy cleanup.)

PHYSICS
AT WORK

Mohan Thomas

DESIGNING AUTOMOBILES
THAT SAVE LIVES

Mo is a Senior Project Engineer at General Motors North American Operation's (NAO) Safety Center and his responsibilities include making sure that different General Motors vehicles meet national safety requirements. Several of the design features that Mo has helped to develop have been implemented into vehicles that are now out on the road.

"This is how it works," he explains. "An engineer for a vehicle comes to us here at the Safety Center and requests technical assistance with design features to help them meet the side impact crash regulations required by the government. You have to analyze the physical forces of an event, which involves one car hitting another car on the side and then the door smashing into the driver," he continues. "We'll study the velocity, acceleration, momentum, and inertia in an event, as well as the materials used in the vehicle itself."

"The initial energy of an impact from one vehicle on another," states Mo, "has to be managed by the vehicle that's getting hit. Our goal is to manage the energy in such a way that the occupant in the vehicle being hit is protected. You take the forces that are coming into the vehicle and you redirect them into areas around the occupant. The framework of the car, therefore, is very important to the design, as well as energy-absorbing materials used in the vehicle."

Mo grew up in Chicago, Illinois, and has always enjoyed math and science, but he was also interested in creative writing. He wanted to combine math and science with creative work and has found that combination in the design work of engineering. "The nice part of being at the Safety Center," states Mo, "is that you know that you are contributing to something meaningful. The bottom line is that the formulas and problems that we are working on are meant to save people's lives."

Chapter 2 Assessment

Your design team will develop a safety system for protecting automobile, airplane, bicycle, motorcycle or train passengers. As you study existing safety systems, you and your design team should be listing ideas for improving an existing system or designing a new system for preventing accidents. You may also consider a system that will minimize the harm caused by accidents.

Your final product will be a working model or prototype of a safety system. On the day that you bring the final product to class, the teams will display them around the room while class members informally view them and discuss them with members of the design team. At this time, class members will generate questions about each others' products. The questions will be placed in envelopes provided to each team by the teacher. The teacher will use some of these questions during the oral presentations on the next day. The product will be judged according to the following:

1. The quality of your safety feature enhancement and the working model or prototype.

2. The quality of a 5-minute oral report that should include:

- **the need for the system**

- **the method used to develop the working model**

- **demonstration of the working model**

- **discussion of the physics concepts involved**

- **description of the next-generation version of the system**

- **answers to questions posed by the class**

3. The quality of a written and/or multimedia report including:

- **the information from the oral report**

- **documentation of the sources of expert information**

- **discussion of consumer acceptance and market potential**

- **discussion of the physics concepts applied in the design of the safety system**

Criteria

Review the criteria that were agreed to at the beginning of the chapter. If they require modification, come to an agreement with the teacher and the class.

Your project should be judged by you and your design team according to the criteria before you display and share it with your class. Being able to judge the quality of your own work before you submit it is one of the skills that will make you a "treasured employee"!

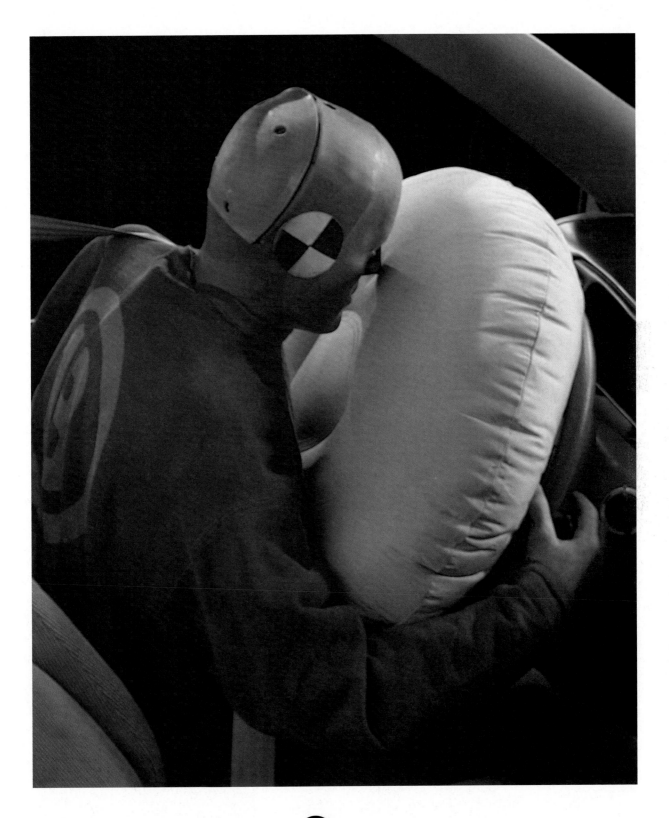

Physics You Learned

Newton's First Law of Motion (inertia)

Pressure (N/m^2)

$$\text{Pressure} = \frac{\text{force}}{\text{area}} \quad (P = \frac{F}{a})$$

Distance vs. time relationships

Time interval =
$$\text{time}_{(final)} - \text{time}_{(initial)} \; (\Delta t = t_f - t_i)$$

Impulse (N × *time*),

Impulse =
force × time interval (Impulse = $F \times \Delta t$)

Stopping distance

Newton's Second Law of Motion, constant acceleration, net force
Force = mass × acceleration
$$F = ma$$

Momentum = mass × velocity ($p = mv$)

Conservation of momentum

Change in momentum is affected by mass and change in velocity

Change in momentum =
mass × change in velocity ($\Delta p = m\,\Delta v$)

Impulse =
change in momentum ($F\Delta t = m\,\Delta v$)

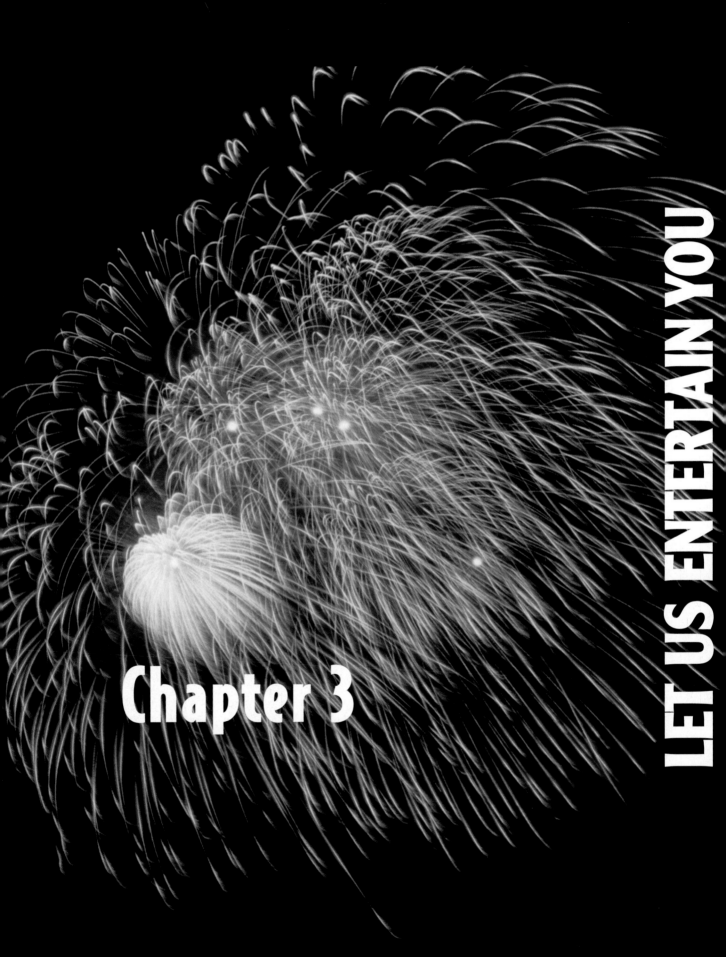

Chapter 3

LET US ENTERTAIN YOU

Scenario

Most entertainment today comes from the communication of sound and light signals. You look forward to television shows, movies, and rock concerts. The sound signals that entertain you come from voices or musical instruments. Light signals make the images you see on TV or in the movies. Specially designed light patterns add to the effect of a rock concert.

Challenge

You have been made part of a committee to design a two- to four-minute sound and light show to entertain other students your age. But unlike the professionals, you have neither the funds nor the technology available to rock stars or MTV™ productions. All the sounds you use must come from musical instruments or sound makers that you build yourself, or from human voices. Some of these sounds may be prerecorded and then played back during your show. If your teacher has a laser and is willing to allow you to use it, you may do so. All other light must come from conventional household lamps.

Criteria

Work with your classmates to agree on the relative importance of the following assessment criteria. Each item in the list has a point value given after it, but your class must decide what kind of grading system to use.

1. The variety and number of physics concepts used to produce the light and sound effects: 30 points

   four or more concepts: 30 points

   three concepts: 25 points

   two concepts: 20 points

   one concept: 10 points

2. Your understanding of the physics concepts: 40 points

   Following your production, you will be asked to:

   a) Name the physics concepts that you used. 10 points
   b) Explain each concept. 10 points
   c) Give an example of something that each concept explains or an example of how each concept is used. 10 points
   d) Explain why each concept is important. 10 points

As a class, you will have to decide if your answers will be in an oral report or a written report.

3. Entertainment value: 30 points

   Your class will need to decide on a way to assign points for creativity. Note that an entertaining and interesting show need not be loud or bright.

   You will have a chance later in the chapter to again discuss these criteria. At that time, you may have more information on the concepts and how you might produce your show. You may want to then propose changes in the criteria and the point values.

Activity 1 Making Waves

GOALS

In this activity you will:

- Observe the motion of a pulse.
- Measure the speed of a wave.
- Observe standing waves.
- Investigate the relationship among wave speed, wavelength, and frequency.
- Make a model of wave motion.

What Do You Think?

On December 26, 2004, one of the largest tsunamis (tidal waves) hit many countries in the Indian Ocean in Southeast Asia, causing massive damage. Some of the waves reached almost 35 meters (100 feet) in height.

- **How does water move to make a wave?**
- **How does a wave travel?**

Record your ideas about these questions in your *Active Physics* log. Be prepared to discuss your responses with your small group and with your class.

For You To Do

1. In an area free of obstacles, stretch out a Slinky® so the turns are a few centimeters apart. Mark the positions of the end of the Slinky by sticking pieces of tape on the floor. Measure the distance between the pieces of tape.

a) Record the distance between the pieces of tape in your log.

2. With the Slinky stretched out to the tape, grab the spring near one end, as shown in the drawing, and pull sideways 20 cm and back. To move it correctly, move your wrist as if snapping a whip. Observe what happens. You have made a transverse pulse.

a) In what direction does the spring move as the pulse goes by?

b) A dictionary definition of *transverse* is: "Situated or lying across." Why is *transverse* a good name for the wave you observed?

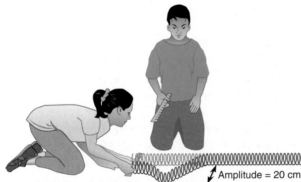

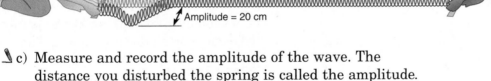

Amplitude = 20 cm

c) Measure and record the amplitude of the wave. The distance you disturbed the spring is called the amplitude. The amplitude tells how much the spring is displaced.

3. After you have experimented with making pulses, measure the speed of the pulse. You will need to measure the time it takes the pulse to go the length of the spring. Take several measurements and then average the values.

a) Record your data in the second and third rows of a table like the one on the following page.

Coordinated Science for the 21st Century

Amplitude	Time for pulse to travel from one end to the other	Average time	Speed = $\dfrac{\text{length of spring}}{\text{average time}}$

4. Measure the speed of the pulses for two other amplitudes, one larger and one smaller than the value used in **Step 3**.

a) Record the results in the table in your log.

b) How does the speed of the pulse depend on the amplitude?

5. Now make waves! Swing one end back and forth over and over again along the floor. The result is called a periodic wave.

a) Describe the appearance of the periodic wave you created.

6. To make these waves look very simple, change the way you swing the end until you see large waves that do not move along the spring. You will also see points where the spring does not move at all. These waves are called standing waves.

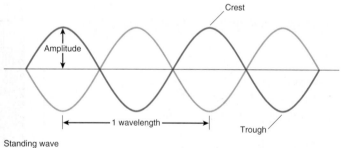

Standing wave

7. The distance from one crest (peak) of a wave to the next is called the wavelength. Notice that you can find the wavelength by looking at the points where the spring does not move. The wavelength is twice the distance between these points. Measure the wavelength of your standing wave.

a) Record the wavelength of your standing wave in your log.

8. You can also measure the wave frequency. The frequency is the number of times the wave moves up and down each second. Measure the frequency of your standing wave. (Hint: Watch the hands of the person shaking the spring. Time a certain number of back-and-forth motions. The frequency is the number of back-and-forth motions of the hand in one second.)

🖊 a) Record the wave frequency in your log. The unit of frequency is the hertz (Hz).

9. Make several different standing waves by changing the wave frequency. Try to make each standing wave shown in the drawings (at right). Measure the wavelength. Measure the frequency.

🖊 a) Record both in a table like the one below.

Wavelength (m/cycle)	Frequency (cycles/s or Hz)	Speed (m/s) wavelength × frequency

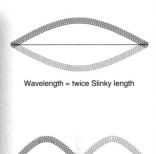

Wavelength = twice Slinky length

Wavelength = Slinky length

🖊 b) For each wave, calculate the product of the wavelength and the frequency. Compare these values with the average speed of the pulse that you found in **Steps 3** and **4** above.

Wavelength = 2/3 Slinky length

10. All the waves you have made so far are transverse waves. A different kind of wave is the compressional (or longitudinal) wave. Have the members of your group stretch out the Slinky between the pieces of tape and hold the ends firmly. To make a compressional wave, squeeze part of the spring and let it go. Measure the speed of the compressional wave and compare it with the speed of the transverse wave.

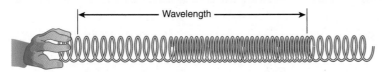

Wavelength

🖊 a) Record your results in a table partly like the one after **Step 3**.

🖊 b) In what direction does the Slinky move as the wave goes by?

Physics Words

periodic wave: a repetitive series of pulses; a wave train in which the particles of the medium undergo periodic motion (after a set amount of time the medium returns to its starting point and begins to repeat its motion).

crest: the highest point of displacement of a wave.

trough: the lowest point on a wave.

amplitude: the maximum displacement of a particle as a wave passes; the height of a wave crest; it is related to a wave's energy.

✎ c) A dictionary definition of *compressional* is: "*a*. The act or process of compressing. *b*. The state of being compressed." A dictionary definition of *longitudinal* is: "Placed or running lengthwise." Explain why *compressional* or *longitudinal wave* is a suitable name for this type of wave.

11. To help you understand waves better, construct a wave viewer by cutting a slit in a file card and labeling it as shown.

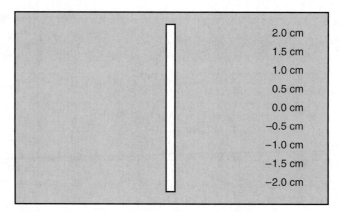

2.0 cm
1.5 cm
1.0 cm
0.5 cm
0.0 cm
−0.5 cm
−1.0 cm
−1.5 cm
−2.0 cm

12. Make a drawing of a transverse wave on a strip of adding machine tape. Place this strip under the wave viewer so you can see one part of the wave through the slit.

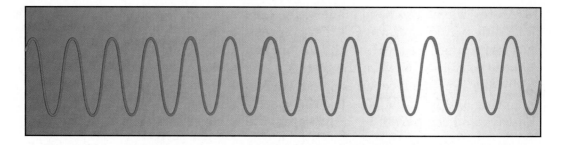

13. With the slit over the tape, pull the tape so that the wave moves. You will see a part of the wave (through the slit) going up and down.

14. Draw waves with different wavelengths on other pieces of adding machine tape. Put these under the slit and pull the adding machine tape at the same speed.

✎ a) Describe what you see.

FOR YOU TO READ

Wave Vocabulary

In this activity, you were able to send energy from one end of the Slinky to the other. You used chemical energy in your muscles to create mechanical energy in your arms that you then imparted to the Slinky. The Slinky had energy. A card at the other end of the Slinky would have moved once the wave arrived there. The ability to move the card is an indication that energy is present. The total energy is transferred but it is always conserved.

Of course, you could have used that same mechanical energy in your arm to throw a ball across the room. That would also have transferred the energy from one side of the room to the other. It would have also moved the card.

There is a difference between the Slinky transferring the energy as a wave and the ball transferring the energy. The Slinky wave transferred the energy, but the Slinky basically stayed in the same place. If the part of the Slinky close to one end were painted red, the red part of the Slinky would not move across the room. The Slinky wave moves, but the parts of the Slinky remain in the same place as the wave passes by. A wave can be defined as a transfer of energy with no net transfer of mass.

Leonardo da Vinci stated that "the wave flees the place of creation, while the water does not." The water moves up and down, but the wave moves out from its center.

In discussing waves, a common vocabulary helps to communicate effectively. You observed waves in the lab activity. We will summarize some of the observations here and you can become more familiar with the terminology.

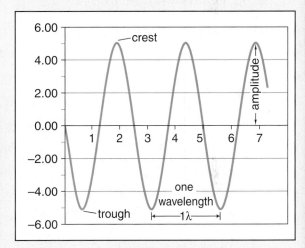

A **periodic wave** is a repetitive series of pulses. In the periodic wave shown in the diagram above, the highest point is called the **crest.** The lowest point is called the **trough.** The maximum disturbance, the **amplitude**, is 5.00 cm. Notice that this is the height of the crest or the height of the trough. It is *not* the distance from the crest to the trough.

The **wavelength** of a periodic wave is the distance between two consecutive points in phase. The distance between two crests is one wavelength or 1 λ. (The Greek letter lambda is used to signify wavelength.) The wavelength of the wave in the diagram is 2.5 cm.

The amplitude of a periodic wave is the maximum disturbance. A large amplitude corresponds to a large energy. In sound, the large amplitude is a loud sound. In light, the large amplitude is a bright light. In Slinkies, the large amplitude is a large disturbance.

The wavelength of the wave in the diagram is 2.5 cm. It is the distance between two crests or the distance between two troughs.

The **frequency** is the number of vibrations occurring per unit time. A frequency of 10 waves per second may also be referred to as 10 vibrations per second, 10 cycles per second, 10 per second, 10 s^{-1}, 10 Hz (hertz). The human ear can hear very low sounds (20 Hz) or very high sounds (20,000 Hz). You can't tell the frequency by examining the wave in the diagram. The "snapshot" of the wave is at an instant of time. To find the frequency, you have to know how many crests pass by a point in a given time.

The **period**, T, of a wave is the time it takes to complete one cycle. It is the time required for one crest to pass a given point. The period and the frequency are related to one another. If three waves pass a point every second, the frequency is three waves per second. The period would be the time for one wave to pass the point, which equals $\frac{1}{3}$ s. If 10 waves pass a point every second, the frequency is 10 waves per second. The period would be the time for one wave to pass the point, which equals $\frac{1}{10}$ second. Mathematically, this relationship can be represented as:

$$T = \frac{1}{f} \text{ or } f = \frac{1}{T}$$

Points in a periodic wave can be "in phase" if they have the same displacement and are moving in the same direction. All crests of the wave shown below are "in phase."

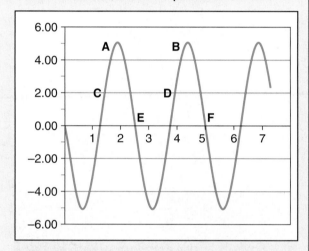

In the wave shown, the following pairs of points are in phase A and B, C and D, E and F.

A **node** is a spot on a standing wave where the medium is motionless. There are places along the medium that do not move as the standing wave moves up and down. The locations of these nodes do not change as the standing wave vibrates. A **transverse wave** is a wave in which the motion of the medium is perpendicular to the motion of the wave. A **longitudinal wave** is a wave in which motion of the medium is parallel to the direction of the motion of the wave.

PHYSICS TALK

Calculating the Speed of Waves

You can find the speed of a wave by measuring the distance the crest moves during a certain change in time.

$$\text{speed} = \frac{\text{change in distance}}{\text{change in time}}$$

In mathematical language:

$$v = \frac{\Delta d}{\Delta t}$$

where v = speed

d = distance

t = time

Suppose the distance the crest moves is 2 m in 0.2 s. The speed can be calculated as follows:

$$v = \frac{\Delta d}{\Delta t}$$

$$= \frac{2 \text{ m}}{0.2 \text{ s}}$$

$$= 10 \text{ m/s}$$

The distance from one crest of a wave to the next is the wavelength. The number of crests that go by in one second is the frequency. Imagine you saw five crests go by in one second. You measure the wavelength to be 2 m. The frequency is 5 crests/second, so the speed is $(5 \times 2) = 10$ m/s. Thus, the speed can also be found by multiplying the wavelength and the frequency.

$$\text{speed} = \text{frequency} \times \text{wavelength}$$

In mathematical language:

$$v = f\lambda$$

where v = speed

f = frequency

λ = wavelength

$\rightarrow$

Physics Words

wavelength: the distance between two identical points in consecutive cycles of a wave.

frequency: the number of waves produced per unit time; the frequency is the reciprocal of the amount of time it takes for a single wavelength to pass a point.

period: the time required to complete one cycle of a wave.

node: a point on a standing wave where the medium is motionless.

transverse pulse or wave: a pulse or wave in which the motion of the medium is perpendicular to the motion of the wave.

longitudinal pulse or wave: a pulse or wave in which the motion of the medium is parallel to the direction of the motion of the wave.

Standing waves happen anywhere that the length of the Slinky and the wavelength have a particular mathematical relationship. The length of the Slinky must equal $\frac{1}{2}$ wavelength, 1 wavelength, $1\frac{1}{2}$ wavelengths, 2 wavelengths, etc. Mathematically, this can be stated as:

$$L = \frac{n\lambda}{2}$$

where L is the length of the Slinky,
λ is the wavelength
n is a number (1, 2, 3...)

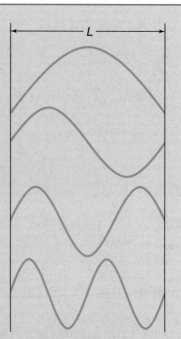

Sample Problem I

You and your partner sit on the floor and stretch out a Slinky to a length of 3.5 m. You shake the Slinky so that it forms one loop between the two of you. Your partner times 10 vibrations and finds that it takes 24.0 s for the Slinky to make these vibrations.

a) How much of a wave have you generated and what is the wavelength of this wave?

Strategy: Draw a sketch of the wave you have made and you will notice that it looks like one-half of a total wave. It is! This is the maximum wavelength that you can produce on this length of Slinky. You can use the equation that shows the relationship between the length of the Slinky and the wavelength.

Givens:

$L = 3.5$ m

$n = 1$

Solution:

$$L = \frac{n\lambda}{2}$$

Rearrange the equation to solve for λ.

$$\lambda = \frac{2\,L}{n}$$

$$= \frac{2\,(3.5\text{ m})}{1}$$

$$= 7.0\text{ m}$$

b) What is the period of vibration of the wave?

Strategy: The period is the amount of time for one vibration. You have the amount of time for 10 vibrations.

Solution:

$$T = \frac{\text{time for 10 vibrations}}{10} = \frac{24.0\text{ s}}{10} = 2.4\text{ s}$$

c) Calculate the wave frequency.

Strategy: The frequency represents the number of vibrations per second. It is the reciprocal of the period.

Given:

$$T = 2.4\text{ s}$$

Solution:

$$f = \frac{\text{number of vibrations}}{\text{time}} \text{ or } f = \frac{1}{T}$$

$$= \frac{1}{2.4\text{ s}}$$

$$= 0.42\text{ vibrations per second}$$

$$= 0.42\text{ s}^{-1} \text{ or } 0.42\text{ Hz}$$

→

153

d) Determine the speed of the wave you have generated on the Slinky.

Strategy: The speed of the wave may be found by multiplying the frequency times the wavelength.

Givens:

$$f = 0.42 \text{ Hz}$$

$$\lambda = 7.0 \text{ m}$$

Solution:

$$v = f\lambda$$

$$= 0.42 \text{ Hz} \times 7.0 \text{ m}$$

$$= 29 \text{ m/s}$$

Remember that Hz may also be written as 1/s so the unit of speed is m/s.

Sample Problem 2

You stretch out a Slinky to a length of 4.0 m, and your partner generates a pulse that takes 1.2 s to go from one end of the Slinky to the other. What is the speed of the wave on the Slinky?

Strategy: Use your kinematics equation to determine the speed.

Givens:

$$d = 4.0 \text{ m}$$

$$t = 1.2 \text{ s}$$

Solution:

$$v = \frac{d}{t}$$

$$= \frac{4.0 \text{ m}}{1.2 \text{ s}}$$

$$= 3.3 \text{ m/s}$$

Reflecting on the Activity and the Challenge

Slinky waves are easy to observe. You have created transverse and compressional Slinky waves and have measured their speed, wavelength, and frequency. For the **Chapter Challenge**, you may want to create musical instruments. You will receive more guidance in doing this in the next activities. Your instruments will probably not be made of Slinkies. You may, however, use strings that behave just like Slinkies. When you have to explain how your instrument works, you can relate its production of sound in terms of the Slinky waves that you observed in this activity.

Physics To Go

1. a) Four characteristics of waves are amplitude, wavelength, frequency, and speed. For each characteristic, tell how you measured it when you worked with the Slinky.
   b) For each characteristic, give the units you used in your measurement.
   c) Which wave characteristics are related to each other? Tell how they are related.

2. a) Suppose you shake a long Slinky slowly back and forth. Then you shake it rapidly. Describe how the waves change when you shake the Slinky more rapidly.
   b) What wave properties change?
   c) What wave properties do not change?

3. Suppose you took a photograph of a wave on a Slinky. How can you measure wavelength by looking at the photograph?

4. Suppose you mount a video camera on a tripod and aim the camera at one point on a Slinky. You also place a clock next to the Slinky, so the video camera records the time. When you look at the video of a wave going by on the Slinky, how could you measure the frequency?

5. a) What are the units of wavelength?
   b) What are the units of frequency?
   c) What are the units of speed?
   d) Tell how you find the wave speed from the frequency and the wavelength.

e) Using your answer to **Part (d)**, show how the units of speed are related to the units of wavelength and frequency.

6. a) What is a standing wave?
   b) Draw a standing wave.
   c) Add labels to your drawing to show how the Slinky moves.
   d) Tell how to find the wavelength by observing a standing wave.

7. a) Explain the difference between transverse waves and compressional waves.
   b) Slinky waves can be either transverse or compressional. Describe how the Slinky moves in each case.

8. a) When you made standing waves, how did you shake the spring (change the frequency) to make the wavelength shorter?
   b) When you made standing waves, how did you shake the spring (change the frequency) to make the wavelength longer?

9. Use the wave viewer and adding machine tape to investigate what happens if the speed of the wave increases. Pull the tape at different speeds and report your results.

10. A Slinky is stretched out to 5.0 m in length between you and your partner. By shaking the Slinky at different frequencies, you are able to produce waves with one loop, two loops, three loops, four loops, and even five loops.
    a) What are the wavelengths of each of the wave patterns you have produced?
    b) How will the frequencies of the wave patterns be related to each other?

11. A tightrope walker stands in the middle of a high wire that is stretched 10 m between the two platforms at the ends of the wire. He is bouncing up and down, creating a standing wave with a single loop and a period of 2.0 s.

a) What is the wavelength of the wave he is producing?
b) What is the frequency of this wave?
c) What is the speed of the wave?

12. A clothesline is stretched 9 m between two trees. Clothes are hung on the line as shown in the diagram. When a particular standing wave is created in the line, the clothes remain stationary.

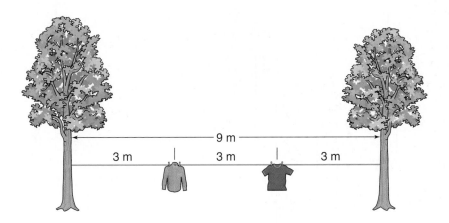

a) What is the term for the positions occupied by the clothes?
b) What is the wavelength of this standing wave?
c) What additional wavelengths could exist in the line such that the clothes remain stationary?

13. During the Slinky lab, your partner generates a wave pulse that takes 2.64 s to go back and forth along the Slinky. The Slinky stretches 4.5 m along the floor. What is the speed of the wave pulse on the Slinky?

14. A drum corps can be heard practicing at a distance of 1.6 km from the field. What is the time delay between the sound the drummer hears ($d = 0$ m) and the sound heard by an individual 1.6 km away? (Assume the speed of sound in air to be 340.0 m/s.)

Activity 2 Sounds in Strings

GOALS

In this activity you will:

- Observe the effect of string length and tension upon pitch produced.
- Control the variables of tension and length.
- Summarize experimental results.
- Calculate wavelength of a standing wave.
- Organize data in a table.

What Do You Think?

When the ancient Greeks made stringed musical instruments, they discovered that cutting the length of the string by half or two-thirds produced other pleasing sounds.

- **How do guitarists or violinists today make different sounds?**

Record your ideas about this question in your *Active Physics* log. Be prepared to discuss your responses with your small group and with your class.

For You To Do

1. Carefully mount a pulley over one end of a table. Securely clamp one end of a string to the other end of the table.

2. Tie the other end of the string around a mass hanger. Lay the string over the pulley. Place a pencil under the

string near the clamp, so the string can vibrate without hitting the table, as shown in the drawing.

3. Hang one 500-g mass on the mass hanger. Pluck the string, listen to the sound, and observe the string vibrate.

✎ a) Record your observations in your log in a table similar to the following:

⚠ **Make sure the area under the hanging mass is clear (no feet, legs). Also monitor the string for fraying.**

Length of vibrating string	Load on mass hanger	Pitch (high, medium, low)

4. Use a key or some other small metal object. Press this object down on the string right in the middle, to hold the string firmly against the table. Pluck each half of the string.

✎ a) Record the result in your table.

5. To change the string length, press down with the key at the different places shown in the diagrams on the next page. Pluck each part of the string.

✎ a) Record the results in your table.

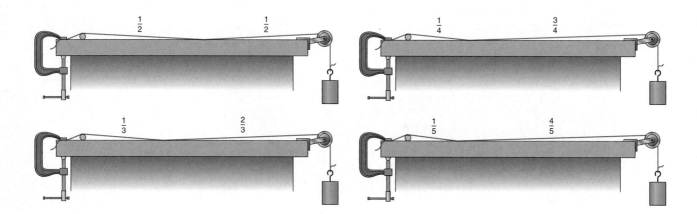

6. When you pluck the string, it does not move at the ends. Look at the drawing under **Step 9** of the **For You To Do** section in **Activity 1**. Measure the length of your string, and find the wavelength of the vibration for each string length.

✎ a) Record the wavelength in your table.

✎ b) Look over the data in your table. Make a general statement about what happens to the pitch you hear as you change the length of the string.

⚠️ **Make sure the string is capable of holding 2 kg.**

7. Remove the key, so the string is its original length. Pluck the string. To investigate the effect of tightening the string, add a second 500-g mass to the mass hanger. Pluck the string again, observe the vibration, and listen to the pitch of the sound.

✎ a) Make up a table to record your data in your log.

✎ b) Add a description of the pitch of the sound to your table. Continue adding weights and observing the sound until the total mass is 2000 g.

✎ c) Look over your data. As the mass increases, the string becomes tighter, and its tension increases. Make a general statement about what happens to the pitch you hear as you change the tension on the string.

FOR YOU TO READ

Changing the Pitch

Sound comes from vibration. You observed the vibration of the string as it produced sound. You investigated two of the variables that affect the sound of a vibrating string.

When you pushed the vibrating string down against the table, the length of the string that was vibrating became shorter. Shortening the string increased the **pitch** (resulted in a higher pitch). In the same way, a guitarist or violinist pushes the string against the instrument to shorten the length that vibrates and increases the pitch.

When you hung weights on the end of the string, that increased the pitch too. These weights tightened the string, so they created more tension in it. As the string tension increased, the pitch of the sound also increased. In tuning a guitar or violin, the performer changes the string tension by turning a peg attached to one end of a string. As the peg pulls the string tighter, the pitch goes up.

Combining these two results into one expression, you can say that increasing the tension or decreasing the length of the string will increase the pitch.

The string producing the pitch is actually setting up a standing wave between its endpoints. The length of the string determines the wavelength of this standing wave. Twice the distance between the endpoints is the wavelength of the sound. The pitch that you hear is related to the frequency of the wave. The higher the pitch, the higher the frequency. The speed of the wave is equal to its frequency multiplied by its wavelength.

$$v = f\lambda$$

where v = speed

f = frequency

λ = wavelength

If the speed of a wave is constant, a decrease in the wavelength will result in an increase in the frequency or a higher pitch. A shortened string produces a higher pitch.

Reflecting on the Activity and the Challenge

Part of the **Chapter Challenge** is to create a sound show. In this activity you investigated the relationship of pitch to length of the string and tension of the string: the shorter the string, the higher the pitch; the greater the tension, the higher the pitch. You also learned that the string is setting up a standing wave between its two ends, just like the standing wave that you created in the Slinky in **Activity 1**. That's the physics of stringed instruments! If you wanted to create a stringed or multi-string instrument for your show, you would now know how to adjust the length and tension to produce the notes you want. If you were to make such a stringed instrument, you could explain how you change the pitch by referring to the results of this activity.

Physics Words

pitch: the quality of a sound dependent primarily on the frequency of the sound waves produced by its source.

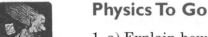

Physics To Go

1. a) Explain how you can change the tension of a vibrating string.
   b) Tell how changing the tension changes the pitch.

2. a) Explain how you can change the length of a vibrating string.
   b) Tell how changing the length changes the sound produced by the string.

3. How would you change both the tension and the length and keep the pitch the same?

4. Suppose you changed both the length and the tension of the string at the same time. What would happen to the sound?

5. a) For the guitar and the piano, tell how a performer plays different notes.
   b) For the guitar and the piano, tell how a performer (or tuner) changes the pitch of the strings to tune the instrument.

6. a) Look at a guitar. Find the tuners (at the end of the neck). Why does a guitar need tuners?
   b) What is the purpose of the frets on a guitar?
   c) Does a violin or a cello have frets?
   d) Why do a violinist and a cellist require more accuracy in playing than a guitarist?

7. a) Using what you have learned in this activity, design a simple two-stringed instrument.
   b) Include references to wavelength, frequency, pitch, and standing waves in your description.
   c) Use the vocabulary of wavelength, frequency, and standing waves from **Activity 1** to describe how the instrument works.

Stretching Exercises

1. Set up the vibrating string as you did in the preceding **For You To Do**. This time, you will measure the frequency of the sound. Set up a frequency meter on your computer. Pick up the sound with a microphone. Investigate how changing the length of the string changes the frequency of the sound. Create a graph to describe the relationship.

2. Set up the vibrating string, computer, and microphone as you did in **Stretching Exercise 1**. This time, investigate how changing the string tension changes the frequency of the sound. Create a graph to describe the relationship.

3. Design an investigation to find how the diameter (thickness) of the string or the type of material the string is made of affects the pitch you hear. Submit your design to your teacher for approval before proceeding to carry out your experiments.

Activity 3　Sounds from Vibrating Air

GOALS

In this activity you will:

- Identify resonance in different kinds of tubes.
- Observe how resonance pitch changes with length of tube.
- Observe the effect of closing one end of the tube.
- Summarize experimental results.
- Relate pitch observations to drawings of standing waves.
- Organize observations to find a pattern.

What Do You Think?

The longest organ pipes are about 11 m long. A flute, about 0.5 m long, makes musical sound in the same way.

• How do a flute and organ pipes produce sound?

Record your ideas about this question in your *Active Physics* log. Be prepared to discuss your responses with your small group and with your class.

For You To Do

1. Carefully cut a drinking straw in half. Cut one of the halves into two quarters. Cut one of the quarters into two eighths. Pass one part of the straw out to one member of your group.

2. Gently blow into the top of the piece of straw.

a) Describe what you hear.

b) Listen as the members of your group blow into their straw pieces one at a time. Describe what you hear.

c) Write a general statement about how changing the length of the straw changes the pitch you hear.

3. Now cover the bottom of your straw piece and blow into it again. Uncover the bottom and blow again.

a) Compare the sound the straw makes when the bottom is covered and then uncovered.

b) Listen as the members of your group blow into their straw pieces, with the bottom covered and then uncovered. Write a general statement about how changing the length of the straw changes the pitch you hear when one end is covered.

4. Obtain a set of four test tubes. Leave one empty. Fill the next halfway with water. Fill the next three-quarters of the way. Fill the last one seven-eighths of the way.

5. Give each test tube to one member of your group. Blow across your test tube.

 **Make sure the outsides of the tubes are dry.**

a) Describe what you hear.

b) Listen as the members of your group blow, one at a time, across their test tubes. Record what you hear.

c) What pattern do you find in your observations?

d) Compare the results of blowing across the straws with blowing across the test tubes. How are the results consistent?

Physics Words

diffraction: the ability of a sound wave to spread out as it emerges from an opening or moves beyond an obstruction.

PHYSICS TALK

Vibrating Columns of Air

The sound you heard when you blew into the straw and test tube was

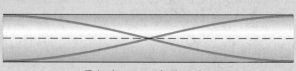

Tube is open at both ends.
1/2 wavelength fits in straw.

produced by a standing wave. If both ends of the straw are open, the air at both ends moves back and forth. The above drawing shows the movement of the air as a standing wave.

When you covered the other end of the straw, you prevented the

Tube is closed at one end.
1/4 wavelength fits in straw.

air from moving at the covered end. This drawing shows the movement of the air as a standing wave.

The velocity of a wave is equal to the frequency multiplied by the wavelength. Therefore,

$$\text{frequency} = \frac{\text{wave speed}}{\text{wavelength}}$$

Using mathematical symbols,

$$f = \frac{v}{\lambda}$$

As the wavelength increases, the frequency decreases. The wavelength in the open straw is half the wavelength in the straw closed at one end. This equation predicts that the frequency of the standing wave in the open straw is twice the frequency of the standing wave in the straw closed at one end.

FOR YOU TO READ
Compressing Air to Make Sound

Sound is a compression wave. The molecules of air bunch up or spread apart as the sound wave passes by.

At the end where the tube is closed, the air cannot go back and forth, because its motion is blocked by the end of the tube. That's why the wave's amplitude goes to zero at the closed end. At the open end, the amplitude is as large as it can possibly be. This back-and-forth motion of air at the open end makes a sound wave that moves from the tube to your ear.

In the compressional Slinky wave, the coils of the Slinky bunched up in a similar fashion when the Slinky wave passed by.

Wave Diffraction

As the sound wave leaves the test tube in this activity, it spreads out. In the same way, when you speak to a friend, the sound waves leave your mouth and spread out. You can speak to a group of friends because the sound leaves your mouth and moves out to the front and to the sides.

This ability of the sound wave to spread out as it emerges from an opening is called **diffraction**. The smaller the opening, the more spreading of the sound. The spreading of the wave as it emerges from two holes can be shown with a diagram.

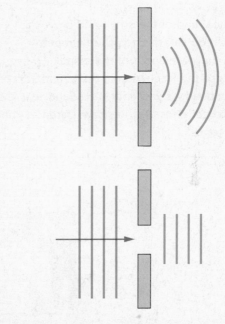

The wave on the top is going through a small opening (in comparison to its wavelength) and diffracts a great deal. The wave on the bottom is going through a large opening (in comparison to its wavelength) and shows little diffraction.

Cheerleaders use a megaphone to limit the diffraction. With a megaphone, the mouth opening becomes larger. The sound wave spreads out less, and the cheering crowd in front of the cheerleader hears a louder sound.

A new musical instrument that you can make uses a straw instead of a test tube.

Take a straw and cut the ends to form a V as show below.

Flatten the V end of the straw and blow this "trumpet." You can shorten the trumpet, decrease the wavelength of the standing wave, and increase the frequency of the sound. Try making a sound. As you emit the sound, use scissors to cut ends off the straw. Listen to the different tones.

You can probably make a trombone by inserting one straw within another.

The sound diffracts from the small opening. You can add a horn to one end of the straw and limit the diffraction. The effect will be that the sound appears louder because it doesn't spread out. You can make a horn out of a piece of paper, as shown.

You may want to adapt this idea of a trumpet and a megaphone and make diffraction a part of your light and sound show.

Reflecting on the Activity and the Challenge

In this activity you have observed the sounds produced by different kinds of pipes. If the pipe is cut to a shorter length, the pitch of the sound increases. Also, when the pipe is open at both ends, the pitch is much higher than if the pipe were open at only one end. You have seen how simple drawings of standing waves in these tubes help you find the wavelength of the sound. If the tube is closed at one end, the air has zero displacement at that end. If the tube is open at one end, the air has maximum displacement there.

For your sound show, you may decide to create some "wind" instruments using test tubes or straws, or other materials approved by your teacher. When it comes time to explain how these work, you can refer to this activity to get the physics right.

Physics To Go

1. a) You can produce a sound by plucking a string or by blowing into a pipe. How are these two ways of producing sound similar?
   b) How are these two ways different?

2. a) For each piece of straw your group used, make a full-sized drawing to show the standing wave inside. Show both the straw closed at one end and open at both ends.
   b) Next to each drawing of the standing waves, make a drawing, at the same scale, of one full wavelength. You may need to tape together several pieces of paper for this drawing.
   c) Frequency times the wavelength is the wave speed. The speed is the same for all frequencies. From your answer to **Part (b)**, what can you predict about the frequencies of the standing waves in the straw pieces?
   d) How well do your predictions from **Part (c)** agree with your observations in this activity?

3. a) What is the length, in meters, of the longest organ pipe?
   b) Assume this pipe is closed at one end. Draw the standing wave pattern.
   c) For this pipe, how long is the wavelength of this standing wave?
   d) Why does a long wavelength indicate that the frequency will be low? Give a reason for your answer.

4. a) Suppose you are listening to the sound of an organ pipe that is closed at one end. The pipe is 3 m long. What is the wavelength of the sound in the pipe?
   b) The speed of sound in air is about 340 m/s. What is the frequency of the sound wave?
   c) Now suppose you are listening to the sound of an organ pipe that is open at both ends. As before, the pipe is 3 m long. What is the wavelength of the sound in the pipe?
   d) What is the frequency of the sound wave?

5. Suppose you listen to the sound of an organ pipe that is closed at one end. This pipe is only 1 m long. How does its frequency compare with the frequency you found in **Question 4, Part (b)**?

6. Waves can spread into a region behind an obstruction.
   a) What is this wave phenomenon called?
   b) Draw a diagram to illustrate this phenomenon.

Stretching Exercises

1. If you have a good musical ear, add water to eight test tubes to make a scale. Play a simple piece for the class.

2. Obtain a 2- to 3- meter-long piece of a 7- to 10-centimeter-diameter plastic pipe, like that used to filter water in small swimming pools. In an area free of obstructions, twirl the pipe overhead. What can you say about how the sound is formed? Place some small bits of paper on a stool. Twirl the pipe and keep one end right over the stool. What happens to the paper? What does that tell you about the air flowing through the pipe? Try to play a simple tune by changing the speed of the pipe as you twirl it.

3. Carefully cut new straw pieces, as you did in **For You To Do**, **Step 1**. This time, you will measure the frequency of the sound. Set up a frequency meter on your computer. Place the microphone near an open end of the straw.
   As before, each person blows into only one piece of straw. Make the sound and record the frequency. Now cover the end of the straw and predict what frequency you will measure. Make the measurement and compare it with your prediction. Repeat the measurements for all of the lengths of straw. Record your results, and tell what patterns you find.

Activity 4   Reflected Light

GOALS

In this activity you will:

- Identify the normal of a mirror.
- Measure angles of incidence and reflection.
- Observe the relationship between the angle of incidence and the angle of reflection.
- Observe changes in the reflections of letters.
- Identify patterns in multiple reflections.

What Do You Think?

Astronauts placed a mirror on the Moon in 1969 so that a light beam sent from Earth could be reflected back to Earth. By timing the return of the beam, scientists found the distance between the Earth and the Moon. They measured this distance to within 30 cm.

- **How are you able to see yourself in a mirror?**
- **If you want to see more of yourself, what can you do?**

Record your ideas about these questions in your *Active Physics* log. Be prepared to discuss your responses with your small group and with your class.

For You To Do

1. Place a piece of paper on your desk. Carefully aim the laser pointer, or the light from a ray box, so the light beam moves horizontally, as shown on the opposite page.

2. Place a glass rod in the light beam so that the beam spreads up and down. Shine the beam on the piece of paper to be sure the beam passes through the glass rod.

3. Carefully stand the plane mirror on your desk in the middle of the piece of paper. Draw a line on the paper along the

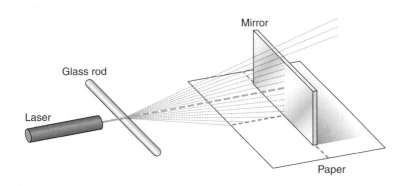

front edge of the mirror. Now remove the mirror and draw a dotted line perpendicular to the first line, as shown. This dotted line is called the **normal**.

4. Aim the light source so the beam approaches the mirror along the normal. Be sure the glass rod is in place to spread out the beam.

✎ a) What happens to the light after it hits the mirror?

5. Make the light hit the mirror at a different angle.

✎ a) What happens now?

✎ b) On the paper, mark three or more dots under the beam to show the direction of the beam as it travels to the mirror. The line you traced shows the incident ray. Also make dots to show the light going away from the mirror. This line shows the reflected ray. Label this pair of rays to show they go together.

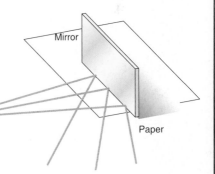

6. Turn the light source so it starts from the same point but strikes the mirror at different angles. For each angle, mark dots on the paper to show the direction of the incident and reflected rays. Also, label each pair of rays.

⚠ **Do not use mirrors with chipped edges. Make sure the ends of the glass rod are polished.**

⚠ **Never look directly at a laser beam or shine a laser beam into someone's eyes. Always work above the plane of the beam and beware of reflections from shiny surfaces.**

Physics Words

normal: at right angles or perpendicular to.

173

7. Most lab mirrors have the reflecting surface on the back. In addition, the light bends as it enters and leaves the glass part of the mirror. In your drawing, the rays may not meet at the mirror surface. Extend the rays until they do meet.

  a) Measure these angles for one pair of your rays.

8. Turn off the light source and remove the paper. Look at one pair of rays. The diagram shows a top view of the mirror, the normal, and an incident and reflected ray. Notice the angle of incidence and the angle of reflection in the drawing. Using a protractor, measure these angles for one pair of rays.

  a) Record your data in a table.

  b) Measure and record the angles of incidence and reflection for all of your pairs of rays.

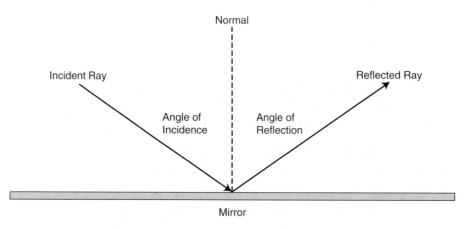

  c) What is the relationship between the angles of incidence and reflection?

  d) Look at the reflected rays in your drawing. Extend each ray back behind the mirror. What do you notice when you have extended all the rays? The position where the rays meet is the location of the image of the light source. All of the light rays leave one point in front of the mirror. The reflected rays all seem to emerge from one point behind the mirror. Wherever you observed the reflection, you would see the source at this point behind the mirror.

  e) Tape a copy of your diagram in your log.

9. Hold the light source, or any object, near the mirror and look at the reflection. Now hold the object far away and again look at the reflection.

✎ a) How is the position of the reflection related to the position of the object?

10. Set up a mirror on another piece of paper, and draw the normal on the paper. Write your name in block capital letters along the normal (a line perpendicular to the mirror). Observe the reflection of your name in the mirror.

✎ a) How can you explain the reflection you see?

✎ b) Which letters in the reflection are closest to the mirror? Which are farthest away?

✎ c) In your log, make a sketch of your name and its reflection.

11. Carefully stand up two mirrors so they meet at a right angle. Be sure they touch each other, as shown in the drawing.

12. Place an object in front of the mirrors.

✎ a) How many images do you see?

✎ b) Slowly change the angle between the mirrors. Make a general statement about how the number of images you see changes as the angle between the mirrors changes.

Physics Words

angle of incidence: the angle a ray of light makes with the normal to the surface at the point of incidence.

angle of reflection: the angle a reflected ray makes with the normal to the surface at the point of reflection.

ray: the path followed by a very thin beam of light.

FOR YOU TO READ

Images in a Plane Mirror

An object like the tip of a nose reflects light in all directions. That is why everybody in a room can see the tip of a nose. Light reflects off a mirror in such a way that the **angle of incidence** is equal to the **angle of reflection**. You can look at the light leaving the tip of a nose and hitting a mirror to see how an image is produced and where it is located. Each **ray** of light leaves the nose at a different angle. Once it hits the mirror, the angle of incidence must equal the angle of reflection. There are now a set of rays diverging from the mirror. If you assume that the light always travels in straight lines, you can extend these rays behind the mirror and find where they "seem" to emerge from. That is the location of the image.

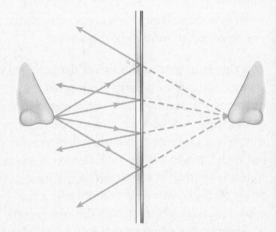

The mirror does such a good job of reflecting that it looks as if there is a tip of a nose (and all other parts of the face) behind the mirror. If you measure the distance of the image behind the mirror, you will find that it is equal to the distance of the nose (object) in front of the mirror. This can also be proved using geometry.

→

175

Diffraction of Light

As you begin to study the reflection of light rays, it is worthwhile to recognize that light is a wave and has properties similar to sound waves.

In studying sound waves, you learned that sound waves are compressional or longitudinal. The disturbance is parallel to the direction of motion of the wave. In sound waves, the compression of the air is left and right as the wave travels to the right. You saw a similar compressional wave using the compressed Slinky.

Light waves are transverse waves. They are similar to the transverse waves of the Slinky. In a transverse wave, the disturbance is perpendicular to the direction of the wave. In the Slinky, the disturbance was up and down as the wave traveled to the right. In light, the fields (the disturbance) are perpendicular to the direction of motion of the waves.

You also read that sound waves diffract—they spread out as they emerge from small openings. You can find out if light waves spread out as they emerge from a small opening. Try this: Take a piece of aluminum foil. Pierce the foil with a pin to create a succession of holes, one smaller than the next. Shine the laser beam through each hole and observe its appearance on a distant wall. You will be able to observe the diffraction of light.

Sample Problem

Light is incident upon the surface of a mirror at an angle of 40°.

a) Sketch the reflection of the ray.

Strategy: The angles of incidence and reflection are always measured from the normal. The Law of Reflection states that the angle of incidence is equal to the angle of reflection. Since the angle of incidence is equal to 40°, the angle of reflection is also 40°.

Given:

$$\theta_i = 40°$$

Solution:

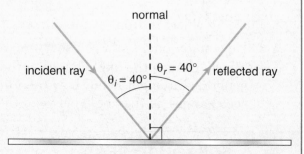

b) At what angle, as measured from the surface of the mirror, did the beam strike the mirror?

Strategy: The angle of incidence is measured from the normal. The question is asking for the complementary angle.

Solution:

$$\theta_i = \theta_r = 40°$$
$$90° - 40° = 50°$$

The angle between the light beam and the mirror is 50°.

Reflecting on the Activity and the Challenge

In this activity you aimed light rays at mirrors and observed the reflections. From the experiment you discovered that the angle of incidence is equal to the angle of reflection. Therefore, you can now predict the path of a reflected light beam. You also experimented with reflections from two mirrors. When you observed the reflection in two mirrors, you found many images that made interesting patterns.

This activity has given you experience with many interesting effects that you can use in your sound and light show. For instance, you may want to show the audience a reflection in one mirror or two mirrors placed at angles. You can probably create a kaleidoscope. You will also be able to explain the physics concept you use in terms of reflected light.

Physics To Go

1. How is the way light reflects from a mirror similar to the way a tennis ball bounces off a wall?

2. a) What is the normal to a plane mirror?

   b) When a light beam reflects from a plane mirror, how do you measure the angle of incidence?

   c) How do you measure the angle of reflection?

   d) What is the relationship between the angle of incidence and the angle of reflection?

3. Make a top-view drawing to show the relationships among the normal, the angle of incidence, and the angle of reflection.

4. a) Suppose you are experimenting with a mirror mounted vertically on a table, like the one you used in this activity. Make a top-view drawing, with a heavy line to represent the mirror and a dotted line to represent the normal.

   b) Show light beams that make angles of incidence of 0°, 30°, 45°, and 60° to the normal.

   c) For each of the above beams, draw the reflected ray. Add a label if necessary to show where the rays are.

5. a) Stand in front of a mirror.

  b) Move your hand toward the mirror. Which way does the reflection move?

  c) Move your hand away from the mirror. Which way does the reflection move?

  d) Use what you learned about the position of the mirror image to explain your answers to **Parts (b)** and **(c)**.

6. Suppose you wrote the whole alphabet along the normal to a mirror in the way you wrote your name in **Step 10** of **For You To Do**.

  a) Which letters would look just like their reflections?

  b) Write three words that would look just like their reflections.

  c) Write three letters that would look different than their reflections.

  d) Draw the reflection of each letter you gave in **Part (c)**.

7. Why is the word *Ambulance* written in an unusual way on the front of an ambulance?

8. Use a ruler and protractor and a ray diagram to locate the image of an object placed in front of a plane mirror. Be careful! You must measure as carefully as you can to obtain the most accurate answer.

⊏⎯⎯⎯⎯⎯⎯⎯⎯⎯⎯⎯⎯⎯⎯⎯⎯⎯⎯⊐ mirror

object ●

9. Locate the image of the lamp shown in the diagram.

10. After reflecting off mirrors A, B, and C, which target will the ray of light hit?

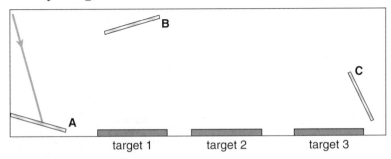

target 1   target 2   target 3

Stretching Exercises

1. Carefully tape together three small mirrors to make a corner reflector. Shine a flashlight down into the corner. Where does the reflected beam go?

2. Build a kaleidoscope by *carefully* inserting two mirrors inside a paper towel holder. You can also use three identical mirrors. Do not force the mirrors into the tube. Tape the edges of the mirrors together, with the mirrored surfaces inside. Describe what you see through your kaleidoscope.

3. Carefully tape together one edge of two mirrors so they can move like a hinge, with the mirrored surfaces facing each other. Place a small object between the mirrors. Investigate how the number of images you see depends on the angle between the mirrors. You will need a protractor to measure this angle. Plot a graph of the results. What mathematical relationship can you find between the angle and the number of images?

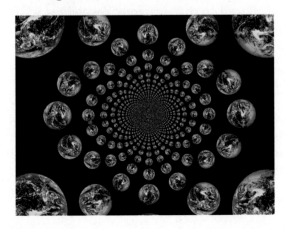

Activity 5   Curved Mirrors

GOALS

In this activity you will:

- Identify the focus and focal length of a curved mirror.

- Observe virtual images in a convex mirror.

- Observe real and virtual images in a concave mirror.

- Measure and graph image distance versus object distance for a convex mirror.

- Summarize observations in a sentence.

What Do You Think?

The curved mirror of the Palomar telescope is five meters across. Mirrors with varying curvatures are used in amusement parks as fun-house mirrors. Store mirrors and car side-view mirrors are also curved.

- **How is what you see in curved mirrors different from what you see in ordinary flat mirrors?**

Record your ideas about this question in your *Active Physics* log. Be prepared to discuss your responses with your small group and with your class.

For You To Do

1. Carefully aim a laser pointer, or the light from a ray box, so the light beam moves horizontally, as you did in the previous activity. Place a glass rod in the light beam so that the beam spreads up and down.

2. Place a convex mirror in the light beam, as shown in the diagram.

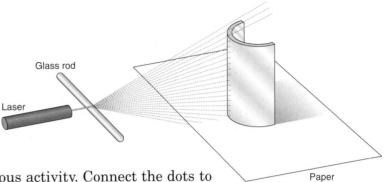

Glass rod

Laser

Paper

3. Shine a beam directly at the center of the mirror. This is the incident beam. Show its path by placing three or more dots on the paper, as you did in the previous activity. Connect the dots to make a straight line. Find the reflected ray and mark its path in a similar way. Label the two lines so you will know they go together.

4. You will move the light source sideways to make a series of parallel beams. To make sure the incident beams are parallel, line up each one with the dots you made to show the incoming beam in **Step 3**. Mark the path of the incoming ray with three dots.

5. Each parallel beam makes a reflected beam. Show the path of each of these reflected rays. Label each incident and reflected beam so you will know that they go together.

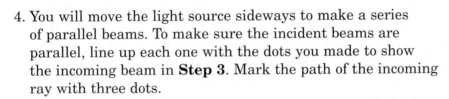

a) Write a sentence to tell what happens to the parallel beams after they are reflected.

b) Make a drawing in your *Active Physics* log to record the path of the light.

6. Remove the mirror. With a ruler, extend each reflected ray backwards to the part of the paper that was behind the mirror.

a) You probably noticed that all the lines converge in a single point. The place where the extended rays meet is called the **focus** of the mirror. The distance from this point to the mirror is called the **focal length**. Measure and record this focal length.

7. Place the concave side of the mirror in the light beam. To help you remember the name *concave*,

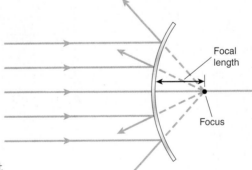

Focal length

Focus

Never look directly at a laser beam or shine a laser beam into someone's eyes. Always work above the plane of the beam and beware of reflections from shiny surfaces.

Physics Words

focus: the place at which light rays converge or from which they appear to diverge after refraction or reflection; also called focal point.

focal length: the distance between the center of a lens and either focal point.

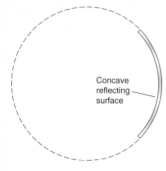

Concave reflecting surface

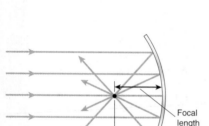

Focal length

Focus

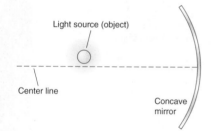

Light source (object)

Center line

Concave mirror

think of the concave mirror as "caving in." Repeat **Steps 3** through **5** for this mirror.

a) Write a sentence to tell what happens to the parallel beams after they are reflected from the concave mirror.

b) Make a drawing in your *Active Physics* log to record the path of the light. The place where the beams cross is called the focus. The distance from the focus to the mirror is the focal length.

c) Measure and record the focal length.

d) How do concave and convex mirrors reflect light differently? Record your answer in your log.

8. Use the concave mirror. Use a 40-W light bulb or a candle as a light source, which will be called the "object." Carefully mount your mirror so it is at the same height as the light source. Place a light bulb about a meter away from the mirror. Put the bulb slightly off the center line, as shown, so that an index card will not block the light from hitting the mirror.

9. Try to find the image of the object on an index card. Move the card back and forth until the image is sharp. The image you found is called a real image because you are able to project it on a card.

a) Record the distance of the bulb from the mirror and of the image on the file card from the mirror. Put your results in the first line of a table like the one below.

Distance of bulb from mirror	Distance of image from mirror

10. Carefully move the mirror closer to the object. Find the sharp image, as before, by moving the index card back and forth.

   a) Record the image and object distances in your table.
   b) Repeat the measurement for at least six object locations.
   c) Draw a graph of the image distance (y-axis) versus the object distance (x-axis).
   d) Write a sentence that describes the relationship between the image distance and the object distance.

11. A mathematical relation that describes concave mirrors is

$$\frac{1}{f} = \frac{1}{D_o} + \frac{1}{D_i}$$

   where

   f is the focal length of that particular mirror

   D_o is the object distance

   D_i is the image distance

   You have measured D_o and D_i. Calculate $\frac{1}{D_o}$ and $\frac{1}{D_i}$. Find their sum for each pair of data.

   a) Record your calculations in your log.
   b) Are your sums approximately equal? If so, you have mathematically found the value of $\frac{1}{f}$ for the mirror you used.

12. A convex mirror cannot form a real image that can be projected onto a screen. It can form an image behind the mirror, like a plane mirror.

   a) Record in your log descriptions of the image in a convex mirror when the mirror is held close and when the mirror is held far from the object.

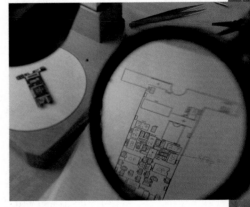

183

Physics Words

real image: an image that will project on a screen or on the film of a camera; the rays of light actually pass through the image.

PHYSICS TALK

Making Real Images

To find how a concave mirror makes a **real image**, you can view a few rays of light. Each ray of light obeys the relation you found for plane mirrors (angle of incidence = angle of reflection). In this case, you choose two easily drawn rays.

Look at the drawing. It shows rays coming into a concave mirror from a point on a light bulb. One ray comes in parallel to the dotted line, which is the axis of the mirror. This ray reflects through the focus. The other ray hits the center of the mirror. This ray reflects and makes the same angle with the mirror axis going out as it did coming in. Where these rays meet is the image of the top of the light bulb.

The next drawing shows the same mirror, but with the object much further from the mirror. Notice how the image in this second drawing is much smaller and much closer to the focus.

As you have seen, the position of the object and image are described by the equation below.

$$\frac{1}{f} = \frac{1}{D_o} + \frac{1}{D_i}$$

Look at the graph of this equation at left. Notice that as the object distance decreases, the image distance becomes very large. As the object distance increases, the image distance moves towards the focal length (f). Also notice that neither the object distance nor the image distance can be less than the focal length.

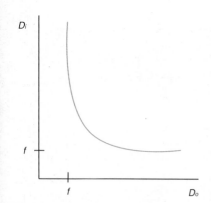

Reflecting on the Activity and the Challenge

You have observed how rays of light are reflected by a curved mirror. You have seen that a concave mirror can make an upside-down real image (an image on a screen). You have also seen that the image and object distances are described by a simple mathematical relationship. In addition, you have seen that there is no real image in a convex mirror, and the image is always smaller than the object.

You may want to use a curved mirror in your sound and light show. You may want to project an image on a screen or produce a reflection that the audience can see in the mirror. What you have learned will help you explain how these images are made.

Since the image changes with distance, you may try to find a way to have a moving object so that the image will automatically move and change size. A ball suspended by a string in front of a mirror may produce an interesting effect. You may also wish to combine convex and concave mirrors so that some parts of the object are larger and others are smaller. Convex and concave mirrors could be shaped to make some kind of fun-house mirror.

Remember that your light show will be judged partly on creativity and partly on the application of physics principles. This activity has provided you with some useful principles that can help with both criteria.

Physics To Go

1. a) Make a drawing of parallel laser beams aimed at a convex mirror.
   b) Draw lines to show how the beams reflect from the mirror.

2. a) Make a drawing of parallel laser beams aimed at a concave mirror.
   b) Draw lines to show how the beams reflect from the mirror.

3. a) Look at the back of a spoon. What do you see?
   b) Look at the inside of a spoon. What do you see?

4. a) If you were designing a shaving mirror, would you make it concave or convex? Explain your answer.
   b) Why do some makeup mirrors have two sides? What do the different sides do? How does each side produce its own special view?
   c) How does a curved side mirror on a car produce a useful view? How can this view sometimes be dangerous?
   d) Why does a dentist use a curved mirror?

D_i (cm)	D_o (cm)
549	15
56	25
20	50
18	91
14	142

5. a) A student found the real image of a light bulb in a concave mirror. The student moved the light bulb to different positions. At each position, the student measured the position of the image and the light bulb. The results are shown in the table on the left. Draw a graph of this data.
   b) Make a general statement to summarize how the image distance changes as the object distance changes.
   c) If the object were twice as far away as the greatest object distance in the data, estimate where the image would be.
   d) If the object were only half as far from the mirror as the smallest object distance in the data, estimate what would happen to the image.

6. A ball is hung on a string in front of a flat mirror. The ball swings toward the mirror and back. How would the image of the ball in the mirror change as the ball swings back and forth?

7. a) A ball is hung on a string in front of a concave mirror. The ball swings toward the mirror and back. How would the image of the ball in the mirror change as the ball swings back and forth?
   b) How could you use this swinging ball in your light show?

8. Outdoors at night, you use a large concave mirror to make an image on a card of distant auto headlights. You make the image on a card. What happens to the image as the car gradually comes closer?

9. The diagram shows a light ray R parallel to the principal axis of a spherical concave (converging) mirror. Point F is the focal point and C is the center of curvature. Draw the reflected light ray.

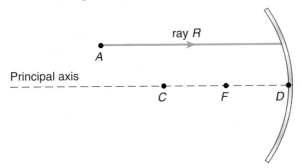

10. The diagram shows a curved mirror surface and a light bulb and its image. In relation to the focal point of the mirror, where is the light bulb (object) most likely located?

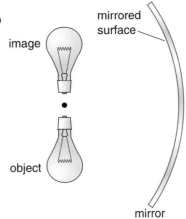

11. A candle is located beyond the center of curvature, C, of a concave spherical mirror having a principal focus, F, as shown in the diagram. Sketch the image of the candle.

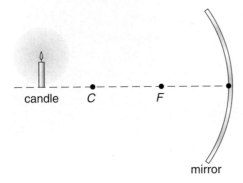

12. The diagram shows four rays of light from object AB incident on a spherical mirror with a focal length of 0.04 m. Point F is the principal focus of the mirror, point C is the center of curvature, and point O is located on the principal axis.

a) Which ray of light will pass through F after it is reflected from the mirror?

b) As object AB is moved from its position toward the left, what will happen to the size of the image produced?

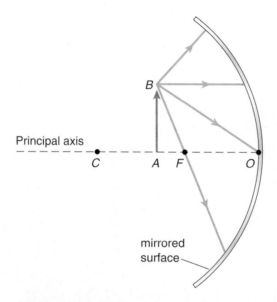

Activity 6   Refraction of Light

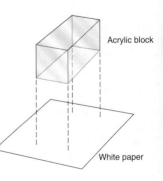

GOALS

In this activity you will:

- Observe refraction.
- Measure angles of incidence and refraction.
- Measure the critical angle.
- Observe total internal reflection.

What Do You Think?

The Hope Diamond is valued at about 100 million dollars. A piece of cut glass of about the same size is worth only a few dollars.

- **How can a jeweler tell the difference between a diamond and cut glass?**

Record your ideas about this question in your *Active Physics* log. Be prepared to discuss your responses with your small group and with your class.

For You To Do

1. Place an acrylic block on a piece of white paper on your desk.

2. Carefully aim a laser pointer, or the light from a ray box, so the light beam moves horizontally, as you

Acrylic block

White paper

⚠ Never look directly at a laser beam or shine a laser beam into someone's eyes. Always work above the plane of the beam and beware of reflections from shiny surfaces.

did in previous activities. Place a glass rod in the light beam so that the beam spreads up and down.

3. Shine the laser pointer or light from the ray box through the acrylic block. Be sure the beam leaves the acrylic block on the side opposite the side the beam enters. Mark the path of each beam. You may wish to use a series of dots as you did before. Label each path on both sides of the acrylic block so you will know that they go together.

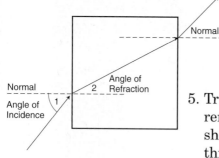

4. The angle of incidence is the angle between the incident laser beam and the normal, as shown in the diagram. Choose two other angles of incidence and again mark the path of the light, as you did in **Step 3**. As before, label each pair of paths.

5. Trace the outline of the acrylic block on the paper and remove the acrylic block. Connect the paths you traced to show the light beam entering the acrylic block, traveling through the acrylic block, and emerging from the acrylic block. Draw a perpendicular line at the point where a ray enters or leaves the acrylic block. Label this line the normal.

6. Measure the angles of incidence (the angle in the air) and refraction (the angle in the acrylic block).

✎ a) Record your measurements in tables like the one shown.

Angle of incidence	Angle of refraction	Sine of angle of incidence	Sine of angle of refraction	$\dfrac{\text{Sin} \angle i}{\text{Sin} \angle R}$

b) Use a calculator to complete the chart by finding the sines of the angles (sin button on calculator).

c) Is the value of $\dfrac{\sin \angle i}{\sin \angle R}$ a constant? This value is called the index of refraction for the acrylic block.

7. Set up the acrylic block on a clean sheet of white paper. This time, as shown in the drawing (next page), aim the beam so it leaves the acrylic block on the side, rather than at the back.

Physics Words

critical angle: the angle of incidence for which a light ray passing from one medium to another has an angle of refraction of 90°.

index of refraction: a property of a medium that is related to the speed of light through it; it is calculated by dividing the speed of light in vacuum by the speed of light in the medium.

Snell's Law: describes the relationship between the index of refraction and the ratio of the sine of the angle of incidence and the sine of the angle of refraction.

8. Make the first angle of incidence (angle 1) as small as possible, so the second angle of incidence (angle 2) will be as large as possible. Adjust angle 1 so that the beam leaves the acrylic block parallel to the side of the acrylic block, as shown. Measure the value of angle 2.

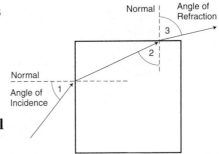

✎ a) Record the value of angle 2. It is called the **critical angle**.

✎ b) What happens to the beam if you make angle 2 greater than the critical angle?

✎ c) What you observed in **(b)** is called "total internal reflection." What is reflected totally, and where?

9. It is possible to bend a long, rectangular acrylic block so the light enters the narrow end of the acrylic block, reflects off one side of the acrylic block, then reflects off the other and back again to finally emerge from the other narrow end. Try to bend an acrylic block rectangle so that the light is reflected as described.

FOR YOU TO READ

Snell's Law

Light refracts (bends) when it goes from air into another substance. This is true whether the other substance is gelatin, glass, water, or diamond. The amount of bending is dependent on the material that the light enters. Each material has a specific **index of refraction**, n. This index of refraction is a property of the material and is one way in which a diamond (very high index of refraction—lots of bending) can be distinguished from glass (lower index of refraction—less bending). The index of refraction is a ratio of the sine of the angle of incidence and the sine of the angle of refraction.

Index of refraction: $n = \dfrac{\sin \angle i}{\sin \angle R}$

This equation is referred to as **Snell's Law**.

As light enters a substance from air, the light bends toward the normal. When light leaves a substance and enters the air, it bends away from the normal. If the light is entering the air from a substance, the angle in that substance may be such that the angle of refraction is 90°. In this special case, the angle in the substance is called the critical angle. If the angle in the substance is greater than this critical angle, then the light does not enter the air but reflects back into the substance as if the surface were a perfect mirror. This is the basis for light fibers where laser light reflects off the inner walls of glass and travels down the fiber, regardless of the bend in the fiber.

Reflecting on the Activity and the Challenge

The bending of light as it goes from air into a substance or from a substance into air is called refraction. It is mathematically expressed by Snell's Law. When light enters the substance at an angle, it bends towards the normal. When light leaves the substance at an angle, it bends away from the normal. As you create your light show for the **Chapter Challenge**, you may find creative uses of refraction. You may decide to have light bending in such a way that it spells out a letter or word or creates a picture. You may wish to have the light travel from air into glass to change its direction. You may have it bend by different amounts by replacing one material with another. Regardless of how you use refraction effects, you can now explain the physics principles behind them.

Physics To Go

1. A light ray goes from the air into an acrylic block. In general, which is larger, the angle of incidence or the angle of refraction?

2. a) Make a sketch of a ray of light as it enters a piece of acrylic block and is refracted.

    b) Now turn the ray around so it goes backward. What was the angle of refraction is now the angle of incidence. Does the turned-around ray follow the path of the original ray?

3. A light ray enters an acrylic block from the air. Make a diagram to show the angle of incidence, the angle of refraction, and the normal at the edge of the acrylic block.

4. Light rays enter an acrylic block from the air. Make drawings to show rays with angles of incidence of 30° and 60°. For each incident ray, sketch the refracted ray that passes through the acrylic block.

5. a) Light is passing from the air into an acrylic block. What is the maximum possible angle of incidence that will permit light to pass into the acrylic block?

    b) Make a sketch to show your answer for **Part (a)**. Include the refracted ray (inside the acrylic block) in your sketch.

6. a) A ray of light is already inside an acrylic block and is heading out. What is the name of the maximum possible angle of incidence that will permit the light to pass out of the acrylic block?

   b) If you make the angle of incidence in **Part (a)** greater than this special angle, what happens to the light?

   c) Make a sketch to show your answer for **Part (b)**. Be sure to show what happened to the light.

7. a) Make a drawing of a light ray that enters the front side of a rectangular piece of acrylic block and leaves through the back side.

   b) What is the relationship between the direction of the ray that enters the acrylic block and the direction of the ray that leaves the acrylic block?

   c) Use geometry and your answer to **Question 2 (b)**, to prove your answer to **Question 7 (b)**.

8. You have seen the colored bands that a prism or cut glass or water produce from sunlight. Light that you see as different colors has different wavelengths. Since refraction makes these bands, what can you say about the way light of different wavelengths refracts?

Stretching Exercises

1. Cover the acrylic block with a red filter. Shine a red laser beam into the acrylic block, as you did in **For You To Do**, **Steps 1** through **3**. What happens? How can you explain what happens?

2. Find some $\frac{1}{2}''$ diameter clear tubing, about 2 m long. Plug one end. Pour clear gelatin in the other end, through a funnel, before the gelatin has had time to set. Arrange the tubing into an interesting shape and let the gelatin set. You may wish to mount your tube on a support or a sturdy piece of cardboard, which can be covered with interesting reflective material. Fasten one end of the tube so laser light can easily shine straight into it. When the gelatin has set, turn on the laser. What do you see? This phenomena is called total internal reflection.

3. Place a penny in the bottom of a dish or glass. Position your eye so you can just see the penny over the rim of the glass. Predict what will happen when you fill the glass with water. Then try it and see what happens. How can you explain the results?

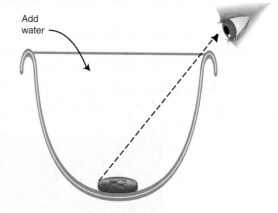

Add water

4. Place an empty, clear drinking glass over a piece of a newspaper. When you look through the side of the glass near the bottom, you can see the printing on the newspaper. What do you think will happen if you fill the glass with water? Try it and see. How can you explain the result? Does it help to hold your fingers over the back of the glass?

Activity 7   Effect of Lenses on Light

GOALS

In this activity you will:

• Observe real images.

• Project a slide.

• Relate image size and position.

What Do You Think?

Engineers have created special lenses that can photograph movie scenes lit only by candlelight.

• **How is a lens able to project movies, take photographs, or help people with vision problems?**

Record your ideas about this question in your *Active Physics* log. Be prepared to discuss your responses with your small group and with your class.

For You To Do

1. Look at the lens your teacher has given you.

✎ a) Make a side-view drawing of this lens in your log. This is a *convex* lens.

2. Point the lens at a window or at something distant

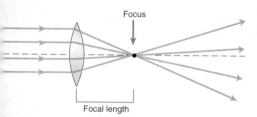

Focus

Focal length

outside. Use a file card as a screen. Look for the image on the file card. Move it back and forth until you see a sharp image of the distant object.

 a) Describe what you see. Is the image large or small? Is it right side up or upside down? Is it reversed left to right? This image is called "real" because you can project it on the screen.

3. Measure the distance between the image and the lens. If the object is very far away, this distance is the focal length of the lens. The position of this image is the focus of the lens. It is the same location at which parallel rays of light would converge.

a) Approximate the object distance.
b) Measure the image distance.
c) Record your object and image distance. Note that the image distance is also the focal length of the lens.

Do not use lenses with chipped edges. Mount lenses securely in a holder. Use only light sources with enclosed or covered electrical contacts. Keep flammables/combustibles away from the candleholder.

4. Set up a 40-W light bulb or a candle to be a light source. Mount the lens at the same height as the light source. If you are using a light bulb, point it right at the lens, as shown.

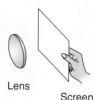

Light

Lens

Screen

5. Place the light bulb about a meter away from the lens. Try to find the image of the light bulb on a screen. The screen can be a file card or a sheet of paper.

a) Record your results in a table, including the distance and appearance of the image.

6. Adjust the position of the object to create a larger image.

a) Describe how the position of the object, the image, and the size of the image have changed. Record the results in a table.

7. Create an object by carefully cutting a hole in the shape of an arrow in an index card. Have someone in your group hold the card close to the light bulb.

a) Can you see the object on the screen? Describe what you see.

b) Have the person holding the object move it around between the light bulb and the convex lens. What happens?

8. Project the object onto the wall. Can you make what you project larger or smaller?

a) In your log indicate what you did to change the size of the image.

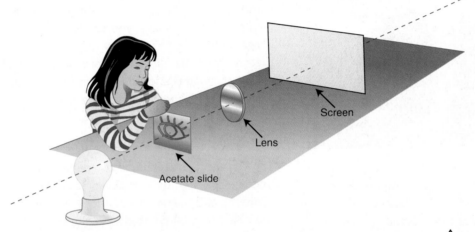

Screen

Lens

Acetate slide

9. Create a slide by drawing with a marking pen on clear acetate. Try placing a 100-W light bulb and the slide in different positions.

a) Describe how you can project a real, enlarged image of your slide onto a screen or wall.

b) How can you use the lens to change the size of the image?

c) Explore the effect of different lenses. In your log, record how you think this effect might be part of your light and sound show.

Caution:
Lamps get very hot. Be careful not to touch the bulb or housing surrounding the bulb.

FOR YOU TO READ

Lens Ray Diagrams

You are probably more familiar with images produced by lenses than you are with images from curved mirrors. The lens is responsible for images of slides, overhead projectors, cameras, microscopes, and binoculars.

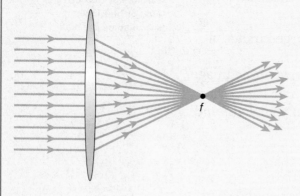

Light bends as it enters glass and bends again when it leaves the glass. The **convex converging lens** is constructed so that all parallel rays of light will bend in such a way that they meet at a location past the lens. This place is the focal point.

If an object is illuminated, it reflects light in all directions. If these rays of light pass through a lens, an image is formed.

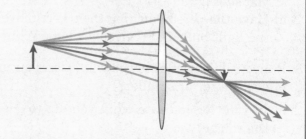

Although all of the light rays from the object help to form the image, you can locate an image by looking at two easy rays to draw— the ray that is parallel to the principal axis and travels through the focal point and the ray that travels through the center of the lens undeflected. (These rays are in red in the diagram.)

You can use this technique to see how images that are larger (movie projector), smaller (camera), and the same size (copy machine) as the object can be created with the same lens.

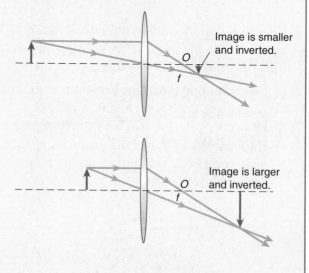

Image is smaller and inverted.

Image is larger and inverted.

If the object is close to the lens (an object distance smaller than the focal distance), then an image is not formed. However, if you were to view the rays emerging, they would appear to have come from a place on the same side of the lens as the object. To view this image, you put your eye on the side of the lens opposite the object and peer through it—it's a magnifying glass!

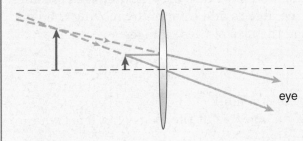

Sample Problem

The diagram shows a lens and an object.

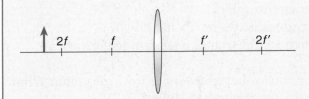

a) Using a ray diagram, locate the image of the object shown.

b) Describe the image completely.

Strategy: Choose a location on the object to be the origin of the rays. A simple choice would be the tip of the arrow. At least two rays must be drawn to locate the image.

Givens:

See the diagram.

Solution:

a)

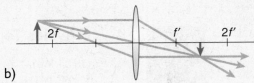

b)
The image is real, reduced, and inverted.

As the object moves closer to the lens, its size will increase. At $d_o = f$ there will be no image and at $d_o < f$ the image will be virtual and upright.

Physics Words

converging lens: parallel beams of light passing through the lens are brought to a real point or focus (if the outside index of refraction is less than that of the lens material); also called a convex lens.

Reflecting on the Activity and the Challenge

You have explored how convex lenses make real images. You have found these images on a screen by moving a card back and forth until the image was sharp and clear, so you know that they occur at a particular place. Bringing the object near the lens moves the image away from the lens and enlarges the image, but if the object is too close to the lens, there is no real image. These images are also reversed left to right and are upside down. You may be able to use this kind of image in your sound and light show. You have also projected images of slides on a wall. You may be able to add interest by moving the lens and screen to change the size of these images.

Physics To Go

1. a) What is the focus of a lens?
   b) If the image of an object is at the focus on a lens, where is the object located?
   c) What is the focal length of a lens?
   d) How can you measure the focal length of a lens?

2. a) You set up a lens and screen to make an image of a distant light. Is the image in color?
   b) Is the image right side up or upside down?
   c) Did the lens bend light to make this image? How can you tell?
   d) A distant light source begins moving toward a lens. What must you do to keep the image sharp?

3. a) You make an image of a light bulb. What can you do to make the image smaller than the light bulb?
   b) What can you do to make the image larger than the light bulb?

4. a) You have two lights, a lens, and a screen, as shown on opposite page. One light is at a great distance from the lens. The other light is much closer. If you see a sharp image of the distant light, describe the image of the closer light.
   b) If you see a sharp image of the closer light, describe the image of the more distant light.

c) Could you see a sharp image of both lights at the same time? Explain how you found your answer.

5. Research how a camera works. Find out where the image is located. Also find out how the lens changes so that you can photograph a distant landscape and also photograph people close up.

6. Using a ray diagram, locate the image formed by the lens below.

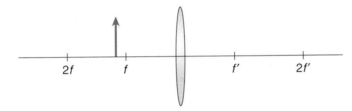

7. An object 1.5 cm tall is placed 5.0 cm in front of a converging lens of focal length 8.0 cm.
   a) Determine the location of the image.
   b) Completely describe the image.

8. A relative wants to show you slides from her wedding in 1972. She brings out her slide projector and screen.
   a) If she puts the screen 2.8 m from the projector and the lens has a focal length of 10.0 cm, how far from the lens will the slide be so that her pictures are in focus?
   b) If each slide is 3.0 cm tall, how big will the image be on the screen?

9. The diagram shows an object 0.030 m high placed at point X, 0.60 m from the center of the lens. An image is formed at point Y, 0.30 m from the center of the lens. Completely describe the image.

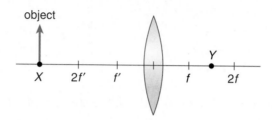

10. The diagram represents an object placed two focal lengths from a converging lens. At which point will the image be located?

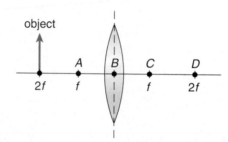

11. The diagram shows a lens with an object located at position A. Describe what will happen to the image formed as the object is moved from position A to position B.

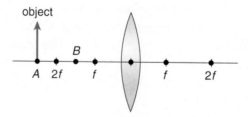

Stretching Exercises

1. To investigate how the image position depends on the object position, find a convex lens, a white card, and a light source. Find the image of the light source, and measure the image and object distance from the lens. Make these measurements for as wide a range of object distances as you can. In addition, make an image of an object outside, such as a tree. Estimate the distance to the tree. The image of a distant object, like the tree, is located very near the focus of the lens. Draw a graph of the results. Compare the graph with the equation

$$\frac{1}{f} = \frac{1}{D_o} + \frac{1}{D_i}$$

2. Find a camera with a shutter that you can keep open (with a bulb or time setting). Place a piece of waxed paper or a piece of a plastic bag behind the lens, where the film would be if you took a picture. Find the image and compare it to the images you made in this activity. Focus the lens for objects at different distances. Investigate how well the object and image location fit the lens equation $\frac{1}{f} = \frac{1}{D_o} + \frac{1}{D_i}$.

   Remember that the focal length of the lens is typically printed on the lens.

3. Research how the concept of "depth of field" is important in photography. Report to the class on what you learn.

Activity 8   Color

GOALS

In this activity you will:

• Analyze shadow patterns.

• Explain the size of shadows.

• Predict pattern of colored shadows.

• Observe combinations of colored lights.

What Do You Think?

When a painter mixes red and green paint, the result is a dull brown. But when a lighting designer in a theater shines a red and a green light on an actress, the actress's skin looks bright yellow.

• **How could these two results be so different?**

• **How are the colors you see produced?**

Record your ideas about these questions in your *Active Physics* log. Be prepared to discuss your responses with your small group and with your class.

For You To Do

1. Carefully cut out a cardboard puppet that you will use to make shadows.

2. Turn on a white light bulb only. Move the puppet around and observe the shadow.

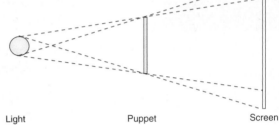

Caution:
Lamps get very hot. Be careful not to touch the bulb or housing surrounding the bulb.

a) Describe the shadow you see.

b) What happens to the shadow if you move the puppet sideways or up and down?

c) What happens to the shadow if you move the puppet close to the screen?

d) What happens to the shadow if you move the puppet close to the bulb?

3. Look at the drawing. It shows a top view of a puppet halfway between the light and the screen.

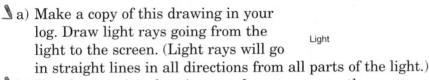

Light   Puppet   Screen

a) Make a copy of this drawing in your log. Draw light rays going from the light to the screen. (Light rays will go in straight lines in all directions from all parts of the light.)

b) Use the top-view drawing you drew to answer these questions: Which part of the screen receives light? Which part receives no light? Which part receives some light?

c) Is the shadow larger or smaller than the puppet? Explain how you found your answer.

d) Now copy the other two top-view drawings and show the path of the light rays.

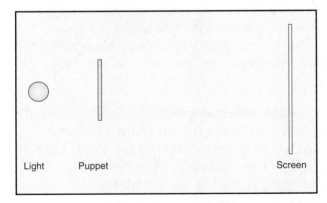

Light   Puppet   Screen

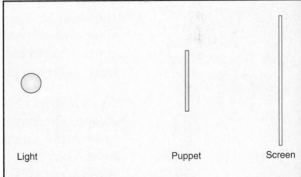

Light   Puppet   Screen

e) On your drawings, show which part of the screen does receive light and which part does not receive light and which part receives some light.

f) For each of these two drawings, tell whether the shadow is larger or smaller than the puppet. For each one, explain how you found your answer.

4. Turn off the white bulb. Turn on red and green bulbs. They should be aimed directly at the center of the screen.

 a) What color do you see on the screen?

b) Predict what color the shadows will be if you bring your puppet between the bulbs and the screen. Record your prediction, and give a reason for it.

c) Make a top-view drawing to show the path of the light rays from the red and green bulbs.

d) On your drawing, label the color you will see on each part of the screen.

5. Turn off the green bulb and turn on a blue one. Repeat what you did in **Step 4**, but with the blue and red bulbs lit.

6. Turn off the red bulb and turn on the green one. Repeat what you did in **Step 4**, but with the blue and green bulbs lit.

7. Turn on the red bulb so all three—red, blue, and green—are lit. Repeat what you did in **Steps 5** and **6**.

Reflecting on the Activity and the Challenge

Different colored lights can combine to make white light. When an object blocks all light, it creates a dark shadow. Since some light comes from all parts of the bulb, there are places where the shadow is black (no light) and places where the shadow is gray (some light reaches this area). An object illuminated by different colored lights can create shadows that prevent certain colors from reaching the wall and allowing other colors to pass by.

In your light show creation, you may choose to use the ideas of colored shadows to show how lights can be added to produce interesting combinations of colors. By moving the object or the lights during the show, you may be able to produce some interesting effects. Lighting design is used in all theater productions. It requires a knowledge and understanding of how lights work, as well as an aesthetic sense of what creates an enjoyable display.

Physics To Go

1. Show how a shadow is created.

2. How can moving the light, the object, and the screen all produce the effect of enlarging the shadow?

3. Explain why a gray halo surrounds a dark shadow made by a light bulb and an object.

4. a) Why is your shadow different at different times of the day?
   b) What is the position of the Sun when your shadow is the longest? The shortest?

5. Why is the gray halo about your shadow so thin when you are illuminated by the Sun?

6. a) Suppose you shine a red light on a screen in a dark room. The result is a disk of red light. Now you turn on a green light and a blue light. The three disks of light overlap as shown. Copy the diagram into your journal. Label the color you will see in each part of the diagram.

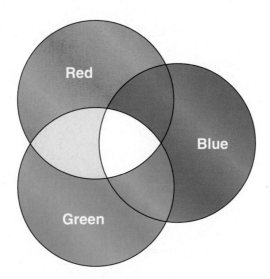

   b) Add the labels "bright," "brighter," and "brightest" to describe what you would see in each part of your diagram.

7. a) Make a drawing of an object in red light. The object casts a shadow on the screen. Label the color of the shadow and the rest of the screen.

   b) Repeat **Part (a)** for an object in green light.

   c) Now make a copy of your drawing for **Part (b)**. Add a red light, as in **Part (a)**. Label the color of all the shadows.

8. List some imaginative ways that you can add colors to your light show.

Stretching Exercises

1. With the room completely dark, shine a red light on various colored objects. Compare the way they look in red light with the way they look in ordinary room light.

2. View 3-D pictures with red and blue glasses. Explain how each eye sees a different picture.

3. Shine a white light and a red light on a small object in front of a screen. What colors are the shadows? How is this surprising?

4. Prepare a large drawing of the American flag but with blue-green in the place of red, and yellow in the place of blue. Stare at the drawing for 30 seconds and then look at a white surface.

PHYSICS AT WORK

Alicja and Dennis Phipps

Alicja Phipps has always been interested in electronics. As a child she wanted to be a television repair person. She now works with her husband, Dennis Phipps, in their company, Light & Sound Entertainment, which designs original content programming in a variety of areas—from rock concerts to the Olympics.

Light & Sound Entertainment got its name because Dennis believes strongly that the two are and should be linked. "It's terrible when the music of a production does not match what you are seeing," he says. "More and more theaters have lighting specialists come in and set up the theater with everything preset on a computerized lighting board. There will be a setting for 'outdoor lighting' and 'nighttime' or 'sunset'. The problem with that is that every production will look the same, when in reality nighttime in Canada looks very different from nighttime in Florida. There are also different lighting considerations depending on your audience. A production in front of a live audience needs different lighting than one being recorded on video. For instance, the human eye picks up shadows a lot better than a camera will. If a production is intended for both a live audience and video, lighting needs to be arranged accordingly."

Dennis continues, "The sound of a production is only as good as its setup, and nothing can replace the actual setting. The Red Rocks Theater in Colorado is terrific, for example, because stone has a very high reverberation rate, which is great for guitars. A huge wooden room like Carnegie Hall also provides a unique sound. However, these spaces and materials are not readily available." The hardest projects, Dennis says, are those in which you cannot control the elements. "Sound elements include the size and shape of the space, reverberation, feedback, and temperature."

"New media has its own set of challenges," explains Alicja who oversees the conversion of live events into various other formats, such as CD-ROMs, virtual reality, and Web sites. "We have to think about how much information (sound and image) we will be able to fit on a disc or on to a Web page and how long it will take to load. If it takes too long, no one will ever see or hear what we've done."

"We enjoy the creative process of every production," claim Alicja and Dennis. "Each one is a unique challenge."

209

Chapter 3 Assessment

With what you learned about sound and light in this chapter, you are now ready to dazzle the world. However, you have neither the funds nor the technology available to professionals. All sounds you use to capture the interest of the class must come from musical instruments that you build yourself, or from human voices. Some of these sounds may be prerecorded and then played in your show. If your teacher has a laser and is willing to allow you to use it, you may do so. All other light must come from conventional household lamps. Gather with your committee to design a two- to four-minute sound and light show to entertain other students your age.

Review the criteria by which you decided that your show will be evaluated. The following suggestions were provided at the beginning of the chapter:

1. The variety and number of physics concepts used to produce the light and sound effects.

2. Your understanding of the physics concepts:
   a) Name the physics concepts that you used.
   b) Explain each concept.
   c) Give an example of something that each concept explains or an example of how each concept is used.
   d) Explain why each concept is important.

3. Entertainment value

At this time you may wish to propose changes in the criteria. Also decide as a class if you wish to modify or keep the point value you established at the beginning of the chapter.

Enjoy the sound and light productions!

Physics You Learned

Compressional and transverse waves

Wave speed = wavelength $\times$ frequency

Standing waves

Pitch and frequency

Sound production in pipes and vibrating strings

Controlling frequency of sounds produced electronically

Angle of incidence and angle of reflection

Location of image in plane and curved mirrors

$$\frac{1}{f} = \frac{1}{D_o} + \frac{1}{D_i} \text{ in curved mirrors}$$

Real images

Angle of incidence and angle of refraction

Lenses and image formation

$$\frac{1}{f} = \frac{1}{D_o} + \frac{1}{D_i} \text{ in lenses}$$

Color addition

$$n = \frac{\sin\angle\,i}{\sin\angle\,R}$$

Chapter 4

Scenario

In this *Active Physics* chapter, you will try to help educate children through the use of toys. With your input, the Homes for Everyone (HFE) organization has developed an appliance package that will allow families living in the "universal dwelling" to enjoy a healthy and comfortable lifestyle. The HFE organization would now like to teach the children living in these homes, and elsewhere, more about electricity and the generation of electricity. They hope that this may encourage interest in children to use electricity wisely, as well as encourage development of alternative sources for electrical energy by future generations.

The HFE organization will work with a toy company to provide kits and instructions for children to make toy electric motors and generators. These toys should illustrate how electric motors and generators work and capture the interest of the children.

In an effort to help others, people often make changes or introduce new products without considering the personal and cultural impact on those whom they are trying to assist. If you ever become involved in a self-help community group, such as HFE, it would be important for you to work together with the people you are assisting to assess their needs, both personal and cultural. Although that is not possible given your limited time in class, you should recognize the need for collaborative teamwork in evaluating the impact of any new product on an established community.

Challenge

Your task is to prepare a kit of materials and instructions that the toy company will manufacture. Children will use these kits to make a motor or generator, or a combination electric motor/generator. It will serve both as a toy and to illustrate how the electric motors in home appliances work or how electricity can be produced from an energy source such as wind, moving water, a falling weight, or some other external source.

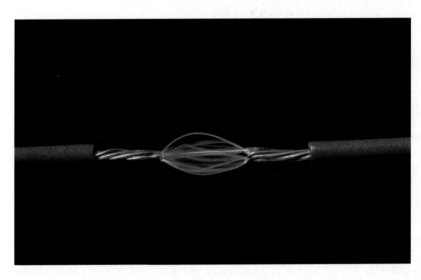

Criteria

Your work will be judged by the following criteria:

- **(30%) The motor/generator is made from inexpensive, common materials, and the working parts are exposed but with due consideration for safety.**

- **(40%) The instructions for the children clearly explain how to assemble and operate the motor/generator device, and explain how and why it works in terms of basic principles of physics.**

  - **(30%) If used as a motor, the device will operate using a maximum of four 1.5-V (volt) batteries (D cells), and will power a toy (such as a car, boat, crane, etc.) that will be fascinating to children.**

  OR

  - **(30%) If used as a generator, the device will demonstrate the production of electricity from an energy source such as wind, moving water, a falling weight, or some other external source and be fascinating to children.**

Activity 1

The Electricity and Magnetism Connection

GOALS

In this activity you will:

• Describe electric motor effect and generator effect in terms of energy transformations.

• Use a magnetic compass to map a magnetic field.

• Describe the magnetic field near a long, straight current-carrying wire.

What Do You Think?

Generators produce electricity. Motors use electricity.

• **What is the significance of motors and generators to your standard of living? That is, how would your life be different if you had no motors or generators?**

Write your answer to these questions in your *Active Physics* log. Be prepared to discuss your ideas with your small group and other members of your class.

For You To Do

1. Set up the equipment as shown in the diagram, or as directed by your teacher.

2. The needle of a compass is a balanced magnet. It can be used as a magnetic field detector. If any magnet is present, the compass will respond. It usually aligns itself with Earth's magnetic field. With no current flowing in the wire, verify that the compass always points in the same direction, north, no matter where it is placed on the horizontal surface.

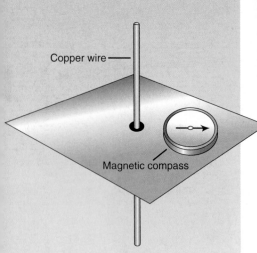

Copper wire —

Magnetic compass —

✎ a) Sketch the compass direction at different places on the horizontal surface in your log.

3. Bring another type of magnet, such as a bar magnet, into the area near the compass needle.

✎ a) Describe your observations in your log.

✎ b) What happens to the dependable north-pointing property of a compass when the compass is placed in a region where magnetic effects, in addition to Earth's magnetic field, exist?

4. You will now make a map of the magnetic field of the bar magnet. Place the magnet on another piece of paper and trace its position. Place the compass at one location and note the direction it points. Remove the compass.

✎ a) Put a small arrow at the location from which you removed the compass to signify the way in which it pointed.

✎ b) Place the compass at a second location about at the tip of the first arrow. Remove the compass and place another small arrow in this location to signify the way in which the compass pointed.

✎ c) Repeat the process at an additional 20 locations to get a map of the magnetic field of a bar magnet. Tape the piece of paper of the map in your log.

⚠ **Do not adjust the power supply settings provided by your teacher.**

5. Return the compass to the horizontal surface surrounding the wire. Observe the orientation of the compass. Send a current through the wire. The direction of the flow of electrons which make up the current in the wire is from the negative terminal of the power supply to the positive terminal. Move the compass to different locations on the horizontal surface, observing the direction in which the compass points at each location. Make observations on all sides of the wire, and at different distances from the wire.

✎ a) Record how the compass was oriented when the bar magnet was removed.

✎ b) Describe any pattern that you observe about how the compass behaves when it is near the current-carrying wire. Use a sketch and words to describe your observations in your log.

✎ c) From your observations, what effect does the electric current appear to have on the wire?

215

6. Reverse the direction of the current in the wire by exchanging the contacts of the power supply. Repeat your observations.

\\a) Describe the results.

\\b) Make up a rule for remembering the relationship between the direction of the current in a wire and the direction of the magnetism near the wire (i.e., when the current is up, the magnetic field . . .). Anyone told your rule should be able to use it with success. Write your rule in your log. Include a sketch. (Hint: One of the rules that physicists use makes use of your thumb and fingers.)

Reflecting on the Activity and the Challenge

This activity has provided you with knowledge about a critical link between electricity and magnetism, which is deeply involved in your challenge to make a working electric motor or generator. The response of the compass needle to a nearby electric current showed that an electric current itself has a magnetic effect which can cause a magnet, in this case a compass needle, to experience force. You have a way to go to understand and be able to be "in control" of electric motors and generators, but you've started along the path to being in control.

Physics To Go

1. If 100 compasses were available to be placed on the horizontal surface to surround the current-carrying wire in this activity, describe the pattern of directions in which the 100 compasses would point in each of the following situations:

   a) no current is flowing in the wire
   b) a weak current is flowing in the wire
   c) a strong current is flowing in the wire

2. If a vertical wire carrying a strong current penetrated the floor of a room, and if you were using a compass to "navigate" in the room by always walking in the direction indicated by the north-seeking pole of the compass needle, describe the "walk" you would take.

3. Use the rule which you made up for remembering the relationship between the direction of the current flowing in a wire and the direction of the magnetic field near the wire to make a sketch showing the direction of the magnetic field near a wire which has a current flowing:

a) downward
b) horizontally

4. Physicists remember the orientation of the magnetic field of a current by placing their left thumb in the direction of the electric current and noting whether the fingers of their left hand curve clockwise or counterclockwise. Copy the following diagrams into your log. Use this rule to sketch the direction of the magnetic field in each case.

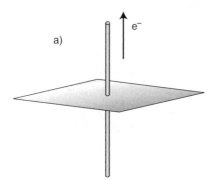

a)

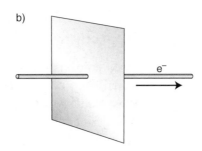

b)

5. Imagine that a second vertical wire is placed in the apparatus used in this activity, but not touching the first wire. There is room to place a magnetic compass between the wires without touching either wire. If a compass were placed between the wires, describe in what direction the compass would point if the wires carry equal currents:

a) which are in opposite directions
b) which are in the same direction

6. A hollow, transparent plastic tube is placed on a horizontal surface as shown in the diagram. A wire carrying a current is wound once around the tube to form a circular loop in the wire. In what direction would a compass placed inside the tube point? (Plastic does not affect a compass; only the current in the wire loop will affect the compass.)

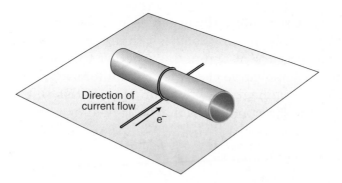

Direction of current flow
e^-

Stretching Exercises

Use a compass to search for magnetic effects and magnetic "stuff." As you know, a compass needle usually aligns in a north-south direction (or nearly so, depending on where you live). If a compass needle does not align north-south, a magnetic effect in addition to that of the Earth is the cause, and the needle is responding to both the Earth's magnetism and some other source of magnetism. Use a compass as a probe for magnetic effects. Try to find magnetic effects in a variety of places and near a variety of things where you suspect magnetism may be present. Try inside a car, bus, or subway. The structural steel in some buildings is magnetized and may cause a compass to give a "wrong" reading. Try near the speaker of a radio, stereo, or TV. Try near electric motors, both operating and not operating.

Do not bring a known strong magnet close to a compass, because the magnet may change the magnetic alignment of the compass needle, destroying the effectiveness of the compass.

Make a list of the magnetic objects and effects that you find in your search.

Activity 2

Electromagnets

GOALS

In this activity you will:

- Describe and explain the magnetic field of a current-carrying solenoid.

- Compare the field of a solenoid to the field of a bar magnet.

- Identify the variables of an electromagnet and explain the effects of each variable.

What Do You Think?

Large electromagnets are used to pick up cars in junkyards.

- **How does an electromagnet work?**

- **How could it be made stronger?**

Write your answer to these questions in your *Active Physics* log. Be prepared to discuss your ideas with your small group and other members of your class.

For You To Do

1. Wind 50 turns of wire on a drinking straw to form a solenoid as shown in the diagram on the next page. Use sandpaper to carefully clean the insulation from a short section of the wire ends to allow electrical connection of the solenoid to the generator.

219

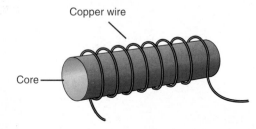

Copper wire

Core

2. Carefully connect the wires from the generator to the wire ends of the solenoid. Bring one end of the solenoid near the magnetic compass and crank the generator to send a current through the solenoid. Observe any effect on the compass needle. Try several orientations of the solenoid to produce effects on the compass needle.

   a) Record your observations in your log.
   b) How can you tell the "polarity" of an electromagnet; that is, how can you tell which end of an electromagnet behaves as a north-seeking pole?

3. Predict what you can do to change the polarity of an electromagnet.

   a) Write your answer in your log.
   b) Test your prediction.

4. Use the solenoid wound on the drinking straw as an electromagnet to pick up paper clips.

   a) Record your observations in your log.

5. Carefully, slip a nail into the drinking straw to serve as a new core. Again, test the effect on a compass needle.

   a) Record your observations in your log.

6. Use the solenoid wound on the nail to pick up paper clips.

   a) Record your observations in your log.
   b) What evidence did you find that the choice of core material for an electromagnet makes a difference?

7. Predict what will happen when you increase the current running through the coiled wire solenoid. This can be done by increasing the speed at which you crank the generator.

   a) Write your answer in your log.
   b) Test your prediction by measuring how many paper clips can be picked up.

8. Predict what will happen when you increase the number of turns of wire forming the solenoid.

   a) Write your answer in your log.
   b) Test your prediction by measuring how many paper clips can be picked up.

Active Physics

Reflecting on the Activity and the Challenge

An electromagnet, often constructed in the shape of a solenoid, and having an iron core, is the basic moving part of many electric motors. In this activity you learned how the amount of current and the number of turns of wire affect the strength of an electromagnet. You will be able to apply this knowledge to affect the speed and strength with which an electric motor of your own design rotates.

Physics To Go

1. Explain the differences between permanent magnets and electromagnets.

2. The diagram below shows an electromagnet with a compass at each end. Copy the diagram and indicate the direction in which the compass needles will be pointing when a current is generated.

3. Which of the following will pick up more paper clips when an electric current is sent through the wire:
   a) A coil of wire with 20 turns, or a coil of wire with 50 turns?
   b) Wire wound around a cardboard core, or wire wound around a steel core?

4. Explain conditions necessary for two electromagnets to attract or repel one another, as do permanent magnets when they are brought near one another.

5. Explain what you think would happen if, when making an electromagnet, half of the turns of wire on the core were made in one direction, and half in the opposite direction.

Stretching Exercises

1. Find out how both permanent magnets and electromagnets are used. Do some library research to learn how electromagnets are used to lift steel in junkyards, make buzzers, or serve as part of electrical switching devices called "relays." For other possibilities, find out how magnetism is used in microphones and speakers within sound systems, or how "super-strong" permanent magnets made possible the small, high-quality, headset speakers for today's portable radio, tape and CD players. Prepare a brief report on your findings.

2. Do some research to find out about "magnetic levitation." "Maglev" involves using super-conducting electromagnets to levitate, or suspend objects such as subway trains in air, thereby reducing friction and the "bumpiness" of the ride.

   a) What possibilities do "maglev" trains, cars, or other transportation devices have for the future?
   b) What advantages would such devices have?
   c) What problems need to be solved? Prepare a brief report on your research.

3. Identify as many variables as you can that you think will affect the behavior of an electromagnet, and design an experiment to test the effect of each variable. Identify each variable, and describe what you would do to test its effects. After your teacher approves your procedures, do the experiments. Report your findings.

Activity 3

Detect and Induce Currents

GOALS

In this activity you will:

• Explain how a simple galvanometer works.

• Induce current using a magnet and coil.

• Describe alternating current.

• Recognize the relativity of motion.

What Do You Think?

In 1820, the Danish physicist Hans Christian Oersted placed a long, straight, horizontal wire on top of a magnetic compass. Both the compass and the wire were resting on a horizontal surface, and both the length of the wire and the compass needle were oriented north-south. Next, Oersted sent a current through the wire, and happened upon one of the greatest discoveries in physics.

• **What do you think Oersted saw?**

Write your answer to this question in your *Active Physics* log. Be prepared to discuss your ideas with your small group and other members of your class.

For You To Do

1. Wrap 10 turns of wire to form a coil that surrounds a magnetic compass. Wrap the wire on a diameter corresponding to the north-south markings of the compass scale, as shown in the diagram. Hold the turns of wire in place with tape, or use the method recommended by your teacher. Use sandpaper to carefully remove the insulation from a short section of the wire ends to allow electrical connection.

2. In **Step 1**, you constructed a galvanometer, a device to detect and measure small currents. Carefully connect a hand generator, a light bulb, and the galvanometer, as shown on the next page (in a series circuit). Rest the galvanometer so that the compass is horizontal, with the needle balanced, pointing north, and free to rotate. Also, turn the galvanometer, if necessary, so that the compass needle is aligned parallel to the turns of wire which pass over the top of the compass.

223

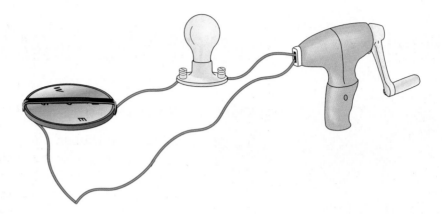

3. Crank the generator to establish a current in the circuit. Think of the compass needle as a meter such as the one in the speedometer of a car. The amount it moves corresponds to the amount of current. The glow of the light bulb verifies that current is flowing.

   a) Does the compass-needle galvanometer also indicate that current is flowing? How? In your log, use words and a sketch to indicate your answer.

4. The amount of current flowing in the circuit can be varied by changing the speed at which the generator is cranked, and the amount of current is indicated by the brightness of the light bulb. Vary the speed at which you crank the generator, and observe the galvanometer.

   a) How does the galvanometer indicate changes in the amount of current? Use words and sketches to indicate your answer.

5. Change the direction in which you crank the generator.

   a) What evidence does the galvanometer provide that changing the direction in which the generator is cranked has the effect of changing the direction of current flow in the circuit? Use words and sketches to give your answer.

6. Carefully connect each wire end of a galvanometer to a wire end of a solenoid wound on a hollow core of non-magnetic

material, such as a cardboard tube. Orient the galvanometer so that it is ready to detect current flow.

7. Hold a bar magnet in one hand and the solenoid steady in the other hand. Rapidly plunge one end of the bar magnet into the hollow core of the solenoid, and then stop the motion of the magnet, bringing the end of the magnet to rest inside the solenoid. Another person should hold the galvanometer in a steady position so that it will not be disturbed if the solenoid is moved. Observe the galvanometer during the sequence. You may need to practice this a few times.

a) Write your observations in your log.

8. Remove the magnet from the solenoid with a quick motion, and observe the galvanometer during the action.

a) Record your observations.

b) A current is produced! How does the direction of the current caused, or induced, when the end of the magnet is entering the solenoid, compare to the direction of the current when the magnet is leaving the solenoid?

c) How can you detect the direction of the current in each case?

225

9. Modify and repeat **Steps 7** and **8** to answer the following questions:

a) What, if anything, about the created or induced current changes if the opposite end of the bar magnet is plunged in and out of the solenoid?

b) How does the induced current change if the speed at which the magnet is moved in and out of the solenoid is changed?

c) What is the amount of induced current when the magnet is not moving (stopped)?

d) What is the effect on the induced current of holding the magnet stationary and moving the solenoid back and forth over either end of the magnet?

e) What is the effect of moving both the magnet and the solenoid?

Reflecting on the Activity and the Challenge

In this activity you discovered that you can produce electricity. A current is created or induced when a magnet is moved in and out of a solenoid. The current flows back and forth, changing direction with each reversal of the motion of the magnet. Such a current is called an alternating current, and you may recognize that name as the kind of current that flows in household circuits. It is frequently referred to by its abbreviated form, "AC." It is the type of current that is used to run electric motors in home appliances. Part of your **Chapter Challenge** is to explain to the children how a motor operates in terms of basic principles of physics or to show how electricity can be produced from an external energy source. This activity will help you with that part of the challenge.

Physics To Go

1. An electric motor takes electricity and converts it into movement. The movement can be a fan, a washing machine, or a CD player. The galvanometer may be thought of as a crude electric motor. Discuss that statement, using forms of energy as part of your discussion.

2. Explain how the galvanometer works to detect the amount and direction of an electric current.

3. How could the galvanometer be made more sensitive, so that it could detect very weak currents?

4. An electric generator takes motion and turns it into electricity. The electricity can then be used for many purposes. The solenoid and the bar magnet, as used in this activity, could be thought of as a crude electric generator. Explain the truth of that statement, referring to specific forms of energy in your explanation.

5. If the activity were to be repeated so that you would be able to see only the galvanometer and not the solenoid, the magnet, and the person moving the equipment, would you be able to tell from only the response of the galvanometer what was being moved, the magnet or the solenoid or both? Explain your answer.

6. Part of the **Chapter Challenge** is to explain how the motor and generator toy works.

   a) Write a paragraph explaining how a motor works.
   b) Write a paragraph explaining how a generator works.

7. In generating electricity in this activity, you moved the magnet or the coil. How can you use each of the following resources to move the magnet?

   a) wind
   b) water
   c) steam

Stretching Exercise

Find out about the 120-V (volt) AC used in home circuits. If household current alternates, at what rate does it surge back and forth? Write down any information about AC that you can find and bring it to class.

Activity 4   AC & DC Currents

GOALS

In this activity you will:

• Describe the induced voltage and current when a coil is rotated in a magnetic field.

• Compare AC and DC generators in terms of commutators and outputs.

• Sketch sinusoidal output wave forms.

What Do You Think?

In the last activity, you used human energy to produce motion to generate electricity.

• **What other kinds of energy can generate electricity?**

Write your answer to this question in your *Active Physics* log. Be prepared to discuss your ideas with your small group and other members of your class.

For You To Do

AC Generator

1. Your teacher will explain and demonstrate a hand-operated, alternating current (AC) generator. During the demonstration, make the observations necessary to gain the information needed to answer these questions:

a) When the AC generator is used to light a bulb, describe the brightness of the bulb when the generator is cranked slowly, and then rapidly. Write your observations in your log.

b) When the AC generator is connected to a galvanometer, describe the action of the galvanometer needle when the generator is cranked slowly, and then rapidly.

2. It is easier to understand the creation of a current if you think of a set of invisible threads to signify the magnetic field of the permanent magnets. The very thin threads fill the space and connect the north pole of one magnet with the south pole of the other magnet. If the wire of the generator is imagined to be a very thin, sharp knife, the question you must ask is whether the knife (the wire) can "cut" the threads (the magnetic field lines). If the wire moves in such a way that it can cut the field lines, then a current is generated. If the wire moves in such a way that it does not cut the field lines, then no current is generated.

a) Look at the diagrams of the magnetic fields shown. In which case, I, II, or III will a current be generated?

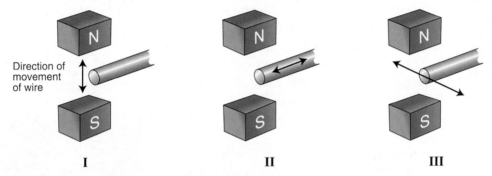

3. The following diagram shows the position of the rotating coil of an AC generator at instants separated by one-fourth of a rotation of the coil. Build a small model of the rectangular coil so that you can move the model to help you understand the drawings. The coil model can be constructed by carefully bending a coat hanger into the shape of the rectangular coil. Rest the coil between two pieces of paper—label the left paper N for the north pole of a magnet; label the right paper S for the south pole of a magnet.

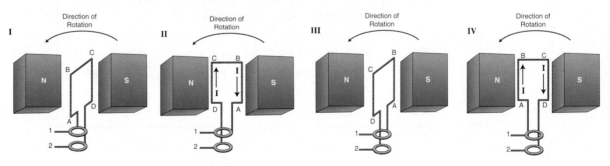

229

4. For the purpose of analyzing the rotating coil figure, the four sides of the rectangular coil of the AC generator will be referred to as sides AB, BC, CD, and DA. Side DA is "broken" to allow extension of the coil to the rings. The "brushes," labeled 1 and 2, make sliding contact with the rings to provide a path for the induced current to travel to an external circuit (not shown) connected to the brushes. The magnetic field has a left-to-right direction (from the north pole to the south pole) in the space between the magnets in the rotating coil figure. It is assumed that the coil has a constant speed of rotation.

a) When the generator coil is in position I shown in the rotating coil, is a current being generated? A current is produced if the wire cuts the magnetic field lines. Record your answer and the reason for your answer in your log.

b) Use a graph similar to the one shown below. Plot a point at the origin of the graph, indicating the amount of induced current is zero at the instant corresponding to the beginning of one rotation of the coil.

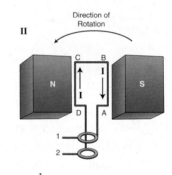

c) One-fourth turn later, at the instant when the rotating coil is in position II, is a current being generated? Record your answer and the reason for your answer in your log.

d) On your graph, plot a point directly above the $\frac{1}{4}$-turn mark at a height equal to the top of the vertical axis to represent maximum current flow in one direction.

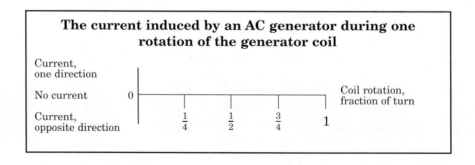

e) One-half turn into the rotation of the coil, at the instant shown in the rotating coil position III, the current again is zero because all sides of the coil are moving parallel to the magnetic field. Plot a point at the $\frac{1}{2}$ mark on the horizontal axis to show that no current is being induced at that instant.

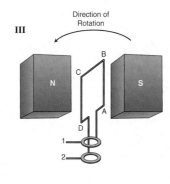

III — Direction of Rotation

f) At the instant at which $\frac{3}{4}$ of the rotation of the coil has been completed, shown by the rotating coil in position IV, the induced current again is maximum because coil sides AB and CD again are moving across the magnetic field at maximum rate. However, this is not exactly the same situation as shown in the rotating coil position II; it is a different situation in one important way: the direction of the induced current has reversed. Follow the directions of the arrows which represent the direction of the current flow in the coil to notice that, at this instant, the current would flow to an external circuit out of brush 2 and would return through brush 1. On your graph, plot a point below the $\frac{3}{4}$-turn mark at a distance as far below the horizontal axis as the bottom end of the vertical axis. This point will represent maximum current in the opposite, or "alternate," direction of the current shown earlier at $\frac{3}{4}$-turn.

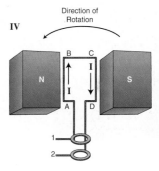

IV — Direction of Rotation

g) The rotating coil in position I is used again to show the instant at which one full rotation of the generator coil has been completed. Again, all sides of the coil are moving parallel to the magnetic field, and no current is being induced. Plot a point on the horizontal axis at the 1-turn mark to show that the current at this instant is zero.

5. You have plotted only 5 points to represent the current induced during one complete cycle of an AC generator.

a) Where would the points that would represent the amount of induced current at each instant during one complete rotation of the generator coil be plotted?

b) What is the overall shape of the graph? Should the graph be smooth, or have sharp edges? Sketch it to connect the points plotted on your graph.

c) What would the graph look like for additional rotations of the generator coil, if the same speed and resistance in the external circuit were maintained?

231

DC Generator

6. Your teacher will explain and demonstrate a hand-operated, direct current (DC) generator. During the demonstration, make the observations needed to answer these questions:

✎ a) When the DC generator is used to light a bulb, describe the brightness of the bulb when the generator is cranked slowly, and rapidly. Write your observations in your log.

✎ b) When the DC generator is connected to a galvanometer, describe the action of the galvanometer needle when the generator is cranked slowly, and rapidly.

7. The diagram shows important parts of a DC generator. As in **Step 3**, build a model of the generator to help you analyze how it works.

8. Use a graph similar to the one shown below. Complete the graph using the same pattern of analysis applied to the AC generator.

✎ a) At the instant shown in the DC generator diagram, the induced current is maximum. The instant corresponds to the rotating coil II. Plot a point on the graph directly above the $\frac{1}{4}$-turn mark at a height equal to the top of the vertical axis to represent maximum current flow at that instant.

✎ b) At the instant $\frac{1}{4}$-turn earlier than the instant shown in the DC generator figure, corresponding to the zero mark of rotation, the current would have been zero because all sides of the coil would have been moving parallel to the direction of the magnetic field. Therefore, plot a point at the origin of the graph.

✎ c) Similarly, the induced current again would be zero at the instant $\frac{1}{4}$-turn later than the instant shown in the DC generator figure; therefore, plot a point on the horizontal axis at the $\frac{1}{2}$-turn mark.

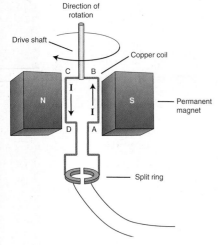

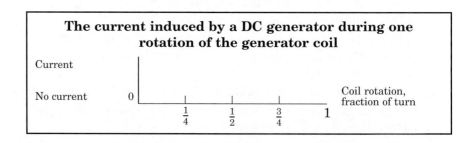

The current induced by a DC generator during one rotation of the generator coil

9. Notice the arrangement used to transfer current from the generator to the external circuit for the DC generator. It is different from the arrangement used for the AC generator. The DC generator has a "split-ring commutator" for transferring the current to the external circuit. Notice that if the coil shown in the DC generator figure were rotated $\frac{1}{4}$-turn in either direction, the "brush" ends that extend from the coil to make rubbing contact with each half of the split ring would reverse, or switch, the connection to the external circuit. Further, notice that the connection to the external circuit would be reversed at the same instant that the induced current in the coil reverses due to the change in direction in which the sides of the coil move through the magnetic field. The outcome is that while the current induced in the coil alternates, or changes direction each $\frac{1}{2}$-rotation, the current delivered to the external circuit always flows in the same direction. Current that always flows in one direction is called direct current, or DC.

a) Plot a point on the graph at a point directly above the $\frac{3}{4}$-turn mark at the same height as the point plotted earlier for the $\frac{1}{4}$-turn mark.

b) As done for the AC generator, find out how to connect the points plotted on this graph to represent the amount of current delivered always in the same direction to the external circuit during the entire cycle.

Reflecting on the Activity and the Challenge

It is time to begin preparing for the **Chapter Challenge**. Now that you know how a generator works, you should begin to think about toys that might generate electricity. You should also think about how you could assemble "junk" into a toy generator, or do some research on homemade generators and motors.

Physics To Go

1. What is the purpose of:

   a) An electric generator?
   b) An electric motor?

2. How does a direct current differ from an alternating current? Use graphs to illustrate your answer.

3. In an electric generator, a wire is placed in a magnetic field. Under what conditions is a current generated?

Stretching Exercises

1. What is the meaning of "hertz," abbreviated "Hz," often seen as a unit of measurement associated with electricity or stereo sound components such as amplifiers and speakers?

2. What does it mean to say that household electricity has a frequency of 60 Hz?

3. Have you ever heard 60 Hz AC being emitted from a fluorescent light or a transformer?

4. Look at a catalog or visit a store where sound equipment is sold, and check out the "frequency response" of speakers— what does it mean?

5. Heinrich Hertz was a 19th-century German physicist. Find out about the unit of measurement named after him, and write a brief report on what you find.

Activity 5   Building an Electric Motor

What Do You Think?

You plug a mixer into the wall and turn a switch and the mixer spins and spins—a motor is operating.

- **How do you think the electricity makes the motor turn?**

Write your answer to this question in your *Active Physics* log. Be prepared to discuss your ideas with your small group and other members of your class.

For You To Do

1. Study the diagram on the following page closely. Carefully assemble the materials, as shown in the diagram, to build a basic electric motor. Follow any additional directions provided by your teacher.

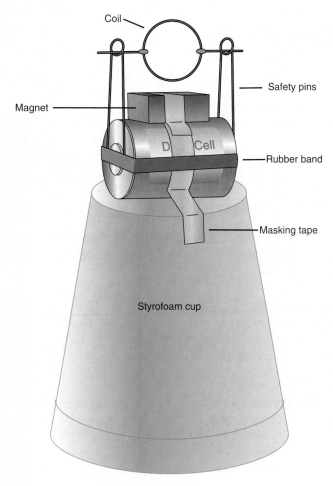

Coil

Safety pins

Magnet

D Cell

Rubber band

Masking tape

Styrofoam cup

2. When your motor is operating successfully, find as many ways as you can to make the motor change its direction of rotation.

   a) Describe each way you tried and identify the ways that were successful.

3. Hold another magnet with your fingers and bring it near the coil from above, facing the original magnet, as the motor is operating.

   a) Describe what happens. Does the orientation of the second magnet make a difference?

4. Replace the single magnet with a pair of attracting magnets on top of the battery.

   a) What is the effect?

5. Think of other ways to change the speed of the motor. With the approval of your teacher, try out your methods.

   a) Describe ways to change the speed of the motor.

6. Use a hand generator as the energy source instead of the battery. You can disconnect the battery without removing it from the structure by placing an insulating material, such as a piece of cardboard, between the safety pin and the battery to open the circuit at either end of the battery. Then clip the wires from the generator to the safety pins to deliver current from the generator to the motor.

   a) Discuss what you find out.

7. Your motor turns! Chemical energy in the battery was converted to electrical energy in the circuit. The electrical energy was then converted to mechanical energy in the motor.

   a) List at least three appliances or devices where the motor spins.

8. The spin of the motor occurs because the current-carrying wire has a force applied to it. You know if something moves, a force must be applied. As you observed, when the battery connection was broken, the motor stopped turning. You know from a previous activity that a current-carrying wire creates a magnetic field. Pause for a bit to remind yourself of the behavior of magnets. Take a bar magnet and place its north pole near a compass. The compass is a tiny bar magnet that can easily turn.

a) Draw a sketch to show the orientation of the compass.

 N S

9. Shift the compass to the south pole of the bar magnet.

a) Draw a sketch to show the orientation of the compass.

10. The north pole of the bar magnet repelled the north pole of the compass. The south pole of the bar magnet attracted the north pole of the compass. This attraction and repulsion is the result of a force on the compass. You can now investigate the force between the poles of a magnet and the magnetic field of a current-carrying wire.

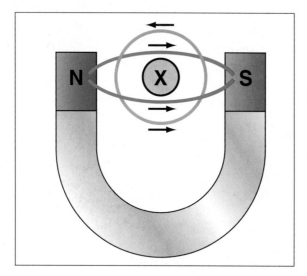

The magnetic field lines are drawn for a horseshoe magnet. The direction of the magnetic field lines is identical to a direction that a compass would point. The compass would point away from the north pole and toward the south pole. The magnetic field of the current-carrying wire is circular, as you investigated in an earlier activity. Compare the direction of this magnetic field to the direction of the magnetic field of the horseshoe magnet.

11. Think of the magnetic field lines above as small compasses.

   a) Write down whether the compasses above the wire attract one another or repel one another.

   b) Write down whether the compasses below the wire attract one another or repel one another.

12. This attraction/repulsion causes the wire to jump. There is a force on the wire. This force on the current-carrying wire is the basis for the electric motor that you built in this activity. The use of the loop of the wire allows the wire to rotate instead of jumping in the way a single wire would.

13. It is the moving electrons in the wire that create the current. In some TV sets, there is an electron beam that shoots the electrons from the back of the TV to the front. There are horseshoe magnets of a sort in the television. The moving electrons experience a force. The electrons' path is affected by the magnetic field. By varying the strength of the magnetic field, the electron beam can hit all parts of the screen and you receive a TV image.

FOR YOU TO READ

The history of science is filled with discoveries that have led to leaps of progress in knowledge and applications. This is certainly true of physics and, in particular, electricity and magnetism. These discoveries "favor" the prepared mind. Oersted's discovery in 1820 of the magnetic field surrounding a current-carrying wire already has been mentioned. Similarly, Michael Faraday

Michael Faraday

discovered electromagnetic induction in 1831. Faraday was seeking a way to induce electricity using currents and magnets; he noticed that a brief induced current happened in one circuit when a nearby circuit was switched on and off. (How would that cause induction? Can you explain it?) Both Oersted and Faraday are credited for taking

advantage of the events that happened before their eyes, and pursuing them.

About one-half century after Faraday's discovery of electromagnetic induction, which immediately led to development of the generator, another event occurred. In 1873, a Belgian engineer, Zénobe Gramme, was setting up DC generators to be demonstrated at an exposition (a forerunner of a "world's fair") in Vienna, Austria. Steam engines were to be used to power the generators, and the electrical output of the generators would be demonstrated. While one DC generator was operating, Gramme connected it to another generator that was not operating. The shaft of the inactive generator began rotation—it was acting as an electric motor! Although Michael Faraday had shown as early as 1821 that rotary motion could be produced using currents and magnets, a "motor effect," nothing useful resulted from it. Gramme's discovery, however, immediately showed that electric motors could be useful. In fact, the electric motor was demonstrated at the very Vienna exposition where Gramme's discovery was made. A fake waterfall was set up to drive a DC generator using a paddle wheel arrangement, and the electrical output of the generator was fed to a "motor" (a generator running "backwards"). The motor was shown to be capable of doing useful work.

Reflecting on the Activity and the Challenge

Decision time about the **Chapter Challenge** is approaching for your group. In this activity you built a very basic, working electric motor. This is an important part of the **Chapter Challenge**. However, knowing how to build an electric motor is only part of the challenge. Your toy must be fascinating to children. You must also be able to explain how it works.

Physics To Go

1. Some electric motors use electromagnets instead of permanent magnets to create the magnetic field in which the coil rotates. In such motors, of course, part of the electrical energy fed to the motor is used to create and maintain the magnetic field. Similarly, electromagnets instead of permanent magnets are used in some generators; part of the electrical energy produced by the generator is used to energize the magnetic field in which the generator coil is caused to turn. What advantages and disadvantages would result from using electromagnets instead of permanent magnets in either a motor or generator?

2. Design three possible toys that use a motor or a generator or both. One of these may be what you will use for your project.

3. The motor/generator you submit for the **Chapter Challenge** must be built from inexpensive, common materials. Make a list of possible materials you could use to construct an electric motor.

4. In the grading criteria for the **Chapter Challenge**, marks are assigned for clearly explaining how and why your motor/generator works in terms of basic principles of physics. Explain how an electric motor or generator operates.

Activity 6   Building a Motor/Generator Toy

GOALS

In this activity you will:

• Design, construct and operate a motor/generator.

What Do You Think?

You may have heard the following expression used before: "The difference between men and boys is the cost of their toys."

• **What characteristics make an item a toy?**

Write your answer to this question in your *Active Physics* log. Be prepared to discuss your ideas with your small group and other members of your class.

For You To Do

1. Confer within your group and between your group and your teacher about whether you will pursue, as a basis for the motor/generator kit for the **Assessment**, the motor design presented in this activity, an alternate design, or both. Whatever design(s) your group chooses to pursue, you are encouraged to be creative.

Most designs can be improved in some way or another by substituting materials or making other changes. There is no single "best way" to go about designing the motor/generator and making it function within a toy or to produce electrical energy from another form of energy. The best way for your group is the way that the group can get the job done.

✎ a) When you have decided on a design, submit your design to your teacher for approval.

2. In your group decide how you will make the motor/generator fascinating to children. You may wish to use some of the ideas you generated in answering the **What Do You Think?** question.

✎ a) Record your ideas in your log.

✎ b) Describe and make a sketch of your final design, and submit it to your teacher for approval.

3. Use the design for a DC motor as shown in the diagram as a basis to begin your construction. It can be adapted, as required, for the **Chapter Assessment**, to power a toy. Also as required, the motor could be adapted to be driven "backwards" by an external energy source to function as a DC generator.

[The motor design shown was adapted from the following public domain work: Educational Development Center, Inc., *Batteries and Bulbs II* (New York: McGraw-Hill, 1971), pp. 85-88.]

cork

clips

thumb tacks

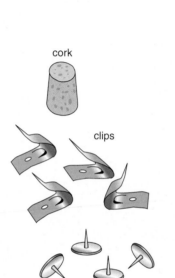

straight pins

magnets

tin can

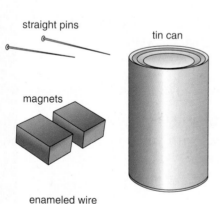

enameled wire

thin stick

wood

masking tape

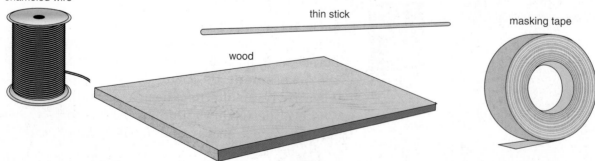

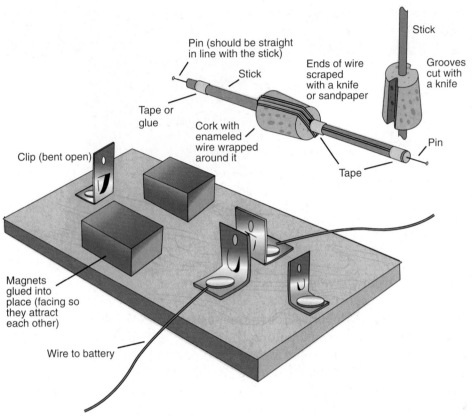

Pin (should be straight in line with the stick)

Stick

Tape or glue

Clip (bent open)

Cork with enameled wire wrapped around it

Magnets glued into place (facing so they attract each other)

Wire to battery

Ends of wire scraped with a knife or sandpaper

Stick

Grooves cut with a knife

Pin

Tape

Reflecting on the Activity and the Challenge

You are now well on your way to completing the **Chapter Challenge**. You have decided on the design for your motor/generator and the toy it will power.

Physics To Go

Your assignment is to prepare to meet the criteria of the **Assessment** of the **Chapter Challenge**.

PHYSICS AT WORK

Uriah Gilmore

HEADED FOR THE STARS

Uriah Gilmore loved to take electric appliances apart when he was growing up. "I couldn't always get them back together," he admits, "but I was so curious I couldn't help myself. I just had to see how they worked." Fortunately, Uriah's parents supported his curiosity.

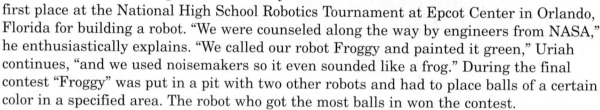

Uriah and his fellow teammates from Cleveland, Ohio's East Technical High School recently won first place at the National High School Robotics Tournament at Epcot Center in Orlando, Florida for building a robot. "We were counseled along the way by engineers from NASA," he enthusiastically explains. "We called our robot Froggy and painted it green," Uriah continues, "and we used noisemakers so it even sounded like a frog." During the final contest "Froggy" was put in a pit with two other robots and had to place balls of a certain color in a specified area. The robot who got the most balls in won the contest.

"In my sophomore year the school I was attending closed and I went to East Technical High School which was the best thing that happened to me." He entered the engineering program and became a member of the engineering team—a team that is more popular than any sports team in his school.

Uriah attended Morehouse College on a NASA scholarship. "But," he states, "it's not enough to be a good student. You also have to be involved with your school and your community." Uriah once led a march on the Cleveland, Ohio, City Hall to protest a law which threatened to fire certain teachers, including one who inspired Uriah and was responsible for the revitalization of East Technical High School.

"My ultimate goal is to travel in space and explore the galaxy," he states. A shorter term goal is to be as involved in college as he has been in high school.

Chapter 4 Assessment

Your task is to prepare a kit of materials and instructions that a toy company will manufacture. Children will use these kits to make a motor or generator, or a combination electric motor/generator. It will serve both as a toy and to illustrate how the electric motors in home appliances work or how electricity can be produced from an energy source such as wind, moving water, a falling weight, or some other external source.

Review and remind yourself of the grading criteria that you and your classmates agreed on at the beginning of the chapter. The following was a suggested set of criteria:

- **(30%) The motor/generator is made from inexpensive, common materials, and the working parts are exposed but with due consideration for safety.**

- **(40%) The instructions for the children clearly explain how to assemble and operate the motor/generator device, and explain how and why it works in terms of basic principles of physics.**

- **(30%) If used as a motor, the device will operate using a maximum of four 1.5-volt batteries (D cells), and will power a toy (such as a car, boat, crane, etc.) that will be fascinating to children.**

OR

- **(30%) If used as a generator, the device will demonstrate the production of electricity from an energy source such as wind, moving water, a falling weight, or some other external source and be fascinating to children.**

Physics You Learned

Motors

Generators

Galvanometers

Magnetic field from a current

Solenoids

Electromagnets

Induced currents

AC and DC generators

Chapter 5

Scenario

Science has enriched the lives of everyone. People no longer fear the movement of the planets. Many enjoy viewing an eclipse. Science and technology have helped feed large numbers of people, and raise the standard of living of many people as well. Science and technology have also complicated lives. New problems have emerged as a result of the technologies that people have decided to use. As people learn more about the natural world through science and technology, they discover that there is more and more to know!

Challenge

Although you have grown up in a society that uses science and technology, it is difficult sometimes to distinguish between science and pseudoscience.

This challenge places you as the head of an institute that provides funding for science research. A number of groups or individuals have submitted proposals to you, all wishing funding from your institute. These include research on:

• force fields
• auras
• telekinesis
• new comets
• failure modes of complex systems
• advent of new diseases
• astrology prediction
• communicating with extraterrestrial beings
• the extinction of dinosaurs
• communication with dolphins
• prediction using biorhythms
• properties of new materials
• dowsing
• earthquake prediction
• election predictions using polling

You will choose two proposals from this list, or invent other proposals to add to the list. One of the proposals will be accepted because of its scientific merit. The other proposal will be denied because it has little or no scientific merit.

You will have to defend your selections in a position paper. You will also write letters to each of the people who submitted these studies for funding.

How will you decide which project to fund and which to deny? As you work through the chapter and think about funding, ask the following questions:

Is the area of study logical?

Is the topic area testable by experiment?

Can any observer replicate the experiment and get the same results?

Is the theory the simplest and most straightforward explanation?

Can the new theory explain known phenomena?

Can the new theory predict new phenomena?

Criteria

Here are the standards by which your work will be evaluated:

- **The selection of proposals reflects an accurate understanding of the nature of science**
- **The selection of proposals reflects an accurate understanding of the role and importance of science in the world**
- **The selection considers all the major differences you've learned about science and pseudoscience**
- **The position paper is clearly written and accurate. Grammar and spelling are correct**
- **The letters explain your reasoning clearly, concisely, and in a businesslike fashion. Grammar and spelling are correct**

Discuss in your small groups and as a class the criteria for this performance task. For instance:

- **How much of the grade should depend on showing the scientific merit of the first idea, or the lack of scientific merit of the second idea?**
- **How much of the grade should depend on quality and clarity of the presentation?**
- **How much should depend on the letters to the hopeful researchers? How should a letter be graded?**
- **What would constitute an "A" for this project?**

Here is a sample grading rubric. You can fill out the descriptions and supply the point values.

Criteria	Excellent max=100%	Good max=70%	Satisfactory max=50%	Poor max=25%
reflects an accurate understanding. . . points				
role and importance of science. . . points				
major differences. . . points				
clearly written. . . points				
letters. . . points				

A=   points   B=   points   C=   points   D=   points

Activity 1   Force Fields

GOALS

In this activity you will:

- Investigate the properties of bar magnets.
- Plot magnetic and electric fields.
- Make a temporary magnet.
- Describe the properties of magnetic and electric fields.
- Make an electromagnet.

 What Do You Think?

Large magnets are able to pick up cars and move them around junkyards.

- **How does a magnet work?**
- **What objects are attracted to magnets and which objects are not attracted to magnets?**

Record your ideas about these questions in your *Active Physics* log. Be prepared to discuss your responses with your small group and the class.

For You To Do

1. Get two bar magnets from your teacher. Use them to answer the following questions:

   ✎ a) Do the magnets exert a force on one another? How do you know?

   ✎ b) Must they touch one another to exert the force?

   ✎ c) Is the force between the two magnets attractive or repulsive? Is it both? Make a drawing that shows how the forces between the magnets act.

   ✎ d) A bar magnet can be described by its ends. These ends are called magnetic poles. They are labeled N and S. Opposite poles attract each other and like poles repel. Are the poles on your magnet marked? If not, get a magnet that is marked, and use it to find out which pole on your magnet is N and which is S.

   ✎ e) Do the magnets exert a force on other objects in the classroom? Find out. Record your findings.

2. The needle of a compass is a small bar magnet. The N pole is usually painted red or shaped like an arrow. Place one of the bar magnets under a sheet of paper. Put the other magnet out of the way by moving it some distance from the paper. Move a small compass back and forth just above the surface of the paper. Be sure to move the compass back and forth in close rows so you don't miss large areas of the paper. As you move the compass, sketch the direction of the N pole of the compass at different points on the paper. The diagram you have made shows the magnetic field around the magnet under the paper. Answer the following questions in your log:

   ✎ a) The N pole of the compass points in the direction of the field. What is the direction of the magnetic field at the N pole of the bar magnet? At the S pole of the bar magnet?

   ✎ b) How does the strength of the magnetic field change with distance from the magnet? Explain your thinking.

   ✎ c) Your plot of the magnetic field is in two dimensions. Do you think that the magnetic field lies in only two dimensions? How could you test your answer?

Coordinated Science for the 21st Century

3. Repeat **Step 2** with a horseshoe magnet.

✎ a) Does the field depend on the shape of the magnet you use?

4. You can map a field around a magnet quickly with iron filings. In a tray place a sheet of stiff paper over the magnet. Sprinkle the filings on the paper and gently tap the paper. The filings will behave like tiny compasses. They will fall along the lines of the magnetic field and make an instant plot of the field.

✎ a) Sketch the magnetic field shown by the iron filings.

5. Get a metal paper clip, sewing needle, or small nail. These objects become magnets when they are stroked by a magnet. Stroke the object with a bar magnet. Stroke in one direction only.

✎ a) Did the object become magnetic? How do you know?

⚠️ **Avoid having the iron filings come into direct contact with the magnet.**

Physics Words

electric field: the region of electric influence defined as the force per unit charge.

magnetic field: the region of magnetic influence around a magnetic pole or a moving charged particle.

FOR YOU TO READ

Magnetic Fields

Some forces act across space. One of these is the force between two magnets. The ancient Greeks first discovered magnetism over 2000 years ago. They found lodestone, an ore that attracted some other rocks. The name *magnet* probably comes from the region in Asia Minor where lodestone was first discovered, Magnesia. By the 10th century, the Chinese and Vikings were using magnetic compasses, which they invented, in boating.

A distinguished English physicist, William Gilbert, who was also a physician to Queen Elizabeth I, studied magnetism in detail in the late 1500s. He published a book called *De Magnete* in 1600. *De Magnete* explained how a magnet had an "invisible orb of virtue" around it. The idea of an "invisible orb of virtue" has been replaced with the theory of **magnetic fields**.

Magnets fill the region of space around themselves with a magnetic field. When another magnet enters the magnetic field, it experiences a force. Magnetic fields can be "seen" when another magnet is brought into the field and the force it experiences is observed. You did this with a magnet and compass. The direction of a magnetic field is the direction of the force on the N magnetic pole of the second magnet.

Gilbert also showed how a small ball of magnetic material could make magnetic compasses brought near them behave like compasses on Earth. His demonstrations led to the idea that the Earth behaves like a giant magnet, an idea that has been shown to be true since his time. The N pole of a magnet is attracted to the magnetic pole of the Earth in the Northern Hemisphere.

The magnetic pole of Earth in the Northern Hemisphere and the geographic North Pole are not in exactly the same place. Today they are about 1500 km from each other. Scientists have evidence that the position of the magnetic poles has changed during the history of the Earth.

All magnets have two poles. If you break a magnet in half, you'll get two magnets, each with a N pole and a S pole. Scientists are trying to break magnets into small enough pieces—atom-size pieces—to make a magnet with just one pole, but they have not been able to do so. If someone does one day, he or she may be on the short list for the Nobel Prize!

The magnetic field is invisible. While it cannot be seen, it can be measured. Anyone can use a compass to find out if a magnetic field exists. Observations of magnetic behavior are predicted by scientific theory. Magicians claim to project the powers of their minds across empty space by filling space with psychic auras. There are no known ways to detect auras. Some people say that they can sense auras. But they cannot demonstrate this "sense" to others. There is no "compass" to detect auras.

Electric Fields

In your previous study of charges, you learned that like charges (+ + or - -) repel and unlike charges (+ - or - +) attract. You also learned that Coulomb's Law describes the force of attraction:

$$F = \frac{kQq}{R^2}$$

You can think of the force on $+q$ as the interaction between the two charges $+Q$ and $+q$.

The force on a charge can also be described with the field concept. The electric field surrounding $+Q$ can be mapped in a way similar to how the magnetic field was mapped.

In mapping the magnetic field of a bar magnet, you observed what would happen to a compass placed near the bar magnet or what would happen to small iron filings (similar to tiny compasses) placed near the bar magnet.

To map the **electric fields** due to a group of charges, you can place small positive charges and observe what will happen. These small positive charges are called "test" charges since they test for the electric field.

If positive test charges were placed near the $+Q$, they would be repelled. You could draw that force of repulsion with little arrows.

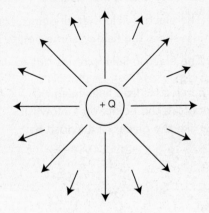

Coordinated Science for the 21st Century

The lengths of the force vectors are intended to show the strength of the force. The test charges placed close to $+Q$ have a larger force than the test charges placed further from $+Q$.

The general pattern of the forces on the test charges is depicted as the electric field of the $+Q$ charge. In this case, the electric field of $+Q$ is:

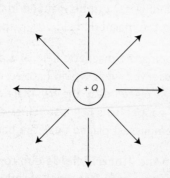

The lines tell you the following properties of the electric field or $E-$ field of $+Q$:

• The direction of the electric field is the direction of the force on a + test charge.

• The electric field is strongest where the lines are close together and weaker where the lines are further apart.

• The electric field is at all points. The lines just show the field at some points.

• The electric field extends out to infinity.

The electric field lines from $+Q$ look like the points on a blowfish or like the pieces of a Koosh ball.

A similar process can be used to find the electric field of a long line of charges.

The test charge will always be pushed away from the line of charge. The force is due to the vector sum of all of the force interactions of the test charge and each charge on the line.

The electric field lines from a line of charge look like the spokes of many bike wheels lined up next to one another. In the diagram below, you don't see the lines coming at you or away from you. The lines radiate out from the line of charge in all directions. As you get further from the line of charge, the electric force gets weaker and the electric field lines spread out as the spokes of a wheel spread out.

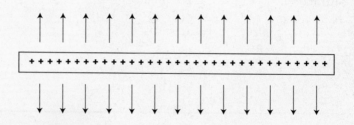

The electric field from a line of charge gets weaker as you get further from the charges, but does not weaken as much as the electric field from a spherical set of charges +Q.

An extremely important electric field for all sorts of electric circuits is the electric field of parallel plates. The top plate has positive charges and the bottom plate has negative charges. The force on a positive test charge will be toward the negative plate and away from the positive plate.

Since the electric field lines do not get closer or further apart within the parallel plates, you know that the force on the positive test charge is a constant force. These parallel plates are also referred to as parallel-plate capacitors.

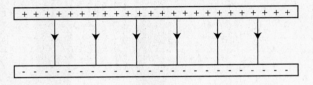

Reflecting on the Activity and the Challenge

You know that magnetic and electric forces act across space. You were able to detect the magnetic field of a bar magnet at different distances from the magnet. The field lines are invisible, yet they can be detected by anyone using tools such as a compass or test charges. Often, people refuse to believe in things that are invisible. Can you convince someone that magnetic and electric forces and fields are real even though they are invisible? Some people believe in psychic auras. These people insist that these are also invisible and should be believed. But ordinary people with everyday tools are not able to detect auras. Psychic auras do not meet the same criteria as do magnetic and electric fields. If you funded research in psychic auras, you would expect the researchers to be able to detect auras.

Physics To Go

1. How do you know that magnets and electric charges act across a distance?

2. Will the N pole of one magnet attract or repel the N pole of another magnet? The S pole of another magnet?

3. Draw each of the magnets shown below. Then draw the magnetic field around each magnet.

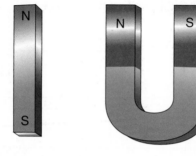

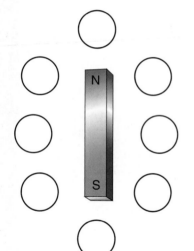

4. Copy the diagram to the left. Each circle represents a magnetic compass. Draw the compass needle for each, as it would point in its position. Use an arrow for the N end of the compass needle.

5. Copy the diagrams of electric charges below. Draw the electric field around each.

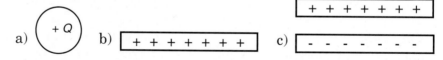

a) $+Q$ b) + + + + + + +   c) - - - - - - - -

6. How does the strength of a magnetic field change with distance from a magnet?

7. A magnet is hanging from a ceiling by strings. Which way will it point?

8. The needle on a magnetic compass points to the magnetic pole of the Earth in the Northern Hemisphere. You know that like poles repel and unlike poles attract. What can you say about the magnetic pole of the Earth in the Northern Hemisphere and the N pole of the magnetic compass?

Stretching Exercises

1. Obtain two or three household flat ("refrigerator") magnets. Use iron filings and paper or a magnetic compass to explore the magnetic field of each magnet. Draw a diagram of your findings.

2. Electricity and magnetism are different forces. However, they are related to one another. Find out how. Carefully wrap a length of wire around an iron nail. Connect each end of the wire to a battery terminal. When you attach the wires to the battery, current starts flowing through the wire. Bring a magnetic compass near the wire coils. Record your observations.

   You've made an electromagnet. Electromagnets are magnetic when current is flowing through the wire. They are the type of magnet used to pick up cars and move them around junk-yards. They have many other uses, too. Research the uses of electromagnets, and report your findings to the class.

Activity 2

Newton's Law of Universal Gravitation

GOALS

In this activity you will:

- Explore the relationship between distance of a light source and intensity of light.

- Graph and analyze the relationship between distance of a light source and intensity of light.

- Describe the inverse square pattern.

- Graph and analyze gravity data.

- State Newton's Law of Universal Gravitation.

- Express Newton's Law of Universal Gravitation as a mathematical formula.

- Describe dowsing and state why the practice is not considered scientific.

What Do You Think?

Astronauts on many Shuttle flights study the effects of zero-gravity. Fish taken aboard the Shuttle react to "zero-gravity" by swimming in circles.

- **How would a fish's life be different without gravity?**

- **Does gravity hold a fish "down" on Earth?**

Record your ideas about these questions in your *Active Physics* log. Be prepared to discuss your responses with your small group and the class.

For You To Do

1. Place a projector 0.5 m from the chalkboard. Insert a blank slide. Turn on the projector.

2. Use chalk to trace around the square of light on the board.

3. Place the photocell in one corner of the light square. Attach it to the galvanometer as directed by your teacher. The photocell and galvanometer measure light intensity. The more light that strikes the cell, the greater the current reading on the galvanometer.

✎ a) Copy the table below in your log. Record the distance to the board, current in galvanometer, and length of a side of the square.

Distance to board (m)	Distance squared	Current in galvanometers (A)	Side of square (cm)	Area of square (cm²)

4. Move the projector to a position 1 m from the board. Adjust the projector so that the original square of light sits in one corner of the new square of light.

✎ a) Enter the data into the table in your log.

5. Repeat **Step 4** with the projector at distances of 1.5 m, 2 m, 2.5 m, and 3 m.

✎ a) Enter the data into the table in your log.

6. Graph the current in the galvanometer versus distance. Label this graph Graph 1.

✎ a) Is Graph 1 a straight line?

✎ b) What does a straight line on the graph tell you?

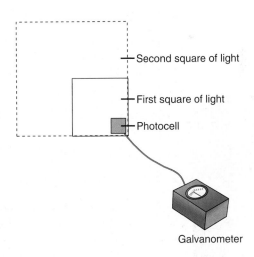

Second square of light

First square of light

Photocell

Galvanometer

Physics Words

acceleration: the change in velocity per unit time.

gravity: the force of attraction between two bodies due to their masses.

Inverse square relation: the relationship of a force to the inverse square of the distance from the mass (for gravitational forces) or the charge (for electrostatic forces).

7. Light intensity decreases with distance as the light from the source spreads out over larger areas. The light is literally spread thin. The light intensity at any one spot increases as the area gets smaller and decreases as it gets larger. This observation is an example of a pattern called the inverse square relation. In an inverse square relation, if you double the distance the light becomes $\frac{1}{2^2}$ or $\frac{1}{4}$ as bright. If you triple the distance, the light becomes $\frac{1}{3^2}$ or $\frac{1}{9}$ as bright. If you increase the distance by 5 times, the light becomes $\frac{1}{5^2}$, or $\frac{1}{25}$ as bright. If you increase the distance by 10 times, the light becomes $\frac{1}{10^2}$, or $\frac{1}{100}$ times as bright.

a) How closely does your data reflect an inverse square relation?

Acceleration Due to the Earth's Gravitational Field at Different Heights	
Height above Sea Level (km)	Acceleration due to Gravity (m/s^2)
0	9.81
3.1	9.76
11	9.74
160	9.30
400 (Shuttle orbit)	8.65
1600	6.24
8000	1.92
16,000	0.79
36,000 (geosynchronous orbit for communications satellite)	0.23
385,000 (orbit of the Moon)	0.003

8. Compute the distances from the center of the Earth (6400 km below sea level). Plot these distances vs. acceleration in a graph. Draw the best possible curve through the points on the graph. Label this graph Graph 2.

a) Does the data form a pattern?

b) Is the pattern familiar to you? Give evidence for your conclusion.

FOR YOU TO READ

An Important Pattern

You've seen one pattern in this activity. But you've seen it in two different ways. In **Steps 1** through **8** you found that light intensity becomes less as the light source is moved further away. In **Step 7,** you've seen that **acceleration** due to **gravity** becomes less as an object moves further from the surface of the Earth. Both are examples of the **inverse square relation.** Although light is not a force, the effect of distance on its behavior in this activity is like that of the effect of distance on the force of gravity. That is, the behavior of light in this activity is analogous to the behavior of gravity. In simple terms for gravity, the inverse square relation says that the force of gravity between two objects decreases by the square of the distance between them.

Mapping the Earth's Gravitational Field

In **Activity 1,** you mapped the magnetic field around a bar magnet using a compass as a probe. You also read about electric fields. In this activity, you used data on acceleration due to gravity to map the Earth's gravitational field. The probe is the acceleration of a falling mass. To see the pattern of Earth's gravitational field, you needed data from satellites. The gravitational field changes very slowly near the surface of the Earth. The pattern is very difficult to see using surface data.

Newton's Law of Universal Gravitation describes the gravitational attraction of objects for one another. Isaac Newton first recognized that all objects with mass attract all other objects with mass.

Experiments show that objects have mass and that the Earth attracts all objects. Newton reasoned that the Moon must have mass, and that the Earth must also attract the Moon. He calculated the acceleration of the Moon in its orbit and saw that the Earth's gravity obeyed the inverse square relation. It is a tribute to Newton's genius that he then guessed that not only the Earth but all bodies with mass attract each other.

Almost 100 years passed before Newton's idea that all bodies with mass attract all other bodies with mass was supported by experiments. To do so, the very small gravitational force that small bodies exert on one another had to be measured. Because this force is very small compared to the force of the massive Earth, the experiments were very difficult. But in 1798, Henry Cavendish, a British physicist, finally measured the gravitational force between two masses of a few kilograms each. He used the tiny twist of a quartz fiber caused by the force between two masses to detect and measure the force between them.

Newton's Law of Universal Gravitation states:

All bodies with mass attract all other bodies with mass.

The force is proportional to the product of the two masses and gets stronger as either mass gets larger.

The force decreases as the square of the distances between the two bodies increases.

Physics and Dowsing: Comparing Forces

Dowsing is a way some people use to locate underground water. It is claimed to work on an apparent "attraction" between running water and a dowsing rod carried by a person. All dowsers claim to feel a force pulling the rod towards water, and many claim to feel unusual sensations when they cross running water. In the 19th century, many dowsers described the force on the rod as an electric force. No evidence supports this idea. In fact, there is no scientific theory to explain any attraction between running water and a dowsing rod.

Despite the skepticism of the scientific community about dowsing, it is widely used in the United States. Even a national scientific laboratory has used dowsers! But the United States Geological Survey has investigated dowsing and finds no experimental evidence for it. Statistics show that the success rate could be a result of random events. Even if experimental evidence supported the success of dowsing, there is no theory to predict its operation. In order to be accepted as scientific, a phenomenon must be reproducible in careful experiments. Its effects must be predictable by a theory. Also, the theory must give rise to other predictions that can be tested by experiments.

PHYSICS TALK

Newton's Law of Universal Gravitation in Mathematical Form

Complex laws like Newton's Law of Universal Gravitation may look easier in mathematical form. Let F_G be the force between the bodies, d be the distance between them, m_1 and m_2 the masses of the bodies and G be a universal constant equal to 6.67×10^{-11} N·m²/kg².

You can express Newton's Law of Universal Gravitation as

$$F_G = \frac{G\, m_1 m_2}{d^2}$$

You can see that the equation says exactly the same thing as the words in a much smaller package.

Reflecting on the Activity and the Challenge

In this activity you determined experimentally how light intensity varies with distance. By plotting measured data, you found that gravity follows an identical pattern. You detect gravity by measuring the acceleration of objects falling at specific locations. Patterns help you understand the world around you. Light follows the inverse square relation and so does gravity. You can detect gravity with masses. You can detect magnetic fields with compasses. But you cannot detect the "attraction" claimed by dowsers. There are no detectors for that! You will be required in the **Chapter Challenge** to differentiate between the measured gravity and its inverse square nature and the dowser's claim of measurement. This activity helped you to understand one difference between science and pseudoscience.

Physics To Go

1. How would the light intensity of a beam from a projector 1 m from a wall change if the projector was moved 50 cm closer to the wall?

2. The gravitational force between two asteroids is 500 N. What would the force be if the distance between them doubled?

3. A satellite sitting on the launch pad is one Earth radius away from the center of the Earth (6.4×10^6 m).

   a) How would the gravitational force between them be changed after launch when the satellite was two Earth radii (1.28×10^7 m) from the center of the Earth?
   b) What would the gravitational force be if it was 1.92×10^7 m from the center of the Earth?
   c) What would the gravitational force be if it was 2.56×10^7 m from the center of the Earth?

4. Why does everyone trust in gravity?

5. Why doesn't everyone trust in dowsing?

6.

a) Which is closer to the Moon, the middle of the Earth or the water on the side of Earth facing the Moon?

b) Use your answer to **a)** to propose an explanation for the uneven distribution of water on Earth's surface, as shown in the diagram.

c) Suggest an explanation for high tides on the side of the Earth facing the Moon.

Stretching Exercises

1. To locate underground water, a dowser uses a Y-shaped stick or a coat hanger bent into a Y. The dowser holds the Y by its two equal legs with the palms up and elbows close to his or her sides. The long leg of the Y is held horizontal. The dowser walks back and forth across the area he or she is searching. When he or she crosses water, the stick jerks convulsively and twists so hard that it may break off in the dowser's hands. Dowsers claim to be unaware of putting any force on the stick. Most observers think that the motion of the stick is probably due to the unconscious action of the dowser.

   According to records of those who believe in dowsing, approximately 1 in 10 people should have the ability to dowse. Do you have dowsing ability? Try this activity to find out. Can you prove that you're a dowser to a classmate? What would constitute proof?

2. Does the inverse square relation apply to magnetic force? Work with your group to plan an experiment to find out. State your hypothesis, and describe the method to test it. If your teacher approves your experimental design, try it. Report your results to the class.

Activity 3   Slinkies and Waves

GOALS

In this activity you will:

- Make a "people wave."
- Generate longitudinal and transverse waves on a Slinky.
- Label the parts of a wave.
- Analyze the behavior of waves on a Slinky.
- Compare longitudinal and transverse waves.
- Define wavelength, frequency, amplitude, and period of a wave.
- Measure the speed of various waves on a Slinky.

What Do You Think?

The Tacoma Narrows Bridge was known as "Galloping Gertie" because light winds caused the bridge's roadway to ripple and oscillate. In 1940 the bridge collapsed. The ripple motion caused the structure to break.

- **Have you ever crossed a bridge that was rippling? How secure did you feel?**
- **If you were an engineer, how would you test the strength of bridges?**

Record your ideas about these questions in your *Active Physics* log. Be prepared to discuss your responses with your small group and the class.

Coordinated Science for the 21st Century

For You To Do

1. With your classmates, make a "people wave," like those sometimes made by fans at sporting events.

   - Sit on the floor about 10 cm apart. At your teacher's direction, raise, then lower your hands. Practice until the class can make a smooth wave.
   - Next, make a wave by standing up and squatting down.

   a) Which way did you move?

   b) Which way did the wave move?

   c) Did any student move in the direction that the wave moved?

   d) What is a wave?

2. Work in groups of three. Get a Slinky® from your teacher. Two members of your group will operate the Slinky; the third will record observations. Switch roles from time to time.

   Sit on the floor about 10 m apart. Stretch the Slinky between you. Snap one end very quickly parallel to the floor. A pulse, or disturbance, travels down the Slinky.

   a) Which way does the pulse travel?

   b) Look at only one part of the Slinky. Which way did that part of the Slinky move as the pulse moved?

   c) Mark a coil on the Slinky by tying a piece of colored yarn around the coil. Send a pulse down the Slinky. Describe the motion of the coil tied with yarn.

   d) Send a pulse down the Slinky. Watch the pulse as it moves. Does the shape of the pulse change?

   e) Sketch the Slinky with a pulse moving through it.

   f) Does the speed of a pulse appear to increase, decrease, or remain the same as it moves along the Slinky?

   g) What happens to a pulse when it reaches your partner's end of the Slinky?

   h) Shake some pulses of different sizes and shapes. Does the speed of a pulse depend on the size of the pulse? Use a stopwatch to time one trip of pulses of different sizes.

3. Instead of sending single pulses down the Slinky, send a wave, a continuous train of pulses, by snapping your hand back and forth at a regular rate.

✎ a) Sketch the wave. Use the diagram below to label the parts of the wave on your sketch.

The crests of this wave are its high points. The troughs are its low points. The wavelength of this wave is the distance between two crests or between two troughs.

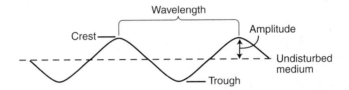

✎ b) How does the number of wave crests passing any point compare to the number of back-and-forth motions of your hand?

4. Lift one end of the Slinky and drop it rapidly. You've sent a vertical pulse down the Slinky. If both ends of the Slinky are lifted and dropped rapidly, the vertical pulses will be sent down the Slinky and will meet in the middle. Try it.

✎ a) What happens when the pulses meet in the middle?

✎ b) Do the pulses pass through each other or do they hit each other and reflect? Perform an experiment to find out.

5. Gather up 7 or 8 coils at the end of the Slinky and hold them together with one hand. Hold the Slinky firmly at each end. Release the group of coils all at once.

✎ a) Describe the pulse that moves down the Slinky.

✎ b) Sketch the pulse.

✎ c) In which direction do the coils move? In which direction does the pulse move?

6. You can use a Slinky "polarizer" to help you determine the direction of oscillation of the transverse wave. A Slinky "polarizer" is a large slit that the Slinky passes through.

Have one student hold the "polarizer" vertically. Have another student send a vertical pulse down the Slinky by moving her arm in a quick up and down motion.

267

✎ a) Describe what happens to the pulse on the far side of the "polarizer."

7. Continue to have one student hold the "polarizer" vertically. Have another student send a horizontal pulse down the Slinky by moving his arm is a quick side-to-side motion.

✎ a) Describe what happens this time to the pulse on the far side of the "polarizer."

8. Predict what might happen if a diagonal pulse were sent toward the vertical "polarizer."

✎ a) Record your prediction.

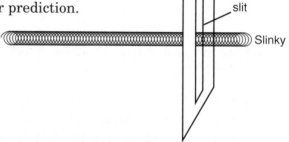
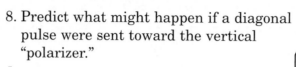

9. Send the diagonal pulse toward the vertical "polarizer."

✎ a) Record your observations.

10. Gather up 7 or 8 coils at the end of the Slinky and hold them together with one hand. Hold the Slinky firmly at each end. Release the group of coils all at once.

✎ a) Did this pulse travel through the "polarizer"?

11. You may have heard of polarizing sunglasses. Your teacher will supply you with two pieces of polarizing film that could be used for sunglasses. Hold each polarizing filter to the light and record your observation. Make a sandwich of the two pieces of polarizing filter and hold it to the light.

✎ a) Record your observations.

12. Rotate one polarizing filter in the sandwich by 90°. Hold it to the light.

✎ a) Record your observations.

13. The Slinky "polarizer" can be used as a way to understand the light polarizers. If a vertical Slinky pulse were sent through two vertical slits, you would expect that the pulse would pass through the first slit and then the second slit.

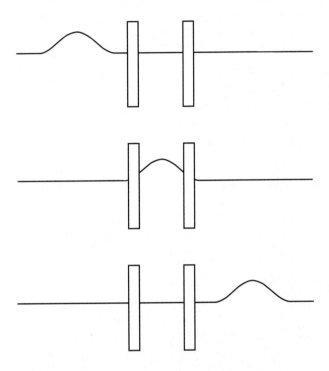

14. Draw a similar set of sketches to show what would happen if the first slit were vertical and the second slit were horizontal.

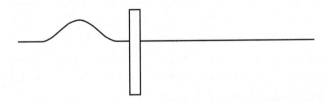

The light behaved in a similar way. When the two polarizers were parallel, the light was able to travel through both. When the two polarizers were perpendicular to each other, the "vertical" light could get through the first polarizer, but that light could not get through the second polarizer.

Physics Words

transverse pulse or wave: a pulse or wave in which the motion of the medium is perpendicular to the motion of the wave.

longitudinal pulse or wave: a pulse or wave in which the motion of the medium is parallel to the direction of the motion of the wave.

wavelength: the distance between two identical points in consecutive cycles of a wave.

frequency: the number of waves produced per unit time; the frequency is the reciprocal of the amount of time it takes for a single wavelength to pass a point.

velocity: speed in a given direction; displacement divided by the time interval; velocity is a vector quantity; it has magnitude and direction.

amplitude: the maximum displacement of a particle as a wave passes; the height of a wave crest; it is related to a wave's energy.

period: the time required to complete one cycle of a wave.

PHYSICS TALK

Describing Waves

You've discovered features of transverse waves and longitudinal waves. In **transverse waves,** the motion of the medium (the students or the Slinky) is perpendicular to the direction in which the wave is traveling (along the line of students or along the Slinky). In **longitudinal,** or **compressional waves,** the medium and the wave itself travel parallel to each other. Four terms are often used to describe waves. They are wavelength, frequency, amplitude, and period.

Wavelength is the distance between one wave and the next. It can be measured from the top part of the wave to the top part of the next wave (crest to crest) or from the bottom part of one wave to the bottom part of the next wave (trough to trough). The symbol for wavelength is the Greek letter lambda (λ). The unit for wavelength is meters.

Frequency is the number of waves that pass a point in one unit of time. Moving your hand back and forth first slowly, then rapidly to make waves in the Slinky increases the frequency of the waves. The symbol for frequency is f. The units for frequency are waves per second, or hertz (Hz).

The **velocity** of a wave can be found using wavelength and frequency. The relationship is shown in the equation

$$v = f\lambda.$$

The **amplitude** of a wave is the size of the disturbance. It is the distance from the crest to the undisturbed surface of the medium. A wave with a small amplitude has less energy than one with a large amplitude. The unit for amplitude is meters.

The **period** of a wave is the time for a complete wave to pass one point in space. The symbol for period of a wave is T. When you know the frequency of a wave, you can easily find the period.

$$T = \frac{1}{f}$$

FOR YOU TO READ

Waves and Media

Waves transfer energy from place to place. Light, water waves, and sound are familiar examples of waves. Some waves need to travel through a medium. Water waves travel along the surface of water; sound travels through the air and other material. As they pass, the disturbances move by but the medium returns to its original position. The wave is the disturbance. At the point of the disturbance, particles of the medium vibrate about their equilibrium positions. After the wave has passed, the medium is left undisturbed.

Light waves, on the other hand, can travel through the vacuum of space and through some media. Radio waves, microwaves, and x-rays are examples of waves that can travel through vacuums and through some media. These waves are transverse waves.

Polarization

Vertical disturbances and transverse waves can travel through a polarizer. Two parallel polarizers will allow these waves to continue to travel. If the two polarizers are perpendicular, then no wave is transmitted through the polarizers. You observed the polarization of light. This is evidence that light is a transverse wave. As you observed with the Slinky, compressional (longitudinal) waves are not affected by polarizing filters.

Reflecting on the Activity and the Challenge

Energy can move from one place to another. For example, throwing a baseball moves energy from the thrower to the catcher. Energy can also move from one place to another without anything moving. In the waves you made with a Slinky, no part of the Slinky moves across the room but the energy gets from one side to the other. This is true of sound waves as well. The activity provides evidence for this "unusual" concept of energy moving without "stuff" moving. Waves in a Slinky, water waves, and sound waves travel through media. Yet, light can travel through empty space. Trust in the wave model lets physicists create a theory for light.

Think about sending thoughts as waves. If a research team proposed this, they would have to explain how they would test this idea. They would also have to show that the study was valid and could produce reliable and repeatable results.

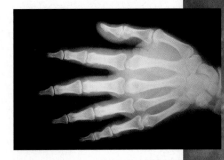

Physics To Go

1. Compare the direction in which people move in a people wave and the way the wave moves.

2. You sent a pulse down a Slinky.
   a) Which way did the pulse move?
   b) Did the shape of the pulse change as it moved?
   c) What happened to the pulse when it reached the end of the Slinky?

3. a) Draw the wave shown above and label the parts of the wave.
   b) What kind of wave is this?

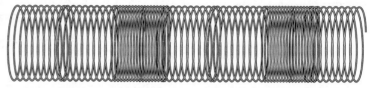

4. a) What kind of wave is this?
   b) Describe the movement of the wave and the movement of the medium.

5. Two pulses travel down a Slinky, each from opposite ends, and meet in the middle.
   a) What do the pulses look like when they meet? Make a sketch.
   b) What do the pulses look like after they pass each other? Make a sketch.

6. In your own words, compare frequency and period.

7. What determines the speed of a wave?

8. Find the velocity of a 2-m long wave with a frequency of 3.5 Hz.

9. Find the period of a wave with the frequency of 3 Hz.

10. Find the frequency of your favorite AM and FM radio stations.

Stretching Exercises

1. Perform this activity and answer the questions. You'll need a basin for water or a ripple tank, a small ball, a ruler, and a pencil.

   - Fill the basin or ripple tank with water. Let the water come to rest so that you are looking at a smooth liquid surface as you begin the activity. If possible, position a "point source" of light above the water basin or tank. The light will help you see the shadows of the waves you produce.
   - Touch the surface of the water with your finger.

   a) Describe the wave you produced and the way it changes as it travels along the water surface.
   - Drop the ball in the water and watch closely.

   b) What happens at the point where the ball hits the water?
   - Drop the ball from different heights and observe the size of the mound of water in the center.

   c) What happens to the size of the mound as the height of the drop increases?

   d) Describe the pattern in which the waves travel.
   - Drop the ruler into the water.

   e) Does the shape or size of the wave maker affect the shape of the wave produced? How?
   - Make waves by dipping one finger into the water at a steady rate.

   f) What is the shape of the wave pattern produced?
   - Now vary the frequency of dipping.

   g) Describe what happens to the distance between the waves as the rate of dipping your finger increases.

   h) Describe what happens to the distance between the waves as the rate of dipping your finger decreases.

   i) Express the results of your observations in terms of wavelength and frequency of the waves.

2. Find or create a computer simulation that will allow you to explore the behavior of waves in slow motion and stop action. If possible, "play" with these simulations in order to get a better sense of the behavior of waves. Demonstrate the simulation to your teacher and others in the class.

Activity 4   Interference of Waves

GOALS

In this activity you will:

- Generate waves to explore interference.
- Identify the characteristics of standing waves.
- Generate and identify parts of water waves.
- Experience interference of sound waves.

What Do You Think?

After a cost of millions of dollars, the Philharmonic Hall in New York City had to be rebuilt because the sound in the hall was not of high enough quality. Now named Avery Fisher Hall, it has excellent acoustics.

- **What does it mean to have "dead space" in a concert hall?**
- **What is the secret to good acoustics?**

Record your ideas about these questions in your *Active Physics* log. Be prepared to discuss your responses with your small group and the class.

For You To Do

1. Work with two partners. Two of you will operate the Slinky and one will record the observations. Switch roles from time to time. Stretch the Slinky to about 10 m. While one end of the Slinky is held in a fixed position, send a pulse down the Slinky by quickly shaking one end.

✎ a) What happens to the pulse when it reaches the far end of the Slinky?

2. Send a series of pulses down the Slinky by continuously moving one of its ends back and forth. Do not stop. Experiment with different frequencies until parts of the Slinky do not move at all. A wave whose parts appear to stand still is called a standing wave.

3. Set up the following standing waves:

   • a wave with one stationary point in the middle
   • a wave with two stationary points
   • a wave with three stationary points
   • a wave with as many stationary points as you can set up

4. You can simulate wave motion using a graphing calculator.

   Follow the directions for your graphing calculator to define a graph and set up the window. Use the following for the Y-VARS, and select FUNCTION. Use the Y= button to enter these values:

   $Y_1 = 4 \sin x$
   $Y_2 = 4 \sin x$
   $Y_3 = Y_1 + Y_2$

   Press GRAPH to view the waves.

✎ a) Describe the two waves you see on the screen.

✎ b) Can you see that Y_3 is equal to $Y_1 + Y_2$?

✎ c) Use the vertical axis on the screen to find the amplitude of the crest of each wave. How do they compare?

   You can edit the Y_1 and Y_2 functions to show waves Y_1 and Y_2 moving from the left to the right, as follows:

   $Y_1 = 4 \sin (x - \pi/4)$
   $Y_2 = 4 \sin (x + \pi/4)$
   $Y_3 = Y_1 + Y_2$

✎ d) How many waves do you see on the screen? Compare the amplitude of the third wave to those of the first two waves.

   Edit again.

$$Y_1 = 4 \sin (x - \pi/2)$$
$$Y_2 = 4 \sin (x + \pi/2)$$
$$Y_3 = Y_1 + Y_2$$

e) Describe the waves you see on the screen. Look for locations on the waves that always remain zero. These locations are called nodes.

f) Draw the waves you see on the screen on graph paper. Label the nodes.

Edit again.

$$Y_1 = 4 \sin (x - 3\pi/4)$$
$$Y_2 = 4 \sin (x + 3\pi/4)$$
$$Y_3 = Y_1 + Y_2$$

g) Describe the waves you see on the screen.

h) Draw the waves you see on the screen on graph paper. Label the nodes.

i) Compare the amplitude of each wave. How does the amplitude of the third wave compare to that of the first and the second wave?

Edit again.

$$Y_1 = 4 \sin (x - \pi)$$
$$Y_2 = 4 \sin (x + \pi)$$
$$Y_3 = Y_1 + Y_2$$

j) Describe the waves you see on the screen.

k) Draw the waves you see on the screen on graph paper. Locate the positions of the nodes.

l) Measure the amplitude of the first wave. What is the amplitude of the second wave? How do they compare?

5. Use a ripple tank to explore what happens when two sources of circular water waves "add together" in the tank.

6. As directed by your teacher, set up two speakers to explore what happens when two identical single tone sounds are broadcast.

7. As directed by your teacher, use a double slit to explore what happens when two beams of laser light are "added."

Physics Words

destructive interference: the result of superimposing different waves so that two or more waves overlap to produce a wave with a decreased amplitude.

constructive interference: the result of superimposing different waves so that two or more waves overlap to produce a wave with a greater amplitude.

standing wave: a stationary wave formed by the superposition of two equal waves passing in opposite directions.

FOR YOU TO READ
Wave Interference

The wave that you sent down the Slinky was reflected and traveled back along the Slinky. The original wave and the reflected wave crossed one another. In the previous activity, you saw that waves can "add" when they pass one another. When waves "add," their amplitudes at any given point also "add." If two crests meet, both amplitudes are positive and the amplitude of the new wave is greater than that of the component waves. If a crest and trough meet, one amplitude is positive and one is negative. The amplitude of the resulting wave will be less than that of the larger component wave. If a wave meets its mirror image, both waves will be canceled out.

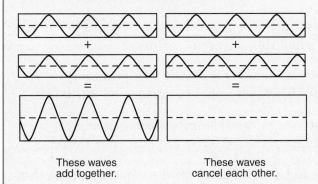

These waves add together.

These waves cancel each other.

In this activity, you created a pattern called a **standing wave.** Two identical waves moving in opposite directions interfere. The two waves are constantly adding to make the standing wave. Some points of the wave pattern show lots of movement. Other points of the wave do not move at all. The points of the wave that do not move are called the nodes. The points of the wave that undergo large movements are called the antinodes.

The phenomena that you have observed in this activity is called wave interference. As waves move past one another, they add in such a way that the sum of the two waves may be zero at certain points. At other points, the sum of the waves produces a smaller amplitude than that of either wave. This is called **destructive interference.** The sum of the waves can also produce a larger amplitude. This is called **constructive interference.** The formation of nodes and antinodes is a characteristic of the behavior of all kinds of waves.

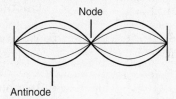

Node

Antinode

Reflecting on the Activity and the Challenge

Imagine you were told that adding one sound to another sound in a space could cause silence. Would you believe that light plus light can create interference fringes, where dark lines are places where no light travels? You might have thought such strange effects are magic. In a Slinky, a wave traveling in one direction and a wave traveling in the opposite direction create points on the Slinky that do not move at all. That is experimental evidence for the interference of waves. Now you know that dead spaces and dark lines can be explained by good science. You can approve funding to study phenomena that appear strange as long as some measurements on which all observers can agree are used in supporting the claims.

Physics To Go

1. What is a standing wave?

2. Describe in your own words how waves can "add."

3. What properties must two pulses have if they are to cancel each other out when they meet on a Slinky?

4. Make a standing wave using a Slinky or a graphing calculator. Draw the wave. Label its nodes and antinodes.

5. What is the distance, in wavelengths, between adjacent nodes in a standing wave pattern? Explain your thinking.

6. In photography, light can scatter off the camera lens. A thin coating is often placed on the lens so that light reflecting off the front of the thin layer and light reflecting off the lens will interfere with each other. How is this interaction helpful to the photographer?

7. Two sounds from two speakers can produce very little sound at certain locations. If you were standing at that location and one of the speakers was turned off, what would happen? How would you explain this to a friend?

8. Makers of noise reduction devices say the devices, worn as headsets, "cancel" steady noises such as the roar of airplane engines, yet still allow the wearer to hear normal sounds such as voices. How would such devices work? What principles of waves must be involved?

Stretching Exercises

An optical hologram is a three-dimensional image stored on a flat piece of film or glass. You have probably seen holograms on credit cards, in advertising displays, and in museums or art galleries. Optical holograms work because of the interference of light. Constructive interference creates bright areas, and destructive interference, dark areas. Your eyes see the flat image from slightly different angles, and your brain combines them into a 3-D image.

Find out how holograms are made. Describe the laboratory setup for making a simple hologram. If your teacher or you have the equipment, make one!

Activity 5   A Moving Frame of Reference

GOALS

In this activity you will:

• Observe the motion of a ball from a stationary position and while moving at a constant velocity.

• Observe the motion of a ball from a stationary position and while accelerating in a straight line.

• Measure motion in a moving frame of reference.

• Make predictions about motion in moving frames of reference and test your predictions.

• Define relativity.

• Define frame of reference and inertial frame of reference.

• Reconcile observations from different frames of reference.

What Do You Think?

As you sit in class reading this line, you are traveling at a constant speed as the Earth rotates on its axis. Your speed depends on where you are. If you are at the Equator, your speed is 1670 km/h (1040 mph). At 42° latitude, your speed is 1300 km/h (800 mph).

• **Do you feel the rotational motion of the Earth? Why or why not?**

• **What evidence do you have that you are moving?**

Record your ideas about these questions in your *Active Physics* log. Be prepared to discuss your responses with your small group and the class.

For You To Do

1. When you view a sculpture you probably move around to see the work from different sides. In this activity, you'll look at motion from two vantage points—while standing still and while moving. Get an object with wheels that is large enough to hold one of your classmates seated as it rolls down the hall. You might use a dolly, lab cart, a wagon, or a chair with wheels.

⚠ **This activity should be done under close supervision of your teacher.**

2. Choose a student to serve as the observer in the moving system. Have the observer sit on the cart and practice pushing the cart down the hall at constant speed. (This will take a little planning. Find a way to make the cart travel at constant speed. Also, find a way of controlling that speed!)

3. Once you can move the cart at a constant speed, give the moving observer a ball. While the cart is moving at constant speed, have the moving observer throw the ball straight up, then catch it.

✎ a) How does the person on the cart see the ball move? Sketch its path as he or she sees it.

✎ b) How does a person on the ground see the ball move? Again, sketch the path of the ball as he or she sees it.

4. With the observer on the cart traveling at constant speed, let a student standing on the ground throw the ball straight up and catch it.

a) How does the moving observer see the ball move? Sketch its path.

b) How does a person on the ground see the ball move? Sketch its path.

5. Work in groups for the following steps of this activity, as directed by your teacher. Get a wind-up or battery-powered car, two large pieces of poster board or butcher paper, a meter stick, a marker, string, and a stopwatch from your teacher.

Use a marker to lay out a distance scale on the poster board. Be sure to make it large enough so that a student walking beside the poster board can read it easily.

Next, lay out an identical distance scale along the side of the classroom or in a hall.

Attach a string to the poster board and practice moving it at a constant speed.

6. Place the toy car on the poster board and let it move along the strip. Measure the speed (distance/time) of the car along the poster board as the board remains at rest. Try the measurement several times to make sure that the motion of the car is repeatable.

a) Record the speed.

7. Move the poster board at constant speed while the car travels on the board. Focus on the car, not on the moving platform. Measure the speed of the car relative to the poster board when the board is moving.

a) Record the speed.

b) Compare the speed of the car when its platform is not moving and when its platform is moving.

c) Do your observations and measurements agree with your expectations?

8. Work with your group to make two simultaneous measurements. Measure the speed of the board relative to the fixed scale (the scale on the floor) and the speed of the car relative to the fixed scale. The second measurement can be tricky. Practice a few times. It may help to stand back from the poster board.

a) Record the measurements.

9. Next, measure the speed of the poster board and the car relative to the fixed scale while moving the board at different speeds. Make and complete a table like the one below.

Speed of the Car Relative to the Board	
Speed of the Board Relative to Fixed Scale	Speed of the Car Relative to Fixed Scale

a) Work with your group to state a relationship between the speed of the car relative to the fixed scale, its speed relative to the board, and the speed of the board relative to the fixed scale. Describe the relationship in your log. Also explain your thinking.

b) What do you think will happen if the car moves in the direction opposite to the direction the board is moving? Record your idea. Now try it. Do the results agree with your predictions?

c) Plan an experiment in which the car is moving along the poster board, the poster board is moving, and the car remains at the same location. Try it. Record the results.

d) What will happen if the car travels perpendicular to the direction in which the board is moving? Record your ideas. Now try it. Do the results agree with your predictions?

e) When the car travels perpendicular to the motion of the frame of reference, does the motion of the board affect your measurement of the car's speed?

FOR YOU TO READ

Frames of Reference

A frame of reference is a coordinate system from which observations and measurements are made. Your usual frame of reference is the surface of the Earth and structures fixed to it.

Have you experienced two frames of reference at once? Many large public spaces have banks of escalators to transport people from one floor to another. If two side-by-side escalators are moving in the same direction and at the same speed, and you and a friend step onto these escalators at the same time, you will seem to be standing still in relation to your friend. From the frame of reference of your friend, you are not moving. From the frame of reference of a person standing at the base of the escalator, you are both moving.

As you saw in the activity, there are other frames of reference. For a person moving at a constant velocity, the vehicle is the local frame of reference. In a moving train or plane or car,

the local frame of reference is the train or plane or car. When you are moving at a constant velocity, the local frame of reference is easier to observe than the frame of reference fixed to the Earth. If you drop an object in front of you while moving at a constant velocity in an airplane, it will fall to the floor in front of you. If the plane is traveling at 300 m/s, how do you explain the motion of the dropped object? Because you and the object are moving at a constant velocity, the object and you act as if you and it were standing still!

Would you be surprised to know that one frame of reference is not "better" than another? No matter what your frame of reference, if you are moving at a constant velocity, the laws of physics apply.

Relativity is the study of the way in which observations from moving frames of reference affect your perceptions of the world.

Relativity has some surprising consequences. For example, you cannot tell if your frame of reference is moving or standing still compared

Physics Words

relativity: the study of the way in which observations from moving frames of reference affect your perceptions of the world.

to another frame of reference, as long as both are moving at constant speed in a straight line. Newton's First Law of Motion states that an object at rest will stay at rest, and an object in motion will stay in motion unless acted on by a net outside force. Newton's First Law holds in each frame of reference. Such a frame of reference is called an **inertial frame of reference.**

If you are in a frame of reference traveling at a constant velocity from which you cannot see any other frame of reference, there is no way to determine if you are moving or at rest. If you try any experiment, you will not be able to determine the velocity of your frame of reference. This is the first postulate in Einstein's Theory of Relativity. Think of it this way: Any observer in an inertial frame of reference thinks that he or she is standing still!

Reflecting on the Activity and the Challenge

Different observers make different observations. As you sit on a train and drop a ball, you see it fall straight down—its path is a straight line. Someone outside the train observing the same ball sees the ball follow a curved path, a parabola, as it moves down and horizontally at the same time. However, a logical relation exists between different observations. If you know what one observer measures, you can determine what the other observer measures. This relation works for any two observers. It is repeatable and measurable. Pseudoscience requires special observers with special skills. No relation or pattern exists between them. Different explanations can be accepted for the same phenomenon and it's still science. Your **Chapter Challenge** is to distinguish between different explanations that are science and different explanations that have no basis and are pseudoscience.

Physics Words

inertial frame of reference: unaccelerated point of view in which Newton's Laws hold true.

Physics To Go

1. A person walking forward on the train says that he is moving at 2 miles per hour. A person on the platform says that the man in the train is moving at 72 miles per hour.

   a) Which person is correct?

   b) How could you get the two men to agree?

2. If you throw a baseball at 50 miles per hour north from a train moving at 40 miles per hour north, how fast would the ball be moving as measured by a person on the ground?

3. You walk toward the rear of an airplane in flight. Describe in your own words how you would find your speed relative to the ground. Explain your thinking.

4. A jet fighter plane fires a missile forward at 1000 km/h relative to the plane.
   a) If the plane is moving at 1200 km/h relative to the ground, what is the velocity of the missile relative to the ground?
   b) What is the velocity of the missile relative to a plane moving in the same direction at 800 km/h?
   c) What is the velocity of the missile relative to a target moving at 800 km/h toward the missile?

5. A pilot is making an emergency air drop to a disaster site. When should he drop the emergency pack: before he is over the target, when he is over the target, or after he has passed the target?

6. Each day you see the Sun rise in the east, travel across the sky, and set in the west.
   a) Explain this observation in terms of your frame of reference.
   b) Compare the observation to the actual motions of the Sun and Earth.

7. How would you explain relativity to a friend who is not in this course. Outline what you would say. Then try it. Record whether or not you were successful.

8. Explain this event based on frame of reference. You are seated in a parked car in a parking lot. The car next to you begins to back out of its space. For a moment you think your car is rolling forward.

FOR YOU TO READ

A Social Frame of Reference

Physics is sometimes a metaphor for life. Just as physicists speak of judging things from a frame of reference, a frame of reference is also used in viewing social issues. For example, a Black American*, one of the authors of this chapter, shared the following story about choosing a career, because his frame of reference conflicted with that of his father.

"I was born in Mississippi in 1942, the place where my parents had spent their entire lives. My father lived most of his life during a period of "separate-but-equal," or legal segregation. He believed that the United States would always remain segregated. So when I was choosing a career, all of his advice was from that frame of reference.

"On the other hand, my frame of reference was changing. To me, the United States could not stay segregated and remain a world power. The time was 1962, about 10 years after the Supreme Court had made its landmark Brown vs. Topeka School Board decision. I reasoned that the opportunities for black people would be greatly expanded.

"Both my parents had encouraged me to get as much education as possible. My mother always said that "the only way to guarantee survival is through good education." I had a master's degree and was teaching in a segregated college. I thought I would need a Ph.D. to stay in my profession, and decided to quit my job and go back to school. That decision brought on an encounter with my father that I shall never forget.

"My father did not say good-bye on the day I left home for graduate school. Our frames of reference had moved very far apart. The possibility of becoming a professor at a white college or university, particularly in the South, was not very high. My father could not understand why I needed a Ph.D. After all, I could have a good life in our segregated system without quitting my prestigious job to return to graduate school.

"As was usual for him, my father eventually supported my decision. At his death in 1989, however, he still had not fully accepted my frame of reference."

* The author uses the term Black American instead of African-American because the use of that term also shows social changes in frames of reference.

Stretching Exercise

1. The famous scientist Albert Einstein is noted for his Theory of Relativity. Research Einstein's life. What kind of a student was he? What was his career path? When did he make his breakthrough discoveries? What were his political beliefs as an adult? What role did he play in American political history? Report your findings to the class.

2. A Social Frame of Reference tells the story of one man's encounter with different ideas about society, or social frames of reference. Write a short story that illustrates what happens when two people operate from different frames of reference. Your story can be based on your own experience, or it can be fiction.

Activity 6   Speedy Light

GOALS

In this activity you will:

• Perform a thought experiment about simultaneous events.

• State Einstein's two Postulates of Special Relativity.

What Do You Think?

It takes light 8 minutes to travel from the Sun to the Earth. If the Sun suddenly went dark, no one would know for 8 minutes.

• **If an event happened on Mars and on Earth at the "same time," what would that mean?**

• **How would a person on Mars report the event to a person on Earth?**

Record your ideas about these questions in your *Active Physics* log. Be prepared to discuss your responses with your small group and the class.

For You To Do

Work with your group to solve these problems. Record your thinking and conclusions in your log.

1. An old, slow-moving man has a large house with a grandfather clock in each room. He has no wristwatch and he cannot carry the clocks from one room to another. He wants to set each clock at 12 noon. He finds that by the time he sets the second clock, it is no longer noon because it takes time to get from the first clock in one room to the second clock in another.

   a) How can he make sure that all the clocks in the house chime the same hour at the same instant?

   b) Would he hear all the chimes at the same instant?

   Imagine that his house is huge—100 km × 100 km × 100 km.

   c) How can he set all the clocks to chime the same hour at the same instant?

   d) Would he hear all the chimes at the same instant?

2. Your group is put in charge of a solar system time experiment. You send clocks to Mercury, Venus, Mars, and Jupiter. These clocks "chime" by sending out radio waves. It takes many hours for the "chime" to travel between planets.

   a) How can you set all the clocks to "chime" at the same hour at the same instant?

   b) Would you "hear" all the clocks at the same instant?

3. Your group gets another mission. You are to send clocks to distant stars. These clocks "chime" by sending out pulses of light. It takes hundreds of years for the light to travel between the different stars.

   a) How can you set all the clocks to "chime" at the same hour at the same instant?

   b) Would you "hear" all the clocks at the same instant?

FOR YOU TO READ

The Theory of Special Relativity

The speed of light is 3.0×10^8 m/s (meters per second) in a vacuum, or 186,000 miles per second. The speed of light is represented by the symbol c. So, $c = 3.0 \times 10^8$ m/s. If light could travel around Earth's equator, it would make over 7 trips each second! The very great speed of light makes it difficult to measure changes in the speed of light caused by motion of frames of reference that are familiar to you on Earth.

In the early part of the 20th century, Einstein showed that light does not obey the laws of speed addition that we have seen in objects on Earth's surface. His theory predicted that light traveled at the same speed in all frames of reference, no matter how fast the frames were moving relative to one another.

To understand exactly how startling this result was, let's use an example. Recall measuring the speed of objects in a moving frame of reference in the previous activity. The following sketch shows a woman standing on a moving cart and throwing a ball forward.

A man watches from the roadside. The speed of the cart relative to the road is 20 m/s.

The speed of the ball relative to the cart is 10 m/s. How fast is the ball traveling according to the man by the side of the road? (20 m/s + 10 m/s = 30 m/s.)

In the second sketch, the ball is replaced by a flashlight.

Once again the cart travels at speed 20 m/s relative to the road. The light travels at speed c relative to the cart. How fast does the light travel according to the man at the side of the road? (Take time to discuss your thinking with your group!)

Imagine the cart traveling at 185,000 mi/s relative to the road. The light travels at 186,000 mi/s. How fast does the light travel according to the man at the side of the road?

As a young clerk in the Swiss patent office, Albert Einstein postulated that the speed of light in a vacuum is the same for all observers. Einstein recognized that light and other forms of electromagnetic radiation (including x-rays, microwaves, and ultraviolet waves) could not be made to agree with the laws of relative motion seen on Earth. Einstein modified the ideas of relativity to agree with the theory of electromagnetic radiation. When he did, he uncovered consequences that have changed the outlook of not only physics but the world.

The basic ideas of Einstein's **Theory of Special Relativity** are stated in two postulates:

- **The laws of physics are the same in all inertial frames of reference. (Remember that inertial frames of reference are those in which Newton's First Law of Motion holds. This automatically eliminates frames of reference that are accelerating.)**
- **The speed of light is a constant in all inertial frames of reference.**

The first postulate adds electromagnetism to the frames of reference discussed. Its implications become clear when you begin to ask questions. Is the classroom moving or standing still? How do you know? Remember that an observer in an inertial frame of reference is sure that he or she is standing still. An observer in an airplane would be convinced that he or she is standing still and that your classroom is moving. The meaning of the first postulate is that there is no experiment you can do that will tell you who is really moving.

The second postulate, however, produces results that seem to defy common sense. You can add speeds of objects in inertial frames of reference. But you cannot add the speed of light to the motion of an inertial frame of reference.

What Are Simultaneous Events?

Like the old man in the **For You To Do** activity, light travels at a finite speed. Although it travels very rapidly, it takes time for light to get from one place to another. Just as the old man had a problem setting his clocks at the same time, physicists have a problem saying when two events happen at the same time.

The speed of light in a vacuum is always the same. Physicists say that two events are simultaneous if a light signal from each event reaches an observer standing halfway between them at the same instant. You can demonstrate this idea in your classroom. An observer standing midway between two books would see them fall at the same instant if their falls were simultaneous. It is a little more difficult to imagine an observer midway between classrooms in two different time zones avidly watching for falling books, but—in principle— the experiment is possible.

An experiment that could be done but would be very difficult to carry out can be replaced by what is called a gedanken, or thought, experiment. Physicists use gedanken experiments to clarify principles. If the principle is called into question, experimenters can always try to conduct the actual experiment, although it may be very difficult to do so. In the activity, you performed a gedanken experiment.

Reflecting on the Activity and the Challenge

Einstein's second postulate is that any observer moving at any speed would measure the speed of light to be 3.0×10^8 m/s. This postulate and his first postulate leads to the idea that simultaneity depends on the observer. You cannot say whether two events in different places occurred at the same time unless you know the position of the observer. For one observer, event A and event B happen at the same time while for a second observer, event A happens before event B. Why should you trust such a strange theory? Why should you trust in new ideas about space and time? You can trust them because they are supported by experimental results.

Can you be in two places at the same time? Should you fund a research project to test this out? If the proposal produces measurements and observations that can be used as evidence, you could fund it. If the proposal requires observations that only certain people are "qualified" to make, or data that cannot be agreed on, you should not fund it.

Physics To Go

1. How long does it take a pulse of light to:

    a) cross your classroom?
    b) travel across your state?

2. Calculate the number of round trips between New York City and Los Angeles a beam of light can make in one second. (New York and Los Angeles are 5000 km or 3000 miles apart.)

3. The fastest airplanes travel at Mach 3 (3 times the speed of sound). If the speed of sound is 340 m/s, what fraction of the speed of light is Mach 3?

4. The Earth is about 150 million kilometers from the Sun. Use 365 days as the length of 1 year, and think of the Earth's orbit as a circle. Find the speed of the Earth in its orbit. What fraction of the speed of light is the Earth's orbital speed?

5. Try this gedanken experiment to clarify the consequences of Einstein's postulates.

   Armin and Jasmin are astronauts. They have traveled far into space and, from our frame of reference, they now pass each other at 90% of the speed of light. Their ships are going in straight lines, in opposite directions, at constant speeds. The astronauts each see their own ship as standing still. (Remember, observers in inertial frames of reference think that they are standing still.)

   a) How does Armin describe the motion of the two ships?
   b) How does Jasmin describe the motion of the two ships?
   c) Is Armin's or Jasmin's description of the motion correct? What is a correct description? (Did you think that your frame of reference is the correct one? Is your frame of reference "better" than Armin's or Jasmin's?)

6. A train is traveling at 70 mph in a straight line. A man walks down the aisle of the train in the direction that the train is traveling at a speed of 2 mph relative to the floor of the train. What is the man's speed as measured by:

   a) the passengers on the train?
   b) a man standing beside the track?
   c) a passenger in a car on a road parallel to the track traveling in the same direction as the train at a speed of 30 mph?
   d) a passenger in a pickup truck on the parallel road traveling in the opposite direction from the train at a speed of 40 mph?

   Each of the above measurements has produced a different result. Who is telling the truth? Explain your answer.

Stretching Exercises

The sound of two radios will reach you at a time depending on your position relative to the radios. The sound will seem simultaneous only when you are at the midpoint between the radios.

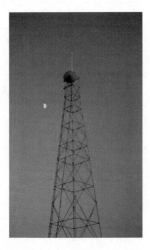 

Place two radios on opposite sides of the room or, if possible, out of doors and far apart and away from traffic. Tune them to the same station. Then move around listening to the two radios until you find a position where you hear the sounds from both radios. Answer the following questions:

a) Is there only one place in the area where the radios are playing the same words at the same time?

b) Describe that place, or those places, in terms of their location(s) compared to the locations of the radios.

c) How would this experiment be different if light from flashlights were used instead of sound from radios?

Activity 7   Special Relativity

What Do You Think?

Einstein's Theory of Special Relativity predicts that time goes more slowly for objects moving close to the speed of light than for you. If you could travel close to the speed of light, you would age more slowly than if you remained on Earth. This prediction doesn't fit our "common sense."

- **Does this prediction make sense to you? Explain your thinking.**

- **What do you mean by "common sense"?**

Record your ideas about these questions in your *Active Physics* log. Be prepared to discuss your responses with your small group and the class.

GOALS

In this activity you will:

- Plot a muon clock based on muon half-life.

- Use your muon clock and the speed of muons to predict an event.

- Identify the ways that special relativity meets the criteria of good science.

For You To Do

1. A muon is a small particle similar to an electron. Muons pour down on you all the time at a constant rate. If 500 muons arrive at a muon detector in one second, then 500 muons will arrive during the next second.

   Muons have a half-life of 2 microseconds. (A microsecond is 1 millionth of a second, or 1×10^{-6} s.) Beginning with 500 muons, after 2 microseconds there will be about 250 muons left. (That is 1 half-life.) After 4 microseconds (2 half-lives) there will be about 125 muons left. After 6 microseconds (3 half-lives) there will be about 62 muons left.

   a) How many muons would be left after 4 half-lives?

2. The half-life of muons provides you with a muon clock. Plot a graph of *the number of muons* versus *time*. Use 500 muons as the size of the sample. This graph will become your clock.

   a) If 125 muons remain, how much time has elapsed?

   b) If 31 muons remain, how much time has elapsed?

   c) If 300 muons remain, how much time has elapsed?

   d) If 400 muons remain, how much time has elapsed?

3. Measurements show that 500 muons fall on the top of Mt. Washington, altitude 2000 m. Muons travel at 99% the speed of light or $0.99 \times 3.0 \times 10^8$ m/s.

   a) Calculate the time in microseconds it would take muons to travel from the top of Mt. Washington to its base.

   b) Use your calculation and the muon clock graph to find how many muons should reach the bottom of Mt. Washington.

4. Experiments show that the actual number of muons that reach the base of Mt. Washington is 400.

   a) According to your muon clock graph, how much time has elapsed if 400 muons reach the base of Mt. Washington?

   b) By what factor do the times you found differ?

   c) Suggest an explanation for this difference.

Physics Words

muon: a particle in the group of elementary particles called leptons (not affected by the nuclear force).

Albert Einstein had an answer. The muon's time is different than your time because muons travel at about the speed of light. He found that the time for the muon's trip (at their speed) should be 0.8 microseconds. That is the time that the muon's radioactive clock predicts.

❧ d) As strange as that explanation may sound, it accurately predicts what happens. Work with your group to come up with another plausible explanation.

FOR YOU TO READ

Special Relativity

Physicists of this century have had a difficult decision to make. They could accept common sense (all clocks and everyone's time is the same), but this common sense cannot explain the data from the **muon** experiment. They could accept Einstein's Theory of Special Relativity (all clocks and everyone's time is dependent on the speed of the observer), which gives accurate predictions of experiments, but seems strange. Which would you choose, and why?

The muon experiment shows that time is different for objects moving near the speed of light. You calculated that muons would take 7 microseconds to travel from the top of Mt. Washington to its base. Experiments show that the muons travel that distance in only 0.8 microseconds. Because of their speed, time for muons goes more slowly than time for you!

Time is not the only physical quantity that takes on a new meaning under Einstein's

theory. The length of an object moving near the speed of light shrinks in the direction of its motion. If you could measure a meter stick moving at 99% of the speed of light, it would be shorter than one meter!

Perhaps the most surprising results of Einstein's theory are that space and time are connected and that energy and mass are equivalent. The relationship between energy and mass is shown in the famous equation $E = mc^2$. Put in simple words, increasing the mass of an object increases its energy. And, increasing the energy of an object increases its mass. This idea has been supported by the results of many laboratory experiments, and in nuclear reactions. It explains how the Sun and stars shine and how nuclear power plants and nuclear bombs are possible.

Meter stick traveling at near
the speed of light, as seen from Earth

Meter stick on Earth's surface

Physics and Pseudoscience

Physics, like all branches of science, is a game played by rather strict rules. There are certain criteria that a theory must meet if it is to be accepted as good science. First, the predictions of a scientific theory must agree with all valid observations of the world. The word *valid* is key. A valid observation can be repeated by other observers using a variety of experimental techniques. The observation is not biased, and is not the result of a statistical mistake.

Second, a new theory must account for the consequences of old, well-established theories. A replacement for the Theory of Special Relativity must reproduce the results of special relativity that have already been solidly established by experiments.

Third, a new theory must advance the understanding of the world around us. It must tie separate observations together and predict new phenomena to be observed. Without making detailed, testable predictions, a theory has little value in science.

Finally, a scientific theory must be as simple and as general as possible. A theory that explains only one or two observations made under very limited conditions has little value in science. Such a theory is not generally taken very seriously.

The Theory of Special Relativity meets all the criteria of good science. When the relative speeds of objects and observers are very small compared to the speed of light, time dilation, space contraction, and mass changes disappear. You are left with the well-established predictions of Newton's Laws of Motion. On the other hand, all the observations predicted by the Theory of Special Relativity have been seen repeatedly in many laboratories.

By contrast, psychic researchers do not have a theory for psychic phenomena. The psychic phenomena themselves cannot be reliably reproduced. Psychic researchers are unable to make predictions of new observations. Thus, physicists do not consider psychic phenomena as a part of science.

Reflecting on the Activity and the Challenge

One of the strangest predictions of special relativity is that time is different for different observers. Physicists tell a story about twins saying good-bye as one sets off on a journey to another star system. When she returns, her brother (who stayed on Earth) had aged 30 years but she had aged only 2 years. Commonsense physicists trust this far-fetched idea because there is experimental evidence that supports it. The muon experiment supports the idea. There is no better explanation for the events in the muon experiment than the Theory of Special Relativity. The theory is simple but it seems to go against common sense. But common sense is not the final test of a theory. Experimental evidence is the final test.

One of the criteria for funding research is whether the experiment can prove the theory false. If muons had the same lifetime when at rest and when moving at high speed, the Theory of Special Relativity would be shown to be wrong. Many theories of pseudoscience cannot be proven false. According to pseudoscientific theories, any experimental evidence is okay. There is no way to disprove the theory. Any evidence that doesn't fit causes the "pseudoscientists" to adjust the theory a bit or explain it a bit differently so that the evidence "fits" the theory.

The proposal you will fund should be both supportable and able to be disproved. The experimental evidence will then settle the matter—either supporting the theory or showing it to be wrong.

Physics To Go

1. Use the half-life of muons to plot a graph of the number of muons vs. time for a sample of 1000 muons.

   a) If 1000 muons remain, how much time has elapsed?
   b) If 250 muons remain, how much time has elapsed?
   c) How many muons are left after 6 half-lives?
   d) How many muons are left after 8 half-lives?

2. If the speed of light were 20 mph . . .

   You don't experience time dilation or length contraction in everyday life. Those effects occur only when objects travel at speeds near the speed of light relative to people observing them. Imagine that the speed of light is about 20 mph. That means that observers moving near 20 mph would see the effects of time dilation and space contraction for objects traveling near 20 mph. Nothing could travel faster than 20 mph. As objects approach this speed, they would become increasingly harder to accelerate.

   Write a description of an ordinary day in this imaginary world. Include things you typically do in a school day. Use your imagination and have fun with the relativistic effects.

Activity 8   The Doppler Effect

GOALS

In this activity you will:

- Describe red shift.
- Sketch a graph.
- Observe changes in pitch.
- Calculate with a formula.

What Do You Think?

You have probably heard the sound of a fast-moving car passing by you.

- **Why is there a change in tone as the car moves by?**

Record your ideas about this question in your *Active Physics* log. Be prepared to discuss your responses with your small group and with your class.

For You To Do

1. Listen to a small battery-powered oscillator. It makes a steady tone with just one frequency. The oscillator is fastened inside a Nerf™ ball for protection.

2. Stand about 3 m away from your partner. Toss the oscillator back and forth between you. Listen to the pitch as the oscillator moves. As you listen, observe how the pitch changes as the oscillator moves.

  ✎ a) How is the oscillator moving when the pitch is the highest?
  ✎ b) How is the oscillator moving when the pitch is the lowest?

3. Stop the oscillator so you can listen to its "at rest" pitch.

  ✎ a) With the oscillator moving, record how the pitch has changed compared to the "at rest" pitch. How has the pitch changed when the oscillator is moving towards you?
  ✎ b) How has the pitch changed when the oscillator is moving away from you?

4. Look at the graph axes shown. The axes show pitch vs. velocity. When the velocity is positive, the oscillator is moving away from you. When the velocity is negative, the oscillator is moving towards you.

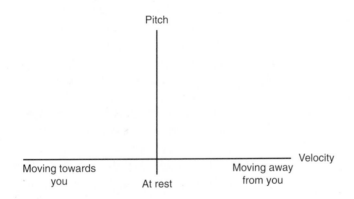

  ✎ a) On a similar set of axes in your log, sketch a graph of your pitch observations. Explain your graph to the other members of your group.

5. You can do an outdoor Doppler lab using the horn of a moving car as the wave source. Tape-record the horn when the car is at rest next to the tape recorder. Then, with the driver of the car maintaining an agreed-upon speed, tape the sound of the horn as the car passes. Have the driver blow the horn continuously, both as the car approaches and as it moves away. Be very careful to stay away from the path of the car.

303

6. You can determine the observed frequency by matching the recorded tone to the output of an oscillator and loudspeaker. Use this formula:

$$f = f_0\left(\frac{s}{(s-v)}\right)$$

f_0 = frequency when car is at rest

v = speed of the car

s = speed of sound = 340 m/s

a) When the car is moving toward you, v is positive. When the car is moving away from you, v is negative. Use the equation to calculate the speed of the car from the data you collected.

FOR YOU TO READ

Measuring Distances Using the Doppler Effect

Astronomers measure distances to stars in two different ways. One way is with parallax, but this method works only for the nearest stars. For all other stars astronomers apply the Doppler effect. They use the Doppler shift of spectral lines. The next-nearest galaxy is Andromeda, more than a million light-years away.

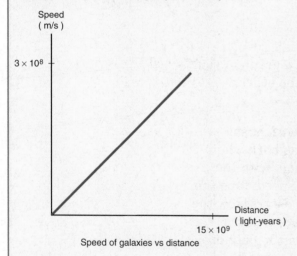

Speed of galaxies vs distance

Astronomers have observed galaxies at far greater distances, up to about 12 billion light-years away. These incredible distances are measured by observation of the absorption lines of light. These lines are consistently Doppler-shifted towards the red end of the spectrum, and the result is called the "red shift."

All the lines are shifted toward longer wavelengths. Since this is a shift towards lower frequencies, the galaxies are moving away from Earth. By measuring the size of the shift, astronomers find the speed of distant galaxies. Different galaxies move away at different speeds, but with a clear pattern. The farther away the galaxy, the faster it is moving away, as shown in the graph.

Astronomers explain this result with the Big Bang theory, which says that the universe began in an explosion about 15 billion years ago. After the explosion, the matter in the galaxy continued to move apart, even after the galaxies formed.

Reflecting on the Activity and the Challenge

You have learned that the pitch of a sound changes if the source of the sound is moving toward you or away from you. This is called the Doppler effect for sound. You also learned that there is a Doppler effect for light where the frequency or color of the light would change if the source of the light were moving. Measurements of sound frequency can be used to determine the speed of the source of sound. Measurements of light frequencies from distant galaxies can be used to determine the speed of the galaxies. The speed of galaxies moving away from Earth has been shown to relate to the distance of the galaxies from Earth. A measurement of light frequency and the Doppler effect can be used to measure distances. Measuring speeds through the use of changes in frequency of sound or light is good science. Some people may say that hearing a person's voice indicates to them whether that person is kind or gentle. If this were to have a scientific basis, you would have to conduct experiments. You should be able to contrast the experimental evidence you have for the Doppler effect and the lack of evidence you have for finding out what a person is like by their voice as you decide the kinds of proposals you may fund.

Physics To Go

1. a) If a sound source is moving towards an observer, what happens to the pitch the observer hears?

   b) If a sound source is moving towards an observer, what happens to the sound frequency the observer measures?

2. a) If a sound source is moving away from an observer, what happens to the pitch the observer hears?

   b) If a sound source is moving away from an observer, what happens to the sound frequency the observer measures?

3. a) If you watch an auto race on television, what do you hear as the cars go by the camera and microphone?

   b) Sketch a graph of the pitch you hear vs. time. Make the horizontal axis of your graph the time, and the vertical axis the pitch. (Hint: Don't put any numbers on your axes. Label the time when the car is going right by you.)

c) Sketch a graph of the frequency you observe vs. time. As in **Part (b)**, label the time when the car is going right by you. (Hint: Don't put any numbers on your axes.)

4. a) In **Question 3** above, what would happen to your graphs if the speed of the racing car doubled? Make a sketch to show the change.

   b) What would happen to your graphs if the speed of the racing car was cut in half? Make a sketch to show the change.

5. a) Red light has a longer wavelength than blue light. Which light has the lower frequency? You will need the equation:

   wave speed = wavelength × frequency

   Show how you found your answer.

   b) When the oscillator moved away from you, was the pitch you heard lower or higher?
   c) When the oscillator moved away from you, was the frequency you heard lower or higher?
   d) If light from a distant galaxy is shifted towards the red, is it shifted to a lower or a higher frequency?
   e) If the light is shifted towards the red, is the galaxy moving away from Earth or towards Earth?

Stretching Exercise

Watch a broadcast of an auto race. Listen closely to the cars as they zoom past the microphone. Use the Doppler effect to explain your observations.

Activity 9   Communication Through Space

GOALS

In this activity you will:

- Calculate time delays in radio communications.
- Express distances in light travel-time.
- Solve distance-rate-time problems with the speed of light.

What Do You Think?

In 1865, Jules Verne wrote *From the Earth to the Moon*. In this book, a team of three astronauts were shot to the Moon from a cannon in Florida. They returned by landing in the ocean. Verne correctly anticipated many of the details of the Apollo missions.

• How well do you think *Star Trek* predicts the future?

Record your ideas about this question in your *Active Physics* log. Be prepared to discuss your responses with your small group and with your class.

For You To Do

1. Alexander Graham Bell's grandson suggested a simple way to talk to Europe long-distance. He recommended placing a long air tube across the bottom of the Atlantic Ocean. He believed that if someone spoke into one end of the tube, someone else at the other end would hear what was said.

   a) Do you think this is practical? Give reasons for your answer.

   b) If the sound could be heard in Europe, how long would it take to send a message? (Hint: The distance to Europe is about 5000 km, and the speed of sound is about 340 m/s.)

   c) Compare this time with the time to communicate with extraterrestrials in the next galaxy using light. The nearest galaxy is Andromeda, which is about two million light-years away. (It takes light about two million years to get from Earth to Andromeda.)

2. The highest speed ever observed is the speed of light, 3.0×10^8 m/s. In addition, a basic idea of Einstein's Theory of Relativity is that no material body can move faster than light. Radio waves also travel at the speed of light. If Einstein is correct, there are serious limitations on communication with extraterrestrials. Look at the table of distances below. These are distances from the Earth.

to the Sun:	1.5×10^{11} m
to Jupiter:	8×10^{11} m
to Pluto:	6×10^{12} m
to the nearest star:	4×10^{16} m
to the center of our galaxy:	2.2×10^{20} m
to the Andromeda galaxy:	2.1×10^{22} m
to the edge of the observable universe:	1.5×10^{26} m

   a) How long would it take to send a message using radio waves to each place?

   b) How long would it take to send this message and get an answer back?

3. A real-life problem occurred when the Voyager spacecraft was passing the outer planets. NASA sent instructions to the spacecraft but had to wait a long time to find out what happened. The ship had to receive the instructions, take data, and send the data back home.

a) If the spacecraft was at Jupiter, how long would it take for the message to travel back-and-forth?

b) If this spacecraft was at Pluto, how long would it take for the message to travel back-and-forth?

4. Make a time-line of Earth history. For the scale of your time-line, make six evenly spaced marks.

a) Label the time-line like the one shown.

100 million years ago   50 million years ago   Present

b) On your time-line, label interesting events in Earth's history that occurred during these times. Possibilities include the end of the last Ice Age (10,000 years ago), the evolution of the modern horse (50 million years ago), the evolution of humans (3 million years ago), the Iron Age (1000 BC), the Stone Age (8000 BC), the Middle Ages in Europe (13th century), the beginning of civilization (3000 BC), and the spread of mammals over the Earth (50 million years ago).(Dates given are approximations.)

5. Many scientists believe that intelligent life would most likely be thousands or millions of light-years away.

a) How would this affect two-way communication?

b) If you asked a question, how long would it be before a response came back? Would you be able to receive the response?

c) What questions would you ask? (Note: Think about the distances involved.)

d) What kind of answers might you expect?

e) What changes have occurred on Earth over this time period?

f) What changes would you expect on Earth before the answer came?

g) Is two-way communication possible over such distances? Is it practical? Is it likely?

309

Reflecting on the Activity and the Challenge

This activity helped you to understand how much time it would take for a light signal to travel from Earth to other places in the solar system, the galaxy, or the edge of the universe. There are many reasons why contact with another life form on another planet would be valuable. You may wish to consider the merits of scientific proposals that seek to communicate with other life forms. You will want to consider whether the proposals take into account the difficulties of sustained communication. You will have to consider the type of communication expected and whether the proposal understands that a response to the simplest question to a life form near the nearest star would take at least six years. If a proposal states that it can communicate faster than the speed of light, they would have to explain how this would be possible since no technique is now known that permits this.

Physics To Go

1. a) The speed of sound is about 340 m/s in air. You and another student take gongs outside about 200 m apart. You hit the gong. After hearing the sound of your gong, the other student hits the other gong. How long is it before you hear the sound of the other gong?
   b) How is this experiment similar to the problem of communicating with extraterrestrial life?

2. a) If extraterrestrial life is probably 1000 light-years away, would it be within this galaxy?
   b) If extraterrestrial life is likely probably several million light-years away, would that be within this galaxy? Could it be in the Andromeda galaxy? (Note: This galaxy has over 100 billion stars.)

3. a) The Moon is 3.8×10^8 m from the Earth. How long does it take a radio wave to travel from the Moon to the Earth?
   b) The Sun is 1.5×10^{11} m from the Earth. How long does it take a light wave to travel from the Sun to the Earth?
   c) Pluto is about 6×10^{12} m from the Sun. How long does it take a light wave to travel from the Sun to Pluto?
   d) The nearest star is 4.3 light-years away from Earth. How

long does it take a radio wave to travel from Earth to the nearest star?

e) This galaxy is about 100,000 light-years across. How long does it take light to go all the way across our galaxy?

f) The nearest galaxy is more than a million light-years away. How long does it take light to reach us from this galaxy?

g) The universe is about 15 billion light-years across. How long does it take light to cross the universe?

4. a) In *Star Trek*, the spaceship can move at "warp speed." This speed is faster than the speed of light. How is warp speed important for space travel?

b) Do you think "warp drive" is likely to be developed? Is it possible? Explain your answers.

5. Suppose your job is to make a plan to send people in space ships to explore nearby galaxies. How would the distances in space affect your plan?

6. a) How would you choose a language for communication with extraterrestrials?

b) Many scientists suggest that a good starting point is to describe the periodic table of the elements. Do you agree? Explain your answer.

c) Is there any evidence that extraterrestrials would observe the same elements, with the same properties, that you observe? Tell what the evidence is.

d) Do you think another advanced civilization would have already discovered the periodic table? Tell why or why not.

e) How would you start to create a language?

f) How would you begin communication?

7. a) Suppose that intelligent extraterrestrial beings exist. Suppose that you are able to communicate with them. Why would you want to?

b) Should you be afraid of extraterrestrial beings?

c) Is it more likely that they would help Earth or enslave Earth? (Note: Consider the distances involved.)

8. a) What is known of the Earth of 2000 years ago?

   b) It takes 2000 years for a spaceship to travel to a star. When the travelers arrive at the star, would their information about the Earth be up-to-date? Explain why or why not.

   c) If the trip to another star took 10,000 years, would such a trip be worthwhile? Explain why or why not.

9. A record was sent into space in an effort to communicate with extraterrestials.

   a) If you were on the team designing the record, what music would you include?

   b) What photographs would you include?

   c) What drawings would you include?

   d) Have you fairly represented the majority of the world with your choices?

10. a) Make a list of movies, books, and TV shows that involve trips to other parts of the galaxy or extraterrestrials visiting the Earth.

    b) Very briefly describe the plot of the story.

    c) How accurately is science represented?

Stretching Exercises

1. Read the Carl Sagan book, *Contact*, or watch the movie. What features of the book and movie have you considered in this chapter? What features have been ignored?

2. Look up the messages that were placed on the Pioneer and Voyager spacecraft. Make a report to the class on how this plaque communicated information about humans.

PHYSICS AT WORK

Loretta Wright

GRANTING WISHES

How would you feel if you could make people's dreams come true? And how would you feel if you knew you could make a major difference in this world? Loretta Wright does that every day. She is a project officer for the Annenberg/CPB-Math and Science Project which distributes over six million dollars a year in grants. The project's goals are to improve math and science education throughout the country. As project officer, Loretta determines whether various grants will be funded.

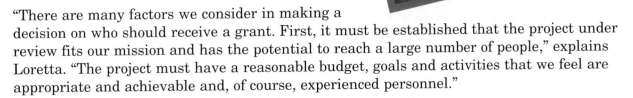

"There are many factors we consider in making a decision on who should receive a grant. First, it must be established that the project under review fits our mission and has the potential to reach a large number of people," explains Loretta. "The project must have a reasonable budget, goals and activities that we feel are appropriate and achievable and, of course, experienced personnel."

One exciting example of an Annenberg/CPB-funded project that Loretta is very proud of is the Tennessee Valley Project. "This is a rural telecommunication project in which teachers and students used the Internet and Web resources along with resources in their communities as tools to learn more about science. For example, students who were learning about water resources went into their community and tested water samples. They then posted their results on the Web and contacted students and scientists all around the country to discuss those results, compare them to the water quality in other communities, and find solutions if water quality was a problem."

"Being a project officer for Annenberg is a tremendous privilege for me. As a youth I never dreamed I would one day work in the corporate world." Loretta graduated from Risley High School in Brunswick, Georgia and went to Fort Valley State College in Georgia. She later received a master's degree in biology from Atlanta University. "I was a science educator in several public school systems for 32 1/2 years," she explains, "and as a teacher and administrator had some experience and success in getting grants from various foundations and agencies for enrichment programs in our school. After all those years spent asking for money, it's nice now to be sitting on the other side of the table."

313

Chapter 5 Assessment

It is now time, if you have not already done so, to choose the proposal you will accept because of its scientific merit, and the one you will deny. For the proposal you chose, can you answer "yes" to the following questions:

Is the topic area testable by experiment? Can experimental evidence be produced to support or refute a hypothesis in this area?

Is the area of study logical?

Is the topic testable by experiment?

Can any observer replicate the experiment and get the same results or will only "believers" in the idea get results that "agree"?

Is the theory the simplest and most straightforward explanation?

Can the new theory explain known phenomena?

Can the new theory predict new phenomena?

Write a letter to defend your selections. Also, write letters to each of the people who submitted proposals for funding.

Before preparing to defend your selection and writing your letter to each person, review the criteria and point value you, your classmates, and teacher agreed upon at the beginning of this chapter.

Physics You Learned

Magnet force fields

Electric force fields

Newton's Law of Universal Gravitation

Inverse Square Relation

Waves

Transverse Waves

Longitudinal (Compression) waves

Wavelength

Amplitude

Frequency

Period

Interference of waves

Frames of reference

Inertial frame of reference

Theory of Special Relativity

Doppler Effect

Red shift

Distances in the universe

Unit 2

Active Chemistry™

Making Connections

You are now going to start a unit of *Active Chemistry*. You investigate the world of chemistry every time you cook a meal. You mix chemicals together, heat them, and new chemicals appear. There are many different foods. How many different chemicals are there? In the *Active Physics* unit you learned about the properties of light waves and sound waves. Now you will explore properties of chemicals and try to make your complex world simpler to understand.

The laws of physics are also the laws of chemistry. In chemistry, you will continue to study forces, energy, and atomic structure. You will learn about these laws of nature from a new perspective.

What Do You Think?

• How do you think that colors are created on the atomic level?

• How is sugar different from salt?

• What makes some chemicals nutritious and others dangerous to eat?

Record your ideas about this in your log. Be prepared to discuss your responses with your small group and the class.

For You To Try

• Look up a recipe for making a cake. When do you think that chemical reactions are taking place?

• Compare the labels on a box of cereal and a cleaning fluid. What similarities and differences are apparent?

Chapter 6

Movie Special Effects

Scenario

You have all been captivated and entertained by special effects in movies. The explosions, makeup, animation, and props in the "blockbuster" features take months and millions of dollars to put together. Many special effects are the result of the application of science and technology.

Your *Active Chemistry* class has been asked to participate in the production of a low-budget movie. However, to make the film exciting, the movie producer would still like to use some awesome special effects. You are being asked to write a simple scene in which you can incorporate some special effects. Of course, in addition to cost, safety is also a major concern.

Chapter Challenge

Your challenge is to create a story line and produce special effects based on the chemistry you have learned in your *Active Chemistry* class. You will need to demonstrate the special effects you created. Your special effects will be evaluated on their quality, interest, entertainment, and the knowledge of chemistry you exhibited in putting them together.

You will need to complete the following tasks:
• Write a script for a simple scene in a movie
• Choose a special effect to include as part of your scene
• Write a procedure on how your special effect is done
• Demonstrate the special effect to the "producer"
• Write an explanation of how the special effect works, including the chemistry behind the demonstration
Using more than one chemical principle will strengthen your presentation.

Criteria

How will your special effect be graded? What qualities should an effective special effect have? How will your demonstration and supporting documentation be graded? Discuss these issues in small groups and with your class. You may decide that some or all of the following qualities should be graded:

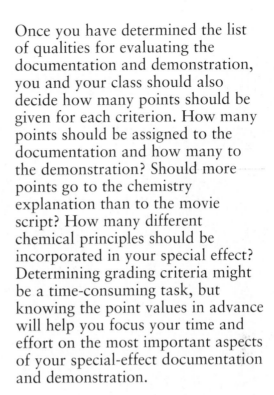

- Demonstration
- Safety
- Quality
- Interest and appeal to audience
- Supporting documentation
- Script — creativity
  - Procedure — clarity, safety, accuracy
    - Chemistry explanation — accuracy and quantity of chemical principles incorporated

Once you have determined the list of qualities for evaluating the documentation and demonstration, you and your class should also decide how many points should be given for each criterion. How many points should be assigned to the documentation and how many to the demonstration? Should more points go to the chemistry explanation than to the movie script? How many different chemical principles should be incorporated in your special effect? Determining grading criteria might be a time-consuming task, but knowing the point values in advance will help you focus your time and effort on the most important aspects of your special-effect documentation and demonstration.

Will each student produce his/her own special effect, will students be required to work in groups, or will both options be offered? Discuss the pros and cons of these possibilities. Keep in mind that if you are going to be working in groups, it is important to discuss before the work begins how each member of the group will be graded. Determine grading criteria that reward each individual in the group for his/her contribution and also reward the group for the final project. You should discuss different strategies and choose the one that is best suited to your situation. Make sure that you understand all the criteria as well as you can before you begin. Your teacher may provide you with a sample rubric to help you get started.

Activity 1   Elements and Compounds

GOALS

In this activity you will:

- Decompose water by electrolysis into the two elements from which it is composed.

- Test the two elements to determine their identities.

- Learn one way to determine the chemical formula of a material.

- Compare characteristic properties of a material to those of its constituent elements.

- Represent materials with chemical formulas using numbers and the symbols of elements.

- Practice safe laboratory techniques with flames and explosions.

What Do You Think?

Matter is the name for all the "stuff" in the universe. Anything that has mass and occupies space is called matter.

- **How many kinds of matter are there in the universe: 1, 10, 100, 1000, 10,000, or more than 10,000?**

- **What makes up matter?**

Record your ideas in your *Active Chemistry* log. Be prepared to discuss your responses with your small group and the class.

Investigate

1. Look around your classroom at the kinds of matter that make up the things you can see.

   a) Make a list of 10 kinds of materials that make up the objects you see. For example, you might list the wood that makes up your pencil, or the glass that makes the windows.

   b) To understand the nature of matter, it helps to know if it is simple or complex. Is it made from only one kind of material or is it a mixture of various materials? Classify each of the 10 materials you listed as pure or mixtures.

320

c) For each material you thought was a mixture, write your best guess about what materials make it up.

2. Sometimes the materials that make up a substance are not obvious. Early scientists thought that water was an element. In other words, they thought that there was only one kind of material in water. They had not discovered a way of breaking it down further. Water, however, can be broken down further.

a) Carefully observe the characteristic properties of water. Record your observations in your log.

3. Put on your safety goggles and apron. Assemble the apparatus for decomposing water as shown in the diagram. Fill two test tubes with water. Submerge them in the water in the beaker and invert them. Make sure you do not allow any air to enter the test tubes. Make sure that the ends of the wires are stripped. Polish them with steel wool. Insert the ends of the wires into the test tubes. Add about 1 to 2 mL of sodium sulfate solution to the water in the beaker.

4. Plug in the battery. Let the reaction run until a test tube is full of gas.

(Your teacher may decide to have you stop the reaction sooner.)

a) Note what happens when the power is turned on. Record your observations in your *Active Chemistry* log.

b) How do the relative amounts of gas formed in the test tubes compare?

5. Disconnect the battery. Place stoppers on the test tubes and remove them from the water.

a) What gas do you think is contained in each test tube? (Hint: You've probably heard that water is H-2-O, written, H_2O.) Record your prediction, and give reasons for your prediction.

b) Observe the physical properties of each gas. Record these properties in your log.

6. You are going to use a lighted wooden splint to identify the gas in each test tube. First, examine the test tube with the smaller volume of gas. Light a wooden splint. Blow out the flame, but leave the splint glowing. Hold the test tube with its mouth up. Remove the stopper. Quickly bring the glowing splint to the mouth of the test tube.

Be certain that the mouth of the test tube is pointed away from everyone.

a) Observe what happens to the splint, and record your observations.

b) What gas do you think was produced in this test tube?

Clean up spills immediately.

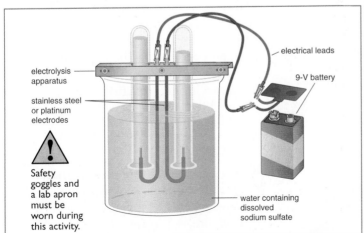

electrical leads

9-V battery

electrolysis apparatus

stainless steel or platinum electrodes

Safety goggles and a lab apron must be worn during this activity.

water containing dissolved sodium sulfate

Safety goggles and a lab apron must be worn during this activity.

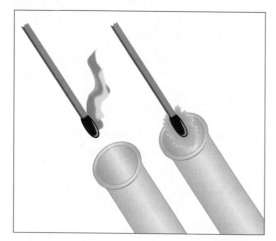

c) In your log record any additional properties of the gas that you discovered.

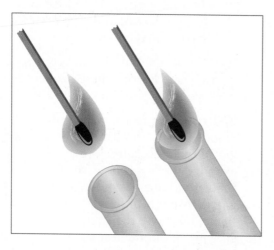

Both the teacher and the students must be shielded.

7. Next, examine the test tube that is full of gas (or contains the larger volume of gas). Light a wooden splint. Remove the stopper. Quickly bring the burning splint to the mouth of the test tube.

a) Record your observations.

b) What gas do you think was produced in this test tube?

c) In your log record any additional properties of the gas that you discovered.

8. Return the splints and equipment as directed by your teacher. Clean up your work station. Wash your hands.

9. Your teacher will demonstrate a way you can use what you investigated to produce a special effect. Your teacher will set up a gas generator, similar to the one shown in the diagram. The test tube contains 10 mL of 6M HCl (hydrochloric acid — a compound of hydrogen and chlorine) and 3 g of mossy zinc.

a) What gas do you think is being produced in the test tube? Give a reason for your answer.

10. An egg has been emptied out. There is a small hole in the top and another in the bottom of the egg. One of the holes is taped. The egg will be placed over the gas generator, and gas will be collected in the egg for several minutes.

11. The egg will then be mounted in an egg holder behind a shield. If you have a video camera available, be prepared to start recording. Your teacher will remove the tape, and light the top of the egg with a burning splint. Begin recording or observing the reaction until it is over.

a) Record your observations in your log.

b) If the gas in the egg was hydrogen and air contains oxygen, what substance do you think may have been created in this process?

Chem Talk

THE STRUCTURE OF MATTER

In this investigation you used electricity to decompose water into two gases. You knew that the gases were different because they reacted differently to the burning and glowing splints. Since water is referred to as H_2O, a first guess would be that hydrogen (H) and oxygen (O) were created in the experiment. The test for hydrogen is a small explosion when exposed to a burning splint. The test for oxygen is igniting a glowing splint. If you look back on the results of the experiment, you find that the hydrogen gas filled one test tube while the oxygen filled half of the other test tube. There was twice as much hydrogen as oxygen. That's where the 2 comes from in the chemical formula H_2O.

Hydrogen and oxygen are elements. An **element** is any material that cannot be broken down into simpler materials by chemical means. You are probably familiar with many elements like hydrogen, oxygen, zinc, gold, or helium. Other elements like strontium and beryllium are more exotic and less likely to be familiar to you. Every kind of matter you observe in your everyday life is made up of the chemical elements. There are only about a hundred different kinds of chemical elements. This is an amazing discovery of chemistry — everything you observe in the world is made of different combinations of a hundred elements. Chemistry is the study of how these elements combine and the characteristics of these combinations.

Chem Words

element: any material that cannot be broken down into simpler materials by chemical means.

Elements are represented by symbols. The symbol is one, two, or three letters that represent the name. It's easier to write O than to write oxygen. It's easier to write H than to write hydrogen. The symbols come from many different sources. However, the same symbols are used for each element in all countries of the world.

Symbols for Some Elements	
Name of Element	Symbol
aluminum	Al
bromine	Br
calcium	Ca
carbon	C
chlorine	Cl
copper	Cu
gold	Au
helium	He
hydrogen	H
iodine	I
iron	Fe
lead	Pb
magnesium	Mg
mercury	Hg
neon	Ne
nickel	Ni
nitrogen	N
oxygen	O
phosphorus	P
potassium	K
silicon	Si
sodium	Na
sulfur	S
tin	Sn
zinc	Zn

When elements combine they form new substances called **compounds**. These compounds have entirely new characteristics. It is like combining the letters of the alphabet to make words. Twenty-six letters can be combined to make thousands of different words.

Water is an example of a compound. A water molecule, H_2O, is composed of two atoms of hydrogen and one atom of oxygen. (For now, think of an atom as the smallest particle of an element and a molecule as the smallest unit of a compound.) In this activity you used electricity to decompose water into its elements, hydrogen and oxygen. This process is called **electrolysis**. You observed that oxygen gas made a glowing splint burst into flame, and that hydrogen gas was explosive. However, to extinguish a burning splint, you could use liquid water. The compound has very different characteristics from the elements from which it is made.

Compounds are represented by **chemical formulas**. A chemical formula shows the symbols of the elements that are combined to make the compound. If there is more than one atom of an element, a subscript is added after the symbol indicating how many atoms of that element there are. For example, as you discovered in this activity, the chemical formula for water is H_2O.

Chem Words

compound: a material that consists of two or more elements united together in definite proportion.

electrolysis: the conduction of electricity through a solution that contains ions or through a molten ionic compound that will induce chemical change.

chemical formula: the combination of the symbols of the elements in a definite numerical proportion used to represent molecules, compounds, radicals, ions, etc.

Examples of Some Chemical Formulas		
Compound	**Common Name**	**Chemical Formula**
calcium carbonate	chalk	$CaCO_3$
carbon dioxide	dry ice	CO_2
hydrochloric acid	muriatic acid	HCl
hydrogen sulfide	rotten-egg gas	H_2S
sodium hydrogen carbonate (or sodium bicarbonate)	baking soda	$NaHCO_3$
sodium chloride	table salt	NaCl
sodium nitrate	fertilizer	$NaNO_3$
sulfuric acid	battery acid	H_2SO_4

From the table of chemical formulas, you can see that carbon dioxide is a compound of carbon and oxygen. There are two atoms of oxygen for every atom of carbon. Sodium hydrogen carbonate (sodium bicarbonate) is a compound of sodium, hydrogen, carbon, and oxygen. There are three atoms of oxygen for every atom of the other elements. Also, there are a total of three atoms in the carbon dioxide formula and a total of six atoms in sodium hydrogen carbonate.

To generate the gas to fill the empty eggshell in this activity (the teacher demonstration), zinc was placed in hydrochloric acid. Zinc is an element. Hydrochloric acid (HCl) is a compound of hydrogen and chlorine. The reaction of the zinc and hydrochloric acid created a gas. Given the explosion you observed, you can guess that the gas produced was hydrogen. The hydrogen gas came from the hydrogen in the hydrochloric acid.

There's much more to the structure of matter than you can discover in just one activity. However, this activity may have raised some new questions in your mind. For example:

• Can all compounds be decomposed into their elements?
• What techniques can be used to decompose compounds?
• What are elements made of?
• What are atoms?
• What are molecules?

These questions and many more will be explored in other *Active Chemistry* activities.

Checking Up

1. In your own words, explain the difference between an element and a compound.

2. Why are symbols useful in describing chemical elements?

3. What are the symbols for the following elements: carbon, copper, gold, and helium?

4. What information does a chemical formula of a compound provide?

Reflecting on the Activity and the Challenge

Part of the problem you are facing in creating a special effect is understanding what matter is made of and how it can change. In this activity you broke a chemical compound down into its component elements using electrolysis. In another part of the activity a compound was made from chemical elements through a fast and noisy reaction. There are only about one hundred elements, but there are many thousand compounds. You should begin thinking of ways in which some of the reactions you observe could be made to appear more dramatic on screen, without making them any larger in real life. You can now use the concepts of elements and compounds to provide the chemistry description of what is occurring.

Chemistry to Go

1. The table on the right contains several common compounds that are probably familiar to you.

   For each compound:

   a) List the names of the elements present.

   b) State the number of atoms of each element present.

   c) Give the total number of atoms present in each compound.

2. Write a chemical formula for nitrous oxide (laughing gas) that is made up of two atoms of nitrogen and one atom of oxygen.

3. Choose one compound from the table in **Question 1**.

   a) Describe the properties of each element in the compound.

Common Name	Formula
sugar	$C_{12}H_{22}O_{11}$
marble	$CaCO_3$
natural gas	CH_4
rubbing alcohol	C_3H_8O
glass	SiO_2

   b) Explain how the property of the compound is different from the property of each element.

Preparing for the Chapter Challenge

In a short paragraph, summarize the difference between an element and a compound and describe how the properties of a compound can be very different from the properties of the elements that make it up. Explain why knowing these differences is important when designing special effects for a movie set.

Inquiring Further

How is electrolysis used in industry?

Use the reference materials available to you to explore how electrolysis is used in industry to produce hydrogen gas and other elements from compounds.

Activity 2   States of Matter: Solid, Liquid, and Gas

GOALS

In this activity you will:

- Create an animation to illustrate the behavior of particles in different phases of matter, and as the material changes phase.

- Observe changes of state of water and describe the process graphically.

- Observe a change of state of carbon dioxide and describe the energy transformations involved.

- Describe the energy transformations and the roles of kinetic and potential energy as heat energy is transferred to or away from a material.

- Describe the behavior of gas particles, based on your observations of how the temperature, pressure, and volume of the gas are affected as heat energy is transferred to or away from the gas.

- Characterize materials by their unique phase-change temperatures.

- Practice safe laboratory techniques in working with temperature extremes.

What Do You Think?

You know that materials can exist as solids, liquids, or gases. Each state of matter has its own characteristics.

- **Draw three circles. In the first circle draw what you think particles of material look like in the solid state. In the next circle draw the particles of the same material as a liquid. In the final circle illustrate the same material as a gas.**

Record your ideas in your *Active Chemistry* log. Be prepared to discuss your responses with your small group and the class.

Investigate

Part A: The Heating Curve of Water

1. Put on your safety goggles and apron. Half-fill a 250-mL beaker with crushed ice.

2. Set up your equipment as shown in the diagram. The hot-plate dial should be in the off position. When you clamp the thermometer into place, be sure that the bulb is in the center of the ice and is not touching the bottom of the beaker.

328

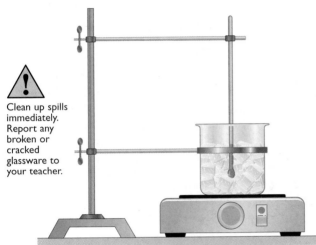

Clean up spills immediately. Report any broken or cracked glassware to your teacher.

Never use a thermometer as a stirring rod.

3. Observe the thermometer closely until it appears to have reached its lowest reading.

   a) Prepare a data table similar to the one shown below. Record the minimum temperature at time 0 min.

4. Turn on the hot plate. Use a medium-low setting, or one suggested by your teacher.

5. Gently stir the ice and water mixture with a stirring rod.

   a) Record the temperature every minute.

6. Continue to record the temperature until the water has been at a full boil for 5 min.

a) Name all the changes of state that you observed. A change of state refers to a change from a solid to a liquid, a liquid to a gas, a solid to a gas, and vice versa.

7. Turn off the hot plate. When the water has cooled discard it. Return all equipment as directed and clean up your station.

8. Use your data to answer the following:

   a) What was the temperature at which all the ice had melted?

   b) What was the boiling point of the water?

   c) Plot a graph of the data with time along the x-axis and temperature on the y-axis. You may wish to use a microcomputer or a calculator to plot your graph.

   d) Describe your graph. Consider: What is happening at the various points along the graph? Heat energy is being continually transferred to the system by the hot plate. At which point is the heat energy causing the temperature to increase? What is the heat energy doing if it is not acting to raise the temperature of the water?

Hot plates may remain hot for some time after they are turned off.

Time (min)	Temperature (°C)	Observations
0		
1		

Coordinated Science for the 21st Century

Part B: Making Particles Shake

1. Your teacher will assemble or have you assemble a small booklet with many blank pages.

2. Use only the right-hand pages of the booklet. At the very bottom right of the last page of the booklet draw a dot. To make the dot appear to move from the top left to the bottom right at a constant speed, draw another dot on the previous page slightly to the left and up. Continue this until you reach the first page of the booklet. Now as you flip through the booklet from the front to the back, the dot will appear to move. The smaller the movements from one page to the next, the smoother the animation effect will appear.

3. Matter can be in a solid state, a liquid state, or a gaseous state. Each state of matter has its own typical motion of particles. The physical properties of each state are a result of how the particles move relative to one another. Use animation to model the movement of particles in each state of matter. Consider using different colors to keep track of the particles.

   a) In a solid the particles stay in the same position but vibrate. Use your flipbook to model the movement of particles in a solid.

   b) In a liquid the particles are about the same distance apart as in the solid, but they can move more freely. Use animation to model a liquid.

4. Gas particles are very, very far apart and they move very quickly.

a) What problems would you have to make a flipbook for gases?

Part C: Volume Changes

1. Draw up a water plug about 1 cm long in a 30-cm long glass tube. Set the glass tubing into a test tube, as shown in the diagram below.

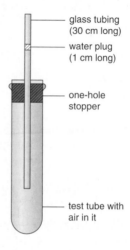

glass tubing (30 cm long)

water plug (1 cm long)

one-hole stopper

test tube with air in it

The water should not be too hot to touch. Glassware is slippery when wet. Handle with care.

a) Observe and describe the movement, if any, of the water plug.

2. Place the test tube into a beaker of warm water.

   a) Record your observations in your *Active Chemistry* log.

   b) How does the warm water affect the volume of the air in the test tube? What evidence did you observe that suggested a volume change?

   c) The warm water was a source of heat energy. What effect did this heat energy have on the volume of air?

3. Place the test tube into a beaker of ice water.

   a) Record your observations in your *Active Chemistry* log.

4. In physics you've learned that for the drop of water to stay in place, the forces must be equal and opposite. The force of gravity pulling down on the water plug must be equal to the upward force of the gas.

When the air in the test tube was heated, the upward force on the water plug must have increased, because the water rose in the tube. Since no additional air molecules were added, the molecules must have moved faster as a result of the additional heat. As the temperature of the air increased, the molecules of air increased their speed and therefore applied a greater force to the drop of water. Once the drop of water rose high enough, the larger volume and fewer air molecules hitting the drop per second compensated for the increased speed of the molecules. The forces of gravity and pressure were equal once again.

   a) Pressure is force per area. What happened to the pressure on each wall of the test tube as you heated up the air (gas)?

   b) Draw a box with a moveable piston, as shown in the diagram at right.

   c) Use animation to show what would happen to the piston if the temperature of the gas inside the

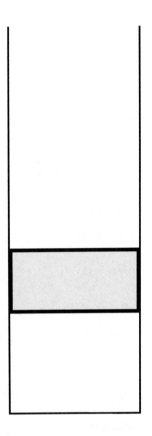

cylinder and below the piston were increased.

Part D: "Special" Ice

1. Your teacher will place a small piece of dry ice (solid carbon dioxide) in an empty beaker.

   a) Record your observations in your *Active Chemistry* log.

   b) Is heat energy being transferred to or away from the dry ice by the surrounding air?

   c) What change of state is taking place?

Chem Talk

CHANGES OF STATE

All matter is made up of tiny particles. Different materials are made of different kinds of particles. These particles are always moving, and there are spaces between them. The more energy the particles have, the faster they move. There are also attractive forces among the particles. The closer the particles are together, the greater are the attractive forces.

Temperature

Chem Words

temperature: the measure of the average kinetic energy of all the particles of the material.

kinetic energy: a form of energy related to the motion of a particle ($KE = \frac{1}{2}mv^2$).

You get an intuitive sense of **temperature** by how hot or cold something feels to your skin. Your body is at 37°C (98.6°F). When something with a higher temperature comes in contact with your skin, you know that it is "hot." When something with a lower temperature comes in contact with your skin, you know that it is "cold." As you observed in the activity, when the temperature of air increased, the drop of water lifted. This drop of water could be a crude thermometer. As the drop rises, you know that the temperature of the air is higher. Liquids like alcohol and mercury expand when they get hot and are used for the thermometers with which you are most familiar.

The movement of the water drop gives you an insight into another interpretation of temperature. The air molecules inside the tube were moving faster as the temperature of the air increased. The temperature of the air is a measure of the speed of the molecules. In physics, you learned that kinetic energy is related to speed. **Kinetic energy** is equal to one-half the mass times the square of the speed of the particles ($KE = \frac{1}{2}mv^2$). Observing the behavior of many gases, scientists have concluded that temperature is a measure of the average kinetic energy of the molecules.

Melting and Boiling Points

In this activity you started with a beaker of crushed ice that was at a temperature less than 0°C. As you heated the ice it did not initially melt, but the temperature of the ice began to rise. As the temperature of any solid increases, the average kinetic energy (energy of motion) of the particles of the material increases.

This motion is mainly a vibration-type motion where the molecules vibrate around a fixed location. As heat energy continued to be transferred, the temperature of the ice increased until it reached 0°C. This is called the **normal melting point** of water. It is the temperature at which water changes from a solid to a liquid state at 1 atm (atmospheric pressure at sea level). It is also the **normal freezing point** of water, when water changes from a liquid to a solid at 1 atm. Each material has its own characteristic normal melting/freezing point.

The temperature then remained at 0°C as the solid water changed to a liquid. Since the temperature remained constant, the average kinetic energy did not change. All the heat energy that was transferred caused a phase change during which the molecules of water were rearranging or decomposing. There was a change in the **potential energy**. If there is a change in kinetic energy you will see a change in temperature and if there is a change in potential energy, the temperature will remain constant while heat energy is transferred to a material.

When all of the water had melted at 0°C, the temperature of the liquid water increased until it reached 100°C. This is called the **normal boiling point** of water at 1 atm.

Chem Words

normal melting point: the characteristic temperature, at 1 atm, at which a material changes from a solid state to its liquid state.

normal freezing point: the characteristic temperature, at 1 atm, at which a material changes from a liquid state to its solid state.

potential energy: stored energy of the material as a result of its position in an electric, magnetic, or gravitational field.

normal boiling point: the temperature at which the vapor pressure of the pure liquid equals 1 atm.

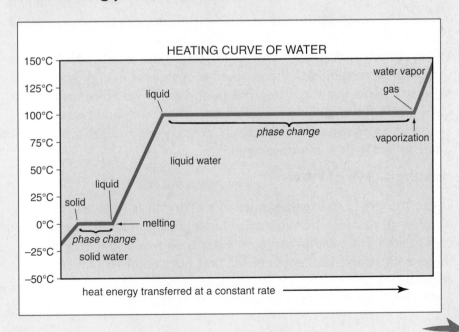

HEATING CURVE OF WATER

If the atmospheric pressure is less than 1 atm, then the water will boil at less than 100°C and is just called the boiling point of the liquid. For example, on Mt. Rainier in Washington State, at an altitude of about 4393 m (14,411 ft), the atmospheric pressure is much less than 1 atm, and you would find that water boils at a lower temperature.

When the water arrived at the boiling point, you again noted that the temperature remained the same, even though heat energy was still being transferred to the water. The temperature would remain the same until all of the liquid is **vaporized.** Then, with additional heat energy, the temperature would again increase and the gas molecules of water would have greater average kinetic energy.

Heating Curve of Water

These changes in the temperature of a material, as heat energy is transferred to it, can be summarized in a graph, similar to the one you constructed. The heating curve of water is shown in the diagram on the previous page. The length of the first horizontal section corresponds to the amount of heat energy required to make the material change from solid to liquid.

Chem Words

vaporization: the change of state from a liquid to a gas.

sublimation: the change of state of a solid material to a gas without going through the liquid state.

Dry ice (solid carbon dioxide) that your teacher used in the demonstration, does not have a normal melting point; instead, it has a normal sublimation point (-78.5°C at 1 atm). **Sublimation** is the process where the solid goes directly to the gaseous state. The changes of state are summarized in the diagram below.

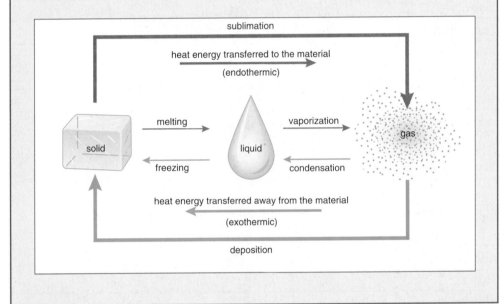

Checking Up

1. What does temperature measure?

2. Describe what is happening to particles of a material when heat energy is transferred to the material and the temperature increases.

3. What happens to the temperature of a material when it is undergoing a change of state?

4. What is the difference between the normal boiling point of water, and the temperature at which water might boil?

Reflecting on the Activity and the Challenge

In this exercise you focused on very simple chemistry, the motions of particles in solids, liquids, and gases. You can use the techniques learned in this activity to animate more complicated chemical systems. Consider how you could illustrate a phase change, like boiling or freezing. With research you might be able to use animation to explain the chemistry you use in staging your special effect.

Coordinated Science for the 21st Century

Chemistry to Go

1. Copy and complete the following table summarizing the changes of state in your *Active Chemistry* log.

Change of State	From	To	Heat (added or removed?)
boiling	liquid	gas	added
condensation	gas	liquid	removed
evaporation			
freezing			
melting			
deposition	gas		
sublimation	solid		
vaporization			

2. Copy and complete the following table in your *Active Chemistry* log.

Definite or Indefinite?	Solid	Liquid	Gas
shape			
volume			

3. The heating curve for water was given in the **ChemTalk** reading section. Create the cooling curve for water and describe each part of the curve.

4. Create a heating curve for water when you have twice the amount of water you used in the investigation. How does this heating curve compare with the original curve?

5. The melting and boiling points of three materials are given on the next page.

   a) Draw a heating curve for each material.

   b) From graphs, explain why, at room temperature (22°C), copper is solid, mercury is liquid, and oxygen is a gas.

6. When water is in a pressure cooker the pressure is greater than 1 atm. What will the boiling point of water be compared to the normal boiling point of water? What information did you use to support your answer?

Material	Melting Point (°C)	Boiling Point (°C)
copper	1083	2336
mercury	- 39	357
oxygen	- 218	- 183

7. A certain material has a normal freezing point of 10°C and a normal boiling point of 70°C. What state is the material in at room temperature (about 22°C)? Explain your answer.

8. The normal boiling point of water is 100°C and the normal boiling point of ethanol is 78.5°C. These two liquids are soluble in each other. What technique would you use to separate these two liquids from the solution?

9. A certain material is white in the solid phase and clear (transparent) in the liquid phase. It has a normal melting point of 146°C. Which of the following materials could this be: water, sugar, or carbon dioxide? Explain your choice. (Hint: You will have to base your argument on concepts you have learned and observations made in this activity, as well as common sense.)

Preparing for the Chapter Challenge

1. Make an animation of 10 water particles changing from the solid state at 0°C to the liquid state at 0°C.

2. Animate the change of state of 10 dry-ice particles placed at room temperature.

Inquiring Further

1. Video animation

Making the leap from flipbook animation to video animation is not difficult. You will need access to a video camera that can record one frame at a time. Set the video camera so that it can record an image on a flat table or desk. To animate the process you can either record a series of drawings on paper, putting one after another on the table and taking a frame or two of each image.

Go to the Internet and see if you can find animations on the Internet. You might find a computer helpful in making the drawings or you could investigate using animation software to make more elaborate animations.

2. Design a thermometer

Use what you learned in this activity to design a thermometer that you could use to measure temperature. If your teacher approves your design, try out your thermometer.

Activity 3 — Solutions, Suspensions, and Colloids

GOALS

In this activity you will:

* Explore different ways that materials can be mixed together to make new materials.

* Test some materials to determine what kinds of mixtures they are.

* Determine why certain kinds of mixtures are manufactured for commercial use in particular situations.

What Do You Think?

One way to get different types of materials is to just mix them together. Lots of different things can happen when materials are mixed. Each kind of mixture has its own characteristics.

* **Is it easier to separate milk from coffee or milk from a bowl of cereal?**

Record your ideas about this question in your *Active Chemistry* log. Be prepared to discuss your responses with your small group and the class.

Investigate

1. Half-fill six large test tubes with water. Number the test tubes.

   a) In your *Active Chemistry* log, prepare a table for your observations. You may wish to use a table similar to the one on the following page.

Safety goggles and a lab apron are required for this activity. Wipe up spills immediately. Report any broken or cracked glassware to your teacher.

No.	Materials mixed with water	Observations before mixing	Observations after mixing	Kind	Filter
#1	water only				
#2	0.5 g sugar				
#3	drops of milk				
#4					

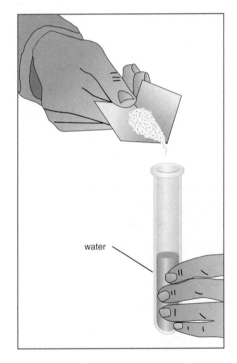

water

2. Add the following materials to the test tubes:

   #1—nothing
   #2—0.5 g sugar
   #3—a few drops of milk
   #4—0.5 g CuSO$_4$ (copper sulfate)
   #5—2 mL olive oil
   #6—0.5 g soil

   a) Describe each material before mixing.

3. Stopper each test tube. Place your finger over the stopper and shake each for several minutes to make a mixture.

   Observe each mixture.

4. Consider the following characteristics of the mixtures:

   • What is the appearance of each mixture after the vigorous mixing? Which ones have visible particles suspended in them? Which ones look totally uniform (homogeneous) throughout?

   • Which mixtures separate (are heterogeneous) after sitting a few moments after vigorous mixing? Which remain mixed?

   • Shine a laser pointer through each mixture. In which mixtures is the laser beam clearly visible? In which mixtures does it pass through with little effect?

   Never look directly at a laser beam.

   • For each mixture place a small beaker below a funnel to catch the filtrate, as shown in the diagram on the next page. Pour the contents of each test tube into a funnel with filter paper. Which mixtures pass through? Which leave part behind on the filter paper?

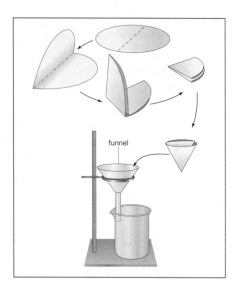

a) Record all your observations in the table in your log.

b) Using your data and the descriptions of the kinds of mixtures in the **ChemTalk** reading section, classify each of the mixtures as a solution, colloid, or suspension.

5. Discard materials and return all equipment as directed by your teacher. Clean up your station and wash your hands.

Chem Talk

CLASSIFYING MIXTURES

In this activity you mixed together water and several different materials to produce different kinds of mixtures. In some cases the materials you used were **pure substances.** A pure substance contains only one kind of particle throughout. For example, sugar is a pure substance. A mixture contains at least two pure substances. You may think of water as a pure substance, however most water found in nature has different materials mixed with it and is in fact a mixture.

Mostly all materials that you find in nature, as well as most human-made materials are mixtures of one or more pure substances. You made one kind of mixture, called a **solution**, when you added sugar to water. In a solution, the particles that dissolve are so tiny they can't be seen with the naked eye. The mixture is said to be homogeneous. The dissolved particles (called the **solute**) remain mixed forever with the solvent. The water, since it is doing the dissolving in this case, is the **solvent**. If a solution is filtered, everything passes through. Light passing through a solution has no special effect.

When you added milk to water, the water appeared cloudy. However, the tiny drops of milk remained suspended in the water and did not

Chem Words

pure substance: a substance that contains only one kind of particle.

solution: a homogeneous mixture of two or more substances.

solute: the substance that dissolves in a solvent to form a solution.

solvent: the substance in which a solute dissolves to form a solution.

settle out over time. You could see the laser beam as it passed through the mixture, and when you filtered the mixture, it all passed through the filter paper. This kind of mixture is a **colloid**. In colloids, the dispersed particles are larger than those in solution and may be visible on close inspection with a microscope. The particles will also stay suspended indefinitely. All parts of the colloid will pass through a filter. When light passes through a colloid it is scattered and you can see where the light beam passes through. This is known as the **Tyndall Effect.**

When you added soil to water, you created a suspension. **Suspensions** have the largest of all the dispersed particles. The particles are visible to the eye and will settle out in time. The suspended particles can be separated by filtration. The mixture is said to be heberageous. A light beam shining through a suspension may be scattered, but the suspension is definitely not transparent.

Chem Words

colloid: a mixture containing particles larger than the solute but small enough to remain suspended in the continuous phase of another component. This is also called a colloidal dispersion.

Tyndall Effect: the scattering of a light beam as it passes through a colloid.

suspension: heterogeneous mixture that contains fine solid or liquid particles in a fluid that will settle out spontaneously. By shaking the container they will again be dispersed throughout the fluid.

Checking Up

1. In your own words describe how you would distinguish among a solution, a colloid, and a suspension.

2. What is the Tyndall Effect?

Reflecting on the Activity and the Challenge

In this activity you made mixtures with solid or liquid solutes and water as the solvent. The same basic principles apply regardless of the states of matter involved. For example, it is possible to have solid solutions. Metal alloys such as brass or bronze are such solutions. Fog, smoke, and clouds are mixtures that show the Tyndall Effect. A common stage effect is to produce smoke or fog to give an eerie setting. Many movies have used the Tyndall Effect to show the path of laser beams or flashlights. One easy way to produce this effect is to use a spray bottle of water to mist the air in a darkened room (problem with allergies/asthma). When a flashlight or laser pointer is directed through the mist, it shows up nicely, thanks to the Tyndall Effect.

Chemistry to Go

1. Classify each of the following as a suspension, colloid, or solution. Explain your reasoning. (Hint: In some cases more than one answer may be possible.)

   a) A mixture is poured through a filter, and the entire mixture passes through.

   b) A mixture is left to stand for a while and small particles settle out.

   c) When viewed under a microscope, small particles are visible in the mixture.

   d) A beam of light passed through the mixture is scattered.

   e) The mixture is blue and transparent.

2. Suggest a method by which you could separate the various materials in each of the following mixtures:

   a) solutions   b) colloids

   c) suspensions

3. On some days it is possible to see rays of light from the Sun distinctly coming through breaks in clouds. What in the atmosphere might account for this difference?

4. The term homogeneous means uniform or the same throughout. Heterogeneous means that different parts of the mixture are clearly visible. Classify each of the mixtures in this activity as homogeneous or heterogeneous.

5. Look in your kitchen at home and choose five products. Make your best guess as to the type of mixture they represent. Elaborate on the evidence that you used to classify the products into the respective category. Explain what would happen to each product that you have chosen if it were in a different kind of mixture. For example, milk would settle if it was a suspension instead of a colloid.

Preparing for the Chapter Challenge

Consider how you could use the properties of a solution, colloid, or suspension to produce a special effect for a movie scene. In a few sentences establish the setting for the scene, and describe the mood you want to create.

In a paragraph or two describe the chemistry required to understand the differences between the different types of mixtures and how they produce the special effect.

Inquiring Further

Cooking and mixtures

Cooking is a practical application of mixtures.

One such application is the recipe for making mayonnaise. Mayonnaise is classified as a colloid. A liquid—liquid colloid is also called an emulsion. If you have access to the following materials, try making mayonnaise at home.

A Recipe for Mayonnaise

Combine in a blender:

Do this only with the approval of a responsible adult.

2 tablespoons beaten egg

1 large egg yolk

$\frac{1}{4}$ teaspoon dry mustard

1 teaspoon lemon juice

Blend the mixture for 20 seconds. While the blender is still running, add 3/4 cup of vegetable oil slowly in the thinnest stream you can manage. Serve immediately or keep for 1 to 2 days in the refrigerator.

Investigate the properties of the ingredients of mayonnaise that allow them to form this colloidal emulsion. What other substances are classified as emulsions? How could emulsions be used in a movie special effect?

Metals in cooking utensils are also examples of mixtures. Investigate the kinds of materials used in making metal cooking utensils. Why are some mixtures more suitable for cooking than others?

Activity 4   Properties of Matter

GOALS

In this activity you will:

- Make modeling dough from common kitchen materials.

- Adjust the properties of the modeling dough by adding another material to it.

- Compare the properties of an emulsion to those of a composite material.

- Recommend whether an emulsion or a composite would be best for certain applications.

- Consider the advantages and disadvantages of using composites for industrial applications.

What Do You Think?

Frequently in movies, models of sets or characters are built and then enlarged on the screen. A giant marshmallow or dough man that appeared to be as tall as a 10-story building was created for a popular movie that used many special effects.

- **Estimate how many marshmallows it would take to build an actual-size giant marshmallow man.**

- **What do you think would be the cost of the marshmallows?**

Record your ideas about these questions in your *Active Chemistry* log. Be prepared to discuss your responses with your small group and the class.

Investigate

1. Use the ingredients and directions on the following page to prepare a batch of modeling dough.

⚠️ Safety goggles and a lab apron must be worn during this activity.

Ingredients:

1 cup flour
$\frac{1}{2}$ cup salt
3 tablespoons oil
2 tablespoons cream of tartar
1 cup water

⚠ Do not overheat the oil.

Directions:

- Warm the oil in a saucepan or large beaker on a hot plate set at medium-low.

- Add the other ingredients and cook 3 to 5 minutes while stirring constantly.

- Using tongs, drop the mixture onto waxed paper or aluminum foil.

- Cool until the modeling dough is easy to handle.

- Knead the dough until the texture is consistent.

- Add a drop of food coloring, if desired.

2. Divide the dough into two samples. Set aside one sample.

   Add a new material (tissue paper or paper towel strips, sand, gravel, pieces of twine, marbles) to the other sample, as directed by your teacher. Knead the mixture well until the new material is well incorporated into the dough.

3. You will now observe and compare the properties of the original sample and the new sample that you made. In your group, discuss how you will test for each of the following properties:

texture,
elasticity,
uniformity,
bounce,
strength,
malleability.

If you are not sure what characteristics you should be testing for, each property is explained in the **ChemTalk** reading section.

a) Record in your log the procedure you will use to test for each property.

b) Make a table in your *Active Chemistry* log to record your observations.

4. When your teacher has approved your procedure, test each property.

a) Record your observations in your log.

b) In a few sentences, describe the differences between the two kinds of mixtures.

c) Speculate on why your new mixture behaved differently from the original modeling dough.

5. Compare the properties of your new mixture with the mixtures of other groups.

a) What new or useful properties did each mixture have?

6. Dispose of materials and return all equipment as directed by your teacher. Clean up your station and wash your hands.

Chem Talk

PHYSICAL PROPERTIES

In this activity you examined a variety of physical properties of a mixture. In addition to the state and color of the mixture, you also observed the **texture** of the material. Texture is defined as the feel or appearance of a surface or substance. For example, if you were looking at a photograph, you may be interested in the texture of the surface of the paper used, or if you were looking at cloth, you might be interested in how the material is woven together. **Uniformity** describes how consistent a material is throughout. Does everything in the material seem to be evenly distributed? **Strength** determines how durable the material is. How well the material withstands the application of a force establishes how strong it is. **Elasticity** determines how well the material will resist deformation and return to its normal size or shape after a force has been applied to it. **Bounce** refers to the material's ability to return to its original position when it is dropped from a given height. Another way to think about bounce is to consider how much it behaves like a ball. **Malleability** determines how easy it is to roll or hammer out the material without breaking it apart. Can the material be reshaped without breaking it apart? Lead is an example of a metal that can easily be shaped into other forms without breaking. Brittle objects shatter easily.

Chem Words

texture: the characteristics of the surface of a material, like how smooth, rough, or coarse it is.

uniformity: the property of how consistent a material is throughout.

strength: the property of how well a material withstands the application of a force.

elasticity: the property of a material to resist deformation and return to its normal size or shape after a force has been applied to it.

bounce: the ability of an object to rebound to its original position when dropped from a given height.

malleability: the property of a material to be able to be hammered into various shapes without breaking.

COMPOSITE MATERIALS

The properties of a mixture can be changed by adding different materials. The modeling dough you made in this activity was an **emulsion**. When you added one of the other materials, you made a **composite**. Composites are heterogeneous mixtures that use the characteristics of the components to make useful substances. Another example of a composite is paper maché.

The composite industry has been growing at a very rapid pace. Think about it. Composites are used as the skin on jets, rotor blades of a helicopter, bulletproof clothing for law enforcement officers, and special armor for tanks. Buildings, cars, boats, trains, and planes all take advantage of different types of composites. If you were trying to define what a composite is, you would probably say that it is a solid that consists of two or more materials. It may be the process by which fibers are embedded within another material. Usually, this will lead to a stronger material and provide the best qualities of both materials. Composites not only help make stronger and lighter materials but they also help in extending the life of materials. Composites are used for materials that cover electrical cables and protect them from chemicals.

A major drawback to composites is the initial cost of research and the use of special raw materials in fabricating the composite. However, overall composites are materials of the future and their manufacture could be a great career choice.

Chem Words

emulsion: a colloid or colloidal dispersion of one liquid suspended in another.

composite: a solid heterogeneous mixture of two or more substances that make use of the properties of each component.

Checking Up

1. Two pieces of cloth have the same color and texture. Name another property of cloth that you might use to distinguish between the two materials.

2. Provide a situation in which each of the following properties of a material might be important:
   a) elasticity
   b) uniformity
   c) strength
   d) bounce

3. What is a composite material?

4. Give an advantage and a disadvantage of using a composite material.

Reflecting on the Activity and the Challenge

One part of making a special effect for a movie is finding materials that could be used for making models. In this activity you learned to make a material that has many characteristics that make it ideal for making models. Consider how each of the properties you investigated is important to a special effect. A malleable material is easy to shape. To construct a large building, you will want to use a composite that is strong. You could also construct a small model of a tall building, and then make a model of a character that can bounce 10 stories high.

Chemistry to Go

1. A composite when tested bounces half as high as the original material. How can you create another composite that bounces only one-quarter as high?

2. Pure, 24-K gold is very malleable. By mixing gold with other metals it becomes less malleable. In what situations would you want to use gold that is not as malleable?

3. Plastics are made of one type of material. Why are they not classified as a composite?

4. When road contractors are laying cement, they first place a series of steel rods in a specific pattern. Then they pour the cement on the rods and allow it to cure. The composite is called reinforced concrete. What is the purpose of the steel rods?

Preparing for the Chapter Challenge

1. Each kind of modeling dough that you made in this activity could be used in making sets for a movie special effect. In your log list the kinds of structures each might be useful in constructing.

2. Give two examples of special effects in movies where each of the following properties is important: strength, elasticity, bounce, uniformity, malleability.

Inquiring Further

1. Other modeling materials

Many other types of modeling materials, such as plaster of Paris or Fimo® dough, are available in crafts and art stores. Research these substances and consider their usefulness as modeling materials for your movie special effect.

2. Different composites

Investigate different composites. Learn about the special properties of these materials. List the items of products that are constructed from composites.

3. Investigating other techniques with modeling dough

Color and appearance of models are crucial to making them look authentic in a movie scene. With the supervision of an adult, try some alternative means of coloring your modeling dough. Try painting the surface of the dough, or using other substances to add color and texture. How do the various techniques affect the characteristics of the dough?

Activity 5   Mass and Volume

What Do You Think?

A piece of steel sinks in water, but a steel boat floats. A tiny rock sinks in water, but a large log floats.

- **Since a kilogram of feathers and a kilogram of lead have the same mass, how do they appear different and why?**

Record your ideas about this question in your *Active Chemistry* log. Be prepared to discuss your responses with your small group and the class.

Investigate

Part A: Mass and Volume of Liquids

1. In your *Active Chemistry* log create a table to record your data for this part of the activity. You may wish to use a table similar to the one on the opposite page.

GOALS

In this activity you will:

- Determine the densities of various liquid and solid materials.
- Make measurements in the laboratory to the precision of the instruments used.
- Learn the difference between accuracy and precision in experimental measurements.
- Retain significant figures in calculations involving experimental measurements.
- Use density measurements to determine the identity of a material.
- Locate sources of the variation in the class's experimental results.

a.

Volume and Mass of Water				
Mass of graduated cylinder (g)	Volume of water (mL)	Mass of graduated cylinder and water (g)	Mass of water (g)	$\frac{Mass}{Volume}$ (g/mL)

2. Measure the mass of an empty, dry, graduated cylinder.

   a) Record the mass of the cylinder in your *Active Chemistry* log.

3. Add 10 mL of water to the graduated cylinder. Remember when reading the volume, take the reading at the lowest part of the meniscus, as shown in the diagram.

a) Record the volume of water in your table. Remember to consider the precision of your measurement when recording your data.

4. Measure the mass of the graduated cylinder and 10 mL of water.

   a) Record the measurement in your log.

   b) Calculate the mass of the water and record this in your table.

5. Add another 10 mL to the graduated cylinder and measure the mass. Calculate the mass of 20 mL of water.

   Repeat this step for 30 mL, 40 mL, 50 mL, and so on up to 100 mL.

   a) Record all your measurements and calculations in the table in your log.

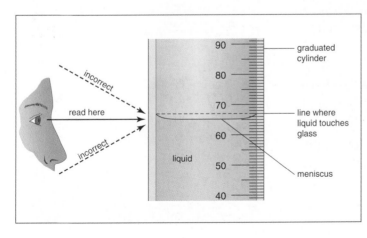

6. Use the data you obtained.

   a) Plot a graph of the mass versus the volume of water. Plot volume on the *x*-axis (horizontal axis) and mass on the *y*-axis (vertical axis).

   b) As the volume of the water increases, what happens to the mass?

   Since the graph you created is a straight line (or close to a straight line), you should draw the best fit line through the data points. Do not connect the points with small segments but draw one line that comes closest to all of the individual points.

   c) From your graph predict the mass of 55 mL of water. What would be the volume of 75 g of water? Predicting values from within a graph is called interpolation.

   d) An important attribute of a straight line graph is its slope. How steep is the graph? Calculate the slope of the graph you plotted. Remember to calculate slope you divide the "rise" by the "run." What does the "rise" of the graph represent? What does the "run" represent?

   e) Divide the mass of each sample of water by the volume. What do you notice about the relationship between the mass and the volume?

   How does the slope of the graph compare to the values you calculated in this step?

7. Your teacher will provide you with a sample of a liquid.

Use the procedure you used to find the mass and corresponding volumes of water to determine the slope of this liquid's mass/volume graph.

   a) Record all your data and calculations in your *Active Chemistry* log.

8. Dispose of your liquid sample as directed by your teacher. Clean up your work station.

Part B: Mass and Volume of Solids

1. Your teacher will provide you with three samples of two different solid materials.

2. As a group, decide on a procedure to calculate the mass/volume ratio and slopes of the graph of each material.

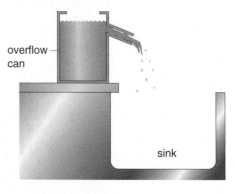

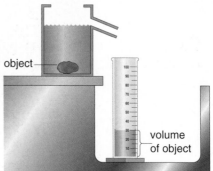

You can consider using either method shown in the diagrams on the preceding page and at right for measuring the volume of each solid. Volume of solids is usually expressed in cubic centimeters. One milliliter is equivalent to one cubic centimeter $(1 \text{ mL} = 1 \text{ cm}^3)$.

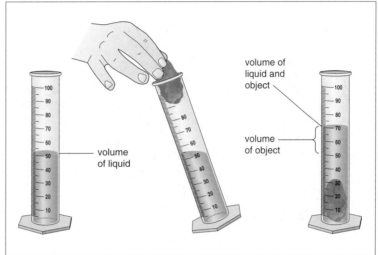

a) Record your procedure in your *Active Chemistry* log. Be sure to include what measurements you need to make, what equipment you will need, what safety precautions you must use, and what calculations you have to do.

3. When your teacher has approved your procedure, carry out your activity.

a) Carefully record all your data.

4. Use the data you collected.

a) Plot a mass versus volume graph for each solid. Plot both solids on the same graph.

b) How do the slopes for the two solids compare? Which solid is more dense?

c) Use the table in the **ChemTalk** reading section to identify the two samples of materials.

5. The mass of a unit volume of a material is called its density. You found the density of water by calculating the slope of the mass versus volume graph. You can also calculate density by dividing the mass of a sample of a material by the volume.

$$\text{Density } (D) = \frac{\text{Mass } (m)}{\text{Volume } (V)}$$

a) Find the densities of water, the other liquid, and the two solid materials.

b) Compare your answers with another lab group.

Part C: Density and Special Effects

1. Your teacher will display a set of four colored liquids that float on one another. The densities of each of the liquids were measured. The top layer has a density of 0.8 g/mL. The next layer has a density of 0.9 g/mL. The following layer has a density of 1.1 g/mL. The bottom layer has a density of 1.3 g/mL.

a) What do you notice about the densities of the liquids and their position in the display?

353

2. Your teacher will drop an ink-pen barrel into the liquids.

   a) What would you predict will happen to the ink-pen barrel? Write your prediction in your *Active Chemistry* log.

   b) Observe the movement of the ink-pen barrel as your teacher places it in the liquid. Record your observation in your log.

3. You will now make an ink-pen barrel float in liquid.

Place the ink-pen barrel in a beaker of ethanol.

Next place the ink-pen barrel in a beaker of distilled water.

Return the ink-pen barrel to the beaker of ethanol. Slowly add distilled water to the ethanol until the ink-pen barrel floats.

   a) Before you begin, predict what you think will happen in each part of this step. Give a reason for your prediction.

Chem Talk

DENSITY

Density as a Property of Matter

If you were to compare a 1 cm³ cube of iron to a 1 cm³ cube of wood, you would probably say that the iron is "heavier." However, if you compared a tree trunk to iron shavings, the tree trunk is obviously heavier. As you discovered in this activity, a "fair" comparison of the "heaviness" of two materials is a comparison of their densities. **Density** is the mass per unit volume of a material. In this activity you measured the density of water and other liquids. You found that each sample of the same liquid had the same density and each different liquid had its own characteristic density. You also found that each solid material you investigated had its own characteristic density. Density can be expressed in grams per milliliter (g/mL) or grams per cubic centimeter (g/cm³). The table on the next page shows the densities of some common liquids and solids.

You used the slope of the mass versus volume graph of a material to calculate density. You also calculated density using the equation:

$$\text{Density } (D) = \frac{\text{Mass } (m)}{\text{Volume } (V)}$$

Chem Words

density: the mass per unit volume of a material.

Density and Flotation

In this activity you further observed that materials with a greater density than a given liquid will sink, and materials with less density than a given liquid will float. In the column of colored liquids, the liquid with the highest density was on the bottom, and the liquid with the lowest density was on the top. The ink-pen barrel sank in ethanol and floated in water. When you added ethanol to the water you created just the right density to have the ink-pen barrel float within the liquid. This position of floating is where the density of the ink-pen barrel is equal to the density of the ethanol/water. The ink-pen barrel "found" the place where the density of the liquid was identical to the density of the ink-pen barrel.

Approximate Densities of Some Common Liquids and Solids	
Material	Density (g/cm³)
wood (balsa)	0.12
wood (birch)	0.66
gasoline	0.69
isopropanol	0.79
vegetable oil	0.92
distilled water	1.00
glycerol	1.26
magnesium	1.70
aluminum	2.70
iron	7.90
copper	8.90
nickel	8.90
silver	10.50
mercury	13.50
gold	19.30

The most famous story about density is when Archimedes jumped out of the bath, ran through the town naked, and shouted "Eureka!" As the story supposedly goes, Archimedes was asked by the king to determine if his crown was solid gold. Archimedes knew the density of gold. He also knew that he could correctly determine if the crown were gold if he knew the density of the crown. The mass of the king's crown was easy to measure. The volume posed a real problem because it had such an unusual shape, and of course the king did not want his crown altered. When Archimedes submerged himself in the bathtub, he realized that the displacement of water would provide him with the volume. *Eureka* is Greek for "I found it."

MAKING MEASUREMENTS AND USING THE MEASUREMENTS TO MAKE CALCULATIONS

Uncertainty of Measurements

Every measurement that you make involves some uncertainty. When you measured the volume of water using a graduated cylinder, you used the division of units marked on the side of the cylinder to make your measurements. Suppose the smallest precision division marked on the graduated cylinder was a milliliter. This means that you can estimate the measure to the nearest tenth of a milliliter, because you can see if the level of the water is at, above, or below the mark. When you record your measurement of volume, you can record it as 10.0 mL, because you can see whether the level of the water is at, above, or below the 10 mL mark.

Remember to always look at the instrument that you are using and determine the smallest precision mark it has. When you make your measurement using the instrument you can only estimate to the next place. If you are using an electric balance to measure mass, it will do the estimating work for you. Most school balances will measure to the tenth, or the hundredth of a gram.

Calculations

When you perform calculations using the measurements that you made in an investigation, you need to express the result of your calculations in a way that makes sense of the certainty of the measurements you made. For example, when calculating the density of a 10.1 mL sample of liquid with a mass of 9.8 g, you may obtain a value of 0.9702770... g/mL using a calculator. This value does not seem reasonable when considering limitations of your measurements.

There are rules that you can use when making your calculations:

Adding and Subtracting

When adding or subtracting numbers, arrange the numbers in columnar form lining up the decimal points. Retain no column that is to the right of a column containing a doubtful digit.

Multiplying and Dividing

In multiplication and division, the result should have no more significant digits than the factor having the fewest number of significant digits. To determine how many digits are significant, count all the digits excluding zeroes at the beginning or end (e.g., 0.00326 and 71800 each have 3 significant digits). (Exception: a zero at the end of a number is significant if the number contains a decimal point — for example, 0.00326 has 3 significant figures, 0.003260 has 4, and 3260 has 3.)

Checking Up

1. Explain the meaning of density.

2. Explain the difference between feathers and lead, using the concept of density.

3. Why is balsa instead of birch wood used in the construction of model airplanes?

4. In an investigation the volume of a material is measured as 80.0 cm³ and its mass is measured as 253 g. Which calculation of density correctly uses the precision rules: 3.1625 g/cm³, 3.163 g/cm³, 3.16 g/cm³, 3.2 g/cm³, or 3 g/cm³?

Reflecting on the Activity and the Challenge

In this activity, you discovered that the ratio of mass to volume (m/V) is a special number associated with each material. The ratio m/V is called the density and is a characteristic property of matter. You can identify whether a piece of metal is gold or gold-plated by measuring the density. You can distinguish one material from another by comparing densities. Objects of greater density than a liquid will sink, while objects of lesser density will float. Objects of the same density will appear suspended. You can make use of these conclusions in your challenge. For instance, you may want to have a movie special effect where a material appears suspended in space. The concept of density will help you create this special effect. You can compare the density of different materials and decide which materials will float and which materials will sink, and also how to make them appear to be suspended somewhere between the top and bottom of a liquid.

Chemistry to Go

1. Look at the table in the **ChemTalk** reading section. Use density to identify the liquid and solid samples you investigated in this activity.

2. Calculate the density of a solid from the following data:

Volume of water	48.4 mL (or cm^3)
Volume of water and solid	62.7 mL (or cm^3)
Mass of solid	123.4 g

3. Determine the density of a liquid from the following data:

Mass of the graduated cylinder	33.79 g
Mass of the cylinder and liquid	40.14 g
Volume of liquid	13.3 mL

4. Methanol has a density of 0.79 g/mL. How much would be the mass of 589 mL of methanol?

5. Copper has a density 8.90 g/cm^3. What would be the volume of a 746 g sample of copper?

6. In a well-known movie, there is a famous scene in which the hero tries to outwit the designers of a trap by replacing a gold statue with a bag of sand of about the same volume.

   a) Given the density of gold is 19.3 g/mL and sand is 3.1 g/mL, does this seem like a scientifically reasonable plan?

   b) In the movie, the hero grabs the gold statue with one hand and appears to handle it quite easily. Given that the volume of the statue appears to be about 1 liter, what would be the mass of the statue?

   c) A mass of 454 g has a gravitational weight of about 4.45 N (newtons) which is about 1 lb. How many pounds would the statue weigh?

   d) One gallon of milk has a mass of 3.7 kg and a weight equivalent of approximately 8 pounds. How many gallons of milk would be equivalent to the gold statue?

7. In each of the following pairs, which has the greater mass?

   a) 1 kg lead or 1 kg feathers?

   b) 1 L gold or 1 L water?

   c) 1 L copper or 1 L silver?

8. Which of the following has the greater volume:

   a) 1 kg lead or 1 kg feathers?

   b) 1 kg gold or 1 kg water?

   c) 1 kg copper or 1 kg silver?

9. Review the measurements you made for mass and volume. How certain were your measurements? If you were to make the measurements again, could you be more certain? Explain your answer.

10. In calculating density you divided the mass of the material by the volume. Review the calculations you made. Adjust the accuracy of your answers using the rule for division given in the **ChemTalk** reading section.

Preparing for the Chapter Challenge

Design a special effect in which an object is suspended in a liquid. Consider the density of the material you will suspend, and the density of the liquid you will use. Show the calculations that you used to make your choice of materials.

Inquiring Further

1. Is it real gold?

The new United States dollar coin has a golden color. Could it be made of real gold? Devise a method to determine if the new golden coin is any of the metals in the table in the **ChemTalk** reading section.

2. Density of gas

Devise an investigation that you could do to determine the density of air or gas.

Activity 6 Metals and Nonmetals

GOALS

In this activity you will:

- Observe some chemical and physical properties of various materials.

- Classify the materials as metals or nonmetals.

- Identify the metals that make up common alloys and learn about some special properties and uses of the alloys.

- Make generalizations about the properties that differentiate metals from nonmetals.

- Explore how heat treatments can alter the properties of metals.

What Do You Think?

If you look around your house you will see hundreds of objects made from dozens of kinds of materials. Have you ever wondered why the manufacturer chose the materials they did for each item?

- **Why are frying pans made of metal and baking dishes often made of glass or ceramic?**

- **Could a baking dish be made of metal? Could a frying pan be made of glass or ceramic?**

Record your ideas about these questions in your *Active Chemistry* log. Be prepared to discuss your responses with your small group and the class.

Investigate

1. Your teacher will provide you with samples of a number of materials: iron, copper, zinc, magnesium, tin, aluminum, brass, solder, chalk, graphite or charcoal, wood, glass, plastic, and concrete.

Choose two of the materials that are obviously different. In your group, brainstorm at least five characteristics or properties of each material.

a) In your *Active Chemistry* log, use the characteristics to describe each material. Could someone else reading your log be able to identify the material?

2. Chemists use specific characteristics or properties to describe and distinguish among materials. You investigated some of these properties in **Activity 4** and **Activity 5**. Additional properties used by chemists are described below.

a) In your log, prepare a table for recording your observations.

3. **Luster:** Is the material shiny or dull in appearance? Does it look more like a mirror or more like mud? If a material has lots of luster it reflects light and you may be able to see images reflected from the surface. Polished metal has high luster. Dull surfaces don't reflect as much light. They appear flat and no images can be reflected.

a) In the table in your log, record whether the material has a high or low luster.

4. **Electrical conductivity:** Test each substance with a conductivity tester or multi-meter. To test each substance, touch the two leads to each end of the sample. Do not allow the leads to touch each other, or it will give a false reading. In the example shown in the diagram, the bulb will glow if the material is conductive. Your teacher will demonstrate how the specific conductivity tester used in your lab works.

a) In the table in your log, record whether the material is conductive. You can use words to reflect this like nonconductive, slightly conductive, or very conductive. If you are using a meter, you can record the reading.

Safety goggles and a lab apron must be worn.

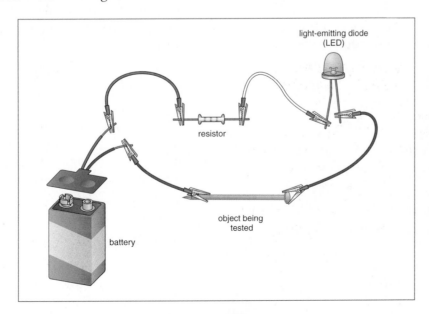

5. **Malleability:** Wrap the material being tested in heavy plastic or a cloth to prevent pieces from flying off the sample. Place the material on a hard, flat surface. Using a hammer, try to pound the material flat. If the sample can be pounded into a flatter shape it is called malleable. If it breaks or doesn't change it is called nonmalleable.

   a) In the table in your log, record whether the material is malleable or nonmalleable.

6. **Reactivity:** Try scraping or sanding a small part of each sample. Is the surface underneath the same in appearance or different? If the surface is different, that means the sample has reacted with the air.

   a) In the table in your log, record whether the material is highly reactive, slightly reactive or nonreactive.

7. **Ductility:** Ductility refers to how easily the substance can be pulled out into a wire or how bendable it is. Try bending each piece to determine how ductile it is.

   a) In the table in your log, record whether or not the material is ductile.

8. **Color**

   a) Record the color of each sample material in the table in your log.

9. Once you have completed the table, compare your list of characteristics of each substance with those recorded by the other students in your class. Have each member of your group pair off with a student from another group. If there is a difference in the results go back to the material and review your observations until everyone agrees on the most accurate list of properties for each material.

   a) Be sure to record any changes you make in your log.

10. Classify the substances into two groups. Use any property you have observed to divide the samples into groups that have the most in common. For example, you could divide the materials into those that do have a luster and those that do not.

    a) Record your classification of the materials in your log.

11. Metals have a shiny and lustrous surface. They conduct electricity and heat. They are malleable and ductile and they are often relatively reactive. Nonmetals have characteristics that are generally opposite to metals in every way. Instead of being lustrous their surfaces are dull in appearance. They are nonconductive, brittle, and nonductile. Now separate your samples into metals and nonmetals. If you have any samples that do not fit clearly as either a metal or a nonmetal, set them aside.

    a) Make a list of the samples in each category in your *Active Chemistry* log.

Chem Talk

METALS, NONMETALS, AND METALLOIDS

In this activity you investigated specific properties of materials. You then used your observations to classify a material as a **metal** or a **nonmetal**. Metals have **luster**. They exhibit **conductivity**. They conduct electricity and heat. They are malleable and **ductile** and they are often relatively **reactive**. Many metals form a compound on their surface that results from reactions with air. When you scrape or sand a piece of metal you are removing that coating of metal compound. Sometimes that natural coating can prevent further reacting and will preserve the metal underneath.

Looking at the drawing at the right, you can see that in solid copper metal the centers of the copper atoms are in fixed locations but they are surrounded by a sea of electrons. If an electric circuit is set up, the electrons are free to move. This is the basis of the metallic property of electrical conductivity.

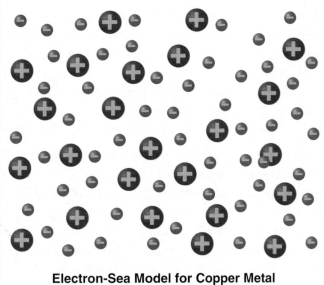

Electron-Sea Model for Copper Metal

Chem Words

metal: classes of materials that exhibit the properties of conductivity, malleability, reactivity, and ductility. Metal elements readily lose electrons to form positive ions.

nonmetal: elements that do not exhibit the properties of conductivity, malleability, reactivity, and ductility. These elements tend to form negative ions. The oxides of the elements are acidic.

luster: the reflection of light from the surface of a material described by its quality and intensity.

conductivity: the property of transmitting heat and electricity.

ductility: a property that describes how easy it is to pull a substance into a new permanent shape, such as, pulling into wires.

reactivity: a property that describes how readily a material will react with other materials.

Coordinated Science for the 21st Century

On the other hand, silicon dioxide is an amorphous solid; you know it as glass. In glass, electrons are fixed into position and are held tightly by each atom due to covalent bonding (sharing of electrons) between silicon and oxygen atoms. Since the electrons are not mobile, the glass does not conduct an electric current like copper metal does. Glass is a nonconductor of electricity.

Preserving metal and preventing its reaction with some of the components in the air is a major task. When metals react with oxygen in the air it is called **oxidation**. This type of reaction is what happens when things rust. Preventing rust is important. While a metal like steel is very strong and makes excellent building material, once it rusts it loses all strength and flakes away. Millions of structures, tools, and vehicles are made primarily of metal. Preventing rust (also called corrosion) is essential if they are to remain in good operating condition. In order to prevent oxidation, metal surfaces can be painted, coated, or combined with another metal to make them less reactive.

Nonmetals have characteristics that are generally opposite to those of metals in every way. Instead of being lustrous, their surfaces are dull in appearance. They are nonconductive, brittle, and nonductile. Over the past 150 years, chemists have developed a chart for classifying and organizing the chemical elements. Elements are classified as metals and as nonmetals. Some other elements are called metalloids. They share some characteristics of metals and some of nonmetals.

Brass and solder are not elements but they are still classified as metals. They are commonly called **alloys**. Alloys are materials that contain more than one metal element and still maintain the characteristic properties of metals. Many metals are not practically useful because they may be too soft and are hard to work with. Gold is a good example of a metal that is too soft for jewelers to work with so they make an alloy of gold, silver, and copper. The alloy is harder and will hold its shape. Iron combined with chromium, nickel, and carbon makes the alloy called steel. This gives it the strength that it needs in construction. The brass that you investigated contains 67% copper and 33% zinc. Solder contains 67% lead and 33% tin. Alloys are classified as metal solutions and if they are uniformly mixed then they are homogeneous mixtures called solution alloys.

Chem Words

oxidation: the process of a substance losing one or more electrons.

alloy: a substance that has metal characteristics and consists of two or more different elements.

Checking Up

1. List five properties of metals and five properties of nonmetals.

2. Why is it important to prevent the oxidation of metals used in construction?

3. Explain the meaning of an alloy.

4. Why are alloys used?

Reflecting on the Activity and the Challenge

As you are creating your movie special effect, you may have to build a stage set, model, or prop. It is important to consider the nature of the materials you choose before you start construction. You need to match the characteristics of the material you choose with the object you are trying to build. You need to decide if the building material should be heavy or light, flexible or rigid. Each of the characteristics of the substance is important. The characteristics required will probably vary with each construction project. Whichever materials you do choose, you will strengthen your report by discussing the specific properties of the materials and classifying them as metals, nonmetals, or metalloids.

Chemistry to Go

1. a) List the names of three metals you are familiar with in your daily life.

   b) For each metal you listed in (a), describe two different uses for each.

2. a) List the names of three nonmetals you are familiar with in your daily life.

   b) For each nonmetal you listed in (a), describe two different uses for each.

3. Examine the following objects or machines that you use every day:

   a) backpack
   b) bicycle
   c) car

   Make a list of the metals and nonmetals that make up the major components of the object. List the function and characteristics of each component.

4. a) List two properties of a material that you can observe using your senses.

   b) List two properties of a material that require tests to observe.

5. Classify each of the following as a metal or nonmetal:

   a) aluminum (Al)
   b) iron (Fe)
   c) oxygen (O)
   d) carbon (C)
   e) mercury (Hg)

6. The way materials are used can change with time. Milk was originally delivered in glass bottles. Now cartons made from wax-coated paper and plastic jugs are used for milk. Snow skis used to be made of wood. Now they are made from fiberglass or graphite. What factors go into decisions about changing what materials should be used when building a product?

Inquiring Further

The effect of heat treatment on the property of a metal

It is possible to change the characteristics of a material by treating them in various ways. Determine the effect of heat on a metal. Obtain a few paper clips or bobby pins from your teacher. As a control, determine how many times it takes to bend the clip or pin back and forth in order to break it. The stress at the point of bending causes the paper clip or bobby pin to break.

Now try some various types of heat treatment to see what effect they have on the metal. Try heating the piece in the flame of a burner until it is red. To do this, hold the piece with tongs or forceps. Allow the piece to cool on its own until it is safe to handle. Then try to break it again, being careful to bend it back and forth exactly as you did with the control. Record your results and try another heating scheme. Consider treatments such as heating to redness and then cooling by plunging into water, or heating and cooling several times, or heating, but not to redness, only heating to a moderate amount.

Record your observations and summarize what seems to be the relationship of heat treatment to the characteristics of the metal.

What characteristic of metals do you think is demonstrated by bending the metals back and forth until breaking?

Activity 7   Polymers

GOALS

In this activity you will:

- Make a polymer-based material that has properties different from other states of matter that you have studied.

- Observe the material's properties and compare them to those of solids and liquids.

- Describe the process of cross-linking in polymeric materials.

- Discuss and invent new commercial uses for water-soluble polymers.

- Compare the viscosities of two non-Newtonian fluids.

What Do You Think?

In ancient times, humans made most of the objects they needed for daily life from natural substances. Relatively little processing was done between harvesting and use. For much of history, people have used metals, cotton, wood, and other natural materials for building, clothing, and tools. In the 1900s things changed dramatically as scientists were able to create new materials.

- **What materials are polyester and rayon made from?**

- **How are they manufactured?**

Record your ideas about these questions in your *Active Chemistry* log. Be prepared to discuss your responses with your small group and the class.

Investigate

1. Measure 50 mL of polyvinyl alcohol (PVA) solution into a beaker. Observe closely the properties of the PVA solution. Refer to the different type of properties you investigated in previous activities to make your observations.

367

a) Record your observations in your *Active Chemistry* log.

2. Measure 10 mL of sodium tetraborate (borax) solution into another beaker. Observe the properties of borax.

a) Record your observations in your *Active Chemistry* log.

3. Add one drop of food coloring to the sodium tetraborate solution.

4. Add the sodium tetraborate solution to the PVA solution while stirring with a wooden stick. Keep stirring until the mixture thickens.

When the mixture has thickened, remove the stick. Place a few paper towels on your desktop and use your hands to remove the thickened mixture.

Mold and stretch the new material while you observe its characteristics.

a) Record your observations in your log.

5. Test your new "slime" and see if it behaves more like a solid or more like a liquid. Try holding the slime in your fingers and dangling it downward. Wait for a few minutes and record what happens.

Place your slime back in the beaker and see what happens as it sits for a few minutes.

Try pulling the slime out slowly and see what happens. Now try the same thing but pull quickly.

Roll the slime into a ball and try bouncing it gently on the tabletop.

a) Record all your observations in your log.

6. It was not difficult to label the PVA solution as a solid or a liquid. Similarly it was easy to label borax. Liquids have very different characteristics from solids. Liquids flow while solids have a rigid shape. Liquids assume the shape of their container and solids have their own shape. Slime blurred the line between the definitions. Liquids spread out when force is applied, but solids break. Liquids splatter when dropped, while solids may bounce.

a) In what ways does your slime behave like a liquid?

b) In what ways does it behave like a solid?

7. The PVA molecules are relatively long, slender molecules. The sodium tetraborate molecules are shorter and can form bonds on both ends.

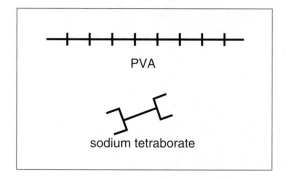

PVA

sodium tetraborate

a) Draw a sketch to represent how the substances act in this reaction. Use the model in the diagram to draw PVA molecules alone, and then draw sodium tetraborate molecules alone. Finally, draw the reaction of sodium tetraborate cross-linking the PVA molecules. Remember your drawing is just a representation of the molecules.

Chem Talk

POLYMERS

In this activity you made "slime." This substance is a **polymer** that has unique characteristics. It is classified as a non-Newtonian liquid. Liquids resist flow. This phenomenon is known as **viscosity**. Newton devised a simple model for fluid flow. You will learn more about this model in later chemistry and physics courses. Liquids like water and gasoline behave according to Newton's model. They are called Newtonian fluids. Ketchup, blood, yogurt, gravy, pie fillings, mud, and slime do not follow the model. They are classified as non-Newtonian liquids.

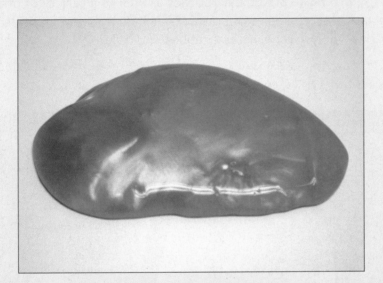

Slime has some characteristics of liquids such as being fluid and taking the shape of its container. On the other hand it bounces and breaks when pulled quickly, as solids do. The unique characteristics of slime are due to the two substances that make it up. The borax solution forms cross-links among the polyvinyl alcohol molecules. These cross-links make the resulting polymer slower to flow and change shape.

The formation of very large molecules from many smaller molecular units is called **polymerization**. Polymers are materials that are made up of many similar small molecules, called monomers, linked together in long chains. These materials have "always" existed in nature, but they have only been produced by industrial processes in the 20th century.

Chem Words

polymer: a substance that is a macromolecule consisting of many similar small molecules (monomers) linked together in long chains.

viscosity: a property related to the resistance of a fluid to flow.

polymerization: a chemical reaction that converts small molecules (monomers) into large molecules (polymers).

Proteins are natural polymers. They are the basic structural unit of plants and animals. There are more than ten thousand different proteins found in nature, yet they are all made up of combinations of about 20 different amino acids. Starch and cellulose are examples of carbohydrates that are polymers. The human digestive system is able to break apart the bonds that form starch molecules, releasing glucose, which the body uses as a source of energy. Humans cannot digest cellulose.

Polyethylene (polyethene) is a polymer made up of many ethane molecules. It is one of many chemicals manufactured from crude oil. It is used to make plastic milk bottles as well as a variety of other objects. Other examples of polymers include polyvinyl chloride, Plexiglas®, polystyrene, natural rubber, and Teflon®.

Checking Up

1. Describe polymers and polymerization in your own words.

2. Name two natural-occurring polymers.

3. Give examples of five polymers that are manufactured using technological processes.

Reflecting on the Activity and the Challenge

What monster movie is complete without slime? Now you know how to make slime. Supplies for making slime are relatively expensive. Keep costs in mind as you plan for your special effect.

Chemistry to Go

1. List three ways you could modify the properties of your slime. If the slime you made in this activity is not quite correct for the special effect you have in mind, how might you modify it for use in your movie?

2. List five other uses that might be possible for the slime substance, in addition to its use as a movie special effect.

3. Thin sheets of solid PVA slime have been used in several new commercial products. Since the sheets are soluble in water, they are ideal recyclable containers. Seed manufacturers have made long tapes of PVA sheets with seeds imbedded inside. When the tapes are planted in the ground the PVA dissolves and the seeds are free to grow. Another use is collecting dirty laundry in hospitals. The solid laundry is placed into PVA bags. When it is time to wash the clothing the entire bag is thrown into the washing machine. It is not necessary for workers to handle the dirty clothes again. What two other uses can you think of for PVA sheet material?

Inquiring Further

1. Self-siphoning slime

Spread your slime on a piece of plastic wrap or waxed paper and let it dry completely. This may take a day or two. Compare the characteristics of the dry slime to the wet slime.

PVA slime has been reported to be "self-siphoning." To siphon most substances, a tube is placed in a liquid and the siphon is started by sucking the liquid out of the container, over the edge and down the tube. The mass of the water flowing out of the tube creates a pressure difference and draws the remainder of the liquid with it. Slime is supposed to have the same effect, but without the tube! Try demonstrating this self-siphoning effect with your slime.

2. Viscosity of non-Newtonian liquids

Ketchup is also classified as a non-Newtonian liquid. Determine which non-Newtonian liquid is the most viscous (resistance to flow).

- Set up a retort stand with a ring to hold a long-stem funnel.

- On the funnel place two marks with a pen at about 2 cm apart.

- Pour one of the liquids into the funnel and allow it to flow through the stem into a beaker.

- Let the liquid level come down to your top mark on the side of the funnel and stop the flow with your finger.

- Using a stopwatch, time how long it takes for the liquid to flow before it reaches the second line after you remove your finger.

371

Activity 8   Identifying Matter

GOALS

In this activity you will:

- Produce colored flames.

- Identify the metal ions present in materials by the colors of light a material gives off when held in a flame.

- Describe how atoms create the colored light.

- Investigate ways of producing new colors not among those produced by the materials you test.

- Practice safe laboratory techniques in working with laboratory burners.

What Do You Think?

You may have read how royalty, suspicious of enemies, had their food tested for poison by having someone else taste it first. That's not a job anyone would enjoy!

• How else could you test for a poisonous substance?

Record your ideas about this question in your *Active Chemistry* log. Be prepared to discuss your response with your small group and the class.

Investigate

1. Your teacher will supply you with seven numbered wooden splints that have been soaked in various solutions. Some are harmful, and some are not.

   a) How could you distinguish one splint from another?

2. Light your Bunsen burner. Immediately make observations about the color of the burner flame.

Safety goggles and a lab apron are required for this activity.

a) Record your observations in your *Active Chemistry* log.

3. Take one of the wooden splints and hold it in the hot part of the flame using forceps, tongs, or a fireproof glove. Note the color of the flame when the solution-soaked wooden splint is heated.

As soon as the flame from the splint is no longer strongly colored, extinguish the splint by placing it in a beaker of water.

a) Organize a table to record your data.

b) Record your observations in your table.

4. Repeat **Step 3** for the remaining six splints.

a) Record all your observations.

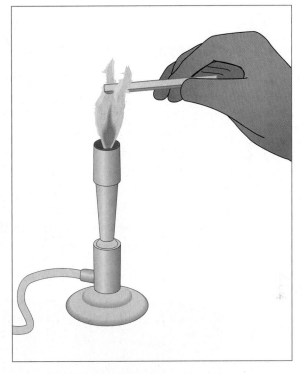

5. Refer to your data table. The salt solutions in which the splints were soaked are metal compounds of the elements lithium, barium, sodium, strontium, potassium, and calcium. One splint was soaked in water.

a) What do you notice that was different about each splint?

b) Your teacher will identify the solution each splint was soaked in. Record this information in your data table.

6. Obtain another splint soaked in an unknown solution.

Hold the wooden splint soaked in the unknown solution in the burner flame. Observe what happens.

a) Record your observations in your log.

b) Use your observations to decide what metal ions are on the splint. Confirm your results with your teacher.

c) Did you correctly identify the metal ion?

d) In your log describe how you decided what the identity of your compound was, and how sure you are that you were correct.

7. Clean up your work station and return all your equipment as directed by your teacher. Wash your hands thoroughly.

Chem Talk

IDENTIFYING ELEMENTS USING ELECTRONS

The process you observed in this activity is not a chemical reaction. When the metal **ions** are placed in the hot flame their **electrons** absorb the energy and move to higher energy levels around the nucleus. When these electrons fall back to their original level they give off the light you see in the **flame test.** Since each atom has a unique arrangement of electrons, each gives a unique color. This experimental technique is called a flame test, and is the basis for identifying the composition of the salt solutions. This same effect is responsible for the glowing light of neon signs and the burst of color in fireworks. In this experiment only the metal ions are changed; the negatively charged ion (chloride) is kept constant.

The flames produced in this activity have interesting colors. It is not obvious where they come from. The metal compounds are made from atoms that have **nuclei** surrounded by electrons. When electrons fall from a higher energy level to a lower energy level, they give off light. In this case the electrons were able to get to a higher energy level

Chem Words

ion: an electrically charged atom or group of atoms that has acquired a net charge, either negative or positive.

electron: a negatively charged particle with a charge of 1.6×10^{-19} coulombs and a mass of 9.1×10^{-31} kg.

flame test: an experimental technique or process in identifying a metal from its characteristic flame color.

nucleus: the very dense core of the atom that contains the neutrons and protons.

by absorbing heat energy from the Bunsen burner flame. Each compound has a different configuration of electrons and electron energy levels. The change in energy levels represents the color of the light given off. If you refer to the higher energy level as E and to the lower energy level as E_l, then you can state:

The energy of the emitted light is equal to $E_h - E_l$. The amount of energy remains the same as it is conserved. The energy the electron gains from the heat energy of the Bunsen burner flame raises the electron to a higher energy state. When the electron drops down to the lower energy state, it gives off that energy as light energy.

Checking Up

1. Explain what is meant by a flame test.

2. Explain how energy is conserved during the flame test.

Reflecting on the Activity and the Challenge

Most of the common fuels are primarily hydrocarbons, which do not contain metal ions. Therefore, you are most familiar with flames that are yellow-orange or occasionally a light blue. The bright colors of the excited metals are a surprise and tend to look "out-of-this-world."

Colors could play an interesting part in a plot line of a movie script. Perhaps an unusually colored flame could be taken as a mystical sign. A sudden change in the color of campfires could be a sign that danger is near. Consider how you might use flame tests to create unusually colored flames for your movie special effect.

Chemistry to Go

1. Compare the colors you observed in your flame tests to the colors you have seen in fireworks displays. Identify what metals are used in producing the different colors in fireworks.

2. Develop a series of sketches showing how an electron of an atom can give off light. Your first sketch should have the electron in the ground state. The next sketch should show the electron excited when you apply energy to the atom and the third sketch should show how light is emitted when the electron falls to a lower energy state.

Preparing for the Chapter Challenge

Name three new colors of light that could be produced by combining some of the metal salts that you tested in the lab, and identify which metal salts you would combine to produce these colors. If time and your teacher permit it, test your predictions.

Inquiring Further

Fireworks

Fireworks have been around for centuries. They were used long before anyone knew why they produced the stunning effects they do. Investigate the manufacturing of fireworks and find the specific compounds used to produce the many colors presented during fireworks displays.

Are all colors equally represented in fireworks displays, or are some colors easier to obtain than others? Research what colors are available and which, if any, do not seem possible given the range of substances readily available.

Activity 9 Organic Substances

GOALS

In this activity you will:

- Combust a material present in fruit rinds.

- Represent the combustion of hydrocarbons as chemical equations.

- Learn the formulas and proper names for simple hydrocarbons.

- Make two-dimensional drawings showing the chemical bonding structure in simple hydrocarbons.

- Classify materials as organic or inorganic.

- Explore various definitions of the term "organic."

What Do You Think?

At a grocery store or market, "regular" apples might cost about 79 cents per pound, while "organic" apples may cost as much as or even more than $1.93 per pound.

- **Why do people buy organic apples if they are so much more expensive?**

- **Does organic mean one thing to shoppers and another to chemists? Explain your answer.**

Record your ideas about these questions in your *Active Chemistry* log. Be prepared to discuss your responses with your small group and the class.

Investigate

1. Obtain a piece of orange peel from your teacher. Make observations of its texture, color, smell, and any other characteristics you notice.

   a) Record your observations in your log.

2. Take one piece of the orange peel and hold it in your hand between your thumb and index finger. Fold the orange peel in half.

Using forceps or tongs, hold it with the smooth fold pointing away from you.

Hold the surface of the orange peel near the flame of a candle. Pinch or squeeze hard.

Repeat using fresh pieces of orange peel.

a) Record your observations in your *Active Chemistry* log.

3. Repeat **Step 2** with other types or fruit and vegetable skins. Use pieces that are the same approximate size and shape as the orange peel.

a) Record your observations in your log.

4. Organic chemistry is the study of molecular compounds of carbon. The gas and liquid that you ignited in the previous steps are carbon compounds. They belong to a class of organic compounds called hydrocarbons.

That means it is a compound composed only of carbon and hydrogen.

a) Draw what the molecules of some simple hydrocarbons might look like. Each carbon molecule can form four chemical bonds. Each hydrogen can form one bond. Carbon can link to another carbon atom or to hydrogen. Draw structures for hydrocarbons containing one through 10 carbons, adding hydrogen as necessary to fill the bonds. The first two hydrocarbons are done as examples.

$$
\begin{array}{ccc}
 & H & \\
H-\!\!\!\!\underset{\substack{| \\ H}}{\overset{H}{C}}\!\!-H & \qquad &
\end{array}
$$

b) In the case of ethylene, a chemical that stimulates the ripening of fruit, the carbons share two bonds between them. Write some formulas for carbon compounds containing two, three and four carbons that contain at least one carbon—carbon double bond. Remember that each carbon can still only have a total of four bonds. If it shares two bonds with another carbon atom that leaves only two others to share.

Chem Talk

THE CHEMISTRY OF MOLECULAR COMPOUNDS OF CARBON

Chemistry attempts to explain macroscopic events through an understanding of microscopic phenomenon. Chemists rationalize how the world they see works, by describing how things they cannot see function. In this activity you found evidence of a gas. This gas is called ethylene and is responsible for the ripening of fruits.

As plants age, they release ethylene gas (C_2H_4) which acts as a plant hormone to begin the ripening process. Green fruit can be made to ripen faster by bathing it in ethylene gas. Absorbing the gas from fruit storage areas can stop the ripening process. This activity demonstrated the presence of the ethylene gas, which is flammable, in the skin of the fruit.

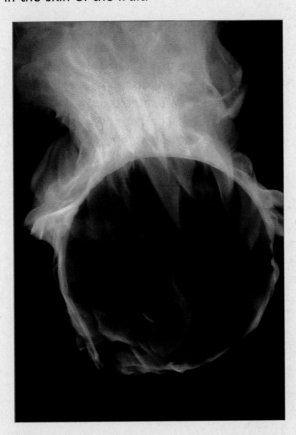

Ethylene is one of many materials classified as organic. To a chemist the term organic means "based on carbon molecules." Everything that is not organic is termed inorganic. In the very early 19th century, compounds were classified into two categories: **organic**, which were those compounds that came from living (or once living) organisms; and **inorganic**, which were obtained from mineral sources. A theory, called "vitalism," proposed that a "vital force"

Chem Words

organic compound: a molecular compound of carbon.

inorganic compound: a compound not based on molecular compounds of carbon.

379

was required to produce organic compounds. This theory was disproved when an organic compound was synthesized in the laboratory. Today, organic chemistry refers to the study of carbon compounds. This is the largest area of study in chemistry. In the natural world alone, plants and animals synthesize millions of carbon compounds. Many manufactured chemicals are facsimiles of these natural products. Chemists have also invented processes that synthesize similar, new chemicals for use in society and technology. Examples of synthesized chemicals include gasoline, artificial sweeteners and flavors, and a variety of medicines.

The popular use of the word, "organic" implies totally natural or not human-made. Chemists do not use the word in this way, and would consider a molecule organic if it came either from a test tube or a tomato plant. They would also argue that all apples are organic, no matter how they are cultivated.

Ethylene belongs to a class of organic compounds called **hydrocarbons**. That means it is a compound composed only of carbon and hydrogen. Most fuels, such as gasoline, kerosene, candle wax, propane, and natural gas are hydrocarbons too. The burning of a hydrocarbon in the presence of oxygen is an example of **combustion**. In combustion, a substance is combined with oxygen and accompanied by the generation of heat and sometimes light.

Chem Words

hydrocarbon: a molecular compound containing only hydrogen and carbon.

combustion: the rapid reaction of a material with oxygen accompanied by rapid evolution of flame and heat.

Checking Up

1. What does a chemist mean by an organic compound?

2. How did the meaning of "organic" change during the 19th century?

3. What is a hydrocarbon?

Reflecting on the Activity and the Challenge

When you held the orange peel near the flame and squeezed it, you forced a natural material made by the orange to be released. The material that comes from the skin of the orange peel is a hydrocarbon fuel. It is similar to the fuel used in barbecues and portable lighters. This material burns very quickly and is quite like the fuels used in movies to make huge explosions and fires. Of course, explosions and fires can be quite dangerous. Often, movie directors will film an explosion on a small scale, by using models, and then make it appear huge on the screen. Think of some ways you could use this very small explosion to represent a larger effect on the screen.

Chemistry to Go

1. Write two or three examples of the use of the word organic to mean natural.

2. Classify each of the following materials as inorganic or organic:

salt (NaCl)	sugar ($C_{12}H_{22}O_{11}$)	vinegar (CH_3COOH)
gasoline (C_8H_{18})	sand (SiO_2)	oxygen (O_2)

3. Draw an example of a simple hydrocarbon that contains five carbon atoms, all with single bonds.

4. a) Draw an example of a hydrocarbon compound containing seven carbons that contain at least two carbon—carbon double bonds.

   b) Is there more than one answer possible in **Part (a)**? Explain your answer.

5. Acetylene (C_2H_2) is also a hydrocarbon. The carbons in this case share three bonds between them. Again, remember that a carbon atom can only have a total of four bonds. Draw the structure for acetylene.

6. In addition to chains, carbon atoms can bond together to form rings and other shapes. Draw a hydrocarbon that is made up of six carbon atoms, all with single bonds, that forms a chain.

7. Some products or processes claim to be "chemical free" or say "no chemicals added." Based on your understanding of chemistry, what are the arguments for and against such statements?

Inquiring Further

Organic produce and federal regulations

The federal government has proposed standards for food products bearing the label "organic." The National Organic Standards Board, of the United States Department of Agriculture, has issued the following definition of an organic food:

"Organic is a labeling term that denotes products produced under the authority of the Organic Foods Production Act.

The principal guidelines for organic production are to use materials and practices that enhance the ecological balance of natural systems and that integrate the parts of the farming system into an ecological whole."

- How does this definition of the term "organic" compare with the chemical definition?

- Investigate the other proposed USDA standards and evaluate them relative to the chemical definition of organic.

Movie Special Effects Assessment

Your Chapter Challenge is to create a story line and produce special effects for a low-budget movie. In this chapter, you have investigated chemistry phenomena which can be adapted to create those special effects.

The following tasks are required:

- Write a script for a simple scene in a movie
- Choose a special effect to include as part of your scene
- Write a procedure on how your special effect is done
- Demonstrate the special effect to the "producer"
- Write an explanation of how the special effect works, including the chemistry behind the demonstration

Using more than one chemical principle will strengthen your presentation.

To begin, you should probably review all of the activities that you have completed. You may wish to construct a chart with three columns. The first column would include a brief description of the activity. The second column would include what kinds of special effects might be done with this activity. The third column would include chemistry principles involved in the activity. As you review the chapter in this manner, you should pay particular attention to the section **Reflecting on the Activity and the Challenge**. This section will help you to see the connection between these three elements — the activity, the special effects show and the chemistry concept. You may also use the end-of chapter summary chart **Chemistry You Learned** as a review tool.

There are two different ways in which to continue — you can create a story that will require some special effects or you can decide which special effects you will use and then make up the story. Special effects do not have to include explosions and fire. Some of the best special effects are more subtle (e.g., where something floats in space or when a young person looks old).

If you choose to write the story line first, you may start the creative process by brainstorming ideas in your group. In a brainstorming process, every team member contributes ideas as the thought arises. After brainstorming for 10 minutes, your team can then choose what may be the most valuable ideas and then put together the story. If your team chooses to decide on the special effects first, you can refer to your completed list of the activities. Which activities do you wish to repeat as your special effects? The **Chapter Challenge** states that more than one chemical principle will strengthen your proposal. Will it be possible to use a few activities together in your special effects presentation?

Compromise will be an important ingredient for your success. You may not be able to include the best special effect because of limitations of time, money or knowledge. You may also have to limit your script ideas for some of the same reasons. A tight and well prepared script and special effects show will take into account these

limitations and still be entertaining, interesting and appealing.

Once you have agreed on the script and the special effects, each team member can be assigned one of the remaining responsibilities. Who will write the detailed instructions on how to complete the procedure? Who will demonstrate the special effect? Who will write the explanation of the chemistry behind the demonstration? Writing the procedure will require a closer review of the specific activity that you are using. Writing an explanation of the chemistry will also require you to carefully review the activity with particular emphasis on the **ChemTalk** section.

If there are 4 members of the team, each team member may be responsible for one special effect and its procedure, demonstration and chemistry. Alternatively, one team member may be writing all of the procedures while another team member writes all of the chemistry explanations. However you divide the work, it will be important for each team member to share the contribution with the entire group before the due date. Be sure to leave some time to provide feedback and improve upon each part before the final project is due.

You should practice your demonstration before the presentation. Does every team member know their part? Is every team member aware of the time limitations? Are safety issues noted and taken into account? Review all of the criteria listed at the beginning of the chapter and the grading criteria agreed to by your teacher and class. Create a checklist and perform a final review of your checklist. Grade yourself before your presentation to ensure that your team is heading for a great grade.

Chemistry You Learned

Elements and symbols	**Physical properties of elements and compounds**	**Temperature**	**Properties of metals and nonmetals**
Chemical formulas		**Solutions**	
Compounds	**Phase changes and properties of matter**	**Suspensions**	**Polymers**
Electrolysis and synthesis of water		**Colloids**	**Hydrocarbons**
Chemical properties of elements and compounds	**Sublimation**	**Composites**	**Organic compounds vs. inorganic compounds**
	Kinetic energy and potential energy	**Tyndall Effect**	
		Density of liquids, solids, and gases	
	Freezing points and boiling points	**Instrumental measurements (precision and accuracy)**	

Chapter 7

The Periodic Table

Scenario

Every time you say you like or don't like something, you are putting it into a category. You have probably developed categories for many things in your life. You may have categories for food you eat for breakfast, as opposed to dinner, or for clothes you wear to school, as opposed to at home. Can you imagine what your life would be like if nothing were sorted into categories? What if you went shopping in a supermarket that displayed milk next to shoe polish, next to oranges, next to oatmeal, next to hams, next to orange juice, next to detergent? Where would you look for yogurt, shoelaces, corn flakes, ground beef, lemonade, and soap?

That kind of supermarket display pretty much describes the state of chemistry in the mid-19th century. By then chemists had identified and isolated a large number of chemical elements, but they needed a way to sort them into categories—much as a supermarket groups milk with yogurt, shoe polish with

shoelaces, oatmeal with corn flakes, ham with ground beef, orange juice with lemonade, and detergent with soap.

Like similar items in a supermarket, some chemical elements were recognized to share similar chemical properties. The first chemist to arrange these elements successfully into a pattern according to their properties was the Russian, Dimitri Mendeleev.

One of the things Mendeleev did was to write down everything that was known about each element on a small card. Then he moved the cards around until he got an arrangement that showed the groups of elements with similar properties.

Dimitri Mendeleev

In Mendeleev's time the periodic table was developed as a way to arrange elements according to their chemical behavior. Surprisingly, it then revealed information about the structure of the atoms of those elements as well.

By writing the properties of the elements onto separate cards and arranging them, Mendeleev created a puzzle, and he solved that puzzle when he arranged the first version of what is now known as the Periodic Table of the Elements. The table was independently created at the same time by the German, Julius Lothar Meyer.

Chapter Challenge

Your challenge in this chapter is to develop a game related to Mendeleev's Periodic Table of the Elements.

How the game is played, whether on a table, with cards, on a computer, or with equipment that only you might choose, is up to you. You might even choose to emphasize some aspects of the periodic table over others, or to focus on some types of information presented by the table rather than others. However, you need to keep in mind the criteria you and your teacher establish.

Criteria

How will your game be graded? What qualities should a good game have? Discuss these issues in small groups and with your class. You may decide that some or all of the following qualities should be graded:

- how well the game shows your understanding of the periodic table
- how well the game enables players to learn about the periodic table
- how interesting the game is to play
- how long the game takes to play
- whether the game is sequential or can be continued

Once you have determined the list of qualities for evaluating the game, you and your class should also decide how many points should be given for each criterion. Make sure that you understand all the criteria as well as you can before you begin. Your teacher may provide you with a sample rubric to help you get started.

Activity 1   Organizing a Store

GOALS

In this activity you will:

- Plan the arrangement of the items for sale in a store.

- Analyze trends in the arrangement of the store.

- Relate the arrangement of items in the store to the arrangement of elements in the periodic table.

What Do You Think?

Some supermarkets now sell books, flowers, and prescription drugs in addition to eggs, meat, and cereal.

- **How many different items do you think that a supermarket has in its inventory?**

Record your ideas about this question in your *Active Chemistry* log. Be prepared to discuss your responses with your small group and the class.

Investigate

1. Suppose that you decided to go into the business of opening and running a supermarket grocery store. In your group, brainstorm a list of between 50 and 100 items you would sell at your supermarket.

   A member of your group should volunteer to record the items suggested by all members of the group. Everyone, including the person serving as recorder, should participate in suggesting items to be sold.

   a) Make a map showing the locations of all of the items in your store. Give some thought to what will be at the front of each aisle, and what will be at the back, and how the store will be arranged from left to right.

b) Keep in mind which items you want shoppers to see as they enter the store and which should be near as they approach the cash register.

Would either of these factors alter your arrangement?

c) Consider the arrangement of items going from left to right across your store. Why did you choose to arrange the items that way?

Reflecting on the Activity and the Challenge

Organizing 50 to 100 items in your store is not unlike the problem faced by Mendeleev when he organized about the same number of chemical elements into the periodic table. In the following activities you will learn about the properties of chemical elements that led Mendeleev to arrange the elements the way he did and the information about them provided by the periodic table. This activity, in which you were asked to organize a group of items familiar in your everyday experience, was designed to acquaint you with some of the problems Mendeleev faced in the hope that you can better appreciate what he did. You may wish to build this experience into the game you design.

Chemistry to Go

1. What is the pattern or arrangement in your store's aisles?

2. Choose one aisle in your store. Describe the arrangement of items going from the front of the store to the back of the store. What is the trend (or general drift) in that aisle?

3. A new item is brought into the store — chocolate covered peanuts. Where would you place this item? Provide an explanation for your decision.

4. Your store decides to sell napkins, plates, and other decorations for Thanksgiving. How will you adapt your store arrangement to accommodate these items?

5. You would like people to purchase a certain item because it gives you a big profit. Where would you place it in your store and why?

6. One of the characteristics of Mendeleev's original periodic table was a series of blank spots. Since As and Se didn't have anything in common with Al and Si, but do with P and S, Mendeleev decided there must be a couple of other elements yet to be discovered. He left spaces for them and put As under P and Se under S where they belong. What would such a "blank" correspond to in your store?

Activity 2

Elements and Their Properties

GOALS

In this activity you will:

- Apply ancient definitions of elements to materials you believe are elements.

- Test some properties of several common chemical elements.

- Classify elements as metals, nonmetals or neither.

- Learn to differentiate between chemical and physical properties of materials.

- Organize a table of the elements you tested based on their properties.

- Practice safe handling of corrosive chemicals in the laboratory.

What Do You Think?

Throughout the ages of history, philosophers and scientists have talked about "the elements." Reference to elements is most frequent today in the field of chemistry.

• What is a chemical element?

Record your ideas about this question in your *Active Chemistry* log. Be prepared to discuss your responses with your small group and the class.

Investigate

1. Work individually first and then in your group.

   a) Make a list of four or more substances you use or encounter in your everyday experience that meet your definition of element.

2. The ancient Greeks believed that the four elements were: earth, air, fire, and water.

   The alchemists of the early Renaissance identified three elements: mercury, sulfur, and salt.

   a) Does each of the above "elements" satisfy your definition of an element? Why or why not?

3. Your teacher will provide a series of jars containing several common chemical elements: aluminum, copper, iodine, iron, magnesium, silicon, sulfur, and zinc.

You will investigate the properties of these elements. By observing common properties, you may gain an insight into how an organizational chart can be created for all of the known elements. Observe the sample of the chemical element in each jar (without removing any).

on the apparatus goes on, that means that a complete circuit is created, and an electric current is passing through both the light bulb and the sample of the element in the jar.

It is important to make sure that the part of the apparatus immersed into the elements stays dry and is not contaminated by any of the other elements it has been immersed in. Also, use steel wool to polish the metal strips before you test them.

Safety goggles and a lab apron must be worn during this activity.

Element	Initial observations	Conducts electricity	Reacts with HCl	Metal or nonmetal
aluminum				
copper				
iodine				
iron				
magnesium				
silicon				
sulfur				
zinc				

a) Record your observations in a table. You may wish to use a table similar to the above to record your observations in this activity.

4. One of the properties of the chemical elements on Mendeleev's cards was the ability of the element to conduct electric current.

Insert the terminals of the electrical conductivity apparatus into the jar containing each element. If the light

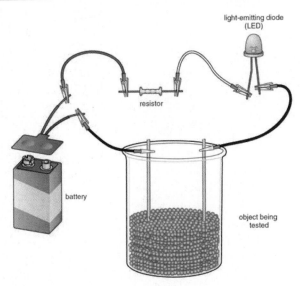

light-emitting diode (LED)

resistor

battery

object being tested

a) Test the samples of each element with the electrical conductivity apparatus and record whether they conduct electric current (yes) or not (no).

5. Another of the properties of each chemical element known to Mendeleev was how it reacts with acid.

You must still be wearing your safety goggles. Pour 5 mL of 1 M hydrochloric acid into each of eight small test tubes. Use a chemical scoop or tongs to remove a small portion of each element from the jar and add it to the hydrochloric acid. It is important to add the hydrochloric acid to the test tube first so that you will not be surprised by a reaction by

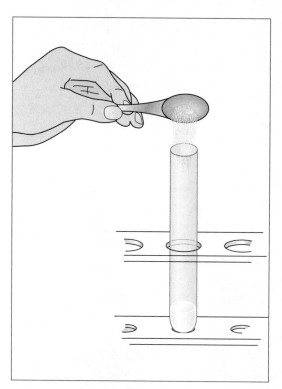

pouring acid over a reactive chemical element. Place a piece of white paper in the background behind the test tube and observe the reaction between the element and hydrochloric acid by looking through the side of the test tube.

a) Test small samples of each element for their reaction with hydrochloric acid and record whether they react with the acid (yes) or not (no).

b) For those that do react, try to determine whether all exhibit the same type of reaction (do they all do the same thing?) and compare the relative vigor of the reactions. If the reaction is vigorous, include a + sign next to your "yes." If the reaction is weak, place a − sign next to your "yes."

6. Dispose of the contents of the test tubes and clean the test tubes as directed by your teacher. Wash your hands.

7. A metal is generally a solid that is shiny, malleable, and a good conductor of heat and electricity. A nonmetal is generally dull, brittle, and is a poor conductor of heat and electricity.

Classify each of the elements you observed as either a metal or nonmetal.

a) Record your observations in the table in your *Active Chemistry* log.

8. Create index cards for each of the elements and find a way to sort them

based on their properties. You may try arranging them and/or color coding them. Your method of sorting will be successful if you can quickly find an element and know from its position whether it:

- conducts electricity
- reacts with HCl
- is metallic or nonmetallic

a) Record your method of sorting the cards in your *Active Chemistry* log.

Chem Talk

PHYSICAL AND CHEMICAL PROPERTIES

Classifying Elements Using Properties

You began this activity by trying to define the meaning of a chemical element. The ancient Greek philosopher Aristotle defined an element as "a body into which other bodies may be analyzed ... and not itself divisible into bodies different in form." The first modern definition of chemical element, which is not much different, is from Robert Boyle: "Bodies, which not being made of any other bodies, or of one another, are the ingredients of which all those ... mixed bodies are ... compounded." We now state that an element is any material that cannot be broken down by chemical means into simpler materials.

Before the mid-19th century, scientists were preoccupied with discovering elements and observing and recording their properties. Then they tried to organize the elements they had discovered in a useful way. At first, they listed the elements alphabetically. However, every time a new element was discovered, the whole list had to be changed. They tried other methods. Could elements be organized by properties like state, color, or taste? None of these methods appeared practical or safe! However, chemists worldwide were sure that elements existed in families that had similar physical and chemical properties. To the Russian scientist, Dimitri Mendeleev (1843-1907), the development of a tool to organize the elements began the same way that so much of science inquiry begins, with a simple question. The question Mendeleev wanted answered was: "What is the relationship of the elements to one another and to the chemical families to which

Coordinated Science for the 21st Century

Checking Up

1. Define a chemical element.

2. What question did Mendeleev use to guide his science inquiry?

3. In your own words, describe the difference between a physical and a chemical property.

they belong?" At that time there were 63 known elements. To help him with his organization, he developed a card game, much the same as you did in this activity. He wrote the properties of each known element on a different card and then spent many hours arranging and rearranging the cards. He was looking for patterns or trends in the data in front of him. Mendeleev, however, had more information on his cards than you presently have. In the following activities, you will look at additional properties of elements that will help you organize your game.

Physical and Chemical Properties

In this activity you observed several properties of the elements you were provided. You probably initially observed the color and the state of the element. You then investigated whether or not the chemical element conducted electricity. You could have also observed the luster, measured the density or the strength, or determined the malleability of each element. In each case, you would not have changed the element itself. In this investigation the element still looked the same in the jar after you removed the electrical conductivity apparatus as it did when you initially inserted it. If measuring a property of a substance does not change the chemical identity of one substance, it is called a **physical property**.

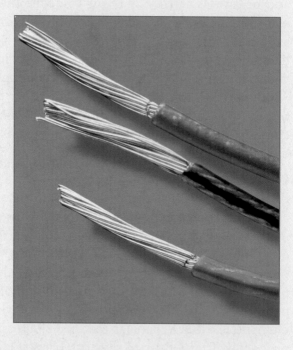

On the other hand, when you observed whether the chemical element reacted with hydrochloric acid, the element clearly changed. A **chemical property** is the kind of reaction that a substance undergoes. Measuring chemical properties changes the chemical composition of a substance.

Chem Words

physical property: a property that can be measured without causing a change in the substance's chemical composition.

chemical property: a characteristic that a substance undergoes in a chemical reaction that produces new substance(s).

Reflecting on the Activity and the Challenge

In this activity you learned not only the definition of a chemical element but also some of the properties of chemical elements. Measuring these properties not only enabled Mendeleev to place the elements in his periodic table but also allowed other chemists to identify the elements. You have tried to sort the cards of elements in the same way that Mendeleev did. Perhaps your periodic table game can have sorting cards as one part of the strategy.

Chemistry to Go

1. Make a list of three more physical properties of a chemical element that you could observe.

2. Make a list of three more chemical properties of a chemical element that you could measure.

3. Why did you want the metals to be clean or polished before you tested them for electrical conductivity?

4. What criteria did you use to differentiate metals from nonmetals in this investigation?

   a) Is this a valid statement of a trend you saw: as the color of the element becomes darker, the element is less metallic? Support your assessment of this statement with evidence that you observed in your investigation.

   b) Is this a valid statement of a trend you saw: the elements react with hydrochloric acid more as you move down a list of the elements listed in alphabetical order? Support your assessment of this statement with evidence that you observed in your investigation.

Preparing for the Chapter Challenge

Prepare a set of index cards for each of the elements with which you are familiar. Record as many properties of each element you know on the card. Use your observations in the following activities and any research you complete on your own to add information to each card.

Activity 3   Atoms and Their Masses

GOALS

In this activity you will:

- Explore the idea of atoms by trying to isolate a single atom.

- Measure how many times greater mass a copper atom has than an aluminum atom.

- Practice careful laboratory technique with measuring masses and filtration.

- Locate sources of the variation in the class's experimental results.

- Compare Dalton's experimental results to the masses of atoms known today.

- See that atoms react in definite proportions of mass when forming a compound.

- Relate the mole concept to real quantities.

- Use scientific notation in calculations.

What Do You Think?

Atoms are the smallest, indivisible part of an element.

- **When did you first hear of atoms? What did they mean to you then, and what do they mean to you now?**

- **Use the following headings: Things I Know about Atoms, Things I Think I Know about Atoms, Things I Would Like to Know about Atoms. Include at least one entry under each heading.**

Record your ideas in your *Active Chemistry* log. Be prepared to discuss your responses with your small group and the class.

Investigate

1. One way to think about an atom is to imagine trying to isolate a single atom from a large number of atoms.

   Take a piece of aluminum foil and cut it in half. Take one of the resulting pieces and cut it in half again. Repeat this process with each successive half until you cannot make another cut.

   a) In your *Active Chemistry* log, record how many cuts you were able to make. How does the size of the smallest piece of aluminum compare to the size of the original piece? Does the smallest piece of aluminum have the same properties as the original piece of aluminum foil?

2. An atom is the smallest part of an element. Since you can still cut the aluminum in pieces, you have not reached the size of a single atom.

Now imagine that there may be a way to cut the smallest piece of aluminum you have into even smaller and smaller pieces.

a) How small can the smallest piece be and still retain the properties of aluminum? Could you cut the piece in half and half again 10 more times? 100 more times? 1000 more times?

Using your imagination of cutting and cutting, you will eventually get to one atom of aluminum.

3. Chemists combine elements to form new substances. By measuring the amounts of elements used and substances formed, they are able to draw conclusions about the properties of the elements involved. You will study the reaction between aluminum metal and a solution of copper (II) chloride.

Read through the procedure below.

a) Make a table in your *Active Chemistry* log for the data you will be collecting. You will need room for measurements (mass) and observations. You can use a table similar to the one provided below.

4. Check your balance to make sure that it reads zero with nothing on it. Then measure the mass of a 50-mL beaker.

a) Record the mass in your *Active Chemistry* log.

Safety goggles and a lab apron must be worn during this activity.

Finding the mass of aluminum	
1. Mass of empty 50-mL beaker	g
2. Mass of beaker and aluminum foil	g
3. Calculate the mass of aluminum from 1 and 2 above.	g
Finding the mass of copper (II) chloride	
4. Mass of weighing paper	g
5. Mass of paper and copper (II) chloride	g
6. Calculate the mass of copper (II) chloride from 4 and 5 above.	g
Finding the mass of the product	
7. Mass of dry filter paper	g
8. Mass of filter paper with product, after drying	g
9. Calculate the mass of the product material from 7 and 8 above.	g

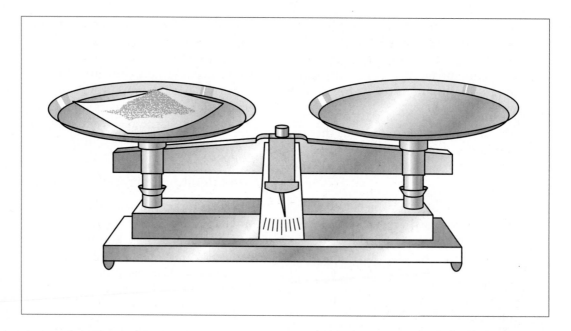

5. Measure out approximately 0.20 g of aluminum foil into the empty beaker. Try to get your mass measurement close to the assigned value.

   If you have a centigram balance, you'll need to adjust the balance to read 0.20 more grams than the beaker alone. Then add pieces of aluminum foil until it rebalances.

   If you have an electronic balance, simply add pieces of aluminum foil until the display indicates 0.20 g more than the empty beaker.

Wipe up any spills immediately.

   a) Record the value that you obtain, even though you might not hit the target value.

6. Measure the mass of a piece of weighing paper.

   Place approximately 2.00 g of copper (II) chloride on the weighing paper. (The chemical elements you would expect to find in copper (II) chloride

are copper and chlorine.) Again, remember that your target value is 2.00 g and that you may be slightly over or under this value.

   a) Record the masses in your *Active Chemistry* log.

7. Add the copper (II) chloride to the beaker.

   Next add water to the beaker until it is approximately half full.

   a) Record your observations in your *Active Chemistry* log. Consider including the following: how the beaker feels when you touch it; what you hear when you listen closely to the beaker; what you see happening in the beaker.

   b) What color forms on the aluminum foil? What do you think is responsible for this color? Where is the color coming from?

398

8. You will now need to find the mass of the substance formed in the chemical reaction. You can filter out this substance and then find its mass.

First measure the mass of a piece of dry, clean filter paper.

a) Record the mass in your *Active Chemistry* log.

9. Set up a filtration system, as shown in the diagram below.

10. Wait until you no longer see or hear any reaction between the aluminum foil and copper (II) chloride and the liquid begins to clear up. Pour the contents of the 50-mL beaker into the funnel. You should rinse the beaker a couple of times with some water to be sure that all of the contents of the beaker are transferred.

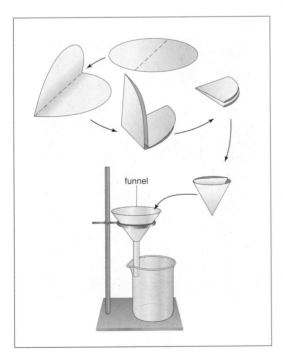

funnel

Remove the filter from the funnel and place it on a piece of folded paper towel and allow it to dry overnight. Label the paper towel so that you can identify your filter paper.

Clean and put away your equipment and dispose of your chemicals as directed by your teacher. Wash your hands.

11. When the filter paper is dry, measure the mass of the filter paper and its contents. Dispose of the filter paper and its contents as directed by your teacher. Wash your hands.

a) Record the mass in your *Active Chemistry* log. Determine and record the mass of the contents of the filter paper.

b) What element is inside the filter paper?

12. The reaction you witnessed is called a single-displacement reaction. In this reaction, a single element (aluminum) replaces another element (copper) in its combined form (copper (II) chloride). As a result of this reaction, the copper leaves its combined form to become an uncombined, or free element. The aluminum leaves its uncombined form to join with the chlorine to form a new compound. It's time to look at your data and see if you can make some sense of the numbers.

a) How many grams of aluminum did you start with? How many grams of copper did you end up with (contents of the dry filter paper)?

b) How many times as great is the mass of the copper as compared to the mass of aluminum you originally used? (What is the ratio of mass of copper to mass of aluminum?)

c) If one atom of copper is released for each atom of aluminum that becomes combined (with chlorine), the masses of copper and aluminum determined in (a) should contain the same number of atoms. What does this tell you about the relative masses of copper and aluminum atoms? The number of atoms in 0.20 g of aluminum EQUALS the number of atoms in 0.71 g of copper. How many times more massive is a copper atom than an aluminum atom? (You may wish to compare

objects that you are familiar with: a dozen bowling balls of mass 60.0 kg can be compared to a dozen eggs of mass 2.4 kg. Since there are a dozen bowling balls and a dozen eggs, you can find that the bowling balls are 25 times as heavy as the eggs (60.0 kg/2.4 kg.)

13. Recall that every group in your class reacted the same mass of aluminum with the same mass of copper (II) chloride. Discuss the similarities and differences in the data and calculations among the groups in the class.

a) Record your thoughts on how and why the results are similar and/or different.

Chem Talk

ATOMIC MASS

Atoms

In **Activity 2**, you defined the term element and explored the properties of some common elements. In this activity, you focused on atoms. An **atom** is the smallest representative part of an element. The ancient Greek philosopher Aristotle did not believe in the existence of atoms. In his thinking, if atoms did exist, there would have to be empty space between them.

Aristotle

Chem Words

atom: the smallest representative part of an element.

Aristotle did not believe it was possible to have empty space. Not everyone agreed with Aristotle. Another ancient Greek named Democritus believed that matter was made up of tiny particles that could not be broken down further. He called the particles atoms, from the Greek word *atomos*, meaning indivisible.

If you could have continued cutting the aluminum foil until it could no longer be cut, by any means, you would have reached one atom of aluminum. A mind-expanding fact is that if you started with 27 g of aluminum, you would find that there are 6.02×10^{23} atoms of aluminum. Nobody has ever counted this nor could they. Scientists have determined this number by other means and are very confident that it is correct.

Masses of Elements and Compounds in a Reaction

By the turn of the 19th century, chemists were combining elements to form new substances. The new substance was called a **compound**, because the atoms of the elements were believed to combine to form what they called a compound atom. The chemists were also particularly interested in measuring the amounts of elements used and substances formed. Their first attempts in determining masses were wrong, possibly due to the equipment that they had available at that time.

John Dalton

John Dalton, an early 19th century chemist who did much to advance the belief in the existence of atoms, expected that atoms combined in the simplest possible relationship. He reported that seven pounds of oxygen reacted with one

Chem Words

compound:
a material that consists of two or more elements united together in definite proportion.

401

pound of hydrogen to form water. Accurate modern experiments give eight pounds to one. We will use modern values rather than historical ones to avoid confusion. Dalton reported that five pounds of nitrogen reacted with one pound of hydrogen to form ammonia. He also reported that seven pounds of oxygen reacted with five pounds of nitrogen to form a compound he called nitrous gas.

In 1809, Joseph Gay-Lussac reported that the hydrogen reacting with oxygen to form water occupied twice as much volume as the oxygen. He also noted that the hydrogen reacting with nitrogen to form ammonia occupied three times as much volume as the nitrogen. Furthermore, he found that equal volumes of nitrogen and oxygen reacted to form nitrous gas (now known as nitric oxide or nitrogen monoxide, NO).

Gay-Lussac's data was inconsistent with Dalton's assumption that water, ammonia, and nitrous gas are formed from one atom of each of the combining elements. This inconsistency was subsequently resolved by Amadeo Avogadro, who furthered the understanding of the correct chemical formulas and atomic masses.

Relative Mass of Atoms

Chem Words

atomic mass unit (amu): a unit of mass defined as one-twelfth of the mass of a carbon-12 atom.

atomic mass: atomic mass is determined by the mass of the protons and neutrons of the atom.

Law of Definite Proportions: the composition of a pure substance is always the same or the elements of the compound always combine in the same proportion by mass.

Eventually, chemists determined a scale of relative masses of atoms through the systematic study of chemical reactions. By measuring the masses of two elements reacting with each other and knowing the formula for the compound that was formed, the relative mass of the two elements was determined. In this way, chemists were able to determine, for example, that one element has twice the mass of a second element. Relative mass does not tell you the exact mass measured in kilograms. It does provide a relative scale. Comparison of many reactions resulted in a scale of relative masses. Atoms of carbon were found to have a mass 12 times greater than the mass of hydrogen atoms, whereas oxygen atoms were found to have a mass 16 times greater than the mass of hydrogen. The units for this scale are called **atomic mass units**, defined in such a way that the mass of one type of carbon (carbon-12) is exactly 12 atomic mass units. The average mass of an atom of a given element in atomic mass units is known as the **atomic mass**. Atoms of hydrogen have an atomic mass of one unit. In addition to the physical and chemical properties of elements, the relative mass (incorrectly called weight) of each element was known and used by Mendeleev as he organized his table. The atomic mass is still one of the most prominent pieces of information provided for each element on the periodic table. The table gives the relative atomic masses of the eight elements that you observed in **Activity 2**.

Element	Relative atomic mass
Aluminum	26.98
Copper	63.55
Iodine	126.90
Iron	55.85
Magnesium	24.31
Silicon	28.09
Sulfur	32.06
Zinc	65.38

Checking Up

1. What is the difference between an element and a compound?

2. What is an atom?

3. How is an atomic mass unit defined?

4. How can the existence of atoms help to explain the Law of Definite Proportions?

The Law of Definite Proportions

Chemists at the beginning of the 19th century noted that eight pounds of oxygen always reacted with one pound of hydrogen to form nine pounds of water. This observation is an example of **The Law of Definite Proportions**. This law, first articulated by Joseph Proust in 1799, states that whenever two elements combine to form a compound, they do so in a definite proportion by mass. Proust based this statement on his observations that 100 pounds of copper, dissolved in nitric acid and precipitated by carbonates of soda (sodium) or potash (potassium), invariably gave 180 pounds of green carbonate. The Law of Definite Proportions is not direct proof of the existence of atoms. However, if you believe in the existence of atoms it does make it easier to explain why the Law of Definite Proportions should hold. The existence of atoms can also help explain why a given mass of aluminum reacting with sufficient copper (II) chloride in solution should always produce the same mass of copper.

Reflecting on the Activity and the Challenge

In this activity you learned how chemists measured elements in chemical reactions to determine the relative masses of atoms and how these masses were assembled into a scale of atomic masses. The atomic mass is one of the most important pieces of information listed for each element on the periodic table. How will you incorporate it into your game about the periodic table? Will you test players' ability to simply identify the atomic mass from a periodic table, or will you require that players understand how the relative scale was determined?

Chemistry to Go

1. John Dalton believed that water was formed from the simplest combination of hydrogen and oxygen atoms—one of each. Observations today show that eight pounds of oxygen react with one pound of hydrogen to form water. We now know that Dalton's values were wrong but his attempt was based on the data that he had available at the time.

   a) Based on these two statements, what conclusion could Dalton draw about the relative masses of oxygen and hydrogen atoms? How many times more massive is an oxygen atom than a hydrogen atom?

   b) The atomic mass of oxygen is 16 and the atomic mass of hydrogen is 1. How do the current atomic masses of oxygen and hydrogen compare to Dalton's?

   c) You know that water molecules are not made from one atom of hydrogen and one atom of oxygen. A water molecule is made up of 2 atoms of hydrogen and one atom of oxygen. If the pound of hydrogen reacting with eight pounds of oxygen is due to twice as many hydrogen atoms as there are oxygen atoms, how many times more massive is an oxygen atom than a hydrogen atom?

   d) Are the values of these revised masses closer to the current atomic masses of oxygen and hydrogen atoms?

2. In Dalton's time, ammonia was observed to be formed when nitrogen reacted with hydrogen. Today's values show that fourteen pounds of nitrogen react with three pounds of hydrogen.

   a) If ammonia were formed from Dalton's simplest formula of one atom of each element, what would he have concluded about the relative masses of nitrogen and hydrogen atoms?

   b) Ammonia molecules are made not from one atom each of hydrogen and nitrogen but from three atoms of hydrogen and one atom of nitrogen. If the three pounds of hydrogen reacting with fourteen pounds of nitrogen is due to three times as many hydrogen atoms as there are nitrogen atoms, how many times more massive is a nitrogen atom than a hydrogen atom?

3. A student uses magnesium instead of aluminum in this activity and obtains the following data:

Mass of beaker: 30.20 g

Mass of beaker + magnesium strip: 30.40 g

Mass of beaker + magnesium + copper (II) chloride: 32.40 g

Mass of beaker + magnesium + copper (II) chloride + water: 57.40 g

Mass of dry filter paper: 0.67 g

Mass of new beaker: 30.50 g

Mass of beaker + wet filter and residue + solution: 58.37 g

Mass of dry filter + residue: 1.19 g

a) How many grams of magnesium did the student use in this experiment?

b) How many grams of copper did the student measure in the dry filter paper?

c) How many times as great is the mass of the copper as the mass of magnesium the student originally used? (What is the ratio of mass of copper to mass of aluminum?)

d) If one atom of copper was released for each atom of magnesium that becomes combined (with chlorine), the masses of copper and magnesium determined in (a) and (b) should contain the same number of atoms. What does this tell you about the relative masses of copper and magnesium atoms? How many times more massive is a copper atom than a magnesium atom?

4. Look at the table in the **ChemTalk** reading section that gives the atomic masses of the eight elements that you observed in **Activity 2**. Add this data to your element cards. Can you now improve upon the way you sorted the cards in the previous activity taking this new information about relative masses into account?

Preparing for the Chapter Challenge

In a paragraph, explain how the relative scale of atomic masses is determined.

Inquiring Further

Avogadro's number and a mole

Chemists are interested in keeping track of quantities of particles. However, the particles are very small so chemists use a particular quantity that is convenient for counting particles. The quantity is called a mole. The quantity of particles in a mole is 602,000,000,000,000,000,000,000. The mole can be represented more easily in scientific notation as 6.02×10^{23}. This is a very large number because many, many small particles (atoms or molecules) make up a mole. The number 6.02×10^{23} is sometimes called Avogadro's number.

Research to find the significance of Avogadro's number and a mole. Record your findings in your *Active Chemistry* log.

Then, to appreciate how huge a mole is, answer the following question:

Imagine you have 7 billion people (7×10^9 people, which is approximately the human population of the world) and they are given the task of dropping $1-bills once every second into a large hole. How long will it take 7×10^9 people to drop 1 mole of dollar bills into the hole? How old will you be when they complete this task?

Activity 4   Are Atoms Indivisible?

GOALS

In this activity you will:

- Observe the behavior of a cathode ray in the presence of a magnet.

- Discuss Thomson's conclusions from 1897 about cathode rays.

- Simulate an experiment from 1911 by Rutherford in which he learned more about the structure of atoms.

- Organize your understanding of some of the different particles that comprise matter.

What Do You Think?

Ever since Democritus from ancient Greece hypothesized the existence of atoms, a major question was how atoms of different elements were different.

- **If you could observe a single atom of gold and a single atom of lead, how do you think they would be different? How could they have something in common?**

Record your ideas about these questions in your *Active Chemistry* log. Be prepared to discuss your responses with your small group and the class.

Investigate

1. Your teacher will demonstrate the behavior of what were called cathode rays a hundred years ago. They were called cathode rays because they were emitted from the negative terminal, or cathode of what was known as a cathode-ray tube, a forerunner of the television or the computer monitor tube.

Do not to try to see the effect of a magnet on an actual television or computer monitor; you may damage the monitor.

a) What happens to the path of the cathode rays when a horseshoe magnet is placed near the tube? Record your observation in your *Active Chemistry* log.

b) Record what you think will happen to the path of the cathode rays when the orientation of the horseshoe magnet is reversed.

c) Observe the path of the cathode rays as your teacher reverses the magnet. Record what does happen.

2. Magnets exert a force on moving electrically charged particles. The effect of the magnet on the cathode rays therefore shows that these rays are moving electrically charged particles. Cathode rays, which have a negative electric charge, are made up of electrons. In 1897, Joseph John (J. J.) Thomson showed that identical rays (electrons) were emitted from the cathode of a cathode-ray tube, regardless of the metal of which the cathode was made.

Discovery of electrons emerging from the atoms of the cathode gave scientists new information about the atom. The atom is not indivisible. It has internal parts, one of which is the electron.

a) In a sentence or two, describe the relationship between cathode rays, the electron, and the structure of atoms.

3. In order to investigate the other components of an atom, you will take part in the following simulation, similar to the game *Battleship*. You will work with a partner for this activity.

You and your classmate should each construct a grid of squares, 8 by 10. Without letting your classmate see your grid, color in a section of ten squares. The squares must touch each other. To make the simulation relatively simple, begin with a fairly compact design. This shape (colored region) represents your target.

You and your partner will try to guess the shape of each other's targets by sending "missiles" onto any of the 80 squares in this array. For the purpose of this description, designate one person to be Player X and the other person to be Player Y. To begin, Player X will tell Player Y the destination (number and letter) of

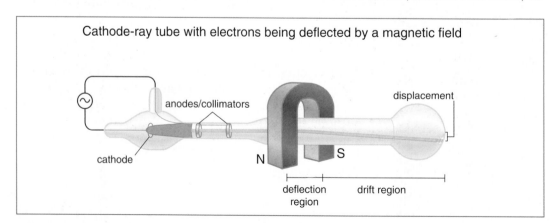

Cathode-ray tube with electrons being deflected by a magnetic field

anodes/collimators

displacement

cathode

N   S

deflection region   drift region

	A	B	C	D	E	F	G	H	I	J
1										
2										
3										
4										
5										
6										
7										
8										

the missile being sent. Player Y will respond, indicating that the missile "hit" or "missed" the target shape. Player X will make note of the response. Then Player Y sends the next missile, noting the response. Continue this process until one player identifies the other player's target.

a) Record the number of turns taken to complete the game.

b) Repeat the game with a target of only 2 adjacent squares. Record the number of turns taken.

4. Now do a thought experiment. The same-size game grid is divided into smaller squares—100 squares across and 100 squares down. There are now 10,000 squares in the same size board as before. A target of only one square is chosen.

a) Record an estimate of how many turns will be required to identify the target square amongst the 10,000 squares in the game grid.

5. Now modify the thought experiment. The same-size grid is now 1000 rows across and 1000 squares down. That is a total of 1,000,000 squares.

a) Record an estimate of how many turns will be required to identify the target square among the 1,000,000 squares in the game grid.

6. An experiment similar to your game of "Battleship" was carried out in 1911 by Lord Ernest Rutherford. Rutherford sought to learn something about the structure of the atom by bombarding gold atoms with energetic particles given off by certain atoms.

In Rutherford's game of "Battleship," it seemed that he was required to send an incredible number of missiles to get a "hit." He concluded that the grid of the atom must be composed of tiny, tiny cells and only one cell contains all of the positive charge of the atom.

a) Explain why you think he concluded this.

Chem Talk

THE CHANGING MODEL OF AN ATOM

J.J. Thomson's Model of an Atom

As you noted in this activity, in the late 1800s J.J. Thomson, an English physicist, found evidence for the existence of negatively charged particles that could be removed from atoms. He called these subatomic particles with a negative charge **electrons.** Using this new information, Thomson then proposed a model of an atom that was a positive sphere, with electrons evenly

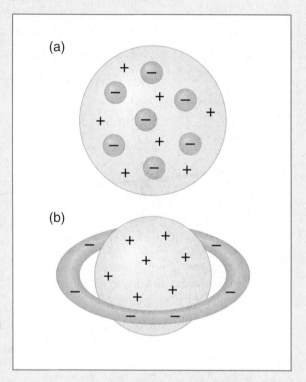

distributed and embedded in it, as shown in the diagram. Using the same evidence, H. Nagaoka, a Japanese scientist, modeled the atom as a large positively charged sphere surrounded by a ring of negative electrons.

Rutherford's Discovery of the Nucleus

For several years there was no evidence to contradict either Thomson's or Nagaoka's atomic models. However, in the early 1900s, Ernest Rutherford, a New Zealand-born scientist, designed experiments to test the current model of an atom. In Rutherford's experiment, alpha particles were sent as "missiles" toward a thin sheet of gold. Gold was used because it is malleable and could be hammered into a thin, thin sheet. Most of the alpha particles went through the sheet and were not deflected. It is as if they missed the target. This was expected since it was assumed that the atom's charge and mass

Chem Words

electron: a negatively charged particle with a charge of 1.6×10^{-19} coulombs and a mass of 9.1×10^{-31} kg.

Ernest Rutherford

was spread evenly throughout the gold. Occasionally one of the alpha particles that "hit" the gold sheet bounced back. This was the big surprise. The conclusion: there must be tiny places containing lots of charge and mass. Since the bouncing back was so unusual, it was assumed that the places where all the charge and mass were concentrated were only 1/100,000 of the area of the gold. Rutherford concluded that almost all the mass and all of the positive charge of the atom is concentrated in an extremely small part at the center, which he called the **nucleus**. He also coined the term **proton** to name the smallest unit of positive charge in the nucleus.

The story of Rutherford's discovery of the atomic nucleus is best told by Rutherford himself. Examining the deflection of high-speed alpha particles as they passed through sheets of gold foil, Rutherford and his student Hans Geiger noticed that some particles were scattered through larger angles than predicted by the existing theory of atomic structure. Fascinated, Rutherford asked Geiger's research student Ernest Marsden to search for more large-angle alpha scattering. Rutherford did not think that any of the alpha particles in his experiment would actually bounce backward. "Then I remember two or three days later Geiger coming to me in great excitement and saying, 'We have been able to get some of the alpha particles coming backwards . . .' It was quite the most incredible event that has ever happened to me in my life. It was almost as incredible as if you fired a 15-inch shell at a piece of tissue paper and it came back and hit you."

Chem Words

nucleus: the very dense core of the atom that contains the neutrons and protons.

proton: a positively charged subatomic particle contained in the nucleus of an atom. The mass of a proton is 1.673×10^{-24}g and it has a charge of $+1$.

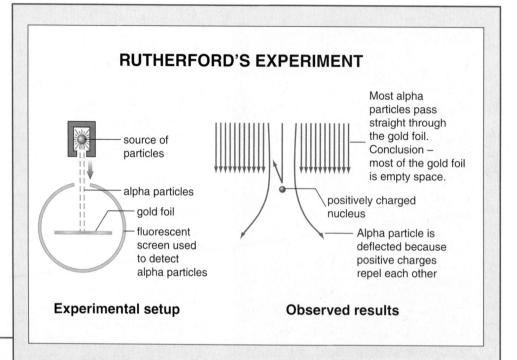

RUTHERFORD'S EXPERIMENT

source of particles

alpha particles

gold foil

fluorescent screen used to detect alpha particles

Most alpha particles pass straight through the gold foil. Conclusion – most of the gold foil is empty space.

positively charged nucleus

Alpha particle is deflected because positive charges repel each other

Experimental setup

Observed results

Checking Up

1. What is an electron?

2. Thomson's model of an atom is sometimes referred to as the "plum-pudding" model. (A plum pudding is a heavy pudding with raisins mixed into it.) Explain why this is an appropriate comparison.

3. Why was Rutherford surprised that some alpha particles bounced back from the gold foil?

4. What is the nucleus of an atom?

A Physics Connection

What was responsible for the wide-angle scattering of the alpha particles and their bouncing back? The force between the positive nucleus and the positive alpha particle is the coulomb force. Positive charges repel one another according to the coulomb force law.

$$F = \frac{kq_1q_2}{d^2}$$

where k is Coulomb's constant
$(k = 9.0 \times 10^9 \text{ N m}^2/\text{C}^2)$,
q is the charge in coulombs, and
d is the distance between the charges.

The closer the alpha particle gets to the nucleus, the larger the force and the larger the deflection of the alpha particle.

Reflecting on the Activity and the Challenge

In this activity you learned of evidence that atoms are made of a positively charged nucleus and negatively charged electrons. The nucleus contains most of the atom's mass and its positive charge is balanced by the combined negative charge of the electrons, resulting in an atom that is electrically neutral. The number of protons of the neutral atom plays a very important role in the periodic table. Called the atomic number, it supports the order in which Mendeleev arranged the elements in his periodic table, long before anything was known about the structure of the atom or atomic number. How will your game reflect your new knowledge about atomic structure and its relationship to the periodic table?

Chemistry to Go

1. Since the electron has a negative electric charge and the nucleus has a positive electric charge, where would you expect to find electrons in atoms?

2. Are atoms indivisible? Support your answer using information from this activity.

3. Construct a chart or diagram to summarize what you have learned in this activity about the particles that make up an atom. Include electric charge and location of the particles.

4. Lead has an atomic number of 82; iron has an atomic number of 26; and copper has an atomic number of 29. How do the charges of the nuclei of these three elements compare?

5. The element chlorine has an atomic number of 17. How many electrons does chlorine have? Support your answer with a logical explanation of how you could arrive at this answer.

6. Sketch the outline of three grids. Pretend that each grid has 100,000 squares.

   a) If the target was 50,000 squares, draw the target.

   b) If the target was 25,000 squares, draw the target.

   c) If the target was only 1 square, draw the target.

   Which grid most closely relates to Rutherford's experiment? Explain your answer.

Inquiring Further

1. An atomic timeline

Construct a timeline that reflects how scientists' views of the atom have changed through the ages. Identify significant scientists, their beliefs, and experimental findings as mentioned in this chapter. You may also wish to consult other resources. Add information to your timeline as you continue to work through this chapter.

2. John Dalton's Atomic Theory

John Dalton, an English scientist, developed his atomic theory in the early 1800s. This theory was based on the Greek concept of atoms and the studies of Joseph Proust's Law of Definite Proportions or Law of Constant Composition. Dalton's atomic theory contained a series of postulates based on the data of his time and his observations:

• Matter consists of small particles called atoms.

• Atoms of one particular element are identical and the properties are identical.

• Atoms are indestructible. In chemical reactions, the atoms rearrange or combine, but they are not destroyed.

• Atoms of different elements have a different set of properties.

• When atoms of different elements combine to form compounds, they combine in a fixed numerical ratio.

From his postulates, the Law of Conservation of Mass would be supported. Since his postulates state that atoms cannot be destroyed but they can be moved around and combine with other atoms to form compounds, then the mass of the compound must be the sum of the atoms of the compound. This law still exists with a slight modification for nuclear reactions. So, you can conclude that if water has a certain mass today, it will have the same mass any other day (unless evaporation occurs).

Investigate whether all of Dalton's postulates are presently accepted or describe how some have been modified based on our current understanding.

Joseph Proust

Activity 5   The Chemical Behavior of Atoms

GOALS

In this activity you will:

- View the spectrum of hydrogen.

- Interpret changes in electron energies in the hydrogen atom to develop an explanation for where the colored light in the hydrogen spectrum comes from.

- Use Bohr's model of the atom to predict parts of the hydrogen atom spectrum.

- Calculate and compare the wavelengths, energies, and frequencies of light of different colors.

- Identify regions in the electromagnetic spectrum.

What Do You Think?

A neon sign uses electricity and a gas-filled tube to produce a colored light. A fluorescent bulb uses electricity, a gas-filled tube and a phosphor coating to produce white light.

- **How is the color produced in a neon sign?**

Record your ideas about this question in your *Active Chemistry* log. Be prepared to discuss your response with your small group and the class.

Investigate

1. In order to directly observe the behavior of atoms, you can observe the spectrum of light given off when atoms are excited by a high-voltage, electric-power supply. You probably have already seen this effect in the familiar red color of neon signs.

   For this demonstration, your teacher will set up a tube of hydrogen gas connected to a high-voltage power supply. This light can be viewed through a spectroscope or a diffraction grating lens, as shown in the diagram on the next page. When the slit at the end of the spectroscope is aimed toward the light, the colors of the spectrum appear separately off to the sides of the slit.

means 10^{-9}. Thus, 656.5 nm is 656.5×10^{-9} m. The frequency of each line can be determined using the equation:

$$f = \frac{c}{\lambda}$$

where,

f is the frequency (waves/s or cycles/s or hertz),

c is the speed of light and is a constant (2.998×10^8 m/s), and

λ is the wavelength measured in nanometers.

Example:

Given red light with a wavelength $\lambda = 656.5 \times 10^{-9}$ m and the speed of light $c = 2.998 \times 10^8$ m/s, calculate the corresponding frequency and energy of the light.

a) What colors do you see in the spectrum of light given off by hydrogen gas?

b) Make a colored diagram in your *Active Chemistry* log of what you see inside the spectroscope. Make sure to draw the colors with the order and spacing between them that you observe.

2. When you observed the spectrum of light given off by hydrogen gas, you probably saw three or four distinct lines, each having a different color. The color of light is determined by its frequency; the greater the energy in the light, the greater the frequency.

The diagram on the right shows the visible lines of the line spectrum of hydrogen.

The wavelengths are measured in nanometers (nm). The prefix *nano*

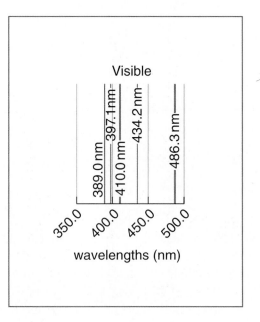

First calculate the frequency.

$$f = \frac{c}{\lambda}$$

$$= \frac{2.998 \times 10^8 \text{ m/s}}{656.5 \times 10^{-9} \text{ m}}$$

$$= 4.567 \times 10^{14} \text{ cycles/s (or Hz)}$$

$$= 4.567 \times 10^{14} \text{ Hz (Hertz)}$$

To calculate the corresponding energy of a photon of light, you can use an equation called Planck's equation that relates the frequency to the energy. A photon is a tiny, indivisible packet of light. Planck's constant is h and has the value of 6.63×10^{-34} J•s.

$$E = hf$$

$$= (6.63 \times 10^{-34} \text{ J•s})$$
$$(4.567 \times 10^{14} \text{ cycles/s})$$

$$= 3.03 \times 10^{-19} \text{ J}$$

a) Calculate the frequencies and energies for the other visible wavelengths of light in the hydrogen spectra. Complete a table like the one below in your *Active Chemistry* log.

3. In 1913, Niels Bohr, a Danish physicist, tried to explain the line spectrum of hydrogen by hypothesizing that the electron in the hydrogen atom is allowed to have only certain amounts of energy. Bohr believed that an electron in a higher energy level would give off light when it fell to a lower energy level, and the amount of energy in the light would be the difference in these energy levels. Bohr pictured the atom to be similar to the solar system, with electrons orbiting around the nucleus in the way that planets orbit around the Sun. If you picture the electron to be in orbit about the nucleus of the hydrogen atom, as Bohr did, allowing the electron to have only certain amounts of energy would mean that the electron could be allowed in orbits of only certain distance from the nucleus. As the electron "jumps" from one energy level to another, it behaves something like a ball falling down a flight of stairs. It's allowed to rest only on one of the steps, nowhere in between.

The light was the result of electron energy jumps. These jumps should be able to produce the corresponding energies of the light. Bohr calculated the lowest energy level to be -2.18×10^{-18} J. Each of the other

Wavelength λ (nm)	Frequency f (cycles/s or Hz)	Energy E
410.3		
434.2		
486.3		
656.5	4.567×10^{14}	3.03×10^{-19} J

Niels Bohr

energy levels can be computed from this first one with the equation:

$$E_n = \frac{1}{n^2} E_1$$

The second energy level:

$$E_2 = \frac{1}{2^2} E_1$$

$$= \frac{1}{4}(-2.18 \times 10^{-18}\,\text{J})$$

$$= -5.45 \times 10^{-19}\,\text{J}$$

The third energy level:

$$E_3 = \frac{1}{3^2} E_1$$

$$= \frac{1}{9}(-2.18 \times 10^{-18}\,\text{J})$$

$$= -2.42 \times 10^{-19}\,\text{J}$$

a) Calculate the energy levels for the fourth, fifth, and sixth levels and record all these values in your log.

4. In Bohr's model, the electrons jump from E_3 to E_2 and give off light with that energy.

$$E_3 - E_2 = (-2.42 \times 10^{-19}\,\text{J}) - (-5.45 \times 10^{-19}\,\text{J})$$

$$= 3.03 \times 10^{-19}\,\text{J}$$

This is equivalent to the energy of the red light. The model is successful, so far.

a) Calculate the energy of light when the electron of hydrogen jumps from:

E_4 to E_2
E_5 to E_2
E_6 to E_2

b) Do these energies correspond to the energies of the light that is emitted from hydrogen that you calculated earlier?

c) If the model is a truly good model, it should be able to predict something unknown. Bohr's model does just this. Calculate the energy of light when the electron of hydrogen jumps from E_2 to E_1.

The light corresponding to this transition was not observed because it is not in the visible range of light. It is in the ultraviolet spectrum. When an ultraviolet detector is used, the wavelength of the ultraviolet light is just as predicted by Bohr's model. The light emitted from transitions of E_3 to E_1 and E_4 to E_1 are also in the ultraviolet.

When the electron jumps from a higher energy level to a lower energy level, light is emitted. The light emitted from a neon sign is based on this phenomenon.

The opposite can also occur. If the electron absorbs light of just the right energy, the electron can jump up to a higher energy level. For instance, an electron in the E_2 energy level can absorb light of 3.02×10^{-19} J and jump up to the E_3 energy level. If enough energy is applied to the hydrogen atom, the electron can be totally removed. This is commonly called the ionization energy; for one electron of hydrogen, the ionization energy is 2.18×10^{-19} J. If the electron in the E_1 level absorbs light of energy 2.18×10^{-19} J, then its new energy will be 0 J. This means that it is no longer bound to the hydrogen nucleus. The atom is ionized (the electron has been removed).

419

ChemTalk

BOHR'S MODEL OF AN ATOM

The Electromagnetic Spectrum

Visible light is only one part of the **electromagnetic spectrum**. You've probably heard of some of the other parts including ultraviolet, infrared, x-rays, gamma rays, microwaves, and radio waves. As you demonstrated in calculations using Bohr's model, the light from some of the transitions is in the ultraviolet. Infrared light is also emitted as the electron jumps from E_4 to E_3 and E_5 to E_3 and other higher energy levels.

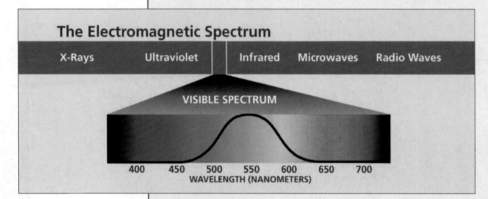

Many people do not think of radio waves as being light waves, but they are part of the electromagnetic spectrum. The problem is that you can only see a very small part of the electromagnetic spectrum. Often, you hear radio announcers say that they are broadcasting at a certain **frequency**. Your FM radio dial may have MHz (megahertz) printed on the side. This tells you that the numbers correspond to frequencies in units of MHz or 10^6 Hz. Frequency tells you the number of cycles or waves that are being produced per second. The unit for frequency is a hertz (Hz). 1 Hz = 1 cycle/s = 1 s^{-1}. Normally, frequency is read as per second and the cycles are dropped from the terminology.

Wavelength (λ), where λ is the Greek letter *lambda*, is the distance from crest to crest of a wave. All light waves travel at the same speed. The speed of electromagnetic radiation is constant and it is called the speed of light (c), $c = 2.998 \times 10^8$ m/s or 3.00×10^8 m/s. From this information you can calculate the frequency of light of a given wavelength. The equation that is used for this is:

$$f = \frac{c}{\lambda}$$

Chem Words

electromagnetic spectrum: the complete spectrum of electromagnetic radiation, such as radio waves, microwaves, infrared, visible, ultraviolet, x-rays, and gamma rays.

frequency: the number of waves per second or cycles per second or hertz (Hz).

wavelength: the distance measured from crest to crest of one complete wave or cycle.

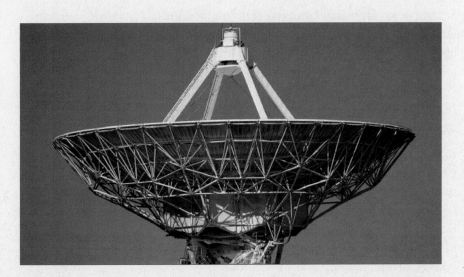

As an example, if the wavelength is 434.2 nm, then the frequency is:

$$f = \frac{2.998 \times 10^8 \text{ m/s}}{434.2 \times 10^{-9} \text{ m}}$$

$$= 6.905 \times 10^{14} \text{ cycles/s or } 6.905 \times 10^{14} \text{ Hz.}$$

As you go across the electromagnetic spectrum you should note that the wavelength continues to get smaller as the frequency increases. Also, you should understand that the energy of the spectrum increases as you go from radio waves to x-rays or gamma rays. Max Planck, a German physicist, found that the energy of a wavelength could be calculated. The equation that he developed was based on measuring the change in energy from one level to another level like you did in the **Investigate** section. The equation he developed is

$E = hf$,

where h is Planck's constant and is 6.63×10^{-34} J · s and f is the frequency.

The corresponding energy of the red light above would be:

$$E = hf$$

$$= (6.63 \times 10^{-34} \text{ J} \cdot \text{s}) (4.567 \times 10^{14} \text{ Hz})$$

$$= 3.03 \times 10^{-19} \text{ J.}$$

So the next time that you are standing around a campfire, you can inform your fellow campers that red light has less energy than blue light and you can also tell them how to calculate these values.

Bohr's Atomic Theory

Niels Bohr, a brilliant Danish physicist, was aware that his theory of electron jumps had incredible success but also raised some problems.

Bohr proposed a "planetary" model of the atom. He theorized that electrons travel in nearly circular paths, called **orbits**, around the nucleus. Each electron orbit has a definite amount of energy, and the farther away the electron is from the nucleus, the greater is its energy. Bohr suggested the revolutionary idea that electrons "jump" between energy levels (orbits) in a quantum fashion. That is, they can never exist in an in-between state. Thus, when an atom absorbs or gives off energy (as in light or heat), the electron jumps to higher or lower orbits. Electrons are the most stable when they are at lower energy levels closer to the nucleus. Bohr's theory could only account for the spectrum of hydrogen and not for the spectra of any other element. Bohr's theory could not explain why only certain orbits were allowed, nor how the electron could jump from one orbit to another. Other scientists improved on Bohr's model as they discovered more about the atom and quantum mechanics.

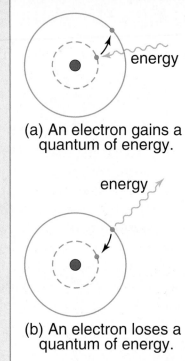

(a) An electron gains a quantum of energy.

(b) An electron loses a quantum of energy.

Chem Words

orbit: the path of the electron in its motion around the nucleus of Bohr's hydrogen atom.

Checking Up

1. How are visible light, ultraviolet light, infrared light, x-rays, gamma rays, microwaves, and radio waves related?

2. Explain the meaning of wavelength.

3. Why is "planetary" model an appropriate name for Bohr's model of the atom?

4. How do the energy levels of different electron orbits compare?

5. Why do elements produce certain color light when heated?

Reflecting on the Activity and the Challenge

In this activity, you've recognized that it is the electrons in an atom that are responsible for the colors of light emitted. Bohr explained the spectrum of light given off by excited hydrogen atoms by hypothesizing that each hydrogen atom's electron was allowed in only certain energy levels. Although Mendeleev knew nothing about electrons or energy levels when he first developed the periodic table (though he did live long enough to know of the electron's discovery by Thomson), the periodic table today is seen as a reflection of the number of electrons in an atom of each element and the energy levels those electrons occupy. Many versions of the periodic table today list the energy levels occupied by each electron in an atom of a given element. You will learn more about this in **Activity 6.** The periodic table is a way of organizing information and knowledge about the atom. Electron structure is a crucial part of that knowledge. Continue to think about how this information could be included in the game you are designing.

Chemistry to Go

1. In this activity you were told that light with greater energy has greater frequency.

   a) Which color in the visible spectrum of hydrogen has the greatest energy? The least energy?

   b) Which color in the visible spectrum of hydrogen has the highest frequency? The lowest frequency?

2. If an electron were to fall down to the E_1 level from the E_3 level, how would its energy compare to one that fell to the E_2 level? Explain.

3. What is the difference measured in energy when an electron falls from E_3 to E_1? How many times greater is this value as compared to the difference of the electron falling from E_3 to E_2?

4. A wavelength of light is 389.0 nm and its frequency is 7.707×10^{14} Hz. Show how this frequency value was calculated.

5. Show that the wavelength of 389.0 nm has an energy of 5.11×10^{-19} J.

6. a) Microwave radiation is absorbed by the water in food. As the heat is absorbed by the water it causes the food to get hot. If the λ of a microwave is 10 cm, calculate the frequency of the microwave and the energy of each photon.

b) The red light you observed in the hydrogen spectrum produced a $\lambda = 656.5$ nm. The energy of the red light was 3.02×10^{-19} J. How many times greater is this value when compared to the energy value that you found for the microwave energy?

Inquiring Further

Balmer, Lyman, and Paschen series

In this activity, you learned that the visible light emitted by the hydrogen lamp resulted when an electron moved from a higher energy level to the second lowest energy level in Bohr's scheme. This series of visible spectral lines is said to comprise the **Balmer series**.

What do you think happens when an electron in a hydrogen atom moves from a higher energy level to the lowest level? This series is known as the **Lyman series** (ultraviolet light).

What if a transition from a higher energy level leads to an electron in the third lowest energy level in Bohr's scheme? How do these energy levels compare with visible light? This series is known as the **Paschen series** (infrared light).

Activity 6   Atoms with More than One Electron

GOALS

In this activity you will:

- View the spectra of various materials.

- Graphically analyze patterns in the amounts of energy required to remove electrons from different kinds of atoms.

- Compare trends in stability of atoms in the periodic table.

- Compare the structure of the periodic table with the patterns of levels and sublevels to which electrons can be assigned.

- Develop a shorthand notation to describe the configuration of electrons in an atom.

What Do You Think?

In **Activity 5** you learned that Niels Bohr was able to explain the spectrum of light emitted by hydrogen using a model that assigned the electron to specific energy levels. Bohr was awarded a Nobel Prize in 1922 for expanding the understanding of atomic structure. He worked with hydrogen, the simplest atom, which contains only one electron. However, the atoms of other elements contain more than one electron.

- **How do you think an increase in the number of electrons would impact the spectrum of an atom?**

- **What modifications in Bohr's model would need to take place to accommodate the extra electrons?**

Record your ideas about these questions in your *Active Chemistry* log. Be prepared to discuss your responses with your small group and the class.

Investigate

1. In **Activity 5** you observed the spectrum of hydrogen gas as its electron moved from a higher energy level to a lower energy level. You also explored a model that used Bohr's theory to explain this spectrum. Now it's time to look at the spectra of some other elements.

   a) Your teacher will connect a tube containing an element other than hydrogen to a high voltage supply. Record the name of the element in your *Active Chemistry* log. Look at the spectrum of light of this element through the spectroscope.

   b) What colors do you see? Make a diagram in your log of the spectrum (pattern of colors) you see inside the spectroscope.

   c) Record how this spectrum is similar to and different from the hydrogen spectrum you observed in **Activity 5**.

   d) Repeat **Steps (a)**, **(b)**, and **(c)** for as many samples as your teacher demonstrates.

2. Although the spectra of such elements as helium and neon are very beautiful, they cannot be explained by Bohr's simple theory for the single electron in the hydrogen atom. The basic idea is still true—light is emitted when electrons jump from a higher energy level to a lower energy level. The energy levels, however, are more complex if there are additional electrons. A more elaborate labeling of electron energy levels is necessary. In this activity you will explore the pattern of electron energy levels in

Atomic Number	Element (Symbol)	1st Ionization Energy J ($\times 10^{-19}$)	2nd Ionization Energy J ($\times 10^{-19}$)
1	H	21.8	
2	He	39.4	87.2
3	Li	8.6	121.2
4	Be	14.9	29.2
5	B	13.3	40.3
6	C	18.0	39.1
7	N	23.3	47.4
8	O	21.8	56.3
9	F	27.9	56.0
10	Ne	34.6	65.6
11	Na	8.2	75.8
12	Mg	12.3	24.1
13	Al	9.6	30.2
14	Si	13.1	26.2
15	P	16.8	31.7
16	S	16.6	37.4
17	Cl	20.8	38.2
18	Ar	25.2	44.3
19	K	7.0	50.7
20	Ca	9.8	19.0
21	Sc	10.5	20.5
22	Ti	10.9	21.8
23	V	10.8	23.5
24	Cr	10.8	26.4
25	Mn	11.9	25.1
26	Fe	12.7	25.9
27	Co	12.6	27.3
28	Ni	12.2	29.1
29	Cu	12.4	32.5
30	Zn	15.1	28.8
31	Ga	9.6	32.9
32	Ge	12.7	25.5
33	As	15.7	29.9
34	Se	15.6	34.0
35	Br	18.9	34.9
36	Kr	22.4	39.0

atoms containing more than one electron.

When multiple electrons are present, some are easier (i.e., require less energy) to remove from the atoms than others. The chart on the left provides information about the amount of energy required to remove the electrons in the two highest energy levels. These are the electrons that are easiest to remove. These energies are called the 1st and 2nd ionization energies, and are given in units of joules. Notice that all values are multiplied by 10^{-19}.

a) Make a graph that shows how the ionization energies vary with atomic number. Since the atomic numbers range from 1 to 36, label the x-axis with atomic numbers from 1 to 36. Since the ionization energies range from 7 to 122, label the y-axis with ionization energies from 0 to 130. Plot the first ionization energy data from the chart in one color, connecting the data points as you go along.

b) Plot the values for the second ionization energies in a different color.

c) Include a title and legend on your graph.

3. Look at the graph of the first ionization energies and answer the following questions:

a) What kinds of patterns do you see? How could you quickly relate the shape of the graph to someone who had not seen it?

b) Where are the ionization energies the largest? The smallest?

c) What happens to the ionization energies as the atomic number increases?

d) Group the elements by their ionization energies into four consecutive "periods." List the range of atomic numbers in each group.

e) Is there any interruption in the general trend of ionization energies as the atomic number increases for a "period"? If so, describe it.

4. Look at the second colored graph line you drew.

a) Describe how the two graphs are alike and/or different. Do you see similarities between the two graphs?

5. If a large amount of energy is needed to remove an electron from an atom, the arrangement of electrons in that atom is considered to be especially stable. Thus, a high first ionization energy means that a lot of energy must be supplied to remove an electron from an atom and that the electron arrangement in that atom is especially stable. Any element that has a larger first ionization energy than its neighboring elements has an electron arrangement in its atoms that is more stable than its neighboring elements.

a) Which element in the first period (atomic numbers 1 and 2) has the most stable arrangements of electrons in its atoms? (Remember, you are looking for elements that have larger ionization energies than their neighbors. In reality you are looking for peaks in your graph, not just those elements with higher values.)

b) Which elements in the second period (atomic numbers 3 through 10) of the periodic table have the most stable arrangements of electrons in their atoms?

c) Which elements in the third period (atomic numbers 11 through 18) of the periodic table have the most stable arrangements of electrons in their atoms?

d) Which elements in the fourth period (atomic numbers 19 through 36) of the periodic table have the most stable arrangements of electrons in their atoms?

6. As mentioned earlier, the Bohr model was not able to account for the spectrum of an element containing more than one electron. A more elaborate model was needed. In this new model, the energy levels are broken down into sublevels. When these sublevels are filled, the atom exhibits a higher degree of stability. In this model, the sublevels are designated by the four letters s, p, d, and f.

The periodic table shows the atomic number, the chemical symbol, and how many electrons in an atom of each element are in each sublevel. The total number of electrons is equal to the atomic number of the element. This is because the atoms are neutral and therefore have a number of electrons equivalent to the number of protons. This arrangement of the electrons in each sublevel will be referred to as the electron assignment or electron configuration of the element. Use this periodic table to answer the following questions:

a) In what sublevel (include number and letter) are the electrons in hydrogen (1 electron) and helium (2 electrons) found?

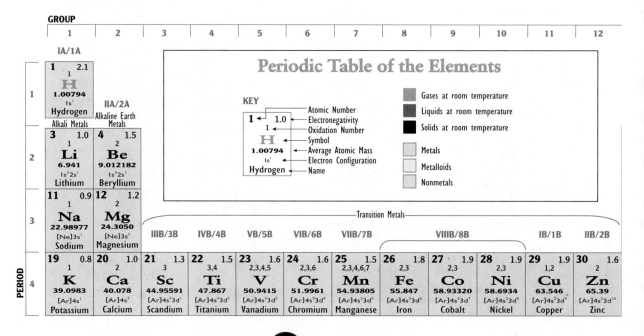

As you move to the second period (second row on the periodic table) each new element has one more proton in its nucleus and one more electron. The electrons must find a place to reside — an energy level and a sublevel within that energy level.

As you move along in the periodic table to increasing atomic numbers, you see that the additional electrons fill the sublevel. A completed sublevel is one that is holding the maximum number of electrons allowed to it before electrons must be placed in the next higher sublevel.

b) In what region of the periodic table are electrons added in an *s* sublevel? What is the greatest number of electrons found in any *s* sublevel?

c) In what region of the periodic table are electrons added in a *p* sublevel? What is the greatest number of electrons found in any *p* sublevel?

d) In what region of the periodic table are electrons added in a *d* sublevel? What is the greatest number of electrons found in any *d* sublevel?

e) In what region of the periodic table are electrons added to an *f* sublevel? What is the greatest number of electrons found in any *f* sublevel?

f) Select a column in the periodic table. (A column of elements on the periodic table is called a family or group.) Look at the electron configuration for each element within the column. Take special note of the last entry, the sublevel to which the last electron in an atom of each element in that column is added. What do all of these sublevels have in common? How many electrons are in these particular sublevels?

g) Mendeleev assigned elements to the same column of the periodic table because the elements had similar properties, both physical and chemical. How, then, does the number and location of the electrons in the outermost sublevel relate to chemical properties? We can now acknowledge that electrons (as opposed to the nucleus) are the key to the chemical properties of elements.

13	14	15	16	17	18
					VIIIA/8A or 0 Noble Gases
					2
					He
					4.002602
					$1s^2$
IIIA/3A	IVA/4A	VA/5A	VIA/6A Chalcogens	VIIA/7A Halides	Helium
5 2.0 3	**6** 2.5 -4,2,4	**7** 3.0 -3,2,3,4,5	**8** 3.5 -2	**9** 4.0 -1	**10**
B	**C**	**N**	**O**	**F**	**Ne**
10.811	12.011	14.00674	15.9994	18.998403	20.1797
$1s^22s^22p^1$	$1s^22s^22p^2$	$1s^22s^22p^3$	$1s^22s^22p^4$	$1s^22s^22p^5$	$1s^22s^22p^6$
Boron	Carbon	Nitrogen	Oxygen	Fluorine	Neon
13 1.5 3	**14** 1.8 2,4	**15** 2.1 -3,3,4,5	**16** 2.5 -2,2,4,6	**17** 3.0 -1,1,3,5,7	**18**
Al	**Si**	**P**	**S**	**Cl**	**Ar**
26.981539	28.0855	30.973762	32.066	35.4527	39.948
[Ne]$3s^23p^1$	[Ne]$3s^23p^2$	[Ne]$3s^23p^3$	[Ne]$3s^23p^4$	[Ne]$3s^23p^5$	[Ne]$3s^23p^6$
Aluminum	Silicon	Phosphorus	Sulfur	Chlorine	Argon
31 1.6 3	**32** 1.8 4	**33** 2.0 -3,3,5	**34** 2.4 -2,4,6	**35** 2.8 -1,1,5,7	**36**
Ga	**Ge**	**As**	**Se**	**Br**	**Kr**
69.723	72.61	74.92159	78.96	79.904	83.80
[Ar]$4s^23d^{10}4p^1$	[Ar]$4s^23d^{10}4p^2$	[Ar]$4s^23d^{10}4p^3$	[Ar]$4s^23d^{10}4p^4$	[Ar]$4s^23d^{10}4p^5$	[Ar]$4s^23d^{10}4p^6$
Gallium	Germanium	Arsenic	Selenium	Bromine	Krypton

429

ChemTalk

A PERIODIC TABLE REVEALED

Ions and Ionization Energy

In the table in the **Investigate** section the amount of energy required to remove an electron from an atom was called **ionization energy**. Atoms are neutral. That is, the number of electrons is equivalent to the number of protons. However, atoms can gain or lose electrons. Atoms that have lost or gained electrons are called **ions** and thus the energy used to remove the electrons is known as the ionization energy. The energy required to remove a single electron from the highest occupied energy level is called the first ionization energy, and the energy needed to remove a second electron from the same atom, after the first one has already been removed, is called the second ionization energy.

Electron Configuration and Energy Levels

As you discovered, the Bohr model was not able to account for the spectrum of an element containing more than one electron. In the new model you investigated, the energy levels are broken down into sublevels. This arrangement of the electrons in each sublevel is called the electron assignment or **electron configuration** of the element. When these sublevels are filled, the atom exhibits a higher degree of stability. The sublevels are designated by the four letters *s*, *p*, *d*, and *f*. The letters come from the words, *sharp, principal, diffuse*, and *fundamental*. The early scientists used these words to describe some of the observed features of the line spectra. They are governed by the following rules:

(i)　The first energy level (corresponding to E_1 in **Activity 5**) has only one type of orbital, labeled 1*s*, where 1 identifies the energy level and *s* identifies the orbital.

(ii)　The second energy level (corresponding to E_2 in **Activity 5**) has two types of orbitals (an *s* orbital and three *p* orbitals) and are labeled as the 2*s* and 2*p* orbitals.

(iii)　The third energy level (corresponding to E_3 in **Activity 5**) has three types of orbitals, (an *s* orbital, three *p* orbitals, and five *d* orbitals) and are labeled as the 3*s*, 3*p*, and 3*d* orbitals.

(iv) The number of orbitals corresponds to the energy level you are considering. For example: E_4 has four types of orbitals (s, p, d, and f); E_5 has five types of orbitals (s, p, d, f, and g).

(v) The maximum number of electrons that can be contained in an orbital is two. Three p orbitals could contain a maximum of six electrons. The number of the type of electrons is indicated by superscript following the orbital designation. For example, $2p^5$ means five electrons in the $2p$ orbitals.

Stability is an important feature for all matter. Remember the excited electron of the hydrogen atom? If the electron were in energy level 3, it would drop down to energy level 2 and give off a specific wavelength of light. Alternatively, the electron in energy level 3, could drop down to energy level 1 and give off a different, specific wavelength of light. The word "excited" is used to describe an electron that has been promoted to a higher energy level, before it falls back down to its original state. The electron in the excited state was unstable and lost energy to go to a more stable form. Particles arranged in an unstable way will move to a more stable arrangement.

The Periodic Table

In previous activities you tried to organize elements by their properties and then by their atomic number. When elements are arranged according to their atomic numbers a pattern emerges in which similar properties occur regularly. This is the periodic law. The horizontal rows of elements in the periodic table are called **periods**. The set of elements in the same vertical column in the periodic table is called a **chemical group**. As you discovered, elements in a group share similar physical and chemical properties. They also form similar kinds of compounds when they combine with other elements. This behavior is due to the fact that elements in one chemical group have the same number of electrons in their outer energy levels and tend to form ions by gaining or losing the same number of electrons.

Chem Words

period: a horizontal row of elements in the periodic table.

chemical group: a family of elements in the periodic table that have similar electron configurations.

Checking Up

1. What is an ion?
2. What is ionization energy?
3. What are the horizontal rows of the periodic table called?
4. Explain the term chemical group.
5. Name three elements in a chemical group.

Reflecting on the Activity and the Challenge

In this activity you learned that electrons in atoms are assigned not only to energy levels but also to sublevels, labeled *s*, *p*, *d*, and *f*. You have also learned that the electron configuration of atoms of all elements in the same column of the periodic table end with the same sublevel and number of electrons in that sublevel. Mendeleev organized elements into columns based on similar chemical properties. Thus, electron energy sublevels are clearly associated with chemical properties of elements and their position on the periodic table. How will you incorporate the information about electron configuration in your game to meet the **Chapter Challenge**?

Chemistry to Go

1. Write the complete sequence of electron energy levels, from 1*s* to 4*f*.

2. Consider the element boron (B) as an example.

   a) What is boron's atomic number?

   b) How many electrons does boron have?

   c) What is the complete electron sequence for boron? (Be sure to include the number and letter of the appropriate sublevels, as well as the number of electrons in each sublevel.)

3. Answer the following questions for the element zinc:

   a) What is zinc's atomic number?

   b) How many electrons does zinc have?

   c) What is the complete electron sequence for zinc? (Be sure to include the number and letter of the appropriate sublevels, as well as the number of electrons in each sublevel.)

   d) What is the last sublevel (number and letter, please) to which electrons are added? How many electrons are in this sublevel?

   e) Where would you expect to find zinc on the periodic table? Support your prediction with your answers from (d).

   f) What other elements might you expect to have chemical properties similar to zinc? Explain your choices.

4. Answer the following questions for the element calcium:

   a) What is calcium's atomic number?

   b) How many electrons does calcium have?

c) What is the complete electron sequence for calcium? (Be sure to include the number and letter of the appropriate sublevels, as well as the number of electrons in each sublevel.)

d) What is the last sublevel (number and letter, please) to which electrons are added? How many electrons are in this sublevel?

e) Where would you expect to find calcium on the periodic table? Support your prediction with your answers from (d).

f) What other elements would you expect to have chemical properties similar to calcium? Explain your choices.

5. A chemist has synthesized a heavy element in the laboratory and found that it had an electron configuration:

$1s^2 2s^2 2p^6 3s^2 3p^6 4s^2 3d^{10} 4p^6 5s^2 4d^{10} 5p^6 6s^2 4f^{14} 5d^{10} 6p^6 7s^2 5f^{14} 6d^8$.

a) What is the number of electrons in this element?

b) What is the atomic number?

c) What might you predict about this element?

6. If the electron configuration is given you should be able to determine what element it is. Identify the following element from its electron configuration:

$1s^2 2s^2 2p^6 3s^2 3p^6 4s^2 3d^6$.

Preparing for the Chapter Challenge

Write a sentence or two to explain in words the pattern you noticed between any group and the electron configurations of the elements belonging to that group.

Inquiring Further

Determining electron configuration

In this activity, you were able to look at the electron configuration for a given element provided in the periodic table. Research other ways that the electron configuration can be determined.

Activity 7

How Electrons Determine Chemical Behavior

THE SODIUM CHLORIDE ROMANCE

GOALS

In this activity you will:

- Investigate more patterns in the electron arrangements of atoms.

- Relate the positions of elements on the periodic table, their electron arrangements, and their distances from the nearest noble gas, to chemical properties of the elements.

- Relate electron arrangements to ionization energies.

- Assign valence numbers to elements and organize the periodic table according to valence numbers.

What Do You Think?

You have learned that electron configurations determine an atom's chemical behavior. You have also learned how these electrons are labeled according to a series of energy sublevels.

- **How does the arrangement of electrons in an atom determine its chemical behavior?**

Record your ideas about this question in your *Active Chemistry* log. Be prepared to discuss your responses with your small group and the class.

Investigate

1. In **Activity 6** you learned that elements with relatively high ionization energies have stable arrangements of electrons. One particular group of elements, located in the column at the extreme right of the table, exhibit high ionization energies and therefore have stable arrangements of electrons. They are called the noble gases.

Look at the periodic table on the inside back cover for the assignment of electrons to energy sublevels for atoms of each of these elements. Focus on the sublevel at the end of the electron arrangement where the last electron is assigned.

a) Make and complete a chart like the one below in your *Active Chemistry* log. An example has been provided for you.

Element	Column A	Column B	Column C	Column D
	Energy level (number) to which the last electron is assigned	Sublevel (letter) to which the last electron is assigned	Number of electrons in the sublevel to which the last electron is assigned	Total number of all electrons of the energy level in Column A
Helium				
Neon				
Argon	3	p	6	8
Krypton				
Xenon				
Radon				

b) Look at the numbers in Column A in your chart. How are these numbers related to the respective rows of the periodic table in which each of the elements is located?

c) What pattern do you notice in Columns B and C?

d) What pattern do you notice in Column D?

2. The chemical behavior of an element can be understood by looking at the electron assignment of an atom of the element as compared to the electron assignments of neighboring noble gas atoms. The chemical inactivity of noble gas atoms reflects the stable arrangement of their electrons, one which other atoms cannot easily disturb. In the following questions you will compare the electron assignments in atoms with those of noble gases.

a) Make and complete a chart like the following one in your *Active Chemistry* log. In this chart you will compare the electron assignments of lithium, beryllium, and boron to the electron assignment of helium ($1s^2$), the closest noble gas. An example has been provided for you.

Element being compared	Number of electrons more than those found in closest noble gas (He)	Energy level (number) to which the last electron is assigned	Energy sublevel (letter) to which the last electron is assigned	Location of element (row) in the periodic table	Location of element (column) in the periodic table
Lithium	1	2	s	Row 2	Column 1
Beryllium					
Boron					

b) Make and complete another chart like the one above in your *Active Chemistry* log. This time you will compare the electron assignments of sodium, magnesium, and aluminum to the electron assignment of neon ($1s^2 2s^2 2p^6$), the closest noble gas.

Element being compared	Number of electrons more than those found in closest noble gas (Ne)	Energy level (number) to which the last electron is assigned	Energy sublevel (letter) to which the last electron is assigned	Location of element (row) in the periodic table	Location of element (column) in the periodic table
Sodium					
Magnesium					
Aluminum					

c) Describe any patterns you notice in the charts in 2(a) and/or 2(b) above.

d) Make and complete a chart like the one at the top of the next page in your *Active Chemistry* log. In this chart you will compare the electron assignments of nitrogen, oxygen, and fluorine to the electron assignment of neon ($1s^2 2s^2 2p^6$), the closest noble gas. An example has been provided for you.

e) Make and complete another chart like the one at the top of the next page in your *Active Chemistry* log. This time you will compare the electron assignments of phosphorus, sulfur, and chlorine to the electron assignment of argon ($1s^2 2s^2 2p^6 3s^2 3p^6$), the closest noble gas.

Element being compared	Number of electrons more than those found in closest noble gas (Ne)	Energy level (number) to which the last electron is assigned	Energy sublevel (letter) to which the last electron is assigned	Location of element (row) in the periodic table	Location of element (column) in the periodic table
Nitrogen	3	2	p	Row 2	Column 5
Oxygen					
Fluorine					

Element being compared	Number of electrons more than those found in closest noble gas (Ar)	Energy level (number) to which the last electron is assigned	Energy sublevel (letter) to which the last electron is assigned	Location of element (row) in the periodic table	Location of element (column) in the periodic table
Phosphorus					
Sulfur					
Chlorine					

f) Describe any patterns you notice in the charts 2 (d) and/or 2(e) above.

3. In your *Active Chemistry* log, draw a simplified periodic table that contains only the first three rows of the periodic table. Your table should have 3 rows and 8 columns, and should contain the elements with atomic numbers 1 through 18.

a) Write the symbol for each element in each appropriate box.

b) In the columns headed by lithium, beryllium, and boron, indicate how many more electrons are found in atoms of those elements than in an atom of the nearest noble gas. Place a plus sign in front of these numbers to indicate that these elements contain more electrons than their nearest noble gas element.

c) In the columns headed by nitrogen, oxygen, and fluorine, indicate how many fewer electrons are found in atoms of those elements than in an atom of the nearest noble gas. Place a minus sign in front of these numbers to indicate that these elements contain fewer electrons than their nearest noble gas element.

d) The carbon column was not listed in the table. How many fewer electrons are found in atoms of these elements than in an atom of the nearest noble gas? How many more electrons are found in atoms of these elements than in an atom of the last noble gas? What can you conclude about the position of the carbon column in respect to the other columns that you have examined?

4. The questions in **Step 2** asked you to compare the electron assignments in atoms of the second and third rows with the electron assignments of the noble gas that was closest in atomic number to those elements. You noted that electrons are added to the *s* and *p* sublevels of the energy level whose number is the same as the row of the periodic table the elements are found in. For instance, lithium has one more electron than its closest noble gas element. That additional electron is added to the *s* sublevel of the second energy level, corresponding to the second row, where lithium is located.

The number of an electron energy level is significant, because the higher the number, the greater the average distance between the nucleus and the electron. The electrons in the energy levels with the highest number are, on average, the farthest from the nucleus. Because differences in electrons located in the outermost level distinguish an atom from its nearest noble gas, these are the electrons responsible for the atom's chemical behavior. These electrons are often called valence electrons. The general definition of valence electrons are those electrons found in the outermost energy level *s* and *p* orbitals. The maximum number of valence electrons that you can have is 8 for the representative elements (2 in the *s* and 6 in *p* sublevel). For example, sodium contains one valence electron since its outermost level is 3 and sublevel is *s*. It does not

have any electrons in the *p* sublevel. Bromine would have seven valence electrons, since its outermost energy level is 4 and it contains 2 electrons in *s* sublevel and 5 electrons in *p* sublevel. Use the assignment of electrons to energy sublevels in the periodic table on the inside back cover and your answers to the questions in **Steps 1** and **2** to answer the following questions:

a) How many valence electrons are in an atom of helium? Neon? Argon? Krypton? Xenon? Radon?

b) How many valence electrons are in an atom of lithium? Sodium? Potassium? Rubidium? Cesium? This family of elements is known as the alkali metals.

c) How many valence electrons are in an atom of beryllium? Magnesium? Calcium? Strontium? Barium? This family of elements is known as the alkaline earth metals.

d) How many valence electrons are in an atom of boron? Aluminum?

e) How many valence electrons are in an atom of carbon? Silicon?

f) How many valence electrons are in an atom of nitrogen? Phosphorus?

g) How many valence electrons are in an atom of oxygen? Sulfur?

h) How many valence electrons are in an atom of fluorine? Chlorine? Bromine? Iodine? This family of elements is known as the halogens.

Chem Talk

THE NOBLE GASES

The Discovery of Argon

Imagine that you prepared two samples of nitrogen, each one liter in volume, and found one to have a mass of 1.250 g (grams) and the other a mass of 1.257 g. You might be tempted to attribute the difference to experimental error. Lord Rayleigh didn't! In 1892 he decomposed ammonia to generate one liter of nitrogen with a mass of 1.250 g. In another preparation method, he isolated one liter of nitrogen with a mass of 1.257 g by removing what he thought were all the other gases from a sample of air.

What accounted for the difference in masses of the two liter samples? Could there be yet another gas in the air that Rayleigh didn't know about? William Ramsay, a colleague of Rayleigh, looked to the experiments conducted by Henry Cavendish a hundred years earlier. Henry Cavendish (the discoverer of hydrogen) had been puzzled by a small bubble of gas remaining after he had chemically absorbed all of a sample of nitrogen he had similarly extracted from the atmosphere.

As Cavendish had done, Ramsay extracted a sample of nitrogen from the atmosphere and then chemically absorbed all of the nitrogen in that sample. He looked at the spectrum of the remaining bubble of gas, just as you have looked at the spectra of various gases in **Activity 4** and **5**. The new spectral lines of color he saw showed him that the bubble of gas was a new element, which today we know as argon. It had previously escaped notice because of its rarity and lack of chemical activity.

A New Family

When Mendeleev formulated the first periodic table argon had not been discovered, and therefore it had not been placed in the periodic table. There was no obvious place for argon. A new column was created. A prediction was therefore made that there would be other elements with similar properties to argon. The other elements of this family (He, Ne, Kr) were subsequently discovered by the end of the 19th century.

Checking Up

1. Why did one liter of nitrogen prepared by Lord Rayleigh appear to have a greater mass than the other liter prepared by a different method?

2. What are two reasons that the noble gases had escaped notice?

439

Chem Words

noble gas (also rare or inert gas): a family of elements (Group 18 or VIIIA) of the periodic table.

Unlike atoms of the other chemical elements, atoms of elements in this column are so stable that they either do not react at all, or they react only in unusual circumstances, with other elements. For this reason, this family has been known as **rare** gases (because they are rare in abundance), **inert** gases (because they are not chemically reactive), or **noble** gases.

Reflecting on the Activity and the Challenge

In this activity you have learned that the electrons in the energy sublevels with the highest number are, on the average, the farthest from the nucleus of an atom. These electrons, also known as valence electrons, determine the atom's chemical behavior. This chemical behavior is best understood in relationship to the arrangement of electrons in energy sublevels in atoms of noble gases, which, by virtue of their chemical inactivity, have a stability which is unmatched by other chemical elements. The key is an atom's excess or deficiency of electrons compared to an atom of the nearest noble gas on the periodic table. This excess or deficiency is readily indicated by the position of an element on the periodic table. How will you include this in your game about the periodic table?

Chemistry to Go

1. From the periodic table on the back inside cover, identify the excess or deficiency of electrons in an atom of a given element relative to an atom of the closest noble gas. Be sure to indicate both the number of electrons and a sign (plus or minus) to indicate whether the electrons are in excess or deficiency.

   a) calcium
   c) potassium

   b) arsenic
   d) iodine

2. Listed below are groups of three elements. For each group determine which two elements have more in common in terms of electron arrangement and therefore exhibit more similar chemical behavior. Give a reason for your selection.

   a) carbon, nitrogen, silicon
   c) sulfur, bromine, oxygen
   e) helium, neon, hydrogen

   b) fluorine, chlorine, neon
   d) sodium, magnesium, sulfur

3. Listed below are pairs of elements. For each pair determine which of the two has the most stable arrangement of electrons. You may refer to the table of first ionization energies in **Activity 6** and the periodic table on the back inside cover. Provide a statement explaining your choice in terms of ionization energy and electron arrangement.

   a) helium and lithium
   c) magnesium and chlorine
   e) neon and krypton

   b) lithium and beryllium
   d) magnesium and argon

4. a) Write the electron configuration for magnesium.

   b) Determine how many valence electrons magnesium contains.

   c) Write the electron configuration for barium.

   d) Determine how many valence electrons barium contains.

   e) How do the number of valence electrons of magnesium compare to the number of valence electrons of barium?

   f) What general statement can you make about the number of valence electrons of each element of the alkaline earth metals?

5. a) Write the electron configuration for cobalt.

   b) How many valence electrons does cobalt contain?

Inquiring Further

1. Valence electron of transition elements

How many valence electrons are there in an atom of iron (Fe, atomic number 26 and called a transition element)? When you look at the periodic table on the back inside cover, you see that iron has all the electrons in an atom of calcium plus six additional electrons in the $3d$ sublevel. Relative to the 2 electrons in the $4s$ sublevel in calcium, do these $3d$ electrons qualify as valence electrons? Explain your thinking. What can you say about the number of valence electrons in the other transition elements, from scandium to copper?

2. Ionization energies of beryllium (Be) atoms

The first three ionization energies of beryllium atoms are as follows:

1st = 1.49×10^{-18} J

2nd = 2.92×10^{-18} J

3rd = 2.47×10^{-17} J

Explain the magnitudes of the energies in terms of electron configurations and from this information determine how many valence electrons are contained in beryllium.

Activity 8

How Atoms Interact with Each Other

GOALS

In this activity you will:

- Relate patterns in ionization energies of elements to patterns in electron arrangements.

- Use your knowledge of electron arrangements and valence electrons to predict formulas for compounds formed by two elements.

- Contrast ionic bonding and covalent bonding.

- Draw electron-dot diagrams for simple molecules with covalent bonding.

What Do You Think?

You have learned that the chemical behavior of an atom is determined by the arrangement of the atom's electrons, specifically the valence electrons. The salt that you put on your food is chemically referred to as NaCl—sodium chloride.

- **How might the valence electrons of sodium (Na) and chlorine (Cl) interact to create this bond?**

Record your ideas about this question in your *Active Chemistry* log. Be prepared to discuss your responses with your small group and the class.

Investigate

1. In **Activity 3** you read that John Dalton assumed that chemical compounds formed from two elements combined in the simplest possible combination—one atom of each element. In **Activity 6** you began to see that an atom's chemical behavior reflects its excess or deficiency of electrons relative to an atom of the closest noble gas on the periodic table. Use the list of ionization energies in **Activity 6** to answer the following questions:

a) Which atoms have the smallest values for first ionization energies? (Remember, the first ionization energy is the amount of energy required to remove the first electron.)

Where are these atoms located on the periodic table?

b) What do you observe about the amount of energy required to remove the second electron from atoms of the elements identified in (a) above?

c) Use your understanding of the arrangement of electrons in this group of elements to suggest a reason for the pattern you noted in (b).

d) Which atoms have the smallest values for second ionization energies? Where are these atoms located on the periodic table?

e) Use your understanding of the arrangement of electrons in this group of elements to suggest a reason for the pattern you noted in (d).

2. Once you recognize the role of an atom's electron arrangement—especially the valence electrons—in an atom's chemical activity, you can often predict formulas for compounds formed by two chemical elements. (Recall that valence electrons are the electrons located in the highest energy level, the levels designated by the sublevel having the highest numbers.)

Sodium (Na) has one valence electron in the 3s sublevel. By losing that electron, the sodium atom becomes a sodium ion and it has the same

stable electron arrangement as neon. What is the electric charge on the resulting Na ion?

Consider a chlorine atom.

a) How many valence electrons does a chlorine atom have?

b) How many electrons does a chlorine atom need to gain to have the same number of electrons as an argon atom?

c) When a chlorine atom gains an electron, a chloride ion is formed. Since the original chlorine atom was electrically neutral and it gained a negative electron to form the ion, what is the electric charge (sign and value) on the resulting ion?

d) Each chlorine atom is capable of accepting one electron. Describe how you think the compound sodium chloride (NaCl) is formed?

3. Consider the reaction between aluminum and zinc chloride (similar to the reaction in **Activity 3**). The zinc atoms in zinc chloride have two valence electrons located in the 4s sublevel. You can note the two valence electrons in the electron arrangement marked on the periodic table.

In order to acquire the electron configuration of argon atoms a rather stable arrangement, the zinc atoms give up their two valence electrons to form zinc ions. Since the original zinc atom was electrically neutral and it lost two negative electrons to form the ion, the resulting ion has a positive charge with a magnitude two times the charge on the electron. It has a plus two charge.

443

a) Each chlorine atom is capable of accepting one electron. How many chlorine atoms are needed to accept the 2 electrons that zinc atoms have to give?

b) When writing the formula for a compound, the number of atoms necessary to balance the loss and gain of electrons can be designated through the use of a subscript, such as the 2 in H_2O. How would you write the formula for the compound zinc chloride?

4. In a reaction between aluminum and zinc chloride, aluminum replaces the zinc in the zinc chloride, forming aluminum chloride and zinc.

a) Consider an atom of aluminum. How many valence electrons does an aluminum atom have?

b) How many electrons does an aluminum atom need to give up to reach the same chemical stability as a neon atom?

c) What are aluminum atoms called after they give up their valence electrons? What is their electric charge (sign and value)?

d) How many chlorine atoms are needed to accept the electrons given up by an aluminum atom?

e) How would you write the formula for the compound aluminum chloride?

Chem Talk

FORMING COMPOUNDS

The Octet Rule

In this activity, you explained the formation of the compounds that you investigated by how the electrons are transferred or shared between atoms. Some scientists explain these observations using the octet rule. The octet rule works well with the representative elements and is stated as follows: atoms tend to gain or lose electrons during chemical reactions so that the atoms have an outer shell configuration of 8 electrons. The exceptions to this are the transition elements. They can form compounds that do not have 8 electrons in their outer shell. For example, when chlorine, one of the **halogens** (Group 17), gains one electron it now has 8 electrons in its outermost s and p sublevels (octet of electrons). Also, you should note that the name of all of the compounds that you have studied always started with the name of the metal and then followed with the nonmetal part. The second thing that you should note is that all of these compounds are binary (meaning two parts). **Binary compounds** always end with the suffix *ide* (except for a few compounds with common names like water and ammonia).

Chem Words

halogens: group VIIA (17) on the periodic table consisting of fluorine, chlorine, bromine, iodine, and astatine.

binary compound: a compound, formed from the combining of two different elements.

ionic bond: the attraction between oppositely charged ions.

covalent bond: when two atoms combine and share their paired electrons with each other.

Covalent and Ionic Bonds

You may have noticed that the column headed by carbon in the periodic table has not received a lot of attention so far. In **Activity 7** you learned that atoms of these elements contain 4 valence electrons. Atoms with a small number of valence electrons give up electrons, and atoms with a large number of valence electrons gain additional electrons to have the same electron arrangement as an atom of the nearest noble gas. Except for helium, with 2 valence electrons, the noble gases each have 8 valence electrons. What do atoms of carbon do: give up their 4 valence electrons, or gain 4 more?

In actuality, atoms of carbon do neither. Instead of giving or taking electrons, carbon atoms share electrons with atoms of other elements. Instead of giving or taking electrons to form what are called **ionic bonds**, carbon atoms share electrons with atoms of other elements to form what are called **covalent bonds**. In fact, all nonmetallic elements whose atoms have four or more valence electrons can form covalent bonds by sharing electrons. This sharing results in a situation in which each atom is associated with 8 valence electrons, as is characteristic of an atom of a noble gas.

Covalent bonding can be illustrated using electron-dot diagrams, in which each valence electron is indicated by a dot (or other appropriate symbol) around the chemical symbol for the element in question. Consider the following covalent compound illustrations:

hydrogen chloride contains one hydrogen atom and one atom of chlorine:	water contains 2 hydrogen atoms and one oxygen atom:	and carbon dioxide contains one atom of carbon and 2 oxygen atoms:
H : C̈l :	H : Ö : H	Ö :: C :: Ö

You can count 8 electrons around the chlorine in hydrogen chloride (note that hydrogen can only have 2 electrons). In water the oxygen has 8 electrons around it and hydrogen again has only 2 electrons. Carbon dioxide shows that oxygen contains 8 electrons and carbon has 8 electrons as well. These shared electrons in these examples all produce stable covalent compounds.

Checking Up

1. When naming a binary compound, which element is named first, the metal or the nonmetal? Give an example to explain your answer.

2. Explain the difference between an ionic and a covalent bond.

3. Draw electron-dot diagrams showing covalent bonding in the following compounds:

   a) water (two atoms of hydrogen, one atom of oxygen). Note: Since the noble gas nearest hydrogen is helium, with only two valence electrons, hydrogen atoms need be associated with only two valence electrons.

   b) methane (four atoms of hydrogen, one atom of carbon).

   c) ammonia (three atoms of hydrogen, one atom of nitrogen).

   d) carbon tetrachloride (four atoms of chlorine, one atom of carbon).

Reflecting on the Activity and the Challenge

In this activity you learned that atoms of two chemical elements will interact with each other in order to achieve a stable electron arrangement like that of nearby noble gases. The way in which atoms interact is based on their excess or deficiency of electrons relative to atoms of the closest noble gas on the periodic table. An atom's excess or deficiency of electrons relative to the closest noble gas is readily indicated by the position of an element on the periodic table. In this way the periodic table can be used to predict the chemical formulas when two elements interact to form a compound. This information can be deduced from the periodic table. Perhaps you can invent a way to make this more explicit as you create your periodic table game. How might you incorporate this information into your game to meet the **Chapter Challenge?**

Chemistry to Go

1. From the periodic table on the back inside cover, identify the excess or deficiency of electrons in an atom of each of the following pairs of elements relative to an atom of the closest noble gas, and then predict the formula of the compound formed when these elements interact:

   a) sodium and oxygen (to form sodium oxide),

   b) magnesium and chlorine (to form magnesium chloride),

   c) aluminum and oxygen (to form aluminum oxide).

2. a) In your *Active Chemistry* log, draw a simplified periodic table that contains only the first three rows of the periodic table. Your table should have 3 rows and 8 columns, and should contain the elements with atomic numbers 1 through 18.

   b) In the box for each element, write the symbol for the element.

   c) In the columns headed by lithium, beryllium, and boron, indicate the charge (sign and value) of the ion formed when atoms of these elements give up electrons to attain a more stable electron arrangement.

   d) In the columns headed by nitrogen, oxygen, and fluorine, indicate how many electrons are gained by atoms of those elements to attain a more stable electron arrangement.

3. You know the formula for sodium chloride (NaCl) and from this knowledge you can also write the formula for potassium chloride, cesium bromide, lithium iodide and sodium fluoride.

a) What tool makes it possible for you to do this, even though you may have never investigated these compounds?

b) Using the periodic table, explain what information you used to help explain how you arrived at the formulas.

4. The formula for calcium chloride is $CaCl_2$ and from this knowledge you should be able to write the formulas of the alkaline earth metals (Group 2) and halogens combining.

a) Write the formula for magnesium bromide, strontium iodide, beryllium fluoride, barium chloride and calcium iodide.

b) What information from the periodic table did you use to support your writing the formulas of these compounds?

5. A compound that was used previously is aluminum chloride ($AlCl_3$). Use your understanding of why elements are grouped together and are called a family, to write the formula for each of the following compounds:

a) boron fluoride,

b) aluminum bromide,

c) gallium iodide,

d) indium (III) chloride,

e) thallium (III) bromide.

6. You have explored how the alkali, alkaline earth metals, and IIIA group (boron, aluminum, gallium, indium, and thallium) combine with the halogens, but how do these metals combine with the group VIA elements (oxygen, sulfur, selenium, tellurium, and polonium)? Look at a few compounds such as sodium oxide that has a formula of Na_2O. Alkali metals like sodium "want" to lose one electron. Oxygen "wants" to gain 2 electrons. Therefore, when the alkali metals combine with oxygen or other elements of group VIA, they should have a similar formula.

a) Write the formula for each of the following compounds:

i) potassium sulfide

ii) rubidium selenide

iii) lithium telluride

iv) sodium sulfide

v) cesium oxide

You can do the same thing with the alkaline earth metals combining with the group VIA elements. Calcium oxide has the formula CaO. The alkaline earth metals want to lose 2 electrons and the group VIA elements want to gain 2 electrons. They will combine in a 1:1 ratio.

b) Write the formula for each of the following compounds:

i) magnesium sulfide

ii) strontium selenide

iii) barium oxide

iv) beryllium telluride

v) calcium sulfide

Magnification of sulfur crystal

Finally, look at the group IIIA elements combining with the group VIA elements. Aluminum oxide has the formula Al_2O_3. The group IIIA elements want to lose 3 electrons and the group VIA elements want to gain 2 electrons. In order to make a whole number exchange 6 electrons must be transferred. (Remember to make sure that your formula has the correct number of electrons being transferred.)

c) Write the formula for the following compounds:

    i) boron sulfide   ii) aluminum selenide

    iii) gallium telluride   iv) indium (III) oxide

    v) thallium (III) sulfide

Inquiring Further

Creating compounds with "inert" gases

For a long period of time the noble gases were called the inert elements, because it was assumed that they were non-reactive and did not want to gain or lose electrons. Research to find out if compounds can be formed with the noble gases. Record your finding in your *Active Chemistry* log. If time permits, share your findings with the class.

Activity 9

What Determines and Limits an Atom's Mass?

GOALS

In this activity you will:

- Investigate the composition of the atom's nucleus.

- Explain why the atomic masses of some elements are not whole numbers.

- Use symbols to represent different isotopes of an element.

- Determine the composition of the nucleus of an atom from its isotope symbol.

- Calculate the average atomic mass of an element from the percent abundance of its isotopes.

What Do You Think?

In **Activity 4** you learned that the structure of an atom includes a nucleus surrounded by electrons. Most of the mass of an atom is concentrated in the small nucleus that has a positive electric charge equal in magnitude to the negative charge of all the electrons surrounding the nucleus.

- **What do you think makes up the nucleus of the atom?**

Record your ideas about this question in your *Active Chemistry* log. Be prepared to discuss your responses with your small group and the class.

Investigate

Part A: What's in the Nucleus?

1. Atomic mass is the average mass of atoms of each element. Atomic number indicates the number of electrons in the atom and the number of protons located in the nucleus needed to produce an electrically neutral atom. Refer to the periodic table to answer the following questions:

449

a) How many protons are there in a hydrogen atom?

b) To the nearest whole number, what is the atomic mass of a hydrogen atom?

c) How many protons are there in a helium atom?

d) Since the mass of an electron is negligible, compared to the nucleus, what would you expect the atomic mass of a helium atom to be? Explain your answer.

e) To the nearest whole number, what is the atomic mass of a helium atom?

2. In **Step 1**, you found that the helium atom has a mass that is four times the mass of a hydrogen atom, while the electric charge on the helium nucleus is only twice that of the hydrogen atom. This suggests the presence of another particle in the nucleus, with about the same mass as the proton but no electric charge. This particle is called a neutron.

Sample:

Boron has atomic number 5. This informs you that there are 5 electrons and that the nucleus contains 5 protons. The average atomic mass of boron is 10.811 atomic mass units. Most boron has 11 atomic mass units and some has 10 atom mass units. Since the mass is the sum of the protons and the neutrons (electrons have very, very little mass) then you can conclude that most boron nuclides have 5 protons and 6 neutrons in the nucleus.

Refer to your table of atomic numbers and atomic masses to answer the following questions:

a) How many protons would you expect to find in the nucleus of a helium atom? (Recall that the number of protons needs to balance the number of electrons.)

b) How many neutrons would you expect to find? (The atomic mass is a combination of the mass of the protons and the mass of the neutrons.)

c) How many protons and neutrons would you expect to find in the nucleus of an atom of each of the following elements?

- lithium
- beryllium
- boron
- carbon
- nitrogen
- oxygen
- fluorine
- neon

3. Refer again to the periodic table.

a) What are the atomic masses of magnesium and chlorine? What are the atomic masses of sodium and fluorine? Which set is closer to whole numbers?

b) We expect protons and neutrons to exist in whole numbers. You cannot have part of a proton in the nucleus. What would you expect the atomic masses of most magnesium, chlorine, sodium, and fluorine atoms to be? Explain your answer.

4. The fact that some atomic masses are not close to whole number multiples of the atomic mass of hydrogen is now explained by the fact that the number of neutrons is not the same in all atoms of a given element. Only the number of protons, the atomic number, is the same in all atoms of a

given element. Atoms of the same element with different number of neutrons in the nucleus are known as isotopes (meaning "same number of protons"). Isotopes are identified by their mass number, the sum of the number of neutrons plus protons.

Sample:

Lithium has an atomic number of 3 and an average atomic mass of 6.941. All lithium atoms have 3 protons in the nucleus. A neutral atom of lithium always has 3 electrons to balance the charge of the three protons. The average atomic mass of a lithium atom is 6.941 atomic mass units, indicating that some lithium atoms have 3 neutrons, to make a total atomic mass of 6 and other lithium atoms have 4 neutrons, to make a total atomic mass of 7. These 2 isotopes are designated lithium-6 and lithium-7. Since there are so many more lithium-7 atoms, the average of all of the atoms is very close to 7.

Refer to your list of atomic masses to answer the following questions:

a) What isotopes (as indicated by their mass numbers) do you expect to account for the known atomic masses of the following elements?

- carbon (carbon-12 atoms with 6 neutrons and carbon-13 atoms with 7 neutrons; more carbon-12 atoms)

- hydrogen
- boron
- magnesium
- beryllium
- sodium

b) In the notation below, the mass number is written at the upper left of the chemical symbol of the element. The atomic number is written at the lower left of the chemical symbol of the element. How many neutrons and protons are present in the following isotopes?

   i) $^{3}_{2}He$ and $^{4}_{2}He$
   ii) $^{6}_{3}Li$ and $^{7}_{3}Li$
   iii) $^{12}_{6}C$ and $^{13}_{6}C$
   iv) $^{14}_{7}N$ and $^{15}_{7}N$

Part B: Forces within the Atom

1. There are two very different forces acting on the electrons, protons, and neutrons in the atom. In order to better understand the atom, you must first understand these forces.

Cut two strips of transparent tape about 12 cm long. Bend one end of each strip under to form a tab. Place one strip sticky-side down on a table and label the tab "B," for "bottom." Place the other strip sticky-side down

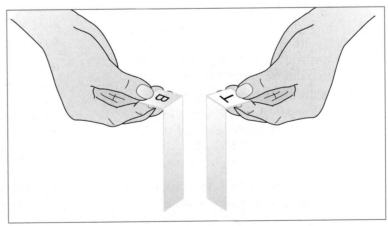

on top of the first strip and label the tab "T," for "top."

Peel off the top strip, using the tab, with one hand and then pick up the bottom strip with the other hand. Hold both strips apart, allowing them to hang down.

Slowly bring the hanging strips toward each other, but do not let them touch.

a) Record your observations.

b) If the strips accelerated toward or away from each other, Newton's Second Law tells you that there must be a force. Is the force between the two strips of tape attractive or repulsive?

2. Make a second set of strips as in **Step 1**.

a) Predict what you think will happen if the two top strips are picked up, one from each set and brought toward each other. Record your prediction in your *Active Chemistry* log.

   Pick up the two top strips by the tabs, allowing both strips to hang down. Slowly bring them toward each other.

b) Record your observations.

c) Was the force attractive or repulsive? Explain.

d) Predict what you think will happen if the two bottom strips of tape are picked up and brought toward each other. Record your prediction.

   Pick up the two bottom strips by the tabs, allowing both strips to

hang down. Slowly bring them toward each other.

e) Record your observations.

f) Was the force attractive or repulsive? Explain.

3. The two different strips of tape have different charges. The top strips have a positive electric charge. They have lost some of their electrons. Since the number of protons has remained the same, the strips are positive. The bottom strips have a negative charge. The bottom strips have gained some electrons. Since the number of protons has remained the same, the strips are negative. The force between the strips is called the electric force.

a) Is the force between two positive strips repulsive or attractive? Use evidence to justify your answer.

b) Is the force between two negative strips repulsive or attractive? Use evidence to justify your answer.

c) When a positive and a negative strip come near each other, is the force attractive or repulsive? Justify your answer.

4. The nucleus has a positive charge due to all of the protons there. The electrons surrounding the nucleus have negative charges.

a) What kind of electric force (attractive or repulsive) exists between the nucleus of an atom and any one of the atom's electrons?

b) What kind of electric force (attraction or repulsion) exists between pairs of protons in the nucleus?

452

5. The nucleus is a very crowded place. The protons in the nucleus are very close to one another. If these protons are repelling each other by an electrostatic force (and they are!), there must be another force, an attractive force, that keeps them there. The attractive force is the nuclear force, also called the strong force. This force is much stronger than the electric force. It acts between pairs of protons, pairs of neutrons, and protons and neutrons. The electron is not affected by the nuclear force.

a) Copy and complete the table below in your *Active Chemistry* log. The first row has been completed for you.

6. If the nucleus were too large, the protons on one side of the nucleus are too far away to attract the protons on the other side of the nucleus. The protons can still repel one another since the coulomb electrostatic force is long-range. The repulsive electrostatic force wins and the nucleus won't form.

A large nucleus will break apart when the electrostatic repulsion between the protons is too great.

The repulsion pushes the fragments of the nucleus apart, releasing a great amount of energy. This process of splitting an atom into smaller atoms is called fission. It occurs in uranium when an additional neutron is added and causes instability.

One example of the fission process can be represented as follows:

$$^{235}_{92}U + ^{1}_{0}n \rightarrow ^{94}_{36}Kr + ^{139}_{56}Ba + 3 \, ^{1}_{0}n + energy$$

a) Is the mass number conserved on both sides of the reaction? What is the total mass number on each side?

b) Is the atomic number conserved on both sides of the reaction? What is the total atomic number on each side?

c) Why does the neutron have a mass number of 1?

d) Why is the atomic number of a neutron equal to 0?

Small nuclei can also combine to form a larger nucleus and release energy. This process is called fusion.

Particles	Coulomb electrostatic force	Strong, nuclear force
electron-proton	*attractive*	*none*
electron-neutron		
proton-proton		
proton-neutron		
neutron-neutron		

Chem Talk

THE NUCLEUS OF AN ATOM

Discovery of the Neutron

The average atomic masses of some elements were known in Mendeleev's time, even though scientists didn't know much about the actual structure of an atom. In **Part A** of this activity you explored the idea of how the atomic mass relates to the atomic number. Mendeleev began organizing his periodic table by listing all the known elements in order of atomic mass. However, he found that organizing the elements in this way did not always make sense in terms of the behavior of the elements. He concluded that his measurements of atomic mass were incorrect and in those situations used the properties of the elements to place them in the table.

As it turned out, Mendeleev's measurements were not necessarily flawed. Although early models of the nucleus included the **proton**, the proton alone could not account for the fact that the mass of a helium atom is four times the mass of a hydrogen atom while the electric charge on the helium nucleus is only twice that of the hydrogen atom. Lord Rutherford (after discovering that atoms had a nucleus) addressed this problem when he suggested that another particle was present in the nucleus, with about the same mass as the proton but no electric charge. He named this particle the **neutron**.

The neutron was actually discovered in 1932 (by Chadwick, a British physicist), adding a great deal to the understanding of the nucleus of the atom. This discovery did not solve all of the mysteries concerning the atomic masses of some elements. Scientists today refer to protons and neutrons as nucleons since they reside in the nucleus and are almost identical in mass. The mass number tells us the number of nucleons.

Isotopes

In **Part A** of this activity you also investigated why the atomic mass of an element is not a whole number. Not all atoms of a given element have the same number of neutrons in the nucleus. Only the number of protons, the atomic number, is the same in all atoms of a given element. Atoms of the same element with different number of

Chem Words

proton: a positively charged subatomic particle contained in the nucleus of an atom. The mass of a proton is 1.673×10^{-24}g and has a charge of $+1$.

neutron: neutral subatomic particle with a mass of 1.675×10^{-24}g located in the nuclei of the atom.

neutrons in the nucleus are known as **isotopes** (meaning "same number of protons"). Isotopes are identified by their mass number, the sum of the number of neutrons plus protons.

You can refer to an element by its name (chlorine), by its atomic symbol (Cl), or by its atomic number (17). All three identifications are equivalent and used interchangeably in chemistry. The same element can have a different number of neutrons in the nucleus. Chlorine, which must have 17 protons in the nucleus, can have 18 or 20 neutrons. Chlorine with 20 neutrons and chlorine with 18 neutrons are the isotopes of chlorine (^{35}Cl and ^{37}Cl).

Electrostatic and Nuclear Forces

In **Part B** of this activity, when you brought the two positive strips near each other, they experienced a repulsive force. This was true for two negative strips as well. When a positive and a negative strip were brought close together, the force was attractive. As you have heard, "opposites attract!"

Inside the nucleus, the protons are repelling one another. Every pair of protons has a repulsive force between them. The force is very large because the distances within the nucleus are very small. The nucleus is between 10,000 and 100,000 times smaller than the atom. The electrical force can be described mathematically.

$$F = \frac{kq_1q_2}{d^2}$$

where F is the force,

k is Coulomb's constant (a number $= 9 \times 10^9$ N m^2/C^2),

q_1 and q_2 are the charges, and d is the distance between the charges.

As the distance between the charges increases the force weakens. Since the distance in the denominator is squared, if the distance triples the electrical force is 9 times (3^2) weaker or one-ninth as strong.

The question then becomes, what holds the protons together in the nucleus? The protons do have an electrical force pushing them apart but they have the larger nuclear force holding them together. The nuclear force is strong at short range. Anywhere beyond a distance of approximately 10^{-14} m (that's less than one 10-millionth of one 10-millionth of a meter), the nuclear force is zero. Neutrons in the nucleus are also attracted to each other and to protons with the nuclear force. Electrons are not affected by the nuclear force. Electrons belong to a different class of particles than protons and neutrons and do not interact with the strong nuclear force.

The nucleus is held together by a new force—the strong nuclear force. The nuclear force:

- is very, very strong at small distances;
- acts only between nucleons (proton-proton, proton-neutron, neutron-neutron);
- is always attractive;
- is very short range (if nucleons are more than 10^{-14} m apart, the nuclear force is zero).

The atom is held together by the electrostatic coulomb force. The electrostatic force:

- is strong at small distances, weak at large distances;
- acts only between charged particles (proton-proton, electron-electron, proton-electron);
- is attractive or repulsive;
- is long range (the force gets weaker at large distances).

All the nucleons are attracted by the nuclear force. The electrostatic force repelling protons in the nucleus is overwhelmed by the attractive nuclear force between these protons.

Unstable Atoms

You might expect to find nuclei of atoms with all sorts of combinations of neutrons and protons. Yet the quantity of isotopes for each element is rather small, and the number of elements is also limited. Moreover, elements do not occur in nature with atomic number greater than 92, and the highest atomic number for an atom created in the laboratory is 117.

There are two stable masses of chlorine, chlorine-35 and chlorine-37. The key word in this statement is "stable." There are other isotopes of chlorine, both heavier and lighter than chlorine-35 and chlorine-37, but they are not stable. The unstable isotopes can convert to a more stable combination of neutrons and protons, and they do so according to a systematic pattern in time. These other isotopes of chlorine are said to be **radioactive**. Understanding why certain elements are radioactive requires a deeper understanding of the structure of the nucleus. Scientists are still trying to fully understand stability of the elements.

If the nucleus of an atom is too large, the protons on one side of the nucleus are too far away to attract the protons on the other side of the nucleus. The protons can still repel one another since the coulomb electrostatic force is long-range. The interaction between the repulsive electrostatic force and the attractive nuclear force is one determining factor on the maximum size of a nucleus.

The stability of an atom varies with the elements. Light elements become more stable as the atomic mass (the number of nucleons) increases. The most stable element is iron (atomic number 26) with an atomic mass of 56. Elements with larger atomic masses become less stable.

In general, elements with nuclear mass much, much less than 56 can combine to gain mass, become more stable, and give off energy. This process is called **fusion**. Elements with nuclear mass much, much greater than 56 can break apart to lose mass, become more stable, and give off energy. This process is called **fission**.

Fusion is the process of small nuclei combining to increase their mass. The best example of fusion processes is what occurs in the Sun and other stars. The fusion process is ideal for supplying safe energy because it releases very large amounts of energy without leaving much dangerous radioactive residue. However, it is very difficult to accomplish this on an industrial level at the present time. In the future we hope scientists will figure out how to harness the energy of nuclear fusion, because it would be an excellent source of energy for society.

The process of splitting an atom into smaller atoms is called fission. This is the process that is used to produce nuclear energy. It is used to power nuclear submarines and to produce electrical energy in nuclear power plants all over the world.

Chem Words

radioactive: an atom that has an unstable nuclei and will emit alpha, positron, or beta particles in order to achieve more stable nuclei.

fusion: nuclei of lighter atoms combining to form nuclei with greater mass and release of a large amount of energy.

fission: the process of breaking apart nuclei into smaller nuclei and with the release of a large amount of energy.

Checking Up

1. Explain the difference between atomic mass and atomic number.

2. What two forces are at work in the nucleus of an atom? Explain how each works.

3. What is an isotope?

4. Why are some isotopes unstable?

5. Construct a table or diagram to compare and contrast the nuclear processes of fission and fusion.

The use of nuclear energy for the production of electricity is quite apparent as you look at the numerous states that depend on nuclear energy. For example, over 40% of Illinois' electricity is produced by nuclear energy. Nuclear fission does create some major problems: (1) Security, (2) Radiation, (3) Removal of spent rods, and (4) Disposal of waste. With these problems, there is a need for continued research. Numerous universities and government facilities are trying to improve the efficiency of nuclear fission and at the same time, trying to develop nuclear fusion for commercial use.

This ongoing research is expensive and depends on the government, industry, and other organizations to continue supporting this research. If we can learn how to harness nuclear fusion we can alleviate our nation's electrical problems while decreasing pollution. The field of nuclear science is going to continue to grow and the future will provide great opportunities for a young scientist like you to get involved.

Reflecting on the Activity and the Challenge

In **Part A** of this activity you learned that the mass of an atom, concentrated in the nucleus, is due to two types of particles, the proton and the neutron. Elements are identified by their atomic number, the number of protons in the nucleus. The atomic mass, the average mass of an atom of a given element, listed on the periodic table is a reflection of the variety of isotopes of a given element that exist. How will you incorporate your expanded understanding of the contents of an atom's nucleus, average atomic masses and isotopes into your game about the periodic table?

In **Part B** of this activity you also learned that only some combinations of neutrons and protons in a nucleus are stable, depending on the balance between the strong force holding the nuclear particles together and the electric force pushing them apart. The nuclear force is a short-range force. Beyond a distance of approximately 10^{-14} m, the nuclear force has no strength. Within that distance, this force between protons and protons, protons and neutrons, and neutrons and neutrons is quite strong. Recognizing the interplay between the electric force in the nucleus and the strong, attractive nuclear force provides an insight into the size of nuclei and the maximum size of a nucleus. These insights can be incorporated into your periodic table game in a creative way.

Chemistry to Go

1. If lithium loses an electron to become Li+, what is the average atomic mass of the lithium ion? Explain how you arrived at your answer.

2. Hydrogen has 3 isotopes with mass numbers of 1, 2, and 3. Write the complete chemical symbol for each isotope.

3. Give the complete chemical symbol for the element that contains 16 protons, 16 electrons, and 17 neutrons.

4. Complete the table below: (Use the periodic table.)

Chemical symbol	$^{39}_{19}$K			
Atomic number		9		
Number of protons			15	
Number of electrons				53
Number of neutrons		10	16	
Atomic mass				127

5. Neutrons can be used to bombard the nucleus of an atom like uranium. Why would it be more difficult to inject the nucleus of uranium with a proton?

6. Complete the following reaction: $^{235}_{92}U + ^{1}_{0}n \rightarrow ^{94}_{38}Sr + \underline{\hspace{1cm}} + 2\,^{1}_{0}n$

7. Radon is a threat to the well-being of people in their homes because it emits radioactive particles at a significant rate. Complete the following radioactive decay equation:

$$^{222}_{86}Rn \rightarrow ^{218}_{84}Po + \underline{\hspace{1cm}}$$

8. Explain why a helium atom is able to exist. What keeps the 2 electrons, 2 protons, and 2 neutrons together?

Inquiring Further

Calculating average atomic mass

If you know the percentages of abundance for the isotopes of a chemical element and the known masses of those isotopes, you can calculate the average atomic mass of that element. The process is similar to calculating the average age of students in your class — add up each person's age and divide by the number of students in your class. However, if you had to average the age of all of the students in your high school, you might choose another route. It would be easier to find out how many students are fourteen, how many are fifteen, and so on. Then you could multiply the number of students in each age group by that age. Then you would add these subsets together and divide by the total number of students.

A similar process is used to average the masses of different isotopes of an element. Consider the element chlorine. There are two stable isotopes of chlorine, chlorine-35 and chlorine-37. Of all the chlorine atoms on Earth, 75.77% of them are the isotope chlorine-35, each having a mass of 34.96885. The other 24.23% of stable chlorine atoms are the isotope chlorine-37, each having a mass of 36.96590. This means that 75.77 out of 100 chlorine atoms have a mass of 34.96885 and 24.23 have a mass of 36.96590. To find the average mass, the number of each isotope is multiplied by that isotope's mass. Then the products are added together. The sum is divided by 100, since the information pertained to 100 chlorine atoms. The result is an average atomic mass of 35.45 for chlorine, the same value stated in the periodic table. The math is shown below:

Chlorine-35 $34.96885 \times 75.77 = 2649.6$

Chlorine-37 $36.96590 \times 24.23 = 895.7$
$3545.3 \div 100 = 35.453$

Magnesium, another isotope you investigated, has three stable isotopes as follows:

mass number	isotopic mass	% abundance
24	23.98504	78.99
25	24.98594	10.00
26	25.98259	11.01

Calculate the average atomic mass for magnesium. Describe how you arrived at your answer. You may use the process described above or challenge yourself to develop your own process.

The Periodic Table Assessment

Your Chapter Challenge is to develop a game related to Mendeleev's Periodic Table of the Elements. The game should be interesting, entertaining and informative. It should demonstrate your understanding of the periodic table while it helps other students learn about the periodic table. When people begin your game, they may have no knowledge of chemistry or the periodic table. After they complete the game, they should be able to report on chemistry principles which are reflected in the periodic table.

You can begin work on the **Chapter Challenge** by reviewing the important features of the periodic table that you want to include in your game. Choose a small section of the periodic table and list what you know about each element including its atomic number, atomic mass, chemical properties, electron configuration, nuclear structure and where and why it is placed where it is in the periodic table. Next to each of these items, you may wish to describe how we know some of these details (e.g., describe the Rutherford scattering experiment for evidence of the nucleus; Law of Definite Proportions for evidence of atoms). You should review the nine activities in the chapter to add to your list. You should pay particular attention to the section **Reflecting on the Activity and the Challenge.** You should also compare your list with that given in the **Chemistry You Learned** summary.

You now have to create the game which will highlight the chemistry

principles that you have investigated in the past few weeks. There are a wide variety of games. There are traditional board games like Monopoly, Parcheesi, Clue and Risk. There are card games like gin rummy, hearts, solitaire, poker and bridge. There are TV games like *Jeopardy, Wheel of Fortune* and *The Price is Right.* There are video games like SimCity, PacMan and Space Invaders. There are also word games like crossword puzzles and anagrams. You have lots of experience with many different kinds of games. The first hurdle for your team will be to decide on the type of game that you will use as the format for communicating your knowledge of the periodic table. Your game should not merely be a periodic table test. It should be an entertaining approach to learning about the periodic table as well as displaying knowledge.

Once you decide and agree upon a game format, you will have to generate a set of rules. These rules will include whether people work individually or in groups, how one earns points or wins, how to get help during the game and suggested strategies for the game.

After the rules of the game are set, you must integrate your expert knowledge of the periodic table into the game. How will people playing the game learn that the atomic number corresponds to the number of protons and electrons in the neutral atom? How will players learn that isotopes of chlorine affect the stated mass? How will players get credit for remembering that the electron configurations are responsible for chemical behavior. The list of chemistry principles that you first completed can now be expanded so that each principle becomes the basis for a part of the game.

After the game structure is completed, you may have to design a board or some computer application so that the game can be played. Some careful preparation here can make a big difference in how well your game is accepted by others.

You should now practice the game as if you and your team members are playing it for the first time. Carefully critique your own game and edit the rules, approaches and chemistry principles so that the game is truly entertaining and educational. Review all of the criteria listed at the beginning of the chapter and the grading criteria agreed to by your teacher and class. Create a checklist and perform a final review of your checklist. Grade yourself before the presentation of your game to ensure that your team is heading for a great grade.

Chemistry You Learned

Elements' chemical and physical properties

Atoms

Single-displacement reactions

Double-displacement reactions

Atomic mass

Compound

Law of Definite Proportions

Avogadro's number and a mole

Electrons

Nucleus

Protons

Line spectrums produced by different elements

Energy, frequency, and wavelength produced by the excitation of electrons

Electromagnetic spectrum

Bohr's Atomic Theory

Orbits of electrons

The spectral lines classified as Balmer series, Lyman series, and Paschen series

Ionization potential of atom's electron

Periodicity of the elements

Atom's energy level and the sublevel energy of the outermost electron

Electron configuration

Family or chemical groups on the periodic table

Group chemical behavior

Noble gas electron configuration and stability

Valence electrons
Ionic compounds and ionic bonding

Covalent compounds and covalent bonds

Chemical formula

Neutrons

Nuclear stability

Isotopes

Fission

Fusion

Electrostatic force

Strong, nuclear force

Chapter 8

1000 mL
± 5%

800

800

700

600

Cool Chemistry Show

Scenario

The fourth- and fifth-grade students at a local elementary school have been studying chemistry in their classes. Because of the students' overwhelming interest, their teachers have asked your class to present a chemistry science show to their students. The elementary teachers have requested that the show be both interesting and informative.

For the chemistry science show, the fourth-grade teachers are asking your class to include demonstrations and explanations about chemical and physical properties and changes. The fifth-grade teacher wants the students to learn more about acids and bases, and about chemical reactions that involve color changes

Chapter Challenge

You and your classmates are being challenged to present an entertaining and informative chemistry science show to fourth- and fifth-grade students.

- The content of the show should meet the needs and interest of your audience. Remember that the fourth-grade teacher has specifically requested that the show address chemical and physical properties and changes. The fifth-grade teacher wants the students to learn more about acids and bases, and about chemical reactions that involve color changes. Your class may choose to add other presentations to enhance the show.

- All presentations must include a demonstration and an audience-appropriate explanation of the chemistry concepts involved.

- You must provide the teachers with a written summary including directions for your chemistry show with explanations of the chemistry.

- As always, safety is a top priority. You and your classmates will wear safety gear, including safety goggles, appropriate to the presentation being conducted. Presentations including flammable or explosive reactions are not appropriate for the elementary audience and may not be included in the show.

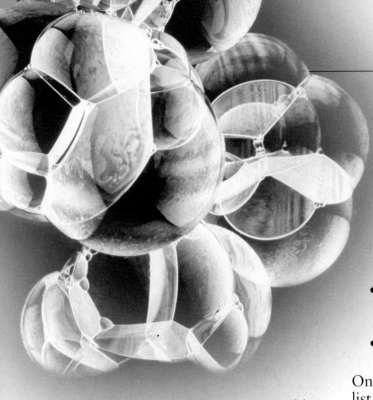

The class, as a whole, is responsible for putting on this Cool Chemistry Show. You will need to coordinate your selection of presentations to provide a show that addresses a variety of chemistry concepts in an entertaining and informative manner.

Criteria

How will your involvement in the Cool Chemistry Show be graded? What qualities should a good presentation have? Discuss these issues in small groups and with your class. You may decide that some or all of the following qualities of your presentation should be graded:

- Chemistry content
  - accuracy
  - meets teacher needs (fourth or fifth grade)
  - number of concepts addressed
- Demonstration
  - carefully planned
  - safety
  - explanation (age-appropriate)

- adherence to assigned time limits
- showmanship
- creativity
- clarity
- organization
- appeal
- written summary
- directions for experiment
- explanation of chemistry
- appropriateness for an elementary school teacher with limited chemistry background
- statements concerning safety needs

Once your class has determined the list of criteria for judging the presentations, you should also decide how many points should be given for each criterion. Determining grading criteria in advance will help you focus your time and effort on the various aspects of the presentation. How many points should be assigned to the content and how many should be assigned to the actual presentation? Will each high school student be involved in only one presentation, or more? For each criterion, you should decide on how excellence is defined and how it compares to a satisfactory effort.

Since you will be working with other students in small groups, you will need to determine grading criteria that reward each individual in the group for his or her contribution and also reward the group for the final presentation. You should discuss different strategies and choose the one that is best suited to your situation. Your teacher may provide you with a sample rubric to help you get started.

Activity 1   Chemical and Physical Changes

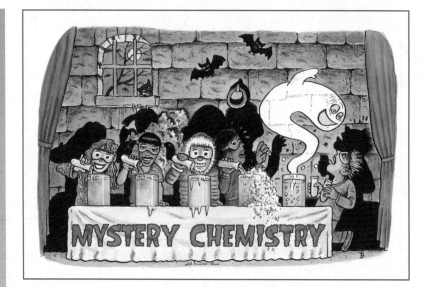

GOALS

In this activity you will:

- Learn to differentiate between chemical and physical changes.

- Make observations and cite evidence to identify changes as chemical or physical.

- Explore the new properties exhibited when new materials are made from combinations of two or more original materials.

- Design an experiment to test properties of different combinations of materials.

What Do You Think?

There are two basic types of changes that matter can undergo: chemical change and physical change. Consider two wooden matches. One is broken in half and the other is ignited by striking it along the side of the matchbox. In both of these instances matter has changed.

- **Which match has undergone a chemical change and which has undergone a physical change? Give specific reasons to support your answer.**

Record your ideas about this question in your *Active Chemistry* log. Be prepared to discuss your responses with your small group and with the class.

Investigate

1. Listed on the following page are 15 opportunities for you to observe changes in matter. Your teacher may choose to do some or all of these as a demonstration or set up stations for you to visit. Notice that the directions call for small amounts of substances.

Safety goggles and a lab apron must be worn during this activity.

Make a data table to organize your observations of the matter before and after any change(s) that may occur. Be detailed in your observations.

• Heat an ice cube in a beaker.

• Boil a small amount of water.

• Melt a small amount of candle wax. Then allow the melted wax to cool.

• Break a wooden splint into several pieces.

• Hold a wooden splint in a flame.

• Add a few drops of lemon juice to a small amount of milk.

• Add a few drops of vinegar to a small amount of baking soda ($NaHCO_3$).

• Add a small amount of table salt to water; stir; boil the solution to dryness; cool and record the result.

• Add several drops of iodine solution to a small amount of starch.

• Add a small piece of zinc to a small amount of hydrochloric acid (0.1 M HCl).

• Add a drop of phenolphthalein indicator solution to a solution of sodium hydroxide (0.1 M NaOH).

• Add two drops of sodium carbonate (0.1 M Na_2CO_3) to two drops of sodium hydrogen sulfate (0.1 M $NaHSO_4$).

• Add a few drops of household ammonia to a small amount of a copper (II) sulfate (0.1 M $CuSO_4$) solution.

• Add a few drops of vinegar to a small piece of chalk.

• Sharpen a pencil and collect the shavings.

Dispose of the materials as directed by your teacher. Clean up your work-station and then wash your hands.

2. Organize the information in your data table.

a) Prepare and complete a chart that organizes your observations into separate columns — one that includes the situations where color changes occurred, one that notes the formation of precipitates (sometimes visible as a cloudy solution), one that includes gas formation (fizz), one to note other changes, and one where no visible change occurred.

3. A physical change involves changes in the appearance of the material but does not involve creation of new materials. A chemical change involves the formation of new materials. Chemical reactions are characterized by a number of changes, including color changes and the formation of a precipitate or gas.

a) Which of the interactions you observed were chemical changes? Explain your answer.

b) Which of the interactions you observed were physical changes? Explain your answer.

c) When you placed the wooden splint into a flame, what other evidence (besides the color change) indicated that a chemical change took place?

d) Imagine a situation where two colorless solutions are mixed

467

together. There is no color change, no precipitate is formed, and no gas is released. However, heat is released as the solutions are mixed. Is this an example of a chemical or physical change? Explain your choice.

4. Each group will be given a piece of disposable diaper. Place the piece in a beaker.

a) Predict how much liquid the diaper will be able to hold. Record your prediction in your log.

b) Design an investigation to measure the amount of liquid that the diaper can absorb. Record your procedure in your log.

c) With the approval of your teacher, carry out your investigation. Record your results.

d) Explain how your prediction compared with your observations.

e) The diaper is made of a material called sodium polyacrylate. When it absorbs water, is this a physical or chemical change? Explain your answer.

5. Your teacher will show you a solution of sodium acetate in a 250-mL flask. Observe the solution carefully.

a) Record your observations in your *Active Chemistry* log.

Your teacher will then add one crystal of sodium acetate to the flask.

b) What happens? Record your observations in your log.

c) Was this a chemical or physical change?

6. In a large throwaway glass jar, mix 100 mL of sodium silicate (sometimes called water-glass solution) and 400 mL of water.

Carefully drop solid-colored crystal compounds of cobalt, copper, nickel, iron, and/or manganese in different locations inside the jar.

a) Is there evidence of a change immediately? In several minutes? In several hours? In several days? In your *Active Chemistry* log, describe the results.

b) Is the phenomenon you see the result of a physical or a chemical change? Explain your answer.

ChemTalk

CHANGES IN MATTER

Physical and Chemical Changes

In this activity you observed a number of situations that involved changes in matter, both physical and chemical. A **physical change** involves changes in the appearance of the material but does not involve creation of new materials. A change of a solid to a liquid is a physical change. When the candle wax melted it may have appeared different, but it was still wax. After it solidified, it had a similar appearance to the initial product. Dissolving is also a physical change. When you added the salt to the water, the salt crystals seemed to disappear as they dissolved in the water. However, they had only spread out into a solution. A **solution** is a homogeneous mixture of at least two different materials. The material being dissolved is called the **solute**, and the material present in the largest amount is called the **solvent**. When you boiled away the solvent, water, the solute, the salt crystals, remained the same as they were originally.

A **chemical change** involves the formation of new materials. The new materials are called **products** and the starting materials are

Chem Words

physical change: a change that involves changes in the state or form of a substance but does not cause any change in chemical composition.

solution: a homogeneous mixture of two or more substances.

solute: the substance that dissolves in a solvent to form a solution.

solvent: the substance in which a solute dissolves to form a solution.

chemical change: a change that converts the chemical composition of a substance into different substance(s) with different chemical composition.

product: the substance(s) produced in a chemical reaction.

Coordinated Science for the 21st Century

Chem Words

reactants: the starting materials in a chemical reaction.

chemical reaction: a process in which new substance(s) are formed from starting substance(s).

precipitate: an insoluble solid formed in a liquid solution as a result of some chemical reactions.

concentration: a measure of the composition of a solution, often given in terms of moles of solute per liter of solution.

saturated solution: a maximum amount of solute that can be dissolved at a given temperature and pressure.

polymer: a substance that is a macromolecule consisting of many similar small molecules (monomers) linked together in long chains.

called **reactants**. The process that brings about a chemical change is called a **chemical reaction**. Chemical reactions are characterized by a number of changes, including color changes, the formation of a **precipitate** or gas, a release of heat or light. Chemical changes are usually not easy to reverse. When you burned the wooden splint you could not put the charcoal together to form the original splint as you could when you simply broke the splint into pieces.

Saturated and Supersaturated Solutions

The solution your teacher used in the demonstration was a supersaturated solution of sodium acetate. Solutions are commonly described in terms of **concentration**. The concentration of a solution is the ratio of the quantity of solute to the quantity of solution. A dilute solution has fewer solute molecules per volume than a concentrated solution. However, in both the dilute and concentrated solutions more solute can dissolve in the solvent. These solutions are unsaturated.

You probably recognize the term "saturated." When something is saturated, it is full. A saturated sponge is full of water; it can't hold any more. A **saturated solution** is one in which no more solute will dissolve under the given conditions. To say that the sodium acetate solution is supersaturated means that it is "over full." A supersaturated solution contains more solute

particles than it normally would under the given conditions. A supersaturated solution can be made using some solutes. If a saturated solution at a high temperature is allowed to cool undisturbed, all the solute may remain dissolved at the lower temperature. The solution is then supersaturated. As you observed in the activity, such solutions are unstable. By introducing a "seed" crystal the extra solute particles "joined" the crystal and came out of the solution.

Polymers

The chemical material that you were working with when you investigated the absorbency of the diaper was sodium polyacrylate. It is a chemical compound called a **polymer**. It is made up of many (poly) repeating units of a smaller group of elements (the monomer called acrylate). This particular polymer has a unique property. It will absorb more than 800 times its own mass in distilled water. The fascinating ability of this polymer (sodium polyacrylate) to absorb large amounts of water has led to its use in a number of commercial endeavors.

Checking Up

1. What is a physical change? Provide two examples.

2. Explain the meaning of a solution, a solute, and a solvent.

3. What is a chemical change? Provide two examples.

4. What "clues" can you look for to determine if a chemical change has occurred?

5. How do you describe the concentration of a solution?

6. Explain the difference between a saturated and a supersaturated solution.

Reflecting on the Activity and the Challenge

Recall that the fourth-grade teacher has specifically requested that your chemistry show addresses chemical and physical properties and changes. You are right on track for the fourth graders. The fifth-grade teacher wants the students to learn more about chemical reactions that involve color changes. You have seen a few of those, too. If you had to conduct the show based on your experiences so far, which activity would you use? What additional information would you need to be able to explain the chemistry to fourth- and fifth-grade students?

Chemistry to Go

1. Which of the following are chemical changes and why?

   a) Toast turns black after being in the toaster too long.

   b) Water condenses on the outside of a glass of iced tea.

c) Green leaves turn orange, yellow, and red in the fall.

d) Green bananas become yellow.

e) Butter melts on a hot summer day.

2. Think back to a recent lunch or dinner. Describe two physical and two chemical changes that were involved in the meal and explain why you think each was a physical or chemical change.

3. Write a paragraph describing a common activity (such as making a cake or driving a car). Underline the physical changes (use one line) and chemical changes (use two lines) taking place within the activity. Select and describe an activity that is sure to have at least two physical changes and two chemical changes.

4. The following information is obtained for the element aluminum. Identify which are physical (use one line) and which are chemical (use two lines) properties.

Aluminum is a shiny silver metal and melts at 660°C. When a strip of aluminum is placed in hydrochloric acid, hydrogen gas is released. The density of aluminum is 2.70 g/cm^3. When polished aluminum is exposed to oxygen over a period of time it forms aluminum oxide (Al_2O_3) on the surface of the metal.

5. How would you determine whether a clear solution in a beaker is saturated sugar water or just water? Remember, you do not taste samples in the laboratory.

Preparing for the Chapter Challenge

Describe how you would demonstrate the difference between a physical and chemical change in a "cool" way.

Inquiring Further

Factors affecting solubility and the rate of dissolving

Understanding the factors that affect how quickly a solute dissolves in a solvent is important in many practical applications in manufacturing. Design an investigation to determine the factors that affect solubility. Consider the following:

• nature of the solute and solvent

• temperature

• agitation (stirring or shaking)

• surface area (for example, try using a sugar cube, granulated sugar, and icing sugar)

• pressure of gases

Remember that your investigation must be controlled, if your results are to be reliable. What will be your independent and what will be your dependent variables?

Activity 2   More Chemical Changes

GOALS

In this activity you will:

• Observe several typical examples of evidence that a chemical change is occurring.

• Make generalizations about the combinations of materials that result in the same evidence.

• Make generalizations about materials that tend to react with everything and materials that tend not to react with anything.

• Practice careful laboratory techniques, such as avoiding contamination of reactants, to ensure that results observed are repeatable and unambiguous.

What Do You Think?

Mix 1 cup flour, 1/3 cup sugar, 1 teaspoon of baking powder with a cup of milk and 1 egg, well-beaten. Place the mixture in an oven for 30 minutes.

Add two drops of sodium carbonate (0.1 M Na_2CO_3) to two drops of sodium hydrogen sulfate (0.1 M $NaHSO_4$).

• **Which of the instructions above will result in a chemical reaction? Why?**

• **Describe one similarity and one difference in the above instructions.**

Record your ideas about these questions in your *Active Chemistry* log. Be prepared to discuss your responses with your small group and with the class.

Investigate

1. Eight solid materials listed on the next page have been dissolved in distilled water to make solutions. You will combine the solutions (one to one) with each other in an organized manner in order to observe their interactions.

473

Safety goggles and a lab apron must be worn during this activity.

- barium nitrate $Ba(NO_3)_2$
- sodium hydroxide (NaOH)
- sodium hydrogen carbonate ($NaHCO_3$)
- copper (II) sulfate ($CuSO_4$)
- potassium iodide (KI)
- silver nitrate ($AgNO_3$)
- iron (III) nitrate $Fe(NO_3)_3$
- hydrochloric acid (HCl)

After mixing two solutions, make notes on your chart of any changes you observe. Don't overlook any color changes, the formation of a precipitate (sometimes observed as a cloudy solution), the formation of a gas (fizzing or bubbles), or a change in temperature.

Using another dropper, continue by adding three drops of the sodium hydrogen carbonate to the second well.

	$Ba(NO_3)_2$	NaOH	$NaHCO_3$	4	5	6	7	8
$Ba(NO_3)_2$								
NaOH								
$NaHCO_3$								
4								
5								
6								
7								
8								

Silver nitrate will stain skin and clothing. Handle with care.

a) Begin by making a chart to record your data. Your chart will require an entire page of your notebook. Allow plenty of room to record your observations. A sample chart has been provided. Notice that some of the blocks in this chart are shaded, indicating there is no need to mix those particular chemicals. Why do you suppose those particular blocks are shaded?

2. Now it is time to mix the solutions.

Begin with barium nitrate. Add three drops of the barium nitrate solution to each of seven wells of a well plate. Add three drops of sodium hydroxide solution to the first well.

It is important that you do not allow the tip of the dropper of one solution to come in contact with another solution. Your attention to this detail will prevent contamination of solutions.

Continue by adding copper (II) sulfate to the third well, and so on.

a) After mixing the pairs of solutions, make note on your chart of any changes you observe.

Continue by putting three drops of sodium hydroxide into each of seven wells and adding the other solutions.

After completing your *entire* chart in this fashion and mixing all possible one-to-one combinations of solutions, clean up your workstation. Your teacher will provide disposal information. Wash your hands.

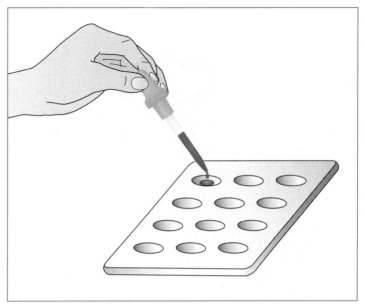

3. Use your chart to answer the following questions:

a) Which combination of reactants seems to produce no reaction when mixed together?

b) Which combination of reactants forms a gas? Can you guess which gas is formed? Try to deduce this from the reactants' names and chemical formulas.

c) Which combination of reactants produces a color change when mixed together?

d) Which combination of reactants forms precipitates quickly? Slowly?

e) Which combination of reactants forms a yellow precipitate? A muddy brown precipitate? A white precipitate? A blue precipitate?

f) Which combination of reactants produces heat? How could you tell?

g) What evidence indicates that a chemical change is occurring?

4. Place the following chemicals in a quart-size resealable plastic bag with a zipper seal:

One teaspoon (scoop)(~28 g) of calcium chloride ($CaCl_2$)

One teaspoon (scoop)(~28 g) baking soda ($NaHCO_3$)

Seal the bag and mix the powders.

a) Record your observations in your *Active Chemistry* log. Did a chemical reaction occur?

Pour 10 mL of phenol red indicator solution into the bag and seal quickly. Make sure the solids come in contact with the indicator solution.

b) Observe the reaction and, in your *Active Chemistry* log, describe what you see.

c) Did a chemical reaction occur in the plastic bag? If so, identify all of the evidence of the chemical change.

d) For this particular reaction, calcium chloride and sodium hydrogen carbonate combined to produce an aqueous solution of sodium chloride and calcium oxide in addition to the carbon dioxide and water.

What are the names of the reactants? What are the names of the products?

5. Your teacher will provide you with a small amount (~25 mL) of limewater, a solution of calcium hydroxide ($Ca(OH)_2$), in a beaker or flask.

Gently blow through a straw into the solution for a minute or so. One end of the straw should be submerged in the solution. You are actually bubbling carbon dioxide through the solution.

a) Did a chemical reaction occur? What is the evidence?

Chem Talk

TESTS FOR CHEMICALS

Chemical Tests for Gases

In this activity you focused on chemical reactions, those processes that result in the formation of new products. You also tested for the presence of some of the new materials. You used **chemical tests** to identify the unknown substances. A chemical test is a form of a diagnostic test. To test for the presence of oxygen, you introduce a glowing splint into a test tube with a small amount of gas. If the splint bursts into a flame, you then know that the gas is oxygen. When you introduce a burning splint into a test tube and heard a loud pop, you assume the gas present to be hydrogen. In this activity you tested for the presence of carbon dioxide. Since carbon dioxide does not burn or support burning, by using a glowing or burning splint, you could not tell if a gas was carbon dioxide.

(If the splint is extinguished you can say that the gas is neither oxygen nor hydrogen and therefore could be carbon dioxide.) The test for carbon dioxide uses limewater, a clear, colorless solution of calcium hydroxide in water. When you blew bubbles into the test tube you were actually blowing carbon dioxide from your lungs into the limewater. The carbon dioxide reacted with the calcium hydroxide forming a precipitate. The precipitate caused the limewater to turn cloudy in appearance.

Indicators for Acids and Bases

When acids and bases are involved in a chemical reaction the appearance of the products is often very similar to the appearance of the reactants. (You will learn more about acids and bases in a later activity.) Therefore, indicators are used to determine the presence of an acid or base. Substances that change color when they react with an acid or a base are called **acid-base indicators**. In this activity you used phenol red, an acid-base indicator that turns yellow in the presence of an acid. Chemists use a great variety of acid-base indicators. You may also have used litmus in previous science classes. It is a very common indicator used in school laboratories.

Checking Up

1. What is a chemical test?
2. Describe how you can use a burning or glowing splint to test for hydrogen or oxygen.
3. Why does a glowing splint test not work with carbon dioxide?
4. What test is used to identify the presence of carbon dioxide?
5. What is a precipitate?
6. What are acid-base indicators and how are they useful?

Chem Words

acid-base indicator: a dye that has a certain color in an acid solution and a different color in a base solution.

Reflecting on the Activity and the Challenge

In this activity you saw evidence of chemical changes taking place when you observed a color change, a change in temperature, a gas being emitted, a precipitate being formed, or light being produced. Which of these chemical reactions would be an exciting or informative addition to your class's **Cool Chemistry Show**? Does your class want to be sure to include a variety of reactions that provide different types of evidence of a chemical reaction, or does your class just want to highlight a few of them? These are decisions your class will need to make as you build your **Cool Chemistry Show**.

Chemistry to Go

1. In both **Activity 1** and **Activity 2** you gathered evidence for chemical changes. However, this evidence does not always indicate a chemical change. For instance, a change in color can be evidence of a chemical change. However, when you add water to a powdered drink mix, the color often changes, but a chemical change has not taken place.

   In each of the following situations indicate whether the evidence suggests a chemical change or not:

   a) An acid is dissolved in water and heat is released.

   b) A burning match produces light.

   c) A "seed" crystal is placed in a supersaturated solution and the extra solute particles "join" the crystal and come out of the solution.

   d) A bottle of a carbonated beverage is opened and carbon dioxide is released.

   e) The glowing filament of a light bulb produces light.

   f) A small piece of metal is placed into an acid and hydrogen is released.

   g) Solutions of sodium hydroxide and copper (II) sulfate are mixed and a blue precipitate appears.

2. Anhydrous copper (II) sulfate ($CuSO_4$) is a white solid. When it is dissolved in water, the solution becomes blue. Is this a chemical change? Explain how you would defend your answer.

3. If a glass of carbonated soda drink is allowed to sit out for a period of time, you find that the drink seems to be flat. Discuss this observation in terms of whether this is a physical or chemical change.

Preparing for the Chapter Challenge

Select one of the reactions you observed in this activity that you thought was pretty cool. Describe how you might incorporate it into a possible event in the **Cool Chemistry Show** you are designing. Would it meet the needs of the fourth-grade teacher, the fifth-grade teacher, or both? What additional information would you need to be able to explain the chemistry to the audience?

Activity 3   Chemical Names and Formulas

GOALS

In this activity you will:

• Predict the charges of ions of some elements.

• Determine the formulas of ionic compounds.

• Write the conventional names of ionic compounds.

• Make observations to determine whether there is evidence that chemical changes occur on combining two ionic compounds.

What Do You Think?

Chemistry is the study of matter and its interactions. A great contribution to the study of chemistry was the discovery that all of the world is composed of elements and that there are approximately 100 elements in nature. The periodic table provides valuable information about each element.

• **What information is provided for the element shown? What significance does this information have?**

Record your ideas about these questions in your *Active Chemistry* log. Be prepared to discuss your responses with your small group and the class.

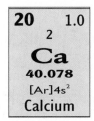

20	1.0
	2

Ca
40.078
[Ar]4s^2
Calcium

Investigate

1. The periodic table lists the elements in order of their atomic number. The atomic number is the number of protons (positively charged particles) in the nucleus of one atom of that element. For a neutral atom, the number of protons also equals the number of electrons (negatively charged particles).

Coordinated Science for the 21st Century

Electrons are found outside the nucleus. A helium atom, with an atomic number of 2, has 2 protons in the nucleus and 2 electrons surrounding its nucleus.

For each of the following elements, write the symbol for the element and indicate the number of protons an atom of that element would have. (Refer to the periodic table.)

a) copper   b) sulfur

c) zinc   d) gold

e) oxygen   f) carbon

g) silver   h) chlorine

i) nitrogen   j) hydrogen

k) magnesium   l) iodine

m) iron   n) calcium

o) aluminum   p) sodium

q) potassium   r) lead

2. Elements can combine to form compounds. A compound results when two or more different elements bond. Some compounds are comprised of positive and negative ions that are bound by their mutual attraction. An ion is an atom that has lost or gained electrons, and therefore is charged because its protons and electrons no longer balance and cancel each other. For example, when a chlorine atom gains one electron, it becomes a chloride ion with a charge of −1 (remember electrons have negative charge). When a sodium atom loses one electron, it becomes a sodium ion with a charge of +1 (because now there is one more proton than the number of electrons). The resulting compound is sodium chloride (NaCl), which you know as table salt and it is an ionic compound.

a) The chemical formula for the compound of potassium and bromine is KBr. Look at where potassium is located on the periodic table (Group 1) and also where bromine is located (Group 17). Each of these has an ionic charge of 1. Potassium is +1, bromine is −1. List four other compounds that are created from elements in Group 1 combining with elements in Group 17.

b) Magnesium forms an ion with a charge of +2 and oxygen forms an oxide ion with a −2 charge. The chemical formula for magnesium oxide is MgO. List four other compounds that are created from elements in Group 2 combining with elements in Group 16.

3. If the values of the charge on a positive ion and a negative ion are the same, the formula of the resulting compound is simply the chemical symbols of each element. If the values of the charge on a positive ion and a negative ion are not the same, subscripts can be used to balance them. For example, aluminum loses 3 electrons to become an ion with a charge of +3. An iodine atom gains only 1 electron to form an ion with a charge of −1. It takes 3 iodine atoms to accept the 3 electrons given up by aluminum. This is reflected in the formula AlI_3. (Note where the 3 is placed for the 3 iodine atoms.) Another example is $CaCl_2$, where 2 chloride ions (each gaining one electron) and one calcium ion (having lost 2 electrons) combine.

Write the chemical formula and name for the compound formed when the following pairs of elements are combined:

a) calcium and oxygen

b) aluminum and fluorine

c) boron and oxygen

d) strontium and nitrogen

e) barium and selenium

4. Some compounds, like baking soda, sodium hydrogen carbonate ($NaHCO_3$), incorporate polyatomic ions. Polyatomic ions are made up of several elements joined together. In the case of baking soda, the sodium (Na) ion has a charge of +1 and the hydrogen carbonate ion (the polyatomic ion HCO_3) has a charge of −1. (Note: hydrogen carbonate ion is also called bicarbonate ion.)

Write the chemical formula for each compound below.

a) potassium nitrate (nitrate: NO_3^{-1})

b) barium sulfate (sulfate: SO_4^{-2})

c) potassium sulfate

d) sodium acetate (acetate: $C_2H_3O_2^{-1}$)

Write the name for each compound below.

e) $(NH_4)_2SO_4$ (ammonium: NH_4^{+1})

f) $Al_2(CO_3)_3$

g) $LiHCO_3$

h) HNO_3

5. You have learned about ionic compounds that are made from positive and negative ions. In another class of compounds, called molecules, the atoms are bound by electrons being mutually attracted to the protons in adjacent atoms. These bonds are called covalent bonds,

because atoms are sharing electrons. It is often useful to imagine, however, that the atoms inside of molecules are charged. These "imagined charges" are called oxidation numbers.

a) The formula for carbon dioxide is CO_2. If you pretend this is an ionic compound, what is the charge (oxidation number) of carbon?

b) Carbon monoxide is CO. What is the oxidation number of carbon now?

c) Explain how you arrived at your answers.

You must wear safety goggles and a lab apron.

6. Do chemical changes occur every time reactants are mixed? Let's find out. Read the directions for this step so you can prepare a data table to record and describe all that you observe.

Put equal amounts of baking soda, crushed Alka-Seltzer™ tablet, and baking powder into three separate test tubes respectively. Be sure to label the test tubes!

Add equal amounts of water to each.

Hold the splint with tongs or wear a heatproof glove. Be sure the mouth of the test tube is pointed away from everyone.

a) Record your observations. (You should now know the chemical formula for baking soda $NaHCO_3$.)

b) Place a glowing splint into the top of each test tube. Make note of what happens. A glowing splint bursts into flames in the presence of oxygen, and a glowing splint is extinguished in the presence of carbon dioxide. Which gases were most likely given off for each reaction?

7. When the reactions have stopped completely, your teacher will put three of the test tubes in a beaker of boiling water. Observe what happens.

   a) Make a note of the results in your *Active Chemistry* log.

8. Repeat **Steps 6** and **7** using clean test tubes, fresh reagents, and instead of water add

   • vinegar   • ammonia.

9. Your teacher will give you a small amount of a white powdered substance that is either baking soda, Alka-Seltzer, or baking powder.

   a) Write down the number of your unknown and determine which of the three substances it is. Provide evidence to support your conclusion.

10. Clean all apparatus and the laboratory bench when you are finished. Dispose of all chemicals as directed by your teacher. Wash your hands.

Chem Talk

FORMING COMPOUNDS

Ionic Compounds

There are certainly more than 100 physically different materials in this world. With approximately 100 elements, how is it possible to have such a variety of materials? How is it possible to invent new materials for clothing, building, and food? Elements can combine to form compounds. A **compound** results when two or more different elements bond.

Some compounds are comprised of positive and negative **ions** that are bound by their mutual attraction. An ion is an atom that has lost or gained electrons, and therefore is charged because its protons and electrons no longer balance and cancel each other. For example, when an iodine atom gains one electron, it becomes a iodide ion with a charge of -1 (remember electrons have negative charge). A negatively charged ion is called an **anion**. When a potassium atom loses one electron, it becomes a potassium ion with a charge of $+1$ (because now there is one more proton than the number of electrons). A positively charged ion is called a **cation**. The resulting **ionic compound** is potassium iodide (KI). Potassium iodide is added to most of the table salt you use. Table salt (NaCl) is another example of an ionic compound.

Chem Words

compound: a material that consists of two or more elements united together in definite proportion.

ion: an electrically charged atom or group of atoms that has acquired a net charge, either negative or positive.

anion: a negatively charged ion.

cation: a positively charged ion.

ionic compound: a compound consisting of positive or negative ions.

If you refer to a periodic table you will notice that elements that form positive ions are on the left side of the table and elements that form negative ions are on the right side. Metals combine with nonmetals to form ionic compounds. Also, when two elements combine the name given the negative ion will end with -ide, and the compound is named with the metal or positive ion first. This is true for binary compounds, for example: sodium chloride, potassium bromide, and magnesium oxide. When dissolved in water, you would find that ionic compounds conduct electricity. In this activity you also investigated some compounds formed with **polyatomic ions**. Polyatomic ions are made up of several elements joined together. For example, Milk of Magnesia™ incorporates the hydroxide ion (OH^{-1}) with the magnesium ion to form magnesium hydroxide, $Mg(OH)_2$. The following table lists some polyatomic ions and their charges.

Polyatomic Ions		
nitrate	NO_3^{-1}	negative one, −1
sulfate	SO_4^{-2}	negative two, −2
hydroxide	OH^{-1}	negative one, −1
carbonate	CO_3^{-2}	negative two, −2
hydrogen carbonate	HCO_3^{-1}	negative one, −1
acetate	$C_2H_3O_2^{-1}$	negative one, −1
ammonium	NH_4^{+1}	positive one, +1

Molecular Compounds

You also learned about **molecular compounds** in this activity. When two atoms of molecular elements come together, neither atom gains nor loses an electron. Instead, the bonding electrons are shared between the two atoms. The mutual attraction of two nuclei for a shared pair of bonding electrons is called a **covalent bond**. Molecular compounds are usually formed by nonmetal-nonmetal combinations. You would find that when dissolved in water, molecular compounds do not conduct electricity.

With covalent bonds you also found that it is often useful to imagine that the atoms inside of molecules are charged. These "imagined

➡

Chem Words

polyatomic ion: an ion that consists of two or more atoms that are covalently bonded and have either a positive or negative charge.

molecular compound: two or more atoms bond together by sharing electrons (covalent bond).

covalent bond: a bond formed when two atoms combine and share their paired electrons with each other.

Checking Up

1. If there are only about 100 elements in this world, why are there so many different materials?

2. What is an ion?

3. How are ionic compounds formed?

4. What is a polyatomic ion? Provide an example of a compound formed with a polyatomic ion.

5. How are molecular compounds formed?

6. Distinguish between an ionic and a covalent bond.

Chem Words

oxidation number:
a number assigned to an element in a compound designating the number of electrons the element has lost, gained, or shared in forming that compound.

charges" are used as a type of bookkeeping and are called **oxidation numbers**. In both ionic compounds and molecular compounds the atoms achieve a stable state, similar to the noble gases.

Reflecting on the Activity and the Challenge

In this activity you have learned how to write the formulas for many compounds and how to name some compounds. You have also investigated both ionic and molecular compounds. As you prepare your presentation for your **Cool Chemistry Show**, you will want to include your knowledge of formulas, the names of compounds, and the different kinds of compounds. Remember that you will be providing the teacher with an explanation of why you included certain demonstrations, and you will also want to include explanations that are grade-appropriate. Think about how much information you will need to provide for each demonstration.

Chemistry to Go

1. Write the chemical formula and name for the compound formed when the following pairs of elements are combined:

   a) sodium and bromine

   b) potassium and sulfur

   c) magnesium and chlorine

   d) cesium and iodine

   e) aluminum and oxygen

2. Write the chemical formula for each of the following:

   a) hydrogen nitrate (nitric acid)   b) ammonium hydroxide

   c) calcium carbonate   d) hydrogen acetate (acetic acid)

3. a) Write the chemical formula for copper (II) sulfate. The (II) indicates that this copper ion has a +2 charge.

   b) Oxygen ions usually have a negative 2 charge (–2). How would formulas for iron (II) oxide differ from iron (III) oxide?

4. You may have noticed that all the elements in the first column of the periodic table (the alkali metals: lithium, sodium, potassium, rubidium, and cesium) have a +1 charge when they combine with negative ions. Another group of positive ions are the alkaline earth metals (beryllium, magnesium, calcium, strontium, and barium), located in the second column of the periodic table. What charge is typical for ions of the alkaline earth metals?

5. The formula for sodium phosphate is Na_3PO_4. What is the charge on the polyatomic phosphate ion? What information did you use to arrive at your answer?

6. When you write the formula for sodium hydroxide, you do not have to put parentheses around the hydroxide polyatomic ion. However, when writing the formula for aluminum hydroxide, you must put parentheses around the hydroxide polyatomic ion.

   a) Write each formula.

   b) Explain why the parentheses are necessary for aluminum hydroxide.

7. a) If the chemical formula for iron (III) chloride is $FeCl_3$, what is the chemical formula for iron (III) nitrate?

   b) If the chemical formula for lead (II) oxide is PbO, what is the chemical formula for lead (II) sulfate?

   c) If the chemical formula for silver chloride is AgCl, what is the chemical formula for silver nitrate?

8. In **Activity 2**, you tested various compounds for chemical changes. (Barium nitrate, sodium hydroxide, sodium hydrogen carbonate, copper (II) sulfate, potassium iodide, silver nitrate, iron (III) nitrate, and hydrochloric acid.) Write the chemical formulas for each of the reactants.

Preparing for the Chapter Challenge

Review any chemical reactions you are considering including in your **Cool** **Chemistry Show**. Write the formulas of any compounds that you plan to use.

Activity 4   Chemical Equations

GOALS

In this activity you will:

- Represent chemical changes using word equations and chemical equations.

- Distinguish between different classes of chemical reactions.

- Predict the possible products of single-displacement and double-displacement reactions.

- Determine whether a reaction has occurred based on evidence observed.

- Use the principle of conservation of matter to balance chemical reactions.

What Do You Think?

In **Activity 1** you mixed zinc metal and hydrochloric acid. You noticed that a gas was produced from the fizzing that occurred.

- **How could you communicate (tell about) the reactants and products of this reaction in a compact way?**

Record your ideas about this question in your *Active Chemistry* log. Be prepared to discuss your responses with your small group and with the class.

Investigate

1. Watch closely as your teacher shows you some cool chemistry.

   a) Record your observations in your *Active Chemistry* log.

2. Here's how the cool chemistry was done. Into each of three beakers that appeared empty, your teacher added about 45 mL of 2.0 M ammonium hydroxide solution.

   Before beginning the demonstration, your teacher had also added the following to each beaker:

Beaker One—20 drops of the indicator phenolphthalein solution;

Beaker Two—15 drops of 1 M magnesium sulfate solution;

Beaker Three—15 drops of 1 M copper (II) sulfate solution.

a) Did a chemical reaction take place in each beaker? What evidence do you have to justify your answer?

3. There are many chemical reactions that can occur. You have already observed some of them. Chemists group most of these reactions into four main categories. They are:

- synthesis reactions
- decomposition reactions
- single-displacement reactions
- double-displacement reactions

a) What do the words synthesis and decomposition mean?

4. In a synthesis reaction two or more chemicals combine to form a compound.

$$A + B \rightarrow AB$$

Here is an example of a synthesis reaction. When magnesium and oxygen react, a white solid, magnesium oxide is formed.

This can be written as a word equation:

Magnesium (solid) and oxygen (gas) produce magnesium oxide (solid).

This can also be written using a chemical equation:

$$?Mg + ?O_2 \rightarrow ?MgO$$

(The subscript communicates the number of atoms in one molecule.

Oxygen is diatomic, that means that it exists as a molecule made up of two atoms.)

a) What do you think are the advantages of writing a reaction using chemical symbols?

Any equation in chemistry must follow scientific laws or principles. The number of atoms of each element must be equal before and after the reaction.

b) In the equation above, how many atoms of oxygen are in the reactants (before the reaction)?

c) How many oxygen atoms are in the product (after the reaction)?

d) What appears to be the problem with the equation? How could you correct this problem?

Write a 2 in front of the MgO.

$$?Mg + O_2 \rightarrow 2MgO$$

(The number in front of a chemical formula, called a coefficient, communicates the number of molecules or formula units that are involved in the reaction. In this equation there are two molecules of magnesium oxide represented. That is, there is a total of two magnesium atoms and two oxygen atoms.)

e) How many magnesium atoms are now represented in the product?

f) How many reactant atoms of magnesium are shown?

g) What now appears to be the problem with the equation? How could you correct this problem?

Write a two in front of the Mg.

$$2Mg + O_2 \rightarrow 2MgO$$

h) The chemical equation above is now balanced. (The number of magnesium and oxygen atoms in the product is equal to the number in the reactant.) In your own words, explain the meaning of a balanced equation. How does the chemical equation communicate what happened in the reaction, and how does it follow the Law of Conservation of Matter?

When writing a chemical equation the states of the reactants and products are also given. The following symbols are used:

• (s) for solid

• (l) for liquid

• (g) for gas

• (aq) for aqueous, meaning in a water solution

The complete balanced chemical equation for the reaction of magnesium and oxygen is:

$$2\ Mg_{(s)} + O_{2(g)} \rightarrow 2\ MgO_{(s)}$$

5. Write a word equation and a balanced chemical equation for each of the following synthesis reactions.

Note that there are eight elements that are diatomic, that means that they exist as a molecule comprised of two atoms. They are hydrogen (H_2), nitrogen (N_2), oxygen (O_2), fluorine (F_2), chlorine (Cl_2), bromine (Br_2), iodine (I_2) and astatine (At_2). If you need to include any of these elements in an uncombined state in a chemical equation, don't forget the 2 as a subscript.

a) Solid carbon (C) burns in air (oxygen gas) to form carbon dioxide gas (CO_2).

b) Hydrogen gas reacts with oxygen gas to form liquid water (H_2O).

c) A piece of solid iron (Fe) over time will react with oxygen to form iron (III) oxide (Fe_2O_3).

d) A piece of solid sodium (Na) is dropped into a container of chlorine gas to produce solid sodium chloride (NaCl).

6. Water can be separated into its elements with an input of energy. The equation for this reaction is:

Water (liquid) and energy produces hydrogen (gas) and oxygen (gas).

$$2H_2O_{(l)} + energy \rightarrow 2\ H_{2(g)} + O_{2(g)}$$

a) Is the equation properly balanced? How did you check?

488

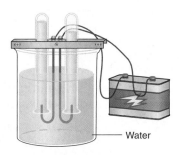

**Decomposition of water
into $H_{2(g)}$ and $O_{2(g)}$ by electrolysis**

When a substance breaks down into its component parts, the process is called a **decomposition** reaction.

form products. Such was the case when solid zinc was dropped into hydrochloric acid in **Activity 1**, forming hydrogen gas and aqueous zinc chloride solution.

The equation for this reaction is:

Zinc (solid) and hydrochloric acid (aqueous) produces hydrogen (gas) and zinc chloride (solution).

$$Zn_{(s)} + 2HCl_{(aq)} \rightarrow H_{2(g)} + ZnCl_{2(aq)}$$

a) Check to ensure that the chemical equation is properly balanced. Complete the following table in your log.

Number of Atoms			
	Before	**After**	**Balanced**
Zn	I	I	yes
H			
Cl			

$AB \rightarrow A + B$

Write word and balanced chemical equations for the following decomposition reactions. (Remember that some elements are diatomic –H_2, N_2, etc.):

b) sodium chloride solid ($NaCl_{(s)}$)

c) potassium iodide solid ($KI_{(s)}$)

d) magnesium bromide solid ($MgBr_{2(s)}$)

7. The reactions mentioned above involve elements combining to form compounds or compounds breaking up to form elements.

There are other reactions that involve elements reacting with compounds to

The reaction with zinc and hydrochloric acid is called a single-displacement reaction because zinc replaces the hydrogen in the acid.

$A + BC \rightarrow B + AC$

Write word and chemical equations for the following:

b) A piece of iron (Fe) is added to an aqueous solution of copper (II) sulfate ($CuSO_4$) and produces iron sulfate ($FeSO_4$) and copper.

c) Solid lead (Pb) is added to an aqueous solution of silver nitrate ($AgNO_3$) and produces lead nitrate $Pb(NO_3)_2$ and silver.

Safety
goggles and
a lab apron
must be
worn during
this activity.

d) Aluminum foil (Al) is placed in a beaker of aqueous copper (II) hydroxide ($Cu(OH)_2$) and produces aluminum hydroxide ($Al(OH)_3$) and copper.

Balance each of the equations, if you have not done so.

8. Another type of reaction is a double-displacement reaction.

$$AB + CD \rightarrow CB + AD$$

Try some double-displacement reactions on your own in your group. Use the chart below to guide your work. The compounds are in water solution.

a) Record your observations of the reactants before you mix them. For example, record your observations of potassium carbonate and silver nitrate before you mix them.

b) Create a chart in your log to record your observations after you mix

the reactants. You may wish to use a chart similar to the one below.

9. Mix three drops of one solution (i.e., potassium carbonate) with three drops of another solution (i.e., silver nitrate), as indicated by the first box on the chart. You can mix these solutions in the well of a spot plate or on a plastic surface. Do not allow the tip of the dropper of one solution to come in contact with another solution. This is important to prevent contamination of solutions.

a) In the chart in your log, record your observations after mixing the reactants.

Continue with the other reactants (i.e., potassium carbonate with copper (II) sulfate; then potassium carbonate with magnesium sulfate; and so on).

b) Record all your observations in your *Active Chemistry* log.

	Silver Nitrate ($AgNO_3$)	Copper (II) Sulfate ($CuSO_4$)	Magnesium Sulfate ($MgSO_4$)	Sodium Hydroxide (NaOH)
Potassium Carbonate (K_2CO_3)	1.	2.	3.	4.
Sodium Hydroxide (NaOH)	5.	6.	7.	8.
Potassium Iodide (KI)	9.	10.	11.	12.
Iron (III) Chloride ($FeCl_3$)	13.	14.	15.	16.

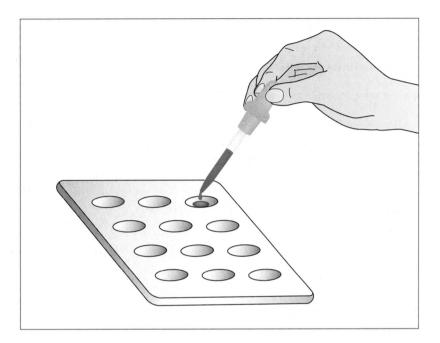

10. Clean all apparatus and the laboratory bench when you are finished. Dispose of all chemicals as directed by your teacher. Wash your hands.

11. Look at the data from the double-displacement reactions that you observed.

    a) Do you think a chemical reaction took place in each case? Explain your answer.

    b) Are you able to predict or identify any of the products that were formed? If so, which ones?

    c) Write word equations and balanced chemical equations for each reaction that you observed. In all cases, assume only two reactants are used and only two products are formed. Use the following formulas to help you write your equations:

copper (II) carbonate ($CuCO_{3(s)}$);	copper (II) hydroxide ($Cu(OH)_{2(s)}$);
copper (II) iodide ($CuI_{2(s)}$);	iron (III) nitrate ($Fe(NO_3)_{3(s)}$);
iron (III) hydroxide ($Fe(OH)_{3(s)}$);	magnesium carbonate ($MgCO_{3(s)}$);
silver carbonate ($Ag_2CO_{3(s)}$);	silver chloride ($AgCl_{(s)}$);
silver hydroxide ($AgOH_{(s)}$);	potassium nitrate ($KNO_{3(aq)}$);
potassium sulfate ($K_2SO_{4(aq)}$);	magnesium hydroxide ($Mg(OH)_{2(s)}$);
silver iodide ($AgI_{(s)}$);	sodium chloride ($NaCl_{(aq)}$);
sodium nitrate ($NaNO_{3(aq)}$);	sodium sulfate ($Na_2SO_{4(aq)}$);

ChemTalk

CHEMICAL REACTIONS

Kinds of Chemical Reactions

A chemical reaction takes place when starting materials (reactants) change to new materials (products). Synthesis, decomposition, single-displacement, and double-displacement reactions are some common kinds of chemical reactions.

Synthesis and Decomposition Reactions

Synthesis means "putting together." In a **synthesis reaction** two or more elements combine to form one or more compounds. In this activity you investigated the reaction of magnesium in oxygen to form magnesium oxide. The opposite kind of reaction is a **decomposition reaction.** In chemical decomposition a compound is separated into its elements.

Single-displacement Reactions

A **single-displacement reaction** is one in which an element reacts with a compound to produce a new element and an ionic compound. For example, a single-displacement reaction occurs when you put a strip of zinc in a copper (II) sulfate solution. The zinc metal exchanges places with the copper cations. You also observed that the free copper metal atoms now plate on the zinc strip.

Activity Series

If you put copper metal in a zinc sulfate solution you would find that no reaction would take place, as expected. Zinc exchanges places with copper, but copper will not exchange places with zinc. You have learned something about a property of copper and zinc. Zinc is more active than copper. If you were to experiment with different metals and metallic solutions, you should be able to create an activity series of metals. The activity series of metals can be put into a table that you can use to predict if a reaction will take place. The table looks like the one on the following page. The table permits you to determine how a metal will react in a metal solution. A metal that is more active than another will dissolve into the metal solution and plate out the less

Chem Words

synthesis reaction: a chemical reaction in which two or more substances combine to form a compound.

decomposition: a chemical reaction in which a single compound reacts to give two or more products.

single-displacement reaction: a reaction in which an element displaces or replaces another element in a compound.

double-displacement reaction: a chemical reaction in which two ionic compounds "exchange" cations to produce two new compounds.

Activity Series of Metals (Most Active to Least Active)	
1. Lithium (Li $\rightarrow$ Li$^+$ + e$^-$)	8. Iron (Fe $\rightarrow$ Fe^{2+} + 2e$^-$)
2. Potassium (K $\rightarrow$ K$^+$ + e$^-$)	9. Lead (Pb $\rightarrow$ Pb^{2+} + 2e$^-$)
3. Calcium (Ca $\rightarrow$ Ca^{2+} + 2e$^-$)	10. Hydrogen (H$_{2(g)}$ $\rightarrow$ 2H$^+$ + 2e$^-$)
4. Sodium (Na $\rightarrow$ Na$^+$ + 1e$^-$)	11. Copper (Cu $\rightarrow$ Cu^{2+} + 2e$^-$)
5. Magnesium (Mg $\rightarrow$ Mg^{2+} + 2e$^-$)	12. Mercury (Hg $\rightarrow$ Hg^{2+} + 2e$^-$)
6. Aluminum (Al $\rightarrow$ Al^{3+} + 3e$^-$)	13. Silver (Ag $\rightarrow$ Ag$^+$ + e$^-$)
7. Zinc (Zn $\rightarrow$ Zn^{2+} + 2e$^-$)	14. Gold (Au $\rightarrow$ Au^{3+} + 3e$^-$)

active metal. Zinc replaced the copper and the copper plated the zinc. For example, let's say that you place a strip of copper in a silver nitrate solution. According to the table, the copper will dissolve into copper ions (Cu^{2+}) and the silver ions will plate out on the copper as silver.

In addition to metals, you will notice that hydrogen gas is also listed in the activity series. In your investigation you found that different metals produced hydrogen gas and metal ions when they reacted with hydrochloric acid. You read above that metals can replace less active metals in metal salt solutions. Metals that are more active than hydrogen can replace the hydrogen from water to form metal hydroxides. As an example, if you were to react potassium metal with water, you would get hydrogen gas and potassium hydroxide solution. The equation is:

$$2K_{(s)} + 2HOH_{(l)} \rightarrow H_{2(g)} + 2\,KOH_{(aq)}$$

Double-displacement Reactions

Double-displacement reactions are different from single-displacement reactions, in that you start with two aqueous phase solutions and when they react they "switch partners." An example of this type of reaction is:

$$Ba(NO_3)_{2(aq)} + Na_2SO_{4(aq)} \rightarrow 2\,NaNO_{3(aq)} + BaSO_{4(s)}$$

Note that the cation of the one compound (Ba^{2+}) exchanged places with the cation ($Na+$) of the other compound. The solid $BaSO_4$ is a

Chem Words

salts: ionic compounds in which the anion is not a hydroxide ion (OH–) and the cation is not a hydrogen proton (H+).

precipitate. When examining double-displacement reactions, you know that a reaction has taken place if you see:

- a precipitate
- a gas
- water

If none of these are present, then no reaction had occurred.

Solubility Rules

A precipitate will form if the compound is not soluble in water. In the example above, barium sulfate was not soluble in water. This was noted in the equation by referring to it as a solid (s). Chemists have created a set of solubility rules for **salts**. Salts are classified as ionic compounds in which the anion is not a hydroxide ion (OH$^-$) and the cation is not a hydrogen proton (H$^+$).

Solubility Rules
1. All salts (defined as ionic compounds) of the alkali (Group 1 on the periodic table) metals and the ammonium ion are soluble in water.
2. All chlorides, bromides, and iodides are soluble with the exception of silver, lead, and mercury halides.
3. All nitrate, chlorate, perchlorate, and acetate salts are soluble.
4. All sulfates are soluble with the exception of calcium, barium, strontium, and lead.
5. All carbonates, phosphates, chromates, hydroxides, and sulfides are insoluble except when they are combined with alkali metals or the ammonium ion.

In the example above, barium sulfate formed a precipitate. Since barium sulfate is insoluble, this agrees with the solubility rules. If you mixed silver nitrate with sodium chloride, would you expect to get a precipitate? The two products that would form are silver chloride and sodium nitrate. The solubility rule #2 tells you that silver chloride is insoluble and solubility rule #3 tells you that sodium nitrate is soluble. Using these rules, you can now predict whether a mixture will produce a precipitate or not.

Checking Up

1. What is a synthesis reaction? Provide an example.

2. What is a decomposition reaction? Provide an example.

3. Distinguish between a single and a double-displacement reaction.

4. What evidence would you look for to determine if a double-displacement reaction has occurred?

5. Will hydrochloric acid react with a clean strip of copper? Explain your answer.

6. Is calcium sulfate soluble in water? Justify your answer.

Reflecting on the Activity and the Challenge

In this activity you have learned about a number of different types of reactions: synthesis, decomposition, single-displacement and double-displacement reactions. Knowing these types of reactions can help you predict the products of some chemical reactions. You'll need to decide if you want the audience for the **Cool Chemistry Show** to learn about these reactions types. You also learned how to write balanced equations for some of the reactions you observed. Think about a creative way of showing how you can explain balancing chemical equation to your elementary school audience. In this activity you have learned how to write the formulas for many compounds and how to name some compounds. You have also investigated both ionic and molecular compounds. As you prepare your presentation for your **Cool Chemistry Show**, you will want to include your knowledge of formulas, the names of compounds, and the different kinds of compounds. Remember that you will be providing the teacher with an explanation of why you included certain demonstrations, and you will also want to include explanations that are grade-appropriate. Think about how much information you will need to provide each.

Chemistry to Go

1. Baking soda ($NaHCO_3$) has been used in several reactions in previous activities. When heat is applied to baking soda, three compounds are produced. Two of the compounds are gases and the other is a solid. If the two gases are water and carbon dioxide, what is the third product? Explain how you arrived at your answer.

2. When solutions of sodium hydroxide and potassium carbonate are mixed together, no apparent reaction takes place. The same is true when you mix sodium hydroxide and potassium iodide together. Explain this observation.

3. If you mix sodium sulfate and barium nitrate solutions together, you get a white precipitate. What is the precipitate that formed? What information did you use to arrive at your answer?

4. There were five combinations that you mixed together that did not form a precipitate. Complete and balance the following equations:

   a) $K_2CO_{3(aq)} + NaOH_{(aq)} \rightarrow$

   b) $KI_{(aq)} + MgSO_{4(aq)} \rightarrow$

   c) $KI_{(aq)} + NaOH_{(aq)} \rightarrow$

   d) $FeCl_{3(aq)} + CuSO_{4(aq)} \rightarrow$

   e) $FeCl_{3(aq)} + MgSO_{4(aq)} \rightarrow$

5. Use the solubility rules to explain why the products in **Question 4** do not form precipitates.

Activity 5   Chemical Energy

GOALS

In this activity you will:

• Make hot packs and cold packs.

• Observe energy changes when matter changes.

• Determine whether energy changes are endothermic or exothermic from a particular point of reference.

What Do You Think?

Two chemicals at room temperature are mixed together and the temperature cools drastically. You may have used cold packs that work like this when you were injured.

• **How do the chemicals in the cold pack lower the temperature?**

Record your ideas about this question in your *Active Chemistry* log. Be prepared to discuss your responses with your small group and with the class.

Investigate

1. To make a cold pack, place 10 g of ammonium nitrate in a quart-size, resealable plastic bag. Add 20 mL water to the bag and seal.

a) Record your observations.

In an endothermic chemical reaction, energy in the form of heat is absorbed in the process. In an exothermic chemical reaction, energy in the form of heat is given off in the process.

Safety goggles and a lab apron must be worn during this activity.

b) Was the cold pack an example of an exothermic or endothermic chemical reaction?

2. Make a hot pack by placing 20 g of sodium carbonate (or calcium chloride) in a quart-size, resealable plastic bag. Add 20 mL of water to the bag and seal.

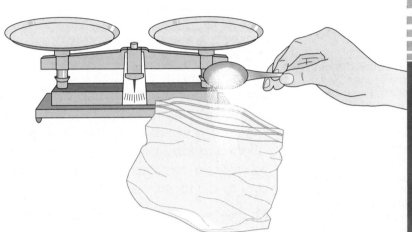

a) Record your observations.

b) Was the reaction exothermic (heat generating) or endothermic (heat absorbing)?

3. To a flask containing 16 g of ammonium thiocyanate, add 32 g of barium hydroxide.

Place a rubber stopper in the mouth of the flask. Shake it vigorously.

Put the stoppered flask on a wood board that has been wet down with puddles of water.

a) Record your observations. Cool chemistry!

b) Was the reaction exothermic (heat generating) or endothermic (heat absorbing)?

4. Using a chemical scoop, transfer a few pellets of sodium hydroxide to a test tube half full of water. Carefully feel the side of the test tube.

a) Record your observations.

b) Was the reaction exothermic (heat generating) or endothermic (heat absorbing)?

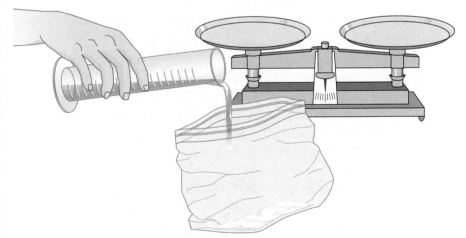

Be careful when working with the sodium hydroxide pellets. Wear rubber gloves and eye protection. If you should accidentally drop a pellet, do not try to pick it up with your bare hands as it may burn them. Use gloved hands to retrieve the pellets.

Chem Talk

ENDOTHERMIC AND EXOTHERMIC REACTIONS

A process is described as **endothermic** when heat energy is absorbed, increasing the internal energy of the system. One example of an endothermic reaction is the cold pack you made with ammonium nitrate. Another example is the decomposition of potassium chlorate. In this reaction, energy must be added to the system in order to cause the decomposition of the potassium chlorate to form the products of oxygen gas and potassium chloride. If you touch a container that holds an endothermic process, it will feel cool to the touch.

An **exothermic** process results when heat energy is released, decreasing the internal energy of the system. If you touch a container that holds an exothermic process, it will feel warm or hot to the touch. One example of an exothermic reaction is the hot pack you made with sodium carbonate. Another example was the combining of sodium hydroxide solution with hydrochloric acid. This reaction produces sodium chloride and water and releases energy to the environment. The terms endothermic and exothermic can be used when describing both physical and chemical changes.

Why is energy so important? In order for a chemical reaction to take place, the particles (reactants) involved in the reaction must interact. Not all collisions result in a chemical reaction. The particles involved must have enough energy to enable them to react with each other. The colliding particles must have enough kinetic energy to break the existing bonds in order for new bonds to be formed. The minimum energy required for a chemical reaction is activation energy. Bond breaking is an endothermic process and requires an addition of energy and bond formation is an exothermic process and requires a release of energy. (Physics reminder: Kinetic energy is the energy of motion $KE = \frac{1}{2}mv^2$.)

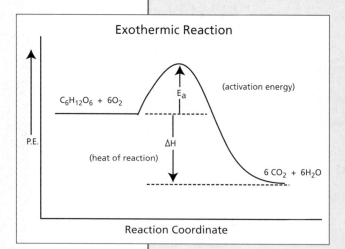

Exothermic Reaction

$C_6H_{12}O_6 + 6O_2$

E_a (activation energy)

P.E.

ΔH

(heat of reaction)

$6 CO_2 + 6H_2O$

Reaction Coordinate

When more energy is released as products form than is absorbed to break the bonds in the reactants, the reaction is said to be exothermic, as shown in the graph. A chemical reaction in which energy is released is called exothermic. The prefix *exo* means exit and *thermo* means heat or energy.

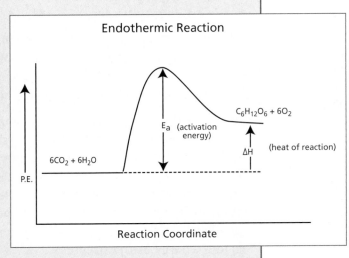

Endothermic Reaction

E_a (activation energy)

$C_6H_{12}O_6 + 6O_2$

ΔH (heat of reaction)

$6CO_2 + 6H_2O$

P.E.

Reaction Coordinate

The cells in your body use glucose to get energy for cellular respiration. Glucose plus oxygen (from breathing) provides you with energy.

$$C_6H_{12}O_6 + 6O_2 \rightarrow 6CO_2 + 6H_2O + energy$$

When more energy is needed to break the bonds in the reactants than is given off by forming the bonds in the products, the reaction is said to be endothermic—a chemical reaction in which energy is absorbed, as shown in the second graph. In some endothermic reactions, energy must be added in order for the reaction to occur.

In the process of photosynthesis, plants use energy from light to break the bonds of carbon dioxide and water in order to form the products glucose and oxygen. The glucose produced makes the plant grow and the reaction also provides animals with oxygen:

$$6CO_2 + 6H_2O + light\ energy \rightarrow$$
$$C_6H_{12}O_6 + 6O_2$$

Conservation of Energy

Energy transfer is an important feature in all chemical changes. Energy is transferred whether the chemical reaction takes place in the human body, like the metabolism of carbohydrates, or in a car, like the combustion of gasoline. In both these cases energy is released into the surroundings.

499

Checking Up

1. Describe an endothermic reaction.
2. Describe an exothermic reaction.
3. Explain whether each of the following is endothermic or exothermic:
   a) bond breaking
   b) bond forming.
4. Distinguish between heat and temperature.

The photosynthesis that occurs in living plants and the decomposition of water into hydrogen and oxygen both remove energy from the surroundings. According to the Law of Conservation of Energy, the energy absorbed from the surroundings or released to the surroundings is equal to the change in the energy of the system.

Energy is the great organizing principle of all science. The conservation of energy allows you to better understand the world around you. Energy exists as light energy, heat energy, sound energy, nuclear energy, kinetic energy, chemical energy as well as other forms. The total energy in any closed system remains the same. In the simplest physical systems, it is quite easy to describe the energy changes. When a bowling ball falls, its original gravitational potential energy becomes kinetic energy as the ball increases its speed. After the ball hits the ground, this kinetic energy is converted to sound energy (you hear the crash) and heat energy (you can measure a temperature rise of the ball and ground) and the compression and vibrations of the ground. Each of these can be measured and it is always found that the total energy before an event is equal to the total energy after the event. When a human being is involved, you notice other energy interactions. A person is able to eat food and digest the food. The energy released from this slow-burn of the food (metabolism) is able to keep the body at about 37°C. As all non-living things in the room cool down to room temperature, humans are able to stay warm in the 22°C room environment. People also use the food energy that they ingest for moving muscles, keeping the heart pumping and operating all human functions. Living organisms are superb energy conversion systems.

Heat and Temperature

Both heat and **temperature** have been mentioned in this activity. It is important to note that heat and temperature are not the same, although they are related. Heat is one form of energy. When two materials of different temperatures interact, they exchange heat energy until they arrive at the same temperature. Temperature is a number associated with how hot or cold something is. On a molecular level, temperature is related to the average kinetic energy of the atoms in the material. All particles in a material are in a constant state of motion. The temperature of the material is a measurement of this molecular motion. If the kinetic energy of the particles increases, the temperature increases. Temperature can be

Chem Words

temperature: the measure of the average kinetic energy of all the particles of the material.

defined as the measure of the average kinetic energy of all the particles of a substance. A thermometer is an instrument that measures temperature. Heat is the transfer of energy, which often results in a change in the kinetic energy of particles—a change in the temperature of the system.

Reflecting on the Activity and the Challenge

Energy is involved in any process that requires the breaking or making of bonds. The process may be a chemical or physical change. Sometimes energy changes are not noticed or measured, but other times an energy change is significant enough to be detected. In this activity you explored both endothermic and exothermic processes. As you select activities to include in the **Cool Chemistry Show,** you must be aware of any heat energy released or absorbed. The audience may be interested to learn about the energy changes that accompany chemical processes. In addition, your awareness will ensure that the presentations are safe for both the presenters and the audience.

Chemistry to Go

1. Identify the following changes as endothermic or exothermic. (Ask yourself whether the reaction requires the addition of heat energy to occur or does it release energy in the form of heat.)

   a) Melting ice.

   b) Lighting a match.

   c) Dry ice subliming into carbon dioxide gas.

   d) Frying an egg.

   e) Burning gasoline.

   f) Explosion of hydrogen gas.

2. The water in a teapot is heated on a stovetop. The temperature of the water increases. Is this an endothermic or exothermic process?

3. If a red-hot piece of iron is dropped into a bucket of water, what type of heat change takes place in reference to the water? What type of heat change takes place in reference to the iron?

4. Explain in terms of energy flow how a cold pack works on a sprained ankle.

5. If ice is at −20°C and you apply some heat by sitting a beaker of the ice on a hot plate, explain why the ice does not appear to be melting.

Preparing for the Chapter Challenge

Review the chemical reactions that you have so far planned to feature in your **Cool Chemistry Show**. Have you included both endothermic and exothermic reactions? In a paragraph describe the difference between the two types of reactions. Be sure to mention why energy transfers are important.

Inquiring Further

1. Commercial cold and hot packs

Research several cold and hot packs. What materials are used in the packs, and how is the chemical reaction activated? You should be aware that there are a variety of commercially available hot and cold packs. In general some are reusable and some are not. Those that are not reusable are typically chemical reactions. Those that are reusable are typically physical changes.

Design a process to make a cold pack, using your research. Have your teacher approve your design before you actually try it out.

2. The colligative property of a solvent

A salt solution will depress the freezing point of water. This is commonly known as the colligative property of the solvent. When you add anti-freeze (ethylene glycol) to water, you find that the freezing point of the water is lowered, which prevents the water in the car radiator from freezing at 0°C. It also elevates the boiling point of the water, which will prevent the water solution from boiling. Design an experiment to demonstrate the colligative property of a solvent. With the approval of your teacher carry out your experiment.

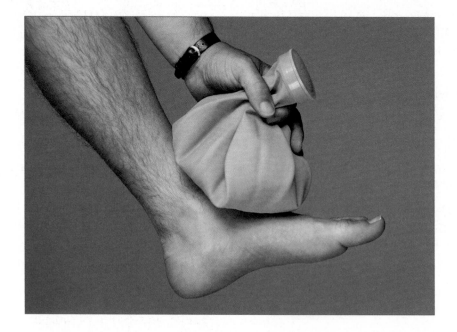

Activity 6   Reaction Rates

What Do You Think?

Have you ever wondered why some chemical reactions, like the burning of a match, take place at a fast rate, while others, like the spoiling of milk, take place slowly? The rate of a chemical reaction is the speed at which the reactants are converted to products.

- **What are some factors that influence the rate of a reaction?**
- **How could you make a reaction take place at a faster rate?**
- **How could you slow a reaction down?**

Record your ideas about these questions in your *Active Chemistry* log. Be prepared to discuss your responses with your small group and with the class.

Investigate

1. One way to study the rate of a reaction is to time the reaction with a stopwatch. Try this!

   Place 20 mL of vinegar into a large test tube.

   To a second test tube, add 10 mL of vinegar and 10 mL of water. Mix well using a stirring rod.

GOALS

In this activity you will:

- Discover conditions that make a reaction proceed faster or slower.
- Discuss explanations for why this happens at the molecular level.

Safety goggles and a lab apron must be worn during this activity.

503

In a third test tube, add 5 mL of vinegar to 15 mL of water. Mix.

Prepare three equal-sized pieces of polished magnesium ribbon. Set your stopwatch so that it is ready to start immediately.

Add a piece of magnesium ribbon to the first test tube, keeping track of the time the reactions take.

a) Record your observations and time on a data table. Repeat for the other two test tubes.

b) In this step, you changed the concentration of one of the reactants (the vinegar). The vinegar was less concentrated (more dilute) in each successive test tube. Did this impact the reaction rate? If so, describe the relationship between the concentration of the reactant and the resulting reaction rate.

2. Repeat the same reaction above using a well plate and smaller amounts of vinegar, water, and magnesium.

a) Record your design. Include the equipment you will need, the amount of reactants, and any safety procedures.

b) With the approval of your teacher, carry out the procedure. Record your data and results.

c) How do the results compare with the reaction in **Step 1**?

3. Place 10 drops of 0.1 M HCl (weak concentration) into one clean well in your well plate and 10 drops of 1.0 M HCl (strong concentration) into a second clean well. Drop a small piece of zinc (equal size) into each well containing HCl.

a) Record your observations.

b) How do these results compare to your earlier results? Do these results support or refute the relationship you stated in **Step 2 (c)** above?

4. Hydrogen peroxide is sold over the counter in pharmacies to be used as a disinfectant for minor injuries. Because hydrogen peroxide decomposes slowly to form oxygen and water, it is also a source of oxygen gas.

$$2H_2O_{2(l)} + light \ energy \rightarrow O_{2(g)} + 2H_2O_{(l)}$$

Pour a small amount of hydrogen peroxide (about 15 mL) into each of two test tubes. Add a small amount of manganese dioxide to one of the test tubes.

a) Record your observations.

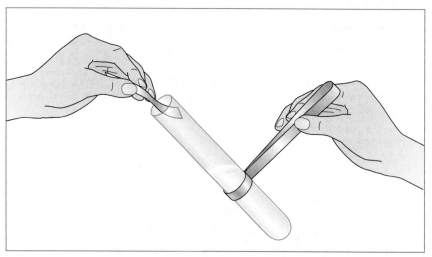

The manganese dioxide did not actually react with the hydrogen peroxide; it simply acted as a catalyst for the decomposition of the hydrogen peroxide. A catalyst is a material that speeds up a reaction without being permanently changed itself. The chemical equation for this reaction is:

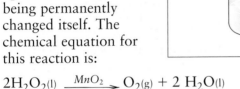

$$2H_2O_{2(l)} \xrightarrow{MnO_2} O_{2(g)} + 2 H_2O_{(l)}$$

5. Design an investigation to prove that the manganese dioxide did not get used up in the reaction.

   Record your design. Include the equipment you will need, the amount of reagents, and any safety procedures.

   a) Record your procedure in your *Active Chemistry* log.

   b) With the approval of your teacher, carry out the experiment.

   c) Record your data and results.

   d) Describe the relationship between the use of a catalyst and the rate of reaction.

6. Pour 200 mL hot water into one beaker and 200 mL ice water into another beaker. Add a tea bag to each.

   a) Record your results.

   b) Repeat the procedure using an Alka-Seltzer tablet in place of the tea bag. Record your observations.

c) Describe the relationship between temperature and the rate of reaction, based on the two situations you studied.

7. Prepare two beakers, each containing equal amounts (about 200 mL) of room temperature water.

   Obtain two Alka-Seltzer tablets. Crush one and leave the other whole.

   Simultaneously add the crushed table to one beaker and the whole tablet to the other beaker.

   a) Record your observations.

   What factor was being studied in **Step 4 (a)**? Describe the relationship between that factor and the rate of reaction.

8. Use the results of this activity to answer the following:

   a) Describe how you could increase the rate of reaction by altering:

   • concentration   • catalyst

   • temperature   • surface area

9. Dispose of all chemicals as directed by your teacher. Clean and put away any

equipment as instructed. Clean up your work station. Wash your hands.

10. Chemical systems that highlight reaction rates can be very interesting. Because time is a factor, these systems are often called clock reactions. Your teacher will do the following reaction as a demonstration.

Your teacher will use the following solutions:

Solution A = 0.1 M potassium iodate (KIO_3)

Solution B = 1% starch solution

Solution C = 0.25 M sodium hydrogen sulfite ($NaHSO_3$)

Two rows of five beakers (all the same size!) will be arranged in front of you.

Each of the beakers in the back row contains 100 mL of Solution A, 50 mL of Solution B, and 100 mL of distilled water.

Each beaker in the front row contains 20 mL of Solution C and 130 mL of water.

Your teacher will add the contents of one beaker from the back row to the contents of one of the beakers in the front row.

Use a stopwatch and stop when a color change occurs.

Your teacher will then combine the next set of beakers as you will again use the stopwatch. She or he will continue down the row.

a) Record the time and observations about the change in each case.

ChemTalk

Chem Words

surface area: changing the nature of the reactants into smaller particles increases the surface exposed to react. Successful reaction depends on collision and increasing the area of the reactant increases the chance of a successful collision taking place. Lighting a log is more difficult than lighting wood shavings. The shavings have a greater surface area and speed up the reaction.

FACTORS AFFECTING THE RATE OF A REACTION

In this activity you investigated several common factors that influence reaction rate. They included surface area, concentration of reactants, temperature, and catalysts. On a molecular level, these factors increase the collision frequency of the particles of the materials involved in the reaction.

Consider **surface area**. In water, a sugar cube dissolves in water at a much slower rate than if the same cube is first crushed. The crushed cube has a greater surface area — more parts of the sugar are in contact with the water. In a fireplace, wood chips burn faster than a pile of logs. In both of these cases, the smaller pieces, with their increased surface area, allow the particles that are reacting to come in

Checking Up

1. List four factors that influence the rate of a reaction.

2. For each of the factors you listed above, describe how the factor increases the collision frequency of the particles of the materials involved in the reaction.

3. How is a catalyst different from the reactants and products of a chemical reaction?

contact with each other more often. This increases the collision frequency.

Another factor that influences reaction rate is the concentration of the reactants. An increase in concentration means an increase in the number of particles in the reaction. This results in an increase in the collision frequency. If a chemist wants to increase the rate of a reaction, an increase in the concentration of one or more of the reactants will do the trick.

Altering collision frequency and efficiency can also be accomplished through temperature changes. According to the kinetic theory, particles move faster at higher temperatures and slower at lower temperatures. The faster motion of the particles increases the energy of the particles and increases the probability that particles will collide. As a result, the reaction rate increases.

Catalysts play an important role in many chemical reactions. A catalyst is a substance that speeds up a reaction without being permanently changed itself. The catalyst lowers the action energy of the reaction. Many commercial reactions make use of catalysts, because the catalysts can be recovered, regenerated, and reused. You are probably familiar with the term catalytic converter, a device used in automobiles to improve the efficiency of unleaded gasoline engine's combustion exhaust. The catalyst in this converter is platinum.

Chem Words

catalyst: a substance that changes the speed of a chemical reaction without being permanently changed itself.

Reflecting on the Activity and the Challenge

In this activity you have observed how different factors affect the rate of a reaction. This knowledge can be applied to the presentations you will make in the **Cool Chemistry Show** demonstrations. If you need to cause a reaction to happen faster or slower, you now know what changes you can make. Of course, it is important for you to check with your teacher before making adjustments to any procedure you might be considering.

Chemistry to Go

1. Explain each of the following in terms of the factors that influence reaction rate:

   a) Which will cook faster: Cookies at 50°C or at 150°C?

   b) A bear (and many other animals) hibernates during the winter months. Scientists claim that the low body temperature slows down the animal's metabolism. Explain.

   c) A sugar cube dissolves slower than the same amount of sugar in granulated form.

   d) Antacid tablets are used to neutralize acid in the stomach. Explain why two tablets are faster than one tablet in neutralizing the acid.

   e) If you tried to burn a sugar cube with a match, you would find it very difficult to get the sugar to burn. However, if you put some cigarette ash on the cube, the cube would then burn when you put a flame to it. Explain the purpose of the cigarette ash in changing the burning of the sugar cube.

   f) Why does powdered aspirin dissolve faster than an aspirin tablet in water?

   g) Sugar dissolves more readily in hot tea than in iced tea. Explain.

2. In most cases, if you increase the temperature, the reaction rate increases. Explain this in terms of the collision theory.

3. Imagine that you purchase a lightstick necklace or wristband at a social event and want to make it last as long as possible. What would you do? Why would it help?

4. Explain why the effervescent antacid tablet did not seem to react as fast when it was put in a more dilute solution of vinegar.

5. Explain in terms of the reaction rate factors that you have studied why it is possible for a person who has been submerged in very cold ice water in some

cases to survive, but individuals who have been submerged in warmer water for the same length of time do not survive.

6. Grain elevators have been known to have explosions because of the production of fine grain powders. Explain in terms of the reaction rate factors that you have studied as to why this could happen.

Preparing for the Chapter Challenge

The factors affecting a reaction can be varied to achieve different reaction rates. How could you use this information in developing a presentation for the **Cool Chemistry Show**? Describe one possible scenario.

Inquiring Further

Quantifying the relationship between temperature and reaction rate

You have seen that temperature is a factor that influences the rate of reaction. In general, if the temperature

increases, the reaction rate increases. When the temperature decreases, the reaction rate decreases. Can this relationship be quantified? Is there a mathematical relationship between temperature and reaction time? To answer these questions, explore the reaction between magnesium ($Mg_{(s)}$) and vinegar ($CH_3COOH_{(aq)}$). Design and conduct an investigation that will use this reaction to show the relationship between temperature and reaction time in a quantitative way. Have your teacher approve your design before you begin. Remember — the point of the investigation is to see if the relationship between temperature and reaction time is quantifiable. You'll need to monitor both the temperature and the time carefully. Plot your data on a graph to make the relationship explicit. In your notes, include the chemical equation for this reaction.

Activity 7

Acids, Bases, and Indicators— Colorful Chemistry

GOALS

In this activity you will:

- Identify common household acids and bases.

- Identify characteristic properties of acids and bases, and learn to tell the difference between acids and bases.

- See how strong acids and bases behave differently from weak acids and bases.

- Make neutral solution by combining an acid and a base by titration.

- Determine the pH of various solutions using indicators.

- Categorize solutions based on the pH scale.

- Use the mathematical definition of pH.

What Do You Think?

When red cabbage is chopped up and added to boiling water and the resulting mixture is allowed to cool, a special bluish colored solution is made. After the solution is separated from the cabbage, it can be used to indicate if other substances are acids or bases. When household vinegar, a common acid, is added to the cabbage-juice water, the solution turns red. When household ammonia, a common base, is added to the cabbage-juice water, the solution turns green.

- **What are some other properties of acids and bases you know about?**

- **How can you tell the difference between an acid and a base?**

Record your ideas about these questions in your *Active Chemistry* log. Be prepared to discuss your responses with your small group and with the class.

Investigate

1. Your teacher will provide you with samples of some of the materials listed below. Place a small amount of each solution in a separate well of a well plate.

   Add a small piece of polished zinc (or magnesium) to each of the solutions.

   a) Make a data table to record your observations.

   hydrochloric acid ($HCl_{(aq)}$)

   lemon or orange juice (citric acid)

   vinegar
   (acetic acid, $CH_3COOH_{(aq)}$)

   sulfuric acid ($H_2SO_{4(aq)}$)

   mineral water

   carbonated beverage such as Sprite® or Seven Up® (contains $H_2CO_{3(aq)}$)

   milk

   dishwashing solution (Ivory®, Palmolive®, Joy®, etc.)

   sodium hydroxide ($NaOH_{(aq)}$)

   Milk of Magnesia® (contains $Mg(OH)_{2(aq)}$)

   apple juice (malic acid)

   potassium hydroxide ($KOH_{(aq)}$)

   calcium hydroxide ($Ca(OH)_{2(aq)}$)

household ammonia
(NH_3 or $NH_4OH_{(aq)}$)

b) Which substances reacted with the metal? How could you tell? What do these substances have in common? (Consider the chemical formulas listed for some of the substances.)

c) Which substances did not react with the metal? What do these substances have in common? (Consider the chemical formulas listed for some of the substances.)

Safety goggles and a lab apron must be worn during this activity.

2. Place small amounts of each solution you used in **Step 1** in a separate well of a well plate.

   Test the solutions with one or more common laboratory indicators. (Your teacher will provide acid-base indicators like blue litmus paper, red litmus paper, phenolphthalein, bromothymol blue, methyl red.) Indicator papers are activated simply by dipping a small piece of the paper into the solution and noting any color change. If the indicator is a solution, add a drop or two to the substance being tested and note any color change. You will need to use fresh test solutions if you want to test with more than one indicator solution.

   a) Make up a chart and record your observations.

3. Use your observations as well as previous experiences to answer the following.

   a) Make a list of some of the observable properties for acids and bases.

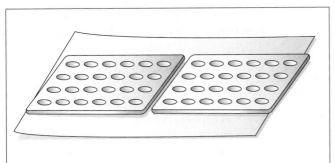

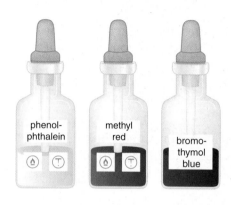

For example:

- How do substances containing acids or bases taste? You should never taste substances in a lab, but you have probably had the opportunity to taste vinegar or lemon juice at home, or you may have accidentally got soap in your mouth.

- How do acids and bases feel. You must be very cautious when handling chemicals both at home and in the lab. However, you've probably had the experience of touching cleaning materials, such as soaps or floor cleaners. Think about vinegar or citrus fruits. How do they feel on a cut on your skin or a canker sore in your mouth?

4. The pH scale can also be used to describe acids and bases. This number scale ranges from 0 to 14. Acid solutions have a pH less than 7. The more acidic a solution is, the lower the pH. Base solutions have a pH greater than 7. The more basic a solution is, the higher the pH. Neutral solutions have a pH of 7.

There are a number of ways you can use to measure pH. You will use pH paper and/or a universal indicator solution. Determine the pH of some of the substances you used in **Step 1.**

a) Make a data table that includes the name of the substance, the pH test, and whether the substance is an acid, a base, or a neutral substance.

b) You may have used both pH paper and universal indicator solution. Both of these are made from a combination of indicators, in order to produce a continuous range of colors throughout the pH scale. Methyl red is a chemical that changes from red to yellow when the pH is between 4.2 and 6.3. Thymolphthalein is a chemical that changes from colorless to blue when the pH is between 9.4 and 10.6. Thymol blue is a chemical that changes from red to yellow when the pH is between 1.2 and 2.8. How could these three chemicals be used to create an indicator scale? What are the limitations of being limited to these three chemicals?

5. Use the pH paper to test additional common household substances in order to determine which are acids and which are bases. (Hint: Try carbonated beverages, tea, coffee, baking powder, mayonnaise, power drinks, pickle juice, window cleaner, stain removers.) Your teacher may give you some pH paper to take home with you.

a) Make a list of common acids and bases found in your school or home. When possible, include both the name and formula for each substance you test.

6. Dispose of all chemicals as directed by your teacher. Clean and put away any equipment as instructed. Clean up your workstation. Wash your hands.

7. Here are two activities that display the characteristics of acids and bases in a colorful way. Your teacher will show you these as demonstrations.

• Paint a message on a large sheet of paper or poster board using phenolphthalein indicator solution. (How about painting a message announcing your **Cool Chemistry Show?**) Allow the message to dry completely and hang the paper/poster board where everyone can see it. Use a window glass cleaner that contains ammonia water and when you are ready to reveal the message, lightly spray the design with the basic solution. (The secret message can also be revealed with a dilute ammonia solution. As the ammonia evaporates, the secret message that has been revealed will disappear again.)

• Rinse a small beaker with a strong acid and label it "A." Rinse another small beaker with a strong base and label it "B." Let both beakers air dry. In another beaker (label it "I") add 20 drops of phenolphthalein indicator solution to about 50 mL of distilled water. When you are ready, pour some of the solution from beaker "I" into beaker "A." Then pour the solution from beaker "A" into beaker "B."

a) Record your observations.

b) Account for the observations in each case.

Caution must be used with these sprays, because they can cause eye damage if they get into the eyes.

ChemTalk

ACIDS AND BASES

Arrhenius' Definition of Acids and Bases

Acids and **bases** were first classified according to their characteristic properties. As you've experienced, acids and bases have different, distinct interactions with indicators (substances that change color with changes in the acidic or basic nature of another material). Some acids react with metals, while bases do not. Bases have a characteristic bitter taste and slippery feel, while acids have a characteristic sour taste. In fact, the term, acid, comes from the Latin word, *acidus*, which means sour. Acids and bases are also good conductors of electricity.

Chem Words

acid: a substance that produces hydrogen ions in water, or is a proton donor.

base: a substance that releases hydroxide ions (OH^-) in water, or is a proton acceptor.

In the 19th century a chemist named Svante Arrhenius attributed the characteristic properties of acids to their ability to produce hydrogen ions when dissolved in water. If you look at the formulas for many common acids (HCl, H_2CO_3, H_2SO_4), you'll notice that they all have H as a common element. When these acids are added to water, a hydrogen atom can be drawn off into the water solution. The hydrogen atom leaves an electron behind, forming a positive hydrogen ion (H+) and a negative ion. Consider the action of hydrochloric acid in solution:

$$HCl_{(g)} \xrightarrow{\text{water}} H^+_{(aq)} + Cl^-_{(aq)}$$

The chemical equation shown above is valuable because of its simplicity. However, in reality, the hydrogen ion (H^+) is simply a proton and readily attaches itself to a water molecule. The result is called a hydronium ion (H_3O^+).

$$\underset{\text{hydrogen ion}}{H^+} + \underset{\text{water}}{H_2O} \rightarrow \underset{\text{hydronium ion}}{H_3O^+_{(aq)}}$$

To be more complete, the chemical equation above could be written as shown below. (Your teacher may allow you to use the simpler form of the equation—using the hydrogen ion as opposed to the hydronium ion).

$$HCl_{(g)} + H_2O \rightarrow H_3O^+_{(aq)} + Cl^-_{(aq)}$$

Arrhenius also addressed bases and their characteristic properties. He defined a base as a substance that produces hydroxide ions (OH^-) when dissolved in water.

Let's look at a base using Arrhenius' definition. When solid sodium hydroxide is dissolved in water, both sodium ions and hydroxide ions are produced, as shown in the chemical equation below:

$$NaOH_{(s)} \xrightarrow{H_2O} Na^+_{(aq)} + OH^-_{(aq)}$$

Over time, scientists have extended their definition of acids and bases beyond Arrhenius' definition to be more inclusive. You will learn more about the contributions of scientists like Johannes Bronsted of Denmark, Thomas Lowry of England, and Gilbert Lewis of the United States in further chemistry courses.

Neutralizing Acids and Bases

When acids and bases react together in solution, the hydrogen ions and hydroxide ions react in a one-to-one ratio to produce water. The remaining ions can join to form a salt. The process of an acid and base reacting to form water and a salt is called **neutralization**. Because the hydrogen ions and hydroxide ions have formed water, the solution is said to be neutral. The process of neutralization is shown in the chemical equations below. The chemical formula for water is actually H_2O. In the equations below the formula is written as HOH, so that you can see where the hydrogen and hydroxide ions end up.

Chem Words

neutralization:
the process of
an acid and base
reacting to form
water and salt.

$$H^+_{(aq)} \quad + \quad OH^-_{(aq)} \quad \rightarrow \quad HOH_{(aq)}$$
hydrogen ion   hydroxide ion   water

$$HCl_{(aq)} \quad + \quad NaOH_{(aq)} \quad \rightarrow \quad HOH_{(aq)} \quad + \quad NaCl_{(aq)}$$
acid   base   water   salt

If a suitable indicator is added to the reaction system, it will change colors when neutralization occurs. The point at which the indicator changes color is called the endpoint.

Consider the reaction of a strong acid (HCl) and a strong base (NaOH), as shown in the equation above. These substances are described as "strong" because they ionize completely in solution. For every HCl molecule, one hydrogen ion is released. For every NaOH, one hydroxide ion is released. These two ions then combine in a one-to-one ratio to form a neutral water molecule. Chemists take advantage of the neutralization process to help determine the concentration of solutions of acids or bases.

The pH Scale

In this activity you observed that one way of describing acids and bases is by examining their effects on indicators. Scientists also use the pH scale to express how acidic or basic a solution is. This number scale ranges from 0 to 14. Acid solutions have a pH less than 7. The more acidic a solution is, the lower the pH. Base solutions have a pH greater than 7. The more basic a solution is, the higher the pH. Neutral solutions have a pH of 7. The pH of a substance can be measured using methods like a pH meter or probe, pH paper, or universal indicator solution.

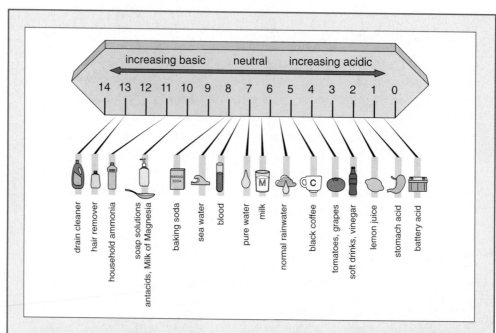

Acid and base indicators are compounds that are sensitive to pH. The color of the indicator changes as the pH of the solution changes. Most indicators are weak acids or weak bases that typically exhibit two different colors under varying pH conditions. The table below shows some common laboratory indicators and the colors they display under different pH conditions.

Common Laboratory Acid-Base Indicators		
Indicator	**Color Change**	**pH Range**
methyl violet	yellow to blue	0.0 to 1.6
thymol blue	red to yellow	1.2 to 2.8
methyl orange	red to yellow	3.2 to 4.4
bromocresol green	orange to violet	3.8 to 6.4
methyl red	red to yellow	4.2 to 6.3
litmus paper	red to blue	5.5 to 8.0
bromothymol blue	yellow to blue	6.0 to 7.6
phenolphthalein	colorless to red	8.2 to 10
thymolphthalein	colorless to blue	9.4 to 10.6
cabbage juice	red to green	2 to 12 (universal)

The pH scale ranges from 0 to 14 and is used to express the concentration of the hydrogen (H+) or hydronium ion (H_3O^+) of a solution at 25°C. Mathematically, it is defined as the negative logarithm of the hydrogen ion concentration. The term **pH** stands for **p**ower of **H**ydrogen ion. It can be written as:

$$pH = -\log_{10}[H^+]$$

where the brackets [ ] stand for "concentration of" (hydrogen ions in solution). Because pH is a logarithmic scale, the concentration of the hydrogen ion [H^+] actually increases or decreases tenfold for each unit on the scale. An acid with a pH of 2 has a [H^+] that is 10 times greater than an acid with a pH of 3 and 100 times the concentration of an acid with pH 4. A base with a pH of 10 has a [H^+] that is 10 times less than a base with a pH of 9.

Chem Words

pH: a quantity used to represent the acidity of a solution based on the concentration of hydrogen ions (pH = – log[H^+]).

Checking Up

1. Use a chart to compare the properties of acids and bases. Be sure to include headings like taste, feel, pH, and reaction with metals.

2. What characteristic property did Arrhenius attribute to acids and bases?

3. Describe the process that occurs when an acid reacts with a base.

4. Why are litmus paper and phenolphthalein particularly useful indicators for distinguishing between acids and bases?

5. What does pH stand for?

6. How much more acidic is a solution of pH 3 than pH 5?

Reflecting on the Activity and the Challenge

In this activity you expanded your knowledge about acids and bases by becoming familiar with many of their characteristics. You learned about Arrhenius' definition of acids and bases. You also learned a bit about pH, another way of expressing the acid or base nature of substances. This information will all come in handy as you plan your presentation for the **Cool Chemistry Show.** Remember that the fifth-grade teacher has specifically asked that your class includes presentations and information about acids and bases.

Chemistry to Go

1. Identify which of the following characteristics relate to acids and which relate to bases:

   a) taste sour

   b) release hydroxide ions ($OH^-_{(aq)}$) when dissolved in water

   c) feel slippery

   d) release hydrogen ions ($H^+_{(aq)}$) when dissolved in water

   e) turn pink in the presence of phenolphthalein

   f) react with metals to produce hydrogen gas

   g) taste bitter

   h) turn red cabbage juice indicator green

2. Use Arrhenius' definition of an acid to help you write a chemical equation that shows the acidic nature of the following:

   a) sulfuric acid (H_2SO_4)

   b) carbonic acid (H_2CO_3)

3. Use Arrhenius' definition of a base to help you write a chemical equation that shows the basic nature of the following:

   a) potassium hydroxide (KOH)

   b) calcium hydroxide ($Ca(OH)_2$)

4. If you prepared the same concentration of two strong acids, sulfuric and hydrochloric, why would the pH of sulfuric be smaller than the hydrochloric acid?

5. Distilled water should have a neutral pH of 7, but water often has a pH less than 7. Suggest a reason for this lowering of the pH.

6. If you bubbled carbon dioxide through water, what would the new pH of the solution be?

7. Lemon juice, curdled milk, vinegar, all taste sour. What other properties would you expect them to have in common?

Preparing for the Chapter Challenge

You have seen a number of interesting color changes using acids, bases, and indicators. Choose one or two different cool activities to demonstrate in your show. Describe the procedure you will use and explain the chemistry involved. You may also wish to include an interesting scenario to accompany your "presto-change-o" demonstrations.

Inquiring Further

1. Titration

Titration is the process whereby a measured amount of solution of known concentration of acid (or base) is added to a known amount of a solution of unknown concentration. Research how chemists perform a titration, and the importance of indicators.

With your teacher's permission, demonstrate titration to your class.

2. The changing definition of an acid and base

The definition of acids and bases has changed through time. You are familiar with the earliest definitions that defined acids and bases in terms of their characteristic properties. The traditional definition has been expanded a number of times to include other substances that behave like acids and bases, but don't fit the traditional definition. Research the expansion through time of the definition of acids and bases. Identify the scientists involved and the changes that were made. Consider researching chemists such as Johannes Bronsted of Denmark, Thomas Lowry of England, or the American chemist Gilbert Lewis.

3. Is it pH balanced?

You may have heard the term "pH balanced" used to describe a shampoo or a deodorant. What does this term mean? What is the pH of most shampoos? Deodorants? Is it important for a shampoo or deodorant to be "pH balanced?" Conduct some research, both in and out of lab, to get answers to these questions. Focus just on shampoos or just on deodorants.

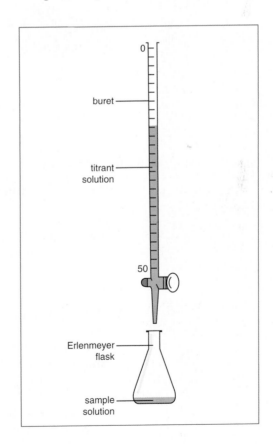

Activity 8

Color Reactions that Involve the Transfer of Electrons

GOALS

In this activity you will:

- Cause different metals to rust by oxidation-reduction (redox) reactions.

- Determine what materials can react with metals, causing the metals to rust.

- Write the word equations and chemical equations for redox reactions.

- Identify the materials that react, and the materials that are simply spectators, in a redox reaction.

- Learn how to impede rusting.

What Do You Think?

What happens to a scratch on a car that is not repaired? What happens to metal barbecue tools that get left out in the rain for a few weeks?

• What is rust and what causes it?

Record your ideas about this question in your *Active Chemistry* log. Be prepared to discuss your responses with your small group and with the class

Investigate

1. Half-fill a test tube with copper (II) sulfate pentahydrate solution ($CuSO_4 \cdot 5H_2O_{(aq)}$). Add a small amount of zinc powder to the test tube. Stopper the test tube, and shake carefully.

   a) Record your observations.

   Dispose of the products as directed by your teacher.

Safety goggles and a lab apron must be worn during this activity.

b) The reaction you just observed was a single-displacement reaction. The zinc displaced the copper. Use this information and your observations to complete the following equations:

zinc + copper (II) sulfate →

_____ + _____

c) Write the word equation as a sentence. Also, explain why this would be classified as a single-displacement reaction. (Refer back to **Activity 4.**)

d) Write the equation using the chemical formulas of the products.

$Zn_{(s)} + CuSO_{4(aq)} →$

_____ + _____

Because the sulfate ion shows on both sides of the equation, it is considered a spectator ion and the equation can be written as shown below:

$Zn_{(s)} + Cu^{2+}_{(aq)} → Cu_{(s)} + Zn^{2+}_{(aq)}$

e) Write a sentence or two describing what happened in the test tube in terms of the zinc and copper.

2. Cut a design or a strip of aluminum from a pie plate or tray. If you cut a strip of aluminum, twist it into an interesting shape.

Place the aluminum in a solution that contains copper (II) ions ($Cu^{2+}_{(aq)}$). Possible solutions include: copper (II) nitrate ($Cu(NO_3)_{2(aq)}$) or copper (II) chloride ($CuCl_{2(aq)}$).

a) Observe and record your results. Dispose of the products as directed by your teacher.

b) What evidence do you have that a chemical reaction occurs? What changes have taken place with the aluminum? With the copper ions?

c) Complete this equation

$Al_{(s)} + Cu^{2+}_{(aq)} →$

_____ + _____

3. Repeat **Step 2** using a different metal such as zinc.

a) Record your observations.

b) Write a chemical equation for the reaction that takes place.

4. Repeat **Step 2** again, this time using a strip of copper in a solution of aluminum nitrate ($Al(NO_3)_{3(aq)}$).

a) Record your observations.

b) Write a chemical equation for the reaction that takes place.

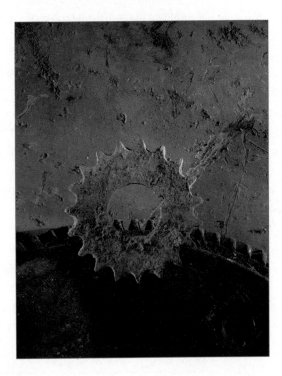

521

ChemTalk

REDOX REACTIONS

When zinc solid reacts with copper ions in solution, a change occurs. Atoms of zinc lose electrons to form zinc ions (Zn^{2+}) that dissolve into the solution. Copper ions (Cu^{2+}) gain the electrons from the zinc atoms to form copper atoms that plate out as a solid. Whenever an atom or ion becomes more positively charged in a chemical reaction, as in the case of zinc atoms forming positive zinc ions, the process is called **oxidation**. Oxidation is the process of losing electrons. Whenever an atom or ion becomes less positively charged in a chemical reaction, as in the case of the copper ions forming copper atoms, the process is called **reduction**. Reduction involves a gain of electrons. The processes of oxidation and reduction happen together and as such are commonly referred to as "redox" reactions. An easy way to remember which is oxidation and which is reduction is by remembering "LiOn GRrr,"— Lose electrons Oxidation; Gain electrons Reduction.

The formation of rust is a redox process. Water and oxygen are necessary for the iron metal to corrode (rust). Iron atoms lose electrons to form mostly Fe^{3+} ions with the help of the moisture in the air and the heat of the Sun. Because the atoms have given up electrons to become more positively charged, oxidation of iron has taken place. Molecules of oxygen gain electrons to form O^{2-} ions. The oxygen has accepted electrons and is said to have been reduced. Corrosion can be prevented by painting a surface of iron to prevent moisture and air from coming in contact with the metal. Let's summarize what you have learned about atoms and ions:

The term atom means that the element is neutral; this means that it has the same number of protons and electrons.

Ions mean that the atom (or ion) has gained or lost electron(s). Examples are:

$Na \rightarrow Na^+ + e^-$ (Sodium atom loses 1 electron and now has a net charge of +1.)

$Cl_2 + 2\,e^- \rightarrow 2\,Cl^-$ (The two chlorine atoms gain 1 electron each and the net charge is −1 for each chloride ion.)

Polyatomic ions like the sulfate ion ($SO_4{}^{2-}$) imply that there are 2 more electrons than protons in the entire structure.

In some cases you will find that an ion can gain or lose an electron and form a new ion. An example of this type is: $Fe^{2+} \rightarrow Fe^{3+} + e^-$ (The iron in the 2+ state loses 1 more electron and now will be in a 3+ state.)

Reflecting on the Activity and the Challenge

Although many colorful chemical reactions involve the use of acids and bases with indicators, there is an entire group of chemical reactions that produce colorful results through the transfer of electrons. In this activity you became familiar with some of the simple concepts behind redox reactions, and you saw several examples of the color changes they can produce. You and/or your classmates may decide to include some redox reactions in the **Cool Chemistry Show.**

Chemistry to Go

1. Aluminum metal can react to form an ion with a charge of +3. Does the aluminum atom gain or lose electrons to form the Al^{+3} ion?

2. A copper ion with a charge of +2 can react to form an atom of copper. Does the copper ion have to gain or lose electrons in this reaction?

3. The element iron can form two different ions. The iron (II) ion (Fe^{+2}) is commonly called a ferrous ion while the iron (III) ion (Fe^{+3}) is called a ferric ion. When ferrous ions undergo a chemical change to become ferric ions, what process has taken place, oxidation or reduction? Explain your answer.

4. In the reaction you did with zinc metal reacting with copper ions, which substance gains electrons? Which loses electrons?

5. What must take place for copper metal to be oxidized?

6. Galvanized iron nails are used to fasten materials that will be exposed to the outdoors. A galvanized nail is a regular iron nail that is coated with zinc.

   a) Why would a zinc coating be an advantage here? What do you think is the purpose of the zinc?

   b) What two reactants could you use to test this in the laboratory? What results would you expect if you were right about the purpose of the zinc?

Inquiring Further

The Statue of Liberty

In the 1980s the Statue of Liberty in New York harbor underwent extensive renovation. Research the involvement of oxidation-reduction reactions in this renovation. Identify what the problem was and its solution.

Cool Chemistry Show Assessment

Your Chapter Challenge is to present an entertaining and informative chemistry science show to fourth- and fifth-grade students. The show for fourth graders should address chemical and physical properties and changes. The show for fifth graders should help students learn about acids and bases, and about chemical reactions that involve color changes. All presentations must include a demonstration and an audience appropriate explanation of the chemistry concepts involved. Finally, the teachers of the fourth- and fifth-grade students need written directions for your chemistry show with explanations of the chemistry concepts included in the show.

Your team should review the activities that you completed. A useful way of completing this review would be to list the activities, the cool chemistry demonstrations that you can perform, whether these demonstrations would be most appropriate for the fourth- or fifth-grade show and the chemistry concepts of the demonstration. You should pay particular attention to the section **Reflecting on the Activity and the Challenge**. You should also compare your list with that given in the **Chemistry You Learned** summary.

Your team's next step would then be to decide on which demonstrations are best suited to your interests and can be the most interesting, informative, creative and entertaining

for the elementary students. Your teacher may wish you to share your decision with the class as a whole so that the full cool chemistry program is not repetitious.

Each demonstration that you plan on producing will be judged along specific criteria. Each demonstration should be carefully planned, should adhere to all safety considerations, should be completed within the assigned time limits and should exhibit showmanship, creativity, clarity and appeal. As you plan and practice your demonstration, you should use these criteria as a checklist to ensure high quality.

Each demonstration must also be accompanied with an explanation of chemistry concepts. You should rehearse your explanation so that it complements the demonstration itself. You may choose to have one team member producing the demonstration, another team member explaining the manipulations and focusing the audience attention on specific observations and a third team member explaining the chemistry concepts. You should decide if your presentation would be strengthened by having some of the chemistry concepts described prior to the demonstration of if the explanation should follow the demonstration.

Your final responsibility is to provide detailed instructions to the fourth- and fifth-grade teachers so that the teacher could repeat the demonstration. This will require a list of materials, step-by-step directions, notes on what should be happening and safety considerations. You must also provide a written explanation of the chemistry concepts illustrated in the **Cool Chemistry Show**. Your chemistry concepts should be more thorough than the oral presentation during the show. Your explanation of chemistry concepts is now targeted to teachers rather than elementary students. You can include more formulas, more background and provide additional experiments which clarify the chemistry concepts.

Your class should generate a grading rubric. The rubric will assign points to the demonstration, the chemistry content and the written summary. It will also assign points to specific aspects of each major category. For instance, how many additional points will be afforded for additional concepts introduced in the demonstration? How many points will be assigned for creativity, showmanship and adherence to time limits of the cool chemistry demonstration? This rubric can be a useful tool for your team to check the quality of your work and to ensure that you have included all necessary parts of the project.

Chemistry You Learned

Chemical and physical changes of matter

Solute, solvent and solution

Saturated and supersaturated solution

Precipitate

Solubility of compounds in water

pH indicators for acids and bases

Cations and anions

Polyatomic ions

Ionic compounds and formulas

Molecular compounds

Chemical equations

Synthesis reactions

Single- and double-displacement reactions

Decomposition reactions

Metal activity series

Endothermic and exothermic reactions

Conservation energy

Kinetic and potential energy

Heat energy

Colligative properties

Reaction rates

Factors that affect the reaction rate:
- Concentration
- Catalyst
- Temperature
- Surface area or nature of the material

Acids and bases

Titration

Oxidation and reduction

Marc Pollack
President, Flix FX

"Everyone talks about 'Movie Magic,'" says Marc Pollack, president of the prestigious Hollywood special effects company Flix FX. "So I guess that makes me a magician." But Pollack is clearly more comedian than magician. The 'magic' he creates for movies like *Blackhawk Down*, *Men In Black* and *Cast Away*, in addition to scores of television commercials, museum installments and Las Vegas casinos, is the product not of mysterious hocus-pocus but rather fundamental principles of science. "One of the most important aspects of our work," he continues, "is to push the limits of how chemicals are designed to be used." Among other things, Pollack and his crew at Flix FX use vacuum-forming thermo-plastics to make tin-based silicon molds for everything from prehistoric creatures to futuristic robots. Through a combination of trial and error experimentation and traditional research science, they've perfected the process. "Silicon is what we call an R.T.V.," Pollack explains. "That stands for room temperature vulcanization. So depending on the type and amount of catalyst we use, the mold will cure at different rates and with slightly differing properties." By manipulating the ratio of silicon to catalysts they can make strong, realistic molds in the most efficient way possible. "Increasing the amount of catalyst will speed up the curing process but too much catalyst will shorten the life of the mold," he says. "Every job is different so determining that balance is one of our many challenges."

Pollack, who is now a master in the art of using chemicals like silicon, polypropylene, urethane and urethane elastimers, is not a chemist by trade. He actually graduated from film school at SUNY Purchase in the hopes of becoming "the next Steven Spielberg." Then, through a twist of fate, he became a special effects nut and eventually founded Flix FX in 1990. "Now," Pollack says, "Spielberg may one day come to me."

Special effects — Pollack creates both mechanical and physical — is an industry in a constant state of transformation. "The industry is always trying new stuff and that's exciting," Pollack says. "For instance, someone just developed a great water-based breakaway glass for stunts called Smash Glass. It's similar to fiberglass without the dangerous elements associated with that material and can be made to break into either large chunks or tiny little pieces. I can't wait to get my hands on it and break it over someone's head. That's part of my job these days and I love it."

Unit 3

Active Biology™

Making Connections

At this point, you have come to the third unit in your *Coordinated Science for the **21st Century*** course. Thus far, you have spent much of your time studying nonliving things. You have been an active participant, as you learned to think about the forces in our world. You have also learned about the building blocks that make up our world. Now you are going to focus on those parts of our world that are considered living. You will start by taking a closer look at the biosphere, the part of Earth where living organisms are found.

What is the difference between living and nonliving things? Historically, this has always been a real puzzle for scientists. Some scientists believed that living things could come from nonliving things. Another group believed that life had a vital force that was neither chemical nor physical.

What Do You Think?

• Is there a difference between the elements found in living things and nonliving things?

• What is the difference between an organic and inorganic chemical?

Record your ideas about this in your log. Be prepared to discuss your responses with your small group and the class.

For You To Try

• Research and report to your class about a career that requires a knowledge of both chemistry and biology.

• What is biophysics? How might what you learned in physics be related to biology? Research and report to your class the connection between physics and biology.

Chapter 9

A VOTE FOR ECOLOGY

A Vote for Ecology

Scenario

Many Americans have begun to realize the importance of ecological issues. Americans now care about problems that were almost unknown a few years ago. Land and water management, pollution, biodiversity, invasive species, and many more concerns are on people's minds.

However, ecological issues cannot be considered on their own. They must be included in the economic and social spheres. The hidden costs of environmental programs are sometimes forgotten. It is necessary to develop a balanced solution to problems. This is the only reasonable way to sustain the environment.

For example, fishing provides food, income, and employment for millions of people. However, fishing has environmental costs. Rare species may be threatened. Marine ecosystems can be disturbed. Also, it is questionable how long the resource will last. Aquaculture presently offers an alternative. It provides a chance to expand the food supply from freshwater or the sea. However, aquaculture can also be ecologically unsound. Natural habitats are lost. The introduction of alien species in an area can pose a threat to the existing ecosystems. The spread of disease from farmed to wild populations is also a concern.

The League of Concerned Voters in your area recognizes the importance of preserving the environment. However, they are also aware that a lack of information about ecological issues could lead to conflict rather than constructive action. That is why they have decided to commission the development of a series of booklets. These booklets are intended to introduce the scientific facts behind current issues.

Chapter Challenge

Your challenge is to create a booklet addressing one current issue. These booklets will be provided to the public. The League hopes that this will produce an informed public, who is better able to decide how to vote on any given issue. Before you begin, you will need to decide which audience you are targeting. You may choose to write your booklet geared to adult voters, teenagers, or a child. Regardless of which you choose, it is assumed that the readers will be non-specialists and the text should be written with this in mind. The booklets should be easily understood by your target audience.

In producing your booklet, you should:

- identify and research one current issue that threats the environment
- provide the relevant data on the issue
- draw attention to areas where data may be weak or lacking
- interpret the data and indicate the limits of the interpretation.

In providing the science behind the issue, you should:

- identify the roles and importance of consumers, producers, and decomposers in an ecosystem
- explain how matter is cycled and energy flows through ecosystems
- provide the meaning and the importance of biodiversity
- describe how changes in an ecosystem are determined and how they can be analyzed
- tell how fluctuations in size of a population are determined by birth, death, immigration, and emigration.

Criteria

How will your booklet be graded? What qualities should a good booklet have? Discuss these matters with your small group and with your class. You may decide some or all of the following qualities are important:

- significance of the issue identified
- completeness and accuracy of the ecology principles presented
- merit of the interpretations suggested
- readability of the booklet
- design and layout of the booklet.

You may have additional qualities that you would like to include. Once you have determined the criteria that you wish to use, you will need to decide on how many points should be given to each criterion. Your teacher may wish to provide you with a sample rubric to help you get started.

Activity 1 Diversity in Living Things

GOALS

In this activity you will:

- Observe a group of diverse organisms.

- Relate the structure of an organism to its adaptation to the environment.

- Describe the organization of the biosphere.

- Define biodiversity and explain its importance.

- Explain the effects of human activity on biodiversity.

- Read about the effects of extinction.

- Practice safe laboratory techniques for handling living organisms.

What Do You Think?

It is estimated that 4% of all living species are found in Costa Rica, even though this country comprises only 0.01% of the area of the Earth.

- **How many species do you think are found in Costa Rica? How many species are found globally?**

- **Why do you think that Costa Rica has such a large number of species?**

Write your answers to these questions in your *Active Biology* log. Be prepared to discuss your ideas with your small group and other members of your class.

For You To Do

This activity provides you with an opportunity to view several very different species of organisms. It should give you an appreciation of the huge diversity of life that fills your world.

Part A: Observing Animal Diversity

1. With your teacher, review the guidelines concerning the proper handling of laboratory animals. Follow these guidelines carefully.

2. In your *Active Biology* log make an enlarged copy of the table shown below. The table should extend across two facing pages. Each of the 13 spaces should allow for several lines of writing.

3. In the *Characteristics* column, copy the words in italics from each of the following questions. The 13th space is for any other observations you make. All the specimens of one animal species and the materials and equipment needed for observing them are arranged at the station. Each team will have a turn at each station. Record only your observations, not what you have read or heard about the organism.

   1 What is the *habitat* of the animal? Does it live in water, on land, or both?

2 Is *body symmetry* radial (symmetry about a center) or bilateral (the left and the right sides of the body are mirror images)?

3 Does the animal have a *skeleton* (a structure that supports the organism's body)? If it does, is it an endoskeleton (on the inside) or an exoskeleton (on the outside)?

4 Is the animal's body *segmented* (divided into sections) or is it *unsegmented*?

Several of the activities that follow involve the use of organisms in water. The water that the organisms are in should be considered a contaminant. Tables, equipment, and hands should be washed carefully so that germs are not inadvertently passed to people.

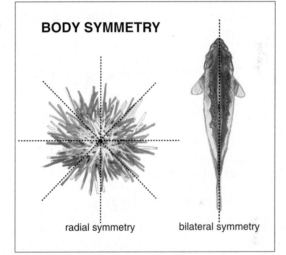

BODY SYMMETRY

radial symmetry   bilateral symmetry

Comparing Animals					
Characteristics	**Hydra**	**Planarian**	**Earthworm**	**Hermit Crab**	**Frog**
1					
2					
3					
13					

5 Which type of *digestive cavity* does the animal have, a sac (only one opening) or a tube (open at both ends)?

6 Does it have *paired appendages*? Are the limbs (arms, legs, fins, wings) found in pairs?

7 How does the animal *obtain oxygen*? Through lungs, gills, skin, or a combination of these?

8 Are any *sense organs* visible? If so, what types and where?

9 How does the animal *move* from one place to another?

10 Does it make any types of *movement* while it remains more or less in one spot?

11 How does the animal *capture* and take in *food*?

12 How does it react when *touched* lightly with a small brush?

Station 1: Observing Hydras

1. Place a single hydra in a small watch glass with some of the same water in which it has been living. Wait until the animal attaches itself to the dish and extends its tentacles. Then slowly add a few drops of a daphnia culture with a dropping pipette.

2. Touch the hydra gently with a soft brush. Observe its reactions.

3. Examine a prepared slide of a lengthwise (longitudinal) section of a hydra under a compound microscope. Try to determine the presence or absence of a skeleton and of a digestive system.

Station 2: Observing Planarians

1. Place one or two planarians in a watch glass containing pond or aquarium water. Add a small piece of fresh raw liver. Observe using a stereomicroscope or hand lens.

2. Use a compound microscope to examine cross sections of a planarian. Examine whole mounts with a stereomicroscope. Determine the presence or absence of a skeleton and a digestive system.

As you move among the stations, keep your hands away from your mouth and eyes. Wash your hands well after the activity.

Station 3: Observing Earthworms

1. Pick up a live earthworm and hold it gently between your thumb and forefinger. Observe its movements. Do any regions on the body surface feel rough? If so, examine them with a hand lens.

2. Place a worm on a damp paper towel. Watch it crawl until you determine its anterior (front) and posterior (back) ends. Use a hand lens to see how the ends differ. Describe.

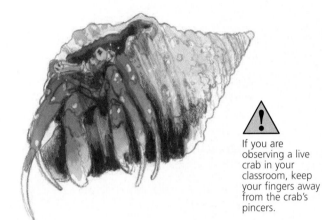

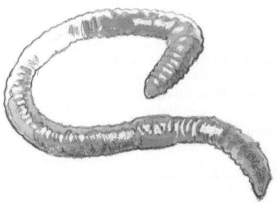

If you are observing a live crab in your classroom, keep your fingers away from the crab's pincers.

2. Place a small piece of food from the food dish in with the hermit crab. Observe how the hermit crab eats.

Station 5: Observing Frogs

1. Observe the breathing movements of a frog while it is not moving.

2. Observe the variety of movements of a live frog.

3. If possible, observe a frog capturing its food and feeding.

3. Place an earthworm on loose soil and observe its movements as it burrows.

4. Examine a model or a diagram of a cross section and lengthwise section of the earthworm's body.

Station 4: Observing Hermit Crabs

1. Observe the movements of the appendages and the pattern of locomotion (movement from one place to another) of a living land hermit crab. Observe the antennae. Touch them gently with a soft brush. Note the animal's reaction.

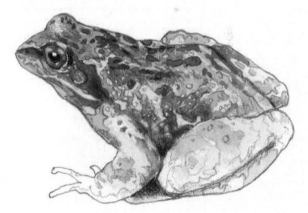

Wash your hands thoroughly before leaving the laboratory.

If you are handling a live frog in the classroom, do not rub your eyes. Wash your hands immediately after handling.

Part B: Animal Adaptations to the Environment

1. Review what you have learned about each of the organisms in **Part A**. By reading across the table, you should be able to compare and contrast the characteristics of the five animals you have studied.

a) For each animal, select two functions it performs as part of its way of life. Describe how its structure enables it to perform these functions.

Bio Talk

BIODIVERSITY

Organization in the Biosphere

The **biosphere** is the area on Earth where living organisms can be found. Most are found in a narrow band where the atmosphere meets the surface of the land and water. Life forms are referred to as the **biotic**, or living, component of the biosphere. The **abiotic**, or nonliving, component is made up of items like rocks, soil, minerals, and factors like temperature and weather.

Bio Words

biosphere: the area on Earth where living organisms can be found

biotic: the living components of an ecosystem

abiotic: the nonliving components of an ecosystem

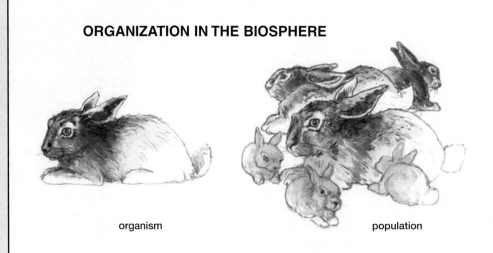

ORGANIZATION IN THE BIOSPHERE

organism   population

Just as you did in this activity, ecologists begin their studies with the **organism**. Their investigations are designed to explore how the individual interacts with its biotic and abiotic environment. However, an organism does not live on its own. It tends to form a group with others of the same **species**. (A species is a group of organisms that can reproduce successfully only with others of the same type.) These groups of species are called **populations**. When more than one population occupies an area, a **community** of organisms is created. The abiotic component as well as the community form a functional unit known as an **ecosystem**.

The Importance of Biodiversity and the Human Threat

In this activity you looked at some very different species of organisms. Scientists have discovered and named close to two million species. That would mean looking at a lot of different organisms. Yet, it may be less than 20 percent of the species that exist! There are thousands of organisms in the world that scientists know very little about. More than 750,000 species of insects have been identified. Yet, it is thought that at least twice that many exist. Biological diversity, or **biodiversity**, is the sum of all the different types of organisms living on Earth.

Bio Words

organism: an individual living thing

species: a group of organisms that can interbreed under natural conditions and produce fertile offspring

population: a group of organisms of the same species occupying a given area

community: all the populations of organisms occupying a given area

ecosystem: a community and the physical environment that it occupies

biodiversity: the sum of all the different types of organisms living on Earth

community

ecosystem

Coordinated Science for the 21st Century

Unfortunately, many organisms are disappearing. This is partly due to alterations of habitats. The result is a decrease in biodiversity. Ecosystems with a large number of different types of organisms are quite stable. Ecosystems with a small number of different organisms are less stable. Humans are partly responsible for this change. As the human population grows it occupies more land. This infringes on or destroys the habitats of many organisms.

The smog created by automobiles and industry is killing many types of trees over a wide area of southern California. The needles of ponderosa pines, for example, gradually turn brown. The tops of palm trees have only small tufts. When this happens, photosynthesis is greatly reduced. The plants die. The Everglades National Park in southern Florida depends on a slowly moving sheet of water. The water flows from north to south. Drainage ditches built at the northern edge of the Everglades have decreased the flow of water over the entire area. As a result, many alligator holes have dried up. These holes helped to contain fires in the Everglades. Now, destructive fires are more frequent in this national park.

Smog is also hazardous to people. This is especially true of those with respiratory problems, the elderly, and children. People have died from the effects of smog.

Bio Words

extinction: the permanent disappearance of a species from Earth

Tropical rainforests are the most diverse ecosystems on Earth. They are home for many different species. Two-thirds of the world's species are located in the tropics and subtropics. The cutting of trees in the rainforests today has grown at a rapid rate. The trees are cut for lumber, grazing land, and other uses. This loss of habitat is destroying many species every day. Nearly half of the Earth's species of plants, animals, and microorganisms will become extinct, be gone forever, or be severely threatened, during the next 25 years.

To find a similar rate of **extinction** (loss of species), you need to go back 65 million years. That was the end of the Cretaceous period when dinosaurs and other organisms disappeared. Because there are more species today than there were then, the absolute number of species lost will be greater now. Hundreds of species of plants and animals are threatened today. They include the whooping crane and some rare pitcher plants. Extinction is a natural process. However, the process has been speeded up because humans have changed whole ecosystems.

In tropical regions, humans are cutting down, burning, or otherwise damaging the rainforests. Extinction of many species as well as change in global climate are some of the effects of this deforestation.

Why is biodiversity important? Why does it matter if whooping cranes and pitcher plants become extinct? One argument comes from genetics. In a field of crop plants planted by humans, all the plants are genetically similar. They have all inherited the same characteristics.

About 90% of the world's food comes from 15 species of plants. Three of them are corn, wheat, and rice. However, there are over 10,000 known species of cereals.

If one individual gets a disease, all the plants may die. In a wild population a vast pool of genetic characteristics are available. This means that some of the plants could resist the disease. Therefore, not all the plants would be destroyed. The extinction of each wild population erases genetic material that could mean healthy crops and animals. Once extinction occurs, the genetic material is gone forever.

A second argument is related to the fact that simple ecosystems are unstable. Think of a field of corn as a simplified ecosystem. Suppose all the corn dies. This means that the whole ecosystem would collapse. The simpler the ecosystem, the easier it is to disrupt its balance. The fewer the species, the easier it is to upset an ecosystem. New species are evolving all the time. However, the process is very slow compared to the rate at which humans are able to cause species to become extinct. Each time a species becomes extinct, the biosphere is simplified a little more. It becomes more difficult to maintain the stable biosphere on which all life depends.

A third argument comes from research on plants. The island of Madagascar, off the east coast of Africa, is the only known habitat of the Madagascar periwinkle. This plant produces two chemicals not produced by other plants. Both of these chemicals are used to fight Hodgkin's disease, a leukemia-like disease. As the human population on Madagascar grew, the habitat for the periwinkle shrank. The periwinkle almost became extinct. Fortunately, botanists collected and grew some of these plants before they were gone forever. The medicines made from the Madagascar periwinkle are worth millions of dollars each year. They also help many people with Hodgkin's disease to live longer. These medicines never would have been known if the plant had become extinct.

Extinction Can Cause a "Domino Effect"

Every organism in an ecosystem is connected to all the other organisms. The reduction in biodiversity caused by the extinction of a single species can cause a "domino effect." The removal of one part from an ecosystem, like the removal of a moving part from a car, can cause the collapse of an entire food chain. If a species acts as a predator, it keeps the population of its prey in check. If a species is prey, it provides an important food source.

For example, sea otters were over-hunted along the Pacific coasts of Asia and North America. This removed the main predator of the sea urchin. Predictably, the number of sea urchins grew rapidly. Sea urchins eat kelp, a form of seaweed. As the number of sea urchins grew, the amount of kelp declined. As a result, the fish that relied on the kelp for habitat and food were reduced in number.

Sea otters very nearly became extinct due to hunting pressure. For humans, killing the sea otters for their fur resulted in a decline in a valuable fishery. Where the sea otter has been reintroduced, sea urchin populations have fallen, kelp beds are being re-established, and the number of fish is increasing.

Restoring the Balance Is a Difficult Task

Introducing the sea otter to the Pacific northwest is an example of an attempt to restore a natural balance. It is not always easy to do. Conservationists have also tried to restore the whooping crane. In spring, whooping cranes fly north to live in the marshes and swamps of the prairies and the Canadian north. There they eat crayfish, fish, small mammals, insects, roots, and berries. Efforts by the United States and Canada have helped increase the population from a low of 14 individuals in 1940 to 183 in 1999. The whooping crane may be a success story, and it may not. Chemical pesticides were the original human threat to the crane. However, it was already struggling.

During the fur trade southern sea otters were hunted to near extinction. They are still a threatened species, and may very well be endangered.

Once widespread throughout North America, the whooping crane wild population dipped to just 15 birds in 1937. Through conservation efforts the whooping crane has begun a slow recovery. However, coastal and marine pollution, illegal hunting, and the draining of wetlands continue to threaten the species.

Cranes must fly a long way between their summer homes in the north and their winter homes on the Gulf of Mexico. Along the way they are vulnerable to hunting and accidents. In addition, the whooping crane reproduces very slowly. Each year females produce two eggs, however, only one will mature. The first fledgling to crawl from the egg kills its brother or sister. This ensures there will be enough food for the survivor. However, it is very difficult for the species to increase its numbers.

Scientists do not understand all the relationships between species ecosystems. They cannot predict what will happen if biodiversity is reduced, even by one species. If one species becomes extinct, it could be disastrous. The extent of the disaster may not be known until later. Sometimes the balance cannot be restored.

Reflecting on the Activity and the Challenge

In this activity you observed several very different living organisms. You then discovered that there are millions of other different organisms alive on Earth. There are reasons why it is important to make sure that these organisms do not disappear forever from the Earth. For your **Chapter Challenge** you may choose to research an issue that relates to the disappearance of a given species. You can now explain why it is important to maintain biological diversity. Whether or not your issue deals with biodiversity, the public still needs to understand why biological diversity should concern them. You need to provide the meaning and importance of biodiversity.

Biology to Go

1. Choose and identify two very different ecosystems.

   a) For each, name some of the populations that might be found in each community.

   b) Describe some of the abiotic factors that could affect each population.

2. What is biodiversity?

3. Explain how humans can influence biodiversity by changing the environment.

4. Why is maintaining biodiversity important?

5. a) Give an example of an ecosystem that has a high biodiversity.

   b) Give an example of an ecosystem that has a low biodiversity.

6. Choose an organism other than one that you studied in this activity. List at least three structures that have helped the organism adapt to its environment. Describe how each helps the organism live in its ecosystem.

Inquiring Further

1. The passenger pigeon and the human influence

Just over a century ago, the passenger pigeon was the most numerous species of bird on Earth. In the Eastern United States they numbered in the billions, more than all other species of North American birds combined. On September 1, 1914, at 1:00 PM the last surviving passenger pigeon died at the age of 29. Research and report on how humans were involved in the extinction of the passenger pigeon.

2. Extinction is forever

Humans were directly responsible for the extinction of passenger pigeons. However, this bird is not the only organism that has been threatened by humans. Research and report on another organism whose existence has been or is endangered by humans.

White rhinos are so large and powerful that in nature they must give way only to the elephant. Yet, humans are a major threat to their existence.

Activity 2

Who Eats Whom?

GOALS

In this activity you will:

- Distinguish between a food chain and a food web.

- Explain the roles of the producers, consumers, and decomposers.

- Understand the meanings of autotroph, heterotroph, herbivore, carnivore, and omnivore.

- Recognize the dependence of organisms on one another in an ecosystem.

What Do You Think?

You have probably heard the question, "If a tree falls in a forest and there is no one there to hear it, does it make a noise?"

- **In ecology you might ask the question, "If a tree falls in a forest and there is no one there to haul it away, what happens to it?"**

Write your answer to this question in your *Active Biology* log. Be prepared to discuss your ideas with your small group and other members of your class.

For You To Do

In this activity you will have an opportunity to explore how organisms in an ecosystem are dependent on one another.

Part A: A Food Chain

1. Look at the organisms in the pictures on the right.

    a) Link the names of the organisms together by the words "is eaten by."

    b) Show the relationship between the organisms as a linked word diagram or chain.

    c) Use arrows to show the direction in which food energy moves in the food chain you constructed in **Part (b)**.

    d) Identify the producer in the food chain.

    e) Identify the consumers in the food chain.

    f) Which consumer is a herbivore (feeds on plants)?

    g) Which consumers are carnivores (feed on other animals)?

    h) What elements are missing from this food chain?

Part B: A Food Web

1. Your teacher will provide you with a card that names an organism, what it does, what it eats, and what it is eaten by. You will also be given a name tag with your organism's name on it.

    Read your card and attach your "name" tag where others can readily identify you.

2. Clear a large area in your classroom, or find another large open area in or near your school. Form a large circle.

snake

grasshopper

frog

green plant

hawk

3. Obtain a big ball of string, about 35 m in length. Give the ball of string to one of the students.

4. The first student will say what organism he/she represents. Also, the student will indicate what the organism eats and what it is eaten by. The ball of string is then directed to one of students who represents the predator or the prey.

   a) As the game progresses, what appears to be forming in the center of the circle?

5. Suppose one organism is removed from the circle. Your teacher will direct you which organism will be removed.

   a) What happens to the web that was created?

   b) How does the removal of an organism impact on the other organisms in the circle?

6. Suppose that your circle has only a few organisms.

   a) What would happen to the web in this case if one of the organisms were removed?

   b) In which situation, a large or small "circle" of organisms, does the removal of an organism have a greater impact?

BioTalk

Food Chains and Webs

A bat ate a mosquito that had bitten a coyote that had eaten a grasshopper that had chewed a leaf. All these living things make up a **food chain**. A food chain is a step-by-step sequence that links together organisms that feed on each other. The story, however, is incomplete. It does not mention that many animals other than coyotes eat grasshoppers and mosquitoes bite other animals. It also does not consider that coyotes and bats eat and are eaten by a great many other living things. When you consider that the kind of plant a grasshopper might eat may also be eaten by various other consumers, you start to build a picture that links together a whole community of living things. Those links resemble a **food web** rather than a food chain. A food web is a series of interconnected food chains or feeding relationships. The diagram shows how members of a community interact in a food web.

Organisms in the Food Web

Autotrophs are organisms that are capable of obtaining their energy (food) directly from the environment. Most autotrophs obtain their energy through the process of photosynthesis. In this process solar energy is converted into a form of energy that can be used by the organism. **Heterotrophs** obtain their energy from autotrophs or other heterotrophs. For this reason autotrophs, the organisms that "make" the food, are called **producers**. In the diagram, grass, vegetables, and trees represent the producers. The heterotrophs are called **consumers**. **Herbivores** are first-order consumers. They feed directly on the plants. These organisms are removed by just one step

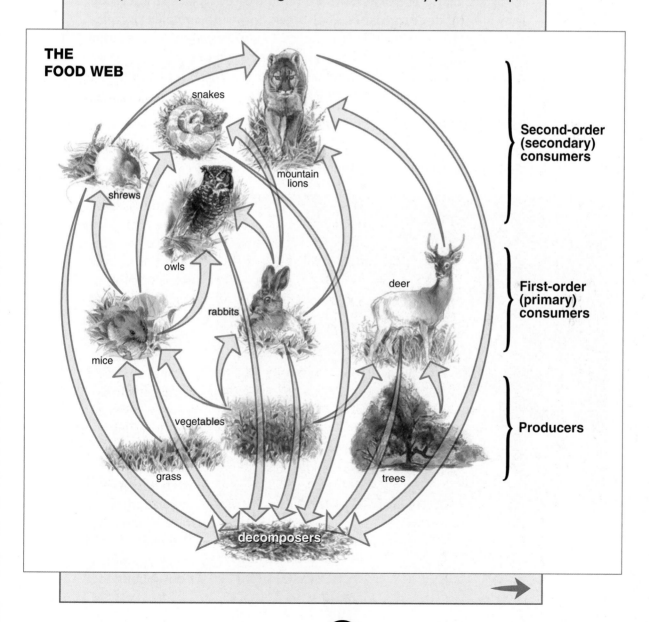

THE FOOD WEB

snakes

mountain lions

shrews

owls

deer

rabbits

mice

vegetables

Producers

grass

trees

decomposers

Second-order (secondary) consumers

First-order (primary) consumers

in the food chain from the producers. In this example, they include mice, rabbits, and deer. **Carnivores** are second-order consumers. They feed on the animals that eat other plants. The owl and the mountain lion are just two examples of carnivores. **Omnivores** eat both plants and animals. A human is an example of an omnivore.

There is another group of organisms in the food web that is so important that these organisms are often treated as a separate group. They are the **decomposers**. They break down the complex organic molecules that are found in the wastes and bodies of other organisms. They do this to obtain food energy for their own use. In the process, they release nutrients back into the ecosystem. Bacteria and fungi make up most of the decomposers.

Alternative Pathways Maintain Stability in Food Webs

The alternative pathways in a food web help maintain the stability of the living community. If the rabbits in some area decrease in number, perhaps because of some disease, the owls might be expected to go hungry. However, this is not the case. The rabbits eat less vegetation. Hence, the greater number of plants produces more fruits, and seeds and furnishes better hiding places for mice. Soon a larger population of mice is present. The owls transfer their attention from rabbits to

The food habits of rabbits vary depending on location, time of year, and species of rabbit. They generally prefer to eat tender, green vegetation. They also eat leaves, bark, seeds, and even fruit of woody plants. Rabbits begin feeding in the evening and continue throughout the night.

mice. This reduces the danger for surviving rabbits, and these primary consumers have a better chance to rebuild their numbers. The greater the number of alternative pathways a food web has, the more stable is the community of living things which make up the web.

Owls are nighttime (nocturnal) birds of prey. Owls feed entirely on living animals. They eat everything from insects to mammals as large as rabbits. The size of the prey is proportional to the size of the owl.

Only a few of the possible offspring of a plant or animal survive to reproduce. Of all the seeds a plant forms, all but a few are eaten by animals. Some die from diseases. Others are killed by poor weather conditions. This can happen either as seeds or somewhat later in life, as young plants that have not yet formed seeds of their own.

Humans are so used to thinking of the welfare of their own species, that they tend to regard as "wasted" all the offspring that do not survive. But there is another side to the picture. For one thing, the world lacks space for so many individuals of any one kind. Also, these individuals are needed as food by a great variety of consumers. Without the fruits, seeds, young plants, and foliage, the primary consumers could not exist. Without the primary consumers, the plants would die. They would become overcrowded or lack nutrients. Without the primary consumers, the secondary consumers would be reduced in numbers because of competition, or would become extinct. Without waste from plants and animals, including dead remains, the decomposers would not be able to get their nutrients. Without decomposers, nutrients that the producers require would not be returned to the soil or water. Through the presence of all these components in the food web, each species is held in check, and the community maintains its stability.

Reflecting on the Activity and the Challenge

In this activity you looked at how every organism is dependent on other organisms and how they are all held together by a food web. You can now begin to understand how the stability of any ecosystem depends on each one of its components. In the **Bio Talk** reading section you were also reintroduced to many terms that are used by ecologists. In discussing your environmental issue, you will be expected to use these terms correctly. You will probably also want to explain the importance of some of these terms in your booklet to educate the public.

How are primary consumers a benefit to plants?

Biology to Go

1. In what ways are living organisms affected by other living organisms?

2. What is the role of decomposers in a biological community?

3. What is the difference between a food chain and food web? Use an example to explain your answer.

4. a) Why are autotrophs called the producers in an ecosystem?

   b) Why are heterotrophs called consumers?

5. Are you a herbivore, carnivore, or omnivore? Explain your answer to show that you understand the meaning of each term.

6. Create a food web that includes you and at least five other organisms. Identify the decomposers, producers, and consumers as you diagram your food web.

7. In which ecosystem would the removal of an organism disrupt stability more, an Arctic ecosystem or a deciduous forest? Explain your answer.

Water makes up the largest part of the biosphere. Aquatic regions, both freshwater and marine, are home to many species of plants and animals. As you inquire further into aquatic food webs, you may be surprised at how many different types of aquatic ecosystems exist.

Inquiring Further

Aquatic food webs

Water covers over two-thirds of the surface of the Earth. Research and construct an aquatic food web. Identify the producers and consumers.

Activity 3   Energy Flow in Ecosystems

GOALS

In this activity you will:

• Infer the loss of energy in the form of heat from the human body.

• Relate the laws of thermodynamics to the transfer of energy in a food chain.

• Calculate the energy lost at a given level in a food web.

• Explain the significance of a pyramid of biomass, a pyramid of numbers, and a pyramid of energy.

What Do You Think?

Heat stroke is caused by a failure of the heat-regulating mechanisms of the body. It can be caused by heavy exercise combined with hot and humid conditions.

• Where does the heat in the body come from?

Write your answer to this question in your *Active Biology* log. Be prepared to discuss your ideas with your small group and other members of your class.

For You To Do

As you work through this activity, consider whether there is any relationship between events like heat stroke and the heat that is stored and lost at each link in a food web.

1. Read through the steps of the activity.

   a) What are you investigating in this activity?

   b) Predict what you think will happen to the water temperatures in the containers.

2. You can now follow the steps to conduct the experiment. Put 600 mL of water in each of three containers. The temperature of the water should be 10°C. You may have to add ice. Remove the ice when the temperature gets to 10°C.

3. Have one student put one hand into the water in container A. Have that student put the other hand into the water in container B. In container A, move the fingers rapidly in the water. Do not move the hand in container B. Keep one hand moving and the other hand still for five minutes.

4. Another student will hold a thermometer in the water in container A. Read the temperature once each minute for 5 minutes.

a) What is the purpose of stirring the water?

b) Record the temperatures in the chart.

 Clean up any spilled water immediately.

Minutes	Temperature Container A (moving hand)	Temperature Container B (still hand)	Temperature Container C (no hand)
1			
2			
3			
4			
5			

a) Record the temperatures in a chart similar to the above.

5. A third student will hold a thermometer and read the temperatures in container B. Also, stir the water in this container using the stirring rod. ⚠ Wash your hands after completing the activity.

6. A fourth student will hold a thermometer in container C. Stir the water in this container. Read the temperature once each minute for five minutes.

a) Why did you have a container that you did not put your hand in?

b) Record the temperature readings in the chart.

7. Make a line graph of the temperature readings for the three containers. You will have three lines on the same graph.

   a) In container B, you held your hand in cold water without

moving it. What happened to the temperature? Does this data support your prediction?

b) In container A, you exercised your hand. How did the temperature of the water change? Do your data support your prediction?

Bio Talk

Pyramids of Mass and Energy

One of the most important abiotic factors that affects relationships in a community is energy. Organisms in an ecosystem are tied together by the flow of energy from one organism to another. The food chain that exists when a herbivore eats a plant and a carnivore eats a herbivore depends on the energy entering the community in the form of sunlight. Without the Sun, there would be no green plants, no herbivores, and no carnivores. (There are a few ecosystems that get their energy from another source.)

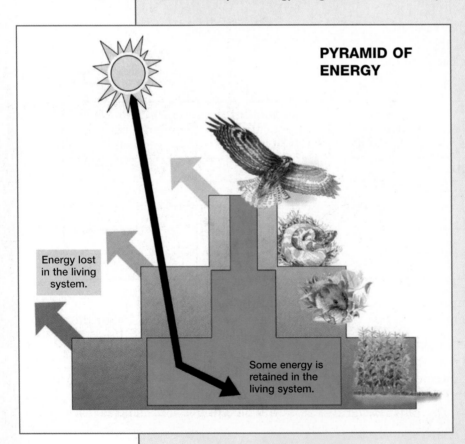

PYRAMID OF ENERGY

Energy lost in the living system.

Some energy is retained in the living system.

The size of a community, therefore, is limited by the amount of energy entering it through its producers. The total amount of chemical energy stored by photosynthesis is the gross primary productivity of the community. Much of that energy is used by the producers to grow and to maintain themselves. The remaining energy, which is available to the consumers as food, is the net primary productivity of the community.

The transfer of energy from producer to primary consumer to secondary consumer, and so on in a food web must follow the laws of **thermodynamics**.

The first law of thermodynamics states that although energy can be transformed, it cannot be created or destroy. Some energy from the Sun is transformed into a form that can be used by living organisms. However, if energy is not destroyed, what happens to it? Why is it necessary to keep adding energy in the form of sunlight? That is where the second law of thermodynamics comes into play. It states that in any energy transformation some energy is lost from the system in an unusable form.

Usually this is in the form of heat. In this activity, you actually measured the temperature increases that resulted from the heat loss from the human body. You noted that with exercise, the heat loss was even greater than without movement.

Among living beings, the transfer of energy in food from "eaten" to "eater" is really quite inefficient, and of course a great deal of the food does not get eaten at all. From grass to sheep the loss is about 90 percent.

It takes about 10 kg of organic matter in the grass to support one kilogram of sheep.

Bio Words

thermodynamics: the study of energy transformations described by laws

Bio Words

pyramid of living matter: a pyramid developed on the basis of the mass of dry living matter at each trophic level

pyramid of energy: a pyramid developed on the basis of the energy at each trophic level

trophic level: the number of energy transfers an organism is from the original solar energy entering an ecosystem; the feeding level of one or more populations in a food web

For the sake of simplification, assume that each consumer lives entirely on one kind of food. Then a person on a lake might live entirely on a given type of fish, for example. To support one kilogram of this person it takes about 10 kg of fish, 100 kg of minnows, 1000 kg of water fleas, and 10,000 kg of algae. This information in graph form is called a **pyramid of living matter**. Mass is a measure of the amount of matter in an object. Because much of the mass of living organisms is water, the producers first must be dried for a truer estimate of their mass when constructing a pyramid of matter. The pyramid shows that the amount of matter is greatest at the producer level.

It is possible to measure the amount of energy available at each level. The **pyramid of energy** that results from graphing these values also

THEORETICAL PYRAMID OF LIVING MATTER

human
100 kg

fish
1000 kg

minnows
10,000 kg

water fleas
100,000 kg

algae
1,000,000 kg

Humans have relied on fish as a source of food throughout history. Most of the fish protein was provided by species caught in the wild.

shows that the energy available is greatest at the producer level and steadily decreases at the other levels. Each step in the pyramid is called a **trophic level** (energy level). Because energy is lost at each transfer, the steps in a pyramid of energy are limited. Usually, there are no more than about five trophic levels in a food chain.

It is also possible to construct a pyramid of numbers by counting the number of organisms in a food chain. Although the largest number of organisms is usually found at the base of the pyramid, this is not always the case. For example, in a meadow there will be many more grass plants than there will be grasshoppers. However, a single tree can sustain many caterpillars.

Reflecting on the Activity and the Challenge

In this activity, you observed the loss of heat from the human body. You then related that to the loss of energy at each step of a food chain. You learned that the further you go up a food chain, the less energy is available. As part of your challenge you are expected to explain how energy flows through an ecosystem. You should also consider how the flow of energy is affected in the environmental issue that you have chosen.

Biology to Go

1. What is the relationship, if any, between the heat energy stored and dissipated at each link in a food web and the heat energy responsible for a heat stroke?

2. In the activity, the student who kept his/her hand in the water may have begun to shiver. Why do you suppose this happened?

3. Explain how the transfer of energy in a food chain follows the laws of thermodynamics.

4. Why is there a limit to the number of trophic levels in an energy pyramid?

5. Why is a pyramid of numbers not always a good example of the flow of energy through a food chain?

6. An energy pyramid illustrates the energy lost at each level of a food web. In general, each level of the pyramid has only 10% of the energy at the level below it. If the producer level (the lowest level) has 10,000 kilocalories available for the rest of the food web, how much energy is available for the other three levels?

7. An energy pyramid illustrates a great loss of energy as you go up the pyramid. When humans eat meat, they act as a top-level consumer. A steer eats a small amount of corn which contains 10 kilocalories. If you were to eat the same amount of corn you would get the same amount of energy from it as the steer. How much energy would you get from that small amount of corn if you ate some hamburger from that steer? Before you calculate this answer think about how this energy pyramid differs from the energy pyramid in the previous questions.

Inquiring Further

Biological amplification

What does biological amplification mean? Use the example of dichloro diphenyl trichloroethane, or DDT, to illustrate how biological amplification and food chains and energy pyramids are related.

The peregrine falcon is a bird of prey at the top of the food chain. As a result of biological amplification, falcons ingested high levels of the pesticide DDT. Falcons contaminated with DDT did not lay eggs or produced eggs with shells that broke.

Activity 4   Factors Affecting Population Size

GOALS

In this activity you will:

- Investigate the factors that affect the size of a population.

- Interpret a graph and make calculations to examine factors affecting fluctuations in populations.

- Calculate the doubling time of the human population.

- Distinguish between an open and closed population.

What Do You Think?

The population of your community may be going up, going down, or remaining the same. The change depends on whether individuals are being added to or taken away from your community.

- **What can take place in your community, or any other community of living things, that can influence the size of the population?**

Write your answer to this question in your *Active Biology* log. Be prepared to discuss your ideas with your small group and other members of your class.

For You To Do

This activity provides an opportunity for you to examine the factors that affect the changes (fluctuations) that occur in a population in an ecosystem.

Part A: Reindeer Population

1. In 1911, 25 reindeer, 4 males and 21 females, were introduced onto St. Paul Island near Alaska. On St. Paul Island there were no predators of the reindeer, and no hunting of the reindeer was allowed. Study the graph shown below and answer the questions in your *Active Biology* log.

a) In 1911 the population was 25 reindeer. What was the size of the population in 1920? What was the difference in the number of reindeer between 1911 and 1920? What was the average annual increase in the number of reindeer between 1911 and 1920?

b) What was the difference in population size between the years 1920 and 1930? What was the average annual increase in the number of reindeer in the years between 1920 and 1930?

c) What was the average annual increase in the number of reindeer in the years between 1930 and 1938?

d) During which of the three periods 1911—1920, 1920—1930, or 1930—1938, was the increase in the population of reindeer greatest?

e) What was the greatest number of reindeer found on St. Paul Island between 1910 and 1950? In what year did this occur?

f) In 1950, only eight reindeer were still alive. What is the average annual decrease in the number of reindeer in the years between 1938 and 1950?

2. In your group, discuss the questions on the next page. Then answer them in your *Active Biology* log.

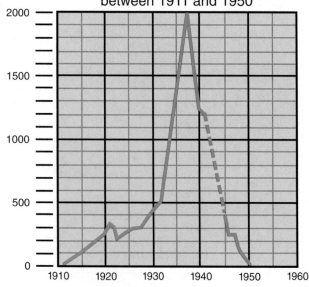

Changes in the
Reindeer Population on St. Paul Island
between 1911 and 1950

a) St. Paul Island is more than 323 km (200 miles) from the mainland. Could leaving or arriving at the island have played a major role in determining the size of the reindeer population? Explain your answer.

b) What might account for the tremendous increase in the population of reindeer between 1930 and 1938, as compared with the rate of growth during the first years the reindeer were on the island?

c) St. Paul Island is about 106 km² (41 square miles). What effect might 2000 reindeer have on the island and its vegetation?

d) Consider all the factors an organism requires to live. What might have happened on the island to cause the change in population size between 1938 and 1950?

e) Beginning in 1911, in which time spans did the reindeer population double? How many years did it take each of those doublings to occur? What happened to the doubling time between 1911 and 1938?

f) If some of the eight reindeer that were still alive in 1950 were males and some females, what do you predict would happen to the population in the next few years? Why?

g) What evidence is there that the carrying capacity (number of individuals in a population that the resources of a habitat can support) for reindeer on this island was exceeded?

h) What does this study tell you about unchecked population growth? What difference might hunters or predators have made?

Part B: Human Population

1. On a piece of graph paper, plot the growth of the human population using the following data.

Human Population Growth between A.D. 1 and 2000			
Date A.D.	Human Population (millions)	Date A.D.	Human Population (millions)
I	250	1930	2070
1000	280	1940	2300
1200	384	1950	2500
1500	427	1960	3000
1650	470	1970	3700
1750	694	1980	4450
1850	1100	1990	5300
1900	1600	2000	6080
1920	1800	2010	?

2. Use your graph to determine the doubling times for the human population between A.D. 1 and 2000.

   a) How much time elapsed before the human population of A.D. 1 doubled the first time?

   b) Is the amount of time needed for the human population to double increasing or decreasing?

   c) What does that indicate about how fast the human population is growing?

3. Extend your graph to the year 2010.

   a) What do you estimate the human population will be in that year?

4. Using the equations below, estimate the doubling time for the current population based on the rate of growth from 1990 to 2000.

   a) In what year will the present population double?

c) In what ways is the Earth as a whole similar to an island such as St. Paul? Does the Earth have a carrying capacity? Explain your answer.

$$\text{Annual rate of growth (in percent)} = \frac{(\text{population in 2000} - \text{population in 1990}) \times 100}{\text{population in 1990} \times \text{number of years}}$$

$$\text{Doubling time} = \frac{70}{\text{annual rate of growth}}$$

5. In your group, discuss the following questions. Then answer them in your *Active Biology* log.

   a) What similarities do you see between the graph of the reindeer population and your graph of the human population?

   b) What are the three or four most important factors required to sustain a population?

d) What might happen to the population of humans if the present growth rate continues?

e) What methods could be used to reduce the growth rate?

f) Suggest several problems in the United States that are related to the human population.

g) What are the most important three or four factors to think about with regard to the world population?

For You To Do

This activity provides you with an opportunity to examine how "life re-establishes itself" after a devastating blow.

1. On August 27, 1883, two volcanoes located on a single island in the Indian Ocean erupted at the same time. The blast was so great that a hole about 250 m deep remained where the peak of the volcano had been. The eruption on the island of Krakatoa has been said to be the loudest noise ever heard on Earth. The blast was heard in Hawaii, several thousands of kilometers away. Hot cinders and lava covered the island. Before the eruption, Krakatoa had been covered with a tropical forest. The eruption completely destroyed life on Krakatoa and two other nearby islands.

2. Two months after the eruption, scientists visited the island of Krakatoa. They found it steaming from a recent rain that had fallen on the lava that was still hot. In some places, the volcanic ash was washing away. In other places the ash was still more than 60 m deep. No life was visible.

3. Scientists visited the island nine months after the explosion, and at later times, to record the living things on Krakatoa. Some of the data recorded is shown in the diagram on the next page. Look for some interesting patterns in the rebirth of life on the island of Krakatoa. Study the plant life. (Reports of the animal life are interesting but too limited to use.)

a) What happens to the number of kinds of plants as the years pass?

b) Is there a change in the number of kinds of plant life?

c) Do the numbers of some kinds of plants change more than the numbers of other kinds?

d) Where do you think these plants might have come from? What reason do you have for your belief?

e) How long a period was needed for the complete recovery of the forest growth?

f) Write a statement that will describe the kinds of changes that have taken place on the island since the eruption.

g) Compare the "rebirth" of plant life on the coastal areas with the rebirth of plant life in the inland areas. How would you explain the difference?

SUCCESSION ON KRAKATOA

COASTAL AREAS

INLAND AREAS

3/4 years since eruption

Only algae and one lone spider found... mostly bare lava.

No plant or animal types found. Ground completely bare.

3 years since eruption

Ground completely covered with grasses. Many ferns, and many tropical seashore plants found. Insects also found.

A few grasses, many ferns and insects found.

13 years since eruption

Completely covered with young coconut trees, horsetail trees, and sugar cane plants. Lizards as well as insects found.

Almost all covered with grasses, orchids, and some horsetail trees. Lizards and insects found.

23 years since eruption

Completely covered as before, but with a greater number of trees.

Completely covered now with grasses, orchids, and groves of horsetail and young coconut trees.

47 years since eruption

By now a dense forest covers the area. All the previously listed plants and animals are found in abundance.

Inland areas now support same amount of plants and animals as the coast.

more wastewater is discharged. Domestic, agricultural, and industrial wastes include the use of pesticides, herbicides, and fertilizers. They can overload water supplies with hazardous chemicals and bacteria. Poor irrigation practices raise soil salinity and evaporation rates. Urbanization of forested areas results in increased drainage of an area as road drains, sewer systems, and paved land replace natural drainage patterns. All these factors put increased pressure on the water equation.

Pollutants that are discharged into the air can also affect the water cycle. Sulfur and nitrous oxides from the burning of fossil fuels, combustion in automobiles, and processing of nitrogen fertilizers enter the atmosphere. They combine with water droplets in the air to form acids. They then return to the surface of the Earth through the water cycle as acid precipitation.

Reflecting on the Activity and the Challenge

In this activity you observed one of the processes that take place in the water cycle. You learned that a great amount of water is transpired by a living plant. You also read about some of the other processes that are involved in the water cycle. The water cycle is very complex, and at any stage humans can have a significant impact. Perhaps the environmental issue you have chosen involves one part of the water cycle.

Biology to Go

1. Name and describe at least four processes that take place in the water cycle.

2. What is the energy source that drives the water cycle?

3. How has the water cycle determined partly where people live in the United States?

4. What would happen to the planet if the hydrologic cycle stopped functioning?

5. Describe three ways in which humans can have a negative effect on the water cycle.

Inquiring Further

Environmental models

Environmental models allow scientists to study what could happen to the plants and animals in an area if changes occurred. Models help check predictions without disrupting a large area.

Build an ecocolumn to research how acid rain affects an ecosystem.

(You will be allowed to use household vinegar as the acid.) An ecocolumn is an ecological model that is especially designed to cycle nutrients.

Record the procedure you will use. Have your teacher approve your procedure before you create your model.

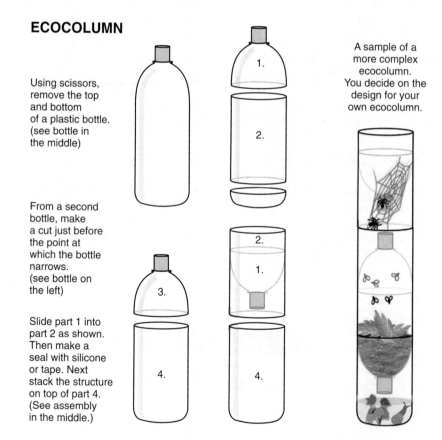

ECOCOLUMN

Using scissors, remove the top and bottom of a plastic bottle. (see bottle in the middle)

From a second bottle, make a cut just before the point at which the bottle narrows. (see bottle on the left)

Slide part 1 into part 2 as shown. Then make a seal with silicone or tape. Next stack the structure on top of part 4. (See assembly in the middle.)

A sample of a more complex ecocolumn. You decide on the design for your own ecocolumn.

Activity 8

Photosynthesis, Respiration, and the Carbon Cycle

GOALS

In this activity you will:

- Learn how oxygen cycles through photosynthesis and respiration.

- Practice safe laboratory techniques for using chemicals in a laboratory situation.

- Describe the cycling of carbon in an ecosystem.

- Speculate how human activities can affect the carbon cycle.

What Do You Think?

Consider the mass of a seed from a giant redwood tree and the tree itself. It is hard to believe that a giant of a tree began as a small seed.

- **From where do the materials come to make up the mass of a mature tree?**

Write your answer to this question in your *Active Biology* log. Be prepared to discuss your ideas with your small group and other members of your class.

For You To Do

In this activity you will investigate what happens when the exchange of carbon dioxide between a leaf and the atmosphere is blocked.

1. Three days before this activity, one plant was placed in the dark. A second plant of the same species was placed in sunlight.

a) Predict what you will find when you test a leaf from each plant for the presence of starch.

Day 1

Isopropyl alcohol is flammable and toxic. Do not expose the liquid or its vapors to heat or flame. Do not ingest; avoid skin/eye contact. In case of spills, flood the area with water and then call your teacher. Make sure you wear goggles, apron, and gloves

2. Remove one leaf from each plant. Use scissors to cut a small notch in the margin of the one placed in sunlight. Using forceps, drop the leaves into a beaker of hot (60°C) tap water.

3. When the leaves are limp, use forceps to transfer the leaves to a screw-cap jar about half full of isopropyl alcohol. Label the jar with your team symbol and store it overnight as directed by your teacher.

4. Select four similar leaves on the plant that has been kept in the dark, but do not remove them from the plant. Using a fingertip, apply a thin film of petroleum jelly to the upper surface of one leaf. Check to be sure the entire surface is covered. (A layer of petroleum jelly, although transparent, is a highly effective barrier across which many gases cannot pass.) Cut one notch in the leaf's margin.

5. Apply a thin film to the lower surface of a second leaf and cut two notches in its margin.

6. Apply a thin film to both upper and lower surfaces of a third leaf and cut three notches in its margin.

7. Do not apply petroleum jelly to the fourth leaf, but cut four notches in its margin; Place the plant in sunlight.

a) What is the purpose of the leaf marked with four notches?

8. Wash your hands thoroughly before leaving the laboratory.

Day 2

9. Obtain your jar of leaf-containing alcohol from Day 1. Using forceps, carefully remove the leaves from the alcohol and place them in a beaker of room-temperature water. (The alcohol extracts chlorophyll from the leaves but also removes most of the water, making them brittle.) Recap the jar of alcohol and return it to your teacher.

10. When the leaves have softened, place them in a screw-cap jar about half full of Lugol's iodine solution.

Lugol's iodine solution is used to test for the presence of small amounts of starch. Starch gives a blue-black color.

11. After several minutes, use forceps to remove both leaves, rinse them in a beaker of water, and spread them out in open Petri dishes of water placed on a sheet of white paper.

a) Record the color of each leaf. Recap the jar of Lugol's iodine solution and return it to your teacher.

b) What was the purpose of the iodine test on Day 2?

c) If you use these tests as an indication of photosynthetic activity, what are you assuming?

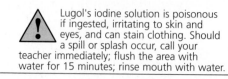

Lugol's iodine solution is poisonous if ingested, irritating to skin and eyes, and can stain clothing. Should a spill or splash occur, call your teacher immediately; flush the area with water for 15 minutes; rinse mouth with water.

12. Wash your hands thoroughly before leaving the laboratory.

Day 4

13. Remove from the plant the four notched leaves prepared on Day 1 and place them on paper towels. To remove the petroleum jelly, dip a swab applicator in the Histoclear™ and gently rub it over the surface of the film once or twice. Then gently use a paper towel to remove any residue of petroleum jelly. Discard the swab applicator and the paper towel in the waste bag.

14. Repeat **Steps 10** and **11**.

   a) Compare the color reactions of the four leaves and record your observations.

   b) In which of the leaves coated with petroleum jelly did photosynthetic activity appear to have been greatest? Least?

15. Wash your hands thoroughly before leaving the laboratory.

 Histoclear is a combustible liquid. Do not expose to heat or flame. Do not ingest; avoid skin/eye contact. Should a spill or splash occur, call your teacher immediately; wash skin area with soap and water.

BioTalk

The Carbon Cycle

You take in carbon in all the foods you eat. You return carbon dioxide to the air every time you exhale. A plant also returns carbon dioxide to the air when it uses its own sugars as a source of energy. When another plant takes in the carbon dioxide during photosynthesis, the cycle of carbon through the community is complete. In this activity you observed what happens when this exchange of carbon dioxide does not take place. However, when the exchange does take place, the plant can use the carbon from the carbon dioxide to live and grow.

Carbon dioxide is also returned to the air by decomposers. When producers or consumers die, decomposers begin their work. As its source of energy, a decomposer uses the energy locked in the bodies of dead organisms. It uses the carbon from the bodies to build its own body. Carbon that is not used is returned to the air as carbon dioxide. Eventually, almost all the carbon that is taken in by plants during photosynthesis is returned to the air by the activity of decomposers.

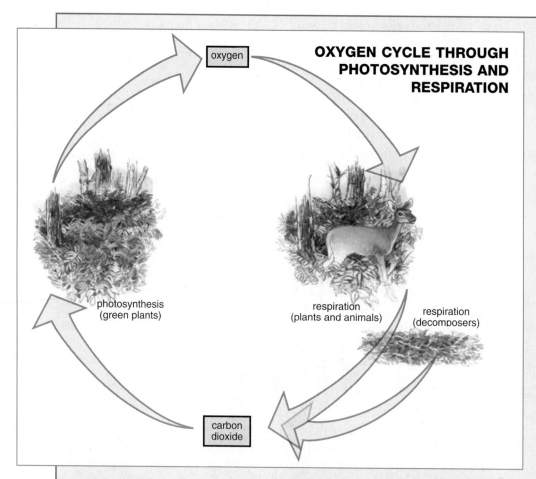

OXYGEN CYCLE THROUGH PHOTOSYNTHESIS AND RESPIRATION

oxygen

photosynthesis (green plants)

respiration (plants and animals)

respiration (decomposers)

carbon dioxide

Hundreds of millions of years ago, many energy-rich plant bodies were buried before decomposers could get to them. When that happened, the bodies slowly changed during long periods of time. They became a source of fuels like coal, oil, and natural gas. Today, when these fuels are burned, energy is released. The carbon in the fuels is returned to the air as carbon dioxide. You can see that even the energy obtained from fuels is a result of photosynthesis. The process in which carbon is passed from one organism to another, then to the abiotic community, and finally back to the plants is called the **carbon cycle**.

The Cycling of Matter

The energy from the Sun flows through the ecosystem in the form of carbon-carbon bonds in organic matter. When respiration occurs, the carbon-carbon bonds are broken and energy is released. The carbon is combined with oxygen to form carbon dioxide. The energy that is released is either used by the organism (to move, digest food, excrete wastes, etc.) or the energy may be lost as heat. In photosynthesis energy is used to combine the carbon molecules from the carbon dioxide, and oxygen is released. This is illustrated in the diagram. All the energy comes from the Sun. The ultimate fate of all energy in ecosystems is to be lost as heat. Energy does not recycle!

However, inorganic nutrients do recycle. They are inorganic because they do not contain carbon-carbon bonds. These inorganic nutrients include the phosphorous in your teeth, bones, and cell membranes. Also, nitrogen is found in your amino acids (the building blocks of protein). Iron is in your blood. These are just a few of the inorganic nutrients found in your body. Autotrophs obtain these inorganic nutrients from the inorganic nutrient pool. These nutrients can usually be found in the soil or water surrounding the plants or algae. These inorganic nutrients are then passed from organism to organism as one organism is consumed by another. Ultimately, all organisms die. They become detritus, food for the decomposers. At this stage, the last of the energy is extracted (and lost as heat). The inorganic nutrients are returned to the soil or water to be taken up again. The inorganic nutrients are recycled; the energy is not.

Bio Words

carbon cycle: the process in which carbon is passed from one organism to another, then to the abiotic community, and finally back to the plants

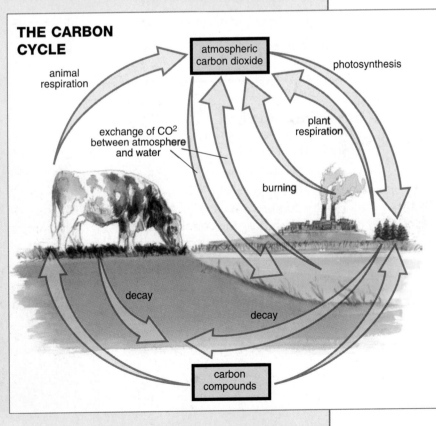

THE CARBON CYCLE

atmospheric carbon dioxide

animal respiration

photosynthesis

exchange of CO_2 between atmosphere and water

plant respiration

burning

decay

decay

carbon compounds

Reflecting on the Activity and the Challenge

In this activity you learned that carbon is the key element in all organic matter. You investigated the process of photosynthesis and then related this process to respiration in the carbon-oxygen cycle. The cycling of matter like carbon is essential to the survival of any ecosystem. You will need to explain this cycle in your booklet.

Biology to Go

1. Explain why photosynthesis and cellular respiration are considered to be paired processes.

2. What is the importance of decomposers in the carbon cycle?

3. What effect does the burning of fossil fuels have on the carbon cycle?

4. Scientists have expressed concerns about the burning of the rainforests to clear the land for the planting of crops.

   a) Explain how the burning of the forests could change oxygen levels.

   b) What impact would the change in oxygen levels have on living things?

Inquiring Further

The greenhouse effect

The term greenhouse effect was coined in the 1930s to describe the heat-blocking action of atmospheric gases. Research and report the connection between the greenhouse effect and the carbon cycle.

Activity 9

The Nitrogen and Phosphorous Cycles

GOALS

In this activity you will:

- Investigate the chemicals that promote and inhibit the growth of plant material.
- Explain the importance of nitrogen and phosphorous to organisms.
- Describe how nitrogen cycles in an ecosystem.
- Describe how phosphorous cycles in an ecosystem.
- Provide examples of how human activities can affect the nitrogen cycle.

What Do You Think?

Nitrogen is essential to all forms of life. Yet, recent studies have shown that excess nitrogen has been introduced into our ecosystems. It has had negative effects on the natural nitrogen cycle.

- **What are the sources of the excess nitrogen?**
- **What are some of the negative effects of too much nitrogen?**

Write your answer to these questions in your *Active Biology* log. Be prepared to discuss your ideas with your small group and other members of your class.

For You To Do

An excessive growth of algae (algal blooms) can make a lake very unappealing. More importantly, it places other organisms in the ecosystem in peril through lack of oxygen. In this activity you will investigate some of the chemicals that promote the growth of algae.

597

Handle all of the liquids and chemicals very carefully. They should all be considered contaminated and toxic. Keep hands away from eyes and mouth during the activity. Wash your hands well after the activity. Clean up any spills immediately.

1. Obtain three 1-L jars. Make sure the jars are rinsed thoroughly, so that there are no leftover traces of any chemicals, including soap. Fill each jar about three-fourths full with distilled water.

   a) Why is it important that the jars be cleaned before beginning this activity?

2. To each jar add a 10-mL sample of pond water. Stir the pond water thoroughly before taking the sample. The pond water will contain algae.

3. Label the jars A through C.

4. To each jar add the following:
   • To Jar A, add 15 gm of detergent.
   • To Jar B, add 15 gm of lawn fertilizer.
   • Do not add anything to Jar C.

   a) Many detergents contain phosphates. Fertilizers contain nitrogen and phosphates. Write as a question what you are investigating in this activity.

   b) What is the purpose of Jar C?

5. Cover each jar with plastic wrap so that dirt will not settle into the jar, but allow for some air to enter the jar.

6. Use a glass marker to mark the water level in each jar.

7. Set all the jars in a well-lighted place, but not in direct sunlight.

   a) Predict in which jar the algal growth will be the greatest? The least? Give reasons for your predictions.

8. Observe the jars each day for about two weeks. As water evaporates from the jars, add distilled water to bring the water back up to its original level. At the end of two weeks, you will pass the water in each jar through a separate filter.

   a) Record your observations every two or three days.

9. Find the mass of each of three pieces of filter paper.

   a) Record the mass of each in a table.

10. Fold the filter paper as shown and insert it into a funnel. Place the funnel in the mouth of another jar to collect the filtrate (the liquid that passes through the filter).

    Filter the liquid in each of the three jars.

11. Allow the filter papers and the algae residue to dry thoroughly.

    Find the mass of each piece of filter paper and algae. Calculate the mass of the algae.

    a) Record your findings in a table.

    b) Did your findings support your predictions? Explain any differences you found.

Place folded filter paper in funnel and soak with water.

Be very careful with the liquid and the algae residue. You should assume that disease organisms have grown in the water during the activity. Be very careful to avoid ingesting any of the water or residue. Dispose of all materials as directed by your teacher when finished.

Bio Talk

THE NITROGEN CYCLE

Nitrogen Fixation

Nitrogen is a basic building block of plant and animal proteins. It is a nutrient essential to all forms of life. Nitrogen is also required to make deoxyribonucleic acid or DNA. DNA is the hereditary material found in all living things. The movement of nitrogen through ecosystems, the soil, and the atmosphere is called the **nitrogen cycle**. Like carbon, nitrogen moves in a cycle through ecosystems. It passes through food chains and from living things to their environment and back again. Life depends on the cycling of nitrogen.

The largest single source of nitrogen is the atmosphere. It is made up of 78 percent of this colorless, odorless, nontoxic gas. With this much nitrogen available, you would think organisms would have no difficulty getting nitrogen. Unfortunately, this is not the case. Nitrogen gas is a very stable molecule. It reacts only under limited conditions. In order to be useful to organisms, nitrogen must be supplied in another form, the nitrate ion (NO_3-).

Bio Words

nitrogen cycle: the movement of nitrogen through ecosystems, the soil, and the atmosphere

nitrogen fixation: the process by which certain organisms produce nitrogen compounds from the gaseous nitrogen in the atmosphere

The nitrogen cycle is very complex. A simplified description is shown in the diagram on the next page. There are two ways in which atmospheric nitrogen can be converted into nitrates. The first method is lightning, and the second is bacteria in the soil. The process of converting nitrogen into nitrates is called **nitrogen fixation**.

A small amount of nitrogen is fixed into nitrates by lightning. The energy from lightning causes nitrogen gas to react with oxygen in the air, producing nitrates. The nitrates dissolve in rain, falling to Earth and forming surface water.

Three processes are responsible for most of the nitrogen fixation in the biosphere: atmospheric fixation by lightning, biological fixation by certain microbes, and industrial fixation. The enormous energy of lightning breaks nitrogen molecules apart. Only about five percent of the nitrates produced by nitrogen fixation are produced by lightning.

The nitrates enter the soil and then move into plants through their roots. Plant cells can use nitrates to make DNA, and they can convert nitrates into amino acids, which they then string together to make proteins. When a plant is consumed by an animal, the animal breaks down the plant proteins into amino acids. The animal can then use the amino acids to make the proteins it needs.

Some bacteria are capable of fixing nitrogen. These bacteria provide the vast majority of nitrates found in ecosystems. They are found mostly in soil, and in small lumps called nodules on the roots of legumes such as clover, soybeans, peas, and alfalfa. The bacteria provide the plant with a built-in supply of usable nitrogen, while the plant

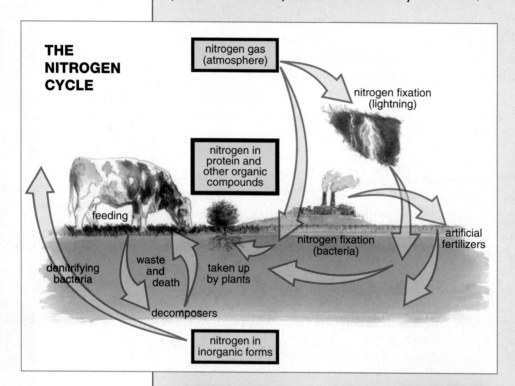

THE NITROGEN CYCLE

nitrogen gas (atmosphere)

nitrogen fixation (lightning)

nitrogen in protein and other organic compounds

feeding

denitrifying bacteria

waste and death

taken up by plants

nitrogen fixation (bacteria)

artificial fertilizers

decomposers

nitrogen in inorganic forms

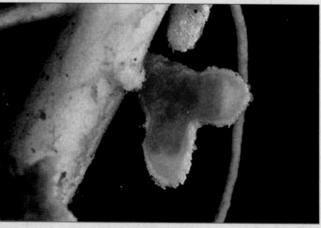

The most familiar examples of biotic nitrogen fixing are the root nodules of legumes, plants like peas, beans, and clover.

supplies the nitrogen-fixing bacteria with the sugar they need to make the nitrates. This plant-bacteria combination usually makes much more nitrate than the plant or bacteria need. The excess moves into the soil, providing a source of nitrogen for other plants. The traditional agricultural practices of rotating crops and mixed plantings of crops, one of which is always a legume, capitalizes on bacterial nitrogen fixation.

All organisms produce wastes and eventually die. When they do, decomposers break down the nitrogen-containing chemicals in the waste or body into simpler chemicals such as ammonia (NH_3). Other bacteria convert ammonia into nitrites, and still others convert the nitrites back to nitrates. These bacteria all require oxygen to function. The nitrates then continue the cycle when they are absorbed by plant roots and converted into cell proteins and DNA.

Farmers and gardeners who use manure and other decaying matter take advantage of the nitrogen cycle. Soil bacteria convert the decomposing protein in the manure into nitrates. Eventually, the nitrates are absorbed by plants.

Denitrification

At various stages in the decay process, denitrifying bacteria can break down nitrates to nitrites, and then nitrites to nitrogen gas. Eventually, the nitrogen gas is released back into the atmosphere. This process

A gardening magazine stated, "grass can actually poison itself as a result of the various chemical processes that occur in the individual grass plants if the grass roots do not have enough air." To what "poison" is the magazine referring?

Coordinated Science for the 21st Century

denitrification: the conversion of nitrates and nitrites to nitrogen gas, which is released into the atmosphere

phosphorous cycle: the cycling of environmental phosphorous through a long-term cycle involving rocks on the Earth's crust, and through a shorter cycle involving living organisms

is called **denitrification**, and is carried out by bacteria that do not require oxygen. Denitrification ensures the balance between soil nitrates, nitrites, and atmospheric nitrogen, and completes the nitrogen cycle.

Older lawns often have many denitrifying bacteria. The fact that denitrifying bacteria grow best where there is no oxygen may help explain why people often aerate their lawns in early spring. By exposing the denitrifying bacteria to oxygen, the breakdown of nitrates to nitrogen gas is reduced. Nitrates will then remain in the soil, and can be used by the grass to make proteins.

THE PHOSPHOROUS CYCLE

The **phosphorous cycle** is different from the water, carbon, and nitrogen cycles because phosphorous is not found in the atmosphere. Phosphorous is a necessary element in DNA, in many molecules found in living cells, and in the bones of vertebrate animals. Phosphorous tends to cycle in two ways: a long-term cycle involving the rocks of the Earth's crust, and a short-term cycle involving living organisms.

In the long cycle living things divert phosphates from the normal rock cycle. Phosphorous is found in bedrock in the form of

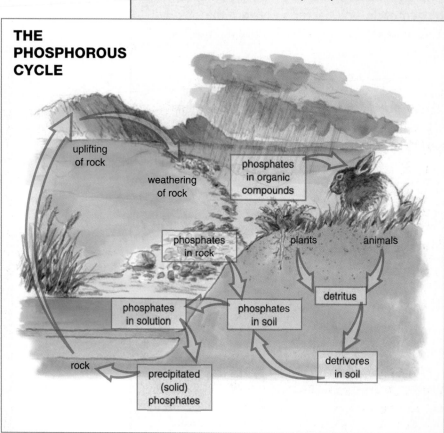

THE PHOSPHOROUS CYCLE

uplifting of rock

weathering of rock

phosphates in organic compounds

phosphates in rock

plants

animals

detritus

phosphates in solution

phosphates in soil

detrivores in soil

rock

precipitated (solid) phosphates

phosphate ions combined with a variety of elements. Phosphates are soluble in water and so can be drawn out of rock as part of the water cycle. Dissolved, phosphates can be absorbed by photosynthetic organisms and so pass into food chains. Phosphates eroded from rock are also carried by water from the land to rivers, and then to the oceans. In the ocean phosphates are absorbed by algae and other plants, where they can enter food chains. Animals use phosphates to make bones and shells. When they die, these hard remains form deposits on the ocean floor. Covered with sediment, the deposits eventually become rock, ready to be brought to the surface again. The cycle can take millions of years to complete. In the short cycle, wastes from living things are recycled by decomposers, which break down wastes and dead tissue and release the phosphates. The short cycle is much more rapid.

AGRICULTURE AND THE NITROGEN AND PHOSPHOROUS CYCLES

The seeds, leaves, flowers, and fruits of plants all contain valuable nutrients, which is why we eat them. However, as crops are harvested, the valuable nitrogen and phosphorous in these plant body parts are removed and do not return to the field or orchard they came from. This diversion of nitrates and phosphate from their cycles would soon deplete the soil unless the farmer replaced the missing nutrients. **Fertilizers** are materials used to restore nutrients and increase production from land. In this activity you investigated the effect fertilizer had on the growth of algae. Some estimates suggest that fertilizers containing nitrogen and phosphates can as much as double yields of cereal crops such as wheat and barley. However, fertilizers must be used responsibly. More is not necessarily better.

The accumulation of nitrogen and phosphate fertilizers produces an environmental problem. As spring runoff carries decaying plant matter and fertilizer-rich soil to streams and then lakes, the nutrients allow aquatic plants to grow more rapidly in what is called an algal bloom. When the plants die, bacteria use oxygen from the water to decompose them. Because decomposers flourish in an environment

Bio Words

fertilizer: a material used to provide or replace soil nutrients

Algae are generally thought of as simple, aquatic plants that do not have roots, stems, or leaves. A recurring problem in many bodies of water is algal bloom. An algal bloom is an abnormal increase of algae in a body of water. The most serious algal blooms are associated with human activities. Algal blooms deplete the water of oxygen and nutrients. In turn, this can kill other species in the water.

with such an abundant food source, oxygen levels in lakes drop quickly, so fish and other animals may begin to die. Dying animals can only make the problem worse, as decomposers begin to recycle the matter from the dead fish, allowing the populations of bacteria to grow even larger, and use still more oxygen.

Reflecting on the Activity and the Challenge

You have now investigated how several different types of matter cycle through ecosystems. You have also had an opportunity to learn about how humans can influence any one of these cycles. Consider how you will describe the importance of each of these cycles to the public. Also, consider whether or not the environmental issue that you have chosen deals specifically with one of these cycles. You will need to examine if any solution you provide will create a problem in any one of these cycles.

Biology to Go

1. Why is nitrogen important to organisms?

2. If plants cannot use the nitrogen in the atmosphere, how do they obtain the nitrogen they need?

3. How do animals obtain their usable nitrogen?

4. Explain why it is a good practice to aerate lawns.

5. Why is phosphorous important to living things?

6. With each harvest, nitrogen is removed from the soil. Farmers have traditionally rotated crops. Wheat, planted one year, is often followed by legumes planted the following year. Because the legumes contain nitrogen-fixing bacteria, nitrogen levels are replenished. The use of nitrogen-rich fertilizers has allowed farmers to not use crop rotation.

   a) What advantages are gained from planting wheat year after year?

   b) New strains of crops have been especially bred to take up high levels of nitrogen and harvests have increased dramatically. Speculate about some possible long-term disadvantages that these crops might present for ecosystems.

7. Before municipal sewers, the backyard outhouse was standard behind homes. They can still be found in some areas. To make an outhouse, a hole was dug in the ground to collect human wastes. Explain why the outhouse poses a risk to neighboring lakes, using information that you have gained about the nitrogen cycle.

Inquiring Further

1. The "new-tank syndrome"

Research to find out what is meant by the "new-tank" syndrome. How is it related to the nitrogen cycle?

2. Too much of a good thing

Which human activities impact on the nitrogen cycle? Choose one and explain how the impact of this activity on the environment could be reduced.

Biology at Work

Christy Todd Whitman

No person on the planet is more qualified to speak about the relationship between politics and environmental issues than Christy Todd Whitman. Whitman, 57, served as Governor of New Jersey from 1993 to 2000 and as the head of the Environmental Protection Agency from 2001 to 2003.

"Growing up on a farm I loved the outdoors. I loved to fish and swim and boat and bike and hike," Whitman says. "After seeing what happened as farms started to develop I got a real sense of the importance of protecting the environment."

Studying international government at Wheaton College in Massachusetts was also a natural extension of her curiosities as a child. "No matter what you're interested in—science or the arts or education—government impacts it in some way so I got interested in politics at an early age."

Whitman's work while EPA administrator forced her to deal with issues on a national level. She says that one of her biggest challenges was to educate the country–both individuals and corporate America–about the perils of ignoring the environment.

"A lot of the issues are very basic and we tried a lot of outreach programs. It's important to explain to people that everything you do has consequences, good and bad. For instance, we all live in a watershed so if you throw something out the window—a cigarette for example—that has an effect. If you change your oil in driveway or over-fertilize your lawn, it all will eventually wash down after a heavy rain. We found that every eight months there is as much oil deposited along the coastline from our everyday activities as was released for the Exxon Valdez spill. So what we're tying to get people to understand is that it

does matter what they do. Everything is cumulative."

And that includes big business. Although, Whitman is careful to point out that economic development does not have to be mutually exclusive to environmental protection.

"When you hear environmentalists yelling about business being bad for the planet or big corporations saying that they can't be profitable and environmentally conscious–you know that neither of those are true. You can have a cleaner, healthier environment, and a thriving economy. In fact, the environment needs the money produced by a healthy economy to invest in new technology. And there isn't a country in the world or a municipality or state in the world that is going to thrive economically if their environment is not good and healthy for the people who live there." Her solution? Incentives. The theory being that if you entice industries to develop environmentally sustainable practices, everybody wins.

Regardless of the fact that she is no longer head of the nations largest environmental group or governor of the 9th most populated state in the union, Whitman is still working to educate the public in an attempt to protect the planet's natural resources. "I am very proud of programs like Energy Star, which identifies energy efficient products such as washing machines, DVD players and other technologies to consumers," Whitman says. "In 2002, purchases from Energy Star saved consumers $7 billion and greenhouse gas emissions equivalent to the removal of 15 million cars from the road." And those are the kind of environmental impact numbers every politician would brag about.

Chapter 10

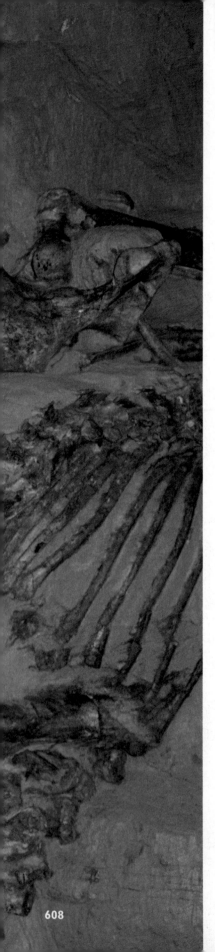

A Highway Through The Past

Scenario

After much study, the State Department of Transportation decided that a stretch of road was needed to connect two very busy state highways. The study had included environmental assessment of the area that the road would be covering. Much to the frustration of the local people, this study had taken over a year to complete.

Now, by state law, before any road construction could begin paleontologists (scientists who study past life) from the state university had been given six months to study the land that would be covered by the new road. The local residents who would use this new road were very upset because of this further delay of six months. They were even more upset when they found out that the findings of the paleontologists would delay the road construction longer than six months, and maybe even indefinitely.

Chapter Challenge

The State Department of Transportation has concluded that in order to allow area residents a chance to express their views, a town-hall meeting should be held to discuss the issue. At this meeting, the paleontologists would also be given an opportunity to provide their findings and explain why they have asked for a delay or even an indefinite postponement of the construction. For your challenge, you may be asked to represent someone who is against building the road, someone who is for building the road, or one of the group of paleontologists. Or, you may be asked to represent a member of one of the levels of government involved in this project to explain why it is necessary to have all these studies

and subsequent delays. Your teacher will act as the chairperson. As you prepare for and participate in role-playing a town meeting, you will be expected to:

- explain why fossils are important

- describe how the age of a fossil can be determined

- indicate how a highway might impact on the natural-selection process of the living organisms in the area

- explain why a great diversity of species is important for the survival of a community.

Criteria

How will your performance at the meeting be graded? Keep in mind that not everyone will be arguing every side of the issue. It is very important that you decide before the work begins how each person will be graded. Discuss this with your small group and then with your class. You may decide some or all of the following qualities are important:

- completeness and accuracy of the science principles presented in your side of the argument

- accuracy of the science principles used to dispute your opponents' positions

- forcefulness or conviction with which you present your argument

- quality of questions you pose to the government officials.

Once you have determined the criteria that you wish to use, you will need to decide on how many points should be given to each criterion. Your teacher may wish to provide you with a sample rubric to help you get started.

Activity 1

Adaptations

GOALS

In this activity you will:

- Explain the meaning of adaptation.

- Speculate how adaptations help an organism survive in their environment.

- Distinguish between structural and behavioral adaptations.

What Do You Think?

Imagine surviving a temperature of −50°C and a blinding snowstorm. Imagine surviving a temperature of 50°C in an extremely dry landscape.

- **How are plants and animals that live daily in these environments adapted for survival?**

Write your answer to this question in your *Active Biology* log. Be prepared to discuss your ideas with your small group and other members of your class.

For You To Do

Part A: Observing Adaptations

An adaptation is an inherited trait or set of traits that improve the chances of survival and reproduction of organisms. In this part of the activity you will look at photographs of animals to observe and speculate about how the different types of adaptations help the organism survive.

1. Look closely at the following photographs. There is a living organism in each picture.

   a) Which organisms are exhibiting camouflage?

   b) How could this adaptation help an organism in capturing prey?

c) How could this adaptation help protect the organism from predators?

d) What other animals can you think of that use this type of adaptation for protection?

2. Some animals are not adapted to disappear into the background, but rather stand out.

Alligator.

Praying Mantis.

Snowshoe hare.

Chameleon.

Hawk moth.

Hawk-moth
caterpillar.

Monarch butterfly.

Viceroy butterfly.

Look at the photographs of the hawk moth and caterpillar.

  a) At first glance, of what animal does each remind you?

  b) Why would birds avoid an animal with large eyes at the front?

  c) What advantage does this adaptation present for the moth and caterpillar?

3. A monarch butterfly stores bad-tasting chemicals in its body that birds hate. The viceroy butterfly also has a bitter taste.

  a) The monarch butterfly is brightly colored. Why do you think that this would be an advantage for the monarch butterfly?

  b) Would the bright colors and bitter taste protect all monarch butterflies? Explain your answer.

  c) Compare the appearance of the monarch and viceroy butterflies. Can you distinguish between them?

  d) How would the viceroy butterfly's coloration be an advantage for its survival?

4. Adaptations are not limited to animals. Look closely at the plants or plant parts shown on the next page for their adaptations to the environment.

a) For each plant shown above, explain the adaptation(s) that you can see. Consider the environment in which the plants live, how they reproduce, and how they get their nutrients when identifying adaptations.

5. Not all adaptations need to be structural. Some adaptations can be behavioral.

a) How is each animal in the photographs adapted to a change in the environmental conditions from summer to winter?

b) How do other animals adapt to an environmental change? Give at least two examples.

c) What type of behavioral adaptation is the plant at right exhibiting?

6. Invent an organism with specific adaptations. Consider one of the following:
   - camouflage
   - mimicry
   - warning coloration

Part B: How Well Adapted Are You?

In this part of the activity you will have an opportunity to examine one of your own adaptations that you probably take for granted.

1. Using masking tape, have your partner tape your thumb to your index finger on each hand. After your thumbs are securely taped, try each of the following activities. Rank the difficulty of each activity on a scale of 1 to 5.
   - picking up and carrying your textbook;
   - writing your name and address on a piece of paper;
   - picking up five coins from the floor and placing them in your pocket;
   - unbuttoning and buttoning a button;
   - tying up a shoe.

   a) Did you find any of the activities impossible?

   b) How did your ratings compare with others in your group and in your class?

   c) Why do you think that an opposable thumb is an important adaptation for humans? (An opposable thumb is an arrangement in which the fleshy tip of the thumb can touch the fleshy tip of all the fingers.)

   d) Do any other animals have opposable thumbs?

Bio Talk

Adaptation

Bio Words

species: a group of organisms that can interbreed under natural conditions and produce fertile offspring

adaptation: an inherited trait or set of traits that improve the chance of survival and reproduction of an organism

Diversity is a striking feature of living organisms. There are countless types of organisms on Earth. They are the result of repeated formation of new **species** and adaptation. There is a type of organism that can live in almost every type of environment on Earth. Living organisms are unique in their ability to adapt. The accumulation of characteristics that improve a species' ability to survive and reproduce is called **adaptation**. Adaptation occurs over long time periods. It is the environment that "selects" the best and most useful inherited variations. In this activity you observed just a few of the large number of adaptations that exist.

Animals Adapt to the Demands of Their Environments

Animals cannot make their own food. Therefore, they must usually seek food. As a result, adaptations that allow animals to move are favorable. Movement is easier if the organism is elongated in the direction of movement. Fish, for example, are streamlined. This reduces water resistance as they swim. It is also easier to move if the sensory organs are concentrated in the head. The organs that detect food, light, and other stimuli should be in a position to meet the environment first. An organism can move more easily if it has a balanced body.

Animals have the type of body plan that is best suited to their lifestyle. The symmetry of an organism gives clues to its complexity and evolutionary development. Higher animals, including humans, are symmetrical along the mid-sagittal plane. This body plan is referred to as **bilateral symmetry**, in which the right and left halves of the organism are mirror images of each other. Some animals, however, are **radially symmetric**, or symmetric about a central axis.

How is body symmetry related to the speed at which an animal moves and to brain development? In general, animals that display radial symmetry are not highly adapted for movement. One explanation for the slower movement can be traced to the fact that no one region always leads. Only bilaterally symmetrical animals have a true head region. Because the head, or anterior region, always enters a new environment first, nerve cells tend to concentrate in this area. The concentration of

Bio Words

bilateral symmetry: a body plan that divides the body into symmetrical left and right halves

radial symmetry: a body plan that is symmetrical about a center axis

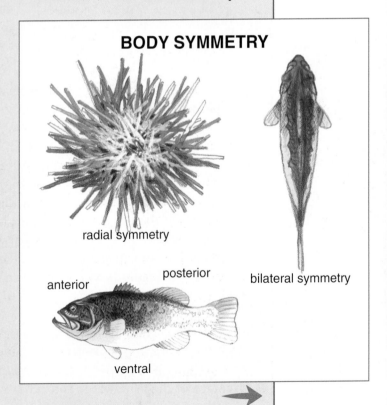

BODY SYMMETRY

radial symmetry

anterior   posterior   bilateral symmetry

ventral

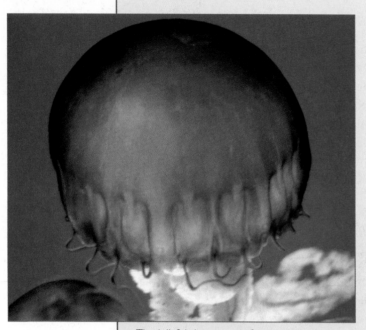

The jellyfish is a group of structurally simple marine organisms. The jellyfish has no head and a nervous system without a brain. The body exhibits radial symmetry.

nerve tissue at the anterior end of an animal's body, is an adaptation that enables the rapid processing of stimuli such as food or danger. Not surprisingly, the faster the animal moves, the more important is the immediate processing of environmental information. Every environment places special demands on the organisms living there. Seawater is fairly uniform. It poses the least stress for animal life. Oxygen is usually adequate. The temperatures and salt content are fairly constant. There is little danger that the organism will dry up. In contrast, the salt and oxygen contents of fresh water vary greatly.

Organisms that live in water have special adaptations. Gills, for example, allow the organisms to use the oxygen found in water. On land, oxygen is plentiful. However, the organisms that live there must protect themselves from the dangers of drying up. These dangers increase greatly because air temperatures change daily and seasonally. Air does not provide the same buoyancy as water. Therefore, large terrestrial, or land-dwelling, animals require good supportive structures. On the other hand, there is less resistance to movement in air than in water. Arms and legs, which would hinder an animal's movement in water, may help on land. Thus, long appendages specialized for locomotion have evolved in terrestrial animals.

Plant Adaptations

Plants lack the ability to move and must survive in the environment in which they are living. A plant must do more than simply survive and grow bigger. It must grow in such a way that it can take the best advantage of the light, water, and other conditions available to it.

Desert plants are an excellent example of adaptation to an environment. Some have a thick waxy coating to prevent them from drying out. Some have long vertical roots enabling a plant to reach water sources beneath the soil. Others develop shallow roots that extend horizontally. This maximizes water absorption at the surface. Many desert plants have small and narrow leaves. This decreases the heating from the Sun.

Even though plants are not able to move, they are still able to disperse. They produce seeds and fruits or other reproductive structures that may be distributed far from the parent plant.

Some plant adaptations are also behavioral. A vine spreads its leaves outward and receives as much light energy as possible. It sends its roots downward and receives more water. Tendrils of a vine touch an object and quickly coil it. This secures the vine in its upward growth. A vine would not live very long if it did not send its roots downward and its stem upward. The manner of plant growth is believed to be governed chiefly by hormones that are produced within the plant. The hormones are produced in response to conditions around the plant such as sunlight and gravity. Thus, the plant can fit itself to the environment in which it lives.

Some plants have even become adapted to feeding on animals. In this activity you looked at the Venus flytrap. Its leaves have been adapted to capture prey. These plants do photosynthesize. However, these plants live in bogs where there is very little nitrogen available. Therefore, they require the nutrients they receive from digesting their prey. Of course, the plant must therefore also be adapted to digest its prey with the secretion of chemicals.

Tendrils are modified stems or leaves that wrap around a support. They enable the plant to achieve fairly extensive horizontal and vertical spread without the use of much energy, since they don't have to support their own weight. Tendrils seem to respond to touch so if the stem or tendril touches an object, it wraps around it. This response is known as thigmotropism.

Reflecting on the Activity and the Challenge

In this activity you had an opportunity to look at adaptations of different organisms. You learned that every environment places various demands on the organisms living there. Organisms have developed special adaptations for living in any given environment. The animals and plants in the area of the highway construction have also adapted to their environment. In an environmental study scientists would have assessed the impact the highway would have had on the animals and plants. You may need to address this issue in the town-hall meeting if you are representing a government employee.

Biology to Go

1. Explain the term adaptation.

2. Distinguish between a structural and a behavioral adaptation.

3. a) How can an animal's structure help it survive in different environments? Give three examples.

   b) How can an animal's behavior help it survive in different environments? Give three examples.

4. Do all animals living in the same environment have similar adaptations? Explain your answer.

5. A cross section represents a cut through the middle of an animal's body. Below are cross sections through an earthworm, sand worm, and a primitive insect.

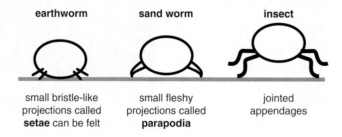

earthworm — small bristle-like projections called **setae** can be felt

sand worm — small fleshy projections called **parapodia**

insect — jointed appendages

   a) The jointed appendages of the insect lift the body from the ground. How does this help the insect move?

   b) What advantages might the fleshy projections of the sand worm have over the bristle-like projections of the earthworm?

   c) Predict which animal would be the fastest and give your reasons.

Inquiring Further

1. Animal adaptations to the arctic

Keeping warm is no easy task in the arctic where frigid weather lasts almost nine months of the year and where temperatures can plunge to −55°C. Even during the brief summer, when the land thaws and the Sun never sets, a sudden snowstorm can freeze everything. What adaptation have animals that live in this region developed?

2. Animal adaptations to the desert

Lack of water creates a survival problem for all desert organisms. However, animals have an additional problem. The biological processes of animal tissue can function only within a relatively narrow temperature range. Fortunately, most desert animals have evolved both behavioral and structural adaptations. Research the adaptations of animals living in desert regions.

Seals are well adapted to a cold environment. Their slick fur sheds water, and a thick layer of blubber beneath the skin keeps them warm in frigid temperatures.

The desert tortoise retreats to its burrow during the hottest times of the summer days. In the cold of winter it hibernates in its underground burrow.

Activity 2

Is It Heredity or the Environment?

GOALS

In this activity you will:

- Observe how an inherited trait can be influenced by the environment.

- Distinguish between a genotype and a phenotype.

- Explain how the environment can influence the development of an inherited characteristic.

What Do You Think?

"You've got your mother's hair and your father's eyes." Almost everyone has heard about heredity at some point.

- **How are personal characteristics passed on from one generation to the next?**

- **Can a personal characteristic be changed?**

Write your answer to these questions in your *Active Biology* log. Be prepared to discuss your ideas with your small group and other members of your class.

For You To Do

In this activity you will use tobacco seeds from parents that carried the characteristics for albinism (no chlorophyll) but did not show it. Your observations will help you to understand that traits are inherited but are also influenced by the environment.

3. Cover two dishes with a lightproof container.

4. Leave the other two dishes exposed to the light.

5. Let the seeds germinate for about a week, adding a few drops of water to the paper every other day or whenever the paper begins to dry.

6. On the tenth day, begin to make entries in your table of results.

   a) Make up tables on which to record your results. You may wish to use tables similar to the ones shown on the next page.

   b) Every day, record how many and what kind of seedlings you observe.

7. When all or most of the seeds have germinated in the darkened dishes (probably the twelfth day) remove the covering. Place these dishes in the light next to the others.

   a) Continue to record the appearance of the seedlings through the thirteenth day.

8. Study all the data you have accumulated.

   a) Try to draw any conclusions that you can from your data.

9. Using your data for the seedlings that were kept in the light all the time, answer the following questions:

   a) How might you explain the differences you observed?

   b) Are these differences caused by heredity or environment?

⚠️ Wash your hands after handling the seeds. If mold forms in the dish, have your teacher dispose of the affected seeds.

1. Place blotting paper in the bottom of each of four Petri dishes. Moisten the paper, but be sure that it is not floating in water. Sprinkle about 40 tobacco seeds evenly over the surface of the paper. Keep the seeds a few seed lengths apart from each other.

2. Replace the covers of the dishes and place the dishes in a well-lighted place, but not in the direct sunlight. The temperature should be approximately 22°C.

Kinds of Leaves from Germinating Tobacco Leaves
(Dishes Continuously Exposed to Light)

	Albino	Green	Percentage of albino each day
10th day			
11th day			
12th day			
13th day			

Kinds of Leaves from Germinating Tobacco Leaves
(Darkened Dishes)

	Albino	Green	Percentage of albino each day
10th day			
11th day			

Kinds of Leaves from Germinating Tobacco Leaves
(Covering Removed from Darkened Dishes)

	Albino	Green	Percentage of albino each day
12th day			
13th day			

10. Consider the seedlings that were kept in darkened dishes.

   a) How do the percentages of albino and green seedlings compare with the percentages of albino and green seedlings that were continuously exposed to light?

   b) What is the environmental factor that is varied in this activity?

   c) Is this difference in percentages of seedlings kept in the light and seedlings kept in the dark due to inherited or environmental factors?

   d) What do you think is causing the differences in the appearance of the seedlings that were in darkened dishes?

11. Consider the seedlings that were first in darkened dishes and then exposed to light.

   a) How do the percentages of green and albino seedlings compare with the percentages of green and albino seedlings in the other situations?

   b) What happened to the appearance of many of the seedlings after the cover was removed?

   c) Does this support your answer to **Step 10 (d)**?

   d) What are the effects of light upon seedlings that carry a certain hereditary characteristic?

Bio Talk

The Importance of Heredity and Environment

Why do offspring resemble their parents? Genetics, a branch of biology, tries to answer these types of questions about inheritance. Geneticists have found that most aspects of life have a hereditary basis. Many traits can appear in more than one form. A **trait** is some aspect of an organism that can be described or measured. For example, human beings may have blond, red, brown, or black hair. They may have tongues that they can roll or not roll. (Try it! Can you roll your tongue? Can your parents?) They may have earlobes that are attached or free. The passing of traits from parents to offspring is called **heredity**.

In *most* organisms, including humans, genetic information is transmitted from one generation to the next by deoxyribonucleic acid (DNA). DNA makes up the **genes** that transmit hereditary traits. Each gene in the body is a DNA section with a full set of instructions. These instructions guide the formation of a particular protein. The different proteins made by the genes direct a body's

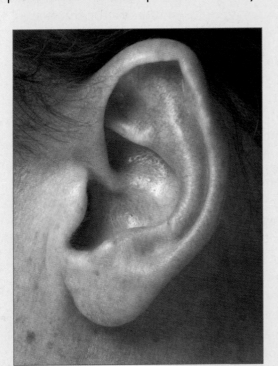

function and structure throughout life.

Chromosomes carry the genes. They provide the genetic link between generations. The number of chromosomes in a cell is characteristic of the species. Some have very few, whereas others may have more than a hundred. You inherit half of your chromosomes from your mother and the other half from your father. Therefore, your traits are a result of the interactions of the genes of both your parents.

Bio Words

trait: an aspect of an organism that can be described or measured

heredity: the passing of traits from parent to offspring

gene: a unit of instruction located on a chromosome that produces or influences a specific trait in the offspring

chromosome: threads of genetic material found in the nucleus of cells

Bio Words

dominant: used to describe the gene that determines the expression of a genetic trait; the trait shows up

recessive: used to describe the gene that is overruled by a dominant gene; the trait is masked

genotype: the genes of an individual

phenotype: the observable traits of an organism that result because of the interaction of genes and the environment

Gregor Mendel was the first person to trace the characteristics of successive generations of a living organism. He was an Augustinian monk who taught natural science to high school students. His origins were humble. However, his work was so brilliant that it took many years for the rest of the scientific community to catch up to it.

The modern science of genetics started with the work of Gregor Mendel. He found that certain factors in a plant cell determined the traits a plant would have. Thirty years after his discovery these factors were given the name genes. Of the traits that Mendel studied, he found that one factor, or gene, could mask the effect of another. This is the principle of dominance. He called the factor that showed up in the offspring **dominant**, and the factor that was masked **recessive**.

Genotype refers to the genes that an organism contains for a particular trait. The **phenotype** is the observable traits of an individual. Phenotype is a product of the interaction between the genotype and the environment.

All genes interact with the environment. Sometimes it is difficult to tell how much of a phenotype is determined by heredity and how much is influenced by the environment. A familiar example of how the environment affects the phenotype is the coloring of Siamese cats. The cats have a genotype for dark fur. However, the special proteins (enzymes) that produce the dark color work best at low temperatures. That is why Siamese cats have dark markings on their ears, nose, paws, and tail. These are all areas that have a low body temperature. Suppose a Siamese cat's tail were shaved and then kept at a higher than normal temperature. It would soon be covered with light-colored fur. However, these changes are temporary and only present if the environmental conditions are met. There are other examples of the influence of the environment on a phenotype. For a fair-skinned person, exposure to sunlight may produce hair that has lightened and a face full of freckles. Primrose plants are red if they are raised at room temperature, but white if they are raised at temperatures about 30°C. Himalayan rabbits are black when raised at low temperatures and white when raised at high temperatures.

The Siamese is a cat in which color is restricted to the points, i.e., nose, ears, legs, and tail. This is known as the Himalayan pattern. This coloration is a result of both hereditary and environmental factors.

Reflecting on the Activity and the Challenge

In this activity you saw that both heredity and environment contribute to the expression of a trait in a plant. You then read about how this also applies to animals. When thinking about how organisms are able to adapt, you must consider both inherited characteristics as well as the influence that the environment has on the organism. You will need to understand this for your **Chapter Challenge**. Also, in the next activity, you will investigate natural selection. Heredity and the environment both play a role in natural selection.

Biology to Go

1. Distinguish between a genotype and a phenotype.

2. What is the difference between a dominant and a recessive gene?

3. A dominant gene for a specific trait is inherited along with a non-dominant gene for the same trait. Which gene's "building instructions" will be used to assemble the specific protein?

4. In guinea pigs black coat is dominant to white. Is it possible for a black guinea pig to give birth to a white guinea pig? Explain your answer.

5. Explain how both heredity and environment contribute to the expression of a trait in plants.

6. Review your observations from the activity. Comment on the following statement: heredity can determine what an organism *may* become, not what it *will* become.

7. Can the environment change the development of an inherited characteristic? Use your observations from this activity to justify your answer.

Inquiring Further

1. Analyzing a genetic condition

What is a genetic condition? Choose a condition from one of the more well-known conditions, such as achondroplasia, cystic fibrosis, hemophilia, Huntington's chorea, Marfan syndrome, dwarfism, Down syndrome, Fragile-X syndrome, Tay-Sachs disease, sickle cell anemia, neurofibromatosis, etc. You may wish to investigate a condition with which you are personally familiar. Construct a hypothetical family tree to do a pedigree analysis of the condition. (A pedigree is used to trace inheritance of a trait over several generations.)

Down syndrome is caused by abnormal cell division in the egg, sperm, or fertilized egg. This results in an extra or irregular chromosome in some or all of the body's cells.

Activity 3   Natural Selection

GOALS

In this activity you will:

- Investigate the process of natural selection.
- Describe the major factors causing evolutionary change.
- Distinguish between the accommodation of an individual to its environment and gradual adaptation of a species.
- Read about the meaning of a theory in science.

What Do You Think?

One hundred rabbits were trapped and introduced to an island with a huge diversity of plants. The rabbits had several noticeable variations. Thirty years later scientists returned to the island. They were amazed that although the number of rabbits was still around 100, the later generations did not vary as much as the earlier rabbits had.

- **What happened to the variations that were evident in the original species?**
- **How would you explain why the variations seemed to have disappeared?**

Write your answer to these questions in your *Active Biology* log. Be prepared to discuss your ideas with your small group and other members of your class.

For You To Do

In this part of the activity you will study the process of natural selection. You will work with a hypothetical population of organisms in a hypothetical environment.

You will use a sheet of newspaper as the environment. You will use paper squares to represent individual prey. You will be given a chance to capture five prey individuals. The remaining prey will reproduce. You will then have another chance to capture the prey.

Part A: Predator and Prey

1. Work in groups of four. One student (the keeper) sets up the environment before each round (generation). The other three in the group act as "predators." They remove prey from the environment.

2. Lay a sheet of newspaper flat on a table or floor.

3. Take at least 50 each of newspaper, white, and red paper squares (150 squares). Keep the three types separate, as each represents a different type of the *same* prey species. Some are brightly colored. The others are not. An example of such different populations is the species *Canis familiaris*, the common dog. Although dogs come in many different colors and sizes, they all belong to the same species. The paper squares represent individuals of different colors, but of the same species.

4. The keeper collects 10 squares from each of the three prey populations. The keeper then mixes them and scatters them on the environment while the predators are not looking. Each predator may look at the environment *only* when it is her or his turn. When it is not your turn, simply close your eyes or turn your back until the keeper indicates that it is your turn. When it is your turn, remove five prey individuals as quickly as you can. Continue in order until each predator has removed five prey individuals.

5. Shake off the individuals left on the environment and count these survivors according to their type. They represent generation 1.

   a) Enter the data for your group in a table similar to the one shown.

   b) Place the data on the chalkboard also, so a class total can be reached.

Generation		Paper-Prey Species		
		Newspaper Individuals	White-Paper Individuals	Red-Paper Individuals
1	Team			
	Class			
2	Team			
	Class			
3	Team			
	Class			
4	Team			
	Class			

6. Analyze your data for the first generation. Record answers to the following questions in your *Active Biology* log:

   a) Does any population have more survivors than the others?

   b) Write a hypothesis that might explain this difference.

   c) Consider your hypothesis. If it is valid, what do you predict will happen to the number of newspaper individuals by the end of the fourth generation? to the red-paper individuals? to the white-paper individuals?

7. The survivors will be allowed to "reproduce" before the next round begins. For each survivor, the keeper adds one individual of that same type. The next generation will then include survivors and offspring. This should bring the total prey number back up to 30.

8. The keeper scatters these 30 individuals on the habitat. Repeat the predation and reproduction procedures for three more generations.

   a) Calculate the change in the number of all three populations after each round.

9. Look at your data and analyze your findings.

   a) Does it take you a longer or shorter period of time to find one prey individual as you proceed through the generations? Give an explanation for this.

   b) How does the appearance of the surviving individuals compare with the environment?

   c) Is your hypothesis and your prediction in question supported, or do they need to be revised?

   d) Were the red-paper individuals suited or unsuited for this environment? Explain.

   e) Would you say this species *as a whole* is better adapted to its environment after several generations of selection by the predators? Explain.

10. Now think of the "real" world.

    a) Is appearance the only characteristic that determines whether an individual plant or animal is suited to its environment? If so, explain. If not, give several other characteristics.

    b) In your own words, what is natural selection? What role does reproduction play in your definition?

11. Now you may test some of your own ideas about natural selection.

    • What would happen if there was a change in the environment, such as a change in color of the habitat?

    • What would be the result if one type of paper square "reproduced" at a faster rate than the others?

Part B: Hypothetical Model

1. Examine the story shown in the pictures on the next two pages. It is purely a hypothetical model and not an actual situation that occurred.

629

2. Discuss the following questions in your small group. Then answer them in your *Active Biology* log.

a) What change took place in the environment of the original moth population?

b) What change was produced in the moth population as a result of this environmental change?

c) Provide evidence that indicates that the change in the moth population is not simply an

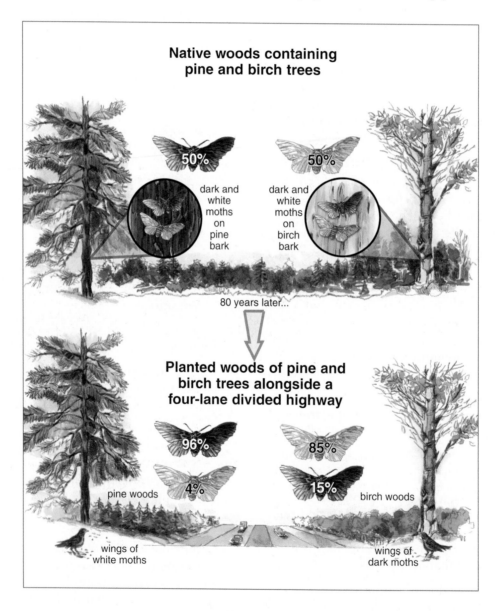

Native woods containing pine and birch trees

50%

50%

dark and white moths on pine bark

dark and white moths on birch bark

80 years later...

Planted woods of pine and birch trees alongside a four-lane divided highway

96%

85%

pine woods

4%

15%

birch woods

wings of white moths

wings of dark moths

effect of the environment, but is really a hereditary population change.

d) Do you think that the change was a result of a change in reproductive capacity of the two kinds of moths? Do you think the change was a result of the survival of the moth best fit for the environment (selection pressure)?

e) What has happened to the frequency of the gene for speckled white color in the moth population now living in the pine woods?

f) What has happened to the frequency of the gene for speckled white color in the moth population living in the birch woods?

g) If environments change over a period of time, what must happen to populations if they are to survive?

h) If natural selection is responsible for the changes in frequency of black and of speckled-white moths in the two types of woods, what comparison can you make between the color of the favored type of moth and the color of the bark of the trees in each woods?

i) Assume there is benefit in protective coloration on the part of moths. What type of predators would you suspect to prey on moths?

j) What special abilities would these predators have to possess if they are really the agents of selection here?

k) Devise an experiment that would test this hypothesis.

Bio Talk

Theories in Science

The popular use and scientific use of the term "theory" are very different. Scientific theories attempt to provide explanations. Scientists make observations and then try to explain them. In popular terms you often hear the expression, "it's just a theory." That usually means that it is a guess. In scientific terms, a **theory** implies that an idea has been strongly supported by observations.

When scientists use the scientific method they often begin with questions from curious observations. They then develop hypotheses that can be tested experimentally. A **hypothesis** is a prediction between an independent (cause) variable and a dependent (result) variable. Hypotheses can either be supported or not, depending on the data collection. A hypothesis is not a guess. You developed and tested hypotheses in this activity. The hypothesis is then tested by further observations and experiments. Over time, if the observations and experiments satisfy the hypothesis, it becomes accepted as a scientific theory.

However, a theory is not the absolute truth. It only provides an explanation. The acceptance of a theory is often measured by its ability to enable scientists to make predictions or answer questions. A good theory provides an explanation that scientists can use to explain other observed events. Theories can be modified as new information becomes available or ideas change. Scientists continually "tinker" with a theory to make it more elegant and concise, or to make it more all encompassing.

Darwin's Hypothesis of Natural Selection

The theory of **evolution** owes much to the work of Charles Darwin. He presented his research in the mid-19th century. However, Darwin never labeled his hypotheses as "evolution." He was interested in how species change and how new species come about. His many years of work led to explanations that have proved to be valid. But Darwin was not the first to think that existing species might evolve into new ones. However, Darwin was a most believable scientist for two reasons. First, he amassed a great deal of evidence. He verified its accuracy and presented it in a convincing way. Second, his hypothesis stated *how* change in organisms might take place, a contribution no one else had made.

On November 24, 1859, the first edition of Darwin's *On the Origin of Species* was published. The book was so popular that its first printing was sold out in one day. There were, of course, many who disagreed with him. The theory of evolution has undergone many changes since Darwin's time. However, Darwin's original thinking still serves as a convenient introduction to the subject.

Here is his analysis:

From 1831 to 1836 Charles Darwin, a British naturalist, served aboard the H.M.S. Beagle on a science expedition around the world. The expedition visited places around the world, and Darwin studied plants and animals everywhere he went, collecting specimens for further study. In South America Darwin found fossils of extinct animals that were similar to modern species. On the Galapagos Islands in the Pacific Ocean he noticed many variations among plants and animals of the same general type as those in South America.

- First, there are many differences among the individuals of every species. In a population, or group, of these individuals, variations occur. Usually it is safe to say that no two individuals are exactly alike. Darwin knew or suspected that many of the individual differences could be inherited.

- Second, the population size of all species tends to increase because of reproduction. One amoeba, for example, divides and produces two. These two divide, and the next generation numbers four. Then there will be 8, 16, 32, and so on.

- Third, this increase in the size of populations cannot go unchecked. If it did, the number of individuals of any species would outgrow the food supply and the available living space.

$\longrightarrow$

633

• Fourth, it is obvious that this huge increase seldom occurs in nature. The number of organisms in a species does not continue to increase over long periods of time. In fact, the sizes of many populations seem to remain nearly the same over time. How can this be explained? Observations of natural populations show that many individuals die before they are able to reproduce.

Why do some individuals die early, but not others? Darwin thought there must be a sort of "struggle for survival." The individuals of a species "compete" for food, light, water, places to live, and other things important for their survival. The "struggle" or "competition" may be either active or passive. That is, sometimes animals actually fight for food or the opportunity to mate. In other cases, there is no direct fight or competition. The first animal that happens to find a suitable living area may settle there. This prevents the area from being used by others. In either case, individuals with certain characteristics, or traits, will survive and produce offspring more often than individuals without them.

While chasing prey, cheetahs often reach speeds of 70 miles per hour. Unfortunately, their great speed may not be enough for this species to survive. Scientists have found that wild cheetahs have virtually no genetic variation. Cheetahs suffer from inbreeding. This lowers their resistance to diseases and also causes infertility and high cub-death rates.

Consider, for example, how the African cheetah came to be such a fast runner. Cheetahs are hunters. They capture their food; mostly antelopes, gazelles, and birds, by first stalking near their prey. Then they run the prey down with a terrific burst of speed over a short distance. In any population of cheetahs, some can run faster than others. Those that run fastest are most successful in getting food. Those that are better at getting food also are more likely to survive.

But survival is not the whole story. The characteristics that make an organism better able to survive in its environment are inherited. Therefore, those who survive are likely to pass on those characteristics to their offspring. For example, the surviving cheetah is likely to produce offspring with long, thin necks and powerful leg muscles, capable of great speed. Over many generations, then, one could expect an increase in the number of individuals that have these traits. The number with less beneficial characteristics would decrease. The organisms with the beneficial characteristics are likely to live longer and produce more offspring. Darwin called this process of survival and reproduction **natural selection**. Darwin thought that several factors were involved in natural selection:

1. The presence of variation among individuals in a population.

2. The hereditary basis of such variable characteristics.

3. The tendency of the size of populations to increase.

4. The "struggle for survival" (or competition for the needs of life).

5. A difference in the inherited characteristics that individuals pass on to succeeding generations.

Bio Words

natural selection: the differences in survival and reproduction among members of a population

Reflecting on the Activity and the Challenge

A change in the environment can have a large impact on the natural selection process. In this activity you investigated two situations. In the first, the "animals" that were best adapted to the environment were the ones to survive. In the second part you saw how a change in an environment could affect the natural selection process. Animals more suited to the changed environment would survive. You will need to explain the process of natural selection as part of your **Chapter Challenge**.

At one time some scientists believed that the necks of giraffes became long as a result of continually stretching to reach high foliage. Using what you know about natural selection, how would you explain the long necks?

Biology to Go

1. What evidence supports the following idea: hereditary differences are important in determining whether or not an individual survives and leaves offspring?

2. What is the difference between natural selection and evolution?

3. What did Darwin emphasize as the major factors in causing evolutionary changes?

4. What did Darwin mean by natural selection?

5. Write a short paragraph expressing your ideas now of what happened to the rabbit population on the island in the **What Do You Think?** section.

6. Comment on the validity of the following statement: Breeders of domestic stock abandon natural selection. Only artificial selection plays an important role in animal-breeding programs.

Inquiring Further

1. Animal-breeding programs

What are the advantages and disadvantages to animal-breeding programs? Research and report on the pros and cons of human intervention in genetic processes.

2. Captive breeding

Captive breeding is one strategy used by governments and non-government organizations to preserve rare and endangered species. What are the advantages and disadvantages of captive breeding?

Activity 4

The Fossil Record

GOALS

In this activity you will:

- Model ways in which fossils are formed.
- Explain the difference between a body fossil and a trace fossil.
- Describe the importance of fossils.
- Predict which animals are more likely to be found in the fossil record.

What Do You Think?

To hold a fossil in the palm of your hand is to have millions of years of history at your grasp. Fossils tell you about history, and like all good history, they help you to understand both the present and the past.

• What is a fossil?

Write your answer to this question in your *Active Biology* log. Be prepared to discuss your ideas with your small group and other members of your class.

For You To Do

In this activity you will have an opportunity to model different ways in which some fossils are formed. You will visit several stations.

Station 1: Preservation in Rock

You will mold a clamshell in plaster to model how it might be preserved in rock.

1. Obtain a large paper cup. Identify the cup with the name of your group. With a paper towel, smear petroleum jelly over the inside of the cup.

2. Mix plaster in another container following the directions on the package. Work quickly to complete the next four steps.

3. Fill the cup half full of plaster.

4. With the paper towel smear some petroleum jelly on both surfaces of a clamshell. Gently press the clamshell into the plaster.

5. Sprinkle a few pieces of confetti over the surface, enough to cover about 50% of the surface.

6. Fill the rest of the container with plaster.

7. Let the plaster harden overnight.

8. In the next class, remove the hardened plaster from the container. Set the plaster on its side and cover it with a towel. With a hammer gently hit the plaster to break it at the layer of confetti.

Wear goggles when using the hammer. Be sure others nearby are also wearing goggles. To contain any bits of plaster, cover the fossil model with cloth or paper before hitting it.

9. Observe your plaster molds and answer the following questions in your *Active Biology* log:

   a) What does the plaster represent?

   b) If you had never seen the clamshell, how would you figure out what the shell looked like by studying the fossil?

   c) Clamshells have two parts to their shell. How many possible imprints could a clamshell form?

   d) Why are fossils most often found in sedimentary rock formations?

Station 2: Preservation in Resin

You will encase a seed in glue to model how it might be preserved in a material like resin.

1. Obtain a small paper plate. Write your group's name on it. Use a paper towel to smear a small amount of petroleum jelly on a spot on the plate.

2. Using a hot-glue gun, put a bead of glue on the greased area of the plate.

3. Using tweezers, place the seed on the bead of glue. Add a few more drops of glue on top of the seed.

4. Let the glue harden overnight.

Wear goggles and be very careful when handling the hot-glue gun. Keep the hot part of the glue gun away from skin and flammable materials. Keep the glue away from skin, cloth, or other materials that may be damaged by it. Work on a surface that will not be damaged by the heat or the glue. Tell the teacher immediately of any accidents, including burns.

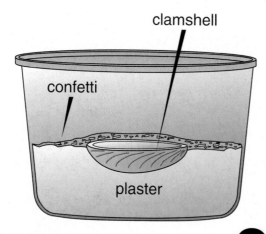

clamshell

confetti

plaster

5. In the next class, remove the bead of glue and observe. Answer the following questions in your *Active Biology* log:

a) Compare your preserved seed with a sample of amber provided by your teacher. How are they different? How are they similar?

b) Explain how a seed might end up being preserved in the resin.

c) Would you ever expect to see a large animal preserved in resin? Explain your answer.

d) Which type of fossil would be easier to identify: one preserved in rock, or one preserved in resin? Explain your answer.

Station 3: Preservation in Ice

You will freeze a small object in a cup of water to model how organisms can be preserved in ice.

1. Your teacher will provide you with a paper cup half full of water that is beginning to freeze. Put your group's name on the cup.

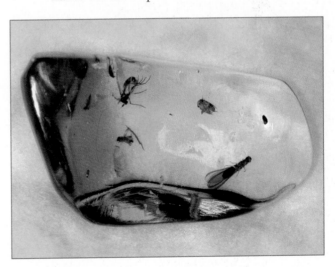

2. Gently push the object under the surface of the ice.

3. Add more water on top of the object.

4. Let the water freeze overnight.

5. In the next class, remove the ice from the paper cup. Answer the following questions in your *Active Biology* log:

   a) How do you think an organism could end up being preserved in ice?

   b) What type of organisms could be preserved in ice?

Do not attempt to remove the object from the ice.

Station 4: Preserving Animal Traces

1. Flatten or roll out a piece of modeling clay to create a flat surface.

2. On the surface of the modeling clay, produce the pathway that an organism might leave in a muddy surface. Use your imagination to produce the pathway. You could represent anything from a worm crawling to a dinosaur trudging.

3. With a paper towel spread a small amount of petroleum jelly over the imprints you left in the modeling clay.

4. Mix plaster in a small plastic bag following the directions on the package. Cut the corner off the plastic bag. Squeeze enough plaster over the impression to fill the area.

5. Let the plaster dry overnight.

6. In the next class, remove the modeling clay from the plaster. Answer the following questions in your *Active Biology* log:

   a) In what kind of ancient environment(s) might you expect to have footprints formed?

   b) Once a set of fresh footprints have been made in the mud, what would have to happen to preserve them as rock?

Bio Talk

THE NATURE OF THE FOSSIL RECORD

Making Models

Scientists often make models to help them understand how living things work. Models can be small-scale structures that simulate what is found in nature. For example, a scientist might reconstruct the climatic conditions of 65 million years ago to uncover what might have happened to the dinosaurs. Another type of model could be nonliving structures that work in a similar fashion. The human heart is often understood from the model of a pump. Recently, scientists have begun using computers to make mathematical models. Unlike the structural models, these models only exist as numbers. In this activity you modeled the formation of fossils.

The Importance of Fossils

What does the fossil record tell you? Among a number of things, it tells you that species are not unchangeable. The species you see around you today are not the ones that have always existed. Fossils provide direct evidence that organisms are continually evolving. However, it is important to note that evidence of evolution is very different from the theories of evolution, which you read about in the previous activity. Fossils tell you that life forms on Earth have changed. The theories attempt to explain how and why these changes took place.

Fossil Formation

Fossils are preserved evidence of ancient life. Some fossils are called **body fossils**. These are the preserved parts of plants and animals. Fossils may also be **trace fossils**. These fossils are traces of the activities of plants and animals, for example, tracks, trails, or scratch marks.

As you investigated in this activity, fossils form as a result of many processes. For example, most animals become fossilized by being buried in sediment. The sediments then accumulate and consolidate to form rock. Molds are fossils formed from the impressions in soft sediment of shells or leaves, for example, or from footprints or tracks. Casts are replicas formed when a hollow mold is subsequently filled with sediment—mud, sand, or minerals. Sometimes an insect might

Bio Words

fossil: any evidence of past life preserved in sediments or rocks

body fossil: a fossil that consists of the preserved body of an animal or plant or an imprint of the body

trace fossil: any evidence of the life activities of a plant or animal that lived in the geologic past (but not including the fossil organism itself)

become trapped in a sticky substance called resin, produced by some types of trees. The resin hardens to form amber. The insect fossil is preserved in amber, often perfectly. At other times natural mummies form when organisms are buried in areas like tar pits and peat bogs or dry environments like deserts or certain caves. Organisms buried in

Body fossil.

Trace fossil.

Cast.

glacial ice also can remain preserved for thousands of years. Finally, the cells and pore spaces of wood and bone can be preserved if filled with mineral deposits, a process called petrifaction.

Not all organisms become fossils. To begin with, very few escape the food chain. They are either eaten by other organisms or are broken down by decomposers. Soft body parts decay very quickly. You know from experience that it takes little time for meat and vegetables to spoil if left out of the refrigerator. More resistant parts, such as the exoskeletons of insects, vertebrate bones, wood, pollen, and

Major Divisions of Geologic Time (boundaries in millions of years before present)		
Era	Period	Event
Cenozoic	Quaternary	modern humans
Cenozoic	Tertiary	abundant mammals
Mesozoic	Cretaceous	flowering plants; dinosaur and ammonoid extinctions
Mesozoic	Jurassic	first birds and mammals; abundant dinosaurs
Mesozoic	Triassic	abundant coniferous trees
Paleozoic	Permian	extinction of trilobites and other marine animals
Paleozoic	Pennsylvanian	fern forests; abundant insects; first reptiles
Paleozoic	Mississippian	sharks; large primitive trees
Paleozoic	Devonian	amphibians and ammonoids
Paleozoic	Silurian	early plants and animals on land
Paleozoic	Ordovician	first fish
Paleozoic	Cambrian	abundant marine invertebrates; trilobites dominant
Proterozoic		primitive aquatic plants
Archean		oldest fossils; bacteria and algae

Boundaries: 1.8, 65, 145, 213, 248, 286, 325, 360, 410, 440, 505, 544, 2500

Time Not to Scale

spores take much longer to decay. Thus, the likelihood of finding these in the fossil record is much greater.

The Fossil Record

Fossils typically form where sediments such as mud or sand accumulate and entomb organisms or their traces. The layers of hardened mud, sand, and other sedimentary materials are like a natural book of the Earth's history. Interpreting each layer is like reading the pages of a book. Unfortunately, there are many surfaces on the Earth where layers are not accumulating or where erosion is removing other layers. Thus, interpreting the layers is like reading a novel that is missing most of its pages. You can read the pages that are preserved and even group them into chapters, but much important information is missing from each chapter. Paleontologists (scientists who study fossils) use **index fossils**. These are fossils of organisms that were widespread but lived for only a short interval of geological time. They use index fossils to divide the fossil record into chapters. For example, dinosaurs are index fossils for the Mesozoic era, the unit of time that runs from roughly 245 million years ago (abbreviated Ma) to 65 Ma. In other

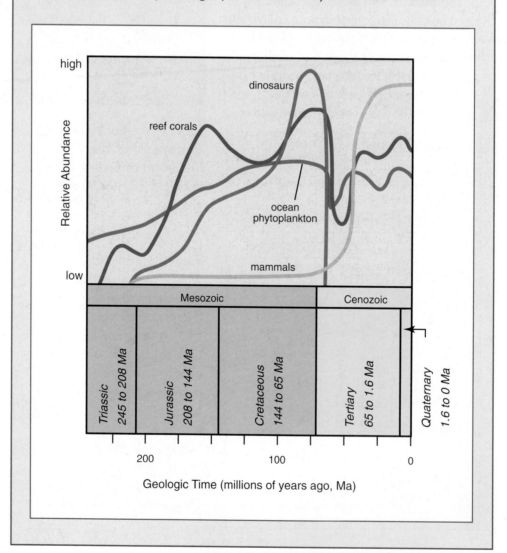

words, all dinosaur species evolved and became extinct during the Mesozoic era. Whales, horses, and many other mammal groups, on the other hand, are index fossils of the Cenozoic era, the unit of time that runs roughly from 65 Ma to the present. Using this and additional fossil evidence, paleontologists infer that the Mesozoic and Cenozoic eras represent two of the major chapters in the history of life.

The graph on the previous page summarizes the distribution of four groups of animals during the Mesozoic and Cenozoic eras. (Please note that this graph is not drawn to scale along the vertical axis; for example, the peak in dinosaur diversity is comparatively a small fraction of present day mammal diversity.) Although the graph is a rough summary of just a small part of the fossil record, paleontologists can get a much more accurate picture of life's history by examining specific pages in the record. You will do this in the next activity.

Reflecting on the Activity and the Challenge

In this activity you modeled some of the ways in which fossils can form. You also read about different types of fossils and the incomplete nature of the fossil record. You may have now developed a sense of the importance of fossils.

You can begin to appreciate what might be lost if fossil records were destroyed or disrupted. You will need to explain this at the town-hall meeting if you are representing one of the paleontologists.

Biology to Go

1. What is the difference between a body fossil and a trace fossil?

2. What are the chances of an organism becoming a fossil? Explain your answer.

3. a) What is an index fossil?

   b) How do paleontologists use index fossils?

4. Use evidence from this activity to explain how the biosphere and the geosphere are connected.

Inquiring Further

1. Carbon-14 dating

How is it possible to determine the age of organic matter using carbon-14? Research to find the physical and chemical principles on which this technique is based. What are the limitations of carbon-14 dating?

Activity 5

Mass Extinction and Fossil Records

GOALS

In this activity you will:

- Investigate fossil data for evidence of mass extinction and adaptive radiation.

- Explain the meaning of mass extinction and adaptive radiation.

- Describe the meaning of niche in an ecosystem.

What Do You Think?

Sixty-five million years ago the curtain came down on the age of dinosaurs when a catastrophic event led to their mass extinction.

- **What type of disastrous event could have led to the extinction of such a large group of animals?**

- **Did any other life forms become extinct at this time in geological history?**

Write your answer to these questions in your *Active Biology* log. Be prepared to discuss your ideas with your small group and other members of your class.

For You To Do

In this activity, you will investigate fossil data from those "pages" that represent the boundary between the Cretaceous and Tertiary periods (about 65 million years ago).

1. Your teacher will divide the class into groups of three or four students. With the other members of your group,

647

examine the six brachiopod fossils. Brachiopods are a group of marine animals.

a) What characteristics might paleontologists use to assign these fossils to different species?

b) What characteristics might paleontologists use to assign these fossils to one group?

2. Now examine **Graph A**. It plots the ranges of 50 different species of brachiopods across 15 m of sedimentary rock at one location in Denmark. This is one of the few places in the world that contains a continuous record of layers that represent the boundary between the

Cretaceous and Tertiary periods. Using a technique known as *magnetostratigraphy*, geologists infer that each meter of this sedimentary sequence represents 0.1 million years of history. The point "0 m" represents the boundary between the Cretaceous and Tertiary systems, 65 Ma (millions of years ago). Paleontologists sampled fossils from the locations shown along the left axis.

a) Which species became extinct at the Cretaceous-Tertiary (K-T) boundary?

b) Which species evolved after the K-T boundary?

c) Which species appear to have become extinct and then reappeared later?

d) What conclusions can you draw from this graph?

e) What are the limitations of the data shown in **Graph A**? (Hint: recall the processes by which fossils are preserved.)

Graph A: Range of Different Species of Brachiopods

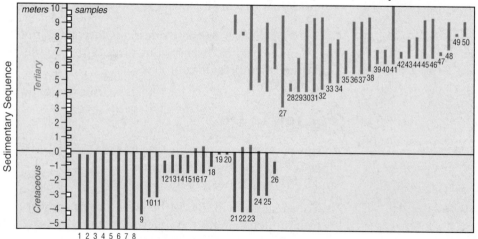

Graph B: Number of Families of Marine Organisms through Time

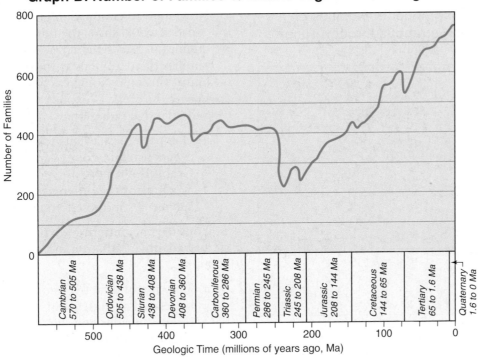

Number of Families (y-axis): 0, 200, 400, 600, 800

Time periods (x-axis):
- Cambrian 570 to 505 Ma
- Ordovician 505 to 438 Ma
- Silurian 438 to 408 Ma
- Devonian 408 to 360 Ma
- Carboniferous 360 to 286 Ma
- Permian 286 to 245 Ma
- Triassic 245 to 208 Ma
- Jurassic 208 to 144 Ma
- Cretaceous 144 to 65 Ma
- Tertiary 65 to 1.6 Ma
- Quaternary 1.6 to 0 Ma

Geologic Time (millions of years ago, Ma): 500, 400, 300, 200, 100, 0

3. Paleontologists have compiled similar data on the ranges of existence of numerous other organisms during the Cretaceous and Tertiary periods. Examine **Graph B**, which shows a set of data assembled to illustrate the number of *families* (groups of closely related species) of marine animals through geological time.

a) When was the number of families the greatest?

b) Has the growth in the number of families been steady? Explain your answer.

c) What do the dips in the graph represent?

d) What inferences can you draw from this graph?

e) What are the limitations of the graph shown in **Graph B**?

4. Now re-examine **Graph B.** Locate the times of the five greatest decreases in the number of families. Discuss why this might represent mass extinctions. Locate the times of the five greatest increases in the number of families. Discuss why this might represent adaptive radiations. (Adaptive radiation describes the rapid changes in a single or a few species to fill many empty functions in an ecosystem.)

a) In your *Active Biology* log, construct a chart that summarizes your findings. Your chart should have two vertical columns, one labeled "Times of Mass Extinction" and the other labeled "Times of Adaptive Radiation." Fill in the chart with the estimated date that each event began and the name of the time period (e.g., "beginning of the Devonian period, roughly 410 Ma").

Graph C: Number of Families of Terrestrial Tetrapods through Time

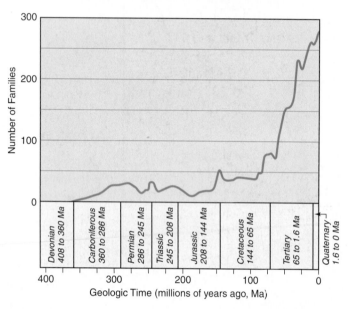

Number of Families

Geologic Time (millions of years ago, Ma)

400   300   200   100   0

Devonian 408 to 360 Ma
Carboniferous 360 to 286 Ma
Permian 286 to 245 Ma
Triassic 245 to 208 Ma
Jurassic 208 to 144 Ma
Cretaceous 144 to 65 Ma
Tertiary 65 to 1.6 Ma
Quaternary 1.6 to 0 Ma

5. Now analyze **Graph C**, a graph constructed to show the number of families of terrestrial tetrapod families (land animals with four limbs) through geological time. Locate the greatest extinction events and adaptive radiations.

   a) Compare these events to the events listed in your chart for **Graph B**. Propose a hypothesis to account for the differences and similarities in these two graphs.

6. Consider the pattern of extinction and adaptive radiation in **Graph B** and C.

   a) How might adaptive radiation be related to mass extinctions? (Hint: consider how life on Earth might be different if dinosaurs still existed.)

BioTalk

Making Inferences in Science

Have you ever wondered how scientists know so much about dinosaurs? No human ever saw a dinosaur eat or run. The huge lizards disappeared from Earth about 65 million years ago. No fossil evidence of the human species, Homo sapiens, appears before 500,000 years ago.

The skeletons of dinosaurs have been reconstructed using fossil records. The skeletons provide indirect evidence of how the dinosaur might have lived. Evidence from the skull of a dinosaur may indicate that the dinosaur might have been a meat eater. The premise that this dinosaur killed other dinosaurs is called an inference. No one ever saw the dinosaur eating meat, the evidence to support this conclusion came from examining the skull shape and the structure of the teeth. Unlike a hypothesis, an inference cannot be tested.

Mass Extinction and Adaptive Radiation

Extinction is the total disappearance of a species. Extinction means that not a single organism of the species lives anywhere on Earth. The fossil record is a virtual graveyard of extinct species. It is strewn with the fossilized remains of millions of extinct species. David Raup, a paleontologist at the University of Chicago, notes that "only about one in a thousand species [that have lived on Earth] is still alive—a truly lousy survival record: 99.9 percent failure!"

Even more striking, however, is the fossil evidence of **mass extinctions**. These are episodes during which large numbers of species became extinct during short intervals of geological time. In geological time a few million years or less is a short period! The extinction of one species often has a domino effect. If one species vanishes, so do many others. Yet mass extinctions can present new opportunities to survivors.

Those best able to survive fill empty **niches**. (An ecological niche is the function a species plays in an ecosystem.) Plants and animals that have the greatest genetic variation are most often best able to fill these empty "spaces." This process is called **adaptive radiation**. In this activity, you investigated evidence of mass extinctions and adaptive radiations by analyzing data from the fossil record. Rapid evolution can also occur when a species moves into a new area. Natural variation within a species makes it easier for the species to adapt to different environments.

One remarkable mass extinction event occurred at the boundary between the Cretaceous and Tertiary periods, roughly 65 million years ago. This boundary separates the age of the reptiles and the age of the mammals. Geologists recognized this event over one hundred years ago when they realized that there was a striking change in the types of fossils deposited on either side of this boundary. This is where the language of science may become difficult to follow. However, no matter how it is said, the concepts are the same. This boundary ➡

Bio Words

extinction: the permanent disappearance of a species from Earth

mass extinction: the extinction of a large number of species during short intervals of geological time

niche: the ecological function of a species; the set of resources it consumes and habitats it occupies in an ecosystem

adaptive radiation: the diversification by natural selection, over evolutionary time, of a species or group of species into several different species that are typically adapted to different ecological niches

Model of a Brachiosaurus.

also separates two eras. These two eras are called the Mesozoic and Cenozoic. Dinosaurs were prevalent during the Mesozoic Era and extinct during the Cenozoic Era. The last segment of the Mesozoic Era is called the Cretaceous Period. The first segment of the Cenozoic Era is called the Tertiary Period. The abbreviation for the boundary between the Cretaceous and Tertiary periods is often referred to as the K-T boundary, where K is the abbreviation for the German form of the word Cretaceous. You may also hear of this time referred to the Mesozoic and Cenozoic boundary. No matter what you call it, there were dinosaurs before and there are no dinosaurs now, and it happened about 65 million years ago!

Moreover, at the end of the Cretaceous period virtually all plant and animal groups were lost from Earth, not just the dinosaurs. Yet, the beginning of the Tertiary period marks the start of the adaptive radiation of mammals.

The ultimate cause of the mass extinction at the Cretaceous/Tertiary boundary is still a debate among scientists. However, more and more evidence suggests that a meteorite impact caused the mass extinction. The impact of the meteorite created a chain of devastating environmental changes for living organisms.

Reflecting on the Activity and the Challenge

In this activity you had an opportunity to see that life forms that dominated the Earth many (geological) years ago are not here now. You also learned that the evidence to support this fact is found in fossil records. The new organisms that evolved to fill the ecological place of the extinct organisms are also part of fossil records. You probably have developed an even further appreciation of the importance of fossil records. You may wish to argue for a delay of construction of the new highway whether you represent the paleontologists or a concerned citizen.

Biology to Go

1. Explain the meaning of adaptive radiation in your own words.

2. What evidence do scientists use to support the idea of mass extinctions?

3. After a mass extinction, which organisms are most likely to survive? Explain your answer.

4. Explain why the extinction of one species can have a domino effect on an ecosystem.

Inquiring Further

1. The Cretaceous and Tertiary boundary event

Research the proposed causes of the mass extinction during this time period.

Provide at least two explanations. Which one do you think is more plausible?

Biology at Work

Pat Holroyd

Paleontologist, University of California Berkeley Museum of Paleontology

Pat Holroyd examines a juvenile sea turtle skull, *Puppigerus camperi.*

"Believe it or not," says Pat Holroyd, "the science of paleontology owes a lot to the oil industry and construction projects in general." This statement may be surprising to you, coming from one of the best paleontologists in the world.

Holroyd's position at the University of California Berkeley Museum of Paleontology (UCBMP) has her in the field looking for rare fossils and overseeing construction projects as much as 12 weeks a year. "Under California law every construction project must produce an environmental impact statement before ever breaking ground," she explains, "and that must include a paleontology component. So, if someone wants to put in a transmission line or a road they must first look at how all the natural resources might be affected."

Yet, according to Holroyd, that rarely means development projects are blocked or even seriously delayed. "Mitigation is the word we like to use," she says. "When excavation begins, a paleontologist will be there to see if anything is coming out of the hole. What usually happens is that the paleontologist will jump in the hole and excavate for as long as she or he has to and then the project continues." As Holroyd explains, "Most important fossil finds do not happen despite development, but rather because of them. Most of the fossils we've found are only found because of construction projects. "The fact is that there are only a few thousand paleontologists in the entire world and they don't have the money to dig a 30-meter hole in the ground."

As a child, she loved playing in the dirt, but thought more about discovering pyramids than fossils. "It wasn't until college that I became interested in paleontology," Holroyd says. After graduating from the University of Kansas, she received a Ph.D. from Duke University in biological anthropology and anatomy. "I usually just say my degree is in paleontology," Holroyd laughs. During graduate school she worked with the U.S. Geological Survey in Denver and continued there for a year after graduation before moving over to UCBMP, where she's been happily digging in the dirt ever since.

In the larger sense, Holroyd's work is the study of small mammals and the effects of global warming. She and other scientists are actively trying to determine what a phenomenon like that does to a whole ecosystem, in an attempt to see what might be happening now. "If there is global warming now, then it's important to look at a period in the past when the globe went through similar changes," she explains. "Almost everything that we, as scientists, look at in terms of the impact humans are having or might have on the environment are things that we can find examples of in the fossil record."

EarthComm®

Making Connections

You have now come to the last unit in your *Coordinated Science for the 21st Century* course. This unit deals with Earth science in your community, *EarthComm®*. Why is it important to look at Earth science in your community? Because Earth is where you live! What happens on Earth affects you.

What is Earth science? Earth science uses physics, mathematics, chemistry, and biology to build an understanding of the Earth and its surroundings. Yes, each of the sciences that you have just studied play an important part in understanding the Earth and its future.

What Do You Think?

• How are physics, chemistry, and biology connected to each of the Earth systems: the atmosphere, the hydrosphere, the cryosphere, the geosphere, and the biosphere?

Record your answer to this in your log. Be prepared to discuss your responses with your small group and the class.

For You To Try

• Research a career in Earth science. Write a short job description of the career of your choice and the education required.

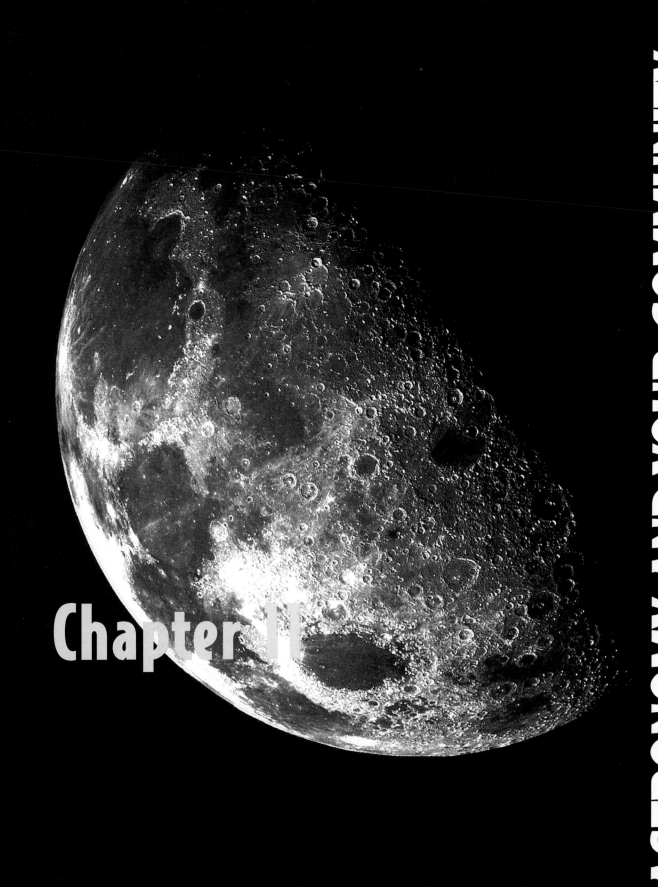

Chapter 11

ASTRONOMY AND YOUR COMMUNITY

Astronomy
...and Your Community

Getting Started

Throughout time, all systems in the universe are affected by processes and outside influences that change them in some way. This includes Earth and the solar system in which it exists. You have years of experience with life on the third planet from the Sun, and you know a lot about your tiny corner of the universe. Think about the Earth in relation to its neighbors in the solar system.

• What objects make up the solar system?

• How far is the Earth from other objects in the solar system?

• Which objects in the solar system can influence the Earth?

• Can you think of any objects or processes outside the solar system that might affect the Earth?

Write a paragraph about Earth and its place in this solar system. After that, write a second paragraph about processes or events in the solar system that could change Earth. Describe what they do and how Earth is, or might be, affected. Try to include answers to the questions above.

Scenario

Scientists recently announced that an asteroid 2-km wide, asteroid 1997XF11, would pass within 50,000 km of Earth (about one-eighth the distance between the Earth and the Moon) in October 2028. A day later, NASA scientists revised the estimate to 800,000 km. News reports described how an iron meteorite blasted a hole more than 1 km wide and 200 m deep, and probably killed every living thing within 50 km of impact. That collision formed Arizona's Meteor Crater some 50,000 years ago. Such a collision would wipe out a major city today. These reports have raised concern in your community about the possibility of a comet or asteroid hitting the Earth. Your class will be studying outer space and the effects that the Sun and other objects in the solar system can have on the Earth. Can you share your knowledge with fellow citizens and publish a booklet that will discuss some of the possible hazards from outer space?

Chapter Challenge

In your publication, you will need to do the following:

- Describe Earth and its place in the universe. Include information about the formation and evolution of the solar system, and about the Earth's distance from and orbit around the Sun. Be sure to mention Earth's place in the galaxy, and the galaxy's place in the universe

- Describe the kinds of solar activities that influence the Earth. Explain the hazardous and beneficial effects that solar activity (sunspots and radiation, for example) have on the planet. Discuss briefly the Sun's composition and structure, and that of other stars

- Discuss the Earth's orbital and gravitational relationships with the Sun and the Moon

- Explain what comets and asteroids are, how they behave, how likely

it is that one will collide with Earth in your lifetime, and what would happen if one did

- Explain why extraterrestrial influences on your community are a natural part of Earth system evolution.

The booklet should have a model of the solar system that will help citizens understand the relative sizes of and distances between solar-system bodies.

Assessment Criteria

Think about what you have been asked to do. Scan ahead through the chapter activities to see how they might help you to meet the challenge. Work with your classmates and your teachers to define the criteria for assessing your work. Record all of this information. Make sure that you understand the criteria as well as you can before you begin. Your teacher may provide you with a sample rubric to help you get started.

Activity 1

The History and Scale of the Solar System

Goals

In this activity you will:

- Produce a scale model of the solar system.

- Identify some strengths and limitations of scale models.

- Calculate distances to objects in the solar system in astronomical units (AU), light-years, and parsecs.

- Explain, in your own words, the nebular theory of the formation of the solar system.

- Explain the formation of the universe.

Think about It

Earth is part of a large number of objects that orbit around a star called the Sun.

- **What objects make up the solar system? Where are they located in relation to Earth?**

What do you think? Record your ideas in the form of a diagram of the solar system in your *EarthComm* notebook. Without looking ahead in this book, draw the Sun and the planets, and the distances from the Sun to the planets, as nearly to scale as you can. Be prepared to discuss your diagram with your small group and the class.

Investigate

1. Use the data in *Table 1* to make a scale model of the solar system. Try using the scale 1 m = 150,000,000 km.

   a) Divide all the distances in the first column by 150,000,000 (one hundred and fifty million). Write your scaled-down distances in your notebook, in meters.

   b) Divide all the diameters in the second column by 150,000,000. Write your scaled-down diameters in your notebook, in meters.

   c) Looking at your numbers, what major drawback is there to using the scale 1 m = 150,000,000 km?

Table I Diameters of the Sun and Planets, and Distances from the Sun		
Object	**Distance from Sun (km)**	**Diameter (km)**
Sun	0	1,391,400
Mercury	57,900,000	4878
Venus	108,209,000	12,104
Earth	149,598,770	12,756
Mars	227,900,000	6794
Jupiter	778,200,000	142,984
Saturn	1,429,200,000	120,536
Uranus	2,875,000,000	51,118
Neptune	4,504,400,000	49,528
Pluto	5,915,800,000	2302

2. Now try another scale: 1 m = 3,000,000 km (three million kilometers).

   a) Divide all the distances in the first column by 3,000,000. Write your scaled-down distances in your notebook in meters.

   b) Divide all the diameters in the second column by 3,000,000. Write your scaled-down diameters in your notebook in meters.

   c) Looking at your numbers, what major drawback is there to using the scale 1 m = 3,000,000 km?

3. Using what you have learned about scaling distances and diameters in the solar system, make models of the Sun and the planets. Each of the planets can be drawn on a different sheet of paper using a ruler to lay out the correct sizes for the different planets and the Sun.

4. To represent the distances from the Sun to the planets you will need to use a tape measure. You may want to measure the size of your stride and use this as a simple measuring tool.

To do this, stand behind a line and take five steps in as normal a way as possible and note where your last step ended. Now measure the distance from where you started to the end. Divide by five to determine how far you walk with each step. Knowing the length of your stride is an easy way to determine distances.

a) Explain the scale(s) you decided to use and your reasons for your choices.

b) Is it possible to make a model of the solar system on your school campus in which both the distances between bodies and the diameters of the bodies are to the same scale? Why or why not?

Reflecting on the Activity and the Challenge

In this activity you used ratios to make a scale model of the solar system. You found out that scale models help you appreciate the vastness of distances in the solar system. You also found out that there are some drawbacks to the use of scale models. Think about how you might use the model you made as part of your **Chapter Challenge**.

Geo Words

astronomical unit: a unit of measurement equal to the average distance between the Sun and Earth, i.e., about 149,600,000 (1.496 x 10^8) km.

light-year: a unit of measurement equal to the distance light travels in one year, i.e., 9.46 x 10^{12} km.

Digging Deeper

OUR PLACE IN THE UNIVERSE

Distances in the Universe

Astronomers often study objects far from Earth. It is cumbersome to use units like kilometers (or even a million kilometers) to describe the distances to the stars and planets. For example, the star nearest to the Sun is called Proxima Centauri. It is 39,826,600,000,000 km away. (How would you say this distance?)

Astronomers get around the problem by using larger units to measure distances. When discussing distances inside the solar system, they often use the **astronomical unit** (abbreviated as AU). One AU is the average distance of the Earth from the Sun. It is equal to 149,598,770 km (about 93 million miles).

Stars are so far away that using astronomical units quickly becomes difficult, too! For example, Proxima Centauri is 266,221 AU away. This number is easier to use than kilometers, but it is still too cumbersome for most purposes. For distances to stars and galaxies, astronomers use a unit called a **light-year**. A light-year sounds as though it is a unit of time, because a year is a unit of time, but it is really the distance that light travels in a year. Because light travels extremely fast, a light-year is a very large distance. For example, the Sun is only 8 light *minutes* away from Earth, and the nearest stars are several light-years away. Light travels at a speed of 300,000 km/s. This makes a light-year 9.46×10^{12} (9,460,000,000,000) km. Light from Proxima Centauri takes 4.21 years to reach Earth, so this star is 4.21 light-years from Earth.

Astronomers also use a unit called the **parsec** (symbol pc) to describe large distances. One parsec equals 3.26 light-years. Thus, Proxima Centauri is 1.29 pc away. The kiloparsec (1000 pc) and megaparsec (1,000,000 pc) are used for objects that are extremely far away. The nearest spiral galaxy to the Milky Way galaxy is the Andromeda galaxy. It is about 2.5 million light-years, or about 767 kpc (kiloparsecs), away.

The Nebular Theory

As you created a scale model of the solar system, you probably noticed how large the Sun is in comparison to most of the planets. In fact, the Sun contains over 99% of all of the mass of the solar system. Where did all this mass come from? According to current thinking, the birthplace of our solar system was a **nebula**. A nebula is a cloud of gas and dust probably cast off from other stars that used to live in this region of our galaxy. More than 4.5 billion years ago this nebula started down the long road to the formation of a star and planets. The idea that the solar system evolved from such a swirling cloud of dust is called the nebular theory.

You can see one such nebula in the winter **constellation** Orion (see *Figure 1*), just below the three stars that make up the Belt of Orion. Through a pair of binoculars or a small, backyard-type telescope, the Orion Nebula looks like a faint green, hazy patch of light. If you were able to view this starbirth region through a much higher-power telescope, you would be able to see amazing details in the gas and dust clouds. The Orion Nebula is very much like the one that formed our star, the Sun. There are many star nurseries like this one scattered around our galaxy. On a dark night, with binoculars or a small telescope, you can see many gas clouds that are forming stars.

Figure I Orion is a prominent constellation in the night sky.

Geo Words

parsec: a unit used in astronomy to describe large distances. One parsec equals 3.26 light-years.

nebula: general term used for any "fuzzy" patch on the sky, either light or dark; a cloud of interstellar gas and dust.

constellation: a grouping of stars in the night sky into a recognizable pattern. Most of the constellations get their name from the Latin translation of one of the ancient Greek star patterns that lies within it. In more recent times, more modern astronomers introduced a number of additional groups, and there are now 88 standard configurations recognized.

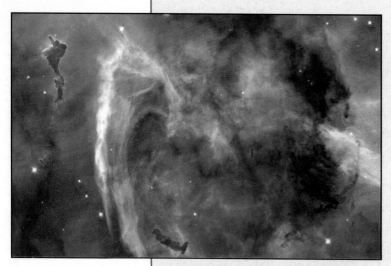

Figure 2 The Keyhole Nebula. Imaged by the Hubble Space Telescope.

In the nebula that gave birth to our solar system, gravity caused the gases and dust to be drawn together into a denser cloud. At the same time, the rate of rotation (swirling) of the entire nebula gradually increased. The effect is the same as when a rotating ice skater draws his or her arms in, causing the rate of rotation to speed up. As the nebular cloud began to collapse and spin faster, it flattened out to resemble a disk, with most of the mass collapsing into the center. Matter in the rest of the disk clumped together into small masses called **planetesimals**, which then gradually collided together to form larger bodies called **protoplanetary bodies**.

At the center of the developing solar system, material kept collapsing under gravitational force. As the moving gases became more concentrated, the temperature and pressure of the center of the cloud started to rise. The same kind of thing happens when you pump up a bicycle tire with a tire pump: the pump gets warmer as the air is compressed. When you let the air out of a tire, the opposite occurs and the air gets colder as it expands rapidly. When the temperature in the center of the gas cloud reached about 15 million degrees Celsius, hydrogen atoms in the gas combined or fused to create helium atoms. This process, called **nuclear fusion**, is the source of the energy from the Sun. A star—the Sun—was born!

Fusion reactions inside the Sun create very high pressure, and like a bomb, threaten to blow the Sun apart. The Sun doesn't fly apart under all this outward pressure, however. The Sun is in a state of equilibrium. The gravity of the Sun is pulling on each part of it and keeps the Sun together as it radiates energy out in all directions, providing solar energy to the Earth community.

The Birth of the Planets

The rest of the solar system formed in the swirling disk of material surrounding the newborn Sun. Nine planets, 67 satellites (with new ones still being discovered!), and a large number of comets and asteroids formed. The larger objects were formed mostly in the flat disk surrounding where the Sun was forming.

Geo Words

planetesimal: one of the small bodies (usually micrometers to kilometers in diameter) that formed from the solar nebula and eventually grew into proto-planets.

protoplanetary body: a clump of material, formed in the early stages of solar system formation, which was the forerunner of the planets we see today.

nuclear fusion: a nuclear process that releases energy when lightweight nuclei combine to form heavier nuclei.

Four of these planets, shown in *Figure 3*—Mercury, Venus, Earth, and Mars—are called the **terrestrial** ("Earth-like") **planets**. They formed in the inner part of our solar system, where temperatures in the original nebula were high. They are relatively small, rocky bodies. Some have molten centers, with a layer of rock called a mantle outside their centers, and a surface called a crust. The Earth's crust is its outer layer. Even the deepest oil wells do not penetrate the crust.

The larger planets shown in *Figure 3*—Jupiter, Saturn, Uranus, and Neptune—consist mostly of dense fluids like liquid hydrogen. These **gas giants** formed in the colder, outer parts of the early solar nebula. They have solid rocky cores about the size of Earth, covered with layers of hydrogen in both gas and liquid form. They lie far from the Sun and their surfaces are extremely cold.

Pluto is the most distant planet from the Sun. Some astronomers do not even classify it as a planet, and there is controversy about whether Pluto should be included among the "official" planets. Pluto is very different from the terrestrial or the gaseous planets. Some scientists think that it may not have been part of the original solar system but instead was captured later by the Sun's gravity. Others think that it was once a moon of an outer planet because it resembles the icy moons of the gas giants.

Figure 3 Composite image of the planets in the solar system, plus the Moon.

All life, as we know it on Earth, is based on carbon and water. Although there is evidence that water may once have been plentiful on Mars, liquid water has not been found on any of the planets in the solar system. Without water, these planets cannot support life as we know it. There is evidence, however, that Jupiter's moon Europa may be covered with ice, and that an ocean may lie beneath the ice. Scientists have found that life on Earth can thrive in even some of the most extreme environments. Powerful telescopes have discovered organic molecules in interstellar gas and have discovered planets around many nearby stars. Although life may not exist on the other planets in the solar system, the possibility of life on Europa and in other places in the universe exists!

Geo Words

terrestrial planets: any of the planets Mercury, Venus, Earth, or Mars, or a planet similar in size, composition, and density to the Earth. A planet that consists mainly of rocky material.

gas giant planets: the outer solar system planets: Jupiter, Saturn, Uranus, and Neptune, composed mostly of hydrogen, helium, and methane, and having a density of less than 2 gm/cm^2.

Geo Words

comet: a chunk of frozen gases, ice, and rocky debris that orbits the Sun.

asteroid: a small planetary body in orbit around the Sun, larger than a meteoroid (a particle in space less than a few meters in diameter) but smaller than a planet. Many asteroids can be found in a belt between the orbits of Mars and Jupiter.

There are trillions of **comets** and **asteroids** scattered throughout the solar system. Earth and other solar-system bodies are scarred by impact craters formed when comets and asteroids collided with them. On Earth, erosion has removed obvious signs of many of these craters. Astronomers see these comets and asteroids as the leftovers from the formation of the solar system. Asteroids are dark, rocky bodies that orbit the Sun at different distances. Many are found between the orbits of Mars and Jupiter, making up what is called the asteroid belt. Many others have orbits outside of the asteroid belt. Comets are mixtures of ice and dust grains. They exist mainly in the outer solar system, but when their looping orbits bring them close to the Sun, their ices begin to melt. That is when you can see tails streaming out from them in the direction away from the Sun. Some comets come unexpectedly into the inner solar system. Others have orbits that bring them close to the Sun at regular intervals. For example, the orbit of Halley's comet brings it into the inner solar system every 76 years.

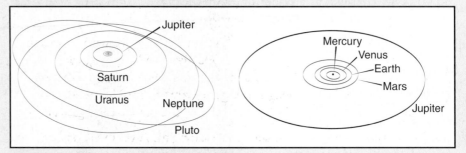

Figure 4 Two diagrams are required to show the orbits of the planets to scale.

Where is the Solar System in Our Galaxy?

Have you ever seen the Milky Way? It is a swath of light, formed by the glow of billions of stars, which stretches across the dark night sky. From Earth, this band of celestial light is best seen from dark-sky viewing sites. Binoculars and backyard-type telescopes magnify the view and reveal individual stars and nebulae. Unfortunately, for those who like to view the night sky, light pollution in densely populated areas makes it impossible to see the Milky Way even on nights when the atmosphere is clear and cloudless.

Galaxies are classified according to their shape: elliptical, spiral, or irregular. Our home galaxy is a flat spiral, a pinwheel-shaped collection of stars held together by their mutual gravitational attraction. Our galaxy shown in *Figure 5* is called the Milky Way Galaxy, or just the galaxy. Our solar system is located in one of the spiral arms about two-thirds of the way out from the center of the galaxy. What is called the Milky Way is the view along the flat part of our galaxy. When you look at the Milky Way, you are looking out through the galaxy parallel to the

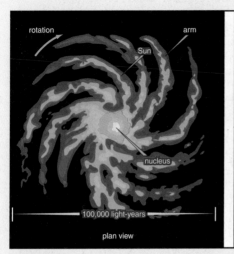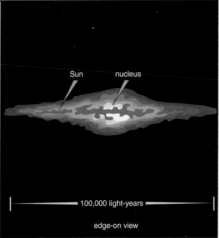

Figure 5 The Milky Way Galaxy. Our solar system is located in a spiral band about two-thirds of the way from the nucleus of the galaxy.

Geo Words

cosmologist: a scientist who studies the origin and dynamics of the universe.

plane of its disk. The individual stars you see dotting the night sky are just the ones nearest to Earth in the galaxy. When you view the Milky Way, you are "looking through" those nearest stars to see the more distant parts of the galaxy. In a sense, you are looking at our galaxy from the inside. In other directions, you look through the nearest stars to see out into intergalactic space!

Our Milky Way Galaxy formed about 10 billion years ago and is one of billions of galaxies in the universe. The universe itself formed somewhere between 12 to 14 billion years ago in an event called the Big Bang. This sounds like the universe began in an explosion, but it did not. In the beginning, at what a scientist would call "time zero," the universe consisted almost entirely of energy, concentrated into a volume smaller than a grain of sand. The temperatures were unimaginably high. Then the universe expanded, extremely rapidly, and as it expanded the temperature dropped and matter was formed from some of the original energy. **Cosmologists** (scientists who study the origin and dynamics of the universe) think that most of the matter in the universe was formed within minutes of time zero! The expansion and cooling that started with the Big Bang continues to this day.

The galaxies and stars are the visible evidence of the Big Bang, but there is other, unseen evidence that it happened. It's called the cosmic background radiation, which is radiation that is left over from the initial moments of the Big Bang. Astronomers using special instruments sensitive to low-energy radio waves have detected it coming in from all directions from the universe. The existence of the cosmic background radiation is generally considered to be solid evidence of how the Big Bang happened.

Check Your Understanding

1. What are the distances represented by a light-year, an astronomical unit, and a parsec?

2. Which of the units in Question 1 would you use to describe each of the following? Justify your choice.
   a) Distances to various stars (but not our Sun)?
   b) Distances to various planets within our solar system?
   c) Widths of galaxies?

3. In your own words, explain the nebular theory for the beginning of our solar system.

4. Briefly describe the origin of the universe.

Understanding and Applying What You Have Learned

1. Using the second scale
   (1 m = 3,000,000 km)
   you used for distance in your
   model of the solar system:

   a) How far away would Proxima
      Centauri be from Earth?
   b) How far away would the
      Andromeda galaxy be on your
      scale, given that Andromeda is
      767 kiloparsecs or 2.5 million
      light-years away?

2. The Moon is 384,000 km from
   Earth and has a diameter of
   3476 km. Calculate the diameter
   of the Moon and its distance from
   the Earth using the scale of the
   model you developed in the
   Investigate section.

3. Refer again to *Table 1*. If the Space
   Shuttle could travel at
   100,000 km/hr, how long would it
   take to go from Earth to each of
   the following objects? Assume that
   each object is as close to the Earth
   as it can be in its orbit.

   a) The Moon?
   b) Mars?
   c) Pluto?

4. What is the largest distance
   possible between any two planets
   in the solar system?

5. Use your understanding of a
   light-year and the distances from
   the Sun shown in *Table 1*.
   Calculate how many minutes it
   takes for sunlight to reach each
   of the nine planets in the solar
   system. Then use the unit "light-
   minutes" (how far light travels
   in one minute) to describe the
   distances to each object.

6. Write down your school address in
   the following ways:

   a) As you would normally address
      an envelope.
   b) To receive a letter from another
      country.
   c) To receive a letter from a friend
      who lives at the center of our
      galaxy.
   d) To receive a letter from a friend
      who lives in a distant galaxy.

Preparing for the Chapter Challenge

Begin to develop your brochure for
the **Chapter Challenge**. In your own
words, explain your community's
position relative to the Earth, Sun, the
other planets in our solar system, and
the entire universe. Include a few
paragraphs explaining what your
scale model represents and how you
chose the scale or scales you used.

Inquiring Further

1. **Solar-system walk**

   Create a "solar-system walk" on your school grounds or your neighborhood. Draw the Sun and the planets to scale on the sidewalk in chalk. Pace off the distances between the Sun and the nine planets at a scale that is appropriate for the site.

2. **Scaling the nearest stars**

   Look up the distances to the five stars nearest to the Sun. Where would they be in your scale model? To show their location, would you need a map of your state? Country? Continent? The world?

3. **Nuclear fusion**

   Find out more about the process of nuclear fusion. Explain how and why energy is released in the process by which hydrogen atoms are converted into helium atoms within the Sun. Be sure to include Albert Einstein's famous equation, $E = mc^2$, in your explanation, and explain what it means.

4. **Star formation**

   Write a newspaper story about star formation. Visit the *EarthComm* web site to find information available on the web sites of the Hubble Space Telescope and the European Southern Observatory to find examples of star-forming nebulae in the galaxy. How are they similar? How are they different? What instruments do astronomers use to study these nebulae?

Activity 2

The Earth–Moon System

Goals

In this activity you will:

- Investigate lunar phases using a model and observations in your community.

- Investigate the general idea of tidal forces.

- Understand the role of the Earth, the Moon, and the Sun in creating tides on Earth.

- Understand the Earth–Moon system and the Moon's likely origin.

- Compare the appearance of the Moon to other solar-system bodies.

Think about It

Think about the last time that you gazed at a full Moon.

- **What happened to make the Moon look the way it does?**

- **What is the origin of the Moon?**

- **How does the Moon affect the Earth?**

What do you think? Record your ideas about these questions in your *EarthComm* notebook. Be prepared to discuss your responses with your small group and the class.

Investigate

Part A: Lunar Phases

1. Attach a pencil to a white Styrofoam® ball (at least 5 cm in diameter) by pushing the pencil into the foam. Set up a light source on one side of the room. Use a lamp with a bright bulb (150-W) without a lampshade or have a partner hold a flashlight pointed in your direction. Close the shades and turn off the overhead lights.

2. Stand approximately 2 m in front of the light source. Hold the pencil and ball at arm's length away with your arm extended towards the light source. The ball represents the Moon. The light source is the Sun. You are standing in the place of Earth.

   a) How much of the illuminated Moon surface is visible from Earth? Draw a sketch of you, the light source, and the foam ball to explain this.

3. Keeping the ball straight in front of you, turn 45° to your left but stay standing in one place.

   a) How much of the illuminated Moon surface is visible from Earth?

   b) Has the amount of light illuminating the Moon changed?

   c) Which side of the Moon is illuminated? Which side of the Moon is still dark? Draw another diagram in your notebook of the foam ball, you, and the light source in order to explain what you see.

4. Continue rotating counterclockwise away from the light source while holding the ball directly in front of you. Observe how the illuminated portion of the Moon changes shape as you turn 45° each time.

   a) After you pass the full Moon phase, which side of the Moon is illuminated? Which side of the Moon is dark?

   b) How would the Moon phases appear from Earth if the Moon rotated in the opposite direction?

   ⚠ Be careful not to poke the sharp end of the pencil into your skin while pushing the pencil into the foam. Use caution around the light source. It is hot. Do not touch the Styrofoam to the light.

Part B: Observing the Moon

1. Observe the Moon for a period of at least four weeks. During this time you will notice that the apparent shape of the Moon changes.

   a) Construct a calendar chart to record what you see and when you see it. Sketch the Moon, along with any obvious surface features that you can see with the naked eye or binoculars.

   b) Do you always see the Moon in the night sky?

   c) How many days does it take to go through a cycle of changes?

   d) What kinds of surface features do you see on the Moon?

   e) Label each phase of the Moon correctly and explain briefly the positions of the Sun, the Earth, and the Moon during each phase.

   ⚠ Tell an adult before you go outside to observe the Moon.

Part C: Tides and Lunar Phases

1. Investigate the relationship between tides and phases of the Moon.

   a) On a sheet of graph paper, plot the high tides for each city and each day in January shown in *Table 1*. To prepare the graph, look at the data to find the range of values. This will help you plan the scales for the vertical axis (tide height) and horizontal axis (date).

   b) On the same graph, plot the Moon phase using a bold line. Moon phases were assigned values that range from zero (new Moon) to four (full Moon).

2. Repeat this process for low tides.

3. Answer the following questions in your *EarthComm* notebook:

   a) What relationships exist between high tides and phases of the Moon?

   b) What relationships exist between low tides and phases of the Moon?

   c) Summarize your ideas about how the Moon affects the tides. Record your ideas in your *EarthComm* notebook.

			Breakwater, Delaware		Savannah, Georgia		Portland, Maine		Cape Hatteras, North Carolina		New London, Connecticut	
Date	**Moon Phase**	**Moon Phase**	**High**	**Low**	**High**	**Low**	**High**	**Low**	**High**	**Low**	**High**	**Low**
1/3/01	First Quarter	2	3.6	0.2	7.3	0.5	8.5	1	2.6	0.2	2.4	0.3
1/6/01	Waxing Gibbous	3	4.5	0	8.2	0.5	9.7	0.1	3.4	−0.4	3	−0.2
1/10/01	Full Moon	4	5.6	−0.9	9.4	−1.5	11.6	−1.9	4.2	−0.8	3.5	−0.7
1/13/01	Waning Gibbous	3	5.1	−0.7	8.8	−0.9	11	−1.4	3.7	−0.6	3	−0.5
1/16/01	Last Quarter	2	4.1	−0.1	7.9	−0.2	9.7	0.1	3	−0.2	2.7	0
1/20/01	Waning Crescent	1	4.3	0.1	7.3	0.2	9.4	0.2	3.2	0	2.8	0
1/24/01	New Moon	0	4.6	0	8.1	−0.1	9.7	−0.1	3.3	−0.1	2.8	−0.1
1/30/01	Waxing Crescent	1	3.7	0.1	7.4	0.1	8.7	0.6	2.6	0	2.4	0.2

Table 1 Heights of High and Low Tides in Five Coastal Locations during January 2001 (All heights are in feet.)

4. *Table 1* shows data from the month of January 2001. At the *EarthComm* web site, you can obtain tidal data during the same period that you are doing your Moon observations. Select several cities nearest your community.

a) Record the highest high tide and the lowest low tide data for each city. Choose at least eight different days to compare. Correlate these records to the appearance of the Moon during your observation period. Make a table like *Table 1* showing high and low tides for each location.

b) What do you notice about the correlation between high and low tides and the appearance of the Moon?

Part D: Tidal Forces and the Earth System
1. Use the data in *Table 2*.

a) Plot this data on graph paper.

Label the vertical axis "Number of Days in a Year" and the horizontal axis "Years before Present." Give your graph a title.

b) Calculate the rate of decrease in the number of days per 100 million years (that is, calculate the slope of the line).

2. Answer the following questions:

a) How many fewer days are there every 10 million years? Every million years?

b) Calculate the rate of decrease per year.

c) Do you think that changes in the number of days in a year reflect changes in the time it takes the Earth to orbit the Sun, or changes in the time it takes the Earth to rotate on its axis? In other words, is a year getting shorter, or are days getting longer? How would you test your idea?

Table 2 Change in Rotation of Earth Due to Tidal Forces		
Period	**Date (millions of years ago)**	**Length of Year (days)**
Precambrian	600	424
Cambrian	500	412
Ordovician	425	404
Silurian	405	402
Devonian	345	396
Mississippian	310	393
Pennsylvanian	280	390
Permian	230	385
Triassic	180	381
Jurassic	135	377
Cretaceous	65	371
Present	0	365.25

Reflecting on the Activity and the Challenge

In this activity you used a simple model and observations of the Moon to explore lunar phases and surface characteristics of the Moon. You also explored the relationship between tides and the phases of the Moon. The tides also have an effect that decreases the number of days in a year over time. That's because tides slow the rotation of the Earth, making each day longer. You now understand that tides slow the rotation of the Earth, and how this has affected the Earth. This will be useful when describing the Earth's gravitational relationships with the Moon for the **Chapter Challenge**.

Geo Words

accretion: the process whereby dust and gas accumulated into larger bodies like stars and planets.

Digging Deeper

THE EVOLUTION OF THE EARTH–MOON SYSTEM

The Formation of the Earth and Moon

Figure I The Moon is the only natural satellite of Earth.

You learned in the previous activity that during the formation of the solar system, small fragments of rocky material called planetesimals stuck together in a process called **accretion**. Larger and larger pieces then came together to form the terrestrial planets. The leftovers became the raw materials for the asteroids and comets. Eventually, much of this material was "swept up" by the newborn inner planets. Collisions between the planets and the leftover planetesimals were common. This was how the Earth was born and lived its early life, but how was the Moon formed?

Scientists theorize that an object the size of Mars collided with and probably shattered the early Earth. The remnants of this titanic collision formed a ring of debris around what was left of our planet. Eventually this material accreted into a giant satellite, which became the Moon. Creating an Earth–Moon system from such a collision is not easy. In computer simulations, the Moon sometimes gets thrown off as a separate planet or collides with the Earth and is destroyed. However, scientists have created accurate models that predict the orbit and composition of both the Earth and Moon from a collision with a Mars-sized object. The Moon's orbit (its distance from the Earth, and its speed of movement) became adjusted so that the gravitational pull of the Earth is just offset by the centrifugal force that tends to make the Moon move off in a straight line rather than circle the Earth. After the Earth–Moon system became stabilized, incoming planetesimals continued to bombard the two bodies, causing impact craters. The Earth's surface has evolved since then. Because the Earth is geologically an active place, very few craters remain. The Moon, however, is geologically inactive. *Figure 2* shows the Moon's pockmarked face that has preserved its early history of collisions.

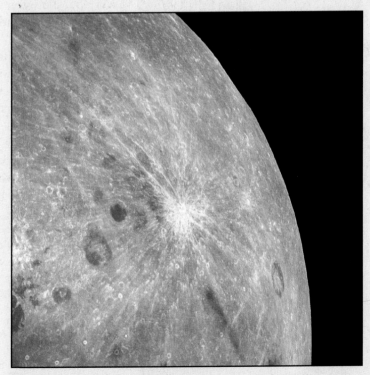

Figure 2 Impact craters on the Moon.

When the Earth was first formed, its day probably lasted only about six hours. Over time, Earth days have been getting longer and longer. In other words, the Earth takes longer to make one full rotation on its axis. On the other hand, scientists have no reason to think that the time it takes for the Earth to make one complete revolution around the Sun has changed through geologic time. The result is that there are fewer and fewer days in a year, as you saw in the **Investigate** section. Why is the Earth's rotation slowing down? It has to do with the gravitational forces between the Earth, the Moon, and the Sun, which create ocean tides.

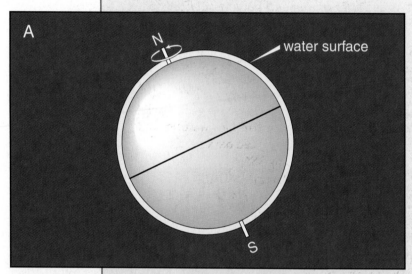

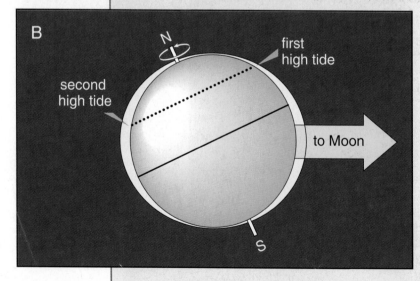

Tides

The gravitational pull between the Earth and the Moon is strong. This force actually stretches the solid Earth about 20 cm along the Earth–Moon line. This stretching is called the Earth tide. The water in the oceans is stretched in the same way. The stretching effect in the oceans is greater than in the solid Earth, because water flows more easily than the rock in the Earth's interior. These bulges in the oceans, called the ocean tide, are what create the high and low tides (see *Figure 3*). It probably will seem strange

Figure 3 Schematic diagram of tides. Diagram A illustrates how the ocean surface would behave without the Moon and the Sun (no tides). Diagram B illustrates, that in the presence of the Moon and the Sun, shorelines away from the poles experience two high tides and two low tides per day.

Geo Words

spring tide: the tides of increased range occurring semimonthly near the times of full Moon and new Moon.

to you that there are two bulges, one pointing toward the Moon and the other away from the Moon. If the tides are caused by the pull of the Moon, why is there not just one bulge pointing toward the Moon? The explanation is not simple. If you are curious, you can pursue it further in the **Inquiring Further** section of this activity.

As the Earth rotates through a 24-h day, shorelines experience two high tides—one when the tidal bulge that points toward the Moon passes by, and once when the tidal bulge that points away from the Moon passes by. The tidal cycle is not exactly 24 h. By the time the Earth has completed one rotation (in 24 h), the Moon is in a slightly different place because it has traveled along about 1/30 of the way in its orbit around the Earth in that 24-h period. That's why the Moon rises and sets about 50 min. later each day, and why high and low tides are about 50 min. later each day. Because there are two high tides each day, each high tide is about 25 min. later than the previous one.

The gravitational pull of the Sun also affects tides. Even though it has much greater mass than the Moon, its tidal effect is not as great, because it is so much farther away from the Earth. The Moon is only 386,400 km away from the Earth, whereas the Sun is nearly 150,000,000 km away. The Moon exerts 2.4 times more tide-producing force on the Earth than the Sun does. The changing relative positions of the Sun, the Moon, and the Earth cause variations in high and low tides.

The lunar phase that occurs when the Sun and the Moon are both on the same side of the Earth is called the new Moon. At a new Moon, the Moon is in the same direction as the Sun and the Sun and Moon rise together in the sky. The tidal pull of the Sun and the Moon are adding together, and high tides are even higher than usual, and low tides are even lower than usual. These tides are called **spring tides** (see *Figure 4*). Don't be confused by this use of the word "spring." The spring tides have nothing to do with the spring season of the year! Spring tides happen when the Sun and Moon are in general alignment and raising larger tides. This also happens at another lunar phase: the full Moon. At full Moon, the Moon and Sun are on opposite sides of the Earth. When the Sun is setting, the full Moon is rising. When the Sun is rising, the full Moon is setting. Spring tides also occur during a full Moon, when the Sun and the Moon are on opposite sides of the Earth. Therefore, spring tides occur twice a month at both the new-Moon and full-Moon phases.

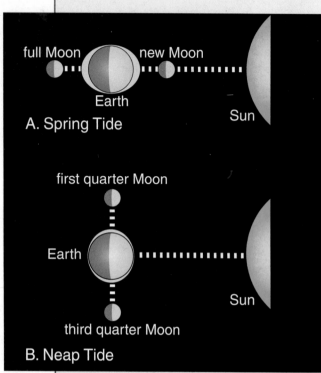

full Moon   new Moon

Earth

A. Spring Tide

Sun

first quarter Moon

Earth

Sun

third quarter Moon

B. Neap Tide

Figure 4 Schematic diagrams illustrating spring and neap tides.

When the line between the Earth and Sun makes a right angle with the line from the Earth to the Moon, as shown in *Figure 4*, their tidal effects tend to counteract one another. At those times, high tides are lower than usual and low tides are higher than usual. These tides are called **neap tides**. They occur during first quarter and third-quarter Moons. As with spring tides, neap tides occur twice a month.

The tide is like a kind of ocean wave. The high and low tides travel around the Earth once every tidal cycle, that is, twice per day. This wave lags behind the Earth's rotation, because it is forced by the Moon to travel faster than it would if it were free to move on its own. That's why the time of high tide generally does not coincide with the time that the Moon is directly overhead. The friction of this lag gradually slows the Earth's rotation.

Another way to look at tides is that the tidal bulges are always located on the sides of the Earth that point toward and away from the Moon, while the Earth with its landmasses is rotating below the bulges. Each time land on the spinning Earth encounters a tidal bulge there is a high tide at that location. The mass of water in the tidal bulge acts a little like a giant brake shoe encircling the Earth. Each time the bulge of water hits a landmass, energy is lost by friction. The water heats up slightly. (This is in addition to the energy lost by waves hitting the shore, which also heats the water by a small fraction of a degree.) Over long periods of time, the tidal bulge has the effect of slowing down the rotation of the Earth, and actually causing the Moon to move away from the Earth. The current rate of motion of the Moon away from the Earth is a few centimeters a year. This has been established by bouncing laser beams off of reflectors on the Moon to measure its distance. Although the Moon's orbit is not circular and is complex in its shape, measurements over many years have established that the

Geo Words

neap tide: the tides of decreased range occurring semimonthly near the times of the first and last quarter of the Moon.

Moon is indeed moving away from the Earth. Special super-accurate clocks have also established that the day is gradually becoming slightly longer as well, because of this same phenomenon, called "tidal friction." The day (one rotation of the Earth on its axis) has gradually become longer over geologic time. As the Earth system evolves, cycles change as well.

In this activity you limited the factors that cause and control the tides to the astronomical forces. These factors play only one part. The continents and their different shapes and ocean basins also play a large role in shaping the nature of the tides. Although many places on Earth have two high tides and two low tides every day (a semidiurnal tide), some places experience only one high tide and one low tide every day (a diurnal tide). There are still other places that have some combination of diurnal and semidiurnal tides (mixed tides). In these places (like along the west coast of the United States) there are two high tides and two low tides per day, but the heights of the successive highs and lows are considerably different from one another.

Figure 5 How do tides affect coastal communities?

Check Your Understanding

1. How did the Moon likely form?

2. Describe the relative positions of the Earth, the Moon, and the Sun for a spring tide and for a neap tide.

3. What effect have tides had on the length of a day? Explain.

Understanding and Applying What You Have Learned

1. Refer back to the graph of the changing length of the day that you produced in the investigation. Think about the causes of tidal friction and the eventual outcome of tidal friction. Predict how long you think the day will eventually be. Explain the reasoning for your prediction.

2. Think about the roles that the Sun and Moon play in causing the ocean tides.

   a) If the Earth had no Moon, how would ocean tides be different? Explain your answer.
   b) How would the ocean tides be different if the Moon were twice as close to the Earth as it is now?
   c) What differences would there be in the ocean tides if the Moon orbited the Earth half as fast as it does now?

3. Look at *Figure 3B* in in the **Digging Deeper** section. Pretend that you are standing on a shoreline at the position of the dotted line. You stand there for 24 h and 50 min, observing the tides as they go up and down.

   a) What differences would do you notice, if any, between the two high tides that day?
   b) Redraw the diagram from *Figure 3B*, only this time, make the arrow to the Moon parallel to the Equator. Make sure you adjust the tidal bulge to reflect this new position of the Moon relative to the Earth. What differences would you now see between the two high tides that day (assuming that you are still at the same place)?
   c) Every month, the Moon goes through a cycle in which its orbit migrates from being directly overhead south of the Equator to being directly overhead north of the Equator and back again. To complicate things, the maximum latitude at which the Moon is directly

overhead varies between about 28.5° north and south, to about 18.5° north and south (this variation is on a 16.8 year cycle). How do you think the monthly cycle relates to the relative heights of successive high tides (or successive low tides)?

4. Return to the tide tables for the ocean shoreline that is nearest to your community (your teacher may provide a copy of these to you).

   a) When is the next high tide going to occur? Find a calendar to determine the phase of the Moon. Figure out how to combine these two pieces of information to determine whether this next high tide is the bulge toward the Moon or away from the Moon.
   b) The tide tables also provide the predicted height of the tides. Look down the table to see how much variation there is in the tide heights. Recalling that the Sun also exerts tidal force on the ocean water, try to draw a picture of the positions of the Earth, Moon, and Sun for:
      i)   The highest high tide you see on the tidal chart.
      ii) The lowest high tide you see on the tidal chart.
      iii) The lowest low tide you see on the tidal chart.

5. The questions below refer to your investigation of lunar phases.

   a) Explain why the Moon looks different in the sky during different times of the month.

b) What advantage is there to knowing the phases of the Moon? Who benefits from this knowledge?

c) It takes 27.32166 days for the Moon to complete one orbit around the Earth. The Moon also takes 27.32166 days to complete the rotation about its axis. How does this explain why we see the same face of the Moon all the time?

Preparing for the Chapter Challenge

Write several paragraphs explaining the evolution of the Earth–Moon system, how mutual gravitational attraction can affect a community through the tides, and how the changing length of the day could someday affect the Earth system. Be sure to support your positions with evidence.

Inquiring Further

1. **Tidal bulge**

   Use your school library or the library of a nearby college or university, or the Internet, to investigate the reason why the tidal bulge extends in the direction away from the Moon as well as in the direction toward the Moon. Why does the Earth have two tidal bulges instead of just one, on the side closest to the Moon?

2. **Tidal forces throughout the solar system**

   Tidal forces are at work throughout the solar system. Investigate how Jupiter's tidal forces affect Jupiter's Moons Europa, Io, Ganymede, and Callisto. Are tidal forces involved with Saturn's rings? Write a short report explaining how tidal friction is affecting these solar-system bodies.

3. **Impact craters**

   Search for examples of impact craters throughout the solar system. Do all the objects in the solar system show evidence of impacts: the planets, the moons, and the asteroids? Are there any impact craters on the Earth, besides the Meteor Crater in Arizona?

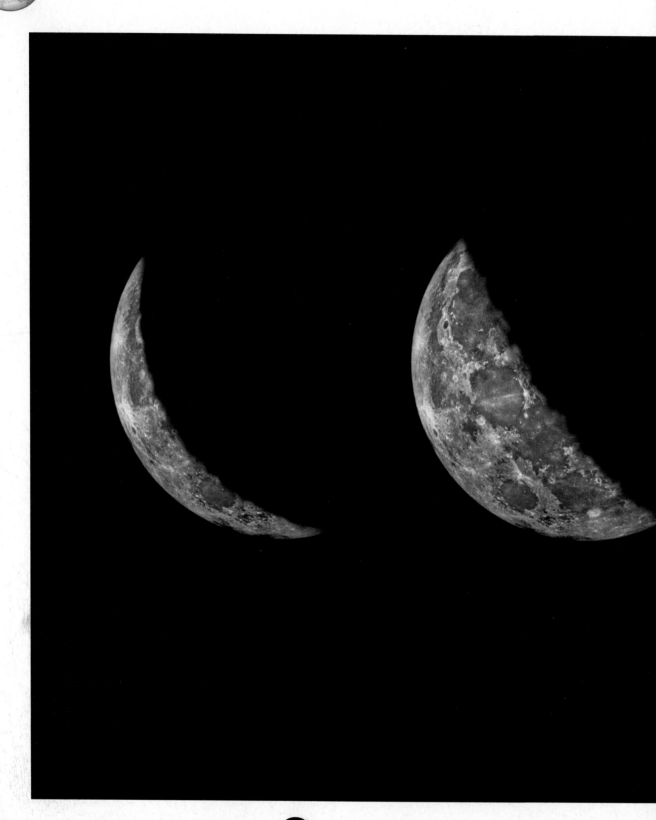

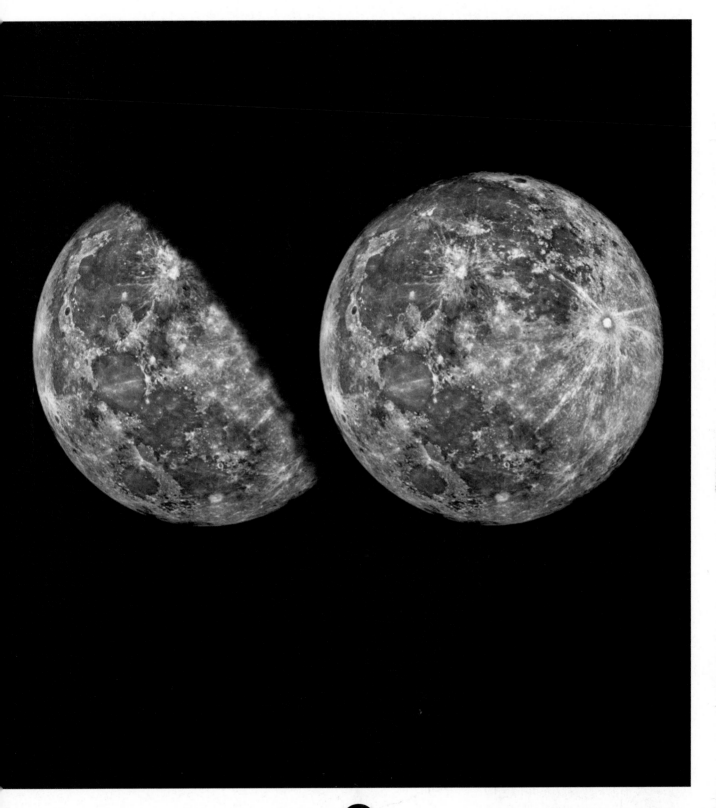

Activity 3   Orbits and Effects

Goals

In this activity you will:

- Measure the major axis and distance between the foci of an ellipse.

- Understand the relationship between the distance between the foci and eccentricity of an ellipse.

- Calculate the eccentricity of the Earth's orbit.

- Draw the Earth's changing orbit in relation to the Sun.

- Explain how the Earth's changing orbit and its rotation rate could affect its climate.

- Draw the orbits of a comet and an asteroid in relation to the Earth and the Sun.

Think about It

The Earth rotates on its axis as it revolves around the Sun, some 150,000,000 km away. The axis of rotation is now tilted about 23.5°.

- **What is the shape of the Earth's orbit around the Sun?**

- **How might a change in the shape of the Earth's orbit or its axis of rotation affect weather and climate?**

What do you think? In your *EarthComm* notebook, draw a picture of the Earth's orbit around the Sun, as seen from above the solar system. Record your ideas about how this shape affects weather and climate. Be prepared to discuss your responses with your small group and the class.

Investigate

1. Draw an ellipse using the following steps:

   - Fold a piece of paper in half.

   - Use a straightedge to draw a horizontal line across the width of the paper along the fold.

   - Put two dots 10 cm apart on the line toward the center of the line. Label the left dot "A" and the right dot "B." unfold the paper

   - Tape the sheet of paper to a piece of thick cardboard, and put two pushpins into points A and B. The positions of the pushpins will be the foci of the ellipse.

   - Tie two ends of a piece of strong string together to make a loop. Make the knot so that when you stretch out the loop with your fingers into a line, it is 12 cm long.

   - Put the string over the two pins and pull the loop tight using a pencil point, as shown in the diagram.

   - Draw an ellipse with the pencil. Do this by putting the pencil point inside the loop and then moving the pencil while keeping the string pulled tight with the pencil point.

   a) Draw a small circle around either point A or point B and label it "Sun."

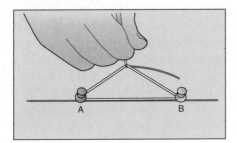

2. Repeat the process using the following measurements and labels:

   - Two points 8 cm apart labeled C and D (1 cm inside of points A and B).

   - Two points 6 cm apart, labeled E and F (2 cm inside of points A and B).

   - Two points 4 cm apart, labeled G and H (3 cm inside points A and B).

   - Two points 2 cm apart, labeled I and J (4 cm inside points A and B).

3. Copy the data table on the next page into your notebook.

   a) Measure the width (in centimeters) of ellipse "AB" at its widest point. This is the major axis, L (see diagram on the next page). Record this in your data table.

   b) Record the length of the major axis for each ellipse in your data table.

   c) Record the distance between the two foci, d (the distance between the two pushpins) for each ellipse in your data table (see diagram).

   d) The eccentricity E of an ellipse is equal to the distance between the two foci divided by the length of the major axis. Calculate the eccentricity of each of your ellipses using the equation $E = d/L$, where d is the distance between the foci and L is the length of the major axis. Record the eccentricity of each ellipse.

   ⚠️ Be sure the cardboard is thicker than the points of the pins. If not, use two or more pieces of cardboard.

Ellipse	Major Axis (L) (cm)	Distance between the Foci (d) (cm)	Eccentricity E = d/L
AB		10	
CD		8	
EF		6	
GH		4	
IJ		2	

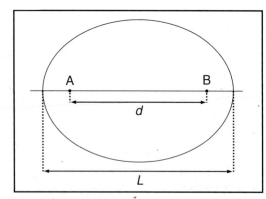

Ellipse with foci AB showing major axis length L and distance between the foci d.

4. Study your data table to find a relationship between the distance between the foci and the eccentricity of an ellipse.

   a) Record the relationship between the distance between the foci and the eccentricity in your notebook.

b) Think of your ellipses as the orbits of planets around the Sun. Does the distance to the center of the Sun stay the same in any orbit?

c) Which orbit has the least variation in distance from the Sun throughout its orbit? Which has the most?

5. Earth's orbit has an eccentricity of about 0.017. Compare this value to the ellipse with the lowest eccentricity of those you drew.

   a) Why does it make sense to describe Earth's orbit as "nearly circular"?

Reflecting on the Activity and the Challenge

In this activity, you explored a geometric figure called an ellipse. You also learned how to characterize ellipses by their eccentricity. The orbits of all nine planets in our solar system are ellipses, with the Sun at one focus of the ellipse representing the orbit for each planet. As you will see, although Earth's orbit is very nearly circular (only slightly eccentric), the shape of its orbit is generally believed to play an important role in long-term changes in the climate. The shape of the Earth's orbit is not responsible for the seasons. You will need this information when describing the Earth's orbital relationships with the Sun and the Moon in the **Chapter Challenge**.

Digging Deeper

ECCENTRICITY, AXIAL TILT, PRECESSION, AND INCLINATION

Eccentricity

Geo Words

eccentricity: the ratio of the distance between the foci and the length of the major axis of an ellipse.

After many years of analyzing observational data on the motions of the planets, the astronomer Johannes Kepler (1571–1630) developed three laws of planetary motion that govern orbits. The first law states that the orbit of each planet around the Sun is an ellipse with the Sun at one focus. The second law, as shown in *Figure 1*, explains that as a planet moves around the Sun in its orbit, it covers equal areas in equal times. The third law states that the time a planet takes to complete one orbit is related to its average distance from the Sun.

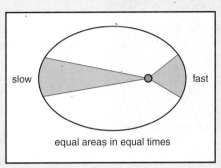

Figure 1 Kepler's Second Law states that a line joining a planet and the Sun sweeps out equal areas in equal intervals of time.

As you saw in the **Investigate** section, the shape of an ellipse can vary from a circle to a very highly elongated shape, and even to a straight line. The more flattened the ellipse is, the greater its eccentricity. Values of **eccentricity** range from zero for a circle, to one, for a straight line. (A mathematician would say that the circle and the line are "special cases" of an ellipse.) The two planets in the solar system with the most elliptical orbits are Mercury, the closest planet, and Pluto, the farthest one. Both have eccentricities greater than 0.2. The orbit of Mars is also fairly elliptical, with an eccentricity of 0.09. In comparison, the Earth's orbit is an ellipse with an eccentricity of 0.017. (This is a much lower value than even the eccentricity of ellipse IJ in the investigation, which looked much like a circle.) If you were to draw an ellipse with an eccentricity of 0.017 on a large sheet of paper, most people would call it a circle. It's eccentric enough, however, to make the Earth's distance from the Sun vary between 153,000,000 km and 147,000,000 km. To make things more complicated, the eccentricity of the Earth's orbit changes over time, because of complicated effects having to do with the weak gravitational pull of other planets in the solar system. Over the course of about 100,000 years, the Earth's orbit ranges from nearly circular (very close to zero eccentricity) to more elliptical (with an eccentricity of about 0.05).

Planetary scientists have found that some solar-system objects have highly elliptical orbits. Comets are a well-known example. As they move closer

Figure 2 The comet's head or coma is the fuzzy haze that surrounds the comet's true nucleus.

to the Sun, the icy mix that makes up a comet's nucleus begins to turn into gas and stream away. The result is a ghostly looking tail and a fuzzy "shroud," that you can see in *Figure 2*. It is called a **coma**, and it forms around the nucleus. When a comet gets far enough away from the Sun, the ices are no longer turned to gas and the icy nucleus continues on its way.

Another good example is the distant, icy world Pluto. Throughout much of its year (which lasts 248 Earth years) this little outpost of the solar system has no measurable atmosphere. It does have a highly eccentric orbit, and its distance from the Sun varies from 29.5 to 49.5 AU. The strength of solar heating varies by a factor of almost four during Pluto's orbit around the Sun. As Pluto gets closest to the Sun, Pluto receives just enough solar heating to vaporize some of its ices. This creates a thin, measurable atmosphere. Then, as the planet moves farther out in its orbit, this atmosphere freezes out and falls to the surface as a frosty covering. Some scientists predict that when Pluto is only 20 years past its point of being closest to the Sun, its atmosphere will collapse as the temperature decreases.

Axial Tilt (Obliquity)

The Earth's axis of rotation is now tilted at an angle of 23.5° to the plane of the Earth's orbit around the Sun, as seen in *Figure 3*. Over a cycle lasting about 41,000 years, the axial tilt varies from 22.1° to 24.5°. The greater the angle of tilt, the greater the difference in solar energy, and therefore temperature, between summer and winter. This small change, combined with other long-term changes in the Earth's orbit, is thought to be responsible for the Earth's ice ages.

Precession

The Earth also has a slight wobble, the same as the slow wobble of a spinning top. This wobble is called the **precession** of the Earth's axis. It is caused by differences in the gravitational pull of the Moon and the Sun on the Earth. It takes about 25,725 years for this wobble to complete a cycle. As the axis wobbles, the timing of the seasons changes. Winter occurs when a hemisphere, northern or southern, is tilted away from the Sun. Nowadays, the Earth is slightly closer to the Sun during winter (January 5) in the Northern Hemisphere. Don't let anybody tell you that winter happens because the Earth is farthest from the Sun! The Earth's orbit is nearly circular! Also, even if a particular hemisphere of the Earth is tilted towards the Sun, it is not significantly closer to the Sun than the other hemisphere, which is tilted away.

Geo Words

coma: a spherical cloud of material surrounding the head of a comet. This material is mostly gas that the Sun has caused to boil off the comet's icy nucleus. A cometary coma can extend up to a million miles from the nucleus.

precession: slow motion of the axis of the Earth around a cone, one cycle in about 26,000 years, due to gravitational tugs by the Sun, Moon, and major planets.

The precession of the Earth's axis is one part of the precession cycle. Another part is the precession of the Earth's orbit. As the Earth moves around the Sun in its elliptical orbit, the major axis of the Earth's orbital ellipse is rotating about the Sun. In other words, the orbit itself rotates around the Sun! These two precessions (the axial and orbital precessions) combine to affect how far the Earth

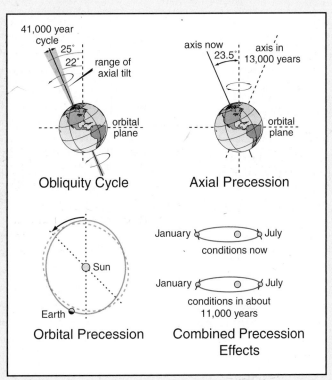

Geo Words

orbital plane: (also called the ecliptic or plane of the ecliptic). A plane formed by the path of the Earth around the Sun.

inclination: the angle between the orbital plane of the solar system and the actual orbit of an object around the Sun.

Figure 3 The tilt of the Earth's axis and its orbital path about the Sun go through several cycles of change.

is from the Sun during the different seasons. This combined effect is called precession of the equinoxes, and this change goes through one complete cycle about every 22,000 years. Ten thousand years from now, about halfway through the precession cycle, winter will be from June to September, when the Earth will be farthest from the Sun during the Northern Hemisphere winter. That will make winters there even colder, on average.

Inclination

When you study a diagram of the solar system, you notice that the orbits of all the planets except Pluto stay within a narrow range called the **orbital plane** of the solar system. If you were making a model of the orbits of the planets in the solar system you could put many of the orbits on a tabletop. However, Pluto's orbit, as shown in *Figure 4*, could not be drawn or placed on a tabletop (a plane, as it is called in geometry). Pluto's path around the Sun is inclined 17.1° from the plane described by the Earth's motion around the Sun. This 17.1° tilt is called its orbital **inclination**.

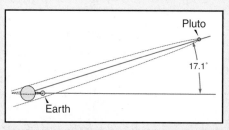

Figure 4 Pluto's orbit is inclined 17.1° to the orbital plane of the rest of the solar system.

What are the orbital planes of asteroids and comets? Both are found mainly in the part of the solar system beyond the Earth. Although some asteroids can be found in the inner solar system, many are found between the orbits of Mars and Jupiter. The common movie portrayal of the "asteroid belt" as a densely populated part of space through which one must dodge asteroids is wrong. The asteroids occupy very little space. Another misconception is that asteroids are the remains of a planet that exploded.

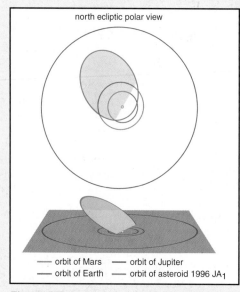

Figure 5 The orbit of the Earth-approaching asteroid 1996 JA, in relation to the Earth.

The orbits of asteroids are more eccentric than the orbits of the planets, and they are often slightly inclined from the orbital plane. As the Earth orbits the Sun, it can cross the orbital paths of objects called Earth-approaching asteroids. There is a great deal of interest in finding Earth-approaching or Earth-crossing asteroids. A collision with an object a few miles across could be devastating, because of its very high velocity relative to the Earth. Astronomers search the skies for asteroids and map their orbits. In this way they hope to learn what's coming toward Earth long before it poses a danger to your community.

Check Your Understanding

1. In your own words, explain what is meant by the eccentricity of an ellipse.

2. For an ellipse with a major axis of 25 cm, which one is more eccentric; the one with a distance between the foci of 15 cm or with a distance between the foci of 20 cm? Explain.

3. How does the precession of the Earth's axis of rotation affect the seasons? Justify your answer.

4. Why is there a danger that a large asteroid might strike the Earth at some time in the future?

Comets are "loners" that periodically visit the inner solar system. They usually originate in the outer solar system. As shown in *Figure 5*, they have very high-inclination orbits—some as much as 30° from the plane of the solar system. In addition, their orbits are often highly eccentric. Astronomers also search the skies for comets. Once a comet is discovered, its orbit is calculated and the comet is observed as it moves closer to the Sun and changes. A collision of a comet's nucleus with the Earth would be serious, but a collision with a comet's tail is much more likely. A collision with the tail would have little, if any, effect on the Earth, because the tail consists mainly of glowing gas with very little mass.

Understanding and Applying What You Have Learned

1. The major axis of the Earth's orbit is 299,200,000 km, and the distance between the foci is 4,999,632 km. Calculate the eccentricity of the Earth's orbit. How does this value compare to the value noted in the **Digging Deeper** reading section?

2. On the GH line on the ellipse that you created for your **Investigate** activity, draw the Earth at its closest position to the Sun and the farthest position away from the Sun.

3. Refer to the table that shows the eccentricities of the planets to answer the following questions:

   a) Which planet would show the greatest percentage variation in its average distance from the Sun during its year? Explain.

   b) Which planet would show the least percentage variation in its average distance from the Sun throughout its year? Explain.

   c) Is there any relationship between the distance from the Sun and the eccentricity of a planet's orbit? Refer to *Table 1* in the **Investigate** section of **Activity 1**.

   d) Why might Neptune be farther away from the Sun at times than Pluto is?

   e) Look up the orbital inclinations of the planets and add them to a copy of the table.

Eccentricities of the Planets	
Planet	**Eccentricity**
Mercury	0.206
Venus	0.007
Earth	0.017
Mars	0.093
Jupiter	0.048
Saturn	0.054
Uranus	0.047
Neptune	0.009
Pluto	0.249

4. Draw a scale model to show changes in the Earth's orbit of about the same magnitude as in nature (a cycle of 100,000 years).

   a) Draw the orbit of the Earth with a perfectly circular orbit at 150,000,000 km from the Sun. Use a scale of 1 cm = 20,000,000 km. Make sure that your pencil is sharp, and draw the thinnest line possible.

   b) Make another drawing of the actual shape of the Earth orbit—an ellipse. This ellipse has 153,000,000 km as the farthest distance and 147,000,000 km as the closest point to the Sun.

   c) Does the difference in distance from the Sun look significant enough to cause much difference in temperature? Explain.

5. Draw the solar system as viewed from the plane of the ecliptic (orbital plane).

a) How will Pluto's orbit look with its 17° inclination?

b) How will the orbits of the other planets look?

c) Draw in the orbits of Earth-crossing asteroids with inclinations of 20° and 30° to the orbital plane.

d) Draw in the orbits of several comets with high inclinations. Some typical high-inclination comets are Comet Halley (162.22°) and Ikeya-Seki (141.86°).

6. Now that you know that the Earth's orbit is elliptical (and the Moon's is too), you can think about a third astronomical factor that controls the nature of tides. Tidal forces are stronger when the Moon is closer to the Earth, and when the Earth is closer to the Sun.

a) Draw a diagram to show the positions of the Moon, Earth, and Sun that would generate the highest tidal ranges (difference in height between high and low tides) of the year.

b) Draw a diagram to show the positions of the Moon, Earth, and Sun that would generate the lowest tidal ranges (difference in height between high and low tides) of the year.

Preparing for the Chapter Challenge

1. In your own words, explain the changes in the Earth's orbital eccentricity, and how it might have affected your community in the past. What effect might it have in the future?

2. Describe the orbits of comets and asteroids and how they are different from those of the planets. What effect could comets and asteroids have on your community if their orbits intersected the Earth's orbit and the Earth and the comet or asteroid were both at that same place in their orbits?

Inquiring Further

1. **The gravitational "slingshot" effect on spacecraft**

The gravitational tug of the Sun and the planets plays a role in shaping the orbits of solar-system bodies. NASA has used the gravitational pull of Jupiter and Saturn to influence the paths of spacecraft like Pioneer and Voyager. Investigate how this gravitational "slingshot" effect works and its role in moving small bodies from one orbit to another.

2. **Investigate the orbits of comets and asteroids**

Look up the orbital information for some typical comets and asteroids. Try to include some with high inclinations and orbital eccentricities.

Activity 4   Impact Events and the Earth System

Goals

In this activity you will:

- Investigate the mechanics of an impact event and make scale drawings of an impact crater.

- Calculate the energy (in joules) released when an asteroid collides with Earth.

- Compare natural and man-made disasters to the impact of an asteroid.

- Understand the consequences to your community should an impact event occur.

- Investigate the chances for an asteroid or comet collision.

Think about It

Meteor Crater in Arizona is one of the best-preserved meteor craters on Earth. It is 1.25 km across and about 4 km in circumference. Twenty football games could be played simultaneously on its floor, while more than two million spectators observed from its sloping sides.

- **How large (in diameter) do you think the meteor was that formed Meteor Crater?**

- **How would the impact of the meteor have affected living things near the crater?**

What do you think? Record your ideas about these questions in your *EarthComm* notebook. Be prepared to discuss your responses with your small group and the class.

Investigate

1. Given the following information, calculate the energy released when an asteroid collides with Earth:

   - The spherical, iron–nickel asteroid has a density of 7800 kg/m³.

   - It is 40 m in diameter.

   - It has a velocity of 20,000 m/s relative to the Earth.

     Note: It is very important to keep track of your units during these calculations. You will be expressing energy with a unit called a "joule." A joule is 1 kg m²/s².

   a) Find the volume of the asteroid in cubic meters. The equation for the volume of a sphere is as follows:

   $$V = \frac{4}{3}\pi r^3$$

   where V is volume of the sphere, and r is the radius of the sphere.

   b) Multiply the volume by the density to find the total mass of the asteroid.

   c) Calculate the energy of the asteroid. Because the asteroid is moving, you will use the equation for kinetic energy, as follows:

   $$KE = \frac{1}{2}mv^2$$

   where KE is kinetic energy, m is the mass, and v is the velocity.

   Express your answer in joules. To do this express mass in kilograms, and velocity in meters per second.

   For some perspective, a teenager uses over 10,000 kJ (kilojoules) of energy each day. (There are 1000 J (joules) in a kilojoule.)

2. The combination of calculations that you just performed can be summarized as:

   $$\text{Energy} = \frac{2}{3}\pi\rho r^3 v^2$$

   where r is the radius,
   ρ is the density, and
   v is the velocity of the object.

   a) Suppose an object makes an impact with the Earth at 10 times the velocity of another identical object. By what factor will the energy of the object increase?

   b) Suppose an object makes an impact with the Earth, and it has 10 times the radius of another object traveling at the same speed. By what factor will the energy of the object increase?

   c) How do these relationships help to explain how small, fast-moving objects can release a tremendous amount of energy as well as larger yet slower-moving objects?

3. The asteroid described in **Step 1** above was the one responsible for Meteor Crater in Arizona.

   a) Copy the following table into your notebook. Enter your calculation for Meteor Crater.

   b) Calculate the energy released by the impacts shown in the table.

Object	Radius (m)	Density (kg/m³)	Impact Velocity (m/s)	Energy (joules)	Richter Scale Magnitude Equivalent
Asclepius	100	3000	30,000		
Comet Swift-Tuttle	1000	1000	60,000		
Chicxulub impactor	5000	3000	32,000		
SL9 Fragment Q	2150	1000	60,000		
Meteor Crater	20	7800	20,000		

Note:

- Asclepius is an asteroid that passed within 690,000 km of Earth in 1989.

- Comet Swift-Tuttle is a future threat to the Earth–Moon system, having passed Earth in 1992 and being scheduled for return in 2126.

- SL9 Fragment Q is a fragment of Comet Shoemaker-Levy that impacted Jupiter in 1994.

- Chicxulub impactor is the name of the asteroid that triggered the extinction of the dinosaurs 65 million years ago.

4. Use the table below to compare the energy from all these events to known phenomena.

   a) In your notebook, explain how the energies of these four impact events compare to some other known phenomena.

Phenomena	Kinetic Energy (joules)
Annual output of the Sun	10^{34}
Severe earthquake	10^{18}
100-megaton hydrogen bomb	10^{17}
Atomic bomb	10^{13}

5. Make a scale drawing of the Chicxulub impactor compared with Earth. The diameter of the Earth is 12,756 km.

   a) If you made the diameter of the Chicxulub impactor 1 mm, what would the diameter of the Earth be?

   b) If you made the diameter of the Chicxulub impactor 0.5 mm, which is probably about as small as you can draw, what would the diameter of the Earth be?

6. How do these impact events compare with the energy released in an earthquake? If you have a calculator capable of handling logarithms, answer the following questions:

a) Calculate the Richter scale equivalent of the energy released by the four impact events. Use the following equation:

$$M = 0.67 \log_{10}E - 5.87$$

where M is the equivalent magnitude on the Richter scale, and E is the energy of the impact, in joules.

b) How do your results compare with the table below, which shows the five greatest earthquakes in the world between 1900 and 1998? Which impacts exceed the world's greatest earthquakes?

Location	Year	Magnitude
Chile	1960	9.5
Prince William Sound, Alaska	1964	9.2
Andreanof Islands, Aleutian Islands	1957	9.1
Kamchatka	1952	9.0
Off the Coast of Ecuador	1906	8.8

Reflecting on the Activity and the Challenge

You have calculated the energy released when asteroids of different sizes hit the Earth's surface, and you have compared these to other energy-releasing events. This comparison will be helpful as you explain the hazards associated with an impact in your **Chapter Challenge** brochure.

Digging Deeper

ASTEROIDS AND COMETS

Asteroids

Asteroids are rocky bodies smaller than planets. They are leftovers from the formation of the solar system. In fact, the early history of the solar system was a period of frequent impacts. The many scars (impact craters) seen on the Moon, Mercury, Mars, and the moons of the outer planets are the evidence for this bombardment. Asteroids orbit the Sun in very elliptical orbits with inclinations up to 30°. Most asteroids are in the region between Jupiter and Mars called the asteroid belt. There are probably at least 100,000 asteroids 1 km in diameter and larger. The largest, called Ceres, is about 1000 km across. Some of the asteroids have very eccentric orbits that cross Earth's orbit. Of these, perhaps a few dozen are larger than one kilometer in diameter. As you learned in the activity, the energy of an asteroid impact event increases with the cube of the radius. Thus, the larger asteroids are the ones astronomers worry about when they consider the danger of collision with Earth.

The closest recent approach of an asteroid to Earth was Asteroid 1994 XM 11. On December 9, 1994, the asteroid approached within 115,000 km of Earth. On March 22, 1989 the asteroid 4581 Asclepius came within 1.8 lunar distances, which is close to 690,000 km. Astronomers think that asteroids at least 1 km in diameter hit Earth every few hundred million years. They base this upon the number of impact craters that have been found and dated on Earth. A list of asteroids that have approached within

Figure 1 Image of the asteroid Ida, which is 58 km long and 23 km wide.

two lunar distances of Earth (the average distance between Earth and the Moon) is provided in *Table 1* on the following page. Only close-approach distances less than 0.01 AU for asteroids are included in this table.

Table 1 Asteroids with Close-Approach Distances to Earth			
Name or Designation	**Date of Close Earth Approach**	**Distance**	
		(AU)	**(LD)**
1994 XM1	1994–Dec–09	0.0007	0.3
1993 KA2	1993–May–20	0.001	0.4
1994 ES1	1994–Mar–15	0.0011	0.4
1991 BA	1991–Jan–18	0.0011	0.4
1996 JA1	1996–May–19	0.003	1.2
1991 VG	1991–Dec–05	0.0031	1.2
1999 VP11	1982–Oct–21	0.0039	1.5
1995 FF	1995–Apr–03	0.0045	1.8
1998 DV9	1975–Jan–31	0.0045	1.8
4581 Asclepius	1989–Mar–22	0.0046	1.8
1994 WR12	1994–Nov–24	0.0048	1.9
1991 TU	1991–Oct–08	0.0048	1.9
1995 UB	1995–Oct–17	0.005	1.9
1937 UB (Hermes)	1937–Oct–30	0.005	1.9
1998 KY26	1998–Jun–08	0.0054	2.1

(AU) – Astronomical distance Unit: 1.0 AU is roughly the average distance between the Earth and the Sun.

(LD) – Lunar Distance unit: 1.0 LD is the average distance from the Earth to the Moon (about 0.00257 AU).

Most (but not all) scientists believe that the extinction of the dinosaurs 65 million years ago was caused by the impact of an asteroid or comet 10 km in diameter. Such a large impact would have sent up enough dust to cloud the entire Earth's atmosphere for many months. This would have blocked out sunlight and killed off many plants, and eventually, the animals that fed on those plants. Not only the dinosaurs died out. About 75% of all plants and animals became extinct. One of the strong pieces of evidence supporting this hypothesis is a 1-cm-thick layer of iridium-rich sediment about 65 million years old that has been found worldwide. Iridium is rare on Earth but is common in asteroids.

Our planet has undergone at least a dozen mass-extinction events during its history, during which a large percentage of all plant and animal species became extinct in an extremely short interval of geologic time. It is likely that at least some of these were related to impacts. It is also likely that Earth will suffer another collision sometime in the future. NASA is currently forming plans to discover and monitor asteroids that are at least 1 km in size and with orbits that cross the Earth's orbit. Asteroid experts take the threat from asteroids very seriously, and they strongly suggest that a program of systematic observation be put into operation to predict and, hopefully avoid an impact.

Geo Words

solar wind: a flow of hot charged particles leaving the Sun.

Comets

Comets are masses of frozen gases (ices) and rocky dust particles. Like asteroids, they are leftovers from the formation of the solar system. There are many comets in orbit around the Sun. Their orbits are usually very eccentric with large inclinations. The orbits of many comets are very large, with distances from the Sun of greater than 20,000 astronomical units (AU). The icy head of a comet (the nucleus) is usually a few kilometers in diameter, but it appears much larger as it gets closer to the Sun. That's because the Sun's heat vaporizes the ice, forming a cloud called a coma. Radiation pressure and the action of the **solar wind** (the stream of fast-moving charged particles coming from the Sun) blow the gases and dust in the coma in a direction away from the Sun. This produces a tail that points away from the Sun even as the comet moves around the Sun. Halley's Comet, shown in *Figure 2*, is the best known of these icy visitors. It rounds the Sun about every 76 years, and it last passed by Earth in 1986.

Figure 2 Halley's Comet last appeared in the night sky in 1986.

Comets have collided with the Earth since its earliest formation. It is thought that the ices from comet impacts melted to help form Earth's oceans. In 1908 something hit the Earth at Tunguska, in Siberian Russia. It flattened trees for hundreds of miles, and researchers believe that the object might have been a comet. Had such an event occurred in more recent history in a more populated area, the damage and loss of life would have been enormous. A list of comets that have approached within less than 0.11 AU of Earth is provided in *Table 2* on the following page.

Geo Words

meteoroid: a small rock in space.

meteor: the luminous phenomenon seen when a meteoroid enters the atmosphere (commonly known as a shooting star).

meteorite: a part of a meteoroid that survives through the Earth's atmosphere.

Table 2 Close Approaches of Comets				
Name	**Designation**	**Date of Close Earth Approach**	**Distance**	
			(AU)	**(LD)**
Comet of 1491	C/1491 B1	1491–Feb–20	0.0094	3.7*
Lexell	D/1770 L1	1770–Jul–01	0.0151	5.9
Tempel-Tuttle	55P/1366 U1	1366–Oct–26	0.0229	8.9
IRAS-Araki-Alcock	C/1983 H1	1983–May–11	0.0313	12.2
Halley	1P/ 837 F1	1837–Apr–10	0.0334	13
Biela	3D/1805 V1	1805–Dec–09	0.0366	14.2
Comet of 1743	C/1743 C1	1743–Feb–08	0.039	15.2
Pons-Winnecke	7P/	1927–Jun–26	0.0394	15.3
Comet of 1014	C/1014 C1	1014–Feb–24	0.0407	15.8*
Comet of 1702	C/1702 H1	1702–Apr–20	0.0437	17
Comet of 1132	C/1132 T1	1132–Oct–07	0.0447	17.4*
Comet of 1351	C/1351 W1	1351–Nov–29	0.0479	18.6*
Comet of 1345	C/1345 O1	1345–Jul–31	0.0485	18.9*
Comet of 1499	C/1499 Q1	1499–Aug–17	0.0588	22.9*
Schwassmann-Wachmann 3	73P/1930 J1	1930–May–31	0.0617	24

* Distance uncertain because comet's orbit is relatively poorly determined.
  (AU) – Astronomical distance Unit: 1.0 AU is roughly the average distance between the Earth and the Sun.
  (LD) – Lunar Distance unit: 1.0 LD is the average distance from the Earth to the Moon (about 0.00257 AU).

Meteoroids, Meteors, and Meteorites

Meteoroids are tiny particles in space, like leftover dust from a comet's tail or fragments of asteroids. Meteoroids are called **meteors** when they enter Earth's atmosphere, and **meteorites** when they reach the Earth's surface. About 1000 tons of material is added to the Earth each year by meteorites, much of it through dust-sized particles that settle slowly through the atmosphere. There are several types of meteorites. About 80% that hit Earth are stony in nature and are difficult to tell apart from Earth rocks. About 15% of meteorites consist of the metals iron and nickel and are very dense. The rest are a mixture of iron–nickel and stony material. Most of the stony meteorites are called chondrites. Chondrites may represent material that was never part of a larger body like a moon, a planet, or an asteroid, but instead are probably original solar-system materials.

Figure 3 Lunar meteorite.

Check Your Understanding

1. Where are asteroids most abundant in the solar system?

2. How might a major asteroid impact have caused a mass extinction of the Earth's plant and animal species at certain times in the geologic past?

3. Why do comets have tails? Why do the tails point away from the Sun?

4. What are the compositions of the major kinds of meteorites?

Understanding and Applying What You Have Learned

1. Look at the table of impact events shown in the **Investigate** section. Compare the densities of the object that formed Meteor Crater and SL9 Fragment Q from the Shoemaker-Levy Comet. Use what you have learned in this activity to explain the large difference in densities between the two objects.

2. If an asteroid or comet were on a collision course for Earth, what factors would determine how dangerous the collision might be for your community?

3. How would an asteroid on a collision course endanger our Earth community?

4. Comets are composed largely of ice and mineral grains. Assume a density of 1.1 g/cm³:

   a) How would the energy released in a comet impact compare to the asteroid impact you calculated in the **Investigate** section? (Assume that the comet has the same diameter and velocity as the asteroid.)

   b) Based upon your calculation, are comets dangerous if they make impact with the Earth? Explain your response.

5. From the information in the **Digging Deeper** reading section, and what you know about the eccentricities and inclinations of asteroid orbits, how likely do you think it is that an asteroid with a diameter of 1 km or greater will hit the Earth in your lifetime? Explain your reasoning. Can you apply the same reasoning to comets?

6. Add the asteroid belt to the model of the solar system you made in the first activity. You will need to think about how best to represent the vast number of asteroids and their wide range of sizes. Don't forget to add in some samples of Earth-approaching asteroids and the orbit of one or two comets.

Preparing for the Chapter Challenge

Assume that scientists learn several months before impact that a large asteroid will hit near your community. Assume that you live 300 km from the impact site. What plans can your family make to survive this disaster? What are some of the potential larger-scale effects of an asteroid impact? Work with your group to make a survival plan. Present your group's plan to the entire class. Be sure to record suggestions made by other groups. This information will prove useful in completing the **Chapter Challenge.**

Inquiring Further

1. **Impact craters on objects other than the Earth**

   In an earlier activity you studied impact sites on the Moon. Look at Mercury, Mars, and the moons of Saturn, Uranus, and Neptune to see other examples of impact craters in the solar system. How are these craters similar to Meteor Crater? How are they different?

2. **Modeling impact craters**

   Simulate an asteroid or comet hitting the Earth. Fill a shoebox partway with plaster of Paris. When the plaster is almost dry, drop two rocks of different sizes into it from the same height. Carefully retrieve the rocks and drop them again in a different place, this time from higher distance. Let the plaster fully harden, then examine, and measure the craters. Measure the depth and diameter and calculate the diameter-to-depth ratio. Which is largest? Which is deepest? Did the results surprise you?

3. **Earth-approaching asteroids**

   Do some research into current efforts by scientists to map the orbits of Earth-approaching asteroids? Visit the *EarthComm* web site to help you get started with your research. How are orbits determined? What is the current thinking among scientists about how to prevent impacts from large comets or asteroids?

4. **Barringer Crater**

   Research the Barringer Crater (Meteor Crater). The crater has been named for Daniel Moreau Barringer, who owned the property that contains the crater. Explain how scientists used Barringer Crater to understand how craters form. Study the work of Dr. Eugene Shoemaker, who was one of the foremost experts on the mechanics of impact cratering.

   ⚠ Wear goggles while modeling impact craters. Work with adult supervision to complete the activity.

Activity 5

The Sun and Its Effects on Your Community

Goals

In this activity you will:

- Explore the structure of the Sun and describe the flow of solar energy in terms of reflection, absorption, and scattering.

- Understand that the Sun emits charged particles called the solar wind, and how this wind affects "space weather."

- Explain the effect of solar wind on people and communities.

- Understand sunspots, solar flares, and other kinds of solar activities and their effects on Earth.

- Learn to estimate the chances for solar activity to affect your community.

Think about It

Every day of your life you are subjected to radiation from the Sun. Fortunately, the Earth's atmosphere and magnetic field provides protection against many of the Sun's strong outbursts.

- **In what ways does solar radiation benefit you?**

- **In what ways can solar radiation be harmful or disruptive?**

What do you think? Record your ideas about these questions in your *EarthComm* notebook. Be prepared to discuss your response with your small group and the class.

Investigate

1. Use the data in *Table 1* to construct a graph of sunspot activity by year.

   a) Plot time on the horizontal axis and number of sunspots on the vertical axis.

   b) Connect the points you have plotted.

   c) Look at your graph. Describe any pattern you find in the sunspot activity.

Table 1 Sunspot Activity 1899 to 1998							
Year	**Number of Sunspots**	**Year**	**Number of Sunspots**	**Year**	**Number of Sunspots**	**Year**	**Number of Sunspots**
1899	12.1	1924	16.7	1949	134.7	1974	34.5
1900	9.5	1925	44.3	1950	83.9	1975	15.5
1901	2.7	1926	63.9	1951	69.4	1976	12.6
1902	5.0	1927	69.0	1952	31.5	1977	27.5
1903	24.4	1928	77.8	1953	13.9	1978	92.5
1904	42.0	1929	64.9	1954	4.4	1979	155.4
1905	63.5	1930	35.7	1955	38.0	1980	154.6
1906	53.8	1931	21.2	1956	141.7	1981	140.4
1907	62.0	1932	11.1	1957	190.2	1982	115.9
1908	48.5	1933	5.7	1958	184.8	1983	66.6
1909	43.9	1934	8.7	1959	159.0	1984	45.9
1910	18.6	1935	36.1	1960	112.3	1985	17.9
1911	5.7	1936	79.7	1961	53.9	1986	13.4
1912	3.6	1937	114.4	1962	37.6	1987	29.4
1913	1.4	1938	109.6	1963	27.9	1988	100.2
1914	9.6	1939	88.8	1964	10.2	1989	157.6
1915	47.4	1940	67.8	1965	15.1	1990	142.6
1916	57.1	1941	47.5	1966	47.0	1991	145.7
1917	103.9	1942	30.6	1967	93.8	1992	94.3
1918	80.6	1943	16.3	1968	105.9	1993	54.6
1919	63.6	1944	9.6	1969	105.5	1994	29.9
1920	37.6	1945	33.2	1970	104.5	1995	17.5
1921	26.1	1946	92.6	1971	66.6	1996	8.6
1922	14.2	1947	151.6	1972	68.9	1997	21.5
1923	5.8	1948	136.3	1973	38.0	1998	64.3

The number of sunspots on the visible solar surface is counted by many solar observatories and is averaged into a single standardized quantity called the sunspot number. This explains the fractional values in the table.

2. *Table 2* contains a list of solar flares that were strong enough to disrupt terrestrial communications and power systems.

   a) Plot the data from *Table 2* onto a histogram.

   b) What pattern do you see in the activity of solar flares?

3. Compare the two graphs you have produced.

   a) What pattern do you see that connects the two?

   b) How would you explain the pattern?

Table 2 Strongest Solar Flare Events 1978–2001			
Date of Activity Onset	**Strength**	**Date of Activity Onset**	**Strength**
August 16, 1989	X20.0	December 17, 1982	X10.1
March 06, 1989	X15.0	May 20, 1984	X10.1
July 07, 1978	X15.0	January 25, 1991	X10.0
April 24, 1984	X13.0	June 09, 1991	X10.0
October 19, 1989	X13.0	July 09, 1982	X 9.8
December 12, 1982	X12.9	September 29, 1989	X9.8
June 06, 1982	X12.0	March 22, 1991	X9.4
June 01, 1991	X12.0	November 6, 1997	X9.4
June 04, 1991	X12.0	May 24, 1990	X9.3
June 06, 1991	X12.0	November 6, 1980	X9.0
June 11, 1991	X12.0	November 2, 1992	X9.0
June 15, 1991	X12.0		

The X before the number is a designation of the strongest flares.
Source: IPS Solar Flares & Space Service in Australia.

Reflecting on the Activity and the Challenge

In this activity you used data tables to plot the number of sunspots in a given year and to correlate strong solar-flare activity with larger numbers of sunspots. You found out that the number of sunspots varies from year to year in a regular cycle and that strong solar flares occur in greater numbers during high-sunspot years. In your **Chapter Challenge**, you will need to explain sunspots and solar flares, their cycles, and the effects of these cycles on your community.

Geo Words

photosphere: the visible surface of the Sun, lying just above the uppermost layer of the Sun's interior, and just below the chromosphere.

chromosphere: a layer in the Sun's atmosphere, the transition between the outermost layer of the Sun's atmosphere, or corona.

corona: the outermost atmosphere of a star (including the Sun), millions of kilometers in extent, and consisting of highly rarefied gas heated to temperatures of millions of degrees.

Digging Deeper

THE SUN AND ITS EFFECTS

Structure of the Sun

From the Earth's surface the Sun appears as a white, glowing ball of light. Like the Earth, the Sun has a layered structure, as shown in *Figure 1*. Its central region (the core) is where nuclear fusion occurs. The core is the source of all the energy the Sun emits. That energy travels out from the core, through a radiative layer and a convection zone above that. Finally, it reaches the outer layers: the **photosphere**, which is the Sun's visible surface, the **chromosphere**, which produces much of the Sun's ultraviolet radiation, and the superheated uppermost layer of the Sun's atmosphere, called the **corona**.

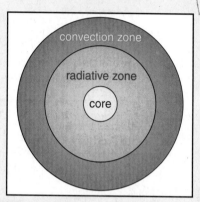

Figure 1 The layered structure of the Sun.

The Sun is the Earth's main external energy source. Of all the incoming energy from the Sun, about half is absorbed by the Earth's surface (see *Figure 2*). The rest is either:

• absorbed by the atmosphere, or
• reflected or scattered back into space by the Earth or clouds.

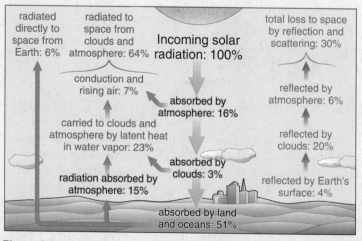

Figure 2 Schematic of Earth's solar energy budget.

Molecules of dust and gas in the atmosphere interfere with some of the incoming solar radiation by changing its direction. This is called scattering, and it explains the blue color of the sky. The atmosphere scatters shorter visible wavelengths of visible light, in the blue range, more strongly than longer visible wavelengths, in the red and orange range. The blue sky you see on a clear day is the blue light that has been scattered from atmospheric particles that are located away from the line of sight to the Sun. When the Sun is low on the horizon, its light has to travel through a much greater thickness of atmosphere, and even more of the blue part of the spectrum of sunlight is scattered out of your line of sight. The red and orange part of the spectrum remains, so the light you see coming directly from the Sun is of that color. The effect is greatest when there is dust and smoke in the atmosphere, because that increases the scattering. The scattered light that makes the sky appear blue is what makes it possible for you to see in a shaded area.

Figure 3 Dust and smoke in the atmosphere enhance the beauty of sunsets.

Most of the sunlight that passes through the atmosphere reaches the Earth's surface without being absorbed. The Sun heats the atmosphere not directly, but rather by warming the Earth's surface. The Earth's surface in turn warms the air near the ground. As the Earth's surface absorbs solar radiation, it re-radiates the heat energy back out to space as infrared radiation. The wavelength of this infrared radiation is much longer than that of visible light, so you can't see the energy that's re-radiated. You can feel it, however, by standing next to a rock surface or the wall of a building that has been heated by the Sun.

Geo Words

albedo: the reflective property of a non-luminous object. A perfect mirror would have an albedo of 100% while a black hole would have an albedo of 0%.

The reflectivity of a surface is referred to as its **albedo**. Albedo is expressed as a percentage of radiation that is reflected. The average albedo of the Earth, including its atmosphere, as would be seen from space, is about 0.3. That means that 30% of the light is reflected. Most of this 30% is due to the high reflectivity of clouds, although the air itself scatters about 6% and the Earth's surface (mainly deserts and oceans) reflects another 4%. (See *Figure 2* in the **Digging Deeper** section.) The albedo of particular surfaces on Earth varies. Thick clouds have albedo of about 0.8, and freshly fallen snow has an even higher albedo. The albedo of a dark soil, on the other hand, is as low as 0.1, meaning that only 10% of the light is reflected. You know from your own experience that light-colored clothing stays much cooler in the Sun than dark-colored clothing. You can think of your clothing as having an albedo, too!

The Earth's Energy Budget

The amount of energy received by the Earth and delivered back into space is the Earth's energy budget. Like a monetary budget, the energy resides in various kinds of places, and moves from place to place in various ways and by various amounts. The energy budget for a given location changes from day to day and from season to season. It can even change on geologic time scales. Daily changes in solar energy are the most familiar. It is usually cooler in the morning, warmer at midday, and cooler again at night. Visible light follows the same cycle, as day moves from dawn to dusk and back to dawn again. But overall, the system is in balance. The Earth gains energy from the Sun and loses energy to space, but the amount of energy entering the Earth system is equal to the amount of energy flowing out, on a long-term average. This flow of energy is the source of energy for almost all forms of life on Earth. Plants capture solar energy by photosynthesis, to build plant tissue. Animals feed on the plants (or on one another). Solar energy creates the weather, drives the movement of the oceans, and powers the water cycle. All of Earth's systems depend on the input of energy from the Sun. The Sun also supplies most of the energy for human civilization, either directly, as with solar power and wind power, or indirectly, in the form of fossil fuels.

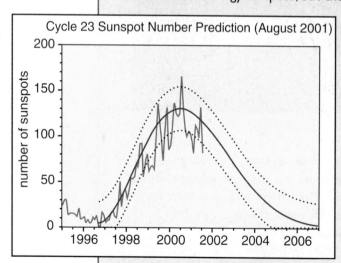

Cycle 23 Sunspot Number Prediction (August 2001)

Figure 4 The jagged line represents the actual number of sunspots; the smooth dark line is the predicted number of sunspots.

Harmful Solar Radiation

Just as there are benefits to receiving energy from the Sun, there are dangers. The ill effects of sunlight are caused by ultraviolet (UV) radiation, which causes skin damage. The gas called ozone (a molecule made up of three oxygen atoms) found in the upper atmosphere shields the Earth from much of the Sun's harmful UV rays. The source of the ozone in the upper atmosphere is different from the ozone that is produced (often by cars) in polluted cities. The latter is a health hazard and in no way protects you. Scientists have recently noted decreasing levels of ozone in the upper atmosphere. Less ozone means that more UV radiation reaches Earth, increasing the danger of Sun damage. There is general agreement about the cause of the ozone depletion. Scientists agree that future levels of ozone will depend upon a combination of natural and man-made factors, including the phase-out, now under way, of chlorofluorocarbons and other ozone-depleting chemicals.

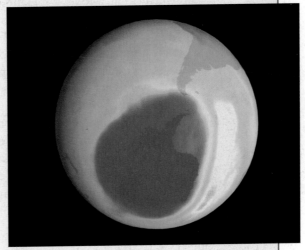

Figure 5 Depletion in the ozone layer over Antarctica. Rather than actually being a hole, the ozone hole is a large area of the stratosphere with extremely low concentrations of ozone.

Sunspots and Solar Flares

Sunspots are small dark areas on the Sun's visible surface. They can be as small as the Earth or as large as Uranus or Neptune. They are formed when magnetic field lines just below the Sun's surface are twisted and poke through the solar photosphere. They look dark because they are about 1500 K cooler than the surrounding surface of the Sun. Sunspots are highly magnetic. This magnetism may cause the cooler temperatures by suppressing the circulation of heat in the region of the sunspot.

Sunspots last for a few hours to a few months. They appear to move across the surface of the Sun over a period of days. Actually, the sunspots move because the Sun is rotating. The number of sunspots varies from year to year and tends to peak in 11-year cycles along with the number of dangerously strong solar flares. Both can affect systems here on Earth. During a solar

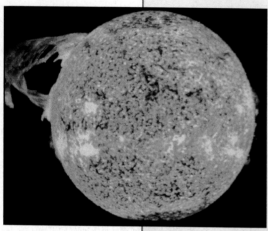

Figure 6 A solar flare.

flare like the one shown in *Figure 6*, enormous quantities of ultraviolet, x-ray, and radio waves blast out from the Sun. In addition, protons and electrons stream from flares at 800 km/hr. These high-radiation events can be devastating to Earth-orbiting satellites and astronauts, as well as systems on the ground. In 1989 a major solar flare created electric currents that caused a surge of power that knocked out a power grid in Canada, leaving hundreds of thousands of people without power. More recently, in 1997, radiation from a flare affected an Earth-orbiting satellite that carried telecommunications traffic. For at least a day people whose beeper messages went through that satellite had no service.

The flow of charged particles (also called a **plasma**) from the Sun is called the solar wind. It flows out from the solar corona in all directions and is responsible for "space weather"—the environment outside our planet. Like severe storms in our atmosphere, space weather can cause problems for Earth systems. Strong outbursts in this ongoing stream of charged particles can disrupt radio signals by disturbing the upper layers of the atmosphere. The sounds of your favorite short-wave radio station or the signals sent by a ham radio operator travel as radio waves (a form of electromagnetic radiation). These signals travel around the Earth by bouncing off the **ionosphere**, a layer of the atmosphere 80 to 400 km above the Earth's surface. The ionosphere forms when incoming solar radiation blasts electrons out of the upper-atmosphere gases, leaving a layer of electrons and of charged atoms, called **ions**. The ionosphere acts like a mirror, reflecting a part of the radio waves (AM radio waves in the 1000 kHz range) back to Earth.

Solar flares intensify the solar wind, which makes the ionosphere thicken and strengthen. When this happens, radio signals from Earth are trapped inside the ionosphere. This causes a lot of interference. As discussed above, solar activity can also be a problem for satellite operations. Astronauts orbiting the Earth and people aboard high-flying aircraft—particularly those who fly polar routes, where exposure to radiation may be greatest—also have cause to worry about space weather. To provide up-to-date information about current solar activity, the United States government operates a Space Environment Center Web site called "Space Weather Now."

At least one effect of space weather is quite wonderful. When the solar wind encounters the Earth's magnetic field, it excites gases in the Earth's atmosphere, causing them to glow. The charged particles from the solar wind end up in an oval-shaped area around the Earth's magnetic poles. The result is a beautiful display called an **aurora**, seen in *Figure 7*. People who live in northern areas see auroras more often than those who live near the Equator do. During periods of heavy solar activity, however, an aurora can be seen as far south as Texas and New Mexico. Auroras are often called the northern lights (aurora borealis) or southern lights (aurora australis). From the ground, they often appear as green or red glows, or shimmering curtains of white, red, and green lights in the sky.

Figure 7 The aurora borealis or northern lights light up the sky in the Northern Hemisphere.

Collecting Data about the Sun

How do astronomers collect data about the Sun? From the ground, they use solar telescopes—instruments outfitted with special sensors to detect the different kinds of solar activity. There are dozens of solar telescope sites around the world. They include the Sacramento Peak Solar Observatory in New Mexico, the McMath Solar telescope in Arizona, and the Mount Wilson solar observatory in California. From space, they study the Sun using orbiting spacecraft like the Yohkoh satellite (*Yohkoh* is the Japanese word for "sunbeam"). Other missions include the Transition Region and Coronal Explorer (TRACE), the Ulysses Solar-Polar mission, the Solar and Heliospheric Observatory, the GOES satellites, and many others. These spacecraft are equipped with detectors sensitive to x-rays, radio waves, and other wavelengths of radiation coming from the Sun. In this way, scientists keep very close track of solar activity and use that information to keep the public informed of any upcoming dangers.

Some scientists theorize that sunspot cycles affect weather on Earth. They think that during times of high sunspot activity, the climate is warmer. During times of no or low sunspot activity, the climate is colder. A sharp decrease in sunspots occurred from 1645 to 1715. This period of lower solar activity, first noted by G. Sporer and later studied by E.W. Maunder, is called the Maunder Minimum. It coincided with cooler temperatures on Earth, part of a period now known as the "Little Ice Age." Similar solar minimums occurred between 1420–1530, 1280–1340, and 1010-1050 (the Oort minimum). These periods preceded the discovery of sunspots, so no correlation between sunspots and temperature is available. Solar astronomers number the solar cycles from one minimum to the next starting with number one, the 1755–1766 cycle. Cycle 23 peaked (was at a maximum) in the year 2000. (See *Figure 4*.) There is still much debate about the connection between sunspot cycles and climate.

Geo Words

aurora: the bright emission of atoms and molecules near the Earth's poles caused by charged particles entering the upper atmosphere.

Check Your Understanding

1. How do solar flares interfere with communication and power systems?

2. In your own words, explain what is meant by the term "solar wind." How does the Sun contribute to "space weather?"

3. Describe the Earth's energy budget.

Coordinated Science for the 21st Century

Understanding and Applying What You Have Learned

1. Study the graph that you made showing sunspot activity. You have already determined that sunspot activity occurs in cycles. Using graph paper, construct a new graph that predicts a continuation of the cycle from 2001 to 2015. Indicate which years you think would see increased solar-flare activity and more dangerous "space weather."

2. The latest sunspot maximum occurred in 2001. Using the data from your sunspot-activity data table, predict the next sunspot minimum.

3. Make lists of the possible consequences of solar flares to the following members of your community: an air traffic controller, a radio station manager, and the captain of a ship at sea. Can you think of other members of your community who would be affected by solar activity?

4. You have read that Earth's albedo is about 0.30.

   a) In your own words, describe what this means.
   b) Is the Earth's albedo constant? Why or why not?
   c) How does changing a planet's albedo change a planet's temperature? Why does this occur?
   d) If Earth's albedo was higher, but Earth was farther from the Sun, could the Earth have the same temperature? Why or why not?

Preparing for the Chapter Challenge

You have been asked to help people in your community to understand how events from outside the Earth affect their daily lives. Write a short paper in which you address the following questions:

1. How has the Sun affected your community in the past?

2. How has the Sun affected you personally?

3. How might the Sun affect your community in the future?

4. What are some of the benefits attained from a study of the Sun?

5. What are some of the problems caused by sunspots and solar flares?

6. Explain how auroras are caused. Explain also why they can or cannot be viewed in your community.

7. Compare the chances of dangerous effects from the Sun with the chances of an impact event affecting the Earth.

Inquiring Further

1. Viewing sunspots

If you have a telescope, you can view sunspots by projecting an image of the Sun onto white cardboard. Never look directly at the Sun, with or without a telescope. Stand with your back to the Sun, and set up a telescope so that the large (front) end is pointing toward the Sun and the other end is pointing toward a piece of white cardboard. You should see a projection of the Sun on the cardboard, including sunspots. If you map the positions of the sunspots daily, you should be able to observe the rotation of the Sun over a couple of weeks. Use the *EarthComm* web site to locate good science sites on the Internet that show daily images of solar activity. Search them out and compare your observations of sunspots to what you see from the large observatories.

⚠ Work with an adult during this activity. Do not look at the bright image for long periods of time.

2. Aurorae

Have people in your community ever seen the northern lights? Even if your community is not very far north, do some research to see if the auroras have ever been spotted from your community.

3. Solar radiation and airplanes

Periods of sunspot maximum increase the dosage of radiation that astronauts and people traveling in airplanes receive. Do some research on how much radiation astronauts receive during sunspot minima and maxima. How much radiation do airplane passengers receive? How do the amounts compare to the solar radiation you receive at the Earth's surface? How do scientists balance safety with the issue of the extra weight that would be added to aircraft, spacecraft, or space suits to provide protection?

4. The hole in the ozone layer

People who live near the South Pole of the Earth are at risk for increased ultraviolet exposure from the Sun. This is due to a thinning in the atmosphere called the ozone hole. Research this ozone hole. Is there a northern ozone hole? Could these ozone holes grow? If so, could your community be endangered in the future?

5. History of science

Research the life of British physicist Edward Victor Appleton, who was awarded the Nobel Prize in physics in 1947 for his work on the ionosphere. Other important figures in the discovery of the properties of the upper atmosphere include Oliver Heaviside, Arthur Edwin Kennelly, F. Sherwood Rowland, Paul Crutzen, and Mario Molina.

Activity 6

The Electromagnetic Spectrum and Your Community

Goals

In this activity you will:

- Explain electromagnetic radiation and the electromagnetic spectrum in terms of wavelength, speed, and energy.

- Investigate the different instruments astronomers use to detect different wavelengths in the electromagnetic spectrum.

- Understand that the atoms of each of the chemical elements have a unique spectral fingerprint.

- Explain how electromagnetic radiation reveals the temperature and chemical makeup of objects like stars.

- Understand that some forms of electromagnetic radiation are essential and beneficial to us on Earth, and others are harmful.

Think about It

Look at the spectrum as your teacher displays it on the overhead projector. Record in your *EarthComm* notebook the colors in the order in which they appear. Draw a picture to accompany your notes.

- **What does a prism reveal about visible light?**

- **The Sun produces light energy that allows you to see. What other kinds of energy come from the Sun? Can you see them? Why or why not?**

What do you think? Record your ideas in your *EarthComm* notebook. Be prepared to discuss your responses with your small group and the class.

Investigate

Part A: Observing Part of the Electromagnetic Spectrum

1. Obtain a spectroscope, similar to the one shown in the illustration. Hold the end with the diffraction grating to your eye. Direct it toward a part of the sky away from the Sun. (**CAUTION:** never look directly at the Sun; doing so even briefly can damage your eyes permanently.) Look for a spectrum along the side of the spectroscope. Rotate the spectroscope until you see the colors going from left to right rather than up and down.

   a) In your notebook, write down the order of the colors you observed.

   b) Move the spectroscope to the right and left. Record your observations.

2. Look through the spectroscope at a fluorescent light.

   a) In your notebook, write down the order of the colors you observed.

3. Look through the spectroscope at an incandescent bulb.

   a) In your notebook, write down the order of the colors you observed.

⚠ Do not look directly at a light with the unaided eye. Use the spectroscope as instructed.

4. Use your observations to answer the following questions:

   a) How did the colors and the order of the colors differ between reflected sunlight, fluorescent light, and the incandescent light? Describe any differences that you noticed.

   b) What if you could use your spectroscope to look at the light from other stars? What do you think it would look like?

Part B: Scaling the Electromagnetic Spectrum

1. Tape four sheets of photocopy paper end to end to make one sheet 112 cm long. Turn the taped sheets over so that the tape is on the bottom.

2. Draw a vertical line 2 cm from the left edge of the paper. Draw two horizontal lines from that line, one about 8 cm from the top of the page, and one about 10 cm below the first line.

3. On the top line, plot the frequencies of the electromagnetic spectrum on a logarithmic scale. To do this, mark off 24 1-cm intervals starting at the left vertical line. Label the marks from 1 to 24 (each number represents increasing powers of 10, from 10^1 to 10^{24}).

4. Use the information from the table of frequency ranges ($\log_{10}$) to divide your scale into the individual bands of electromagnetic radiation. For the visible band, use the entire band, not the individual colors.

Frequency Range Table			
EMR Bands	**Frequency Range (hertz)**	**Log_{10} Frequency Range (hertz)**	**10^{14} Conversions**
Radio and Microwave	Near 0 to 3.0×10^{12}	0 to 12.47	.
Infrared	3.0×10^{12} to 4.6×10^{14}	12.47 to 14.66	.
Visible	4.6×10^{14} to 7.5×10^{14}	14.66 to 14.88	4.6×10^{14} to 7.5×10^{14}
Red	4.6×10^{14} to 5.1×10^{14}	14.66 to 14.71	4.6×10^{14} to 5.1×10^{14}
Orange	5.1×10^{14} to 5.6×10^{14}	14.71 to 14.75	5.1×10^{14} to 5.6×10^{14}
Yellow	5.6×10^{14} to 6.1×10^{14}	14.75 to 14.79	5.6×10^{14} to 6.1×10^{14}
Green	6.1×10^{14} to 6.5×10^{14}	14.79 to 14.81	6.1×10^{14} to 6.5×10^{14}
Blue	6.5×10^{14} to 7.0×10^{14}	14.81 to 14.85	6.5×10^{14} to 7.0×10^{14}
Violet	7.0×10^{14} to 7.5×10^{14}	14.85 to 14.88	7.0×10^{14} to 7.5×10^{14}
Ultraviolet	7.5×10^{14} to 6.0×10^{16}	14.88 to 16.78	.
X-ray	6.0×10^{16} to 1.0×10^{20}	16.78 to 20	.
Gamma Ray	1.0×10^{20} to...	20 to ...	.

5. To construct a linear scale, you will need to convert the range of frequencies that each band of radiation covers for the logarithmic scale. This will allow you to compare the width of the bands of radiation relative to each other. Convert the frequency numbers for all bands (except visible) to 10^{14} and record them in the table.

   Example: 10^{17} is 1000 times greater than 10^{14}, so $2.5 \times 10^{17} = 2500 \times 10^{14}$.

6. On the lower horizontal line, mark off ten 10-cm intervals from the vertical line. Starting with the first interval, label each mark with a whole number times 10^{14}, from 1×10^{14} to 10×10^{14}. Label the bottom of your model "Frequency in hertz." Plot some of the 10^{14} frequencies you calculated on the bottom line of your constructed model. Plot the individual colors of the visible spectrum and color them.

a) Compare the logarithmic and linear scales. Describe the differences.

7. Look at the range of ultraviolet radiation.

a) How high do the ultraviolet frequencies extend (in hertz)?

b) Using the same linear scale that you constructed in **Step 6** (10 cm = 1×10^{14} Hz), calculate the width (in centimeters) of the ultraviolet electromagnetic radiation band.

c) Using this same scale what do you think you would need to measure the distance from the beginning of the ultraviolet band of the electromagnetic radiation to the end of the ultraviolet band of the electromagnetic radiation?

d) Using your calculations above and the linear scale you created in **Steps 5** and **6**, how much wider is the ultraviolet band than the entire visible band? How does this compare to the relative widths of these two bands on the log scale you created in **Steps 1-4**?

8. X-rays are the next band of radiation.

a) Using the same linear scale (10 cm = 1×10^{14} Hz) calculate the distance from the end of the ultraviolet band to the end of the x-ray band. Obtain a map from the Internet or use a local or state highway map to plot the distance.

b) Based on your results for the width of the x-ray band, what would be your estimate for the width of the gamma-ray band of radiation? What would you need to measure the distance?

Part C: Using Electromagnetic Radiation in Astronomy

1. Astronomers use electromagnetic radiation to study objects and events within our solar system and beyond to distant galaxies. In this part of the activity, you will be asked to research a space science mission and find out how astronomers are using the electromagnetic spectrum in the mission and then report to the rest of the class. The *EarthComm* web site will direct you to links for missions that are either in development, currently operating, or operated in the past. A small sampling is provided in the table on the next page. Many missions contain multiple instruments (it is very expensive to send

instruments into space, so scientists combine several or more studies into one mission), so you should focus upon one instrument and aspect of the mission and get to know it well. The **Digging Deeper** reading section of this activity might help you begin your work.

Questions that you should try to answer in your research include:

- What is the purpose or key question of the mission?

- How does the mission contribute to our understanding of the origin and evolution of the universe or the nature of planets within our solar system?

- Who and/or how many scientists and countries are involved in the mission?

- What instrument within the mission have you selected?

- What wavelength range of electromagnetic radiation does the instrument work at?

- What is the detector and how does it work?

- What does the instrument look like?

- How are the data processed and rendered? Images? Graphs?

- Any other questions that you and your teacher agree upon.

2. When you have completed your research, provide a brief report to the class.

Descriptions of Selected Missions	
Mission/Instrument	**Description**
Hubble – NICMOS Instrument	Hubble's Near Infrared Camera and Multi-Object Spectrometer (NICMOS) can see objects in deepest space—objects whose light takes billions of years to reach Earth. Many secrets about the birth of stars, solar systems, and galaxies are revealed in infrared light, which can penetrate the interstellar gas and dust that block visible light.
Cassini Huygens Mission to Saturn and Titan	The Ultraviolet Imaging Spectrograph (UVIS) is a set of detectors designed to measure ultraviolet light reflected or emitted from atmospheres, rings, and surfaces over wavelengths from 55.8 to 190 nm (nanometers) to determine their compositions, distribution, aerosol content, and temperatures.
SIRTF	The Space InfraRed Telescope Facility (SIRTF) is a space-borne, cryogenically cooled infrared observatory capable of studying objects ranging from our solar system to the distant reaches of the Universe. SIRTF is the final element in NASA's Great Observatories Program, and an important scientific and technical cornerstone of the new Astronomical Search for Origins Program.
HETE-2 High Energy Transient Explorer	The High Energy Transient Explorer is a small scientific satellite designed to detect and localize gamma-ray bursts (GRB's). The primary goals of the HETE mission are the multi-wavelength observation of gamma-ray bursts and the prompt distribution of precise GRB coordinates to the astronomical community for immediate follow-up observations. The HETE science payload consists of one gamma-ray and two x-ray detectors.
Chandra X-Ray Observatory	NASA's Chandra X-ray Observatory, which was launched and deployed by Space Shuttle Columbia in July of 1999, is the most sophisticated x-ray observatory built to date. Chandra is designed to observe x-rays from high-energy regions of the universe, such as the remnants of exploded stars.

Reflecting on the Activity and the Challenge

The spectroscope helped you to see that visible light is made up of different color components. Visible light is only one of the components of radiation you receive from the Sun. In the second part of the activity, you explored models for describing the range of frequencies of energy within electromagnetic radiation.

Finally, you researched a space mission to learn how astronomers are using electromagnetic radiation to understand the evolution of the Earth system. Radiation from the Sun and other objects in the universe is something you will need to explain to your fellow citizens in your **Chapter Challenge** brochure.

718

Digging Deeper

ELECTROMAGNETIC RADIATION

The Nature of Electromagnetic Radiation

In 1666, Isaac Newton found that he could split light into a spectrum of colors. As he passed a beam of sunlight through a glass prism, a spectrum of colors appeared from red to violet. Newton deduced that visible light was in fact a mixture of different kinds of light. About 10 years later, Christiaan Huygens proposed the idea that light travels in the form of tiny waves. It's known that light with shorter wavelengths is bent (refracted) more than light with longer wavelengths when it passes through a boundary between two different substances. Violet light is refracted the most, because it has the shortest wavelength of the entire range of visible light. This work marked the beginning of **spectroscopy**—the science of studying the properties of light. As you will learn, many years of research in spectroscopy has answered many questions about matter, energy, time, and space.

In your study of the Sun you learned that the Sun radiates energy over a very wide range of wavelengths. Earth's atmosphere shields you from some of the most dangerous forms of electromagnetic radiation. You are familiar with the wavelengths of light that do get through and harm you—mostly in the form of sunburn-causing ultraviolet radiation. Now you can take what you learned and apply the principles of spectroscopy to other objects in the universe.

In the **Investigate** section, you used a **spectroscope** to study the Sun's light by separating it into its various colors. Each color has a characteristic wavelength. This range of colors, from red to violet, is called the **visible spectrum**. The visible spectrum is a small part of the entire spectrum of **electromagnetic radiation** given off by the Sun, other stars, and galaxies.

Electromagnetic radiation is in the form of electromagnetic waves that transfer energy as they travel through space. Electromagnetic waves (like ripples that expand after you toss a stone into a pond) travel at the speed of light (300,000 m/s). That's eight laps around the Earth in one second. Although it's not easy to appreciate from everyday life, it turns out that electromagnetic radiation has properties of both particles and waves. The colors of the visible spectrum are best described as waves, but the same energy that produces an electric current in a solar cell is best described as a particle.

Geo Words

spectroscopy: the science that studies the way light interacts with matter.

spectroscope: an instrument consisting of, at a minimum, a slit and grating (or prism) which produces a spectrum for visual observation.

visible spectrum: part of the electromagnetic spectrum that is detectable by human eyes. The wavelengths range from 350 to 780 nm (a nanometer is a billionth of a meter).

electromagnetic radiation: the energy propagated through space by oscillating electric and magnetic fields. It travels at 3×10^8 m/s in a vacuum and includes (in order of increasing energy) radio, infrared, visible light (optical), ultraviolet, x-rays, and gamma rays.

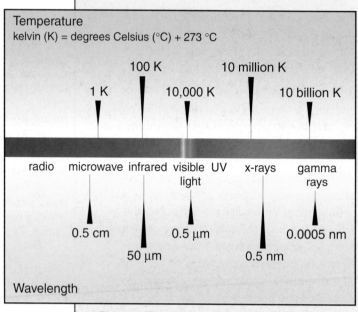

Temperature
kelvin (K) = degrees Celsius (°C) + 273 °C

100 K

10 million K

1 K

10,000 K

10 billion K

radio   microwave   infrared   visible   UV   x-rays   gamma
                                  light   rays

0.5 cm   0.5 μm   0.0005 nm

50 μm   0.5 nm

Wavelength

Figure I The electromagnetic spectrum. Wavelengths decrease from left to right, and energy increases from left to right. The diagram shows that a relationship exists between the temperature of an object and the peak wavelength of electromagnetic radiation it emits.

Figure I summarizes the spectrum of energy that travels throughout the universe. Scientists divide the spectrum into regions by the wavelength of the waves. Long radio waves have wavelengths from several centimeters to thousands of kilometers, whereas gamma rays are shorter than the width of an atom.

Humans can see only wavelengths between 0.4 and 0.7 μm, which is where the visible spectrum falls. A micrometer (μm) is a millionth of a meter. This means that much of the electromagnetic radiation emitted by the Sun is invisible to human eyes. You are probably familiar, however, with some of the kinds of radiation besides visible light. For example, **ultraviolet** radiation gives you sunburn. **Infrared** radiation you detect as heat. Doctors use x-rays to help diagnose broken bones or other physical problems. Law-enforcement officers use radar to measure the speed of a motor vehicle, and at home you may use microwaves to cook food.

Astronomy and the Electromagnetic Spectrum

Humans have traveled to the Moon and sent probes deeper into our solar system, but how do they learn about distant objects in the universe? They use a variety of instruments to collect electromagnetic radiation from these distant objects. Each tool is designed for a specific part of the spectrum. Visible light reveals the temperature of stars. Visible light is what you see when you look at the stars through telescopes, binoculars, or your unaided eyes. All other forms of light are invisible to the human eye, but they can be detected.

Radio telescopes like the Very Large Array (VLA) and Very Large Baseline Array (VLBA) in New Mexico are sensitive to wavelengths in the radio range. Radio telescopes produce images of celestial bodies by recording the different amounts of radio emission coming from an area of the sky

Geo Words

ultraviolet: electromagnetic radiation at wavelengths shorter than the violet end of visible light; with wavelengths ranging from 5 to 400 nm.

infrared: electromagnetic radiation with wavelengths between about 0.7 to 1000 μm. Infrared waves are not visible to the human eye.

radio telescope: an instrument used to observe longer wavelengths of radiation (radio waves), with large dishes to collect and concentrate the radiation onto antennae.

observed. Astronomers process the information with computers to produce an image. The VLBA has 27 large dish antennas that work together as a single instrument. By using recorders and precise atomic clocks installed at each antenna, the signals from all antennas are combined after the observation is completed.

Geo Words

x-ray telescope: an instrument used to detect stellar and interstellar x-ray emission. Because the Earth's atmosphere absorbs x-rays, x-ray telescopes are placed high above the Earth's surface.

The galaxy M81 is a spiral galaxy about 11 million light-years from Earth and is about 50,000 light-years across. The spiral structure is clearly shown in *Figure 2*, which shows the relative intensity of emission from neutral atomic hydrogen gas. In this pseudocolor image, red indicates strong radio emission and blue weaker emission.

Figure 2 The galaxy M81.

The orbiting Chandra **x-ray telescope** routinely detects the highly energetic radiation streaming from objects like supernova explosions, active galaxies, and black holes. The Hubble Space Telescope is outfitted with a special infrared instrument sensitive to radiation being produced by star-forming nebulae and cool stars. It also has detectors sensitive to ultraviolet light being emitted by hot young stars and supernova explosions.

A wide array of solar telescopes both on Earth and in space study every wavelength of radiation from our nearest star in minute detail. The tools of astronomy expand scientists' vision into realms that human eyes can never see, to help them understand the ongoing processes and evolution of the universe.

Figure 3 Astronauts working on the Hubble Space Telescope high above the Earth's atmosphere.

Geo Words

peak wavelength: the wavelength of electromagnetic radiation with the most electromagnetic energy emitted by any object.

Using Electromagnetic Radiation to Understand Celestial Objects

The wavelength of light with the most energy produced by any object, including the Sun, is called its **peak wavelength**. Objects that are hot and are radiating visible light usually look the color of their peak wavelength. People are not hot enough to emit visible light, but they do emit infrared radiation that can be detected with infrared cameras. The Sun has its peak wavelength in the yellow region of the visible spectrum. Hotter objects produce their peaks toward the blue direction. Very hot objects can have their peaks in the ultraviolet, x-ray, or even gamma-ray range of wavelength. A gas under high pressure radiates as well as a hot solid object. Star colors thus reflect temperature. Reddish stars are a "cool" 3000 to 4000 K (kelvins are celsius degrees above absolute zero, which is at minus 273°C). Bluish stars are hot (over 20,000 K).

One of the most important tools in astronomy is the spectrum—a chart of the entire range of wavelengths of light from an object. Astronomers often refer to this chart as the spectrum of the star. These spectra come in two forms: one resembles a bar code with bright and dark lines (see *Figure 4*), and the other is a graph with horizontal and vertical axes (see *Figure 5*). Think of these spectra as "fingerprints" that reveal many kinds of things about an object: its chemical composition, its temperature and pressure, and its motion toward or away from us.

Figure 4 One of the forms of the spectrum of a star. The data encoded here tell astronomers that this star is bright in some elements and dim in others.

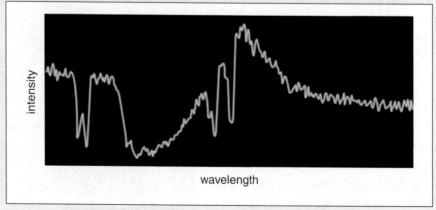

Figure 5 The spectrum of a star can also be represented as a graph with horizontal and vertical axes.

Each chemical element in the universe has its own unique spectrum. If you know what the spectrum of hydrogen is, you can look for its fingerprint in a star. If you suspect that a star may have a lot of the elements helium or calcium, for example, you can compare the spectrum of the star with the known spectra of helium or calcium. If you see bright lines in the stellar spectrum that match the patterns of bright lines in the helium or calcium spectra, then you have identified those elements in the star. This kind of spectrum is known as an **emission spectrum**. If you look at the star and see dark lines where you would expect to see an element—especially hydrogen—it is likely that something between you and the star is absorbing the element. This kind of spectrum is known as an **absorption spectrum**.

The positions of lines in a spectrum reveal the motion of the star toward or away from Earth, as well as the speed of that motion. You have experienced the effect yourself, when a car, a truck, or a train passes by you with its horn blowing. The pitch of the sound increases as the object approaches you, and decreases as the object passes by and moves away from you. That is because when the object is coming toward you, its speed adds to the speed of the sound, making the wavelength of the sound seem higher to your ear. The reverse happens when the object is moving away from you. The same principle applies to the spectrum from a distant object in space, which might be moving either toward Earth or away from Earth.

Geo Words

emission spectrum: a spectrum containing bright lines or a set of discrete wavelengths produced by an element. Each element has its own unique emission spectrum.

absorption spectrum: a continuous spectrum interrupted by absorption lines or a continuous spectrum having a number of discrete wavelengths missing or reduced in intensity.

Check Your Understanding

1. What are the colors of the spectrum of visible sunlight, from longest wavelength to shortest? Are there breaks between these colors, or do they grade continuously from one to the next? Why?

2. Which wavelengths of light can be more harmful to you than others? Why?

3. What tools do astronomers use to detect different wavelengths of light?

4. How can the speed of a distant object in space be measured?

Understanding and Applying What You Have Learned

1. Imagine that you are on a distant planet. Name two parts of the electromagnetic spectrum that you would use to investigate Earth. Explain the reasoning for your choices.

2. Refer to *Figure 1* to answer the questions below.

    a) Describe the relationship between wavelength and energy in the electromagnetic spectrum.

    b) Based upon this relationship, why do astronomers use x-ray telescopes to study supernova explosions and black holes?

3. The Sun looks yellow, can warm the surface of your skin, and can also give you a bad sunburn. Explain these three everyday phenomena in terms of the electromagnetic spectrum and peak wavelength.

Preparing for the Chapter Challenge

Recall that your challenge is to educate people about the hazards from outer space and to explain some of the benefits from living in our solar system. Electromagnetic radiation has both beneficial and harmful effects on life on Earth. Use what you have learned in this activity to develop your brochure.

1. Make a list of some of the positive effects of electromagnetic radiation on your community. Explain each item on the list.

2. Make a list of some of the negative effects of electromagnetic radiation on your community. Explain each item on the list.

3. Make a list of celestial radiation sources and any effects they have on Earth systems. What are the chances of a stellar radiation source affecting Earth?

Inquiring Further

1. **Using radio waves to study distant objects**

   Radio waves from the Sun penetrate the Earth's atmosphere. Scientists detect these waves and study their strength and frequency to understand the processes inside the Sun that generate them. Do some research on how these waves are studied.

2. **Detecting electromagnetic radiation**

   Investigate some of the instruments that astronomers use to detect electromagnetic radiation besides light. Where are you likely to find ultraviolet detectors? Describe radio telescope arrays.

3. **Technologies and the electromagnetic spectrum**

   Research some of the technologies that depend on the use of electromagnetic radiation. These might include microwave ovens, x-ray machines, televisions, and radios. How do they work? How is electromagnetic radiation essential to their operation? What interferes with their operation?

Activity 7

Our Community's Place Among the Stars

Goals

In this activity you will:

- Understand the place of our solar system in the Milky Way galaxy.

- Study stellar structure and the stellar evolution (the life histories of stars).

- Understand the relationship between the brightness of an object (its luminosity) and its magnitude.

- Estimate the chances of another star affecting the Earth in some way.

Think about It

When you look at the nighttime sky, you are looking across vast distances of space.

- **As you stargaze, what do you notice about the stars?**

- **Do some stars appear brighter than others? Larger or smaller? What about their colors?**

What do you think? Record your impressions and sketch some of the stars in your *EarthComm* notebook. Be prepared to discuss your thoughts with your small group and the class.

Investigate

Part A: Brightness versus Distance from the Source

1. Set a series of lamps with 40-, 60-, and 100-W bulbs (of the same size and all with frosted glass envelopes) up at one end of a room (at least 10 m away). Use the other end of the room for your observing site. Turn all the lamps on. Close all of the shades in the room.

   a) Can you tell the differences in brightness between the lamps?

2. Move the lamp with the 40-W bulb forward 5 m toward you.

   a) Does the light look brighter than the 60-W lamp?

   b) Does it look brighter than the 100-W lamp?

3. Shift the positions of the lamps so that the 40-W lamp and the 100-W lamp are in the back of the room and the 60-W lamp is halfway between you and the other lamps.

   a) How do the brightnesses compare?

4. Using a light meter, test one bulb at a time. If you do not have a light meter, you will have to construct a qualitative scale for brightness.

   a) Record the brightness of each bulb at different distances.

5. Graph the brightness versus the distance from the source for each bulb (wattage).

   a) Plot distance on the horizontal axis of the graph and brightness on the vertical axis. Leave room on the graph so that you can extrapolate the graph beyond the data you have collected. Plot the data for each bulb and connect the points with lines.

   b) Extrapolate the data by extending the lines on the graph using dashes.

6. Use your graph to answer the following questions:

   a) Explain the general relationship between wattage and brightness (as measured by your light meter).

   b) What is the general relationship between distance and brightness?

   c) Do all bulbs follow the same pattern? Why or why not?

   d) Draw a light horizontal line across your graph so that it crosses several of the lines you have graphed.

   e) Does a low-wattage bulb ever have the same brightness as a high-wattage bulb? Describe one or two such cases in your data.

   f) The easiest way to determine the absolute brightness of objects of different brightness and distance is to move all objects to the same distance. How do you think astronomers handle this problem when trying to determine the brightness and distances to stars?

7. When you have completed this activity, spend some time outside stargazing. Think about the relationship between brightness and distance as it applies to stars.

   a) Write your thoughts down in your *EarthComm* notebook.

   ⚠️ Do not stare at the light bulbs for extended periods of time.

Part B: Luminosity and Temperature of Stars

1. An important synthesis of understanding in the study of stars is the Hertzsprung-Russell (HR) diagram. Obtain a copy of the figure below. Examine the figure and answer the following questions:

   a) What does the vertical axis represent?

   b) What does the horizontal axis represent?

   c) The yellow dot on the figure is the Sun. What is its temperature and luminosity?

   d) Put four more dots on the diagram labeled A through D to show the locations of stars that are:

      A. hot and bright
      B. hot and dim
      C. cool and dim
      D. cool and bright

2. Obtain a copy of the *Table 1* and the HR diagram that shows the locations of main sequence stars, supergiants, red giants, and white dwarfs.

   a) Using the luminosity of the stars, and their surface temperatures, plot the locations of stars shown in *Table 1* on a second HR diagram.

3. Classify each of the stars into one of the following four categories, and record the name in your copy of the table:

   — Main sequence
   — Red giants
   — Supergiants
   — White dwarfs

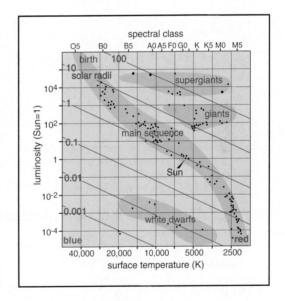

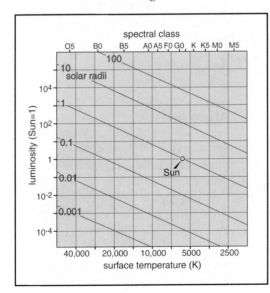

Table 1 Selected Properties of Fourteen Stars							
Star	**Surface Temperature (K)**	**Luminosity (Relative to Sun)**	**Distance (Light-Years)**	**Mass (Solar Masses)**	**Diameter (Solar Diameters)**	**Color**	**Type of Star**
Sirius A	9100	22.6	8.6	2.3	2.03	Blue	
Arcturus	4300	115	36.7	4.5	31.5	Red	
Vega	10300	50.8	25.3	3.07	3.1	Blue	
Capella	5300	75.8	42.2	3	10.8	Red	
Rigel	11000	38,679	733	20	62	Blue	
Procyon A	6500	7.5	11.4	1.78	1.4	Yellow	
Betelgeuse	2300	4520–14,968 (variable)	427	20	662	Red	
Altair	7800	11.3	65.1	2	1.6	Yellow	
Aldebaran	4300	156–171 (variable)	65	25	51.5	Red	
Spica	25300	2121	262	10.9	7.3	Blue	
Pollux	4500	31	33	4	8	Red	
Deneb	10500	66,500	1600	25	116	Yellow	
Procyon B	8700	0.0006	11.2	0.65	0.02	White	
Sirius B	24000	0.00255	13.2	0.98	0.008	Blue-white	

Note: Mass, diameter, and luminosity are given in solar units. For example, Sirius A has 2.3 solar masses, has a diameter 2.03 that of the Sun, and has luminosity 22.6 times brighter than the Sun.

1 solar mass = 2×10^{30} kg = 330,000 Earth masses; 1 solar diameter = 700,000 km = 110 Earth diameters.

Reflecting on the Activity and the Challenge

Measuring the apparent differences in brightness of the light bulbs at different distances helps you to see that distance and brightness are important factors in helping you understand the objects in our universe. When you look at the stars at night, you are seeing stars at different distances and brightnesses. In your **Chapter Challenge** you will be telling people about the effects of distant objects on the Earth. When you assess danger from space, it is important to understand that stars in and of themselves don't pose a danger unless they are both relatively nearby and doing something that could affect Earth. The spectral characteristics of stars help you to understand their temperature, size, and other characteristics. In turn, that helps you to understand if a given star is or could be a threat to Earth. The light from distant stars can also be used to understand our own star, and our own solar system's makeup and evolution.

Digging Deeper

EARTH'S STELLAR NEIGHBORS

Classifying Stars

You already know that our solar system is part of the Milky Way galaxy. Our stellar neighborhood is about two-thirds of the way out on a spiral arm that stretches from the core of the galaxy. The galaxy contains hundreds of billions of stars. Astronomers use a magnitude scale to describe the brightness of objects they see in the sky. A star's brightness decreases with the square of the distance. Thus, a star twice as far from the Earth as an identical star would be one-fourth as bright as the closer star. The first magnitude scales were quite simple—the brightest stars were described as first magnitude, the next brightest stars were second magnitude, and so on down to magnitude 6, which described stars barely visible to the naked eye. The smaller the number, the brighter the star; the larger the number, the dimmer the star.

Today, scientists use a more precise system of magnitudes to describe brightness. The brightest star in the sky is called Sirius A, and its magnitude is −1.4. Of course, the Sun is brighter at −27 and the Moon is −12.6! The dimmest naked-eye stars are still sixth magnitude. To see anything dimmer than that, you have to magnify your view with binoculars or telescopes. The best ground-based telescopes can detect objects as faint as 25th magnitude. To get a better view of very faint, very distant objects, you have to get above the Earth's atmosphere. The Hubble Space Telescope, for example can detect things as dim as 30th magnitude!

Figure 1 This NASA Hubble Space Telescope near-infrared image of newborn binary stars reveals a long thin nebula pointing toward a faint companion object which could be the first extrasolar planet to be imaged directly.

Perhaps you have seen a star described as a G-type star or an O-type star. These are stellar classifications that depend on the color and temperature of the stars. They also help astronomers understand where a given star is in its evolutionary history. To get such information, astronomers study stars with spectrographs to determine their temperature and chemical makeup. As you can see in the table below, there are seven main categories of stars:

Stellar Classification	Temperature (kelvins)
O	25,000 K and higher
B	11,000–25,000 K
A	7500–11,000 K
F	6000–7500 K
G	5000–6000 K
K	3500–5000 K
M	less than 3500 K

The Lives of Stars

Astronomers use the term **luminosity** for the total rate at which a star emits radiation energy. Unlike apparent brightness (how bright the star appears to be) luminosity is an intrinsic property. It doesn't depend on how far away the star is. In the early 1900s Ejnar Hertzsprung and Henry Norris Russell independently made the discovery that the luminosity of a star was related to its surface temperature. In the second part of this activity, you worked with a graph that shows this relationship. It is called the Hertzsprung-Russell (HR) diagram in honor of the astronomers who discovered this relationship. The HR diagram alone does not tell you how stars change. By analogy, if you were to plot the IQ versus the weight of everyone in your school, you would probably find a very poor relationship between these two variables. Your graph would resemble a scatter plot more than it would a line. However, if you plotted the height versus weight for the same people, you are more likely to find a strong relationship (data would be distributed along a trend or line). The graph doesn't tell you why this relationship exists — that's up to you to determine. Similarly, the HR diagram shows that stars don't just appear randomly on a plot of luminosity versus temperature, but fall into classes of luminosity (red giants, white dwarfs, and so on).

The life cycle of a star begins with its formation in a cloud of gas and dust called a **molecular cloud**. The material in the cloud begins to clump

Geo Words

luminosity: the total amount of energy radiated by an object every second.

molecular cloud: a large, cold cloud made up mostly of molecular hydrogen and helium, but with some other gases, too, like carbon monoxide. It is in these clouds that new stars are born.

together, mixing and swirling. Eventually the core begins to heat as more material is drawn in by gravitational attraction. When the temperature in the center of the cloud reaches 15 million kelvins, the stellar fusion reaction starts up and a star is born. Such stars are called main-sequence stars. Many stars spend 90% of their lifetimes on the main sequence.

Newborn stars are like baby chickens pecking their way out of a shell. As these infant stars grow, they bathe the cloud surrounding them in strong ultraviolet radiation. This vaporizes the cloud,

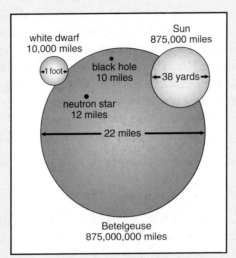

Figure 2 Scaling stars to 10,000 miles to one foot reveals the relative sizes of various stars.

creating beautiful sculpted shapes in the cloud. In the photograph in *Figure 3*, the Hubble Space Telescope studied a region of starbirth called NGC 604. Notice the cluster of bright white stars in the center "cavern" of the cloud of gas and dust. Their ultraviolet light has carved out a shell of gas and dust around the stellar newborns.

Figure 3 The starforming region NGC 604 in the galaxy M33.

Figure 4 The Orion Nebula is an example of a molecular cloud, from which new stars are born.

How long a star lives depends on its mass (masses of selected stars are shown in *Table 1* in the **Investigate** section of the activity). Stars like our Sun will live about 10 billion years. Smaller, cooler stars might go on twice that long, slowly burning their fuel. Massive supergiant stars consume their mass much more quickly, living a star's life only a few tens of millions of years. Very hot stars also go through their fuel very quickly, existing perhaps only a few hundred thousand years. The time a star spends on the main sequence can be determined using the following formula:

Time on main sequence = $\dfrac{1}{M^{2.5}}$ × 10 billion years

where *M* is the mass of the star in units of solar masses.

Even though high-mass stars have more mass, they burn it much more quickly and end up having very short lives.

In the end, however, stars of all types must die. Throughout its life a star loses mass in the form of a stellar wind. In the case of the Sun this is called the solar wind. As a star ages, it loses more and more mass. Stars about the size of the Sun and smaller end their days as tiny, shrunken remnants of their former selves, surrounded by beautiful shells of gas and dust. These are called planetary nebulae. In about five billion years the Sun will start to resemble one of these ghostly nebulae, ending its days surrounded by the shell of its former self.

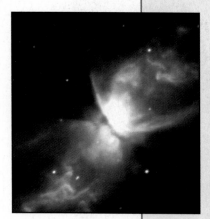
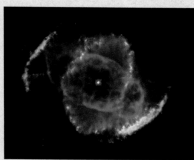
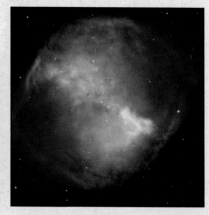

Figures 5A, 5B, 5C Three examples of the deaths of stars about the size of the Sun. **A:** The Butterfly Nebula. **B:** The Cat's-Eye Nebula. In both cases at least the dying star lies embedded in a cloud of material exhaled by the star, as it grew older. **C:** The Dumbbell Nebula. European Southern Observatory.

Massive stars (supergiants tens of times more massive than the Sun) also lose mass as they age, but at some point their cores collapse catastrophically. The end of a supergiant's life is a cataclysmic explosion called a **supernova**. In an instant of time, most of the star's mass is hurled out into space, leaving behind a tiny remnant called a **neutron star**. If the star is massive enough, the force of the explosion can be so strong that the remnant is imploded into a **stellar black hole**—a place where the gravity is so strong that not even light can escape.

The material that is shed from dying stars (whether they end their days as slowly fading dwarf stars, or planetary nebulae, or supernovae) makes its way into the space between the stars. There it mixes and waits for a slow gravitational contraction down to a new episode of starbirth and ultimately star death. Because humans evolved on a planet that was born from a recycled cloud of stellar mass, they are very much star "stuff"—part of a long cycle of life, death, and rebirth.

Astronomers search the universe to study the mechanics of star formation. Star nurseries and star graveyards are scattered through all the galaxies. In some cases, starbirth is triggered when one galaxy collides with (actually passes through) another. The clouds of gas and dust get the push they need to start the process.

Scientists also search for examples of planetary nebulae. They want to understand when and how these events occur. Not only are these nebulae interesting, but also they show scientists what the fate of our solar system will be billions of years from now.

What would happen if there were a supernova explosion in our stellar neighborhood sometime in the future? Depending on how close it was, you could be bombarded with strong radiation and shock waves from the explosion. The chances of this happening are extremely small—although some astronomers think that a supernova some five billion years ago may have provided the gravitational kick that started our own proto–solar nebula on the road to stardom and planetary formation.

Figure 6 The Crab Nebula is the remnant of a supernova explosion first observed in the year AD 1054.

Geo Words

supernova: the death explosion of a massive star whose core has completely burned out. Supernova explosions can temporarily outshine a galaxy.

neutron star: the imploded core of a massive star produced by a supernova explosion.

stellar black hole: the leftover core of a massive single star after a supernova. Black holes exert such large gravitational pull that not even light can escape.

Check Your Understanding

1. How do astronomers classify stars?
2. Write a brief outline of how stars are born.
3. What determines the way a star dies?

Understanding and Applying What You Have Learned

1. Using an astronomy computer program or a guidebook to the stars, make a list of the 10 nearest stars, and their distances, magnitudes, and spectral classes. What do their classes tell you about them?

2. What is mass loss and how does it figure in the death of a star? Is the Sun undergoing mass loss?

3. What happens to the material left over from the death of a star?

4. Two identical stars have different apparent brightnesses. One star is 10 light-years away, and the other is 30 light-years away from us. Which star is brighter, and by how much?

5. Refer to *Table 1* to answer the questions below:

   a) Calculate how long the Sun will spend on the main sequence.
   b) Calculate how long Spica will spend on the main sequence.
   c) Relate your results to the statement that the more massive the star, the shorter they live.

6. Explain the relationships between temperature, luminosity, mass, and lifetime of stars.

Preparing for the Chapter Challenge

You are about to complete your **Chapter Challenge**. In the beginning you were directed to learn as much as you could about how extraterrestrial objects and events could affect the Earth and your community. In order to do this you have explored the stars and planets, looking at all the possibilities. By now you have a good idea about how frequently certain kinds of events occur that affect Earth. The Sun is a constant source of energy and radiation.

In this final activity you learned our solar system's place in the galaxy, and you read about how stars are born and die. Because the birth of our solar system led directly to our planet, and the evolution of life here, it's important to know something about stars and how they come into existence.

You now know that the solar system is populated with comets and asteroids, some of which pose a threat to Earth over long periods of time.

The evolution of the Earth's orbit and its gravitational relationship with the Moon make changes to the Earth's climate, length of year, and length of day. The solar system is part of a galaxy of other stars, with the nearest star being only 4.21 light-years away. The Sun itself is going through a ten-billion-year-long period of evolution and will end as a planetary nebula some five billion years in the future. Finally, our Milky Way galaxy is wheeling toward a meeting with another galaxy in the very, very distant future. Your challenge now that you know and understand these things is to explain them to your fellow citizens and help them understand the risks and benefits of life on this planet, in this solar system, and in this galaxy.

Inquiring Further

1. **Evolution of the Milky Way galaxy**

   The Milky Way galaxy formed some 10 billion years ago, when the universe itself was only a fraction of its current age. Research the formation of our galaxy and find out how its ongoing evolution influenced the formation of our solar system.

2. **Starburst knots in other galaxies**

   Other galaxies show signs of star birth and star death. You read about a starbirth region called NGC 604 in the **Digging Deeper** reading section of this activity. Astronomers have found evidence of colliding galaxies elsewhere in the universe. In nearly every case, such collisions have spurred the formation of new stars. In the very distant future the Milky Way will collide with another galaxy, and it's likely that starburst knots will be formed. Look for examples of starbirth nurseries and starburst knots in other galaxies and write a short report on your findings. How do you think such a collision would affect Earth (assuming that anyone is around to experience it)?

Earth Science at Work

ATMOSPHERE: *Astronaut*
There is no atmosphere in space;
therefore, astronauts must have
pressurized atmosphere in their
spacecraft cabins. Protective suits
protect them when they perform
extra-vehicular activities.

BIOSPHERE: *Exobiologist*
"Did life ever get started on
Mars?" By learning more about the
ancient biosphere and
environments of the early Earth,
exobiologists hope that they may
be able to answer such questions
when space missions return with
rocks gathered on Mars.

CRYOSPHERE: *Glaciologist*
Ice is abundant on the Earth's
surface, in the planetary system,
and in interstellar space.
Glaciologists study processes at or
near the base of glaciers and ice
sheets on Earth and other planets.

GEOSPHERE: *Planetary geologist*
By researching Martian volcanism and
tectonism, or the geology of the icy
satellites of Jupiter, Saturn, and Uranus,
planetary geologists hope to develop a
better understanding of our place in the
universe.

HYDROSPHERE: *Lifeguard*
Surfers and other water-sport
enthusiasts rely on lifeguards to inform
them of the time of high and low tides.
Low tides or high tides can create
dangerous situations.

How is each person's work related to the Earth system, and to Astronomy?

Chapter 12

CLIMATE CHANGE
...and Your Community

Getting Started

The Earth's climate has changed many times over geologic history.

- What kinds of processes or events might cause the Earth's climate to change?

What do you think? Write down your ideas about these questions in your *EarthComm* notebook. Be prepared to discuss your ideas with your small group and the class.

Scenario

Your local newspaper would like to run a series of articles about global warming. However, the newspaper's science reporter is unavailable. The newspaper has come to your class to ask you and your classmates to write the articles. These feature articles and an editorial will be run in the Science and Environment section of the newspaper. The newspaper editor wants to give the readers of the paper a thorough scientific background to understand the idea of global climate change.

Chapter Challenge

Article 1: How Has Global Climate Changed Over Time?

Many people are not aware that the Earth's climate has changed continually over geologic time. This article should contain information about:

- the meaning of "climate," both regional and global
- examples of different global climates in the geologic past
- how geologists find out about past climates, and
- a description of your community's present climate and examples of past climates in your part of the country.

Article 2:
Causes of Global Climate Change

Some people might not be aware that human production of greenhouse gases is not the only thing that can cause the Earth's climate to change. There are many different factors that may affect how and when the Earth's climate changes. This article should include information about:

- Milankovitch cycles
- plate tectonics
- ocean currents, and
- carbon dioxide levels.

Article 3:
What is "Global Warming" and How Might It Affect Our Community?

Although almost everyone has heard the terms "greenhouse gases" and "global warming," there is a lot of confusion about what these terms actually mean. This article should contain information about:

- greenhouse gases
- how humans have increased the levels of carbon dioxide in the atmosphere
- why scientists think increased carbon dioxide might lead to global warming
- possible effects of global warming, focusing on those that would have the greatest impact on your community, and
- why it is difficult to predict climate change.

Editorial

The final piece is not an article but rather an editorial in which the newspaper expresses its opinion about a particular topic. In the editorial, you should state:

- whether your community should be concerned about global warming and why, and
- what steps, if any, your community should take in response to the possibility of global warming.

Assessment Criteria

Think about what you have been asked to do. Scan ahead through the chapter activities to see how they might help you to meet the challenge. Work with your classmates and your teachers to define the criteria for assessing your work. Record all of this information. Make sure that you understand the criteria as well as you can before you begin. Your teacher may provide you with a sample rubric to help you get started.

Activity 1

Present-Day Climate in Your Community

Goals

In this activity you will:

- Identify factors of the physical environment.

- Use a topographic map to gather data about elevation and latitude, and physical features.

- Interpret data from a climate data table.

- Compare and contrast climate information from two different parts of the United States.

- Understand how physical features can influence the climate of an area.

Think about It

A friend e-mails you from Italy to ask what your environment and climate are like. You plan to e-mail her a reply.

- **How would you describe the physical environment of your community?**

- **How would you describe the climate of your community?**

What do you think? Record your ideas about these questions in your *EarthComm* notebook. Be prepared to discuss your responses with your small group and the class.

Investigate

Part A: Physical Features and Climate in Your Community

1. Depending on where your community is located and how large it is, you might wish to expand your definition of "community" to include a larger area, like your county or state. For example, your town does not have to be right on the ocean to have its climate influenced by the ocean.

a) Write a "definition" of the area that you will examine as your "community."

2. Use the climate data tables provided on the following pages and topographic maps of your "community" to describe the climate in your community.

Climatic Zones of North America (Mercator Projection)

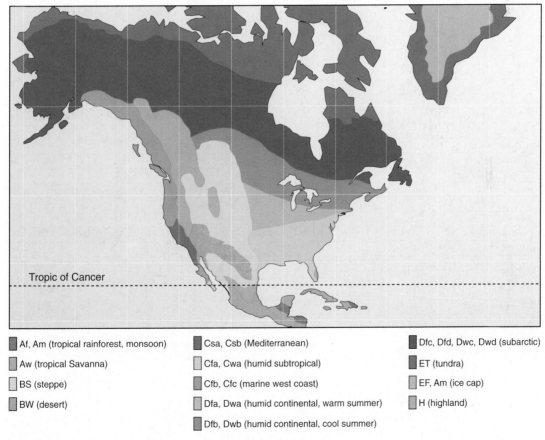

Tropic of Cancer

Af, Am (tropical rainforest, monsoon)	Csa, Csb (Mediterranean)	Dfc, Dfd, Dwc, Dwd (subarctic)
Aw (tropical Savanna)	Cfa, Cwa (humid subtropical)	ET (tundra)
BS (steppe)	Cfb, Cfc (marine west coast)	EF, Am (ice cap)
BW (desert)	Dfa, Dwa (humid continental, warm summer)	H (highland)
	Dfb, Dwb (humid continental, cool summer)	

Abbreviations for climate zones are based upon the Koppen classification system.

Temperatures in the United States															
Country and Station United States (Conterminous):	Latitude	Longitude	Elevation	Record Length	Temperature									Extreme	
					Average Daily										
					January		April		July		October				
					Max.	Min.	Max.	Min.	Max.	Min.	Max.	Min.	Max.	Min.
	′ ″	′ ″	Feet	Yrs.	°F	°F	°F	°F	°F	°F	°F	°F	°F	°F
Albuquerque, NM	35 03N	106 37W	5311	30	46	24	69	42	91	66	71	45	105	−17
Asheville, NC	35 26N	82 32W	2140	30	48	28	67	42	84	61	68	45	100	−16
Atlanta, GA	33 39N	84 26W	1010	30	52	37	70	50	87	71	72	52	105	− 8
Austin, TX	30 18N	97 42W	597	30	60	41	78	57	95	74	82	60	109	− 2
Birmingham, AL	33 34N	86 45W	620	30	57	36	76	50	93	71	79	52	106	−10
Bismarck, ND	46 46N	100 45W	1647	30	20	0	55	32	86	58	59	34	114	−45
Boise, ID	43 34N	116 13W	2838	30	36	22	63	37	91	59	65	38	112	−28
Brownsville, TX	25 54N	97 26W	16	30	71	52	82	66	93	76	85	67	106	12
Buffalo, NY	42 56N	78 44W	705	30	31	18	53	34	80	59	60	41	99	−21
Cheyenne, WY	41 09N	104 49W	6126	30	37	14	56	30	85	55	63	32	100	−38
Chicago, IL	41 47N	87 45W	607	30	33	19	57	41	84	67	63	47	105	−27
Des Moines, IA	41 32N	93 39W	938	30	29	11	59	38	87	65	66	43	110	−30
Dodge City, KS	37 46N	99 58W	2582	30	42	20	66	41	93	68	71	46	109	−26
El Paso, TX	31 48N	106 24W	3918	30	56	30	78	49	95	69	79	50	112	− 8
Indianapolis, IN	39 44N	86 17W	792	30	37	21	61	40	86	64	67	44	107	−25
Jacksonville, FL	30 25N	81 39W	20	30	67	45	80	58	92	73	80	62	105	7
Kansas City, MO	39 07N	94 36W	742	30	40	23	66	46	92	71	72	49	113	−23
Las Vegas, NV	36 05N	115 10W	2162	30	54	32	78	51	104	76	80	53	117	8
Los Angeles, CA	33 56N	118 23W	97	30	64	45	67	52	76	62	73	57	110	23
Louisville, KY	38 11N	85 44W	477	30	44	27	66	43	89	67	70	46	107	−20
Miami, FL	25 48N	80 16W	7	30	76	58	83	66	89	75	85	71	100	28
Minneapolis, MN	44 53N	93 13W	834	30	22	2	56	33	84	61	61	37	108	−34
Missoula, MT	46 55N	114 05W	3190	30	28	10	57	31	85	49	58	30	105	−33
Nashville, TN	36 07N	86 41W	590	30	49	31	71	48	91	70	76	49	107	−17
New Orleans, LA	29 59N	90 15W	3	30	64	45	78	58	91	73	80	61	102	7
New York, NY	40 47N	73 58W	132	30	40	27	60	43	85	68	66	50	106	−15
Oklahoma City, OK	35 24N	97 36W	1285	30	46	28	71	49	93	72	74	52	113	−17
Phoenix, AZ	33 26N	112 01W	1117	30	64	35	84	50	105	75	87	55	122	16
Pittsburgh, PA	40 27N	80 00W	747	30	40	25	63	42	85	65	65	45	103	−20
Portland, ME	43 39N	70 19W	47	30	32	12	53	32	80	57	60	37	103	−39
Portland, OR	45 36N	122 36W	21	30	44	33	62	42	79	56	63	45	107	− 3
Reno, NV	39 30N	119 47W	4404	30	45	16	65	31	89	46	69	29	106	−19
Salt Lake City, UT	40 46N	111 58W	4220	30	37	18	63	36	94	60	65	38	107	−30
San Francisco, CA	37 37N	122 23W	8	30	55	42	64	47	72	54	71	51	109	20
Sault Ste. Marie, MI	46 28N	84 22W	721	30	23	8	46	30	76	54	55	38	98	−37
Seattle, WA	47 27N	122 18W	400	30	44	33	58	40	76	54	60	44	100	0
Sheridan, WY	44 46N	106 58W	3964	30	34	9	56	31	87	56	62	33	106	−41
Spokane, WA	47 38N	117 32W	2356	30	31	19	59	36	86	55	60	38	108	−30
Washington, D.C.	38 51N	77 03W	14	30	44	30	66	46	87	69	68	50	104	−18
Wilmington, NC	34 16N	77 55W	28	30	58	37	74	51	89	71	76	55	104	0

Source: NOAA

Country and Station United States (Conterminous):	Record Length	Average Precipitation												
	Years	Jan. In.	Feb. In.	Mar. In.	Apr. In.	May In.	Jun. In.	Jul. In.	Aug. In.	Sep. In.	Oct. In.	Nov. In.	Dec. In.	Year In.
Albuquerque, NM	30	0.4	0.4	0.5	0.5	0.8	0.6	1.2	1.3	1.0	0.8	0.4	0.5	8.4
Asheville, NC	30	4.2	4.0	4.8	4.0	3.7	3.5	5.9	4.9	3.6	3.1	2.8	3.6	48.1
Atlanta, GA	30	4.4	4.5	5.4	4.5	3.2	3.8	4.7	3.6	3.3	2.4	3.0	4.4	47.2
Austin, TX	30	2.4	2.6	2.1	3.6	3.7	3.2	2.2	1.9	3.4	2.8	2.1	2.5	32.5
Birmingham, AL	30	5.0	5.3	6.0	4.5	3.4	4.0	5.2	4.9	3.3	3.0	3.5	5.0	53.1
Bismarck, ND	30	0.4	0.4	0.8	1.2	2.0	3.4	2.2	1.7	1.2	0.9	0.6	0.4	15.2
Boise, ID	30	1.3	1.3	1.3	1.2	1.3	0.9	0.2	0.2	0.4	0.8	1.2	1.3	11.4
Brownsville, TX	30	1.4	1.5	1.0	1.6	2.4	3.0	1.7	2.8	5.0	3.5	1.3	1.7	26.9
Buffalo, NY	30	2.8	2.7	3.2	3.0	3.0	2.5	2.6	3.1	3.1	3.0	3.6	3.0	35.6
Cheyenne, WY	30	0.5	0.6	1.2	1.9	2.5	2.1	1.8	1.4	1.1	0.8	0.6	0.5	15.0
Chicago, IL	30	1.9	1.6	2.7	3.0	3.7	4.1	3.4	3.2	2.7	2.8	2.2	1.9	33.2
Des Moines, IA	30	1.3	1.1	2.1	2.5	4.1	4.7	3.1	3.7	2.9	2.1	1.8	1.1	30.5
Dodge City, KS	30	0.6	0.7	1.2	1.8	3.2	3.0	2.3	2.4	1.5	1.4	0.6	0.5	19.2
El Paso, TX	30	0.5	0.4	0.4	0.3	0.4	0.7	1.3	1.2	1.1	0.9	0.3	0.5	8.0
Indianapolis, IN	30	3.1	2.3	3.4	3.7	4.0	4.6	3.5	3.0	3.2	2.6	3.1	2.7	39.2
Jacksonville, FL	30	2.5	2.9	3.5	3.6	3.5	6.3	7.7	6.9	7.6	5.2	1.7	2.2	53.6
Kansas City, MO	30	1.4	1.2	2.5	3.6	4.4	4.6	3.2	3.8	3.3	2.9	1.8	1.5	34.2
Las Vegas, NV	30	0.5	0.4	0.4	0.2	0.1	*	0.5	0.5	0.3	0.2	0.3	0.4	3.8
Los Angeles, CA	30	2.7	2.9	1.8	1.1	0.1	0.1	*	*	0.2	0.4	1.1	2.4	12.8
Louisville, KY	30	4.1	3.3	4.6	3.8	3.9	4.0	3.4	3.0	2.6	2.3	3.2	3.2	41.4
Miami, FL	30	2.0	1.9	2.3	3.9	6.4	7.4	6.8	7.0	9.5	8.2	2.8	1.7	59.9
Minneapolis, MN	30	0.7	0.8	1.5	1.9	3.2	4.0	3.3	3.2	2.4	1.6	1.4	0.9	24.9
Missoula, MT	30	0.9	0.9	0.7	1.0	1.9	1.9	0.9	0.7	1.0	1.0	0.9	1.1	12.9
Nashville, TN	30	5.5	4.5	5.2	3.7	3.7	3.3	3.7	2.9	2.9	2.3	3.3	4.2	45.2
New Orleans, LA	30	3.8	4.0	5.3	4.6	4.4	4.4	6.7	5.3	5.0	2.8	3.3	4.1	53.7
New York, NY	30	3.3	2.8	4.0	3.4	3.7	3.3	3.7	4.4	3.9	3.1	3.4	3.3	42.3
Oklahoma City, OK	30	1.3	1.4	2.0	3.1	5.2	4.5	2.4	2.5	3.0	2.5	1.6	1.4	30.9
Phoenix, AZ	30	0.7	0.9	0.7	0.3	0.1	0.1	0.8	1.1	0.7	0.5	0.5	0.9	7.3
Pittsburgh, PA	30	2.8	2.3	3.5	3.4	3.8	4.0	3.6	3.5	2.7	2.5	2.3	2.5	36.9
Portland, ME	30	4.4	3.8	4.3	3.7	3.4	3.2	2.9	2.4	3.5	3.2	4.2	3.9	42.9
Portland, OR	30	5.4	4.2	3.8	2.1	2.0	1.7	0.4	0.7	1.6	3.6	5.3	6.4	37.2
Reno, NV	30	1.2	1.0	0.7	0.5	0.5	0.4	0.3	0.2	0.2	0.5	0.6	1.1	7.2
Salt Lake City, UT	30	1.4	1.2	1.6	1.8	1.4	1.0	0.6	0.9	0.5	1.2	1.3	1.2	14.1
San Francisco, CA	30	4.0	3.5	2.7	1.3	0.5	0.1	*	*	0.2	0.7	1.6	4.1	18.7
Sault Ste. Marie, MI	30	2.1	1.5	1.8	2.2	2.8	3.3	2.5	2.9	3.8	2.8	3.3	2.3	31.3
Seattle, WA	30	5.7	4.2	3.8	2.4	1.7	1.6	0.8	1.0	2.1	4.0	5.4	6.3	39.0
Sheridan, WY	30	0.6	0.7	1.4	2.2	2.6	2.6	1.2	0.9	1.2	1.1	0.8	0.6	15.9
Spokane, WA	30	2.4	1.9	1.5	0.9	1.2	1.5	0.4	0.4	0.8	1.6	2.2	2.4	17.2
Washington, D.C.	30	3.0	2.5	3.2	3.2	4.1	3.2	4.2	4.9	3.8	3.1	2.8	2.8	40.8
Wilmington, NC	30	2.9	3.4	4.0	2.9	3.5	4.3	7.7	6.9	6.3	3.0	3.1	3.4	51.4

Average Precipitation in the United States

Source: NOAA

Average Snowfall in the United States

Country and Station United States (Conterminous): Data through 1998	Record Length Years	Average Snowfall (includes ice pellets)											
		Jan. In.	Feb. In.	Mar. In.	Apr. In.	May In.	Jun. In.	Jul. In.	Aug. In.	Sep. In.	Oct. In.	Nov. In.	Dec. In.	Year In.
Albuquerque, NM	59	2.5	2.1	1.8	0.6	0	T	T	T	T	0.1	1.2	2.6	10.9
Asheville, NC	34	5	4.3	2.8	0.6	T	T	T	T	0	T	0.7	2	15.4
Atlanta, GA	62	0.9	0.5	0.4	T	0	0	0	0	0	T	0	0.2	2
Austin, TX	57	0.5	0.3	T	T	T	0	0	0	0	0	0.1	T	0.9
Birmingham, AL	55	0.6	0.2	0.3	0.1	T	T	T	0	T	T	T	0.3	1.5
Bismarck, ND	59	7.6	7	8.6	4	0.9	T	T	T	0.2	1.8	7	7	44.1
Boise, ID	59	6.5	3.7	1.6	0.6	0.1	T	T	T	T	0.1	2.3	5.9	20.8
Brownsville, TX	59	T	T	T	0	0	0	0	T	0	0	T	T	T
Buffalo, NY	55	23.7	18	11.9	3.2	0.2	T	T	T	T	0.3	11.2	22.8	91.3
Cheyenne, WY	63	6.6	6.3	11.9	9.2	3.2	0.2	T	T	0.9	3.7	7.1	6.3	55.4
Chicago, IL	39	10.7	8.2	6.6	1.6	0.1	T	T	T	T	0.4	1.9	8.1	37.6
Des Moines, IA	57	8.3	7.2	6	1.8	0	T	T	0	T	0.3	3.1	6.7	33.4
Dodge City, KS	56	4.3	3.9	5	0.8	T	T	T	T	0	0.3	2.1	3.6	20
El Paso, TX	57	1.3	0.8	0.4	0.3	T	T	T	0	T	0	0.9	1.6	5.3
Indianapolis, IN	67	6.6	5.6	3.4	0.5	0	T	0	0	0	0.2	1.9	5.1	23.3
Jacksonville, FL	57	T	0	0	T	0	T	0	0	0	0	0	0	T
Kansas City, MO	64	5.7	4.4	3.4	0.8	T	T	T	0	T	0.1	1.2	4.4	20
Las Vegas, NV	48	0.9	0.1	0	T	0	0	0	T	0	T	0.1	0.1	1.2
Los Angeles, CA	62	T	T	T	0	0	0	0	0	0	0	0	T	T
Louisville, KY	51	5.4	4.6	3.3	0.1	T	T	T	0	0	0.1	1	2.1	16.6
Miami, FL	56	0	0	0	0	T	0	0	0	0	0	0	0	T
Minneapolis, MN	60	10.2	8.2	10.6	2.8	0.1	T	T	T	T	0.5	7.9	9.4	49.7
Missoula, MT	54	12.3	7.3	6	2.1	0.7	T	T	T	T	0.8	6.2	11.3	46.7
Nashville, TN	56	3.7	3	1.5	0	0	T	0	T	0	0	0.4	1.4	10
New Orleans, LA	50	0	0.1	T	T	T	0	0	0	0	0	T	0.1	0.2
New York, NY	130	7.5	8.6	5.1	0.9	T	0	T	0	0	0	0.9	5.4	28.4
Oklahoma City, OK	59	3.1	2.4	1.5	T	T	T	T	T	T	T	0.5	1.8	9.3
Phoenix, AZ	61	T	0	T	T	T	0	0	0	0	T	0	T	T
Pittsburgh, PA	46	11.7	9.2	8.7	1.7	0.1	T	T	T	T	0.4	3.5	8.2	43.5
Portland, ME	58	19.6	16.9	12.9	3	0.2	0	0	0	T	0.2	3.3	14.6	70.7
Portland, OR	55	3.2	1.1	0.4	T	0	T	0	T	T	0	0.4	1.4	6.5
Reno, NV	54	5.8	5.2	4.3	1.2	0.8	0	0	0	0	0.3	2.4	4.3	24.3
Salt Lake City, UT	70	13.8	10	9.4	4.9	0.6	T	T	T	0.1	1.3	6.8	11.7	58.6
San Francisco, CA	69	0	T	T	0	0	0	0	0	0	0	0	0	T
Sault Ste. Marie, MI	55	29	18.4	14.7	5.8	0.5	T	T	T	0.1	2.4	15.8	31.1	117.8
Seattle, WA	48	2.9	0.9	0.6	0	T	0	0	0	0	0	0.7	2.2	7.3
Sheridan, WY	58	11	10.4	12.6	9.9	2	0.1	T	0	1.3	4.6	9.2	10.9	72
Spokane, WA	51	15.6	7.5	3.9	0.6	0.1	T	0	0	T	0.4	6.3	14.6	49
Washington, D.C.	55	5.5	5.4	2.2	T	T	T	T	T	0	0	0.8	2.8	16.7
Wilmington, NC	47	0.4	0.5	0.4	T	T	T	T	0	0	0	T	0.6	1.9

Trace (T) is recorded for less than 0.05 inch of snowfall.

Last updated on 5/25/2000 by NRCC.

Source: The National Climatic Data Center/NOAA

Frost-Free Days in the United States

City/State	Last Frost Date	First Frost Date	No. of Frost-Free Days per Year
Albany, NY	May 7	September 29	144 days
Albuquerque, NM	April 16	October 29	196 days
Atlanta, GA	March 13	November 12	243 days
Baltimore, MD	March 26	November 13	231 days
Birmingham, AL	March 29	November 6	221 days
Boise, ID	May 8	October 9	153 days
Boston, MA	April 6	November 10	217 days
Charleston, SC	March 11	November 20	253 days
Charlotte, NC	March 21	November 15	239 days
Cheyenne, WY	May 20	September 27	130 days
Chicago, IL	April 14	November 2	201 days
Columbus, OH	April 26	October 17	173 days
Dallas, TX	March 18	November 12	239 days
Denver, CO	May 3	October 8	157 days
Des Moines, IA	April 19	October 17	180 days
Detroit, MI	April 24	October 22	181 days
Duluth, MN	May 21	September 21	122 days
Fargo, ND	May 13	September 27	137 days
Fayetteville, AR	April 21	October 17	179 days
Helena, MT	May 18	September 18	122 days
Houston, TX	February 4	December 10	309 days
Indianapolis, IN	April 22	October 20	180 days
Jackson, MS	March 17	November 9	236 days
Jacksonville, FL	February 14	December 14	303 days
Las Vegas, NV	March 7	November 21	259 days
Lincoln, NB	March 13	November 13	180 days
Los Angeles, CA	None likely	None likely	365 days
Louisville, KY	April 1	November 7	220 days
Memphis, TN	March 23	November 7	228 days
Miami, FL	None	None	365 days
Milwaukee, WI	May 5	October 9	156 days
New Haven, CT	April 15	October 27	195 days
New Orleans, LA	February 20	December 5	288 days
New York, NY	April 1	November 11	233 days
Phoenix, AZ	February 5	December 15	308 days
Pittsburgh, PA	April 16	November 3	201 days
Portland, ME	May 10	September 30	143 days
Portland, OR	April 3	November 7	217 days
Richmond, VA	April 10	October 26	198 days
Salt Lake City, UT	April 12	November 1	203 days
San Francisco, CA	January 8	January 5	362 days
Seattle, WA	March 24	November 11	232 days
St. Louis, MO	April 3	November 6	217 days
Topeka, KS	April 21	October 14	175 days
Tulsa, OK	March 30	November 4	218 days
Washington, D.C.	April 10	October 31	203 days
Wichita, KS	April 13	October 23	193 days

Include the following important climatic factors in your description of the climate in your community:

- Average daily temperatures in the winter and summer.
- Record high and low temperatures in the winter and summer.
- Average monthly precipitation in the winter and summer.
- Average winter snowfall.
- Growing season (number of days between last spring frost and first autumn frost).

3. Inspect a topographic map of your town.

   a) What is the latitude of your town?

   b) What is the elevation of your school?

   c) What is the highest elevation in your town?

   d) What is the lowest elevation in your town?

   e) Which of the following physical features can be found in or fairly near your community: mountains, rivers, valleys, coasts, lakes, hills, plains, or deserts? Specify where they are in relation to your community.

   f) Describe some of the ways that the physical features of your community might influence the climate.

Part B: Physical Features and Climate in a Different Community

1. Select a community that is in a part of the United States that is very different from where you live. For example, if you live in the mountains, pick a community on the plains. If you live near an ocean, pick a community far from a large body of water.

   a) Record the community and the reason you chose that community in your *EarthComm* notebook.

2. Describe the climate in this community.

   a) Include information for the same climatic factors that you used to describe your own community.

3. Inspect a topographic map of this community.

   a) Describe the same physical features that you did for your community.

4. Compare the physical features and climates of the two communities.

   a) In what ways might the physical features influence climate in the two places?

Part C: Heating and Cooling of Land versus Water

1. How do the rates at which rock and soil heat and cool compare to the rate at which water heats and cools? How might this affect climate in your community?

   a) Write down your ideas about these two questions.

   b) Develop a hypothesis about the rate at which rock or soil heat and cool compared to the rate at which water heats and cools.

2. Using materials provided by your teacher, design an experiment to investigate the rates of cooling and heating of soil or rock and water. Note the variable that you are

manipulating (the independent variable), the variable that you are measuring (the dependent variable), and the controls within your experiment.

a) Record your design, variables, controls, and any safety concerns in your notebook.

3. When your teacher has approved your design, conduct your experiment.

4. In your *EarthComm* notebook, record your answers to the following questions:

a) Which material heated up faster?

b) Which material cooled more quickly?

c) How did your results compare with your hypothesis?

d) How does this investigation relate to differences in climate between places near a body of water, versus places far from water?

⚠ Have your teacher approve your design before you begin your experiment. Do not touch any heat source. Report any broken thermometers to your teacher. Clean up any spills immediately.

Reflecting on the Activity and the Challenge

In this activity, you learned about the physical features and climate of your community. You also compared the physical features and climate of your community to that of another community in the United States. This helped you begin to see ways in which physical features influence climate. This will help you explain the meaning of the term "climate" and describe the climate of your community in your newspaper article.

Digging Deeper

WEATHER AND CLIMATE

Factors Affecting Climate

Weather refers to the state of the atmosphere at a place, from day to day and from week to week. The weather on a particular day might be cold or hot, clear or rainy. **Climate** refers to the typical or average weather at a place, on a long-term average. For example, Alaska has a cold climate, but southern Florida has a tropical climate. Minnesota's climate is hot in the summer and cold in the winter. Western Oregon has very rainy winters. Each of these regions has a definite climate, but weather that varies from day to day, often unpredictably.

The climate of a particular place on Earth is influenced by several important factors: latitude, elevation, and nearby geographic features. ➡

Geo Words

weather: the state of the atmosphere at a specific time and place.

climate: the general pattern of weather conditions for a region over a long period of time (at least 30 years).

Latitude

Latitude is a measure of the distance of a point on the Earth from the Equator. It is expressed in degrees, from zero degrees at the Equator to 90° at the poles. The amount of solar energy an area receives depends upon its latitude. At low latitudes, near the Equator, the Sun is always nearly overhead in the middle of the day, all year round. Near the poles, the Sun is low in the sky even in summer, and in the winter it is nighttime 24 hours of the day. As a result, regions near the Equator are much warmer than regions near the poles. Assuming a constant elevation, temperatures decrease by an average of about one degree Fahrenheit for every three degrees latitude away from the Equator.

Elevation

Elevation, the height of a point on the Earth's surface above sea level, also affects the physical environment. Places at high elevations are generally cooler than places at low elevations in a given region. On average, temperatures decrease by about 3.6°F for every 1000 ft. (300 m) gain in elevation. In many places at high elevation, **glaciers** form because the summers are not warm enough to melt all of the snow each year. The mountains in Glacier National Park in Montana, which are located at high elevation as well as fairly high latitude, contain glaciers, as shown in *Figure 1*.

Figure 1 Mountains at high elevation and high latitudes often contain glaciers.

Geographic Features

Geographic features, like mountain ranges, lakes, and oceans, affect the climate of a region. As shown in *Figure 2*, mountains can have a dramatic effect on precipitation in nearby areas. The **windward** side of a mountain chain often receives much more rainfall than the leeward side. As wind approaches the mountains, it is forced upwards. When the air rises, it cools, and water vapor condenses into clouds, which produce precipitation. Conversely, the **leeward** side of a mountain range is in what is called a **rain shadow**. It often receives very little rain. That is because the air has already lost much of its moisture on the windward side. When the air descends the leeward slope of the mountain, it warms up as the greater air pressure compresses it. That causes clouds to evaporate and the humidity of the air to decrease. The deserts of the southwestern United States have low rainfall because they are in the rain shadow of the Sierra Nevada and other mountain ranges along the Pacific coast.

Geo Words

heat capacity: the quantity of heat energy required to increase the temperature of a material or system; typically referenced as the amount of heat energy required to generate a 1°C rise in the temperature of 1 g of a given material that is at atmospheric pressure and 20°C.

Figure 2 The rain shadow effect. Most North American deserts are influenced by this effect.

Large bodies of water can also affect climate dramatically. The ocean has a moderating effect on nearby communities. Temperatures in coastal communities vary less than inland communities at similar latitude. This is true both on a daily basis and seasonally. The effect is especially strong where the coast faces into the prevailing winds, as on the West Coast of the United States. Kansas City's average temperature is 79°F in July and 26°F in January. San Francisco's average temperature is 64°F in July and 49°F in January. On average, New York City's January temperatures vary only 11°F during a day, whereas Omaha's January temperatures vary 20°F during a day. In each of these two cases the difference between the two cities' climates is too great to be related to their latitude, which is only different by less than about 1.5°. Instead, the differences in climate in these two places are because water has a much higher **heat capacity** than soil and rock. That means that much more heat is needed to raise the temperature of water than to raise the

Geo Words

lake-effect snow: the snow that is precipitated when an air mass that has gained moisture by moving over a relatively warm water body is cooled as it passes over relatively cold land. This cooling triggers condensation of clouds and precipitation.

global climate: the mean climatic conditions over the surface of the Earth as determined by the averaging of a large number of observations spatially distributed throughout the entire region of the globe.

Little Ice Age: the time period from mid-1300s to the mid-1800s AD. During this period, global temperatures were at their coldest since the beginning of the Holocene.

Check Your Understanding

1. What is the difference between weather and climate?

2. What is the difference between regional climate and global climate?

3. Compare the climate of a city along the Pacific coast with that of a city with a similar latitude but located inland.

4. Explain how there can be snow on the top of a mountain near the Equator.

5. What is the "Little Ice Age"?

temperature of soil or rock. In the same way, water cools much more slowly than soil and rock. Land areas warm up quickly during a sunny day and cool down quickly during clear nights. The ocean and large lakes, on the other hand, change their temperature very little from day to day. Because the ocean absorbs a lot of heat during the day and releases it at night, it prevents daytime temperatures in seaside communities from climbing very high and prevents nighttime temperatures from falling very low (unless the wind is blowing from the land to the ocean!). By the same token, oceans store heat during the summer and release it during the winter, keeping summers cooler and winters warmer than they would be otherwise.

Lake-effect snow is common in late autumn and early winter downwind of the Great Lakes in north–central United States. Cold winds blow across the still-warm lake water, accumulating moisture from the lake as they go. When they reach the cold land, the air is cooled, and the water precipitates out of the clouds as snow. The warm oceans also supply the moisture that feeds major rainstorms, not just along the coast but even far inland in the eastern and central United States.

Global Climate

Climates differ from one region to another, depending on latitude, elevation, and geographical features. However, the entire Earth has a climate, too. This is called **global climate**. It is usually expressed as the year-round average temperature of the entire surface of the Earth, although average rainfall is also an important part of global climate. Today, the average temperature on the surface of the Earth is about 60°F. But the Earth's climate has changed continually over geologic time. During the Mesozoic Era (245–65 million years ago), when the dinosaurs roamed the Earth, global climate was warmer than today. During the Pleistocene Epoch (1.6 million–10,000 years ago), when mastodons and cave people lived, global climate seesawed back and forth between cold glacial intervals and warmer interglacial intervals. During glacial intervals, huge sheets of glacier ice covered much of northern North America. Just a few hundred years ago, the climate was about 3°F cooler than now. The time period from about the mid-1300s to the mid-1800s is called the **Little Ice Age**, because temperatures were generally much colder than today, and glaciers in many parts of the world expanded. Global temperatures have gradually increased since then, as the Earth has been coming out of the Little Ice Age.

Over the next several activities, you will be looking at what causes these changes in global climate, and how human activity may be causing global climate change.

Understanding and Applying What You Have Learned

1. What is the nearest body of water to your community? In what ways does it affect the physical environment of your community?

2. Identify one physical feature in or near your community and explain how it affects the climate. You may need to think regionally. Is there a mountain range, a large lake, or an ocean in your state?

3. How would changing one feature of the physical environment near your community affect the climate? Again, you may need to think on a regional scale. Name a feature and tell how the climate would be different if that feature changed. How would this change life in your community?

4. Can you think of any additional reasons to explain the differences in climate between your community and the other community you looked at in the **Investigate** section?

Preparing for the Chapter Challenge

1. Clip and read several newspaper articles about scientific topics.

2. Using a style of writing appropriate for a newspaper, write a few paragraphs in which you:

   - explain the term climate;
   - describe the climate of your community (including statistics about seasonal temperatures, rainfall, and snowfall);
   - explain what physical factors in your community or state combine to produce this climate;
   - explain the difference between regional climate and global climate, and
   - describe the Earth's global climate.

Inquiring Further

1. **Weather systems and the climate of your community**

   Investigate how weather systems crossing the United States affect the climate in your community. In many places, weather systems travel in fairly regular paths, leading to a somewhat predictable series of weather events.

2. **Jet stream**

   Do some research on the jet stream. How does its position affect the climate in your community?

Activity 2   Paleoclimates

Goals

In this activity you will:

- Understand the significance of growth rings in trees as indicators of environmental change.

- Understand the significance of ice cores from glaciers as indicators of environmental change.

- Investigate and understand the significance of geologic sediments as indicators of environmental change.

- Examine the significance of glacial sediments and landforms as evidence for climate change.

- Investigate and understand the significance of fossil pollen as evidence for climate change.

Think about It

The cross section of a tree trunk shows numerous rings.

- **What do you think the light and dark rings represent?**
- **What might be the significance of the varying thicknesses of the rings?**

What do you think? Record your ideas about these questions in your *EarthComm* notebook. Be prepared to discuss your responses with your small group and the class.

Investigate

Part A: Tree Rings

1. Examine the photo that shows tree growth rings from a Douglas fir. Notice the arrow that marks the growth ring that formed 550 years before the tree was cut.

a) Where are the youngest and the oldest growth rings located?

b) Not all the growth rings look identical. How are the rings on the outer part of the tree different from those closer to the center?

c) Mark the place on a copy of the picture where the change in the tree rings occurs.

2. Using the 550-year arrow as a starting point, count the number of rings to the center of the tree. Now count the number of rings from the arrow to where you marked the change in the way the rings look.

a) Record the numbers.

b) Assuming that each ring represents one year, how old is the tree?

c) Assume that the tree was cut down in the year 2000. What year did the tree rings begin to look different from the rings near the center (the rings older than about 550-years old)?

3. Compare the date you calculated in **Step 2 (c)** to the graph that shows change in temperature for the last 1000 years.

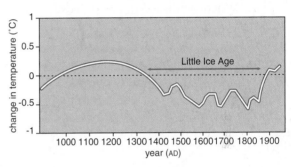

a) What was happening to the climate during that time?

b) From these observations, what would you hypothesize is the correlation between the thickness of tree rings and climate?

4. Examine the diagram that shows temperature change for the last 150,000 years.

a) How does the duration of the Little Ice Age compare to the duration of the ice age (the time between the two interglacial periods) shown in this figure?

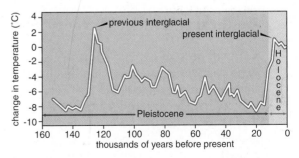

Part B: Fossil Pollen

1. Using blue, red, green, and yellow modeling clay, put down layers in a small container. You may put them down in any order and thickness. The container represents a lake or pond, and the clay represents sediment that has settled out over a long time.

2. When you have finished laying down your "sediments," use a small pipe to take a core. Push the pipe straight down through all of the layers. Then carefully pull the pipe back up. Use a thin stick to push the core of sediments out of the pipe.

⚠️ Place folded paper towels or other padding between your hand and the upper end of the pipe before pushing it into the clay.

   a) Draw a picture of the core in your notebook. Note which end is the top.

   b) Measure and record the thickness of each layer of sediment to the nearest tenth of a centimeter.

3. The different colors of clay represent sediments that have settled out of the lake water at different times. For this exercise, imagine that each centimeter of clay represents the passage of 1000 years.

   a) How many years does your core represent from top to bottom?

   b) How many years does each layer represent?

4. Imagine that the different colors of clay represent the following:

- Blue: sediments containing pollen from cold-climate plants like spruce and alder trees.

- Red: sediments containing pollen from warm-climate plants, like oak trees and grasses.

- Green: sediments containing mostly spruce and alder pollen, with a little oak and grass pollen.

- Yellow: sediments containing mostly oak and grass pollen, with a little spruce and alder pollen.

   a) How do you think the pollen gets into the lake sediments?

   b) Describe what the climate around the lake was like when each layer of sediment was deposited.

   c) Write a paragraph describing the climate changes over the period of time represented by your core. Make sure you say at how many years before the present each climate change occurred. Note whether transitions from one type of climate to another appear to have happened slowly or quickly.

Reflecting on the Activity and the Challenge

You will need to explain some of the ways that geologists know about climates that existed in the geologic past. The activity helped you understand two of the ways that geologists find out about ancient climates. Geologists study tree growth rings. They relate the thickness of the tree rings to the climate. Geologists also collect cores from layers of sediments and study the kinds of pollen contained in the sediments. The pollen shows what kinds of plants lived there in the past, and that shows something about what the climate was like.

Digging Deeper

HOW GEOLOGISTS FIND OUT ABOUT PALEOCLIMATES

Direct Records and Proxies

A **paleoclimate** is a climate that existed sometime in the past: as recently as just a few centuries ago, or as long as billions of years ago. For example, in the previous activity you learned that the world was warmer in the Mesozoic Era (245 – 65 million years ago) and experienced periods of glaciation that affected large areas of the Northern Hemisphere continents during the Pleistocene Epoch (1.6 million–10,000 years ago). At present the Earth is experiencing an interglacial interval—a period of warmer climate following a colder, glacial period. The Earth today has only two continental ice sheets, one covering most of Greenland and one covering most of Antarctica.

The last retreat of continental glaciers occurred between about 20,000 years ago and 8000 years ago. That was before the invention of writing, so there is no direct record of this change. Systematic records of local weather, made with the help of accurate weather instruments, go back only about 200 years. A global network of weather stations has existed for an even shorter time. Historical accounts exist for individual places, most notably in China. For certain places in China records extend back 2000 years. They are useful, but more extensive information is required to understand the full range of climate variability. **Paleoclimatologists** use a variety of methods to infer past climate. Taken together, the evidence gives a picture of the Earth's climatic history.

Unfortunately, nothing gives a direct reading of past temperature. Many kinds of evidence, however, give an indirect record of past temperature. These are called **climate proxies**. Something that represents something else indirectly is called a proxy. In some elections, a voter can choose another person to cast the vote, and that vote is called a proxy. There are many proxies for past climate, although none is perfect.

Fossil Pollen

Pollen consists of tiny particles that are produced in flowers to make seeds. Pollen is often preserved in the sediments of lakes or bogs, where it is blown in by the wind. For example, a layer of sediment may contain a lot of pollen from spruce trees, which grow in cold climates. From that you can infer that the climate around the lake was cold when that layer of sediment was being deposited. Geologists collect sediment from a succession of

Geo Words

paleoclimate: the climatic conditions in the geological past reconstructed from a direct or indirect data source.

paleoclimatologist: a scientist who studies the Earth's past climate.

climate proxy: any feature or set of data that has a predictable relationship to climatic factors and can therefore be used to indirectly measure those factors.

sediment layers. They count the number of pollen grains from different plants in each layer. Then they make charts that can give an idea of the climate changes that have taken place. (See *Figure 1*.) Pollen is easy to study because there is so much of it. Geologists also study fossil plants and insects to reconstruct past climates.

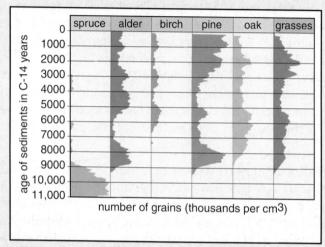

	spruce	alder	birch	pine	oak	grasses

age of sediments in C-14 years: 0, 1000, 2000, 3000, 4000, 5000, 6000, 7000, 8000, 9000, 10,000, 11,000

number of grains (thousands per cm3)

Figure 1 Changes over time in the relative amounts of different types of pollen from various trees and grasses give clues as to how climate has changed in the past.

Ice Cores

Figure 2 Scientists are able to obtain clues into past climatic conditions from air bubbles trapped in ice cores.

In recent years, study of cores drilled deep down into glaciers, like the one shown in *Figure 2*, has become a very powerful technique for studying paleoclimate. Very long cores, about 10 cm (4 in.) in diameter, have been obtained from both the Greenland and Antarctic ice sheets. The longest, from Antarctica, is almost 3400 m (about 1.8 mi.) long. Ice cores have been retrieved from high mountain glaciers in South America and Asia.

Glaciers consist of snow that accumulates each winter and does not melt entirely during the following summer. The snow is gradually compressed into ice as it is buried by later snow. The annual layers can be detected by slight changes in dust content. The long core from Antarctica provides a record of climate that goes back for more than 400,000 years. See *Figure 3*.

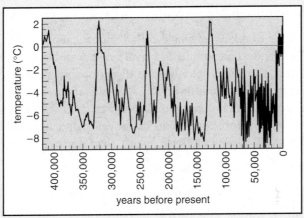

Figure 3 Temperature variation over the past 420,000 years, relative to the modern surface temperature at Vostok (−55.5°C).

Geo Words

isotope: one of two or more kinds of atoms of a given chemical element that differ in mass because of different numbers of neutrons in the nucleus of the atoms.

Bubbles of air trapped in the ice contain samples of the atmosphere from the time when the snow fell. Paleoclimatologists study the oxygen in the water molecules in the ice. Almost all of the oxygen atoms in the atmosphere are in two forms, called **isotopes**. The two isotopes are oxygen-16 (written ^{16}O) and oxygen-18 (written ^{18}O). They are the same chemically, but they have slightly different weights. ^{18}O is slightly heavier than ^{16}O. The proportion of these two isotopes in snow depends on average global temperatures. Snow that falls during periods of warmer global climate contains a greater proportion of ^{18}O, and snow that falls during periods of colder global climate contains a smaller proportion. The ratio of ^{18}O to ^{16}O can be measured very accurately with special instruments. Another important way of using the glacier ice to estimate global temperature is to measure the proportions of the two naturally occurring isotopes of hydrogen: ^{1}H, and ^{2}H (which is called deuterium).

The air bubbles in the ice contain carbon dioxide. The amount of carbon dioxide in the glacier ice air bubbles depends on the amount of carbon dioxide in the air at that time. The amount of carbon dioxide in the atmosphere can be correlated to global temperatures. During times when the paleoclimate is thought to have been warm, the ice core record shows relatively higher levels of atmospheric carbon dioxide compared to times of interpreted colder climate. Measurements of carbon dioxide taken from the cores give a global picture, because carbon dioxide is uniformly distributed in the global atmosphere.

A third component of the ice that yields clues to paleoclimates is dust. During colder climates, winds tend to be stronger. The stronger winds erode more dust, and the dust is deposited in small quantities over large areas of the Earth.

Climate Change

Geo Words

foraminifera: an order of single-celled organisms (protozoans) that live in marine (usually) and freshwater (rarely) environments. Forams typically have a shell of one or more chambers that is typically made of calcium carbonate.

loess: the deposits of wind-blown silt laid down over vast areas of the mid-latitudes during glacial and postglacial times.

Check Your Understanding

1. How do preserved tree rings indicate changes in climate?

2. List three ways that sediments in the ocean help scientists understand ancient climates.

3. Imagine an ice core taken from the Antarctic ice sheet. A layer of ice called "A" is 100 m below the surface. A layer of ice called "B" is 50 m below the surface. Explain why layer "A" represents the atmospheric conditions of an older climate than layer "B."

Deep-Sea Sediments

Sand-size shells of a kind of single-celled animal called **foraminifera** ("forams," for short) accumulate in layers of ocean-bottom sediment. During warm climates, the shells spiral in one direction, but during cold climates, the shells spiral in the opposite direction. Also, the shells consist of calcium carbonate, which contains oxygen. Geologists can measure the proportions of the two oxygen isotopes to find out about paleoclimates. The shells contain more ^{18}O during colder climates than during warmer climates.

Glacial Landforms and Sediments

Glaciers leave recognizable evidence in the geologic record. Glacial landforms are common in northern North America. Glaciers erode the rock beneath, and then carry the sediment and deposit it to form distinctive landforms. Cape Cod, in Massachusetts, and Long Island, in New York, are examples of long ridges of sediment deposited by glaciers. Similar deposits are found as far south as Missouri.

Fine glacial sediment is picked up by the wind and deposited over large areas as a sediment called **loess**. There are thick deposits of loess in central North America. The loess layers reveal several intervals of glaciation during the Pleistocene Epoch.

Glaciers also leave evidence in the ocean. When glaciers break off into the ocean, icebergs float out to sea. As the icebergs melt, glacial sediment in the icebergs rains down to the ocean bottom. The glacial sediment is easily recognized because it is much coarser than other ocean-bottom sediment.

Tree Rings

Paleoclimate is also recorded in the annual growth rings in trees. Trees grow more during warm years than during cold years. A drawback to tree rings is that few tree species live long enough to provide a look very far in the past. Bristlecone pines, which can live as long as 5000 years, and giant sequoias, which are also very long-lived, are most often used.

Figure 4 A glacier carries ground-up pieces of rock and sediment into lakes and oceans.

Understanding and Applying What You Have Learned

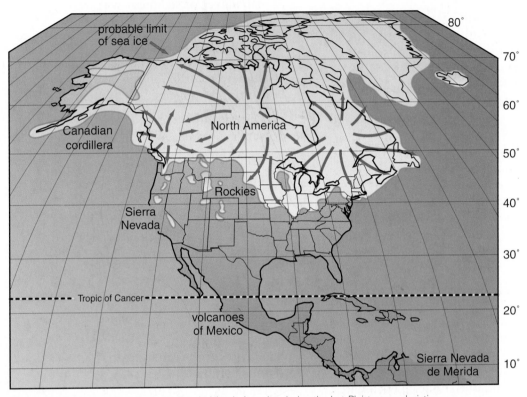

The approximate extent of the ice sheets in North America during the last Pleistocene glaciation.

1. Using evidence from glacial landforms and sediments, geologists have pieced together the maximum advance of glaciers about 18,000 years ago, which is shown in the figure above.

   a) Was your community located under ice during this time period? If not, describe how far your part of the country was from the ice sheet.

   b) What do you think the climate was like in your part of the country during the glacial maximum?

2. Use colored pencils and a ruler to draw a hypothetical series of lake-bottom sediments representing the sequence of climates given on the following page. Use the same colors as you used in the **Investigate** section to represent layers containing different kinds of pollen. Again, assume that it takes 1000 years to deposit 1 cm of sediment.

759

- 7000–5000 years ago: Warm climate supporting grasses and oaks.
- 5000–3000 years ago: Moderate climate supporting mostly grasses and oaks, with some spruce and alder.
- 3000–2000 years ago: Colder climate supporting mostly spruce and alder, with some grasses and oaks.
- 2000–1500 years ago: Moderate climate supporting mostly grasses and oaks, with some spruce and alder.
- 1500 years ago to the present: Warm climate supporting grasses and oaks.

a) Describe what the climate around the lake was like when each layer of sediment was deposited.

b) From this data, what time period marks the coldest climate recorded in the lake bottom sediments?

c) Did the climate cool at the same rate as it warmed?

d) Think of a hypothesis that might explain your answer to **Part (c)**. What additional observations might help you test this hypothesis?

Preparing for the Chapter Challenge

Using a style appropriate for a newspaper article, write a paragraph or two about each of the following topics, explaining how geologists use them to find out about paleoclimates:

- deep-sea sediments
- glacial landforms and sediments
- ice cores from Antarctica
- pollen studies
- tree rings

Inquiring Further

1. **GISP2 (Greenland Ice Sheet Project)**

   Research GISP2 (Greenland Ice Sheet Project), a project that is collecting and analyzing ice cores. What are the most recent discoveries? How many ice cores have been collected? How many have been analyzed? How far into the past does the data currently reach? Visit the *EarthComm* web site to help you start your research.

2. **Dating deep-sea sediments**

   Investigate some of the techniques geologists use to date deep-sea

sediments: carbon-14 dating, isotope dating of uranium, fission-track dating of ash layers, and geomagnetic-stratigraphy dating.

3. **Paleoclimate research in your community**

   Investigate whether any of the paleoclimate techniques discussed in this activity have been used near your community or in your state to research paleoclimates.

Activity 3

How Do Earth's Orbital Variations Affect Climate?

Goals

In this activity you will:

- Understand that Earth has an axial tilt of about 23 1/2°.

- Use a globe to model the seasons on Earth.

- Investigate and understand the cause of the seasons in relation to the axial tilt of the Earth.

- Understand that the shape of the Earth's orbit around the Sun is an ellipse and that this shape influences climate.

- Understand that insolation to the Earth varies as the inverse square of the distance to the Sun.

Think about It

When it is winter in New York, it is summer in Australia.

- **Why are the seasons reversed in the Northern and Southern Hemispheres?**

What do you think? Write your thoughts in your *EarthComm* notebook. Be prepared to discuss your responses with your small group and the class.

Investigate

Part A: What Causes the Seasons? An Experiment on Paper

1. In your notebook, draw a circle about 10 cm in diameter in the center of your page. This circle represents the Earth.

   Add the Earth's axis of rotation, the Equator, and lines of latitude, as shown in the diagram and described below. Label the Northern and Southern Hemispheres.

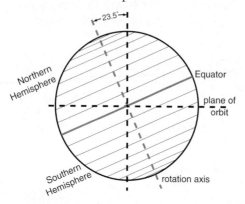

   Put a dot in the center of the circle. Draw a dashed line that goes directly up and down from the center dot to the edge of the circle. Use a protractor to measure 23 1/2° from this vertical dashed line. Use a blue pen or pencil to draw a line through the center of the "Earth" at 23 1/2° to your dashed line. This blue line represents the Earth's axis of rotation. Use your protractor to draw a red line that is perpendicular to the axis and passes through the center dot. This red line represents the Equator of the Earth. Label the Northern and Southern Hemispheres. Next you need to add lines of latitude. To do this, line your protractor up with the dot in the center of your circle so that it is parallel with the Equator.

   Now, mark off 10° increments starting from the Equator and going to the poles. You should have eight marks between the Equator and pole for each quadrant of the Earth. Use a straight edge to draw black lines that connect the marks opposite one another on the circle, making lines that are parallel to the Equator. This will give you lines of latitude in 10° increments so you can locate your latitude fairly accurately. Note that the lines won't be evenly spaced from one another because latitude is measured as an angle from the center of the Earth, not a linear distance.

2. Imagine that the Sun is directly on the left in your drawing. Draw horizontal arrows to represent incoming Sun rays from the left side of the paper.

3. Assume that it is noon in your community. Draw a dot where your community's latitude line intersects the perimeter of the circle on the left. This dot represents your community.

   a) Explain why this represents noon.

   b) At any given latitude, both north and south, are the Sun's rays striking the Northern Hemisphere or the Southern Hemisphere at a larger angle relative to the plane of orbit?

   c) Which do you think would be warmer in this drawing—the Northern Hemisphere or the Southern Hemisphere? Write down your hypothesis. Be sure to give a reason for your prediction.

   d) What season do you think this is in the Northern Hemisphere?

4. Now consider what happens six months later. The Earth is on the opposite side of its orbit, and the sunlight is now coming from the right side of the paper. Draw horizontal arrows to represent incoming Sun's rays from the right.

5. Again, assume that it is noon. Draw a dot where your community's latitude line intersects the perimeter of the circle on the right.

a) Explain why the dot represents your community at noon.

b) Are the Sun's rays striking the Northern Hemisphere or the Southern Hemisphere more directly?

c) Which do you think would be warmer in this drawing—the Northern Hemisphere or the Southern Hemisphere? Why?

d) What season do you think this is in the Northern Hemisphere?

Part B: What Causes the Seasons? An Experiment with a Globe

1. Test the hypothesis you made in **Part A** about which hemisphere would be warmer in which configuration. Find your city on a globe. Using duct tape, tape a small thermometer on it. The duct tape should cover the thermometer bulb, and the thermometer should be over the city. With a permanent black marking pen, color the duct tape black all over its surface.

2. Set up a light and a globe as shown.

⚠️ Use only alcohol thermometers. Place a soft cloth on the table under the thermometer in case it falls off. Be careful not to touch the hot lamp.

3. Position the globe so that its axis is tilted 23 1/2° from the vertical, and the North Pole is pointed in the direction of the light source.

4. Turn on the light source.

a) Record the initial temperature. Then record the temperature on the thermometer every minute until the temperature stops changing.

5. Now position the globe so that the axis is again tilted 23 1/2° from the vertical but the North Pole is pointing away from the light source. Make sure the light source is the same distance from the globe as it was in **Step 4**. Turn on the light source again.

a) Record the temperature every minute until the temperature stops changing.

6. Use your observations to answer the following questions in your notebook:

a) What is the difference in the average temperature when the North Pole was pointing toward your "community" and when it was pointing away?

b) What caused the difference in temperature?

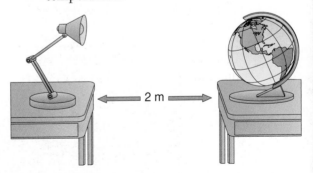

Part C: What Would Happen if the Earth's Axial Tilt Changed? An Experiment on Paper

1. Repeat the experiment you did in **Part A**, except this time use an axial tilt of 10°.

   a) Compared to an axial tilt of 23 1/2°, would your hemisphere experience a warmer or colder winter?

   b) Compared to an axial tilt of 23 1/2°, would your hemisphere experience a hotter or cooler summer?

2. Repeat the experiment you did in **Part A**, except this time use an axial tilt of 35°.

   a) Compared to an axial tilt of 23 1/2°, would your hemisphere experience a warmer or colder winter?

   b) Compared to an axial tilt of 23 1/2°, would your hemisphere experience a hotter or cooler summer?

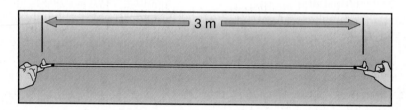

Part D: The Earth's Elliptical Orbit around the Sun

1. Tie small loops in each end of a rope, as shown in the diagram above.

2. Pick a point in about the middle of the floor, and put the two loops together over the point. Put a dowel vertically through the loops, and press the dowel tightly to the floor.

3. Stretch out the rope from the dowel until it is tight, and hold a piece of chalk at the bend in the rope, as shown in the diagram below. While holding the chalk tight against the rope, move the chalk around the dowel.

   a) What type of figure have you constructed?

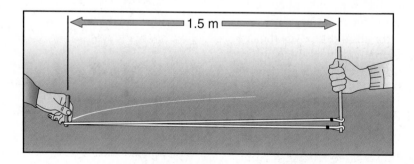

4. Draw a straight line from one edge of the circle that you just made to the opposite edge, through the center of the circle. This line represents the diameter of the circle. Mark two points along the diameter, each a distance of 20 cm from the center of the circle. Put the loops of the rope over the two points, hold them in place with two dowels, and use the chalk to draw a curve on one side of the straight line. Move the chalk to the other side of the line and draw another curve.

a) What type of figure have you constructed?

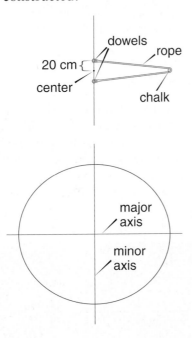

5. Using different colors of chalk, make a few more curved figures in the same way. Choose sets of dowel points that are farther and farther away from the center.

a) Describe the shapes of the figures you constructed. Make sketches in your *EarthComm* notebook.

b) What would be the shape of the curve when the dowel points are spaced a distance apart that is just equal to the length of the rope between the two loops?

Part E: How Energy from the Sun Varies with Distance from the Sun

1. Using scissors and a ruler, cut out a square 10 cm on a side in the middle of a poster board.

2. Hold the poster board vertically, parallel to the wall and exactly two meters from it.

3. Position a light bulb along the imaginary horizontal line that passes from the wall through the center of the hole in the poster board. See the diagram.

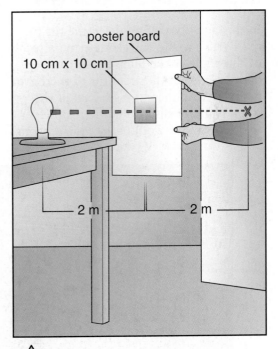

⚠️ Avoid contact with the hot light bulb. Do not look directly into the light.

4. Turn on the light bulb, and turn off the lights in the room. If the room is not dark enough to see the image on the wall, close any curtains or shades, or cover the windows with dark sheets or blankets.

5. With the chalk, trace the edge of the image the hole makes on the wall.

   a) Measure and record the length of the sides of the image you marked with the chalk.

   b) Divide the length of the image on the wall by the length of the sides of the square in the poster board. Now divide the distance of the light bulb from the wall by the distance of the poster board from the wall. What is the relationship between the two numbers you obtain?

   c) Compute the area of the image on the wall, and compute the area of the square hole in the poster board. Divide the area of the image by the area of the hole. Again, divide the distance of the light bulb from the wall by the distance of the poster board from the wall. What is the relationship between the two numbers you obtain?

6. Repeat **Part E** for other distances.

   a) What do you notice about the relationship between the area of the image and the area of the hole?

Reflecting on the Activity and the Challenge

In this activity you modeled the tilt of the Earth's axis to investigate the effect of the angle of the Sun's rays. You discovered that the axial tilt of the Earth explains why there are seasons of the year. You also discovered that if the tilt were to vary, it would affect the seasons. You also modeled the Earth's elliptical orbit around the Sun. This will help you to understand one of the main theories for explaining why the Earth's climate varies over time. You will need to explain this in your newspaper articles.

Digging Deeper

THE EARTH'S ORBIT AND THE CLIMATE

The Earth's Axial Tilt and the Seasons

The Earth's axis of rotation is tilted at about 23 1/2° away from a line that is perpendicular to the plane of the Earth's orbit around the Sun (*Figure 1*). This tilt explains the seasons on Earth. During the Northern Hemisphere summer, the North Pole is tilted toward the Sun, so the Sun shines at a high angle overhead. That is when the days are warmest, and days are longer than nights. On the summer solstice (on or about June 22) the Northern Hemisphere experiences its longest day and shortest night of the year. During the Northern Hemisphere winter the Earth is on the other side of its orbit. Then the North Pole is tilted away from the Sun, and the Sun shines at a lower angle. Temperatures are lower, and the days are shorter than the nights. On the winter solstice (on or about December 22) the Northern Hemisphere experiences its shortest day and longest night of the year.

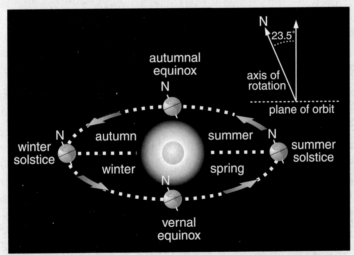

Figure 1 The tilt of the Earth's axis explains the seasons. Note that the Earth and Sun are not shown to scale.

How Do Earth's Orbital Variations Affect Climate?

In **Activity 2** you saw that paleoclimatologists have developed a good picture of the Earth's climatic history. The advances and retreats of the continental ice sheets are well documented. Nevertheless, questions remain.

Geo Words

eccentricity: the ratio of the distance between the foci and the length of the major axis of an ellipse.

obliquity: the tilt of the Earth's rotation axis as measured from the perpendicular to the plane of the Earth's orbit around the Sun. The angle of this tilt varies from 22.5° to 24.5° over a 41,000-year period. Current obliquity is 23.5°.

precession: slow motion of the axis of the Earth around a cone, one cycle in about 26,000 years, due to gravitational tugs by the Sun, Moon, and major planets.

orbital parameters: any one of a number of factors that describe the orientation and/or movement of an orbiting body or the shape of its orbital path.

insolation: the direct or diffused shortwave solar radiation that is received in the Earth's atmosphere or at its surface.

inverse-square law: a scientific law that states that the amount of radiation passing through a specific area is inversely proportional to the square of the distance of that area from the energy source.

Why does Earth's climate sometimes become cold enough for ice sheets to advance? Why does the climate later warm up and cause ice sheets to retreat? The answers to these questions are not yet entirely clear. Most climatologists believe that variations in the geometry of the Earth's orbit around the Sun are the major cause of the large variations in climate. These variations have caused the advance and retreat of ice sheets in the past couple of million years.

If the Earth and the Sun were the only bodies in the solar system, the geometry of the Earth's orbit around the Sun and the tilt of the Earth's axis would stay exactly the same through time. But there are eight other known planets in the solar system. Each of those planets exerts forces on the Earth and the Sun. Those forces cause the Earth's orbit to vary with time. The Moon also plays a role. The changes are slight but very important. There are three kinds of changes: **eccentricity**, **obliquity**, and **precession**. These three things are called the Earth's **orbital parameters**.

Eccentricity

The Earth's orbit around the Sun is an ellipse. The deviation of an ellipse from being circular is called its eccentricity. A circle is an ellipse with zero eccentricity. As the ellipse becomes more and more elongated (with a larger major diameter and a smaller minor diameter), the eccentricity increases. The Earth's orbit has only a slight eccentricity.

Even though the eccentricity of the Earth's orbit is very small, the distance from the Earth to the Sun varies by about 3.3% through the year. The difference in **insolation** is even greater. The word insolation (nothing to do with insUlation!) is used for the rate at which the Sun's energy reaches the Earth, per unit area facing directly at the Sun. The seasonal variation in insolation is because of what is called the **inverse-square law**. What you found in **Part E** of the investigation demonstrates this. The area of the image on the wall was four times the area of the hole, even though the distance of the wall from the bulb was only twice the distance of the hole from the bulb. Because of the inverse-square law, the insolation received by the Earth varies by almost 7° between positions on its orbit farthest from the Sun and positions closest to the Sun.

Because of the pull of other planets on the Earth–Sun system, the eccentricity of the Earth's orbit changes with time. The largest part of the change in eccentricity has a period of about 100,000 years. That means that one full cycle of increase and then decrease in eccentricity takes 100,000 years. During that time, the difference in insolation between the date of the shortest distance to the Sun and the date of the farthest distance to the Sun ranges from about 2° (less than now) to almost 20° (much greater than now!).

The two points you used to make ellipses in **Part D** of the investigation are called the foci of the ellipse. (Pronounced "FOH-sigh". The singular is focus.) The Sun is located at one of the foci of the Earth's elliptical orbit. The Earth is closest to the Sun when it is on the side of that focus, and it is farthest from the Sun when it is on the opposite side of the orbit. (See *Figure 2*.) Does it surprise you to learn that nowadays the Earth is closest to the Sun on January 5 (called **perihelion**) and farthest from the Sun on July 5 (called **aphelion**)? That tends to make winters less cold and summers less hot, in the Northern Hemisphere.

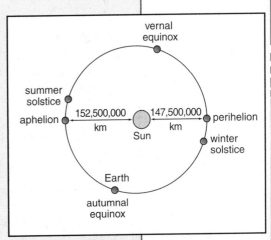

Figure 2 Schematic diagram showing occurrence of the aphelion and perihelion.

Obliquity

The tilt of the Earth's axis relative to the plane of the Earth's orbit is called the obliquity. The axis is oblique to the plane rather than perpendicular to it. In the investigation you discovered that a change in the obliquity would cause a change in the nature of the seasons. For example, a smaller obliquity would mean warmer winters and cooler summers in the Northern Hemisphere. This might result in more moisture in the winter air, which would mean more snow. In cooler summers, less of the snow would melt. You can see how this might lead to the buildup of glaciers.

Geo Words

perihelion: the point in the Earth's orbit that is closest to the Sun. Currently, the Earth reaches perihelion in early January.

aphelion: the point in the Earth's orbit that is farthest from the Sun. Currently, the Earth reaches aphelion in early July.

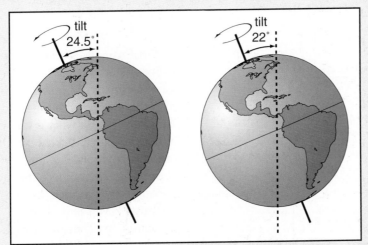

Figure 3 The angle of tilt of the Earth's axis varies between about 24.5° and 22°, causing climatic variations over time.

Coordinated Science for the 21st Century

Geo Words

Milankovitch cycles: the cyclical changes in the geometric relationship between the Earth and the Sun that cause variations in solar radiation received at the Earth's surface.

axial precession: the wobble in the Earth's polar axis.

orbital precession: rotation about the Sun of the major axis of the Earth's elliptical orbit.

Again, because of the varying pull of the other planets on the Earth–Sun system, the Earth's obliquity changes over a period of about 40,000 years. The maximum angle of tilt is about 24 1/2°, and the minimum angle is about 22°. At times of maximum tilt, seasonal differences in temperature are slightly greater. At times of minimum tilt angle, seasonal differences are slightly less.

Precession

Have you ever noticed how the axis of a spinning top sometimes wobbles slowly as it is spinning? It happens when the axis of the top is not straight up and down, so that gravity exerts a sideways force on the top. The same thing happens with the Earth. The gravitational pull of the Sun, Moon, and other planets causes a slow wobbling of the Earth's axis. This is called the Earth's **axial precession**, and it has a period of about 26,000 years. That's the time it takes the Earth's axis to make one complete revolution of its wobble.

There is also another important kind of precession related to the Earth. It is the precession of the Earth's orbit, called **orbital precession**. As the Earth moves around the Sun in its elliptical orbit, the major axis of the Earth's orbital ellipse is rotating about the Sun. In other words, the orbit itself rotates around the Sun! The importance of the two precession cycles for the Earth's climate lies in how they interact with the eccentricity of the Earth's orbit. This interaction controls how far the Earth is from the Sun during the different seasons. Nowadays, the Northern Hemisphere winter solstice is at almost the same time as perihelion. In about 11,000 years, however, the winter solstice will be at about the same time as aphelion. That will make Northern Hemisphere winters even colder, and summers even hotter, than today.

Milankovitch Cycles

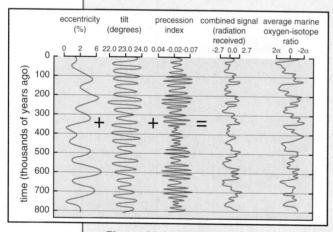

Figure 4 Interpretation of Milankovitch cycles over the last 800,000 years.

Early in the 20th century a Serbian scientist named Milutin Milankovitch hypothesized that variations in the Earth's climate are caused by how insolation varies with time and with latitude. He used what is known about the Earth's orbital parameters (eccentricity, obliquity, and precession) to compute the variations in insolation. Later scientists have refined the computations. These insolation cycles are now called **Milankovitch cycles**. (See *Figure 4*.)

Climatologists now generally agree that Milankovitch cycles are closely related to the glacial–interglacial cycles the Earth has experienced in recent geologic time. *Figure 3* in **Activity 2**, based on a long Antarctic ice core, shows how temperature has varied over the past 420,000 years. There is a clear 100,000-year periodicity, which almost exactly matches the eccentricity cycle. Temperature variations seem to have been controlled by the 100,000-year eccentricity cycle during some time intervals but by the 41,000-year obliquity cycle during other time intervals.

Climatologists are trying to figure out how Milankovitch cycles of insolation trigger major changes in climate. The Milankovitch cycles (the "driver" of climate) are just the beginning of the climate story. Many important climate mechanisms must be taken into account. They involve evaporation, precipitation, snowfall, snowmelt, cloud cover, greenhouse gases, vegetation, and sea level. What makes paleoclimatology difficult (and interesting!) is that these factors interact with one another in many complicated ways to produce climate.

Check Your Understanding

1. Explain why the days are longer than the nights during the summer months. Include a diagram to help you explain.

2. What are the three factors in Milankovitch cycles?

3. Explain how Milankovitch cycles might cause changes in global climate.

Understanding and Applying What You Have Learned

1. You have made a drawing of winter and summer in the Northern Hemisphere showing the tilt of Earth.

   a) Make a drawing showing Earth on or about March 21 (the vernal equinox). Indicate from which direction the Sun's rays are hitting the Earth.

   b) Explain why the daytime and nighttime last the same length of time everywhere on the Earth on the vernal equinox and on the autumnal equinox.

2. In **Part E** of the **Investigate** section you explored the relationship between energy from the Sun and distance to the Sun. How would you expect the area of light shining on the wall to change if the light source was moved farther away from the wall (but the cardboard was left in the same place)?

3. The tilt of the Earth varies from about 22° to about 24.5° over a period of 41,000 years. Think about how the solar radiation would change if the tilt was 24.5°.

   a) What effect would this have on people living at the Equator?

   b) What effect would this have on people living at 30° latitude?

   c) What effect would this have on people living at 45° latitude?

   d) What modifications in lifestyle would people have to make at each latitude?

Preparing for the Chapter Challenge

Use a style of writing appropriate for a newspaper to discuss the following topics:

- How does the tilt of the Earth's axis produce seasons?

- How does a variation in this tilt affect the nature of the seasons?
- How might a variation in the tilt affect global climate?
- How does the shape of the Earth's orbit influence climate?

Inquiring Further

1. **Sunspots and global climate**

   Do sunspots affect global climate? There is disagreement over this in the scientific community. Do some research to find out what sunspots are and how the activity of sunspots has correlated with global climate over time. Why do some scientists think sunspot activity affects global climate? Why do some scientists think that sunspot activity does NOT affect global climate?

2. **Milutin Milankovitch**

   Write a paper on Milutin Milankovitch, the Serbian scientist who suggested that variations in the Earth's orbit cause glacial periods to begin and end. How were his ideas received when he first published them?

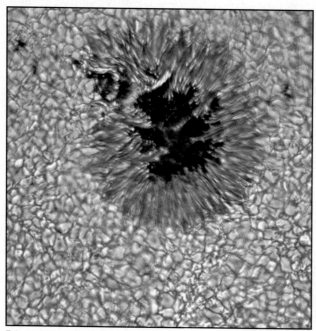

False color telescope image of a sunspot.

Activity 4

How Do Plate Tectonics and Ocean Currents Affect Global Climate?

Goals

In this activity you will:

- Model present and ancient land masses and oceans to determine current flow.

- Explain how ocean currents affect regional and global climate.

- Understand how ocean currents are affected by Earth's moving plates.

- Understand the relationship between climate and Earth processes like moving plates, mountain building, and weathering.

Think about It

Ocean currents help to regulate global climate by transferring heat and moisture around the globe.

- **How would a change in the position of a land mass influence global climate?**

What do you think? Record your ideas about this question in your *EarthComm* notebook. Be prepared to discuss your responses with your small group and the class.

Investigate

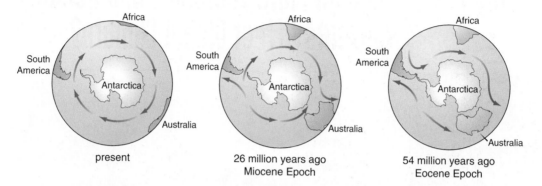

present

26 million years ago
Miocene Epoch

54 million years ago
Eocene Epoch

1. Divide your class into three groups. Each group will investigate the flow of water at one of these three periods of time:

   Group 1: The present.

   Group 2: During the Miocene Epoch, 26 million years ago.

   Group 3: During the Eocene Epoch, 54 million years ago.

2. Obtain a copy of the map for your assigned time period. Put this map under a clear plastic container.

3. Using clay, construct the correct land masses inside the container using the map as a template. Make the land masses at least 3 cm high.

4. Remove the map and add water to the container up to the level of the land masses.

5. Obtain a blue-colored ice cube. Place this as close to the South Pole as possible.

   a) As the ice cube melts and the cold water flows into the clear water, draw arrows on the map to record the direction of flow.

   ⚠ Use only food coloring to dye the ice. Clean up spills immediately. Dispose of the water promptly.

   b) Write a paragraph in which you describe your observations.

6. Present your group's data to the class.

   a) Compare the direction of current flow on your map to those presented by the other groups. How does the current change as the land masses change from the Eocene Epoch to the Miocene Epoch to the present?

7. Use the results of your investigation to answer the following:

   a) Based on the map that shows the position of Australia in the Eocene Epoch and your investigation, what tectonic factors do you think are most important to consider when contemplating Australia's climate in the Eocene Epoch? How might Australia's Eocene Epoch climate have been different from its climate today?

   b) What tectonic factor(s) (for example, the position of the continents, occurrence of major ocean currents, mountain ranges, etc.) is/are most important in affecting the climate in your community today?

8. Examine the map of ocean surface currents carefully.

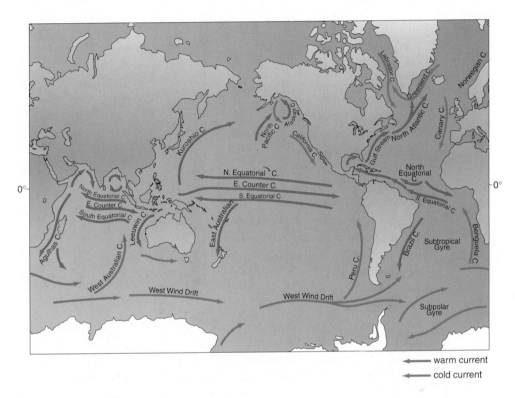

a) How do you think surface ocean currents modify the climatic patterns in the United States?

b) What changes in the surface-current patterns might arise if North America and South America split apart from one another, leaving an open passageway from the Atlantic Ocean to the Pacific Ocean?

c) How might the climate of the East Coast of the United States change if tectonic forces changed the positions of the continents so that the Gulf Stream no longer flowed north?

d) How might some other parts of the world (including your community) be affected if the Gulf Stream stopped flowing north?

Reflecting on the Activity and the Challenge

Ocean currents play a large role in global climate. This activity helped you see how ocean currents change in response to movements of the Earth's lithospheric plates.

Coordinated Science for the 21st Century

thermohaline circulation: the vertical movement of seawater, generated by density differences that are caused by variations in temperature and salinity.

Digging Deeper

CHANGING CONTINENTS, OCEAN CURRENTS, AND CLIMATE

How Ocean Currents Affect Regional Climates

A community near an ocean has a more moderate climate than one at the same latitude inland because water has a much higher heat capacity than rocks and soil. Oceans warm up more slowly and cool down more slowly than the land. Currents are also an important factor in coastal climate. A coastal community near a cold ocean current has cooler weather than a coastal community near a warm ocean current. For example, Los Angeles is located on the Pacific coast near the cold California Current. The city has an average daily high temperature in July of 75°F. Charleston, South Carolina, is located at a similar

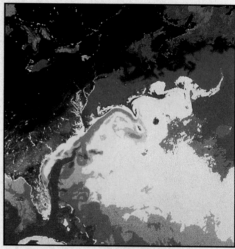

Figure I This thermal infrared image of the northwest Atlantic Ocean was taken from an NOAA satellite. The warm temperatures (25°C) are represented by red tones, and the cold temperatures (2°C) by blue and purple tones.

latitude, but on the Atlantic coast near the warm Gulf Stream. (See *Figure 1*.) Charleston's average daily high temperature in July is 90°F.

How Ocean Currents Affect Global Climate

Patterns of ocean circulation have a strong effect on global climate, too. The Equator receives more solar radiation than the poles. However, the Equator is not getting warmer, and the poles are not getting colder. That is because oceans and winds transfer heat from low latitudes to high latitudes. One of the main ways that the ocean transfers this heat is by the flow of North Atlantic Deep Water (abbreviated NADW). It works like this: In the northern North Atlantic, the ocean water is cold and salty, and it sinks because of its greater density. It flows southward at a deep level in the ocean. Then at low latitudes it rises up toward the surface as it is forced above the even denser Antarctic Bottom Water. Water from low latitudes flows north, at the ocean surface, to replace the sinking water. As it moves north, it loses heat. This slow circulation is like a "conveyor belt" for transferring heat. This kind of circulation is usually called **thermohaline circulation**. (*thermo* stands for temperature, and *haline* stands for saltiness.)

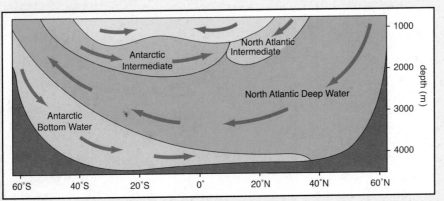

Figure 2 Circulation cell formed by the movement of deep-water masses in the ocean.

Geo Words

lithospheric plate: a rigid, thin segment of the outermost layer of the Earth, consisting of the Earth's crust and part of the upper mantle. The plate can be assumed to move horizontally and adjoins other plates.

plate tectonics: the study of the movement and interaction of the Earth's lithospheric plates.

When this conveyor belt is disturbed, the entire global climate is affected. For example, about 12,000 years ago, glaciers were melting rapidly, because the Earth was coming out of a glacial age. The melting glaciers discharged a lot of fresh water into the North Atlantic in a short time. The fresh water decreased the salinity of the ocean water thus reducing its density. This decrease was so much that the production of NADW was decreased. This seems to have plunged the world back into a short cold period, which lasted about 1000 years.

How Plate Tectonics Affects Global Climate

The positions of the continents on the Earth change as the Earth's **lithospheric plates** move. (**Plate tectonics** is the study of the movement and interaction of the Earth's lithospheric plates.) The arrangement of the continents has a strong effect on the Earth's climate. Think about the requirements for the development of large continental ice sheets. Glaciers form only on land, not on the ocean. For an ice sheet to develop there has to be large land areas at high latitudes, where snow can accumulate to form thick masses of ice. Where oceans occupy polar areas, accumulation of snow is limited by melting in the salty ocean waters. Polar oceans, like the Arctic Sea, around the North Pole, are mostly covered by pack ice. This ice is no more than several meters thick.

Today, the two continental landmasses with permanent ice sheets are Antarctica, in the Southern Hemisphere, and Greenland, in the Northern Hemisphere. The continent of Antarctica has not always been centered on the South Pole. About two hundred million years ago, all of the Earth's continents were welded together. They formed a single continent,

777

Geo Words

Pangea: Earth's most recent supercontinent, which was rifted apart about 200 million years ago.

called the supercontinent of **Pangea**. Pangea was eventually rifted apart into several large pieces. One of the pieces, the present Antarctica, moved slowly southward. Eventually it moved close enough to the South Pole for ice sheets to form. In recent geologic time, Antarctica has been

Figure 3 An ice-core station in Antarctica.

directly over the South Pole, so the Antarctic ice sheet has remained in existence even during interglacial periods. We know that because otherwise global sea level would have been much higher during the interglacial periods of the past million years.

At present, most of the Earth's continental land area is in the Northern Hemisphere. Much of North America and Eurasia is at a high latitude. Ice sheets can form during the parts of Milankovitch cycles that are favorable for decreased global temperatures. During times of increased global temperatures, the North American and Eurasian ice sheets have melted away completely. The picture is very different in the Southern Hemisphere. Except for Antarctica, there is not enough continental land area at high latitudes for large continental glaciers to form.

Figure 4 Mt. St. Helens is an example of a volcano associated with a plate boundary.

Plate tectonics affects climate in other ways besides changing the positions of the continents. Volcanoes like the one shown in *Figure 4* from along active plate margins. Increased activity at these margins causes increased volcanic activity. Volcanoes release carbon dioxide, which is a gas that traps heat in the atmosphere. (You'll learn more about carbon dioxide in the following activity.) In this way, plate tectonics might cause global climate to warm. However, volcanic eruptions also add dust to the atmosphere, which blocks out some solar radiation. This tends to decrease global temperatures.

Continent–continent collisions create huge mountain ranges. The Himalayas and the Alps are modern examples. Many scientists believe that the weathering of such mountain ranges uses up carbon dioxide from the Earth's atmosphere because some of the chemical reactions that break down the rock use carbon dioxide from the atmosphere. This causes global climate to cool. For example, the collisions between continents that produced the supercontinent Pangea resulted in high mountain ranges like the one at the present site of the Appalachian Mountains. The Appalachians were much taller and more rugged when they first formed—perhaps as tall as the Himalayas shown in *Figure 5*. Three hundred million years of erosion have given them their lower and well-rounded appearance. On a global scale, all that weathering (which uses up carbon dioxide) may have contributed to the period of glaciation that began about 300 million years ago and ended about 280 million years ago.

Figure 5 The geologically young Himalayan mountains formed when India collided with Asia.

Check Your Understanding

1. Explain how North Atlantic Deep Water circulates.

2. Why do glaciers form only on continents and not in oceans?

3. Explain how plate tectonics can affect global climate.

Understanding and Applying What You Have Learned

1. In the **Investigate** section, you made models designed to demonstrate how ocean currents were different during the Eocene, Miocene, and today.

   a) Assuming that the maps were accurate, in what ways was your model helpful in exploring possible differences in oceanic currents?

   b) What are some of the drawbacks or problems with your model (how is a model different than the "real world"?)

   c) What improvements could you make to the model so that it would behave in a more accurate way?

779

2. Increased weathering of rocks uses up carbon dioxide. Decreasing carbon dioxide in the atmosphere contributes to global cooling. Reconstructions of the collision between India and Asia suggest that India first collided with Asia during the Late Eocene but that most of the mountain building took place during the Miocene and later.

   a) When would you expect to observe the greatest changes in weathering rate?

   b) Why?

3. Melting glaciers discharged a lot of fresh water into the North Atlantic in a short period of time about 12,000 years ago. Adding fresh water "turned off" the North Atlantic Deep Water current for about 1000 years.

   a) What other changes could disturb the NADW?

   b) In the event that this happened, what effect would it have had on global climate?

   c) How would this change affect your community? Even if you don't live near the Atlantic Ocean, your physical environment might still be greatly affected.

Preparing for the Chapter Challenge

Using a style of writing appropriate for a newspaper, write several paragraphs containing the following material:

- Explain how the locations of the continents on the Earth affect global climate

- Explain how ocean currents affect global climate
- Explain how moving continents change ocean currents.

Inquiring Further

1. **Modeling North Atlantic Deep Water flow**

   Make a physical model of the flow of North Atlantic Deep Water. Experiment with several ideas. Think of how you might do it using actual water, and how you might do it using other materials. If your model idea is large and expensive, draw a diagram to show how it would work. If your model idea is small and simple, construct the model and see if it works.

Activity 5

How Do Carbon Dioxide Concentrations in the Atmosphere Affect Global Climate?

Goals

In this activity you will:

- Compare data to understand the relationship of carbon dioxide to global temperature.

- Evaluate given data to draw a conclusion.

- Recognize a pattern of information graphed in order to predict future temperature.

- Understand some of the causes of global warming.

Think about It

"What really has happened to winter?" You may have heard this type of comment.

- **What causes "global warming"?**

What do you think? Write down your ideas to this question in your *EarthComm* notebook. Be prepared to discuss your responses with your small group and the class.

Investigate

Part A: Atmospheric Carbon Dioxide Concentrations over the Last Century

Data on 10-year Average Global Temperature and Atmospheric Carbon Dioxide Concentration		
time interval	average global temperature (°F)	atmospheric carbon dioxide (ppm)
1901–1910	56.69	297.9
1911–1920	56.81	301.6
1921–1930	57.03	305.19
1931–1940	57.25	309.42
1941–1950	57.24	310.08
1951–1960	57.20	313.5
1961–1970	57.14	320.51
1971–1980	57.26	331.22
1981–1990	57.71	345.87
1991–2000*	57.87	358.85

*carbon dioxide data only through 1998.

1. Graph the concentration of carbon dioxide in the atmosphere from 1900 to 2000. Put the year on the *x* axis and the CO_2 levels (in parts per million) on the *y* axis.

2. On the same graph, plot the global average temperature for the same period. Put another *y* axis on the right-hand side of the graph and use it for global average temperature.

    a) Is there a relationship between carbon dioxide concentration and global average temperature? If so, describe it.

    b) What do you think is the reason for the relationship you see?

Part B: Atmospheric Carbon Dioxide Concentrations over the Last 160,000 Years

1. Look at the figure showing data from an ice core in Antarctica. The graph shows changes in concentrations of carbon dioxide and methane contained in trapped bubbles of atmosphere within the ice, and also temperature change over the same

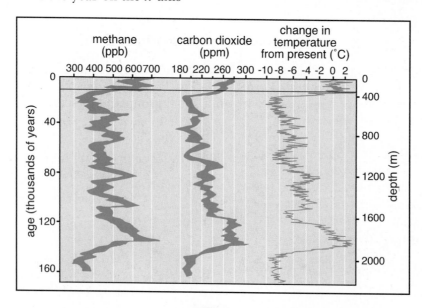

time interval. Data was obtained from the study of an ice core from the Antarctic Ice Sheet. The core was approximately 2200 m long. It was analyzed for methane concentrations (in parts per billion—left graph), carbon dioxide concentrations (in parts per million—middle graph), and inferred change in temperature from the present (in °C—right graph), over the last 160,000 years.

2. Obtain a copy of the graph. Use a straightedge to draw horizontal lines across the three maximum temperatures and the three minimum temperatures.

   a) Describe the correlation between these temperature events and changes in levels of carbon dioxide.

   b) Label likely glacial intervals (low temperatures) and interglacial intervals (higher temperatures).

   c) When did the most recent glacial interval end, according to these graphs?

Part C: The Greenhouse Effect

The phrase "greenhouse effect" is used to describe a situation in which the temperature of an environment (it could be any environment like a room, a car, a jar or the Earth) increases because incoming solar energy gets trapped because heat energy cannot easily escape. The incoming energy easily enters into the environment, but then, once it has been absorbed and is being re-radiated, it is harder for the energy to escape back out of the environment.

1. Work as a group to design an experiment to demonstrate the greenhouse warming in the atmosphere. The experiment should be simple in design, include a control element, and be performed in a short period of time (for example, a class period). The experiment will be presented to the community as a way to show the greenhouse effect.

   a) Record your design in your *EarthComm* notebook. Remember to include a hypothesis. Be sure to also include any safety concerns.

2. Decide on the materials you will use. The materials should be inexpensive and easy to get. The following is a possible list:

   • two identical 2-L plastic bottles with labels removed and tops cut off or two identical beakers

   • water

   • a clear plastic bag

   • a thermometer

   • ice cubes

   • a sunny windowsill or two similar lamps

   a) Record your list in your *EarthComm* notebook.

3. Decide on the measurements that you will make.

   a) Prepare a data table to record your observations.

4. With the approval of your teacher, conduct your experiment.

⚠️ Have the design of your experiment checked carefully by your teacher for any safety concerns.

5. Use the results of your experiment to answer the following questions:

a) How did this experiment demonstrate (or fail to demonstrate) the greenhouse effect?

b) How can this experiment serve as an analogy for atmospheric greenhouse effects?

c) Was there any difference observed between the greenhouse experiment and the control?

d) If there was a difference (or differences) describe it (them) in both qualitative and quantitative terms.

e) How did the data in each case change through time during the experiment?

f) Did the experiment reach a point of equilibrium where continuing changes were no longer observed? (Note: To answer this question, it may take longer than the class period, or, alternatively, you could hypothesize an answer to this question based on the trends of the data that you were able to gather.)

Reflecting on the Activity and the Challenge

In this activity you designed an experiment to demonstrate the greenhouse effect. You also examined the concentration of atmospheric carbon dioxide to see if it is correlated with changes in global average temperature. You discovered that an increase in carbon dioxide seems to be correlated with an increase in global average temperature. You will need this information to begin writing your article on "What is Global Warming?"

Geo Words

correlation: a mutual relationship or connection.

Digging Deeper

CARBON DIOXIDE AND GLOBAL CLIMATE

Correlation Studies

The relationship between carbon dioxide and global climate was mentioned in previous activities. When there is more carbon dioxide in the atmosphere, global temperatures are higher. When there is less carbon dioxide in the atmosphere, temperatures are lower. A scientist would say that there is a **correlation** between carbon dioxide concentration and global temperature. You might think, "Oh, that's because carbon dioxide concentration affects global temperature." And you might be right—but you might be wrong.

It is important to keep in mind always that a correlation does not, by itself, prove cause and effect. There are three possibilities: (1) carbon dioxide affects temperature; (2) temperature affects carbon dioxide; and (3) both are affected by a third factor, and are independent of one another! Any one of these three possibilities is consistent with the observations. It is the scientists' job to try to figure out which is the right answer. There are good reasons to think that the first possibility is the right one. That is because carbon dioxide is a "greenhouse gas."

What Are Greenhouse Gases?

The reason that the Earth is warm enough to support life is that the atmosphere contains gases that let sunlight pass through. Some of these gases absorb some of the energy that is radiated back to space from the Earth's surface. These gases are called **greenhouse gases**, because the effect is in some ways like that of a greenhouse. Without greenhouse gases, the Earth would be a frozen wasteland. Global temperatures would be much lower. Water vapor is the most important contributor to the greenhouse effect. Other greenhouse gases include carbon dioxide, methane, and nitrogen oxides.

How do greenhouse gases work? Most solar radiation passes through the clear atmosphere without being absorbed and is absorbed by the Earth's surface (unless it's reflected back to space by clouds first). There is a law in physics that states that all objects radiate electromagnetic radiation. The wavelength of the radiation depends on the objects' surface temperature. The hotter the temperature, the shorter the wavelength. The extremely hot surface of the Sun radiates much of its energy as visible light and other shorter-wavelength radiation. The much cooler surface of the Earth radiates energy too, but at much longer wavelengths. Heat energy is in the infrared range (*infra-* means "below," and the color red is associated with the longest wavelength in the color spectrum). See *Figure 1*.

Geo Words

greenhouse gases: gases responsible for the greenhouse effect. These gases include: water vapor (H_2O), carbon dioxide (CO_2), methane (CH_4), nitrous oxide (N_2O), chlorofluorocarbons (CF_xCl_x), and tropospheric ozone (O_3).

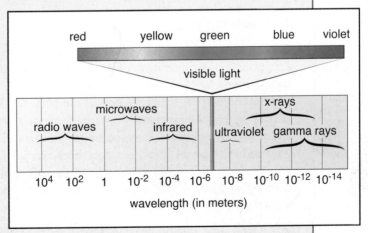

Figure 1 The spectrum of electromagnetic radiation.

785

Greenhouse gases are those that absorb some of the outgoing infrared radiation. None of them absorb all of it, but in combination they absorb much of it. They then re-radiate some of the absorbed energy back to the Earth, as shown in *Figure 2*. That is what keeps the Earth warmer than if there were no greenhouse gases.

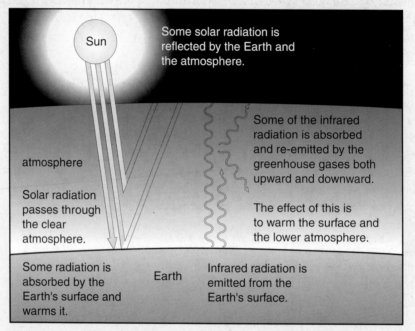

Sun

Some solar radiation is reflected by the Earth and the atmosphere.

atmosphere

Solar radiation passes through the clear atmosphere.

Some of the infrared radiation is absorbed and re-emitted by the greenhouse gases both upward and downward.

The effect of this is to warm the surface and the lower atmosphere.

Some radiation is absorbed by the Earth's surface and warms it.

Earth

Infrared radiation is emitted from the Earth's surface.

Figure 2 Schematic diagram illustrating how the greenhouse effect works.

The Carbon Cycle

Carbon dioxide is put into the atmosphere in two main ways: during volcanic eruptions, and by oxidation of organic matter. Oxidation of organic matter happens naturally in the biosphere. It occurs when plant and animal tissue decays. The organic matter is converted back to carbon dioxide and water. It also happens when animals breathe (and when plants respire too!). When you breathe, you take in oxygen, which you use to oxidize organic matter—your food. Then you breathe out carbon dioxide. Organic matter is also oxidized (more rapidly!) when it is burned. Carbon dioxide is released into the atmosphere whenever people burn wood or fossil fuels like gasoline, natural gas, or coal.

Plants consume carbon dioxide during photosynthesis. It is also consumed during the weathering of some rocks. Both land plants and algae in the ocean use the carbon dioxide to make organic matter, which acts as a storehouse for carbon dioxide. Carbon dioxide is constantly on the move from place to place. It is constantly being transformed from one form to another. The only way that it is removed from the "active pool" of carbon dioxide at or near the Earth's surface is to be buried deeply with sediments. Even then, it's likely to reenter the Earth–surface system later in geologic time. This may be a result of the uplift of continents and weathering of certain carbon-rich rocks. This transfer of carbon from one reservoir to another is illustrated in the carbon cycle shown in *Figure 3*.

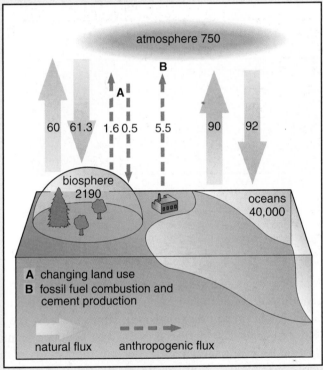

Figure 3 Global carbon cycle. Fluxes are given in billion metric tons per year and reservoirs in billion metric tons.

Carbon Dioxide and Climate

It appears that the more carbon dioxide there is in the environment, the warmer global temperatures are. Scientists have determined this from geologic data like the kind you worked with in the investigation. To what extent is this because carbon dioxide in the atmosphere acts as a greenhouse gas?

It is valuable to look at this question on two different time scales. On a scale of hundreds of thousands of years, carbon dioxide and global temperature track each other very closely. This correlation occurs through several glacial–interglacial cycles (*Figure 4*). It is not easy to develop a model in which carbon dioxide is the cause and global temperature is the effect. It's much more likely that variations are due to Milankovitch cycles. They may well explain the variation in both global temperature and carbon dioxide. On a scale of centuries, however, the picture is different. It seems very likely that the increase in carbon dioxide has been the cause of at least part of the recent global warming.

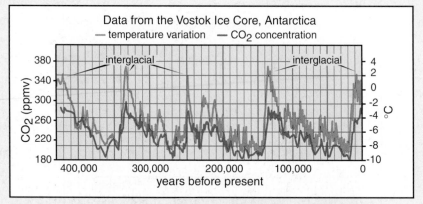

Figure 4 Variations in temperature and carbon dioxide (CO_2) concentration in parts per million by volume (ppmv) over the past 420,000 years interpreted from Antarctic ice cores. Temperature change is relative to the modern surface temperature at Vostok (–55.5°C).

Human emissions of greenhouse gases contribute significantly to the total amount of greenhouse gases in the atmosphere. For a long time humans have been adding a lot of carbon dioxide to the atmosphere by the burning of fossil fuels. This has especially increased in the past couple of centuries. Before the Industrial Revolution, carbon dioxide concentrations in the atmosphere were approximately 300 ppm (parts per million). As of 1995, carbon dioxide concentrations were almost 360 ppm. Scientists are concerned that the temperature of the Earth may be increasing because of this increasing concentration of carbon dioxide in the atmosphere.

Many nations have a commitment to reduce the total amount of greenhouse gases produced. It is their effort to reduce the risk of rapid global temperature increase. The trouble is that the size of the effect is still uncertain. Some people take the position that the increase in carbon dioxide should be reversed. They believe this is necessary even though the size of the contribution to global warming is not certain. It is their belief that the consequences would be very difficult to handle. Other people take a different position. They consider that it would be unwise to disrupt the world's present economy. They consider the future danger to be questionable. The big problem is that no one is certain that rapid global warming will take place. If it does, it may be too late to do anything about it!

Not all of the carbon dioxide released by burning of fossil fuels stays in the atmosphere. Carbon dioxide is also dissolved in ocean water. As carbon is put into the atmosphere, some of it is absorbed by the oceans. That lessens the impact of burning of fossil fuels on climate. Some people have even suggested that enormous quantities of carbon dioxide should be pumped into the oceans. That would tend, however, to just postpone the problem until later generations. Carbon dioxide is also stored by **reforestation**. Reforestation is the growth of forests on previously cleared farmland. Did you know that there is a lot more forested land in the eastern United States now than at the time of the Civil War? The Civil War took place almost 150 years ago. By some estimates, the United States is a sink, rather than a source, for carbon dioxide. Extensive reforestation is occurring east of the Mississippi, despite the continuing expansion of suburbs and shopping malls!

Figure 5 Clear-cut forest area in Olympic National Forest, Washington.

Check Your Understanding

1. List four greenhouse gases. Which gas contributes most to the greenhouse effect?

2. Explain how greenhouse gases make it possible for humans to live on Earth.

3. What are two ways in which carbon dioxide is put into the Earth's atmosphere?

Understanding and Applying What You Have Learned

1. Which of the following activities produce carbon dioxide? Which consume carbon dioxide? Explain how each can influence global climate.

   a) cutting down tropical rainforests
   b) driving a car
   c) growing shrubs and trees
   d) breathing
   e) weathering of rocks
   f) volcanic eruptions
   g) burning coal to generate electricity
   h) heating a house using an oil-burning furnace.

2. Describe the carbon cycle in your community. List the ways that carbon dioxide is produced and used up and the organisms responsible for cycling.

3. What are some difficulties involved with predicting concentrations of atmospheric carbon dioxide into the future?

4. Examine the graph your group prepared. You have gathered data through the year 2000. You have seen that this data has changed over time. Using additional graph paper, try to continue this pattern for the next 10 years.

5. The United States has a population of about 280 million people (according to the 2000 census) and uses about 70 billion gigajoules of energy a year. India has a population of about 835 million people (1990) and uses about 7 billion gigajoules of energy a year.

   a) Divide the United States' total yearly energy use by its population to find out the yearly energy use per person.
   b) Calculate the yearly energy use per person for India.
   c) Give as many reasons as you can to explain the difference.
   d) Do you think you use more or less energy than the typical American? Explain.
   e) If you wanted to use less energy, what would you do?
   f) Why is how much energy you use important when considering how much carbon dioxide is in the air?

6. Determine one source of greenhouse gas emission in your community.

   a) What gas is being produced?
   b) How is it produced?
   c) Can you think of a way to determine the level of the gas that is being produced by your community?
   d) Propose a means for limiting emissions of this gas.

Preparing for the Chapter Challenge

1. Using a newspaper style of writing, write several paragraphs in which you:

   - explain how humans have increased the concentration of carbon dioxide in the atmosphere
   - explain why scientists think that increased carbon dioxide levels might lead to global climate change.

2. Clip and read several newspaper articles containing quotations.

3. Interview a member of your community about global warming. Is this person concerned about global warming? What does he or she think people should do about it? Look over your notes from your interview. Pick out several quotations from the community member that might work well in a newspaper article.

Inquiring Further

1. **Intergovernmental Panel on Climate Change (IPCC)**

   The Intergovernmental Panel on Climate Change (IPCC) is a group of more than 100 scientists and economists from many countries that is investigating the possibility of global warming and proposing ways that the nations of the world should respond. Do some research on the IPCC and what they have reported.

2. **Earth Summit**

   Investigate the 1997 United Nations Earth Summit in New York. What did the world's nations agree to at the Summit? Have the nations stuck to their promises?

Activity 6

How Might Global Warming Affect Your Community?

Goals

In this activity you will:

- Brainstorm the ways that global warming might influence the Earth.

- List ways that global warming might affect your community.

- Design an experiment on paper to test your ideas.

- Explain some of the effects of global warming that computer models of global climate have predicted.

- Understand positive and negative feedback loops and their relationship to climate change.

- Evaluate and understand the limitations of models in studying climate change through time.

Think about It

Some scientists think that the average global temperature may increase by several degrees Fahrenheit by the end of the 21st century.

- **How do you think global warming could affect your community?**

What do you think? List several ideas about this question in your *EarthComm* notebook. Be prepared to discuss your ideas with your small group and the class.

Investigate

1. In small groups, brainstorm as many effects of higher global temperatures (a few degrees Fahrenheit) as you can. Each time you come up with a possible result of global warming, ask yourselves what effect that result might have. For example, if you think glaciers will recede, ask yourself what the implications of that would be. At this point, do not edit yourself or criticize the contributions of others. Try to generate as many ideas as possible. Here are a few ideas to get your discussion going.

   How might higher temperatures affect the following processes?

   - evaporation
   - precipitation
   - glacial activity
   - ocean circulation
   - plant life
   - animal life.

   a) List all the ideas generated.

2. As a group, review your list and cross off those that everyone in the group agrees are probably incorrect or too far-fetched. The ideas that remain are those that the group agrees are possible (not necessarily proven or even likely, but possible). It's okay if some of the ideas are contradictory. Example: More cloud cover might block out more solar radiation (a cooling effect) vs. More cloud cover might increase the greenhouse effect (a warming effect).

   a) Make a poster listing the ideas that remain. Organize your ideas on the poster using the following headings:

   - geosphere
   - hydrosphere
   - atmosphere
   - cryosphere
   - biosphere.

   b) On a separate piece of paper, write down how each of the possible results might affect your community.

3. Imagine that your group is a group of scientists who are going to write a proposal asking for grant money to do an experiment. Pick one of the ideas on your poster that you would like to investigate.

   a) On paper, design an experiment or project to test the idea. Choose ONE of the following:

   - Design an experiment that you could do in a laboratory that would model the process. Draw a diagram illustrating the model. Tell what materials you would need and how the model would work. Describe what the results would mean. Tell which parts of the experiment would be difficult to design or run, and explain why.

   ⚠️ If you plan to perform your experiments do so only under careful supervision by a knowledgeable adult.

- Design a project in which you would gather data from the real world. Include a diagram or sketches illustrating how you would gather data. Tell what kind of data you would gather, how you would get it, how frequently and how long you would collect it, and how you would analyze it. Tell which parts of the project would be difficult to design or carry out, and explain why.

4. Present your poster and your proposal for an experiment or project to the rest of the class.

Reflecting on the Activity and the Challenge

In this activity, you brainstormed ways in which an increase in global temperatures might affect the geosphere, the hydrosphere, the atmosphere, the cryosphere, the biosphere, and your community. Then you designed an experiment or a project for how you might test one of your ideas. This process modeled the way in which scientists begin to think about how to investigate an idea. This will help you to explain which possible effects of global warming would have the greatest impact on your community, and also why it is difficult for scientists to accurately predict climate change.

Digging Deeper

EFFECTS OF GLOBAL WARMING IN YOUR COMMUNITY

Problems with Making Predictions

Many scientists believe that the world's climate is becoming warmer as a result of the greenhouse gases (carbon dioxide, methane, and nitrogen-oxide compounds) that humans are adding to the atmosphere. Because the world's climate naturally experiences warmer years and colder years, it is hard to say for sure whether global average temperature has been increasing. Nowadays, remote sensing of the land and ocean surface by satellites makes it easy to obtain a good estimate of global temperature. The problem is that such techniques didn't exist in the past. Therefore, climatologists have to rely on conventional weather records from weather stations. That involves several problems. Thermometers change. The locations of the weather stations themselves often have to be changed. The move is usually away from city centers. As urban areas have been developed, they become warmer because of the addition of pavement and the removal of cooling vegetation. That is called the **urban heat-island effect**. Climatologists try to make corrections for these effects. The consensus is that global average temperature really is increasing. The questions then become: how much of the warming is caused by humankind, and how much is natural? It is known that there have been large variations in global temperature on scales of decades, centuries, and millennia long before humankind was releasing large quantities of carbon dioxide into the atmosphere (*Figure 1*).

Geo Words

feedback loops: the processes in which the output of a system causes positive or negative changes to some measured component of the system.

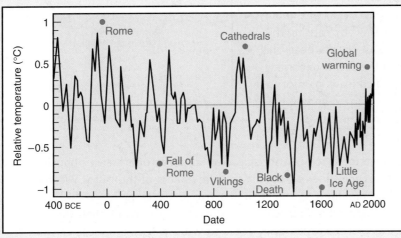

Figure 1 Relative global temperature from 400 BCE to the present.

Scientists who study global warming use very complicated computer models to try to predict what might happen. These computer models do not simply look at carbon dioxide and global temperature. They also do calculations based on many other factors that might be involved in climate change— everything from how much moisture is held in the Earth's soils to the rate at which plants transpire (give off water vapor). The physics of clouds is an especially important but also especially uncertain factor. The workings of the Earth's atmosphere are still much more complex than any computer model. Climatologists are hard at work trying to improve their models.

Drawbacks to the Computer Models

As you have seen, many factors influence the climate on Earth: carbon dioxide and other greenhouse gases, Milankovitch cycles, ocean currents, the positions of continents, weathering of rocks, and volcanic activity. Many of these factors interact with each other in ways that scientists do not fully understand. This makes it hard to make accurate predictions about how the atmosphere will respond to any particular change. The ways that different factors interact in global climate change are called **feedback loops**.

795

Feedback may be positive or negative. Positive feedback occurs when two factors operate together and their effects add up. For example, as the climate cools, ice sheets grow larger. Ice reflects a greater proportion of the Sun's radiation, thereby causing the Earth to absorb less heat. This results in the Earth becoming cooler, which leads to more ice forming. Ice cover and global cooling have a positive feedback relationship.

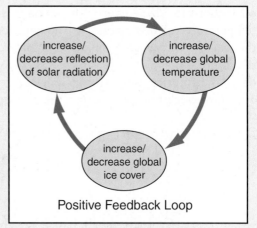

Figure 2 An example of a positive feedback loop.

In a negative feedback relationship, two variables operate in opposition with each other. Each tends to counteract the effects of the other. Weathering and carbon dioxide are one such negative feedback pair. Weathering uses up carbon dioxide, which causes the temperature to drop. When the temperature drops, weathering rates slow down, using up less carbon dioxide, and slowing the rate of temperature decrease. In this sense, weathering acts as a negative feedback for global cooling. The Earth's global climate involves many feedback loops like these. Understanding of such feedback loops is extremely important in understanding how the Earth's physical environment changes through time.

How many feedback loops are there? How do they work? These are questions that scientists are working on every day. The uncertainty about feedback loops is one thing that makes it hard to predict how the Earth's climate might respond to an increase in carbon dioxide. Another major unknown in global warming is clouds. With warmer temperatures, there will be more evaporation and therefore more clouds. Clouds reflect incoming solar radiation into space—a cooling effect. But clouds also act like a blanket to hold in heat—a warming effect. Which effect predominates? Scientists aren't yet sure.

Figure 3 How can clouds influence climate?

What Do the Computer Models Say?

Scientists continue working on their computer models of global climate. They learn as they go and make improvements all the time. Because some of the models have been used for years, scientists can test some of the predictions of past years against weather data collected recently. This helps them make changes to the computer models to make them work better.

Computer climate models have come up with some possible scenarios that may result from the increased concentration of greenhouse gases in the atmosphere. Remember, however, that these scenarios are theoretical outcomes, not certainties.

Changes in Precipitation

Warmer temperatures lead to more clouds. More clouds lead to more rain. Some models predict more rain with global warming. Others predict a change in rainfall patterns—more precipitation in the winter and less in the summer, for example. Some areas of the world would receive more rain, others less. In that respect, some countries would be winners, and others, losers. With the increase in evaporation brought on by warmer temperatures, an increase in extreme events (stronger hurricanes and winter snowstorms) might be likely.

Figure 4 Global warming may cause an increase in the number of extreme winter snowstorms.

Coordinated Science for the 21st Century

Changes in Sea Level

Glaciers around the world have been shrinking in recent years. If the Earth's climate continues to warm, more and more glacier ice sheets will melt. Meltwater is returned to the ocean. This would result in a worldwide rise in sea level. Some models predict a sea-level rise of as much as a meter by 2100.

Changes in Agriculture

In the Northern Hemisphere, where most of the world's cropland is located, warmer temperatures would cause a northward shift of the regions where certain crops are grown. Agriculture would also be affected by changes in rainfall patterns. Some regions might become too dry to support present crops. Other places might become too wet to support present crops. Many areas might continue growing traditional crops but experience declines in productivity. In other words, farmers might still grow corn in Iowa but produce fewer bushels per year. An example of a change that might decrease crop productivity is an increase in nighttime temperatures. Corn and some other grain crops do best when the temperature drops below 70°F at night. Another change that could reduce productivity (or increase costs) is a switch to wetter winters and drier summers.

Changes in Ocean Circulation

The addition of fresh meltwater from glaciers into the North Atlantic could disturb the production of North Atlantic Deep Water. The same thing happened 12,000 years ago (see **Activity 4**). The circulation of North Atlantic Deep Water helps distribute heat from solar radiation evenly around the globe. If this flow is disturbed, there might be far-reaching effects on global climate.

Check Your Understanding

1. Explain how ice cover and global cooling work as a positive feedback loop.
2. How might global warming lead to increased precipitation?
3. Why is it hard to predict how the global climate might react to an increase of carbon dioxide in the atmosphere?

Understanding and Applying What You Have Learned

1. Using the information in **Digging Deeper**, add to the poster you made in the **Investigate** section.

2. For each new item you added to your poster, hypothesize how your community would be affected by that outcome.

3. Using what you learned in **Digging Deeper**, modify the experiment or model you proposed in the **Investigate** section OR design another experiment or model.

Preparing for the Chapter Challenge

1. Using a style of writing appropriate for a newspaper, write a paragraph on each of the following possible effects of global warming. Make sure you make it clear that these are only possible scenarios, not certainties:

   - changes in rainfall patterns
   - increase in extreme events
   - changes in sea level
   - changes in ocean currents
   - changes in agriculture.

2. Write an editorial about how you think your community should respond to global warming. Should your community wait for further research? Should your community take action? What kind of action should be taken? Would these actions benefit your community in other ways, in addition to slowing global warming?

Inquiring Further

1. **Community energy use**

   Make a plan for calculating how much energy your school uses for heating, air conditioning, lights, and other electrical uses. Make a plan for calculating the energy used in gasoline for students, teachers, staff, and administrators to travel to and from school each day. How could you test your estimates to see how accurate they are? What are some ways your school could reduce its energy use? How can a reduction in energy use influence climate?

2. **"CO_2-free" energy sources**

   Investigate some sources of energy that do not produce carbon dioxide, like solar and wind power.

3. **Climate change and crops**

   Call your state's cooperative extension service and find out what are the top three crops grown in your state. Visit the *EarthComm* web site to determine if your state's cooperative extension service has a web site. What are the optimal climatic conditions for maximizing productivity of these crops? How might climate changes due to global warming affect farmers who grow these crops in your state?

Earth Science at Work

ATMOSPHERE: *Plant Manager*
Some companies are providing their workers with retraining to use new equipment that has been designed to reduce the emission of greenhouse gases and control global warming.

BIOSPHERE: *Farmer*
Agriculture is an area of the economy that is very vulnerable to climate change. Climate change that disturbs agriculture can affect all countries in the world. However, there are also steps that farmers can take to reduce the amount of carbon dioxide and other greenhouse gases.

CRYOSPHERE: *Mountaineering Guide*
Regions where water is found in solid form are among the most sensitive to temperature change. Ice and snow exist relatively close to their melting point and frequently change phase from solid to liquid and back again. This can cause snow in mountainous areas to become unstable and dangerous.

GEOSPHERE: *Volcanologist*
Volcanoes can emit huge amounts of carbon dioxide gas as well as sulfur dioxide gas into the atmosphere with each eruption. Both of these substances can have adverse effects on the atmosphere, including global warming as a possible result.

HYDROSPHERE: *Shipping Lines*
Even under "normal" climate conditions, ocean circulation can vary. A changing climate could result in major changes in ocean currents. Changes in the patterns of ocean currents and storm patterns will have important consequences to shipping routes.

How is each person's work related to the Earth system, and to Climate Change?

Chapter 13

Energy Resources
...and Your Community

Getting Started

As the population of a community increases, so does the demand for energy resources for transportation, electricity, and heating fuels. Ensuring a supply of energy to meet the growing needs of a community requires careful analysis and planning.

Think about all of the ways that you have used energy resources in your day so far, starting with when you got up in the morning until you arrived at this class.

• What was the source of energy for each of the activities?

What do you think? Sketch out some of your ideas about energy resources on paper. First, make a list of all the ways that energy resources have been used in your day so far, starting with the moment you woke up (the alarm clock, for instance) and ending with your present classroom. There should be at least five items on your list. Next, make a column for the energy resource responsible for the activity (coal burned to produce electricity for the alarm clock, for instance). Make another column for the source of this energy resource (coal deposit in New Mexico, for example). Finally, make a column for energy alternatives for each activity. For instance, one energy alternative to taking the bus to school is to ride your bike to school. An energy alternative to taking a ten-minute hot shower is to take a five-minute hot shower. The alternatives do not necessarily have to reduce the energy required. They could have other benefits, such as reducing an environmental hazard. Be prepared to discuss your table with your small group and the class.

Scenario

Community leaders often depend upon experts to outline a 10-year plan that addresses the impact of an increase in population on energy consumption and supply. Your community has called upon your *EarthComm* classroom to evaluate energy consumption and use in your community. Community leaders need realistic solutions or alternatives for energy use in your community, to help prevent possible energy shortages while maintaining the quality of the environment.

Chapter Challenge

Your challenge is to produce a report that is written for a general audience. In the report, you must critically analyze energy use in your community on the assumption of a population growth of 20%, and to provide realistic solutions to avoid an energy-supply shortage. Your report needs to help community members understand the origin, production, and consumption (use) of energy resources from an Earth System perspective. You need to address the following points:

- What are the current energy uses and consumption rates in your community?

- How will energy needs change if the population of your community increases by 20%?

- Will your community be able to meet these needs, and at what costs?

- What solutions or alternatives exist that may help to avoid a potential energy crisis?

Your report should explain at least one example for each of the following questions:

- What is the difference between renewable and nonrenewable energy resources?

- How are energy resources used to do work?

- How are energy resources formed, discovered, and produced?

- How much energy conservation is possible in your community? What steps can be taken?

- Does the production and consumption of energy affect the environment?

Assessment Criteria

Think about what you have been asked to do. Scan ahead through the chapter activities to see how they might help you to meet the challenge. Work with your classmates and your teacher to define the criteria for assessing your work. Record all of this information. Make sure that you understand the criteria as well as you can before you begin. Your teacher may provide you with a sample rubric to help you get started.

Activity 1

Exploring Energy Resource Concepts

Goals

In this activity you will:

- Investigate heat transfer by the processes of conduction, convection, and radiation.

- Investigate the conversion of mechanical energy into heat.

- Learn about the Second Law of Thermodynamics and how it relates to the generation of electricity.

Think about It

A car moving along a mountain road has energy. It has energy due to its motion (kinetic energy), energy due to its position in a gravity field (potential energy), and energy stored as fuel in its gas tank (chemical energy).

- **Classify each item below as having kinetic energy, potential energy, or chemical energy:**

    a) **a rock balanced at the edge of a cliff**
    b) **a piece of coal**
    c) **a landslide**
    d) **a roller-coaster car**
    e) **a diver on a 10-m platform**
    f) **a car battery**
    g) **tides**

What do you think? Record your ideas in your *EarthComm* notebook. Be prepared to discuss your responses with your small group and the class.

Investigate

Energy can neither be created nor destroyed (except in nuclear reactions), but it can be changed from one form into another. The following activities will help you to explore basic concepts that govern the use of energy.

Part A: Heat Transfer
Station 1

1. Put your hand close to a 100-W light bulb and notice the heating that occurs in your hand. This is similar to the heat generated from direct sunlight.

   a) Describe what happens to the temperature of your hand as you move it slowly toward and away from the bulb.

   b) Hold a piece of paper between your hand and the light bulb. Describe and explain the change in temperature of your hand.

   c) Compare and explain the temperature difference of your hand when you hold it above the light bulb versus holding it near the side of the bulb.

 Be careful not to touch the hot bulb.

Station 2

1. Which cup will keep the water hot for a longer amount of time, a metal cup or a Styrofoam® cup? Why?

   a) Write down your hypothesis.

2. Design an experiment to test your hypothesis.

   a) Record your experimental design in your notebook.

b) With the approval of your teacher, carry out your experiment, and record your observations.

3. Five minutes after you fill the cup, place your hand around each of the cups.

   a) Which one feels hotter? Why?

 Be sure your teacher approves your design before you begin. The water should not be hot enough to scald. Wipe up spills immediately. Use alcohol thermometers only.

Station 3

1. Set up two solar cookers as shown in the diagrams below. One is a standard solar cooker and the other is an identical solar cooker inside an insulated box.

   a) What differences do you expect in the temperature inside the two solar cookers over time? Write down your hypothesis in your notebook.

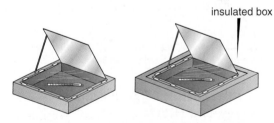

insulated box

2. Design an investigation to test your hypothesis. Your design should include a plan to measure the temperature in each solar cooker and to record data every minute for at least 25 minutes.

   a) Set up a table to record your data.

3. Place a thermometer in each solar cooker and close the lids. You will want to be able to read the thermometer without blocking the path of solar energy and without opening the boxes.

   a) Record and graph the data.

4. Use the evidence that you have collected to answer the following questions:

   a) How did your results compare with your hypothesis?

   b) What heating mechanism causes the cookers to heat up in the first place?

   c) What are the different heat transfer mechanisms that are taking place in the cookers? Use diagrams to record your ideas in your notebook.

   d) What mechanism keeps the heat from escaping?

   e) What improvements could be made to the cooker if you had to do it over again?

⚠️ Be careful when you touch items after they have been in the solar cooker. They will be hot.

Part B: Kinetic Energy, Potential Energy, and Heat

1. The following is a thought experiment. The graph shows the path of a small lump of modeling clay that is thrown into the air.

   a) Copy the graph onto a sheet of graph paper.

2. Imagine that you had thrown the clay into the air so that it landed on a tabletop. In your group, discuss and record your ideas about the following:

   a) How does the kinetic energy of the piece of clay change over time? When is it highest? When is it lowest?

   b) How does the potential energy of the lump of clay change over time? When is it highest? When is it lowest?

   c) How was kinetic energy transformed into potential energy? When did this happen?

   d) How was kinetic energy transformed into heat? When did this happen?

   e) Find a way to represent the changes in these three forms of energy over time. Record your ideas on the sheet of graph paper that shows the path of the modeling clay.

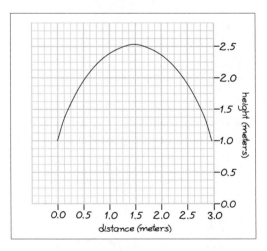

Part C: Energy Units and Conversions

1. Look at the conversion table. (In this activity you will record all your data in metric units. The table gives both metric and English equivalents to all the units that you will be using in this activity. Refer to this table whenever necessary.)

a) Begin a concept map to show how the units are interconnected. Complete the concept map as you work through this part of the activity.

Energy Conversion Table
Heat
I kcal (kilocalorie) = the heat needed to raise the temperature of one kilogram of water from 14.5°C to 15.5°C I Btu (British thermal unit) = the heat needed to raise the temperature of one pound of water from 60°F to 61°F I kcal = 1000 cal = 3.968 Btu
Force, mass, and velocity
I kg = 0.069 slug acceleration of gravity (g) = 9.8 m/s² = 32 ft/s² I N (newton) = I J/m (joule per meter) = 0.225 pounds I m/s = 3.28 fps (feet per second) = 2.24 mph (miles per hour)
Energy and work (the mechanical equivalent of heat)
I kcal = 1000 cal = 4184 J (joules) I Btu = 252 cal = 777.9 ft-lb (foot-pounds) = 1055 J I kWh = 3,600,000 J = 3413 Btu I quad (Q) = 10^{15} Btu
Power (the rate at which work is done)
I W (watt) = I J/s (joules per second) I hp (horsepower) = 550 ft-lb/s = 746 W

2. Do you think that you can produce power equal to that of a 100-W light bulb? Obtain and weigh a steel ball.

a) Record the weight of the ball in newtons. As shown by the conversion tables, a newton is a unit of force. The weight of the ball is the same as the force exerted on the ball by the pull of gravity. Show your work in your *EarthComm* notebook.

3. Work is defined as the product of a force times the distance through which the force acts. The work needed to lift the steel ball a certain vertical distance is the force (weight of the ball, in newtons) times the vertical distance, or

$$W = F \bullet d,$$

where W is work in joules (J), F is force in newtons (N), and d is the height it is raised in meters (m).

a) In order for an object to obtain kinetic energy, work must be done on it. Calculate the work necessary to lift the steel ball to a height of 2 m.

4. Power is the rate at which work is done. The power you produce when you lift the ball is equal to the work divided by the time it took to lift the ball. If you lift the ball a number of times in a certain time period, the average power you produce is equal to the work of each lift, times the number of lifts, divided by the total time it took to do all of the lifting. Remember that the work is measured in joules and the time is measured in seconds.

$$P = W/t$$

where P is power in watts (W), W is work in joules (J), and t is time in seconds (s).

a) Calculate the power produced by lifting the ball 10 times in one minute. Note from the table that the unit for power is the watt. One watt = one joule per second.

5. In your group, discuss what a person would have to do to produce as much power as a 100-W bulb. Examples include running (how fast?) or climbing stairs (how fast?). Do this as a "thought experiment", one that you will describe (with calculations) but not conduct.

a) Record your thought experiment. Show your calculations.

6. The energy it took to produce the power to the ball came from chemical energy. In this case, the chemical energy was energy stored in the food you ate for breakfast. Assume that your body was 100% efficient (all of the stored energy is converted into kinetic energy).

a) Calculate the number of times you could lift the ball to equal a 200 Calorie candy bar (use the table and remember that one food calorie = 1 kilocalorie or 1000 calories).

b) In nature, no energy change is 100% efficient. Some energy is lost to the environment. In the case of lifting the ball, what form does the lost energy take?

7. As a class discuss the question of whether a person can produce as much power as a 100-W light bulb.

a) Record the results of your discussion.

Reflecting on the Activity and the Challenge

In **Part A** of this activity you looked at different ways that heat transfer occurs. **Part B** helped you to understand the concepts of potential and kinetic energy. In **Part C** you explored concepts of work, power, and units of energy. You also completed calculations to determine whether or not the exertion required to lift a ball can be equivalent to the power produced by a 100-W light bulb. These activities will help you think about how energy is transformed into a form that you can use. It will also help you think about ways to conserve energy resources so that your community can meet its growing energy needs.

Digging Deeper

HEAT AND ENERGY CONVERSIONS

Heat Transfer

Heat is really the **kinetic energy** of moving molecules. **Temperature** is a measure of this motion. The term **heat transfer** refers to the tendency for heat to move from hotter places to colder places. Many of the important aspects of heat transfer (see *Figure 1*) that you observed with the solar cooker had to do with heat conduction, which is one of the processes of heat transfer. All matter consists of atoms. At temperatures above **absolute zero** (about −273°C, the coldest anything can be!), the atoms vibrate. You sense those vibrations as the temperature

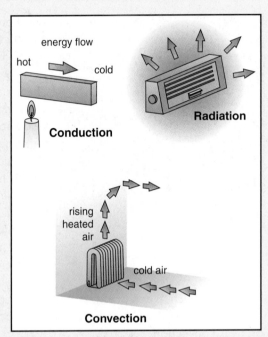

Figure I Three types of heat transfer.

of the material. The stronger the vibration, the hotter the material. When a hotter material is in contact with a colder material, collisions between adjacent vibrating atoms in the two materials cause the energy of the vibrations to even out, cooling the hot material and warming the cold material.

Conduction is the type of heat transfer you experience when you take a hot bath, when you heat a piece of metal, or when the air cools a cup of hot coffee left on top of a table. For instance, when you put a metal pot on the stove, only the bottom of the pot is in contact with the burner, yet the heat flows through the entire pot all the way to the handle. Materials differ greatly in how well they conduct heat. In **thermal insulators**, like Styrofoam, crumpled paper, or a down jacket, the heat flows slowly. Thermal insulators like these contain a large amount of trapped air. Air is a poor conductor because the air molecules are not in constant contact. Metals, on the other hand, are very good conductors of heat. Heat conduction is very important in your community. Keeping your home warm in the winter ➔

Geo Words

heat: kinetic energy of atoms or molecules associated with the temperature of a body of material.

kinetic energy: a form of energy associated with motion of a body of matter.

temperature: a measure of the energy of vibrations of the atoms or molecules of a body of matter.

heat transfer: the movement of heat from one region to another.

absolute zero: the temperature at which all vibrations of the atoms and molecules of matter cease; the lowest possible temperature.

conduction: a process of heat transfer by which the more vigorous vibrations of relatively hot matter are transferred to adjacent relatively cold matter, thus tending to even out the difference in temperature between the two regions of matter.

thermal insulator: a material that impedes or slows heat transfer.

Geo Words

convection: motion of a fluid caused by density differences from place to place in the fluid.

convection cell: a pattern of motion in a fluid in which the fluid moves in a pattern of a closed circulation.

electromagnetic radiation: the movement of energy, at the speed of light, in the form of electromagnetic waves.

would be very difficult (and expensive) without the insulating properties of the walls and the roof. Improving the insulation of your home by using insulating materials like those shown in *Figure 2* can greatly reduce the amount of energy needed to heat or cool your home.

Figure 2 Thermal insulation helps to keep your home warm. It conserves energy needed for space heating.

Another form of heat transfer is **convection**, which is important in liquids and gases. When a liquid or a gas is heated, its density decreases. That causes it to rise above its denser surroundings. In a room heated with a wood stove or a steam or hot water radiator, for instance, a natural circulation pattern is developed. The hot air from the stove rises towards the ceiling and cooler air travels down the walls and across the floor towards the stove. That kind of circulation is called a **convection cell**. Heat convection is also very important to your community. Many of the features of weather, such as sea breezes and thunderstorms, are caused by convection. Also, the way that you heat or cool your home depends strongly on heat convection.

Figure 3 The Sun emits electromagnetic radiation that warms the surface of the Earth.

A third form of heat transfer is **electromagnetic radiation**. Everything emits electromagnetic radiation. Examples of electromagnetic radiation are radio and television waves, visible light, ultraviolet light, and x-rays. Hotter materials emit more energy of electromagnetic radiation than colder materials. The warmth you feel from a hot fire, the Sun, or a light bulb is due to electromagnetic radiation traveling (at the speed of light!) from the hot object to you.

Radiation is important to the community for many reasons. Solar radiation causes things in the community to be heated. Solar radiation heated the solar cooker. It also heats someone standing in the sunshine on a cold winter day, or a parked car in the Sun in the summer with all its windows closed. If a building is designed appropriately, the heat from the Sun can substitute for heat from other energy resources for space heating and hot water. Using insulation or light-colored reflective materials reduces solar heating in warmer months when heat is not desired.

Energy, Work, and Power

In the investigation, you dealt with four forms of energy: energy of motion, called kinetic energy; energy of position, called **potential energy**; energy stored in the chemical bonds of a substance, called **chemical energy**, and heat. Kinetic energy and potential energy together are called **mechanical energy**. You know that objects in motion have energy, because of what they can do to you when they hit you. The energy of motion is called kinetic energy. The more mass the body has, and the faster it is moving, the more kinetic energy it has. When you threw (or imagined throwing) the lump of modeling clay up in the air, you gave it kinetic energy. The kinetic energy was gradually converted to potential energy. When the lump reached its highest point, its kinetic energy was at a minimum. On the way down, the lump regained its kinetic energy. When it hit the table, all of its kinetic energy was changed to heat. The change in temperature was so small that you would need a very sensitive thermometer to measure it. That's an example of how kinetic energy is changed to heat energy by **friction**. When you rub your hands together to keep them warm, you are converting kinetic energy to heat by friction. Of course, you are always resupplying your hands with kinetic energy by the action of your arm muscles.

Figure 4 A coal-powered train is an example of how chemical energy stored in coal is converted into heat energy that in turn is converted to mechanical energy.

Geo Words

potential energy: mechanical energy associated with position in a gravity field; matter farther away from the center of the Earth has higher potential energy.

chemical energy: energy stored in a chemical compound, which can be released during chemical reactions, including combustion.

mechanical energy: the sum of the kinetic energy and the potential energy of a body of matter.

friction: the force exerted by a body of matter when it slides past another body of matter.

Coordinated Science for the 21st Century

Geo Words

work: the product of the force exerted on a body and the distance the body moves in the direction of that force; work is equivalent to a change in the mechanical energy of the body.

force: a push or pull exerted on a body of matter.

power: that time rate at which work is done on a body or at which energy is produced or consumed.

watt: a unit of power.

horsepower: a unit of power.

biomass: the total mass of living matter in the form of one or more kinds of organisms present in a particular habitat.

In physics, the term **work** has a very specific meaning. Work is equal to the **force** you exert on some object multiplied by the distance you move the object in the direction of the force. The importance of work is that it causes a change in the mechanical energy (kinetic and/or potential) of the object. When you threw the lump of modeling clay up in the air, your hand did the work. It exerted an upward force on the clay for a certain distance to give it its kinetic energy.

Power is the term used for the rate at which work is done or at which energy is produced or used. Think once more about the now-famous lump of modeling clay. You could have given it its upward kinetic energy by swinging your arm upward slowly for a long distance, generating low power but for a long time. Or, you could have swung your arm upward fast over only a short distance, generating high power but for only a short time. Whenever your muscles move your own body or some other object, you are generating power. The **watt** is the unit of power that is commonly used to describe the power of electrical devices. **Horsepower** is the unit of power that is often used to describe the power of other mechanical devices.

Converting Heat into Mechanical Energy

Figure 5 Coal is fed by a conveyor into a combustion chamber, where it is burned.

You have explored the idea that mechanical energy always tends to be converted into heat by friction. Nothing on Earth is completely frictionless, although some things, like air-hockey pucks, involve very little friction. Only in the emptiness of outer space can bodies move without friction. But how about energy conversion in the opposite direction: from heat to mechanical energy?

The conversion of heat into mechanical energy is central to most of the processes for producing electricity from energy resources. These resources include coal, natural gas, petroleum, sunlight, **biomass**, and nuclear energy. In these processes, water is heated to produce steam. When water boils (at atmospheric pressure) it undergoes about a thousand-fold increase in volume. The pressure of the steam exerts a force that does work to increase the kinetic energy of a turbine. The steam pressure is used to turn a turbine that generates electricity.

The **Second Law of Thermodynamics** states that you can never completely convert heat into mechanical energy. In fact, in converting any form of energy into another, there is always a decrease in the amount of "useful" energy. Stated in general terms, the **efficiency** of a machine or process is the ratio of the desired output (work or energy) to the input:

$$\% \text{ Efficiency} = \frac{\text{useful energy or work out}}{\text{energy or work in}} \times 100$$

Electrical power plants have efficiencies of about 30%. An efficiency of 33% means that for every three trainloads of coal that are burned to produce electricity, the chemical heat energy from only one of those trainloads is converted to electricity.

Some methods for generating electricity are not based on the conversion of heat to mechanical energy. Hydropower and wind power are examples. In hydropower, the mechanical energy of the falling water is converted directly to the mechanical energy of the rotating turbine. The efficiency of hydropower is only about 80% rather than 100%, however, because of friction and the incomplete use of available mechanical energy. Similarly, the wind already has mechanical energy. The efficiency of wind power is no greater than about 60%, mainly because some of the wind goes around the turbine without adding to its rotation. Actual efficiencies of most wind turbines range from 30% to 40% (The windmills shown in *Figure 6* have an efficiency of only 16%.) By comparison, the efficiency of a normal automobile engine is about 22%.

Figure 6 The efficiency of these windmills is only about 16%. Modern wind turbines have efficiencies between 30% and 40%.

Check Your Understanding

1. What are some of the methods for generating energy that are based on the conversion of heat to mechanical energy?

2. Describe the three processes of heat transfer.

3. In your own words define mechanical energy.

4. Why can't the efficiency of a device be more than 100%?

5. Why is the efficiency of a device always less than 100%?

Understanding and Applying What You Have Learned

1. a) Explain how all of the different parts of the solar cooker work in terms of different heat transfer processes.

   b) How would you adapt your solar cooker to make it more effective and efficient?

2. Describe how a one-liter and two-liter container of water in the same oven differ in their heat and their temperature.

3. Describe how you think heat is transferred in the following situations:

   a) A cold room becomes warm after turning on a hot-water radiator.

   b) Your hand is heated as you grasp the handle of a heated pan on the stove.

   c) The bottom of a pan is heated when placed on an electric burner.

   d) A cold room becomes warm after window drapes are opened on a sunny day.

4. If the energy input of a system is 2500 cal and the energy output is 500 cal, what is the efficiency of the system?

5. A 300 hp engine is equivalent to how many foot pounds per second? In your own words state what this means.

6. When you drive in a car, energy is not lost, even though gasoline is being used up. Use what you have learned in this activity to explain what happens to this energy.

Preparing for the Chapter Challenge

Your **Chapter Challenge** is to help community members think about how they will meet their growing energy needs. Draft an introduction to your report. Use what you have learned in this activity to explain how energy resources are used to do work. Help people to understand how mechanical energy is converted to heat in the devices they use in their everyday lives. You might also begin to think about steps that community members might take to improve their energy efficiency.

Inquiring Further

1. **Perpetual-motion machines**

   The United States Patent Office receives many applications for perpetual-motion machines. All the applications are turned down. What is a perpetual-motion machine, and why can no one get a patent for one?

2. **Improving efficiencies of electricity generation**

Innovative methods for power generation are now being developed to improve the efficiency of generating electricity from energy resources. What are some new methods for generating electricity from coal, natural gas, or oil that have improved efficiencies? Visit the *EarthComm* web site to help you find this information.

3. **History of science**

Research the work of James Prescott Joule. A Scottish physicist, Joule conducted a famous experiment to observe the conversion of mechanical energy to heat energy. How did the experiments help Joule to conclude that heat is a form of energy?

4. **Solar cooking applications**

In your investigation, you explored a model of one kind of solar food cooker. Research:

• How people are using solar cookers and reducing the consumption of wood and fossil fuels for cooking food.

• Where are solar cookers most commonly used?

• Are they a suitable energy alternative for your community?

• How does the use of a solar cooker reduce the effect on the biosphere?

Activity 2   Electricity and Your Community

Goals

In this activity you will:

- Compare energy resources used to generate electricity in the United States to other countries.

- Identify the major energy sources used to produce electricity in the United States and your state.

- Identify trends and patterns in electricity generation.

- Understand the difference between electric power and electric energy.

- Be able to describe commonly used methods of generating electric power.

Think about It

Electricity is a key part of life in the United States. Factories, stores, schools, homes, and most recreational facilities depend upon a supply of electricity. The unavailability of electricity almost always makes the news!

- **What is electric energy?**
- **What are some consequences of not having electricity when it is needed?**

What do you think? Record your ideas in your *EarthComm* notebook. Be prepared to discuss your responses with your small group and the class.

Investigate

Part A: Global and United States Electricity Generation

1. Use the data table showing world net electricity generation by type to answer the questions on the following page.

World Net Electricity Generation by Type, 1998 (in billion-kilowatt hours)					
Region Country	Fossil Fuels	Hydro	Nuclear	Geothermal and Other[1]	Total
North America					
Canada	148.7	328.6	67.7	6.1	551.1
Mexico	134.2	24.4	8.8	5.4	172.8
United States	2550.0	318.9	673.7	75.3	3617.9
The Caribbean & South America					
Bolivia	2.0	1.4	0.0	0.1	3.5
Brazil	15.6	288.5	3.1	9.7	316.9
Puerto Rico	17.0	0.3	0.0	0.0	17.3
Venezuela	21.6	52.5	0.0	0.0	74.0
Western Europe					
France	52.9	61.4	368.6	2.8	485.7
Spain	93.2	33.7	56.0	3.5	186.4
Switzerland	2.3	33.1	24.5	1.1	61.1
United Kingdom	235.3	5.2	95.1	6.2	341.9
Eastern Europe & Former U.S.S.R.					
Bulgaria	20.2	3.3	15.5	0.0	39.0
Romania	27.5	18.7	4.9	0.0	51.1
Russia	530.1	157.9	98.3	0.0	786.3
Middle East					
Cyprus	2.8	0.0	0.0	0.0	2.8
Saudi Arabia	116.5	0.0	0.0	0.0	116.5
Africa					
Angola	0.5	1.4	0.0	0.0	1.9
Egypt	47.1	12.1	0.0	0.0	59.2
Ethiopia	0.0	1.6	0.0	0.0	1.6
Morocco	11.6	1.7	0.0	0.0	13.4
Far East & Oceania					
China	880.2	202.9	13.5	0.0	1096.5
Japan	571.3	91.6	315.7	23.8	1002.4
Malaysia	52.5	4.8	0.0	0.0	57.3
Mongolia	2.5	0.0	0.0	0.0	2.5
Nepal	0.1	1.1	0.0	0.0	1.2
Total	**5535.7**	**1645.1**	**1745.4**	**134**	**9060.3**

[1] Geothermal and Other consists of geothermal, solar, wind, wood, and waste generation.
Source: Adapted from the Energy Information Administration/International Energy Annual 1999 report.

a) List the three countries that generate the most electricity.

b) What type of electricity generation is used most by these areas? Which is used least?

c) List the three countries that generate the least electricity.

d) What type of electricity generation is used most by these areas? Which is used least?

e) How do the resources for electricity generation differ between the top and bottom regions? How do you account for the differences?

2. Consider electricity generation by fuel type.

a) Without including the United States, rank global electricity generation by fuel type from highest to lowest.

b) Rank United States electricity generation by fuel type from highest to lowest.

c) How does the electricity generation in the United States match global electricity generation? Why do you think this is so?

3. Devise a way to represent the data for the United States graphically (either a pie chart or a histogram plot).

a) Complete your graph in your *EarthComm* notebook.

Part B: Investigating Electricity Generation in your State

1. Go to the *EarthComm* web site to visit the Department of Energy's Energy Information Administration web page. Click on your state to obtain your "state electricity profile." Use the information at the site to answer the following questions:

a) Is your state a net importer or net exporter of electricity? What does this mean?

b) What energy resources are used to generate electricity in your state?

c) Which energy resource does your state depend upon the most to generate electricity?

d) What are the trends in energy resource use for electric power generation in your state over time?

e) How has the generation of electricity changed in your state over time? How much has it grown in the most recent 10-year period? What is the annual rate of growth?

f) How does the growth in electricity generation in your state compare to the rate of population growth in your state? How might you explain the relationship?

g) How do the types of resources used by your state for electricity generation compare with the averages for the world and the United States, which you looked at in **Part A** of the investigation?

h) What sector (residential, commercial, industrial, or other) consumes the most energy for electricity in your state? What sector consumes the least?

i) What resources does your state use most for residential and commercial purposes? for transportation? for industry?

2. Review the data and figures for the electricity profile for your state.

   a) How do they help you to meet your **Chapter Challenge**?

Reflecting on the Activity and the Challenge

In this activity, you looked at the different types of resources that are used to generate electricity. You also investigated the trends in energy resource use and electricity generation in your state. Knowing where and how your state uses energy resources to generate electricity will help you to think about how to meet the energy needs of an expanding population in the **Chapter Challenge**.

Digging Deeper

MEETING ELECTRICITY NEEDS

Energy Resources

As the world's population increases and there is continued comparison to the current western European, Japanese, and North American living standards, there is likely to be demand for more energy. Several different energy resources can be used either for heating or cooling, electricity generation, industrial processes, and fuels for transportation. Energy resources available in the world include coal, **nuclear fission**, hydroelectric, natural gas, petroleum, wood, wind, solar, refuse-based, biomass, and oceanic (tides, waves, vertical temperature differences). In addition, nuclear fusion has been proposed as the long-term source, although research progress toward making it a reality has been very slow.

Geo Words

nuclear fission: the process by which large atoms are split into two parts, with conversion of a small part of the matter into energy.

Figure 1 Nuclear power can come from the fission of uranium, plutonium, or thorium or the fusion of hydrogen into helium.

Geo Words

electric power: power associated with the generation and transmission of electricity.

electric energy: energy associated with the generation and transmission of electricity.

fossil fuel: fuel derived from materials (mainly coal, petroleum, and natural gas) that were generated from fossil organic matter and stored deep in the Earth for geologically long times.

geothermal energy: energy derived from hot rocks and/or fluids beneath the Earth's surface.

photovoltaic energy: energy associated with the direct conversion of solar radiation to electricity.

turbine: a rotating machine or device that converts the mechanical energy of fluid flow into mechanical energy of rotation of a shaft.

Generating Electric Energy

Energy resources are used to generate electricity, an energy source with which you are very familiar. **Electric power** is the rate at which electricity does work—measured at a point in time. The unit of measure for electric power is a watt. The maximum amount of electric power that a piece of electrical equipment can accommodate is the capacity or capability of that equipment. You can check the tags or labels on electrical appliances for this information, such as a "1200-W hair dryer" or "40-W stereo receiver". **Electric energy** is the amount of work that can be done by electricity. The unit of measure for electric energy is the watt-hour. A 1200-W hair dryer used for 15 min would require (theoretically) 300 Wh of electric energy.

Fossil fuels supply about 70% of the energy sources for electricity in the United States. Coal, petroleum, and natural gas are currently the dominant fossil fuels used by the electrical power industry. When fossil fuels are burned to generate electricity, a variety of gases and particulates are formed. If these gases and particulates are not captured by some pollution control equipment, they are released into the atmosphere. Other sources of energy can also be converted into electricity, including **geothermal energy**, solar thermal energy, **photovoltaic energy**, and biomass. These alternative sources of energy have many advantages over fossil fuels, as you will explore in **Activity 5**.

Fossil Fuels and Nuclear Energy

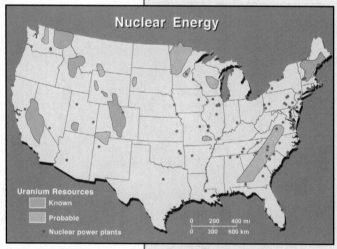

Figure 2 The location of nuclear power plants and uranium resources in the continental United States.

Most of the electricity in the United States is produced in steam **turbines**. A turbine converts the kinetic energy of a moving fluid (liquid or gas) to mechanical energy. In a fossil-fueled steam turbine, the fuel is burned in a boiler to produce steam. The resulting steam then turns the turbine blades that turn the shaft of the generator to produce electricity. In a nuclear-powered steam turbine, the boiler is replaced by a reactor containing a core of nuclear fuel (primarily enriched uranium). Heat produced in the reactor by fission of the uranium is used to make steam. The steam is then passed through the turbine generator to produce electricity, as in the fossil fueled steam turbine. *Figure 2* shows the nuclear-power plant locations and uranium resources available in the United States.

Hydroelectric Power

Water is the leading **renewable energy source** used by electric utilities to generate electric power. **Hydroelectric power** is the result of a process in which flowing water is used to spin a turbine connected to a generator. Hydroelectric plants operate where suitable waterways are available. Many of the best of these sites have already been developed. (*Figure 3* shows the hydroelectric plants developed in the United States.) Seventy percent of the hydroelectric power in the United States is generated in the Pacific and Rocky Mountain States. The two basic types of hydroelectric systems are those based on falling water and natural river current. In a falling-water system, water accumulates in reservoirs created by dams. This water then falls through conduits and applies pressure against the turbine blades to drive the generator to produce electricity. In the second system, called a run-of-the-river system, the force of the river current (rather than falling water) applies pressure to the turbine blades to produce electricity. Because they do not store water, these systems depend upon seasonal changes and stream flow.

Geo Words

renewable energy source: an energy source that is powered by solar radiation at the present time rather than by fuels stored in the Earth.

hydroelectric power: electrical power derived from the flow of water on the Earth's surface.

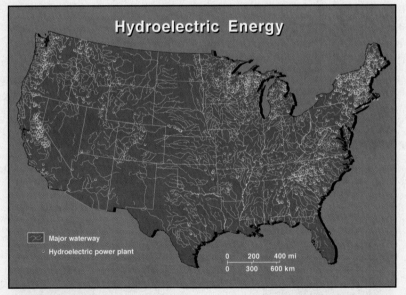

Figure 3 The location of hydroelectric power plants relative to major waterways in the continental United States.

Generating electricity using water has several advantages. The major advantage is that water, a renewable resource, is a source of cheap power. In addition, because there is no fuel combustion, there is little air pollution in comparison with fossil fuel plants and limited thermal pollution compared with nuclear plants. They can start quickly because they do not need to wait for water to be heated into steam. Also, the flow of water can be adjusted to make quick changes in power output during peak demands for electricity. Like other energy sources, the use of water for generation has limitations, including environmental impacts caused by damming rivers and streams, which affects the habitats of the local plants and animals. Another disadvantage to some hydroelectric power plants is that they depend upon the flow of water which varies with seasons and during droughts.

Other Resources Used to Generate Electricity

Currently, renewable resources (other than water) and geothermal energy supply less than 1% of the electricity generated by electric utilities. They include solar, wind, and biomass (wood, municipal solid waste, agricultural waste, etc.). Although geothermal energy is a nonrenewable resource, we will not run out of it. Geothermal power comes from heat energy buried beneath the surface of the Earth. Most of this heat is at depths beyond current drilling methods. In some areas of the country, magma—the molten matter under the Earth's crust from which igneous rock is formed by cooling—flows close enough to the surface of the Earth to produce steam. That steam can then be harnessed for use in conventional steam turbine plants. *Figure 4* shows the locations of geothermal plants within the United States. Most are found in the western United States where magma is close enough to the surface to supply steam.

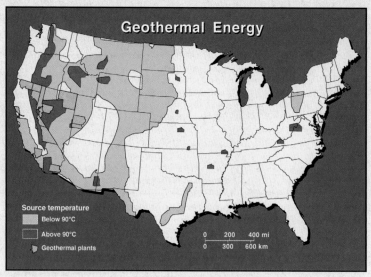

Figure 4 The location of geothermal power plants and source temperatures in continental United States.

Check Your Understanding

1. What is the difference between electric energy and electric power?

2. Compare and contrast steam turbine versus hydroelectric power generation. How are they similar? How are they different?

3. From an Earth systems perspective, what are the advantages and disadvantages of hydroelectric power?

4. In your own words, explain why biomass and wind are called renewable energy sources.

Understanding and Applying What You Have Learned

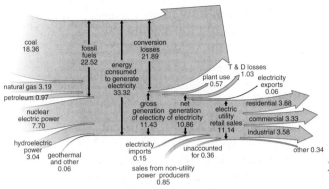

Electricity flow - electric utilities 1999
(quadrillion Btus)

1. Use the diagram above to answer the following questions about the generation of electricity in the United States:

a) Which of the energy sources listed above are nonrenewable? What percentage of energy consumed to produce electricity in 1999 did they represent?

b) Which of the energy sources listed above are renewable? What percentage of energy consumed in 1999 did they represent?

c) In **Activity 1** you learned about energy efficiency. Compare the value for energy consumed to produce electricity versus net generation of electricity in 1999.

What is the overall efficiency of electricity generation systems in the United States?

d) Conversion losses accounted for 21.89 quads of the energy consumed to generate electricity. Explain what this means in relation to your answer to **Question 1(c)** above.

2. Converting one form of energy into another comes with a loss in usable energy (waste heat). The "net efficiency" of a multi-step process is the product of the efficiencies of each step. For example, if each step in a two-step process was 50% efficient, the net efficiency would be $0.50 \times 0.50 = 0.25 = 25\%$.

a) Use the information below to calculate the net efficiency of a home water heater.

b) The lowest efficiency of this process is the generation of electricity. Why do you think this is so?

3. Each year, the United States uses more than 71 trillion Btus of solar energy. One million Btu equals 90 lb. of coal or eight gallons of gasoline.

Step	Efficiency (in percent)
Production of coal	96
Transporting coal to power plant	97
Generating electricity	33
Transmission of electricity (from plants to homes)	85
Heating efficiency of the hot water heater	70
Net Efficiency	**?**

a) How many tons of coal per year does this equate to?

b) How many gallons of gasoline does this equate to?

c) If the United States increased its use of solar energy 5% annually, how many tons of coal would this equate to during the next 10 years?

Preparing for the Chapter Challenge

Write a short paper in which you review the energy resources used in your community and how their use affects your community. Be sure to include information on how energy resources are used to generate electricity and fuel in your community, which of the current means of producing power your community relies on most, and which seems the least important. This paper will help you to think about what resources your community is using before the population increase.

Inquiring Further

1. **Storage of solar energy**

   Electricity from solar energy offers a clean, reliable, and renewable source of energy for home use. Batteries are used to store electrical energy for use at night or when bright sunlight is not available. Research recent advances in technologies used to store solar energy. How do car batteries perform in comparison to solar batteries in solar systems? Present your report to the class.

2. **Energy from the oceans**

   Three novel ways have been proposed to use the energy of the oceans for electrical generation: tides, waves, and vertical temperature differences. Choose one of these, and research the theory and techniques involved in its use for generating electricity.

3. **Other methods of generating electricity**

   Research one of the following methods of generating electricity. How do they work? Where are they being used? What are the ideal conditions for using these alternatives? What are the advantages of these methods over fossil fuels? What are the disadvantages or limitations? Visit the *EarthComm* web site for sources of information.

   - photovoltaics
   - geothermal
   - solar thermal

Activity 3   Energy from Coal

Goals

In this activity you will:

- Classify the rank of coal using physical properties.

- Interpret a map of coal distribution in the United States.

- Understand that fossils fuels represent solar energy stored as chemical energy.

- Understand what coal is made of and how coal forms.

- Be able to explain in your own words why coal is a nonrenewable resource.

Think about It

Coal is the largest energy source for electricity in the United States. Known coal reserves are spread over almost 100 countries. At current production levels, proven coal reserves are estimated to last over 200 years.

- **How does coal form?**
- **Why is coal referred to as "stored solar energy"?**

What do you think? Record your ideas in your *EarthComm* notebook. Be prepared to discuss your responses with your small group and the class.

Investigate

Part A: Types of Coal

1. Obtain a set of samples of four different types of coal. Examine the samples. Look for evidence of plant origin, hardness, luster (the way light is reflected off the surface of the sample), cleavage (tendency to split into layers), and any other characteristics that you think distinguish the four types of coal. Discuss similarities and differences with members of your group.

   a) In your *EarthComm* notebook, create a data table and record your observations.

2. Use your completed data table, to answer the following questions:

   a) How might the samples be related?

   b) Put the samples in order from least compacted to most compacted.

   c) Which sample do you think contains the most stored energy? Why do you think so?

3. The next step of this activity will be a demonstration. You will observe your teacher igniting a small piece of coal held over a Bunsen burner. The sample will be removed from the flame and you will observe how the sample burns. Four types of coal will be tested.

   a) Note the speed of ignition, color of the flame, speed of burning, and odor. Note any other characteristic that you think distinguishes the samples. Summarize your observations in a data table or chart.

   b) In your *EarthComm* notebook, summarize the major differences you observed.

   c) Review your answer to **Step 2(c)**. Have your ideas changed? Explain.

   ⚠️ Wear goggles. Clean up all loose pieces of coal. Wash your hands after you are done.

Part B: Coal Resources

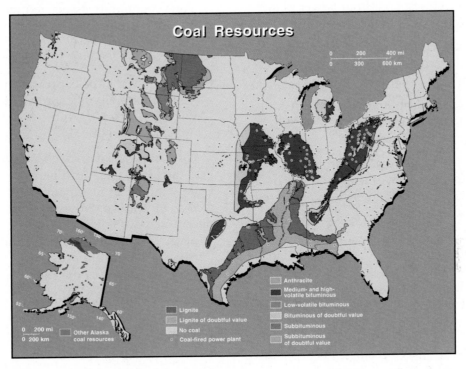

1. Examine the map of coal resources in the United States.

   a) Summarize the major trends, patterns, or relationships in the distribution of coal and the locations of coal-fired power plants.

2. Refer to the map to answer the following questions about coal resources in the United States. Record your responses in your *EarthComm* notebook.

   a) How many states contain coal deposits? Which states are they?

   b) What are the main types of coal present in the following regions: east-central, southeast, and west-central?

   c) Can you infer from the map which state has or produces the most coal? Why or why not?

   d) Measure how far your community is from a source of coal.

   e) Look at the distribution of coal-fired power plants. Why do you think coal is used in these plants rather than petroleum or natural gas?

   f) What type of coal is closest to your community?

3. Refer to the map to answer the following questions about coal-fired power plants in the United States. Record your responses in your *EarthComm* notebook.

   a) Is there a coal-fired power plant in your state? If so, where is it located?

b) How far from your community is the closest coal-fired power plant? Where is the plant located?

c) In what part of the country are the most coal-fired power plants found? Why do you think this is so?

d) Identify the states that have coal-fired power plants but do not have coal deposits.

e) How might distance from coal deposits affect the cost of electricity produced by a coal-fired power plant?

4. Examine a geologic map of your community or state. A geologic map shows the distribution of bedrock at the Earth's surface. The bedrock shown on the map might be exposed at the surface, or it might be covered by a thin layer of soil or very recent sediment. Every geologic map contains a legend that shows the kinds of rocks that are present in the area.

a) Write down or note the names of the different rock types present in your community, the names of the rock units, their ages, and their locations.

5. Compare the geologic map to the map that shows the distribution of coal deposits in the United States.

a) Are any of the rocks present in your community associated with coal deposits? What kinds of rocks are they? How old are the rocks?

b) Are any of the rocks present in your community likely to yield coal deposits in the future? Which ones? How do you know?

6. Go to the *EarthComm* web site to visit the Department of Energy's Energy Information Administration web page. Click on your state to open a new web page. Use the "Energy Consumption, Prices, and Expenditures" profiles page to answer the following questions:

a) In your state, approximately what percentage of primary energy consumption for residential purposes comes from coal?

b) In your state, what percentages of primary energy consumption for commercial, industrial, and transportation purposes come from coal?

c) How would an increase in the population of your community affect this usage?

Reflecting on the Activity and the Challenge

In this activity you described four types of coal and observed how the different types of coal react when burned. You then examined maps to determine the distribution of coal deposits in the United States and also to determine if coal deposits are found near your community. You looked at the location of coal-fired power plants relative to coal deposits and thought about how the distance between the two can affect the cost of electricity production. These activities will help you to determine how coal is used as an energy resource in your community and the impacts of its use.

Digging Deeper

COAL AS A FOSSIL FUEL

There are only four primary sources of energy available for use by humankind. They are solar radiation, the Earth's interior heat, energy from decay of radioactive material in the Earth, and the tides. You can assume that starlight, moonlight, and the kinetic energy of meteorites hitting the Earth are so small that they can be neglected. The energy in coal is energy from solar radiation that is stored as chemical energy in rock.

The energy in coal originates as solar energy. Plants in the biosphere store this solar energy by a process called **photosynthesis**. During photosynthesis, green plants convert solar energy into chemical energy in the form of **organic (carbon-based) molecules**. Only 0.06% of the solar energy that reaches the Earth is stored through photosynthesis, but the amount of energy that is stored in the Earth's vegetation is enormous. Photosynthesis yields the carbohydrate glucose (a sugar) and water, as shown in the equation below:

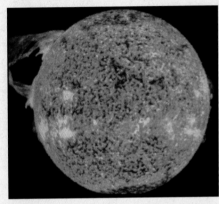

Figure 1 Solar radiation is one of the primary sources of energy on Earth.

Geo Words

photosynthesis: the process by which plants use solar energy, together with carbon dioxide and nutrients, to synthesize plant tissues.

organic (carbon-based) molecules: molecules with the chemical element carbon as a base.

oxidation: the chemical process by which certain kinds of matter are combined with oxygen.

respiration: physical and chemical processes by which an organism supplies its cells and tissues with oxygen needed for metabolism.

decomposition: the chemical process of separation of matter into simpler chemical compounds.

Sunlight used

$$6CO_2 \quad + \quad 6H_2O \quad \longrightarrow \quad C_6H_{12}O_6 \quad + \quad 6O_2$$

Carbon dioxide removed from atmosphere and biosphere	Water removed from hydrosphere	Solar energy stored as chemical energy	Glucose (sugar) for biosphere	Oxygen released to atmosphere

The energy stored in glucose (a sugar) and other organic molecules can be released when broken in a reaction with oxygen (oxidized). **Oxidation** occurs through **respiration, decomposition**, or combustion (burning).

Chemical energy released

$$6O_2 \quad + \quad C_6H_{12}O_6 \quad \longrightarrow \quad 6H_2O \quad + \quad 6CO_2$$

Oxygen removed from atmosphere and biosphere	Glucose from biosphere	Heat released to organism by respiration or to environment by decomposition or combustion	Water returned to hydrosphere	Carbon dioxide returned to atmosphere and biosphere

Coordinated Science for the 21st Century

Geo Words

sediments: loose particulate materials that are derived from breakdown of rocks or precipitation of solids in water.

lithosphere: the rigid outermost shell of the Earth, consisting of the crust and the uppermost mantle.

nonrenewable resource: an energy source that is powered by materials that exist in the Earth and are not replaced nearly as fast as they are consumed.

You can see from the reactants (on the left sides of the two equations) and the products (on the right sides of the two equations) that photosynthesis and oxidation are the reverse of each other.

The energy that enters the biosphere by photosynthesis is nearly equal to the energy lost from the biosphere by oxidation, as shown in *Figure 2*. Most of the carbon in the biosphere is soon returned to the atmosphere (or to the ocean) as carbon dioxide, but a very small percentage is buried in **sediments**. It is protected from oxidation and becomes part of the **lithosphere** (in the form of peat, coal, and petroleum) for future use as fossil fuels. The rate of storage of energy used by organisms for photosynthesis is very slow. However, these processes have been active throughout much of Earth's history. Therefore, the amount of energy stored as fossil fuels is enormous (32×10^{20} Btus). Fossil fuels are consumed far faster than they form, and therefore they are classified as **nonrenewable resources**.

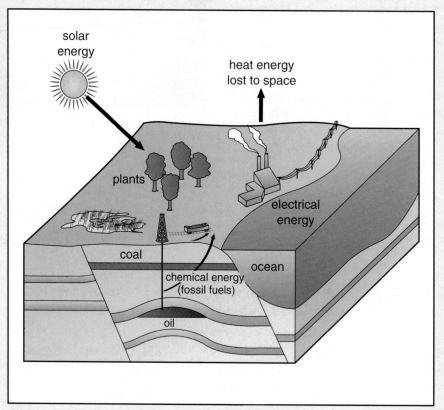

Figure 2 The flow of energy from the Sun, to plants, to storage in fossil fuels, and loss back into space.

The Formation of Coal

Most coal starts out as **peat**. Peat is an unconsolidated and porous deposit of plant remains from a bog or swamp. Structures of the plant matter, like stems, leaves, and bark can be seen in peat. When dried, peat burns freely, and in some parts of the world it is used for fuel. Today, most peat comes from peat bogs that formed during the retreat of the last ice sheets, between about ten thousand and twenty thousand years ago.

Coal, by definition, is a combustible rock with more than 50% by weight of carbonaceous material. Coal is formed by compaction and hardening of plant remains similar to those in peat. The plant remains are altered physically and chemically through a combination of bacterial decay, compaction, and heat. Most coal has formed by lush growth of plants in coastal fresh-water swamps, called coal swamps, in low-lying areas that are separated from any sources of mud and sand. Plants put their roots down into earlier deposits of plant remains, and they in turn die and serve as the medium for the roots of even later plants. In such an environment, the accumulation of plant debris exceeds the rate of bacterial decay of the debris. The bacterial decay rate is reduced because the available oxygen in organic-rich water is completely used up by the decay process. Thick deposits of almost pure plant remains build up in this way. Coal swamps are rare in today's world, but at various times in the geologic past they were widespread.

Geo Words

peat: a porous deposit of partly decomposed plant material at or not far below the Earth's surface.

coal: a combustible rock that had its origin in the deposition and burial of plant material.

For the plant material to become coal, it must be buried by later sediment. Burial causes compaction, because of the great weight of overlying sediment. During compaction, much of the original water that was in the pore spaces of the plant material is squeezed out. Gaseous products (methane is one) of alteration are expelled from the deposit. The percentage of the deposit that consists of carbon becomes greater and greater. As the coal becomes enriched carbon, the coal is said to increase in rank. (See *Figure 3*.) The stages in the rank of coal are in the following order: peat, lignite, sub-bituminous coal, bituminous coal, anthracite coal, and finally graphite (a pure carbon mineral). It is estimated that it takes many meters of original peat material to produce a thickness of one meter of bituminous coal.

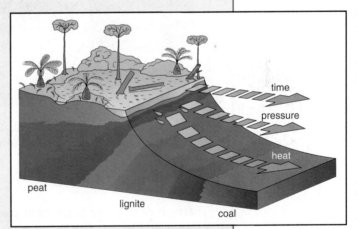

Figure 3 Increasing time, pressure, and heat result in the formation of progressively higher ranking coal.

Geo Words

sedimentary rock: a rock, usually layered, that results from the consolidation or lithification of sediment.

As shown in *Figure 4*, coal is always interbedded with other **sedimentary rocks,** mainly sandstones and shales. Environments of sediment deposition change with time. An area that at one time was a coal swamp might later have become buried by sand or mud from some nearby river system. Eventually, the coal swamp might have become reestablished. Upon burial, the plant material is converted to coal, and the sand and mud form sandstone and shale beds. In this way, there is an alternation of other sedimentary rock types with the coal beds. Some coal beds are only centimeters thick, and are not economically important. Coal beds that are mined can be up to several meters thick.

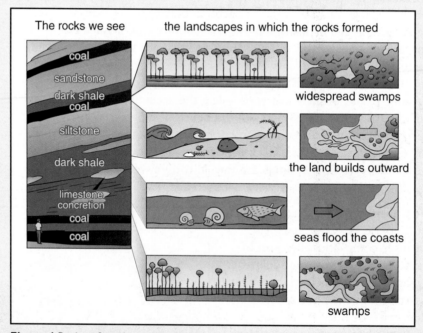

Figure 4 Rocks reflect the environments of the past.

Types of Coal

The type of coal that is found in a given region depends partly on the composition of the original plant material (together with any impurities that were deposited in small quantities and at the same time). But mainly the type of coal depends on the depth and temperature of later burial. Coal that is buried very deeply attains high rank, because of the high pressures and temperatures associated with deep burial. Low-rank coals (lignite and sub-bituminous) have not been buried as deeply.

Coal varies greatly in composition. One of the most important features of coal composition is sulfur content. Sulfur is important because it is released into the atmosphere as sulfur dioxide when the coal is burned. The sulfur dioxide then combines with water in the atmosphere to form sulfuric acid, causing acid rain. The sulfur content of coal can range from a small fraction of 1% to as much as 5%, depending mainly on the sulfur content of the original plant material. The carbon content of coal increases with increasing rank of coal, from lignite to anthracite, as illustrated in *Table 1*. The heat content is also an important feature of coal. The greater the heat content, the smaller the mass of coal that needs to be burned to produce the needed heat. The heat content of coal increases with the rank of the coal, because it depends mainly on the carbon content.

Table 1 Percentages of Carbon and Volatile Matter in Coal			
Coal Rank	**Carbon Content (%)**	**Volatile matter (%)**	**Btus per pound**
Lignite	25–35	up to ~50%	4000–8300
Sub-bituminous	35–45	up to ~30%	8300–13,000
Bituminous	45–86	less than 20%	10,500–15,000
Anthracite	86–98	less than 15%	15,000

The carbon content of coal supplies most of its heat energy per unit weight. The amount of energy in coal is expressed in British thermal units (Btu) per pound. A Btu is the amount of heat needed to raise the temperature of one pound of water one degree Fahrenheit. Peat can be used as a source of fuel, but it has a very low heat content per pound of the fuel burned. Lignite (also called brown coal), the least buried and usually the youngest type of coal, is used mainly for electric power generation. Sub-bituminous coal is a desirable heat source because of its often low sulfur content. Sub-bituminous coal is found in the western United States and Alaska. Bituminous coal is the most abundant coal in the United States, with a large deposit in the Appalachian Province of the East. Bituminous coal is used mainly for generating electricity and making coke for the steel industry. Anthracite, found in a very small supply in the eastern United States, has been used mainly for home heating.

Some of the constituents of coal are not combustible. The part of coal that is not consumed by burning is called ash. Most of the ash consists of sand, silt, and clay that was deposited in the coal swamp along with the vegetation. The purest coal has only a small fraction of 1% ash. The ash content of usable coal can be much higher. Some of the ash remains in the combustion chamber, and some goes up the flue. Ash in coal is undesirable because it reduces the heat content slightly. It also must be removed from the combustion chamber and discarded.

Check Your Understanding

1. Explain how solar energy is stored and released by the processes of photosynthesis and oxidation.

2. Describe, in your own words, the formation of coal, from plant matter to anthracite.

3. Why is the heat content of coal (in Btu/pound) important in determining how the coal is used?

4. What is the origin of the sulfur content of coal, and why is sulfur an undesirable constituent of coal?

Understanding and Applying What You Have Learned

1. How much impact do you think the proximity of your community or state to a coal deposit has on the use of coal for energy generation? Support your response with the information you collected in the investigation.

2. Discuss what happens to the carbon content of peat if it is allowed to decay in the presence of oxygen.

3. Use what you have learned about coal in this activity to provide two reasons why it would be a better idea to burn anthracite in a home fireplace than lignite.

4. What are the advantages to using coal to meet energy needs? What are the disadvantages?

Preparing for the Chapter Challenge

You have learned how coal deposits form, and you looked at the distribution of coal deposits in the United States and your community. You also considered how reliant your community is on coal as an energy resource. Write a paper in which you consider how the number of coal deposits in the United States, along with the time required to produce a coal deposit, shape the way that your community uses coal today and how it will use coal in the future, in light of projected population growth.

Inquiring Further

1. **Model the formation of coal**

   Line a plastic shoebox (or two-liter bottle with the top cut off) with plastic wrap. Pour water into the container to a depth of four inches. Spread about two inches of sand on the bottom. Drop small leaves, sticks, and pieces of fern on the sand. Let it stand for two weeks. Record what you observe as changes in color and decomposition occur. Gently sift fine sand or mud on top of the plant layer to a depth of two inches. Wait two weeks and drain any remaining water. Let it sit and dry for another two weeks. Remove the "formation" out of the container. Slice it open to see how you have simulated coal where the plants were, and fossil imprints from the plant leaves.

2. **Plants associated with coal**

   Research the ancient plants whose remains formed the United States coal deposits.

!\ Wash your hands after handling any material in this activity. Wear goggles. Complete the activity under adult supervision.

Activity 4   Coal and Your Community

Goals

In this activity you will:

- Investigate the production and consumption of coal in the United States.

- Investigate how coal sources are explored.

- Understand methods of coal mining.

- Determine how coal is used in your state.

- Evaluate possible practices to conserve coal resources.

Think about It

Coal production in the United States decreased in 1999 and 2000, but coal consumption continues to increase.

- **Why do you think that the production of coal has decreased?**

- **Why do you think that the consumption of coal has increased?**

- **Given that coal is a finite, nonrenewable source of energy, what are some ways to extend the supply of coal?**

What do you think? Record your ideas in your *EarthComm* notebook. Be prepared to discuss your responses with your small group and the class.

Investigate

Part A: Trends in Coal Production and Consumption

Year	Appalachian	Interior	Western	Total
Table 1 United States Coal Production, by Region 1991–2000 (in million short tons[1])				
1991	457.8	195.4	342.8	996.0
1992	456.6	195.7	345.3	997.5
1993	409.7	167.2	368.5	945.4
1994	445.4	179.9	408.3	1033.5
1995	434.9	168.5	429.6	1033.0
1996	451.9	172.8	439.1	1063.0
1997	467.8	170.9	451.3	1089.9
1998	460.4	168.4	488.8	1117.5
1999	425.6	162.5	512.3	1100.4
2000	420.9	144.7	509.9	1075.5

Source: United States Energy Information Administration, 1996, 1998, 2000
[1] One short ton = 2000 pounds.

1. *Table 1* gives recent trends in coal production in the United States during the last 10 years. Using these values, summarize the trends in coal production for the three major coal-producing regions, and the trend in total coal production.

   a) Appalachian

   b) Interior

   c) Western

   d) Total

2. Make one graph showing coal production in the three regions. Leave room at the end of the graph to project coal production for the next 50 years.

   a) Include your graph in your *EarthComm* notebook.

3. Extrapolate the trends in the data to the year 2050. To do this, you will have to produce a best-fit line through the data points and estimate what you think to be the trend in coal production. There is a little guesswork involved in determining how much of the data you think represents the trend. Is it the last four years or the last eight years?

   a) On the basis of your projections only (you are ignoring all other factors that might influence future production), which coal-producing region do you predict will be the first to exhaust its supply of coal? In what year will this happen?

   b) In your group, identify at least three factors that might affect actual coal production. Record your ideas in your notebook.

4. Examine the data in *Table 2*. Summarize the major trends in coal consumption for each sector from 1991 to 2000, and the total coal consumption.

a) Electric power

b) Coke plants (steel manufacturing)

c) Other industrial plants

d) Residential and commercial users

e) Total

f) What percentage of total coal consumption did electrical power make in 1991? In 2000?

Table 2 United States Coal Consumption by Sector, 1991–2000 (in million short tons[1])					
	Consumption by Sector (million short tons)				
Year	Electric Power	Coke Plants	Other Industrial Plants	Residential and Commercial Users	Total
1991	777.2	33.9	75.4	6.1	892.5
1992	786	32.4	74.0	6.2	898.5
1993	820.8	31.3	74.9	6.2	933.9
1994	826.7	31.7	75.2	6.0	939.6
1995	849.8	33.0	73.1	5.8	961.7
1996	896.9	31.7	70.9	6.0	1005.6
1997	922.0	30.2	71.5	6.5	1030.1
1998	937.8	28.2	67.4	4.9	1038.3
1999	946.8	28.1	65.5	4.9	1045.3
2000	970.7	29.3	65.5	4.9	1070.5

Source: United States Energy Information Administration, 1996, 1998, 2000
[1] One short ton = 2000 pounds.

5. Using the data from *Table 2*, make a graph that shows coal consumption for electrical power.

a) Include the graph in your *EarthComm* notebook.

6. Extrapolate the trend in the data to the year 2050, as in **Step 3**.

a) On the basis of recent trends, how much coal will be needed for electrical power generation in the year 2020? 2050?

7. Coal consumption for electrical power increased at an average rate of 1.65% per year between 1996 and 2000.

a) On the basis of this average, predict coal consumption for electrical power for the years 2010 and 2020 (Hint: Begin by multiplying the value for the year 2000 by 1.0165. This gives you a prediction for the year 2001.)

837

8. Assume that by conserving electricity the average rate of growth in consumption of coal is cut in half, to 0.825% per year.

a) Predict the amount of coal consumed in 2010 and 2020.

9. Draw a new graph that shows your prediction of coal production and coal consumption for the period 2000 to 2050. Do this by superimposing the curves you fitted in **Steps 3** and **6**. Also, in a third curve, take into account the prediction you made in **Step 8**, on the assumption of savings by conservation.

a) Include your graph in your *EarthComm* notebook.

b) Do you predict that consumption will exceed production? If so, how do you think that the shortfall in production will be made up? Or do you predict that production will exceed consumption? If so, do you think that that is a reasonable or likely scenario?

c) In what ways do you think that production and consumption are related? Does production drive consumption? If so, how and to what extent? Or does consumption drive production? If so, how, and to what extent?

Table 3 Logs for Core Holes (elevations are in feet above sea level; ms = mudstone, ss = sandstone).									
Location 1		**Location 2**		**Location 3**		**Location 4**		**Location 5**	
3930–4000	ms	3990–4000	ms	3890–4020	ss	3960–3990	ss	3980–4030	ss
3920–3930	coal	3970–3990	ss	3750–3890	ms	3900–3960	ms	3710–3980	ms
3870–3920	ms	3890–3970	ms	3660–3750	ss	3895–3900	ss	3675–3710	coal
3835–3870	coal	3885–3890	coal	3620–3660	ms	3720–3895	ms	3665–3675	ms
3660–3835	ms	3855–3885	ms	3570–3620	coal	3715–3720	ss	3610–3665	coal
3590–3660	coal	3850–3855	coal	3500–3570	ms	3690–3715	ms	3500–3610	ms
3500–3590	ms	3760–3850	ms		coal	3650–3690	coal		
	ms	3710–3760	coal		coal	3645–3650	ms		
		3700–3710	ms			3590–3645	coal		
		3630–3700	coal			3500–3590	ms		
		3600–3630	ms						

Part B: Coal Exploration

1. In 1998, the Western Region overtook the Appalachian Region as the largest coal-producing region in the United States with 488.8 million short tons produced, up by 8.3% over 1997. The low-sulfur Powder River Basin coal fields in Wyoming dominated growth in coal production in the Western Region. The diagram on the following page shows a cross section of five core holes drilled along an east–west line. *Table 3* provides drilling results for these wells. The results include the elevations and types of rock units in the core holes.

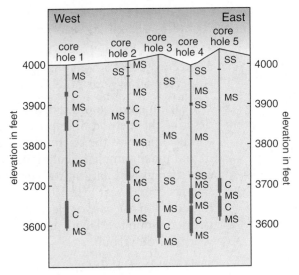

a) Complete the cross section in the diagram. Match up the rock units as you imagine them to be connected in the subsurface. (Hint: The mudstone at the bottom of all five wells is the same unit.)

b) Compare your results with those from other members of your group. How do they compare? How do you explain the differences?

c) Where would you drill your next well in order to determine whether or not your interpretation of the cross section is correct? Explain.

d) Is the information in the cross section sufficient to determine the coal seam with the greatest volume of coal? Why or why not?

e) What is the average thickness of coal in the lower coal seam?

f) Assume that the lower coal seam covers an area of 300,000 acres and that each acre-foot yields 1770 short tons of coal. (An acre-foot is the volume of a 1-foot thick layer that covers an area of 1 acre.) How many short tons of coal does the lower coal seam contain?

Part C: Conserving Coal Resources

1. A 100-W bulb burning for 10 h uses 1 kWh (kiloWatt-hour) of electricity (the same as ten 100-W bulbs burning for one hour).

   a) Calculate the kiloWatt-hours of electricity used in one year for a 100-W bulb running continuously.

   b) Assume that electricity costs $0.06/kWh. (Your teacher may give you a more accurate figure for your community.) Calculate the yearly cost of running the 100-W bulb continuously.

   c) Assume that one pound of good-quality coal can produce about 1 kWh of electricity. Calculate the amount of coal required per year to keep the 100-W bulb running continuously.

   d) How much coal is required each year to keep a 100-W bulb burning continuously in 20 million households?

   e) What do you think are the environmental consequences of burning this much coal?

   f) The average electricity bill for a family of four in the United States is about $50 per month (this is for homes where cooking and heating are by natural gas or oil). Estimate the yearly amount of electricity (in kiloWatt-hours) that this is equivalent to. How many tons of coal are needed?

2. A variety of researchers are currently seeking methods to reduce our consumption of electricity, much of which is produced from coal. In your group decide on five ways to make

homes (or offices) in your community more energy efficient. Go to the *EarthComm* web site for useful links that will help you to explore ways to conserve energy resources. Suggestions are provided below.

- Water heaters — How do you ensure the best energy efficiency of an electric home water heater and reduce energy costs?

- Major home appliances — What are the most efficient ways to use major home appliances and how can you improve their efficiency?

- Home tightening — How can you slow the escape of heat energy from buildings, saving money and making them more comfortable?

- Insulation — How can insulation be used to reduce energy consumption?

- Home cooling — What strategies can you suggest for keeping a building cool in summer and improving the efficiency of air conditioning units?

a) Record the methods that you decide on in your notebook.

3. For each method you decided on, describe the method of improving energy efficiency, the science behind it, and make sample calculations of cost savings and natural resource savings over 1 year and 10 years (based upon the cost of electricity in your community). Divide up the work in your group.

a) How much electricity would be saved?

b) How much coal would be conserved over the course of a year? Over 10 years?

4. Conservation (reduction in use) is another way to reduce energy usage.

a) In your group, decide on 10 ways to conserve energy.

b) Calculate the savings.

Source: Energy Information Administration, Electric Power Monthly March 2001.

Figure 1 Percentage share of United States electric power industry net generation.

petroleum and other 5%
hydro 7%
coal 52%
gas 16%
nuclear 20%

Digging Deeper

COAL EXPLORATION AND MINING

In 2000, over 52% of electricity in the United States was produced from coal (see *Figure 1*). There was a fall in coal production around the middle of the last century. This was in part due to the decline of the steel industry, which uses coal in steel production. In addition, oil and gas have largely replaced coal for transportation and home heating. Since the early 1960s, however, there has been a steady increase in the production of coal. This is mainly because of the increasing demand for electricity. There has also been a development of major coalfields in the western United States and Canada.

Of all fossil fuels, the largest reserves are contained in coal deposits (92%). Three major factors determine which coals are currently economical for mining. One factor is the cost of transportation to areas where the coal is utilized. Another factor is the environmental concern associated with the mining and use of coal. From a geologic perspective, the quality, thickness, volume, and depth of coal are important in determining whether or not a coal is mined.

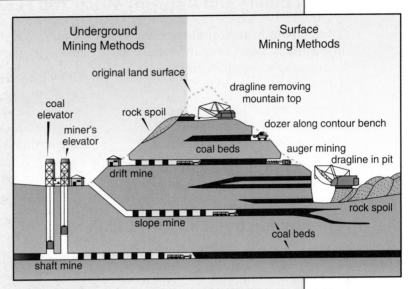

Figure 2 Different methods used to mine coal from the ground.

Figure 2 illustrates underground and surface mining methods (Note: You would not typically see all these mining methods used in one location.) Underground mining methods include drift, slope, and shaft mining.

Surface mining methods include area, contour, mountaintop removal, and auger mining. (An auger is a tool used for boring a hole.) *Figure 3* shows a surface coal mine.

Figure 3 A mine high wall showing layered sandstone and mudstone in a Gulf Coast coal mine. The haul truck provides a sense of scale.

Check Your Understanding

1. What are the major factors that determine whether a coal seam is economically and geologically suitable for mining?

2. What are the major factors that have determined the trends in coal consumption during the 20th century?

Coordinated Science for the 21st Century

Understanding and Applying What You Have Learned

1. Think about historical differences in electricity use. Compare your community's electricity use 10 years ago and today.

2. Estimate the total energy use of your community. How much could your community reasonably conserve?

3. Compare electricity usage of a single college student vs. a family of four. Which group uses more electricity per person?

4. Think about regional differences in electricity use. Which part of the country do you think would be the lowest? Which part of the country do you think would be the highest? What are the reasons for your opinions?

Preparing for the Chapter Challenge

Assume that your community is expected to grow 10% in the next 10 years. If electricity usage remains the same, this means that the electrical capacity required to meet this increase must also increase. This may be undesirable and involve the cost of building a new power plant or an increased environmental hazard.

Determine specific steps that could be taken by the community to conserve electricity. How would the reduced demand for electricity offset the growth in the community? What incentives would be given to encourage the community to implement these proposed changes?

Inquiring Further

1. **Investigating a coal mine**

   Investigate the coal mine closest to your community.

   a) What mining method do they use?
   b) What is their annual production?
   c) How has the mine influenced the economy of the community?
   d) Has the mine impacted the environment?
   e) What steps has the mining company taken to reduce the environmental impact of the mine?

   Have your teacher arrange a field trip to this mine and discuss the trip with your class.

2. **Electricity usage in other countries**

   Find out information about electricity usage in other countries. What can you learn from other countries about energy conservation? (For instance, Japan has a high gross national product (GNP) yet very low energy use. How do they manage this?)

3. **Personal electricity usage**

   Learn how to read your electricity meter, and with your family conduct some conservation experiments for a few days (or weeks) and report the results of these experiments to the class.

Activity 5

Environmental Impacts and Energy Consumption

Goals

In this activity you will:

- Examine one of the environmental impacts of using coal.

- Understand how the use of fossil fuels relates to one of Earth's major geochemical cycles—the carbon cycle.

- Understand the meaning of pH.

- Understand how weather systems and the nature of bedrock geology and soil affect the impact of acid rain.

- Determine ways that energy resources affect your community.

- Analyze the positive and negative effects of energy use on communities.

Think about It

Sulfur dioxide and nitrogen oxides are the primary causes of acid rain. In the United States, about two-thirds of all sulfur dioxide and one-quarter of all nitrogen oxides comes from electric power generation that relies on burning fossil fuels like coal.

- **How does acid rain affect the environment?**

- **What can be done to reduce the amount of sulfur dioxide and nitrogen oxides released into the atmosphere?**

What do you think? Record your ideas in your *EarthComm* notebook. Be prepared to discuss your responses with your small group and the class.

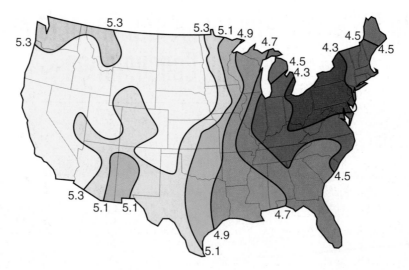

Investigate

1. The map shows the acidity of rainfall across the United States. Normal rainfall is slightly acidic (pH 5.6, where 7.0 is neutral). Carbon dioxide in the atmosphere reacts with water to form a weak acid called carbonic acid. The lower the pH, the more acidic is the rainwater.

   a) What part of the United States has the most acidic rainwater?

   b) How does the pattern of rainwater pH correlate with the locations of coal-producing regions and coal-fired power plants? (See **Activity 3**.)

   c) What parts of the United States have the least acidic rainwater? Why?

   d) What can you infer about the direction in which wind and weather move across the United States?

2. Your teacher will pour an "acid rain" solution (pH of about 4.5) through samples of crushed limestone and granite. Containers below the crushed rock contain pH indicator solution. A change in the color of the pH indicator reveals a change in the pH.

3. Note your answers to the following questions in your *EarthComm* notebook:

   a) Describe any color changes in the collection containers.

   b) What was the pH of the liquids after passing through the two types of crushed rock?

   c) Which type of rock—limestone or granite—neutralized more of the acid?

   d) Where would you expect a greater environmental impact from acid rain—in a region with granite bedrock or one with limestone bedrock? Explain.

e) What does this investigation suggest about how areas may differ in their sensitivity to acid rainwater?

4. Your teacher will repeat the investigation using crushed samples of rock from your community.

a) Record all your observations and explain your results.

Reflecting on the Activity and the Challenge

In this activity you examined one of the major environmental impacts of using coal to produce energy. You explored how the extent to which acid rain affects a region depends upon several factors, like the nature of the rock and soil, weather patterns, and the location of coal-fired power plants. This will help you to understand the benefits of understanding how the growing consumption of energy resources impacts Earth systems.

Digging Deeper

ENERGY RESOURCES AND THEIR IMPACT ON THE ENVIRONMENT

Fossil Fuels and the Carbon Cycle

When a fossil fuel is burned, its carbon combines with oxygen to yield carbon dioxide gas (CO_2). Fossil fuels supply 84% of the primary energy consumed in the United States. They are responsible for 98% of United States emissions of carbon dioxide. The amount of carbon dioxide produced depends on the carbon content of the fuel. For each unit of energy produced, natural gas emits about half, and petroleum fuels about three-quarters, of the carbon dioxide produced by coal. Important questions that scientists are trying to answer are what is happening to all of this carbon dioxide, and how does it affect the Earth system?

According to the **First Law of Thermodynamics**, the amount of energy always remains constant. The amount of chemical energy consumed when a fossil fuel is burned equals the amount of heat energy released. Scientists think of matter in a similar way. Matter cannot be created or destroyed. The **Law of Conservation of Matter** helps you to understand what happens when one type of matter is changed into another type. For example, fossil fuels consist mainly of hydrocarbons, which are made up of hydrogen

Geo Words

First Law of Thermodynamics: the law that energy can be converted from one form to another but be neither created nor destroyed.

Law of Conservation of Matter: that law that in chemical reactions, the quantity of matter does not change.

and carbon. When a fossil fuel is burned (to heat homes or power cars), the carbon-containing compounds are changed chemically into carbon dioxide. The amount of carbon consumed equals the amount of carbon produced. The carbon never "goes away", but moves from one reservoir within the Earth system to another.

The idea that many processes work together in a global movement of carbon from one reservoir to another is known as the **carbon cycle** (see *Figure 1*). You have already learned about several important reservoirs of carbon. The atmospheric reservoir of carbon is carbon dioxide gas. Biomass contains carbon (mostly carbon in plants and soil). In the geosphere, carbon is found in solids (coal, limestone), liquids (petroleum hydrocarbons), and gas (methane). The ocean is the largest reservoir of carbon (exclusive of carbon-bearing bedrock), found in the form of bicarbonate ions.

Photosynthesis and respiration dominate the movement (flux) of carbon dioxide between the atmosphere, land, and oceans (the wide blue arrows show this flux in *Figure 1*). The smaller red arrows indicate the flux of carbon related to human (**anthropogenic**) activities. Natural processes like photosynthesis can remove some of the net 6.6 billion metric tons of anthropogenic carbon dioxide emissions produced each year. This means that an estimated 3.3 billion metric tons of this carbon is added to the atmosphere annually in the form of carbon dioxide.

Figure 1 Global carbon cycle (billion metric tons).

Geo Words

carbon cycle: the global cycle of movement of carbon, in all of its forms, from one reservoir to another.

anthropogenic: generated or produced by human activities.

By measuring concentrations of carbon dioxide in the atmosphere over time, scientists have learned that levels of carbon dioxide in the atmosphere have increased about 25% in the last 150 years (see *Figure 2*). Scientists believe that human activity has caused this growth. This coincides with the beginning of the industrial age. The increase is due largely to the burning of fossil fuels and to deforestation (*Figure 1*). How does this affect the Earth system? Carbon dioxide is one of several **greenhouse gases** (gases that slow the escape of heat energy from the Earth to space). Some scientists believe that the rapid addition of carbon dioxide to the atmosphere is changing the energy budget of the Earth, causing global climate to warm.

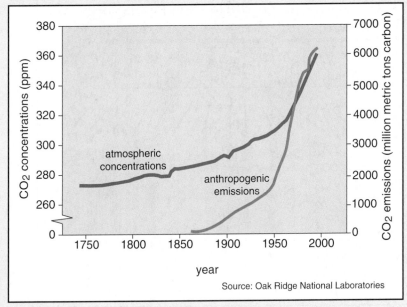

Figure 2 Atmospheric CO_2 concentrations and anthropogenic CO_2 emissions over time.

Coal and Acid Rain

Acidity is a measure of the concentration of hydrogen ions (H^+) in an aqueous (water) solution. A solution with a high concentration of hydrogen ions is acidic. A solution with a low concentration of hydrogen ions is basic. The concentration of hydrogen ions is important, because hydrogen ions are very reactive with other substances. Mixing **acids** and **bases** can cancel out their effects, much as mixing hot and cold water evens out the water temperature. That's because the concentration of hydrogen ions in the mixture lies between that of the two original solutions. A substance that is neither acidic nor basic is said to be **neutral**.

Geo Words

greenhouse gases: gases in the Earth's atmosphere that absorb certain wavelengths of the long-wavelength radiation emitted to outer space by the Earth's surface.

acid: a compound or solution with a concentration of hydrogen ions greater than the neutral value (corresponding to a pH value of less than 7).

base: a compound or solution with a concentration of hydrogen ions less than the neutral value (corresponding to a pH value of greater than 7).

neutral: having a concentration of hydrogen ions that corresponds to a value of pH of 7.

The pH scale shown in *Figure 3* measures how acidic or basic a substance is. It ranges from 0 to 14. A pH of 7 is neutral, a pH less than 7 is acidic, and a pH greater than 7 is basic. Each whole pH value below 7 is 10 times more acidic than the next higher value, meaning that the concentration of hydrogen ions is 10 times as great. For example, a pH of 4 is 10 times more acidic than a pH of 5 and 100 times (10 times 10) more acidic than a pH of 6. The same holds true for pH values above 7, each of which is 10 times more alkaline (another way to say basic) than the next lower whole value. For example, a pH of 10 is 10 times more alkaline than a pH of 9.

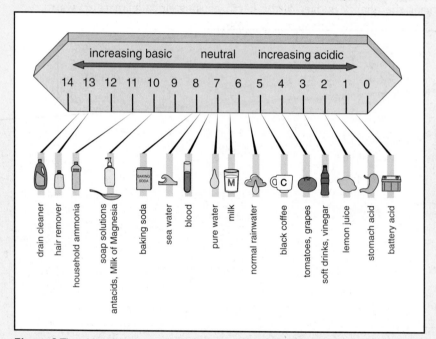

Figure 3 The pH scale.

Pure water is neutral, with a pH of 7.0. Natural rainwater, however, is mildly acidic, with a pH of about 5.7. The reason is that carbon dioxide in the atmosphere dissolves in rainwater, and some of the dissolved carbon dioxide reacts with the water to form a weak acid, called carbonic acid, H_2CO_3. As the carbon dioxide content of the atmosphere has gradually increased in recent decades, because of the burning of fossil fuels, natural rainwater has become slightly more acidic, even aside from the serious problem of acid rain caused by sulfur dioxide.

You learned in a preceding activity that coal contains as much as 5% sulfur. When the coal is burned, the sulfur is emitted as sulfur dioxide gas, SO_2. The sulfur dioxide then reacts with water in the atmosphere to form sulfuric acid, a strong acid. The last reaction in this process is written as follows:

$$SO_3 + H_2O \rightarrow H_2SO_4$$

Some of the sulfuric acid then dissociates into hydrogen ions and sulfate ions in solution in the water. This reaction can be broken down into two parts, as follows:

$$H_2SO_4 \rightarrow H^+ + HSO_3^- \rightarrow 2H^+ + SO_4^{2-}$$

The H^+ ions reach the Earth's surface dissolved in raindrops, either over land or over the ocean. Burning of fuels also produces nitrogen oxide gases, which react with water to form nitric acid, another strong acid. Some of the nitric acid breaks down to release hydrogen ions, by a reaction similar to that of sulfuric acid. Burning of coal is not the only source of sulfuric acid and nitric acid: petroleum-fueled power plants, smelters, mills, refineries, and motor vehicles produce these acids as well.

Acid rain is especially damaging to lakes. If the pH of lake water becomes too acidic from acid rain, it kills fish, insects, aquatic plants, and plankton. Environmental scientists describe lakes that have suffered heavily from acid rain as "dead," because the entire food web of the lake has been disrupted to the point where little is left alive in the lake.

Acid rain is a serious problem locally in many areas of the United States, but the biggest problem is in the Northeast. The reason is that there are a large number of coal-fueled electric power plants in the Midwest, especially in the heavily populated areas of Ohio, Michigan, and Illinois, and these plants burn mostly the relatively high-sulfur coal that is mined in the Interior coal province (see **Activity 3**). Winds blow mostly from west to east in the mid-latitudes of North America, and that tends to bring the acids to the Northeast, where large amounts fall with rain.

Figure 4 This marble gravestone from 1921 is being slowly dissolved by acid rain.

Geo Words

scrubbing: removal of sulfur dioxide, ash, and other harmful byproducts of the burning of fossil fuels as the combustion products pass upward through a stack or flue.

Another reason for the impact of acid rain in the northeastern United States has to do with the types of rock and soil in that region. In the activity, you learned that limestone has a greater ability to neutralize acid than granite. If the soil upon which acid rain falls is rich in limestone or if the underlying bedrock is either composed of limestone or marble, then the acid rain may be neutralized. This is because limestone and marble are more alkaline (basic) and produce a higher pH when dissolved in water. The higher pH of these materials dissolved in water offsets or buffers the acidity of the rainwater, producing a more neutral pH.

If the soil is not rich in limestone or if the bedrock is not composed of limestone or marble, then no neutralizing effect takes place, and the acid rainwater accumulates in the bodies of water in the area. This applies to much of the northeastern United States where the bedrock is typically composed of granite and similar rocks. Granite has no neutralizing effect on acid rainwater. Therefore, over time more and more acid precipitation accumulates in lakes and ponds.

Coal supplies about half of all electricity generated in the United States. In recent years, as the problem of acid rain has grown, various technologies have been developed to reduce the emission of sulfur dioxide from coal-fueled electric power plants. The general term for these processes is flue-gas desulfurization, also called **scrubbing**.

Several kinds of scrubbers are in use. The most common is the wet scrubber. In wet scrubbing, the flue gas from the power plant is sprayed with a calcium carbonate solution in the form of a slurry. The SO_2 is oxidized to form calcium sulfate. Scrubbers of this kind can remove up to 95% of the SO_2 that is emitted from the power plant. Scrubbers can have a useful byproduct. Certain kinds of scrubbers are now in operation that produce pure calcium sulfate (gypsum) as a byproduct. This gypsum can be used industrially, to make plaster and wallboard, so that gypsum does not have to be mined and processed.

Advantages and Disadvantages of Energy Resources

Energy resources affect the community in positive and negative ways. Energy resources allow you to maintain a very high quality of life. Energy resources also indirectly provide employment for a large sector of the community. Use of energy resources, however, can have obvious and not-so-obvious negative impacts on the community. One of the most obvious negative impacts is a deterioration of air quality due to the burning of fossil fuels. Not-so-obvious impacts include acid rain, possible global warming, ground-water contamination, and the financial burden on households. It is important to realize that you, both personally and as a member of the community, can control how you use energy resources to potentially reduce the negative impacts.

Check Your Understanding

1. Describe the carbon cycle in relation to the consumption of fossil fuels.

2. Explain what is meant by the phrase "The Earth is a closed system with respect to matter." How does this statement relate to the Law of Conservation of Matter?

3. In your own words, explain the concept of pH.

4. Why is acid rain a greater problem in the Northeast than elsewhere in the United States?

5. How is sulfur dioxide removed from the gases emitted from coal-fueled electric power plants?

Understanding and Applying What You Have Learned

1. What are some of the positive impacts associated with the use of energy resources?

2. What are some of the negative impacts associated with the use of energy resources?

3. The unit used to measure the amount of SO_2 gas emitted by a coal-fired power plant is pounds of SO_2 per million Btu (lb. SO_2/MBtu).

    a) Assume a 1000-megawatt power plant emits 2.5 lb. SO_2/MBtu. Calculate the amount of SO_2 gas emitted in one year, assuming that the plant operates on average at 75% capacity.

    b) Assume that the coal used has an energy output of 10,000 Btu/pound and use a heat-to-electricity efficiency of 30%. How many tons of coal does the plant use each year?

4. At the present time the United States has about 470 billion tons of coal reserves and the average annual production is about 1.05 billion tons.

    a) Calculate how many years the reserves will last at the present production rate.

    b) What would cause the reserves to run out sooner than this number?

    c) What would cause the reserves to run out later than this number?

5. Assume that you live in a small community in the midwest. The electricity for the community is provided by a 1000-megawatt power plant, and the electricity is produced entirely by the burning of coal. The power plant is located in the community and provides employment to about 75 employees. The coal comes from an underground coal mine about 50 miles away. The coal mine produces about one million tons of coal a year and employs about 150 people. Many of the people that work in the mine also live in your community. The sulfur content of the coal is about 4.5% and the ash content is about 11%. The amendments to the Clean Air Act require that by the year 2010 the power plant must reduce SO_2 emissions by 60%. Also, it is possible that by the year 2010 a "carbon tax" will be levied on electricity produced from coal to address global warming issues. There are a number of ways that this can be done, including retrofitting the power plant to burn natural gas (natural gas emits half the CO_2 gas and almost no SO_2 gas), install a wet scrubber, conserve electricity, etc.

    Answer the following questions:

    a) How will the implementation of environmental laws affect your community?

    b) What alternatives does your community have to address changes that could be made to minimize the impact?

    c) How will the community gain from the environmental laws?

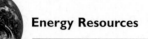

Preparing for the Chapter Challenge

Your challenge is to prepare a plan that will help your community to meet its growing energy needs. Part of your plan must address how the use of energy resources impacts the environment. Based upon what you have learned in this activity, write an essay about ways to reduce the impact of acid rain on communities.

Inquiring Further

1. **The Clean Air Act and environmental regulation**

   Look up information on the amendments to the Clean Air Act having to do with SO_2 emissions and report your findings to the class.

2. **Acid rain and other Earth systems**

   How has acid rain affected your community? How do the processes that lead to acid rain also work to lower the pH of ground water? The latter is a challenge that community officials must face in areas that are mined for coal and some kinds of minerals. Extend your investigation into acid rain by examining acid mine drainage. The *EarthComm* web site will help you to extend your investigations.

3. **Coal mine**

   Investigate the coal mine closest to your community. What are the percent sulfur, weight of SO_2 per million Btu, and percent ash for the coal? What reclamation steps are taken at the mine?

4. **Power plant**

   Investigate the power plant closest to your community. If possible, have your teacher arrange a field trip to this plant.

   a) Which fuels are used at the power plant?
   b) What are the sources of these fuels?
   c) What is the power plant's capacity, in megawatts?
   d) How many people are employed at the power plant?
   e) What is done with the ash from the power plant (i.e., the solid materials that are caught as they pass up through the flue of the plant)?

Activity 6   Petroleum and Your Community

Goals

In this activity you will:

- Recognize the overwhelming dependence of today's society on petroleum.

- Graph changes in domestic oil production and foreign imports to predict future needs and trends.

- Understand why oil and gas production have changed in the United States over time.

- Describe the distribution of oil and gas fields across the United States.

- Understand the origin of petroleum and natural gas.

- Investigate the consumption of oil and natural gas in your community.

Think about It

Oil accounts for nearly all transportation fuel. It is also the raw material for numerous products that you use.

- **What percentage of oil used every day in the United States is produced in the United States?**
- **Where are oil and natural gas found in the United States?**

What do you think? Record your ideas in your *EarthComm* notebook. Be prepared to discuss your responses with your small group and the class.

Investigate

Part A: Trends in Oil Production

Table 1 United States Petroleum Production, Imports, and Total Consumption from 1954 to 1999				
Year	Total U.S. Wells (thousands)	Total U.S. Production (thousand barrels/day)	Total Foreign Imports (thousand barrels/day)	Total U.S. Total Consumption (thousand barrels/day)
1954	511	7030	1052	8082
1957	569	7980	1574	9559
1960	591	7960	1815	9775
1963	589	8640	2123	10,763
1966	583	9580	2573	12,153
1969	542	10,830	3166	13,996
1972	508	11,180	4741	15,921
1975	500	10,010	6056	16,066
1978	517	10,270	8363	18,633
1981	557	10,180	5996	16,176
1984	621	10,510	5437	15,947
1987	620	9940	6678	16,618
1990	602	8910	8018	16,928
1993	584	8580	8620	17,200
1996	574	8290	9478	17,768
1999	554	7760	10,551	18,311

Source: U.S. Energy Information Agency

1. *Table 1* shows statistics for petroleum every three years from 1954 to 1999. Copy the data table into your notebook. If you have Internet access, you can download the complete set of data (annual data since 1954) at the *EarthComm* web site as a spreadsheet file.

   a) Construct a graph of United States (domestic) petroleum production, foreign petroleum imports, and petroleum consumption. Leave room at the right side of your graph to extrapolate the data for 20 years.

2. What are the major trends in petroleum needs during the last 45 years? Use your graph and the data table to answer the following questions:

   a) Describe how domestic production has changed during the 45-year period. About when did it peak?

   b) Describe how petroleum imports have changed during that period.

   c) Describe how total petroleum consumption has changed over the period.

d) In what year did the United States begin to import more petroleum than it produced?

e) What percentage of the total petroleum consumption was met by domestic production in 1954? In 1999? How does this compare to your answer to the first **Think about It** question at the start of this activity?

f) In 1954, the estimated population of the United States was 163 million. In 1999, the estimated population was 273 million. How many barrels of petroleum per person per day were needed in the United States in 1954? In 1999?

g) A barrel of oil is equal to 42 gallons. Convert your answers in **Step 2(f)** to gallons per person per year needed in the United States in 1954 versus 1999.

h) Has the consumption of petroleum changed at the same rate as the growth in population? What might account for the change?

i) Use the data in the table to calculate the average well yield for oil production in the United States.

3. What do you predict about future needs for and sources of petroleum in the United States?

a) Extrapolate the three curves to the year 2020. Explain your reasoning behind your extrapolations for each of the three curves.

b) On the basis of your extrapolations, how much petroleum will be produced domestically, imported into the United States, and consumed in the year 2010? In 2020?

c) Identify several factors that might make the actual curves for production, import, and consumption different from what you have predicted. Record your ideas in your *EarthComm* notebook.

4. Share your results from **Step 3**.

a) Set up a data table to record the predictions from each person in the class for the years 2010 and 2020. An example is provided below.

b) Calculate the minimum, maximum, and average for the class.

c) How do you explain the differences in predictions?

Sample Data Table. Results of Class Predictions of United States Oil Production, Imports, and Total Production in 2010 and 2020.						
	United States Production		Imports		Total	
	2010	2020	2010	2020	2010	2020
Student 1						
Student 2						
Student...						
Class Average						

Part B: United States Oil and Gas Resources

In **Part A** you learned that petroleum production has declined in the United States. The questions below will help you to investigate where oil and gas are found in the United States, and where these natural resources are refined into products that communities depend upon.

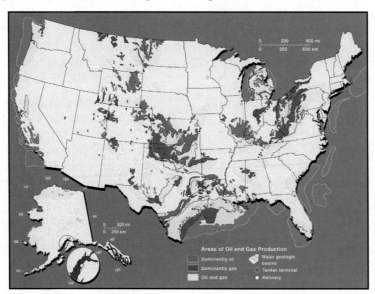

1. Use the map of oil and gas production to explore general trends and patterns in petroleum and gas fields.

   a) Are oil and gas deposits distributed evenly across the country?

   b) Which states have no oil or gas production?

   c) What patterns or trends do you see in oil-producing and gas-producing regions?

   d) Are oil and gas always found together? Give at least two examples.

   e) The map also shows the locations of major sedimentary basins. Sedimentary basins are depressions of the Earth's crust where thick sediments have accumulated. What is the relationship between oil and gas production and sedimentary basins?

2. Refer to the map to investigate where oil and gas are found, refined, and distributed in relation to your community.

   a) How far is your community from the nearest oil field?

   b) How far is your community from the nearest gas field?

   c) How far from your community is the nearest refinery located?

   d) Do you think the factors above might affect the price of petroleum products (gasoline, heating oil, or propane) in your community? Explain your answer.

Part C: Oil and Gas Resource Use in Your Community

1. Investigate trends in petroleum and natural gas production, consumption, and distribution in your state. Go to the *EarthComm* web site to obtain links to the data sets for your state available at the Energy Information Agency. This will allow you to examine several sources of data that will help you with the **Chapter Challenge**.

2. Develop your own questions as you examine the data and review the **Chapter Challenge**.

   a) How have oil and gas consumption in your state changed during the last 40 years? How can you use the data to predict future needs?

   b) How much petroleum is consumed per capita in your state? How much natural gas is consumed? How can you use this information to predict how much oil and gas your community will need 20 years from now if the population grows by 20%?

   c) How do oil and gas reach your community? If switching to another type of fuel is part of your plan for your community's energy future, how will these resources get to your community if they are not local?

Reflecting on the Activity and the Challenge

In this activity you graphed the recent history of oil consumption in the United States. You used these trends to predict what the future needs might be. You examined a map of oil and gas fields in the United States and learned that these resources are not distributed evenly throughout the country. In **Part C**, you investigated trends in oil and gas resource use in your community. Think about how your work in this activity will help you to solve the **Chapter Challenge** on the energy future of your community.

Geo Words

petroleum: an oily, flammable liquid, consisting of a variety of organic compounds, that is produced in sediments and sedimentary rocks during burial of organic matter; also called crude.

crude oil: (see petroleum).

natural gas: a gas, consisting mainly of methane, that is produced in sediments and sedimentary rocks during burial of organic matter.

source rocks: sedimentary rocks, containing significant concentrations of organic matter, in which petroleum and natural gas are generated during burial of the deposits.

seal: an impermeable layer or mass of sedimentary rock that forms the convex-upward top or roof of a petroleum reservoir.

reservoir: a large body of porous and permeable sedimentary rock that contains economically valuable petroleum and/or natural gas.

Digging Deeper

PETROLEUM AND NATURAL GAS AS ENERGY RESOURCES

The Nature and Origin of Petroleum and Natural Gas

Petroleum, also called **crude oil**, is a liquid consisting mainly of organic compounds that range from fairly simple to highly complex. **Natural gas** consists mainly of a single organic compound, methane (with the chemical formula CH_4). Oil and natural gas reside in the pore spaces of some sedimentary rocks.

The raw material for the generation of oil and gas is organic matter deposited along with the sediments. The organic matter consists of the remains of tiny plants and animals that live in oceans and lakes and settle to the muddy bottom when they die. Much of the organic matter is oxidized before it is permanently buried. But, enormous quantities are preserved and buried along with the sediments. As later sediments cover the sediments more and more deeply, the temperature and pressure increase. These higher temperatures and pressures cause some of the organic matter to be transformed into oil and gas. Petroleum geologists use the term

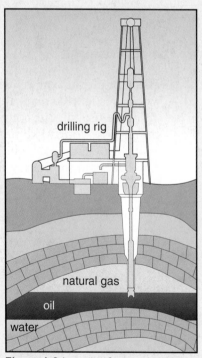

Figure I Schematic of a petroleum trap.

"maturation" for this process of change in the organic matter. They call the range in burial depth that is appropriate for generation of oil and gas the "oil window" or "gas window" and the regions rich in organic matter subjected to these depths "the kitchen." Depending upon the details of the maturation process, sometimes mostly oil is formed, and sometimes mostly gas.

The mudstone and shale that form the **source rocks** for oil and gas are impermeable. Fluids can flow through them very slowly. The oil and gas rise very slowly and percolate through the source rocks, because they are less dense than water, which is the main filler of the pore spaces. Once they reach fractures and much more permeable rocks like some kinds of sandstone and limestone, however, they can migrate much more rapidly. Much of the oil and gas rise all the way to the Earth's surface, to form oil and gas seeps, and escape to the atmosphere or produce tar mats. If the rocks are capped deep in the Earth by an impermeable layer called a **seal**, the oil and gas are detained in their upward travel. A large volume of porous rock containing oil and gas with a seal above is called a **reservoir**. The oil and gas can be brought to the surface by drilling deep wells into the reservoir, as shown in *Figure 1*. Petroleum reservoirs around the world, in certain favorable geological settings, range in age from more than a billion years to geologically very recent, just a few million years old. Oil and gas are nonrenewable resources, however, because the generation of petroleum operates on time scales far longer than human lifetimes.

Geo Words

feedstocks: raw materials (for example, petroleum) that are supplied to a machine or processing plant that produces manufactured material (for example, plastics).

Oil and natural gas are useful fuels for several reasons. They have a very high heat content per unit weight. They cost less than coal to transport and are fairly easy to transport. Oil and gas can also be refined easily to form many kinds of useful materials. Petroleum is used not only for fuel. It is used as the raw material (called **feedstock**) for making plastics and many other synthetic compounds like paints, medicines, insecticides, and fertilizer.

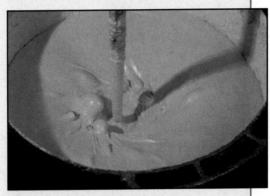

Figure 2 Petroleum is found in many common products, including paint.

The History of Oil Production in the United States

The United States has had several basic shifts in its energy history (see *Figure 3*). The overall pattern of energy consumption in the United States has been that of exponential growth. The burning of wood to heat homes and

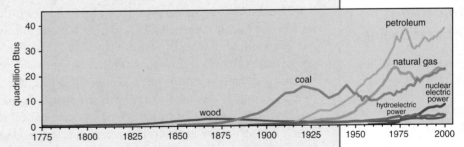

Figure 3 Energy consumption, by source, in the United States since 1775.

provide energy for industry was the primary energy source in about 1885. Coal surpassed wood in the late 1800s, only to be surpassed by petroleum and natural gas in the 1950s. Hydroelectric power appeared in 1890 and nuclear electric power appeared in about 1957. Solar photovoltaic, advanced solar thermal, and geothermal technologies also represent recent developments in energy sources. Of all the recently developed energy sources, petroleum and natural gas have had the greatest impact.

Figure 4 This early 1900's drilling rig was used to prospect for oil. Drilling rigs have advanced greatly since then.

The first successful oil-drilling venture was Colonel Titus Drake's drilling rig in 1859 in northwestern Pennsylvania. Drake reached oil at a depth of 70 ft. Petroleum got major boosts with the discovery of Texas's vast Spindletop Oil Field in 1901. With the advent of mass-produced automobiles, consumption of oil began to grow more rapidly in the 1920s. In the years after World War II, coal lost its position as the premier fuel in the United States. Trucks that ran on petroleum took business once dominated by railroads, which began to switch to diesel locomotives themselves. Labor troubles and safety standards drove up the cost of coal production. Natural gas, which began as a source for lighting (gas lanterns and streetlights), replaced coal in many household ranges and furnaces. By 1947 consumption of petroleum and natural gas exceeded that of coal and then quadrupled in a single generation. No source of energy has ever become so dominant so quickly.

United States Petroleum Production

In 1920, the United States produced more than two-thirds of the world's oil. By 1998, the United States supplied only 12% of the world's oil needs. Oil and gas are found in **sedimentary basins** in the Earth's crust. To find oil, geologists search for these basins. Most of the basins in the United States have already been explored and exploited. In fact, the United States is the most thoroughly explored country in the world. Some people estimate that the United States has less than 35% of its original oil remaining. There is not nearly enough easily recoverable petroleum in the United States to meet the demand, although domestic production will continue for years to come.

Oil and gas resources are not evenly distributed throughout the country. The top ten oil fields account for 33% of all remaining oil in the United States (including a field in California that was discovered in 1899!). Texas has been the largest producing state since the late 1920s, when it surpassed California. In the late 1980s, Alaska rivaled Texas.

Check Your Understanding

1. What is the origin of oil and natural gas?

2. Why do some people refer to the present times as the "petroleum age"?

3. What are the advantages of petroleum and natural gas as fuels?

4. Why are oil and gas considered nonrenewable resources?

Understanding and Applying What You Have Learned

1. What is the world's future supply of petroleum relative to world demand?

2. Is the future decrease in petroleum production likely to be abrupt or gradual? Why?

3. Look at *Table 1* from the investigation.

   a) Calculate the average yield per well in barrels of petroleum per day for each year in the table.

   b) Graph the total number of wells producing and the average well yield over time.

   c) Describe any relationships that you see. How is well yield changing over time?

4. Think back to your projections of future oil production, foreign imports, and consumption.

   a) How might tax incentives for switching to renewable energy sources affect oil and gas consumption rates in the future?

   b) How might a change in the cost of fuels (gasoline, jet fuel, etc.) affect consumption rates in the future?

   c) How might new discoveries of oil affect domestic production?

   d) Identify two other factors that you think might affect one of your three projections.

Inquiring Further

1. **United States oil and gas fields**
   Research one of the top 10 United States oil and gas fields. Go the *EarthComm* web site for links that will help you to research the geology and production history of the top 10 United States oil fields. Why are these fields so large? What is the geologic setting? When was oil or gas discovered? What is the production history of the field? How much oil and gas remains?

Oil Fields
Alaska (Prudhoe Bay; Kuparuk River), California (Midway-Sunset; Belridge South; Kern River; Elk Hills), Texas (Wasson; Yates; Slaughter), Gulf of Mexico (Mississippi Canyon Block 807).

Gas Fields
New Mexico (Blanco/Ignacio-Blanco; Basin), Alaska (Prudhoe Bay), Texas (Carthage; Hugoton Gas Area [also in Kansas and Oklahoma); Wyoming (Madden), Alabama (Mobile Bay), Colorado (Wattenberg); Utah (Natural Buttes); Virginia (Oakwood).

Activity 7   Oil and Gas Production

Goals

In this activity you will:

- Design investigations into the porosity and permeability of an oil reservoir.

- Understand factors that control the volume and rate of production in oil and gas fields.

- Explore the physical relationships between natural gas, oil, and water in a reservoir.

- Understand why significant volumes of oil and gas cannot be recovered and are left in the ground.

- Use this knowledge to understand the variability in estimates of remaining oil and gas resources.

- Appreciate the importance of technological advances in maximizing energy resources.

Think about It

When oil and gas are discovered, petroleum geologists calculate the volume of the oil or gas field. They also need to estimate how much they can recover (remove). This helps them to determine the potential value of the discovery.

- **When oil is discovered, what percentage of it can actually be recovered and brought to the surface?**

- **When will global oil production begin to decline? When will global gas production begin to decline?**

What do you think? Record your ideas in your *EarthComm* notebook. Be prepared to discuss your responses with your small group and the class.

Investigate

Part A: Reservoir Volume

1. In your group, discuss the questions below. Record your ideas in your notebook.

   a) Which material can hold more oil: gravel or sand? Why do you think this is so?

   b) How might this knowledge be useful in oil and gas exploration?

2. Develop a hypothesis related to the question in **Step 1 (a)**.

   a) Record your hypothesis in your notebook.

3. Using the following materials, design an investigation to test your hypothesis:
   two 500-mL (16-oz.) clear plastic soda bottles (with bottoms removed)
   fine cheesecloth
   electrical tape
   stands with clamps to hold the plastic bottles
   sand
   aquarium gravel
   50-mL graduated cylinder
   200 mL of vegetable oil
   water

   a) Record your procedure. Note any safety factors.

4. When your teacher has approved your design, conduct your investigation. (Note: At the end of your investigation, do not empty out the sediments and oil. You will need them for **Part B**.)

   a) Record your results.

5. Use your results to answer the following questions:

   a) What is the porosity (percentage of open space) of your gravel sample?

   Porosity is the ratio of pore space to total volume of solid material and pore space, usually expressed as a percentage by multiplying the ratio by 100. 1 mL = 1 cm^3 (cubic centimeter). For example, suppose you poured 25 mL of sediment into a container and found that 10 mL of water completely filled the spaces between the grains. One milliliter is equal to one cubic centimeter. Expressed as a percentage, the porosity of the sediment would be:

   $$(10 \text{ cm}^3/25 \text{ cm}^3) \times 100 = 40\% \text{ porosity}$$

   b) What is the porosity of your sand sample?

   c) Imagine that a geologist discovers an oil field in a sandy reservoir (in petroleum geology, a reservoir is a porous material that contains oil or gas). The total volume of the reservoir (sand plus oil) is equivalent to 60 million barrels of oil. How many barrels of oil are in the reservoir?

   d) If the reservoir described above were gravel, how many barrels of oil would it contain?

   ⚠ Wear goggles. Clean up spills. Wash your hands when you are done.

Part B: Recovery Volume

1. Suppose you were to remove the cap at the bottom of the plastic bottle of gravel and oil.

   a) What volume of oil do you think would drip out of the container if you left it out overnight? Explain your answer.

2. Make sure that the cheesecloth is secured to the bottle with electrical tape. Unscrew the cap from the bottom of the container of gravel and oil. Allow the oil to seep through the container and drain into a collection device overnight.

3. Record your answers to the following questions:

   a) What is the volume of oil recovered?

   b) Calculate the percentage of oil recovered.

   c) How do you explain your results?

   d) If your model represented a discovery of 100 million barrels of oil, how many barrels of oil would actually be removed from the reservoir?

4. Brainstorm about ways to improve your results.

   a) How can you recover a greater percentage of oil? If your situation permits, test your method of removing more oil.

Part C: Factors that Affect Oil and Gas Production Rates

1. In your group, discuss your ideas about the following question: How does the grain size of sediment affect the ease with which fluids flow through them?

2. Develop a hypothesis about the question above. You will be testing the rate of flow of water through gravel, coarse sand, fine sand, and silt/clay. Your hypothesis should include a prediction (what you expect) and a reason for your prediction.

   a) Record your hypothesis.

3. Write down your plan for conducting the investigation. An idea about how to set up the equipment is shown in the diagram below.

   a) What is your independent variable?

   b) What is your dependent variable?

   c) How will you control other variables that might affect the validity of your results?

   d) How will you record and keep track of your results?

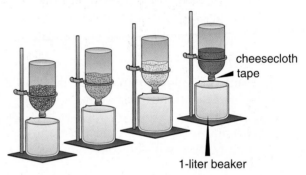

cheesecloth
tape

1-liter beaker

4. When you are ready, conduct your investigation.

   a) Describe your results. Was your hypothesis correct?

   b) Describe the relationship between grain size and permeability.

   c) How do your results relate to the production of oil?

5. If time permits, develop and test a hypothesis about the relationship between the viscosity of a fluid and its rate of flow through material. Viscosity is a term used to describe a fluid's internal friction or resistance to flow. In this case, you might wish to compare the speed at which oil, water, and a third fluid flow through a particular type of sediment.

Part D: Properties of Fluids in Reservoirs

1. Pour water into a 500 mL clear plastic soda bottle until the bottle is about half full. Add vegetable oil until the bottle is three-fourths full. Screw the cap onto the bottle. The water, vegetable oil, and air in this model represent water, crude oil, and natural gas. The bottle represents an oil and gas reservoir.

2. Slowly turn the bottle upside down (if you shake the bottle or turn it quickly, you will create bubbles in the oil, which will interfere with your observations).

   a) In your notebook, sketch a diagram to represent the relationship between the oil, water, and air.

   b) Why do you think that the material stacks in the order that it does?

   c) If a well is drilled into an oil and gas reservoir, what material would it encounter first? Second? Third?

3. Tilt the bottle at a 45° angle and hold it there. Repeat this with a 10° angle. Look at the bottle from the side, and then from the top.

   a) Draw a diagram of the side view of the bottle when it is tilted at a 45° angle versus a 10° angle.

   b) How does the surface area covered by the oil change with the angle of the oil reservoir (bottle)?

   c) Imagine drilling a vertical well through the bottle. This represents a well being drilled into reservoirs that are sloping at different angles in the subsurface. Would it penetrate a greater thickness of oil when the reservoir is at a 10° angle or when the reservoir is at a 45° angle? Why?

4. Look at the data table below. It shows the results of drilling of four wells. The elevation of the top of each well was 5000 ft. above sea level.

Feature	Well A	Well B	Well C	Well D
Elevation of top of VanSant Sandstone	3650 ft.	3850 ft.	3950 ft.	3900 ft.
Elevation of base of VanSant Sandstone	3450 ft.	3700 ft.	3825 ft.	3800 ft.
Result	Water (dry hole)	Oil at 3850 ft.; water at 3800 ft.	Gas at 3950 ft.; oil at 3875 ft.	Gas at 3900 ft.; oil at 3875 ft.

The locations of the wells are shown on the map below. The map shows the elevation of top of the VanSant Sandstone. Symbols on the map show the results of drilling.

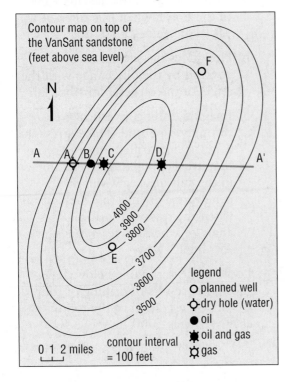

Contour map on top of the VanSant sandstone (feet above sea level)

N

A   A B C   D   A'

4000
3900
3800
3700
3600
3500

E

legend
○ planned well
⬦ dry hole (water)
● oil
✸ oil and gas
✿ gas

0 1 2 miles   contour interval
= 100 feet

a) Draw a cross section across the oil field along the east–west line labeled A–A'. Plot the top of the VanSant Sandstone and the base of the VanSant Sandstone.

b) Use the results of drilling to show the level of gas, oil, and water in the cross section.

c) Use the results and your cross section to map the areal extent of the oil and gas. Color the gas red and the oil green on the map.

d) An example from another oil and gas field is provided below.

A geologist proposes to drill two more wells at locations E and F. Use your map and cross section to decide whether or not you would support the plan.

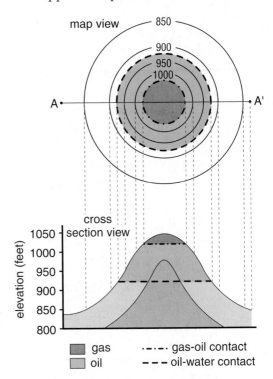

map view   850
             900
             950
             1000

A   A'

cross
section view

1050
1000
950
900
850
800

elevation (feet)

▨ gas   ·—·—· gas-oil contact
▢ oil   – – – oil-water contact

Reflecting on the Activity and the Challenge

In this activity, you explored the percentage of open space in various kinds of sediment. You saw that materials hold different amount of oil and that it is not easy to get all of the oil out of the material. You also explored the rate at which fluids flow through different materials. In the last part of the activity, you investigated the relationship between gas, oil, and water, and used the relationships to make predictions about where oil and gas would be in an oil and gas field. As you read further, you will see how the factors you have explored relate to the ability to predict how much oil and gas can be produced to meet future energy needs.

Digging Deeper

PETROLEUM RECOVERY

Porosity and Permeability

Most sedimentary rocks have open spaces in addition to the solid materials. These open spaces are called pores. Sedimentary rocks are the most porous rocks. Think about what a sandstone looks like, on the inside, when it is first deposited by flowing water. The sand particles are in contact with one another in the form of a framework. This framework is much like the way oranges are stacked in the supermarket, except that the sand particles usually have a much less regular shape than oranges. As you learned in the investigation, the **porosity** is defined as the volume of pore spaces divided by the bulk volume of the material, multiplied by 100 to be expressed as a percentage.

The porosity of loose granular material like sand can be as much as 30%–40%. If the sand particles have a wide range of sizes, however, the porosity is less, because the smaller particles occupy the spaces between the larger particles. When the sand slowly becomes buried in sedimentary basins to depths of thousands of meters, where temperatures and pressures are much higher, new mineral material, called **cement**, is deposited around the sand grains. The cement serves to make the sediment into a rock. The cement in sedimentary rocks is not the same as the cement that's used to make concrete, although it serves the same purpose. The addition of the cement reduces the porosity of the rock, sometimes to just a few percent.

Geo Words

porosity: the ratio of pore space to total volume of a rock or sediment, multiplied by 100 to be expressed as a percentage.

cement: new mineral material precipitated around the particles of sediment when it is buried below the Earth's surface.

Geo Words

permeability: the ease with which a fluid can be forced to flow through a porous material by a difference in fluid pressure from place to place in the material.

The most productive petroleum reservoir rocks are those in which the pore spaces are mostly connected rather than isolated. When the pore spaces are all connected to one another, fluid can flow slowly through the rock. Fluid flows through a porous rock when the fluid pressure differs from one place to another. Here's a simple "thought experiment" to show the concept. Attach a long horizontal pipe to the base of a large barrel. Stuff the pipe with sand, and put a screen over the downstream end of the pipe. When you fill the barrel with water, the water flows through the sand in the pipe, because the pressure at the bottom of the barrel is high but the pressure at the open end of the pipe is low. The water finds its way down the pipe through the connected pore spaces in the sand. The rate of flow through the porous material in response to a pressure difference is related to the **permeability** of the material. For a rock to have a high permeability, it generally has to have high porosity, but the pores must be fairly large and also well connected with one another. Good petroleum reservoir rocks must have high porosity, to hold the oil or gas, and also high permeability, so that the oil or gas can flow through the rock to the bottom of a well. Some reservoir rocks are highly fractured, and the oil and gas can move along open fractures as well as through pore spaces.

Fluids also flow in porous rocks because of differences in density. Most sedimentary rocks at depth in the Earth have their pore spaces filled with water. When droplets of oil or bubbles of gas are formed in the rock during maturation of the organic matter in the rock, they tend to rise upward through the pore water. They do this because both oil and gas are less dense than water. Only when they are trapped and accumulate beneath a seal do they form a petroleum reservoir.

For a mass of rock to form a seal, it has to be impermeable. In its simplest form, it also has to have a convex-upward shape, like an umbrella. Of course, it is usually much more complicated in its geometry than an umbrella. Just imagine raindrops falling upward into the inside of your umbrella, rather than downward to be shed off the umbrella! The vertical distance from the crest or top of the seal to its lower edge, where the petroleum can spill out, is called the closure of the reservoir. Giant petroleum reservoirs can have closures of many tens of meters, as well as horizontal dimensions of thousands of meters. Reservoirs that large can hold hundreds of millions of barrels of oil.

Recovery of Petroleum

Oil and gas are removed from reservoirs by drilling deep wells into the reservoir. At greater depths, the wells are only a few inches in diameter, but they can extend down for as much as several thousand meters. The technology of drilling has become very advanced. Wells can be drilled vertically downward for some distance and then diverted and angled off to the side at precise orientations. Sometimes the oil is under enough pressure to flow out of the well to the ground surface without being pumped, but usually it has to be pumped out of the well. The pumping lowers the pressure in the well, and the oil flows slowly into the well from the surrounding reservoir.

Figure 1 Pumps are used to extract oil.

Much of the oil in a reservoir remains trapped in place after years of production. The main reason is that oil adheres to the walls of the pores in the form of coatings. The coatings are especially thick around the points of contact of sediment particles, where the pore spaces narrow down to small "throats." Several techniques have been developed to recover some of the remaining petroleum. Use of such techniques is called **secondary recovery**. One of the most common methods of secondary recovery is to inject large quantities of very hot steam down into a well, which mobilizes some of the remaining petroleum by causing it to flow more easily.

Estimates of Petroleum Reserves

Petroleum geologists attempt to estimate the future recoverable reserves of petroleum in the world. Estimates of this kind are very important, because they are needed to predict when world petroleum production will start to level off and then decline. This is a very difficult task, because it has to include regions of the world which have not yet been explored in detail but which seem to be similar, geologically, to areas already explored. Estimates of recoverable reserves have increased steadily with time, as petroleum has been discovered in regions that were not believed to contain petroleum. That can't go on forever, of course, because there is only a finite volume of petroleum in the Earth. The most recent estimate from the United States Geological Survey gives a good idea of recoverable reserves. Rather than being a single-value estimate, it is based on probability. The 95% probable value is about 2250 billion barrels of oil. That means that the probability that recoverable reserves are greater than 2250 billion barrels is 95%. The 50% estimate is about 3000 billion barrels, and the 5% estimate is about 3900 billion barrels (a one-in-twenty chance). You can see from this great range in values how indefinite such estimates are.

Geo Words

secondary recovery: the use of techniques to recover oil still trapped among sediment particles after years of production.

Figure 2 A drilling platform offshore.

The world's supply of petroleum is finite. Most of the land areas of the world have been intensively explored already, and the major new areas for petroleum exploration and production (called "plays," in the language of the oil business) lie in deep offshore areas of the oceans. It's generally agreed among petroleum geologists that such areas are the last major exploration frontier.

Gradually, over the next few decades, petroleum production will level off and start to decline, although only slowly. Because demand is likely to keep on rising, there will be a growing shortfall in petroleum. That shortfall will have to be made up by use of various alternative energy sources. Solid organic matter contained in certain sandstones, called tar sands, and in certain shales, called oil shales, are already becoming economically favorable to produce, and they will certainly be a major factor in future energy production. Canada is especially rich in tar sands, and there are abundant oil shales in the western United States. Developing fuel-efficient vehicles and giving greater attention to conservation of fuel will also help communities to address the shortfall of petroleum.

Check Your Understanding

1. In your own words, describe the difference between porosity and permeability.

2. What causes the permeability of a sediment deposit to decrease as it is buried deeply and turned into a sedimentary rock?

3. Why is it not possible to extract all of the oil from a reservoir?

4. What are the benefits of predicting oil and gas reserves?

Figure 3 Heavy oil found in tar sands at Vernal, Utah.

Understanding and Applying What You Have Learned

1. a) Why does the permeability of a fine-grained sedimentary rock tend to be less than that of a coarse-grained rock?

   b) Why does the permeability of a rock with particles of approximately equal size tend to be greater than a rock with particles of a wide range of sizes?

2. Why does a porous and permeable sedimentary rock need to have an upper seal to have the potential to be a petroleum reservoir?

3. What do you think replaces the oil in the pore spaces of a reservoir rock when the oil is pumped out? Why?

4. It is very expensive to drill wells and recover oil and gas. Use what you have learned about porosity and permeability to explain why geologists evaluate these features of reservoirs.

Preparing for the Chapter Challenge

Communities depend upon a fairly constant supply of the fuels and other products that are developed from oil and gas. Use what you have learned about oil and gas exploration and production to write a brief essay. The essay should outline the basics of oil and gas exploration and production. It should also help community members to understand the scientific basis for differences in estimates about the future supply of oil and gas.

Inquiring Further

1. **Secondary recovery methods**

   Research modern methods of secondary recovery:

   - carbon dioxide flooding
   - nitrogen injection
   - horizontal drilling
   - hydraulic fracturing

2. **Protection of sensitive environments**

   Investigate how oil and gas exploration and production companies are using advanced technologies to obtain resources while minimizing environmental impacts. Examples include:

   - The use of slimhole rigs (75% smaller than conventional rigs) to reduce the effect on the surface environment.
   - Using directional drilling technology to offset a drilling rig from a sensitive environment.
   - Using downhole separation technology to reduce the volume of water produced from a well.

3. **Careers in the petroleum industry**

   Investigate careers in the oil and gas industry. No single industry employs as many people as the petroleum industry.

4. **Conserving transportation fuels**

   Investigate methods of reducing consumption of transportation fuels.

Activity 8

Renewable Energy Sources— Solar and Wind

Goals

In this activity you will:

- Construct a solar water heater and determine its maximum energy output.

- Construct a simple anemometer to measure wind speeds in your community.

- Evaluate the use of the Sun and the wind to reduce the use of nonrenewable energy resources.

- Understand how systems based upon renewable energy resources reduce consumption of nonrenewable resources.

Think about It

The solar energy received by the Earth in one day would take care of the world's energy requirements for more than two decades, at the present level of energy consumption.

- **How can your community take advantage of solar energy to meet its energy needs?**

- **Is your community windy enough to make wind power feasible for electricity generation?**

What do you think? Record your ideas in your *EarthComm* notebook. Be prepared to discuss your responses with your small group and the class.

Investigate

Part A: Solar Water Heating

1. Set up a solar water heater system as shown in the diagram. Put the source container above the level of the heater so that water will flow through the tubing by gravity. Do the experiment on a sunny day and face the surface of the solar water heater directly toward the Sun. This will enable you to calculate maximum output.

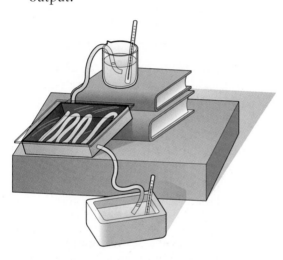

2. In its simplest form, a solar water heater uses the energy of the Sun to increase the temperature of water as the water flows through a coil of tubing.

    A Btu (British thermal unit) is the amount of heat energy required to raise the temperature of one pound of water one degree Fahrenheit. You need to determine the number of Btus of heat generated. If you divide this value by the surface area of the heater and by the time it takes to heat the water, you will have the Btu/ft²/min

output of your solar water heater. It will help to know that 1000 mL of water weighs about 1 kg, or 2.205 pounds.

   a) Decide how you will do the investigation and record the data.

   b) Prepare a data table to record relevant data.

   c) Think about the conversions that you will need to make in order to end up with Btu/ft²/min.

3. Run about one liter of cold water through the heater. Refill the source container with two liters of cold tap water and run the experiment.

4. Answer the following questions:

   a) What is the output of your solar water heater in Btu/ft²/min?

   b) Convert the output of your solar water heater from Btu/ft²/min to Btu/ft²/hour.

   c) Assume that you could track the Sun (automatically rotate the device so that it faces the Sun all day long). Assuming an average of 12 hours of sunlight per day, what would its output be per day?

   d) Convert this value to kWh/m²/day by dividing by 317.2.

5. Examine the maps on the following page of average daily solar radiation for the United States. The first map shows the values for January, and the second one shows the values for July. The maps represent 30-year averages and assume 12 hours of sunlight per day and 30 days in a month.

a) How does the average daily solar radiation per month for January compare with that of July? How might you explain the difference?

b) Find your community on the map. Record the values for average daily solar radiation for January and July.

c) How does the daily output of your solar water heater compare to the average daily solar radiation for the two months shown?

d) Percent energy efficiency is the energy output divided by the energy input, multiplied by 100. Calculate the estimated efficiency of your solar water heater. If you have Internet access, you can obtain the average values for the month in which you do this activity. Otherwise, calculate the efficiency for whatever month is closest—January or July.

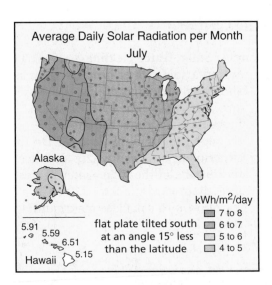

Average Daily Solar Radiation per Month — July

6. In your group, brainstorm about ways to increase the efficiency of your solar water heater. Variables that you might explore include the role of insulation, diameter of the tubing, type of material in the tubing, and flow rate of the water through the system. If time and materials are available, conduct another test.

Part B: Harnessing Wind Energy

1. Set up an anemometer as shown in the diagram below.

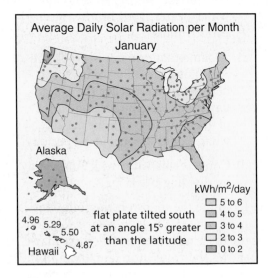

Average Daily Solar Radiation per Month — January

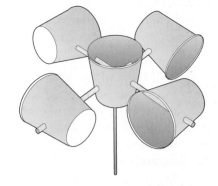

2. To find the wind speed, count the number of revolutions per minute. Next, calculate the circumference of the circle (in feet) made by the rotating paper cups. Multiply the revolutions per minute by the circumference of the circle (in feet per revolution), and you will have the velocity of the wind in feet per minute. (This will be an underestimate, because the cups do not move quite as fast as the wind.)

3. Take your anemometer home during a weekend. This will give you a chance to make measurements at different times of the day and over two days.

4. Make a map of your house and the yard around the house. Plan to place the anemometer at different locations and make velocity measurements at different times of the day.

5. Using a watch, count the number of times the colored cups spins around in one minute. Use your calibration to convert this to wind velocity.

   a) Keep a record of the wind speeds you are measuring for the next few days.

6. Measure the wind speed at different times of the day.

   a) Is the wind speed the same in the morning; the afternoon; the evening?

   b) Move your anemometer to another location. Is it windier in other places?

   c) Do trees or buildings block the wind?

7. Back in class, determine the best location for a wind turbine at your home.

   a) On the basis of the data that you have taken, make a plot of wind velocity throughout the day. You will assume that the wind will blow the same way every day. (Is this a good assumption?)

8. Following the example given below, calculate the electricity that can be produced at your home for a wind turbine with a 2-m blade (the cost of this wind turbine is about $3000).

   a) What percentage of your household electricity needs does this represent?

   b) How much wind power could be produced with a 3-m blade? (Cost about $5000.)

Example calculation of wind power: Wind power per square meter of turbine area is equal to 0.65 times the cube of the wind velocity. For instance, if a wind turbine has a blade 2 m long, then the area that the turbine sweeps is $\pi(1 \text{ m})^2 = 3.14 \text{ m}^2$.

If a constant wind of 10 m/s (22 mph) is blowing, the power is $(10^3)(0.65)(3.14) = 2000$ W. The efficiency of wind turbines is about 40%, so the actual power that can be produced is about 800 W. If the wind blew a constant 10 m/s every day of the year, this wind turbine could produce 7000 kWh of electricity.

9. Get together as a class and discuss this activity.

   a) How do your calculations compare with those of other students?

   b) What are the characteristics of the locations of homes that have the highest wind velocities?

   c) What are the characteristics of those with the lowest wind velocities?

   d) Overall, how well suited is your community for wind power?

   e) Which locations in your community are optimally suited for wind turbines?

   f) How realistic is the assumption that two days' worth of measurements can be extrapolated to yearly averages?

   g) For how long should measurements of wind velocity be taken to get an accurate measure of wind velocities in your community?

Reflecting on the Activity and the Challenge

Water heating can consume as much as 40% of total energy consumption in a residence. In this investigation, you determined the energy output of a flat-plate collector, a device used to heat water using solar energy. You also built a simple anemometer in the classroom and used it to measure wind speed around your home at different times of the day. Finally, you calculated whether it would be feasible and cost-effective for you to generate part of your family's electricity needs using wind power. You may wish to consider the use of different forms of solar energy as an alternative energy source for your community.

Digging Deeper

SOLAR ENERGY

Forms of Solar Energy

Forms of solar energy include direct and indirect solar radiation, wind, photovoltaic, biomass, and others. Tidal energy is produced mainly at the expense of the Earth–Moon system, but the Sun contributes to the tides as well. The three forms of solar energy with the most potential are solar-thermal (direct solar radiation to generate electricity), wind power, and photovoltaic cells.

Solar energy is produced in the extremely hot core of the Sun. In a process called nuclear fusion, hydrogen atoms are fused together to form helium atoms. A very small quantity of mass is converted to energy in this process. The ongoing process of nuclear fusion in the Sun's interior is what keeps the Sun hot. The surface of the Sun, although much cooler than the interior, is still very hot. Like all matter in the universe, the surface radiates electromagnetic energy in all directions. This electromagnetic radiation is mostly light, in the visible range of the spectrum, but much of the energy is also in the ultraviolet range (shorter wavelengths) and in the infrared range (longer wavelengths). The radiation travels at the speed of light and reaches the Earth in about eight minutes. Only an extremely small fraction of this energy reaches the Earth, because the Earth is small and very far away from the Sun. It is far more than enough, however, to provide all of the Earth's energy needs, if it could be harnessed in a practical way. The Sun has been producing energy in this way for almost five billion years, and it will continue to do so for a few billion years more.

Capturing the Sun's energy is not easy, because solar energy is spread out over such a large area. In any small area it is not very intense. The rate at which a given area of land receives solar energy is called **insolation**. (That's not the same as insulation, which is material used to keep things hot or cold.) Insolation depends on several factors: latitude, season of the year, time of day, cloudiness of the sky, clearness of the air, and slope of the land surface. If the Sun is directly overhead and the sky is clear, the rate of solar radiation on a horizontal surface at sea level is about 1000 W/m^2 (watts per square meter). This is the highest value insolation can have on the Earth's surface except by concentrating sunlight with devices like mirrors or lenses. If the Sun is not directly overhead, the solar radiation received on the surface is less because there is more atmosphere between the Sun and the surface to absorb some of the radiation. Note also that insolation decreases when a surface is not oriented perpendicular to the Sun's rays. This is because the surface presents a smaller cross-sectional area to the Sun.

Geo Words

insolation: the rate at which a given area of land receives solar energy.

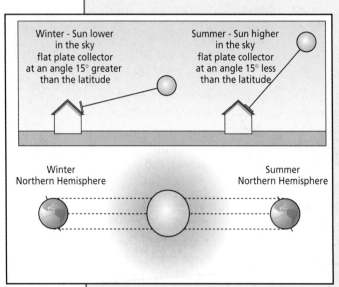

Winter - Sun lower
in the sky
flat plate collector
at an angle 15° greater
than the latitude

Summer - Sun higher
in the sky
flat plate collector
at an angle 15° less
than the latitude

Winter
Northern Hemisphere

Summer
Northern Hemisphere

Figure 1 The tilt of the Earth on its axis during winter and during summer.

The Earth rotates on its axis once a day, and it revolves in an almost circular but slightly elliptical orbit around the Sun once a year. The Earth's rotation axis is tilted about 23° from the plane of the Earth's orbit around the Sun, as shown in *Figure 1*. It points toward the Sun at one end of its orbit (in summer) and away from the Sun at the other end (in winter). In summer, when the Earth's axis points toward the Sun, the Sun traces a path high in the sky, and the days are long. In winter, the Sun's path across the sky is much closer to the horizon, and the days are short. This has important effects on seasonal insolation and on building designs.

Solar Heating

Home heating is one of the main uses of solar energy. There are two basic kinds of solar heating systems: active and passive. In an active system, special equipment, in the form of a solar collector, is used to collect and distribute the solar energy. In a passive system, the home is designed to let in large amounts of sunlight. The heat produced from the light is trapped inside. Passive systems do not rely on special mechanical equipment, but they are not as effective as active systems.

Water heating is another major use of solar energy. Solar water heating systems for buildings have two main parts: a solar collector and a storage tank. The solar energy is collected with a thin, flat, rectangular box with a transparent cover. As shown in *Figure 2*, the bottom of the collector box is a plate that is coated black on the upper surface and insulated

glazing frame

flow tubes

outlet connection

glazing

inlet connection

enclosure

absorber plate

insulation

Figure 2 Flat plate collector.

on the lower surface. The solar energy that strikes the black surface is converted to heat. Cool water is circulated through pipes from the hot collector box to a storage tank. The water is warmed as it passes through the collector box. These collectors are not very expensive, and in sunny regions they can provide most or all of the need for hot water in homes or swimming pools.

Many large commercial buildings such as the one shown in *Figure 3*, can use solar collectors to provide more than just hot water. A solar ventilation system can be used in cold climates to preheat air as it enters a building. The heat from a solar collector can even be used to provide energy for cooling a building, although the efficiency is less than for heating. A solar collector is not always needed when using sunlight to heat a building. Buildings can be designed for passive solar heating. These buildings usually have large, south-facing windows with overhangs. In winter, the Sun shines directly through the large windows, heating the interior of the building. In summer, the high Sun is blocked by the overhang from shining directly into the building. Materials that absorb and store the Sun's heat can be built into the sunlit floors and walls. The floors and walls then heat up during the day and release heat slowly at night. Many designs for passive solar heating also provide day lighting. Day lighting is simply the use of natural sunlight to brighten up the interior of a building.

By tapping available renewable energy, solar water heating reduces consumption of conventional energy that would otherwise be used. Each unit of energy delivered to heat water with a solar heating system yields an even greater reduction in use of fossil fuels. Water heating by natural gas, propane, or fuel oil is only about 60% efficient, and although electric water heating is about 90% efficient, the production of electricity from fossil fuels is generally only 30% to 40% efficient. Reducing use of fossil fuel for water heating not only saves stocks of the fossil fuels but also eliminates the air pollution and the emission of greenhouse gases associated with burning those fuels.

Figure 3 Solar collectors on the roofs of these buildings can be used to provide energy for heating water and regulating indoor climates.

Photovoltaics

Generation of electricity is another major use of solar energy. The most familiar way is to use photovoltaic (PV) cells, which are used to power toys, calculators, and roadside telephone call boxes. PV systems convert light energy directly into electricity. Commonly known as solar cells, these systems are already an important part of your lives. The simplest systems power many of the small calculators and wristwatches. More complicated systems provide electricity for pumping water, powering communications equipment, and even lighting homes and running appliances. In a surprising number of cases, PV power is the cheapest form of electricity for performing these tasks. The efficiency of PV systems is not high, but it is increasing as research develops new materials for conversion of sunlight to electricity. The use of PV systems is growing rapidly.

Wind Power

People have been using wind power for hundreds of years to pump water from wells. Only in the past 25 years, however, have communities started to use wind power to generate electricity. The photograph in *Figure 4* is taken from an electricity-generating wind farm near Palm Springs, California. The triple-blade propeller is one of the most popular designs used in wind turbines today. Wind power can be used on a large scale to produce electricity for communities (wind farms), or it can be used on a smaller scale to meet part or all of the electricity needs of a household.

Figure 4 Wind turbines on a wind farm in California.

Figure 5 shows typical wind velocities in various parts of the country. California leads the country in the generation of electricity from wind turbines (in 2000, California had a wind-power capacity of 1700 MW, with 20,000 wind turbines). Many other areas in the country have a high potential for wind power as well. These areas include the Rocky Mountains, the flat Midwest states, Alaska, and many other areas. Commercial wind turbines can have blades with a diameter as large as 60 m.

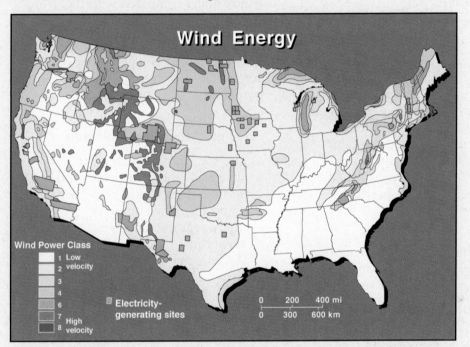

Figure 5 Location of electricity-generating wind-power plants relative to wind-power classes in the United States.

The cost of producing electricity from wind has dropped significantly since the early 1990s. This is due mainly to design innovations. In 2000, the cost of wind power was less than six cents per kilowatt-hour, which makes it competitive with electricity produced by coal-fired plants. The price is expected to decrease below five cents per kilowatt-hour by 2005.

Check Your Understanding

1. How does the Sun produce solar energy?

2. What factors determine the rate at which solar energy is received by a given area of the Earth's land surface?

3. What are the differences between passive solar heating systems and active solar heating systems?

4. Why must solar energy systems occupy a much larger area, per unit of energy produced, than conventional systems that burn fossil fuels?

Understanding and Applying What You Have Learned

1. Suppose that you live in a region that requires 1 million Btus (or 293 kWh) per day to heat a home. An average solar water-heating system with an output of 720 Btu/ft²/day costs $50 per square foot to install.

   a) What size of system would be required to meet these heating requirements?
   b) What would the system cost?
   c) Find out how much it costs to heat a home for five months in the winter where you live (November through March). Calculate how many years it would take to recover the cost of the solar water-heating system.

2. Why is metal used for tubing (rather than the plastic tubing that you used in the activity) in flat plate solar collectors? Use what you have learned about heat conduction to explain.

3. Why is wind power more efficient than generating electricity from all the methods that involve heating water to make steam?

4. Why is wind power less efficient than hydropower?

5. A wind turbine is placed where it is exposed to a steady wind of eight mph for the entire day except for two hours when the wind speed is 20 mph. Calculate the wind energy generated in the two windy hours compared with the 22 less windy hours.

6. Looking at the United States map of wind velocities, where does your community fall in terms of wind potential?

7. Assume that you live in a community in eastern Colorado.

   a) Use the wind map above to determine the number of wind turbines that would be needed to produce the electricity equivalent to a large electric power plant (1000 MW). Assume that each of the wind turbines has a blade diameter of 50 feet.
   b) If these wind turbines are replacing a coal-fired power plant that burns coal with a ratio of pounds of SO_2 per million Btu of 2.5, calculate the reduction in the amount of SO_2 emitted per year.
   c) What are some problems that might be encountered with the wind farm?

Preparing for the Chapter Challenge

Your challenge is to find a way to meet the growing energy needs of your community. You have also been asked to think about how to reduce the environmental impacts of energy resource use. Using what you have learned, describe how some of the energy needs of your community could be accommodated by wind or solar power.

Inquiring Further

1. **Solar-thermal electricity generation**

   Research how a solar thermal power plant, like the LUZ plant in the Mojave Desert in California, produces electricity. Diagram how such a system works. What are the kiloWatt-hour costs of producing electricity using this method? What does the future hold for power plants of this kind?

2. **Photovoltaic electricity**

   Investigate photovoltaic electricity generation and discuss the results of your investigation with the class.

3. **History of wind energy or solar energy**

   Investigate how people in earlier times and in different cultures have harnessed the Sun or wind energy. What developments have taken place in the past hundred years? How is wind energy being used today? Include diagrams and pictures in your report.

4. **Wind farms**

   Prepare a report on how electricity is generated on wind farms. Describe types of wind generators, types and sizes of wind farms, the economics of electricity production on wind farms, and the locations of currently operating wind farms in the United States. Include diagrams.

Earth Science at Work

ATMOSPHERE: *Air-Monitoring Technician*
Sophisticated equipment is used by government agencies to monitor air quality. Reports are issued that inform the public as to the type and amount of pollutants present on any given day.

BIOSPHERE: *Environmental Scientist*
Companies are concerned about the impact they may have on the environment. Environmental impact studies are completed before new projects are undertaken.

CRYOSPHERE: *Snowmobile Dealers*
There are approximately 1.6 million registered snowmobiles in the United States. They are often used for recreation, but in some areas, they are the preferred forms of transportation during winter months.

GEOSPHERE: *Exploration Geologist*
Before a company initiates any drilling operations, extensive surveys are completed. Geologists construct field maps, examine rock samples, and study seismic data to search for oil and gas deposits.

HYDROSPHERE: *Turbine Manufacturer*
In the generation of hydroelectricity, falling water strikes a series of blades attached around a shaft, causing a turbine to rotate.

How is each person's work related to the Earth system, and to Energy Resources?

Chapter 14

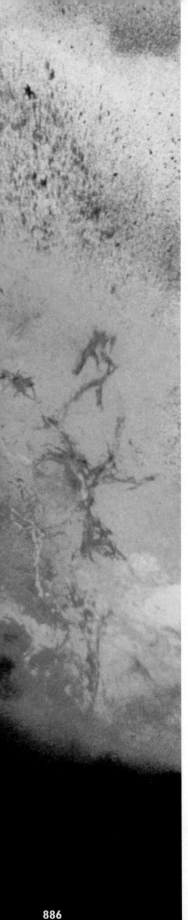

Volcanoes
...and Your Community

Getting Started

In 1883, on the island of Krakatoa in the East Indies, one of the most violent eruptions of recorded time took place. Half of the island was blown away by a volcanic eruption. Over a cubic mile of rock was hurled into the air. The sound of the explosion was heard in Australia, over 2000 miles away.

• Can a volcano that erupts on the other side of the world affect your community?

What do you think? Look at the Earth systems at the front of this book. In your notebook, draw a picture to show one way that a volcanic eruption changes an Earth system. Then, think about how that change might cause a change in another Earth system. Add this to your drawing. Continue until you have connected the volcanic eruption to your community. Be prepared to discuss your pictures with your small group and the class.

Scenario

"The clouds became thicker, and it was increasingly difficult to see as we struggled up the narrow, steep path toward the summit. The ground was hot under our feet, but the moisture from the clouds kept us cold and damp, and made the ash stick to our hair and eyelashes. We began to see larger volcanic rocks, some as large as two feet across. Suddenly we came across a large fissure, about one foot wide and 60 feet long. As I leaned over it, a hot blast of sulfur-smelling air scorched my nostrils. Then, like a warning growl from a watchdog, came a rumble from deep within. This was the moment we had been anticipating with dread..."
Many motion pictures are based on exciting geologic events.

Can you use your knowledge of volcanoes to make a thrilling, yet informative and scientifically correct movie?

Chapter Challenge

Your challenge is to write a screenplay or story, set in your community, that would help audiences understand volcanoes. You need to teach them about volcanic hazards. You also need to help them see volcanoes as part of the Earth system, and to realize that volcanoes affect all communities in some way. Can you use Earth science to create an exciting story and help others understand the hazards and the benefits of volcanoes?

Think about how your story will address the following items:
- Locations of volcanoes closest to your community
- Evidence that shows past or recent volcanic activity
- Types of volcanic hazards

- How volcanoes change the atmosphere, hydrosphere, and other Earth systems.

Assessment Criteria

Think about what you have been asked to do. Scan ahead through the chapter activities to see how they might help you to meet the challenge. Work with your classmates and your teachers to define the criteria for assessing your work. Record all this information. Make sure that you understand the criteria as well as you can before you begin. Your teacher may provide you with a sample rubric to help you get started.

Activity 1

Where are the Volcanoes?

Goals

In this activity you will:

- Find the latitude and longitude of volcanoes nearest your community when given a map of historically active volcanoes.

- Search for and describe patterns in the global distribution of volcanoes.

- Make inferences about possible locations of future volcanic activity.

- Understand that most volcanism occurs beneath the ocean.

- Understand that map projections distort regions near the poles and eliminate some data.

Think about It

Volcanoes are one of nature's most feared, yet spectacular activities.

- **Can volcanoes form anywhere on Earth? Why or why not?**

What do you think? Record your ideas about this question in your *EarthComm* notebook. Be prepared to discuss your responses with your small group and the class.

Investigate

1. "Thought experiments" are experiments that scientists dream up and then run in their imagination, rather than in a real laboratory. They are a useful way to develop new ideas and insights into scientific problems. Here's a thought experiment for you to do, to help you understand the Mercator map projection. It wouldn't even be too difficult to do this in real life, if you could obtain the right materials.

Visualize a large, see-through plastic ball. Poke holes on opposite sides, and stick a wooden dowel or a chopstick through to make the North Pole and the South Pole. Install a bright light directly at the center of the ball, somehow. With a felt-tipped pen, draw a fake continent on the ball. Make the continent extend from near the Equator to near the North Pole. Now wrap a clear sheet of stiff plastic around the globe, to make a tight-fitting cylinder that's parallel to the Earth's axis. See the figure on the right for how to arrange this. Turn the light on, and observe how the border of your continent projects onto the plastic cylinder. Trace that image on the cylinder with the felt-tipped pen. Unwrap the cylinder from the globe, and lay it flat on the table. You now have a map with what's called a Mercator projection of your continent!

a) Describe how the image of your continent is changed in shape (distorted) when it is projected onto the cylinder.

b) If you drew a short east–west line with a certain length near the southern end of your continent, and another east–west line with the same length near the northern end of your continent, how would the lengths of the lines compare when they are projected onto the cylinder map?

c) If you drew a short north–south line with a certain length near the southern end of your continent, and another north–south line with the same length near the northern end of your continent, how would the lengths of the lines compare when they are projected onto the cylinder map?

d) How would the image of a continent that is centered on the North Pole project onto the cylinder map?

e) Which part of your map shows the least distortion?

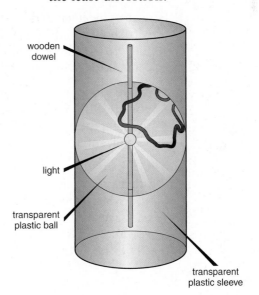

wooden dowel

light

transparent plastic ball

transparent plastic sleeve

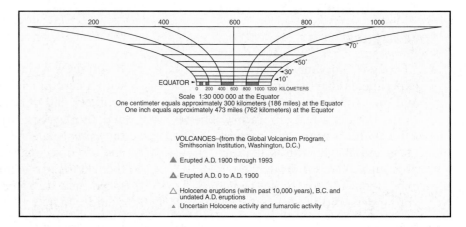

Scale 1:30 000 000 at the Equator
One centimeter equals approximately 300 kilometers (186 miles) at the Equator
One inch equals approximately 473 miles (762 kilometers) at the Equator

VOLCANOES–(from the Global Volcanism Program,
Smithsonian Institution, Washington, D.C.)

▲ Erupted A.D. 1900 through 1993

▲ Erupted A.D. 0 to A.D. 1900

△ Holocene eruptions (within past 10,000 years), B.C. and
undated A.D. eruptions

▴ Uncertain Holocene activity and fumarolic activity

2. Obtain the USGS map called *This Dynamic Planet*. Look at the map key, also shown above, to learn the meaning of the various symbols and how to use the map scale.

   a) What do each of the four kinds of triangles represent?

   b) What do the solid red lines represent?

   c) Describe how the scale of the map changes with latitude.

   d) Does the map cover the entire Earth? Why or why not?

3. For each time interval of volcanic activity shown on the map, find the latitude and longitude of three volcanoes closest to your community.

   a) Make a data table to record your results. When complete, the data table should list 12 volcanoes.

   b) Compare your data with that of other groups in your class. Did your class agree on the locations of the nearest historically active volcanoes? How did you resolve any differences?

4. Obtain a copy of a world map. Use this map to summarize any patterns in the global distribution of volcanoes.

   a) When volcanoes follow a linear pattern, draw a thick line on the world map. Use the string of volcanoes within the Aleutian Islands and southern Alaska as an example.

   b) For the red lines that appear on the USGS map, draw thin lines on your copy of the map. See the examples in the Pacific Ocean near Oregon and Washington.

   c) Where volcanoes are less concentrated, outline (circle) the area that they cover. Try to be as accurate as possible. See the group of volcanoes in the Cascades as an example.

5. When your map is complete, answer the following questions in your notebook:

   a) Are most volcanoes found in random places or do they show a trend or pattern? Explain.

b) Does the USGS map show volcanoes that have not erupted during the last 10,000 years?

c) Does the USGS map show eruptions after 1993, or new volcanoes?

d) Does the USGS map show any volcanoes associated with the red lines in the ocean basins?

e) What information does the map give about the size or hazard of the volcanoes?

f) Suppose that tomorrow a volcano forms somewhere in the United States. Could it form in or near your state? Support your answer with evidence from this activity.

g) What are some limitations of the evidence you used?

Reflecting on the Activity and the Challenge

By looking at a world map of recent volcanic activity you found patterns in the data. This helped you to make inferences about the possible location of the next volcanic eruption in the United States. The data you looked at are incomplete. This may limit the conclusions you can draw. However, you now have some knowledge that will help you decide where in the U.S. you might "stage" a volcanic eruption.

Digging Deeper

THE GLOBAL DISTRIBUTION OF VOLCANOES

Volcanoes beneath the Sea

The USGS map *This Dynamic Planet* shows historical volcanic activity throughout the world. It tells a story about how our dynamic planet releases its internal storehouse of energy. No single source of data tells the whole story, but a map is a great place to begin.

On average, about 60 of Earth's 550 historically active volcanoes erupt each year. Geologists have long known that volcanoes are abundant along the edges of certain continents. The presence of volcanic rocks on the floors of all ocean basins indicates that volcanoes are far more abundant under water than on land.

Coordinated Science for the 21st Century

Geo Words

mid-ocean ridge: a continuous mountain range extending through the North and South Atlantic Oceans, the Indian Ocean, and the South Pacific Ocean.

rift valley: the deep central cleft in the crest of the mid-oceanic ridge.

magma: naturally occurring molten rock material. generated within the Earth

All of the Earth's ocean basins have a continuous mountain range, called a **mid-ocean ridge** extending through them. These ridges, over 80,000 km long in total, are broad rises in the ocean floor. They are usually in water depths of 1000 to 2000 m. *Figure 1* shows a vertical cross section through a mid-ocean ridge. At the crest of the ridge there is a steep-sided **rift valley**. **Magma** (molten rock) from deep in the Earth rises up into the rift valley to form submarine volcanoes. These volcanoes have even been observed by scientists in deep-diving submersibles. All of the floors of the oceans, beneath a thin layer of sediments, consist of volcanic rock, so we know that volcanoes form all along the mid-ocean ridges, at different times at different places. At a few places along the mid-ocean ridges, as in Iceland, volcanic activity is especially strong, and volcanoes build up high enough to form islands.

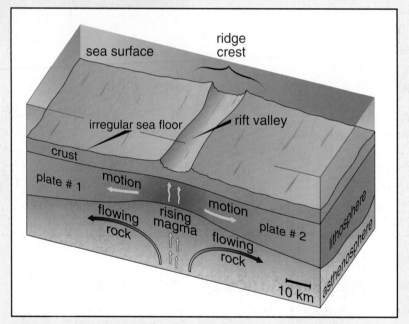

Figure 1 Cross section through a mid-ocean ridge.

Volcanoes on Land

Volcanoes that erupt on land are much more dangerous than volcanoes beneath the ocean. Eruptions along the western edge of the United States have formed the Cascades volcanic mountain range. They also form island chains, like the Aleutians in Alaska. Volcanoes like these are common in a narrow belt all around the Pacific Ocean. Geologists call this the "Ring of Fire." A famous example of an eruption along the Ring of Fire was the dramatic eruption of Mt. St. Helens in Washington in 1980.

Ring of Fire

Around the edges of the Pacific Ocean, the plates of the Pacific Ocean slide down beneath the continents. Look at *Figure 2* to see an example. The Nazca Plate, moving eastward from the East Pacific Ridge, slides down beneath the west coast of South America. The plate is heated as it sinks into the much hotter rocks of the deep Earth. The heat causes fluids, especially water, to leave the plate and rise into overlying hot rocks. The added water lowers the melting point of the solid rock. If enough water is added the rock melts and magma is formed. The magma rises upward, because it is less dense than the rocks. It feeds volcanoes on the overlying plate. Nearly four-fifths of volcanoes on land form where one plate slides beneath another plate.

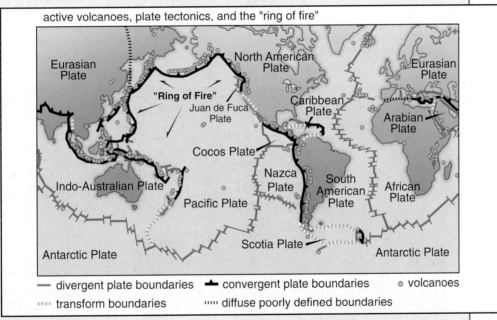

active volcanoes, plate tectonics, and the "ring of fire"

— divergent plate boundaries ▲— convergent plate boundaries ⊙ volcanoes
····· transform boundaries ⁗⁗ diffuse poorly defined boundaries

Figure 2 The plates of the Earth, and the "Ring of Fire" around the Pacific. The circles show active volcanoes.

Volcanoes Formed by Rifting on the Continents

Volcanoes in the East African rift valley form where two parts of the African continent are moving apart from each other. The process is very similar to what happens at mid-ocean ridges. The continental plate is stretched and broken. One of the breaks becomes the main one, and opens up to form the rift valley, as shown in *Figure 3*.

Geo Words

lava: molten rock that issues from a volcano or fissure.

hot spot: a fixed source of abundant rising magma that forms a volcanic center that has persisted for tens of millions of years.

map projections: the process of systematically transforming positions on the Earth's spherical surface to a flat map while maintaining spatial relationships.

Mercator projection: a map projection in which the Equator is represented by a straight line true to scale, the meridians by parallel straight lines perpendicular to the Equator and equally spaced according to their distance apart at the Equator, and the parallels by straight lines perpendicular to the meridians and the same length as the Equator. There is a great distortion of distances, areas, and shapes at the polar regions.

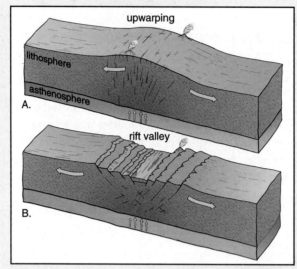

Figure 3 Formation of a rift valley on a continent.

In the United States, continental rifting long ago formed the rocks that make up the tall cliffs on the western bank of the Hudson River. These rocks formed when magma intruded the crust during this rifting. The rocks are seen for more than 80 km along the bank of the Hudson River and can be up to 300 m thick! Other evidence of magma formed during this rifting is found in many states along the East Coast.

Figure 4 Mount Kilimanjaro is a famous example of volcanism at a continental rift. Many other volcanoes in the East African rift valley have erupted in historic times.

Volcanoes at Hot Spots

Volcanoes discussed so far occur near the edges of plates. However, a small percentage of volcanoes occur in the interior of a plate. The Hawaiian Islands, shown in *Figure 5*, are an example. Studies of volcanic rock show that the islands get older to the northwest. Only the youngest island, the "big island" of Hawaii, has active volcanoes.

Here's how geologists explain the pattern of the Hawaiian Islands. Deep beneath Hawaii, there is a fixed source of abundant rising magma, called a **hot spot.** As the Pacific Plate moves to the northwest, away from the East Pacific Ridge, it passes over the fixed hot spot. Magma from the hot spot punches its way through the moving plate to form a chain of islands. The sharp bend in the chain was formed when the direction of movement of the plate changed abruptly at a certain time in the past. Far to the northwest the chain consists of **seamounts**.

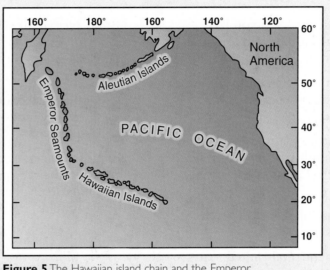

Figure 5 The Hawaiian island chain and the Emperor seamount chain.

Geo Words

seamount: a peaked or flat-topped underwater mountain rising from the ocean floor.

Map Projections

There is always a big problem in drawing a map of the world, because you have to try to show the curved surface of the Earth's globe on a flat sheet of paper. Many different ways of doing this, called **map projections**, have been developed, but they all have some kind of distortion. The USGS map uses a **Mercator projection.** As you move away from the Equator, the map becomes more and more distorted. For example, it makes all lines of latitude look like they are equal in length. This makes it difficult to measure distances on the map. Another problem is that the USGS map stops at 70° north and south latitude, because of the Mercator projection. This keeps you from seeing all of the data. For example, the USGS map cuts off the mid-ocean ridge north of Iceland. The scale of the map also presents a problem. The larger the area covered by the map, the less detail the map can show. In this case, the triangular symbols that represent volcanoes often overlap in areas with many volcanoes. This makes them difficult to count.

Check Your Understanding

1. What evidence do geologists have that volcanoes occur on the ocean floor?
2. What is the Ring of Fire, and where is it located?
3. Where do most volcanoes on land form?
4. How are rift valleys formed?
5. What are hot spots? Provide an example of a hot spot on Earth.
6. Why does the horizontal scale of a Mercator projection increase with latitude?

Understanding and Applying What You Have Learned

1. What difficulties did you have finding the latitude and longitude of volcanoes?

2. Where on Earth do most volcanoes occur? Explain your answer.

3. Are most volcanoes on land caused by the Earth's plates moving away from each other or moving toward each other? Explain your answer.

4. In your own words, describe the likely cause of historically active volcanoes in:

   a) The continental United States
   b) The Aleutian Islands and southern Alaska
   c) The Hawaiian Islands.

5. Based on your results from this investigation, list the five states that you think are most likely to experience the next volcanic eruption. Explain each choice.

6. Of the average of 60 volcanoes that erupt in any given year, how many are likely to erupt along the Ring of Fire?

7. Why did the Mercator projection not show volcanoes near the Earth's poles?

8. Do most volcanoes on land occur in the Northern Hemisphere or the Southern Hemisphere? Explain why you think this is so.

Preparing for the Chapter Challenge

Think about how you can help the audience understand why you chose the probable location of the volcanic eruption for your story. Explain the map that you made for this activity. Note the volcanic eruptions that are closest to your area. Explain where most volcanoes occur in the United States. You should also note where they have not happened recently.

Inquiring Further

1. **Eruptions near your community**

   Find out more about the historical eruptions at the volcanoes nearest your community. The Volcano World web site lists hundreds of historically active volcanoes. (Consult the AGI *EarthComm* web site for current addresses.) Your data table of latitudes and longitudes will help you to identify them.

2. **Volcanoes and the water on Earth (the hydrosphere)**

   Research to find answers to the following questions, and any other questions which you have formed:

   - How do volcanoes at mid-ocean ridges affect the temperature of seawater?

   - How do volcanoes change the chemistry of seawater?

   - How does seawater affect the composition of the volcanic rock that is formed at the mid-ocean ridge?

   - Would volcanoes affect a small body of seawater, such as the Red Sea, the same way as a large ocean like the Atlantic?

   - Can a change in the volume of volcanic rock formed at mid-ocean ridges change sea level?

Activity 2   Volcanic Landforms

Goals

In this activity you will:

- Make a topographic map from a model.

- Understand the meanings of contour line, contour interval, and relief.

- Interpret topographic maps.

- Recognize volcanic landforms on a topographic map, and predict where lava would flow on them.

- Understand basic relationships between magma composition and type of volcano formed.

Think about It

When most people think about volcanoes, they probably have in mind a steep-sided cone. Many volcanoes, however, have very gentle slopes.

- **Why do different volcanoes have different shapes?**

What do you think? Record your ideas about this question in your *EarthComm* notebook. Include a quick sketch for each question. Be prepared to discuss your responses with your small group and the class.

Investigate

1. Use a piece of paper and tape to make a model of a volcano. The model should be small enough to fit into a shoebox.

2. Draw horizontal curves on the model at regular heights above the table. To help you draw the lines, attach a strip of stiff cardboard at right angles to a centimeter ruler at the 1-cm mark, as shown in the diagram. Hold the ruler upright on the table, with the "zero" end down, and move it around the model so that the cardboard strip is near the surface of the model. Make a series of small dots on the model at this 1-cm height, and then connect the dots to form a horizontal curve. Repeat this with the cardboard strip attached at the 2-cm mark. Continue increasing the height above the table by 1 cm until you reach the top of the model.

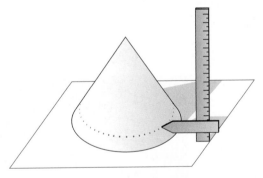

3. Place the model into a shoebox.

4. Clip an overhead transparency onto a clear clipboard. Lay the clipboard on the box.

5. Look straight down into the top of the shoebox at the lines you drew on the mountain. With a grease pencil or marker, trace the lines onto the transparency. Be sure to keep looking straight down, whenever you're tracing the lines! Also, it might help to keep one eye closed. These are contour lines, or lines of equal elevation above sea level.

6. Remove the transparency from the box. Write an elevation on each line of the transparency. Let each centimeter in height on the model represent 100 m in elevation on the map. The numbers should increase toward the center of the transparency.

7. Compare your map to the map of Mt. St. Helens shown below. Answer the following questions in your notebook and on your map:

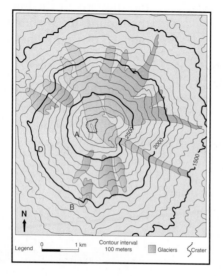

Legend 0 1 km Contour interval 100 meters ▨ Glaciers ⌇ Crater

a) Describe two similarities between the maps.

b) Note the legend on the map of Mt. St. Helens. Add a legend to your map. Include a scale, north arrow, and contour interval.

c) What do the shaded regions on the map of Mt. St. Helens represent?

d) Why do the shaded regions cross the contour lines at right angles?

e) Which part of Mt. St. Helens is steeper, the slope between 1500 m and 2000 m, or the slope between 2000 m and 2500 m? Explain your answer.

f) What are the lowest and highest elevations on the map of Mt. St. Helens? What is the difference in elevation between these two points?

g) Note the locations A, B, C, and D on the map. If lava erupted at point A, would it flow to point B, C, or D? Explain why.

Reflecting on the Activity and the Challenge

You made a map from a model of a volcano. The map showed lines of equal elevation. The lines are contour lines, and the map is a topographic map. You can use topographic maps to predict volcanic hazards. Gravity pulls the lava erupted from volcanoes downhill. A topographic map shows the paths the lava might take. It might also help you to guess whether a certain region has volcanoes.

Digging Deeper

TOPOGRAPHY OF VOLCANIC REGIONS

Topographic Maps

Topographic maps have **contour lines**. These are curves that connect all points at the same elevation. The **contour interval** is the difference in elevation between adjacent contour lines. A **topographic map** shows how steep or gentle a slope is. It also shows the elevation and shape of the land. **Relief** is the difference in elevation between the highest and lowest points on the map.

The following are some important points to consider when interpreting topographic maps:

- Contour lines never cross (but two or more can run together, where there is a vertical cliff).
- The closer together the contour lines, the steeper the slope.
- Contour lines for closed depressions, such as a volcanic crater, are marked with "tick marks" (short lines at right angles to the contour line) pointing downslope, into the depression.
- On most topographic maps, every fifth contour line on a map is darker and its elevation is always marked.

Magma Composition

Volcanoes are often pictured as cone-shaped mountains. However, volcanoes come in many shapes and sizes. Ice, wind, and rain can change the shape of a volcano, both between eruptions and after the volcano becomes dormant. A large eruption or giant landslide can remove the top or side of a volcano. The chemical composition of magma can have an even greater effect on the shape the volcano takes as it forms.

Magma is a mixture of liquid, melted rock, and dissolved gases. The most abundant chemical elements in magma are silicon and oxygen. As the magma cools, minerals form. Silicon and oxygen are the building blocks of the most common minerals, called silicate minerals, that form from magmas. One silicon atom and four oxygen atoms become tightly bonded together to form an ion, called the silicate ion. These combine with ions of other elements, mainly aluminum, iron, calcium, sodium, potassium, and magnesium, to form silicate minerals.

When geologists make a chemical analysis of an **igneous rock**, a rock that formed when molten materials became solid, they express the results as percentages of several "oxides," such as SiO_2, Al_2O_3, or CaO. In one

Geo Words

contour lines: a line on a map that connects points of equal elevation of the land surface.

contour interval: the vertical distance between the elevations represented by two successive contour lines on a topographic map.

topographic map: a map showing the topographic features of the land surface.

relief: the physical configuration of a part of the Earth's surface, with reference to variations of height and slope or to irregularities of the land surface.

igneous rock: rock or mineral that solidified from molten or partly molten material, i.e., from magma.

Geo Words

silica: material with the composition SiO$_2$.

shield volcano: a broad, gently sloping volcanic cone of flat-dome shape, usually several tens or hundreds of square miles in extent.

way, this is a fake, because real oxide minerals are a very small part of most igneous rocks. It's just a generally accepted practice. Because silicon and oxygen are the most abundant elements in magmas, the "oxide" SiO$_2$, called **silica**, is the most abundant "oxide." The percentage of silica in magma varies widely. This is important to know for two reasons. First, magmas rich in silica tend to have more dissolved gases. Second, silica content affects how easily magma flows. Magmas that are rich in silica do not flow nearly as easily as magmas that are poor in silica. Because of this, silica-rich magmas are more likely to remain below the Earth's surface, at shallow depths, rather than flowing freely out onto the surface. These two factors combine to make eruptions of silica-rich magmas likely to be dangerously explosive. Here's why: As the magma rests below the surface, the dissolved gases gradually bubble out, because the pressure on the magma is much less than it was down deep in the Earth, where the magma was formed. It's just like what happens when you pour a can of soda into a glass: the carbon dioxide dissolved in the soda gradually bubbles out of solution. Unlike your soda, however, the magma is so stiff that the bubbles can't readily escape. Instead they build up pressure in the magma, and that often leads to a catastrophic explosion. The table in *Figure 1* shows how magma properties relate to magma composition.

Properties of Magma as They Relate to Magma Composition			
Magma Property	**Magma Composition (percent silica content)**		
	Low Silica	**Medium Silica**	**High Silica**
Silica content (% SiO$_2$)	~ 50	~ 60	~ 70
Viscosity	lowest	medium	highest
Tendency to form lava	highest	medium	lowest
Tendency to erupt explosively	lowest	medium	highest
Melting temperature	highest	medium	lowest
Volume of an eruption	highest	medium	lowest

Figure 1   Adapted from *Earth Science*, 7th Edition, Tarbuck and Lutgens, 1994

Types of Volcanic Landforms

When low-silica magma erupts, lava tends to flow freely and far. If it erupts from a single opening (vent) or closely spaced vents, it forms a broad **shield volcano**, as shown in *Figure 2*.

Figure 2 Volcanoes such as these are called shield volcanoes because they somewhat resemble a warrior's shield. They are formed when low-silica magma erupts.

Figure 3 The eruption of low-silica magma along long, narrow openings in the Columbia Plateau flowed over a vast area. The result was a broad lava plateau that makes up the cliffs.

Silica-rich magmas are far less fluid. They often stop moving before they reach the surface. If they do reach the surface, they ooze slowly, like toothpaste squeezed out of a vertical tube. The thick, stiff lava forms volcanic domes with steep slopes, as shown in *Figure 4*. If the volcano's vent gets plugged, gases cannot escape and pressure builds up. The pressure can be released in a violent eruption that blasts pieces of lava and rock (pyroclastics) into the atmosphere.

903

Geo Words

composite cone (stratovolcano): a volcano that is constructed of alternating layers of lava and pyroclastic deposits.

caldera: a large basin-shaped volcanic depression, more or less circular, the diameter of which is many times greater than that of the included vent or vents.

Figure 4 Silica-rich magma does not flow readily and often forms a volcanic dome such as the one shown in this photograph.

A **composite cone**, as shown in *Figure 5*, forms by many eruptions of material with medium or high silica content. They erupt violently when pressure builds up in the magma. After the explosion, gooey (viscous) lava oozes out of the top. The volcano becomes quiet. Over time, pressure may build up and repeat the cycle. Composite volcanoes are tall and have steep slopes because the lava does not flow easily.

Figure 5 Composite cones include the beautiful yet potentially deadly Cascades in the northwestern United States (Shasta, Rainier, Mt. St. Helens, etc.).

When a very large volume of magma is erupted, the overlying rocks may collapse, much like a piston pushing down in a cylinder. The collapse produces a hole or depression at the surface called a **caldera**, shown in *Figure 6.* A caldera is much larger than the original vent from which the magma erupted.

Figure 6 Calderas are deceptive volcanic structures. They are large depressions rather than conical peaks. Oregon's Crater Lake, formed nearly 7000 years ago, is an example of this type of volcano.

Check Your Understanding

1. Explain in your own words the meaning of a contour line, contour interval, relief, and topographic map.

2. Arrange corn syrup, water, and vegetable oil in order of low to high viscosity.

3. What is the silica content of magma that has a low viscosity?

4. Why do silica-poor magmas produce broad volcanoes with gentle slopes?

5. Why does high-silica magma tend to form volcanic domes with steep sides?

6. How is a caldera formed?

Understanding and Applying What You Have Learned

1. What is the contour interval on the topographic map of Mt. St. Helens?

2. Sketch a contour map of a volcano that shows:

   a) a gentle slope
   b) a steep slope
   c) a nearly vertical cliff
   d) a crater or depression at the top

3. Imagine that your paper model was a real volcano. Lava begins to erupt from the top. Shade your topographic map to show where a stream of lava would flow. Explain your drawing.

4. For the volcanoes shown in *Figures 2* and *5*, sketch a topographic map. Show what the volcano would look like from above. Apply the general rules for interpreting topographic maps. Include a simple legend.

5. Use the topographic map below, or obtain a topographic map of your state or region, to answer the questions.
   a) Record the contour interval, and the highest and lowest elevations. Calculate the relief.

b) Identify areas that look like the volcanic landforms you explored in this activity. Describe possible paths of lava flows.

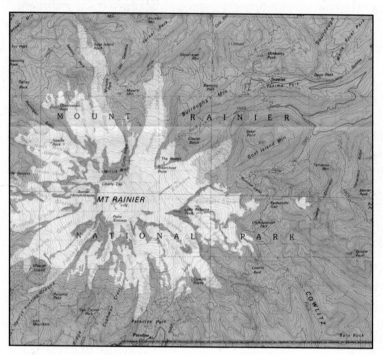

Preparing for the Chapter Challenge

In your story or play decide how you will convey to the audience the importance of using topographic maps to identify volcanic landforms. Indicate how the maps can also help geologists predict the paths of lava.

Inquiring Further

1. **Cascade volcano in your community**

   Build a scale model of a Cascade volcano and a scale model of your community. To do so, find a topographic map of a Cascade volcano. Trace selected contours on separate sheets of paper. Cut and glue each contour level onto pieces of card or foam board. Stack the board to make a three-dimensional model. Do the same using a topographic map of your community. Make sure that the scales of the maps match.

   ⚠️ Be cautious when cutting foam board.

Activity 3   Volcanic Hazards: Flows

Goals

In this activity you will:

- Measure and understand how volume, temperature, slope, and channelization affect the flow of fluid.

- Apply an understanding of factors that control lava flows, pyroclastic flows, and lahars (mudflows).

- Apply understanding of topographic maps to predict lahar flow (mudflow) patterns from a given set of data.

- Describe volcanic hazards associated with various kinds of flows.

- Become aware of the benefits of Earth science information in planning evacuations and making decisions.

- Show understanding of the nature of science and a controlled experiment.

Think about It

Only one person in the entire city of St. Pierre, on Martinique in the Caribbean, survived the hot ash and rock fragments that swept over the city from the explosive eruption of Mt. Pelée in 1902. He was a prisoner in a dungeon deep underground.

• How do volcanoes affect the biosphere?

What do you think? Record your ideas about this question in your *EarthComm* notebook. Be prepared to discuss your responses with your small group and the class.

907

Investigate

Part A: Area of Lava Flow

1. Suppose a volcano produces twice the amount of lava than in a prior eruption. Write a hypothesis based upon the following question: What is the relationship between the volume of an eruption and the size of the area it covers?

   a) Record your hypothesis in your notebook.

2. Check your hypothesis to see if it could be disproved. A hypothesis must be a prediction that can be falsified. The statement "Some stars will never be discovered," cannot be disproved. Therefore, it is not a hypothesis.

3. In this investigation, you will use liquid soap to simulate flow during a volcanic eruption. Volcanic flows include lava, gases, and mixtures of solid particles and gases.

   a) In your notebook, set up a data table. The table should help you record the relationship between the volume of liquid soap and the surface area that the soap covers. You will do trials with 0.5, 1, 2, 4, 8, and 16 cm^3 (cubic centimeters) of liquid soap.

4. Place an overhead transparency of a square grid on a flat surface.

5. Pour 0.5 cm^3 of liquid soap onto the transparent graph paper.

6. When the soap stops flowing, measure the area of the flow.

   a) Record the area of the flow in your data table.

7. Wipe the surface clean. Repeat the trials using 1, 2, 4, 8, and 16 cm^3 of liquid soap.

   a) Record your data in your table. Look for patterns.

8. Develop a hypothesis and design a test for one of the following questions related to the flow of fluids. Remember that during scientific inquiry, you can return to the materials or your data and revise your procedures as needed.

   - What effect does temperature have on resistance to flow (viscosity)?

   - What happens to fluid when slope changes from steep to gentle?

   - What effects would you see if fluids moved through narrow channels?

   a) Write down your hypothesis.

   b) Record your procedure in your notebook.

   c) Describe the variables you investigated.

9. Present your procedure to your teacher for approval. Then run your test.

   a) Record your data.

   b) Summarize your conclusions.

   c) Was your hypothesis correct?

⚠ Heat sources can cause burns. Hot objects and liquids look like cool ones. Feel for heat at a distance before touching.

⚠ Clean up any spills immediately. Liquids being used can cause floors and equipment to be sticky or slippery.

Part B: Travel Time of Lahars

1. Examine the table of expected travel times of lahars (mudflows) triggered by a large eruption of Mt. St. Helens. The values in the table come from computer simulations and actual behavior of mudflows in the 1980 eruption.

Expected Travel Times for Lahars Triggered by a Large Eruption of Mt. St. Helens (USGS)		
Distance (via river channels) from Mt. St. Helens (km)	Estimated travel time (hours:minutes)	
	North Fork Toutle River	South Fork Toutle River, Pine Creek, Muddy River, Kalama River
10	0:37	0:11
20	1:08	0:30
30	1:37	0:54
40	2:16	1:21
50	2:53	1:49
60	3:27	2:20
70	3:48	2:53
80	4:43	3:31
90	6:36	4:18
100	8:50	5:12

2. Convert the travel times into minutes.

   a) Record the times in your notebook.

3. Make a graph of travel time (in minutes on the vertical axis) versus distance (in kilometers on the horizontal axis) for both data sets.

   a) Plot both data sets on the same graph.

   b) Connect the data points so that you can compare the data.

   c) Calculate an average velocity for mudflows along each fork of the Toutle River.

4. Answer the following questions in your notebook:

   a) Which area (North Fork or South Fork) is more likely to have a steeper gradient? Use the results of your investigation in **Activity 2** to support your answer.

   b) Explain the evidence in your graphs that suggests the gradients are not constant?

   c) Based on the information in the table, explain whether or not you think that a community located 50 km from Mt. St. Helens along either of these river valleys would have time to evacuate in the event of an unexpected massive eruption.

Reflecting on the Activity and the Challenge

In this activity, you found that temperature, volume, channels, and slope affect the flow of liquids. Analyzing data from a computer model, you predicted the flow of volcanic fluids down river valleys near Mt. St. Helens. You can now describe the volcanic hazards associated with various kinds of flows and factors which affect the flows. In your movie you may wish to locate a town in the path of a potentially dangerous flow.

Geo Words

lava flow: an outpouring of molten lava from a vent or fissure; also, the solidified body of rock so formed.

viscosity: the property of a fluid to offer internal resistance to flow.

pyroclastic flow: a high-density mixture of hot ash and rock fragments with hot gases formed by a volcanic explosion or aerial expulsion from a volcanic vent.

Digging Deeper

FLOW-RELATED HAZARDS

Lava

Lava flows are streams of molten rock that come from vents and fissures in the Earth's crust. Lava flows destroy almost everything in their path. However, most lava flows move slowly enough for people to move out of the way. Slope and cooling affect the flow of lava. Lava flows faster on a steeper slope. As lava cools, it flows less and less easily. The term **viscosity** is used to describe a fluid's resistance to flow or internal friction. As lava cools, it becomes more viscous.

Lava that is low in silica is less viscous. (See *Figure 1* in the previous activity that shows the properties of magma as they relate to magma composition.) Flows of low-silica lava can travel tens

Figure 1 Lava tubes form when the surface of a flow cools and crusts over, but the interior of the flow is still fluid.

of kilometers from the source. Sometimes it sets up an internal "plumbing system." The surface may cool, crust over, and insulate the interior. This keeps the lava at a higher temperature as it moves away from the source. Evidence of this is found in the lava tubes, as shown in *Figure 1*, found in flows of low-silica lava. When lava breaks out of the leading edge of a flow, the lava can drain out. A hollow tube remains behind.

Basalt flows can move at speeds of up to 10 km/h (kilometers per hour) on steep slopes. On a shallow slope, basalt flows typically move less than one kilometer per hour. Basaltic lava flows confined within channels or lava tubes can travel at speeds of 45 km/h. Basaltic lava flows can cover a considerable area. The largest lava flow in recent history occurred in 1783 at Laki in Iceland. Lava erupted from the Laki fissure covered 500 km^2, an area roughly equal to 100,000 soccer fields.

Since the start of the eruption in 1983, lava flows erupted from the Kilauea volcano in Hawaii and entered communities repeatedly. The flows destroyed more than 180 homes, a visitor center in a national park, highways, and historical and archaeological sites. The village of Kalapana was buried in 1990 by 15–25 m of lava erupted during a period of seven months. See *Figure 2*.

It is sometimes possible to control the flow of lava. In 1973, lava flows at Heimaey, Iceland threatened to cut off a vital harbor. Citizens sprayed water onto the lava from ships in the harbor. This stopped the flow. Lava flows can also be diverted away from populated areas. Workers must carve a new channel or pathway through the landscape for the lava to follow.

Andesitic lava is cooler and has a higher silica content than basaltic lava. It moves only a few kilometers per hour. Andesitic lava rarely flows beyond the base of the volcano. Dacitic and rhyolitic lavas are even higher in silica and are even more viscous. Their lava usually forms steep mountains called lava domes, which extend only short distances from the vent.

Pyroclastic Flows

Topography plays a role in two other types of volcanic flows: pyroclastic flows and lahars. **Pyroclastic flows** are high-density mixtures of hot ash and rock fragments with hot gases. Pyroclastic flows occur in explosive eruptions. They move away from the vent at speeds up to 350 km/h. They often have two parts. A lower flow of coarse fragments moves along the ground. A turbulent cloud of ash rises above the lower flow. Both parts ride upon a cushion of air. This enables the material to move

Figure 2 The former village of Kalapana was buried by lava flows.

911

rapidly. The more dense material follows the topography in a twisting path downslope. Pyroclastic flows are extremely dangerous. They destroy everything in their path. The pyroclastic flow produced by the Mt. St. Helens eruption, as shown in *Figure 3*, was impressive, but it was small compared to pyroclastic flows in prehistoric times.

Lahar

A **lahar** is a wet mixture of water, mud, and volcanic rock fragments, with the consistency of wet concrete, that flows down the slopes of a volcano and its river valleys. Lahars can carry rock debris ranging in size from clay, to gravel, to boulders more than 10 m in diameter.

Figure 3 Flow of pyroclastic materials from Mt. St. Helens destroyed everything in its path.

Check Your Understanding

1. Name two factors that influence the viscosity of a lava flow.

2. Describe two ways in which lava flows can be controlled.

3. What is a pyroclastic flow?

4. What is a lahar?

5. How are lahars formed?

6. Explain how topography influences volcanic flows.

Eruptions may trigger lahars. Heat from the eruption may melt snow and ice, or the eruption may displace water from a mountain lake or river. Lahars sometimes form when the erupted material dams the mountain's drainage, causing a lake to form. The lake may spill over the loose volcanic material and send water and debris down valley. Lahars are also formed when rain soaks the loose volcanic debris during or after an eruption, causing it to start to flow. As a lahar flows downstream, it poses a risk to everyone in the valley downstream. When a lahar finally comes to a stop, it can bury an entire village under many meters of mud.

Figure 4 On the left side of the photograph, the dark region extending down the side of Mt. St. Helens is an example of a lahar flow.

Understanding and Applying What You Have Learned

1. How does the volume of an eruption affect the area? Describe any mathematical pattern in your data.

2. When the Mauna Loa volcano erupted in 1984, lava flowed toward Hilo, Hawaii. It is an excellent example of how scientists used their understanding of the factors that control the flow of lava to predict where lava would flow and decide whether to evacuate residents. The map shows the path of a series of lava flows from Mauna Loa. Each flow is given a letter (A through G) in the order it happened.

   a) Look at flow D on the map. What is the elevation of the top of flow D, and what is the elevation of the Kulani Prison?

   b) How close did flow D get to the prison?

   c) Do you think that the prison was put on alert?

   d) Look at flow E on the map. The flow was channeled. Do you think it moved swiftly or slowly? Explain.

   e) Lava from flow E crossed an important road. It headed straight for the city of Hilo. The lava then broke through walls of the channel. What do you think happened to the width of the flow after it broke through the channel? How do you think this changed the speed of the flow?

3. Refer back to the reading that described the lava flow at Heimaey, Iceland.

   a) Why did spraying the lava flow with water slow it down?

   b) This was a very unusual circumstance. What factors made this effort successful?

4. Why might a lahar (mudflow of volcanic debris and water) affect a community more severely than a lava flow?

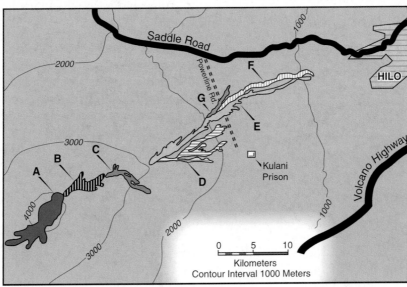

Source: USGS

Preparing for the Chapter Challenge

Think about what you have learned about volcanic flows. Prepare a one-page information sheet to raise awareness of how flows affect communities. Focus on three or more of the following:

a) How local topography controls where lava would flow.

b) Major roads that must be protected to ensure evacuation.

c) Natural and developed areas most likely to be affected.

d) Areas least likely to be affected, and why.

e) Living things that would not escape advancing flows.

f) Ways that flows might be controlled (diverting the flow, using water, and so on).

Consider how you can creatively work this information into your story line.

Inquiring Further

1. **Research a famous lava flow**

   Search the web for information about the Columbia River basalt group in the northwest. Prepare a report to the class about the members of this famous basalt group in relation to largest, longest, thickest, cooling characteristics, effects on ancient topography, and cause.

2. **Lava and the biosphere**

   How have lava flows at Mauna Loa and Kilauea volcanoes affected Hawaiian communities? How does the lava that enters the Pacific Ocean in Hawaii affect coastal ecosystems? What kinds of organisms develop and thrive at the "black smokers" along mid-ocean ridges? Research the 1783 Laki fissure flow in Iceland. It was 40 km long and covered 500 km^2. How did it affect vegetation and livestock?

3. **Lava and the cryosphere**

   What happens when lava erupts from an ice-capped or snow-capped volcano? This is an issue in the Cascade volcanoes. Mt. Rainier, which overlooks Seattle, has 27 glaciers. Some insights might be gained from exploring the recent eruption at Grimsvotn in Iceland.

Activity 4

Volcanic Hazards: Airborne Debris

Goals

In this activity you will:

- Understand why ash from volcanic eruptions can affect a much larger region than lava, pyroclastic flows, or lahars.

- Define tephra and describe some of the hazards it creates.

- Interpret maps and graph data from volcanic eruptions to understand the range in scale of volcanic eruptions.

- Understand that the explosive force of a volcano is not the only factor that determines its potential to cause loss of life and property.

- Interpret maps of wind speed and direction to predict the movement of volcanic ash.

Think about It

Volcanic ash put into the stratosphere from the great eruption of Krakatoa in Indonesia (see **Getting Started** on page two) caused spectacular sunsets all around the world for many months.

- **Could material from a volcanic eruption ever reach your community? Explain your ideas.**

What do you think? Record your ideas about this question in your *EarthComm* notebook. Be prepared to discuss your responses with your small group and the class.

Investigate

1. Look at the map of the 1980 eruption of Mt. St. Helens. It shows the pattern of ash. Use the map to answer the following questions:

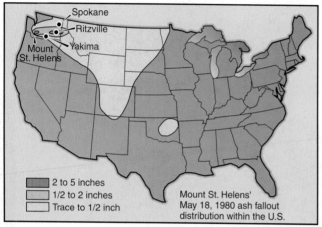

Distribution of ash from Mt. St. Helens eruption.

a) How many states showed at least a trace of volcanic ash?

b) In what direction did the ash move?

c) Was Canada affected by ash from Mt. St. Helens? Why or why not?

d) Would you consider this a small, medium, large, or gigantic eruption? Explain your choice.

2. Make a bar graph of the data shown in the table.

a) Plot the name of each volcano on the horizontal axis.

b) Plot the volume of volcanic eruption on the vertical axis. Arrange the volumes in order from least to greatest.

Volumes of Volcanic Eruptions		
Volcano	Date	Volume (cubic kilometers)
Ilopango, El Salvador	300	40
Krakatoa, Indonesia	1883–84	2.4
Long Valley, California (Bishop Tuff)	740,000 years ago	500
Mazama, Oregon	4000 B.C.	75
Mt. Pelée, Martinique	1902	0.5
Mt. St. Helens, Washington	1980	1.25
Nevado del Ruiz, Colombia	1985	0.025
Pinatubo, Philippines	1991	10
Santorini, Greece	1450 B.C.	60
Tambora, Indonesia	1815	150
Valles, New Mexico	1.4 million years ago	300
Vesuvius, Italy	79	3
Yellowstone, Wyoming (Lava Creek Ash)	600,000 years ago	1000

Note: Volumes are approximate.

3. Use your graph and the table to answer the following questions. Record your answers in your notebook.

   a) Can you group the eruptions by size (small, medium, and so on)? Mark the groups on your plot. Explain how you chose the groups.

   b) What group does the 1980 eruption of Mt. St. Helens fit into?

   c) Suppose you wanted to predict the area that would be covered with ash by each eruption. What other information (besides volume erupted) would help you to predict how far the ash would go?

4. The following map shows the areas covered by five of the eruptions in the data table. Use the map, data table, and your bar graph to do the following:

   a) Rank the area of eruptions in order from smallest to largest. Record your rankings.

   b) Compare the areas to the volumes. Describe any relationships.

   c) Compare the location of each volcano to the path of the ash. Describe any patterns. What might explain any patterns you see?

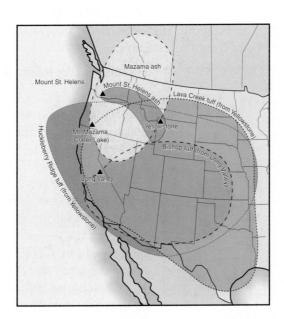

Reflecting on the Activity and the Challenge

This activity gave you a chance to explore factors that affect the movement of volcanic ash. How did this change your ideas about the areas that can be affected by an erupting volcano? Be sure to include the effect of airborne volcanic hazards into your story line.

Geo Words

tephra: a collective term for all the particles ejected from a volcano and transported through the air. It includes volcanic dust, ash, cinders, lapilli, scoria, pumice, bombs, and blocks.

volcanic bomb: a blob of lava that was ejected while viscous and received a rounded shape (larger than 64 mm in diameter) while in flight.

lapilli: pyroclastics in the general size range of 2 to 64 mm.

ash: fine pyroclastic material (less than 2 mm in diameter).

Digging Deeper

AIRBORNE RELEASES

Particle Types

Tephra is a term for pieces of volcanic rock and lava that are ejected into the air. It ranges from less than 0.1 mm to more than one meter in diameter. Tephra is classified by size. Names for sizes of tephra include **volcanic bombs** (greater than 64 mm), **lapilli** (between 2 and 64 mm), and **ash** (less than 2 mm). Bombs and lapilli usually fall to the ground on or near the volcano. Ash can travel hundreds to thousands of kilometers. (See *Figure 1*.) The height of the ash and the wind speed control how far the ash travels.

Distribution

A volcanic eruption can send ash many kilometers into the atmosphere. Ash from the 1980 eruption of Mt. St. Helens reached a height of 19 km. Winds carried the ash to the east. Five days after the eruption, instruments in New England detected ash. An eruption at Yellowstone 2 million years ago produced 1000 times as much ash. An area of 1000 km^2 received one meter of ash. Ten centimeters of ash covered an area of 10,000 km^2. You could look at it this way: if the ash from Mt. St. Helens filled a shoe box, the ash from Yellowstone would fill a bedroom to a depth of a meter!

Hazards of Volcanic Ash

Volcanic ash presents many kinds of hazards. Ash that falls on homes, factories, and schools can collapse roofs. More than 300 people died after the 1991 eruption of Mt. Pinatubo in the Philippines. Most of these deaths were caused by roof collapse. At ground level, fine ash causes breathing problems in humans and animals. It can also damage automobile and truck engines. Ash that coats the leaves of plants interferes with photosynthesis. Ash injected higher into the atmosphere can damage aircraft. In the last 15 years, 80 commercial aircraft have been damaged as they flew through volcanic ash. The only death outside the immediate area of Mt. St. Helens occurred from the crash of a small plane that was flying through the ash. Ash that falls on the slopes of a volcano poses great risk. When soaked by rain, loose ash can form lahars. Years after the eruption, lahars remain a source of concern to communities at the base of Mt. Pinatubo.

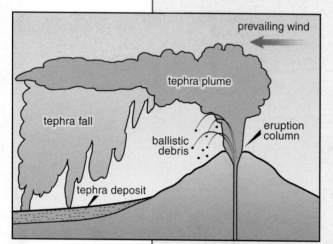

Figure 1 Ballistic debris refers to volcanic bombs and lapilli that fall on or near the volcano. Ash can travel much further.

Figure 2 Volcanic ash landing on buildings can result in death. Volcanic ash is also a hazard to airplanes on the ground as well as in the air.

Volcanic Explosivity Index

The table in *Figure 3* is a scale of eruption magnitude. The scale is known as **Volcanic Explosivity Index**, or VEI. The VEI is based on the volume of erupted material and the height it reaches. The size of an eruption depends upon several factors. Two important factors are the composition of the magma and the amount of gas dissolved in the magma. The viscosity of a magma depends on two things: the temperature of the magma, and its chemical composition. The higher the silica content of magma, the more viscous it is. The more viscous it is, the more likely it is for gas pressure to build. High-silica volcanoes, like Yellowstone, erupt extremely violently, but on a scale of tens or hundreds of thousands of years. Volcanoes with intermediate silica contents, like Mt. St. Helens, commonly produce violent eruptions with a frequency of hundreds or thousands of years. Silica-poor magmas, like those erupted at Kilauea, feed less explosive eruptions that occur more often.

Geo Words

Volcanic Explosivity Index: the percentage of pyroclastics among the total products of a volcanic eruption.

Volcanic Explosivity Index (VEI)				
VEI	**Plume Height**	**Volume**	**How often**	**Example**
0	<100 m	1000's m^3	Daily	Kilauea
1	100–1000 m	10,000's m^3	Daily	Stromboli
2	1–5 km	1,000,000's m^3	Weekly	Galeras, 1992
3	3–15 km	10,000,000's m^3	Yearly	Ruiz, 1985
4	10–25 km	100,000,000's m^3	10's of years	Galunggung, 1982
5	>25 km	1 km^3	100's of years	St. Helens, 1980
6	>25 km	10's km^3	100's of years	Krakatoa, 1883
7	>25 km	100's km^3	1000's of years	Tambora, 1815

Figure 3

Check Your Understanding

1. Review any words from the **Digging Deeper** sections of the previous activities that are used in this section but may still be unfamiliar to you. Briefly explain the meaning of each of the following terms: lahar, pyroclastic flow, caldera.

2. In your own words explain the meaning of tephra and how volcanic bombs, lapilli, and ash relate to tephra.

3. Name two factors that can affect the distance that volcanic ash can travel.

4. How does the silica content of magma affect how explosive a volcano can be?

5. a) What does VEI represent?

   b) Is VEI on its own a good indicator of the dangers involved with a volcanic eruption? Explain your answer.

It might seem that the number of deaths caused by an eruption should always increase as the VEI increases. The table of VEI and deadliest eruptions in *Figure 4* shows that this is not the case. For example, mudflows after the 1985 eruption of Nevado del Ruiz (Colombia) killed more than 25,000 people. This was the worst volcanic disaster since Mount Pelée in 1902. However, both eruptions had a VEI below five. Of the seven most deadly eruptions since 1500 A.D., only Tambora and Krakatoa erupted with greater explosive force (VEI above 5).

Volcanic Explosivity Index (VEI) of the Deadliest Eruptions since 1500 A.D.			
Eruption	**Year**	**VEI**	**Casualties**
Nevado del Ruiz, Colombia	1985	3	25,000
Mount Pelée, Martinique	1902	4	30,000
Krakatoa, Indonesia	1883	6	36,000
Tambora, Indonesia	1815	7	92,000
Unzen, Japan	1792	3	15,000
Lakagigar (Laki), Iceland	1783	4	9000
Kelut, Indonesia	1586	4	10,000

Figure 4

Tambora erupted in 1815. It had a VEI of 7. Pyroclastic flows streamed down its slopes. Ash rose 44 km. About 150 km³ of ash were erupted (about 150 times more than the 1980 eruption of Mt. St. Helens). A caldera formed when the surface collapsed into the emptying magma chamber. The Tambora caldera is 6 km wide and 1.1 km deep. Tephra fall, a tsunami (a giant sea wave caused by the explosion), and pyroclastic flows killed about 10,000 people. More than 82,000 people died from famine. It is thought that the ash shortened the growing season.

Understanding and Applying What You Have Learned

1. In your own words, compare the sizes of the areas affected by lava, pyroclastic flows, and ash falls.

2. Is volcanic ash a concern only in the western United States? Explain your answer.

3. Why do eruptions in Hawaii differ from the Mt. St. Helens eruption?

4. Look at your list of the three volcanoes that are closest to your community. Go to the AGI *EarthComm* web site to find out how to simulate the eruption of one of your three volcanoes on the Internet. Simulate the eruption of one of your three volcanoes.

   a) Print out and describe the paths of the ash from the simulation.

   b) What do the maps tell you about the prevailing wind directions for your community?

   c) Do the prevailing wind directions change seasonally in your area? If yes, how would this affect the pattern of ash fallout?

Preparing for the Chapter Challenge

In **Activity 1** you were asked to describe to your audience the place that you thought a volcanic eruption was most likely to occur. Did you leave the impression in your story that this was the only area to be affected by the eruption? Think about how your study of the movement and hazards of volcanic ash has changed your ideas. Be sure to include this information in your story.

Inquiring Further

1. **Make a model of tephra transport**

   Build a model of a volcano that has exploded (Mt. St. Helens). Run a tube up through the vent of the volcano. Mix a small amount of baby powder with some sand. Use a funnel to pour the sand mixture down the other end of the tube. Attach a bicycle pump to pump the sand out of the volcano. Use a fan or hair dryer to simulate winds.

   Devise a method to outline the distribution of material when there is no wind, weak wind, and strong wind. Compare how far the sand travels and how far the baby powder travels. Consider the factors of particle size, wind speeds, wind direction, and topography. As part of your **Chapter Challenge** you may wish to include a presentation and explanation of your model.

⚠ Use eyewear whenever non-water liquids and/or particles such as sand are used. If this activity is done indoors use a large, clear area. Clean the area well when you are done. Sand on the floor can be slick.

Activity 5

Volcanoes and the Atmosphere

Think about It

Following the eruption of Tambora in Indonesia in 1815, snow fell in New England during each of the summer months that year!

- **What else escapes from a volcano besides lava, rock, and ash?**

What do you think? Record your ideas about this question in your *EarthComm* notebook. Be prepared to discuss your responses with your small group and the class.

Goals

In this activity you will:

- Measure the amount of dissolved gas in a carbonated beverage.

- Understand that volcanoes emit gases such as water vapor, carbon dioxide, and sulfur dioxide.

- Describe how volcanoes are part of the hydrosphere and water cycle.

- Demonstrate awareness of how volcanoes can affect global temperatures.

- Recognize that volcanoes are part of interactive systems on Earth.

Investigate

1. Use a can of your favorite carbonated soft drink to explore the quantity of gas that can be dissolved in a liquid under pressure.

   a) How many milliliters (mL) of liquid are in the can of soda?

   b) Predict how many milliliters of gas (carbon dioxide) a can of soda contains. Record and explain your prediction.

2. Obtain these materials: heat source, 1-liter Pyrex® beaker, water, rubber tubing (about 50 cm), smaller beaker or bottle, plastic container (shoebox size), modeling clay, safety goggles.

3. Devise a way to use the materials to measure the gas that escapes from the can of soda. Note: You will need to heat the soda after you have opened it. To do this safely, put the can in a water bath (container of water) and heat the water bath.

   a) Draw a picture of how you will set up your materials.

   b) Write down the procedures you will follow. Include the safety precautions you will take.

4. After your teacher has approved your design, set up your materials. Run your experiment.

   a) Record your results.

   b) How do your results compare to your prediction?

   c) Describe anything that might have affected your results.

 Plan your activity carefully in detail to avoid potential hazards.

Reflecting on the Activity and the Challenge

You worked with a material that resembled volcanic products. When you opened the can of soda, you lowered the pressure inside the can.

This allowed carbon dioxide (the dissolved gas) to come out of solution. Dissolved gases emerge from the Earth's interior in much the same way.

Digging Deeper

VOLCANOES AND THE ATMOSPHERE

Volcanic Gases

Gases that escape in greatest abundance from volcanoes are water vapor, carbon dioxide, hydrogen chloride, and nitrogen. These and certain other gases have played an important role in the Earth system throughout the long span of geologic time, and they continue to do so at the present time.

The atmosphere of the Earth early in its history contained abundant carbon dioxide but no oxygen. After primitive algae made their appearance partway through Earth history, the carbon dioxide emitted by volcanoes was gradually converted to oxygen by photosynthesis.

Carbon dioxide is more dense than air and sometimes accumulates in a low spot near a volcanic eruption. High concentrations of carbon dioxide are hazardous, because they cause people and animals to suffocate.

Water vapor is an essential component of the Earth system. It is especially important for human communities, because it sustains life. When you think of the water cycle, do you think of volcanoes? Volcanoes release abundant water vapor. Most of the Earth's surface water seems to have been released from the Earth's interior by volcanoes throughout the Earth's history.

Some volcanoes emit sulfur dioxide gas in great abundance. Sulfur dioxide combines with water vapor and oxygen to form sulfuric acid. The sulfuric acid is washed out of the atmosphere by rain, over large areas downwind of the eruption. Rain that contains sulfuric acid, and certain other acids as well, is called acid rain. It is produced not only by volcanoes but also by power plants that burn coal containing sulfur. Acid rain damages plants both on land and in lakes.

Volcanoes and Climate Change

Volcanoes illustrate the complexity of Earth's systems, because the gases from volcanic eruptions can contribute both to global cooling and to global warming.

How do volcanoes affect climate? If the Earth system were simple, the task of answering that question might be easy. Suppose that volcanic activity is the independent variable. This is the variable that, when changed, causes a change in something else (the dependent variable). In a simple model, climate would be the dependent variable. You could plot volcanic activity over time and compare it to temperature (an aspect of climate that can be measured) over time. Temperature changes that follow volcanic events would allow you to make inferences about the effects of eruptions on climate. However, the Earth system is complex. Records of climate and volcanic activity are imperfect. Some volcanic products should warm the atmosphere (carbon dioxide, a greenhouse gas). Others should cool the climate (dust, which reduces sunlight). The task of understanding climate change is obviously very complicated. The evidence at hand, however, suggests that major volcanic eruptions can lower the average temperature of the Earth's surface by a few tenths of a degree Celsius for as long as a few years.

It is often thought that volcanic eruptions increase or cause rainfall near or downwind of the eruption. Volcanoes put dust into the air. Water droplets in clouds form around small dust particles. Eruptions can also heat the local atmosphere. This should increase convection, or circulation, of the atmosphere. Finally, some volcanic eruptions release great quantities of water vapor needed to form clouds and rain. However, a number of studies show that an increase in rainfall is rare after an eruption. The major eruption of Krakatoa in 1883 did not increase rainfall, and it occurred during the wet (monsoon) season. It seems that conditions in the atmosphere near a volcanic eruption have to be just right for rainfall to increase just because of the eruption.

Enormous quantities of sulfur dioxide gas from a volcanic eruption can be put all the way into the stratosphere (the upper layer of the atmosphere, above the weather). It then slowly reacts with water to form tiny droplets of sulfuric acid, less than a thousandth of a millimeter in diameter. Unlike in the troposphere (the lowest layer of the atmosphere), these sulfur dioxide droplets are not affected by the water cycle. They stay suspended in the stratosphere for as long as a few years. The sulfur dioxide droplets, as well as the large quantities of very fine volcanic ash particles that also reach the stratosphere during major volcanic eruptions, reflect sunlight and are thought to cause the global cooling that is often observed for a few years after a major volcanic eruption. For example, following the eruption of Tambora in Indonesia in 1815, many areas in the United States and Canada had unusually cold summer weather. In New England, 1815 was called the "year without a summer."

Check Your Understanding

1. What gases escape from volcanoes?

2. Why does the emission of carbon dioxide pose a threat near volcanic eruptions?

3. How are volcanoes connected to the water cycle?

4. a) How is acid rain formed?

   b) Are volcanoes the only source of acid rain?

5. Do volcanic eruptions increase or decrease the temperature of the Earth? Explain your answer.

Understanding and Applying What You Have Learned

1. Think about the air you are breathing. How much of it came from some distant volcano?

2. If a volcano erupted huge amounts of ash, would you expect global temperatures to go up or down? Why?

3. If warm air rises, why would hot gases from a volcano be a threat to people in the valley below? (Hint: think about volume's effect in your work with the lava flow lab.)

4. If a system consists of many parts that affect each other, how are volcanoes part of systems on Earth?

Preparing for the Chapter Challenge

Use the information in this activity to argue that volcanoes have affected virtually every community.

Consider ways in which you can include these arguments in your story line.

Inquiring Further

1. **Cascades eruptions**

   Examine the figure showing the eruptions of Cascade volcanoes during the last 4000 years.

   a) Which volcano has been most active? Which volcano has been least active? Explain.

   b) What three volcanoes do you think are most likely to erupt next?

   c) Visit the AGI *EarthComm* web site for a link to the USGS Cascades Volcano Observatory web site. Find out about their monitoring efforts.

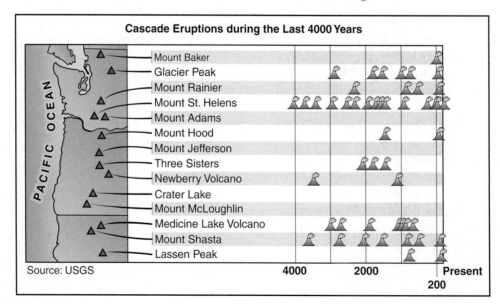

Cascade Eruptions during the Last 4000 Years

Source: USGS

Activity 6   Volcanic History of Your Community

Goals

In this activity you will:

- Demonstrate awareness of the knowledge used to construct geologic maps.

- Examine and identify several common igneous rocks.

- Identify the distribution of active volcanoes on one map and rock types on a given geologic map

- Recognize that volcanic rocks indicate a history of volcanic activity.

Think about It

Geologists can recognize volcanic eruptions dating back to early Earth history, over 3 billion years ago.

- **If your community once had a volcanic eruption long ago, what evidence would be left behind?**

What do you think? Record your ideas about this question in your *EarthComm* notebook. Be prepared to discuss your responses with your small group and the class.

Investigate

Part A: Rock Samples

1. Examine a set of rocks collected near a volcano.

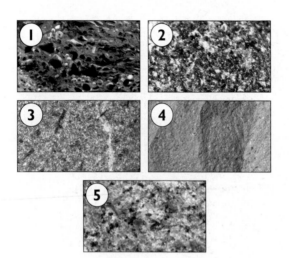

a) Record the sample number and briefly describe the physical characteristics of each rock.

b) Sort them into different groups by color (light, medium, dark).

c) Sort them again by texture (fine grains, large grains, mixture of sizes of grains, bubble holes).

2. Examine a set of rocks collected in your community.

a) Observe each rock, noting the size of the grains, color, and any other distinguishing features. Briefly describe each rock.

b) Note which ones you think might be volcanic.

c) Are all the rocks in the collection originally local? Explain.

⚠ Soft rock material can come off rock samples as grit. Avoid contact with eyes.

Part B: Geologic Maps

1. Examine the geologic map of your state or region. Find your community on the map. An example of a geologic map is shown.

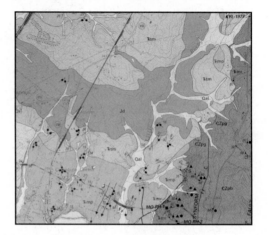

2. Look for patterns in the distribution of map colors (different rock types).

3. Compare the names of the rocks in the formations on the map with the names of rocks found in a reference book or field guide. Formation names followed by Basalt, Andesite, Dacite, Rhyolite, Ash, or Tuff indicate volcanic origin. Also, use any rock columns provided with the map and look for symbols that indicate volcanic rocks.

a) Record in your notebook the geologic age of any volcanic rocks you find on the map.

4. Using a measuring device and the map scale, determine how far your community is from the nearest volcanic rock unit.

a) Record this distance in your notebook.

Reflecting on the Activity and the Challenge

You have seen evidence that rocks formed by volcanic eruptions look different from those formed in other ways. They also vary greatly in their chemical composition. You can try to use what you have learned about the ages of the rocks in your area to decide on the probability of a volcano erupting in your community. Consider how you can share this knowledge with your audience.

Digging Deeper

IGNEOUS ROCKS

Introduction

Igneous rocks crystallize from cooling magma and lavas. Some igneous rocks form from magma that has cooled slowly beneath Earth's surface. The slow cooling allows crystals to form and grow, yielding coarse-grained igneous rocks such as granite. We know from laboratory work on melted rock that it takes extremely long times for large crystals to grow from a cooling magma. Such slow cooling can happen only deep within the Earth. These **intrusive igneous rocks**, also called **plutonic igneous rocks**, are made of crystals large enough to be seen with the naked eye. The sizes and shapes of bodies of intrusive igneous rocks vary greatly, from human scale to whole mountain ranges. Because intrusive igneous rocks form underground, they can only be seen where uplift and erosion have removed the overlying rocks.

Other igneous rocks, called **extrusive igneous rocks**, form from magma that is brought to the Earth's surface. Magma at or near the Earth's surface cools more rapidly, and crystals do not have time to grow to a large size. Sometimes, lava cools so fast that no crystals have a chance to form. Instead, the lava forms a kind of glass, called obsidian, as shown in *Figure 3*. Extrusive rocks are only one kind of volcanic rock. Remember from an earlier activity that pyroclastic volcanic rocks are also important.

Geo Words

intrusive igneous rock (plutonic igneous rock): igneous rock formed at considerable depth by the crystallization of magma.

extrusive igneous rock: an igneous rock that has formed by eruption of lava onto the surface of the Earth.

Figure 1 Granite is one kind of intrusive igneous rock formed by slow cooling of magma below Earth's surface.

Usually, the color of igneous rocks reflects the composition of the magma from which they form. Rocks from magmas high in silica (rhyolite and granite) tend to be lighter in color because their minerals are lighter in color, such as quartz, muscovite, and feldspar. These minerals are relatively poor in magnesium and iron, which are chemical elements that tend to make minerals dark.

Figure 2 Scoria is an extrusive igneous rock with a frothy texture.

Igneous rocks from magmas low in silica (basalt and gabbro) are darker in color, because they contain a large percentage of dark-colored minerals like pyroxenes, amphiboles, and dark micas. These minerals are relatively rich in magnesium and iron.

Figure 3 Obsidian is an extrusive igneous rock with a glassy texture.

Volcanic rocks of intermediate composition are mixtures of light and dark minerals that give the rock an intermediate color. A good example is andesite. It is named after the Andes Mountains, a mountain chain in South America that has many volcanoes.

In summary, igneous rocks crystallize from melted rock. They are divided into two types:

- Intrusive (plutonic) rocks are coarse-grained (>1 mm) and composed of crystals large enough to be seen with the naked eye. This implies slow cooling at depth.

- Extrusive (volcanic) rocks are fine-grained (<1 mm) and are composed mainly of crystals that are too small to be seen without a magnifying glass. This implies rapid cooling at or near the surface. Some may even be glassy, and many are filled with bubble holes, called vesicles.

Classification of Igneous Rocks				
Color	**Light**	**Intermediate**	**Dark**	**Dark**
Mineral composition	quartz (≥5%) plagioclase feldspar potassium feldspar iron-magnesium rich minerals (≤15%)	quartz (<5%) plagioclase feldspar potassium feldspar iron-magnesium rich minerals (15-40%)	no quartz plagioclase feldspar (~50%) no potassium feldspar iron-magnesium rich minerals (~40%)	nearly 100% iron magnesium rich minerals
Texture Crystals >10 mm	granite pegmatite	diorite pegmatite	gabbro pegmatite	
Crystals 1–10 mm	granite	diorite	gabbro	peridotite
Crystals <1 mm	rhyolite	andesite	basalt	
Glassy	obsidian		obsidian	
Frothy	pumice		scoria	

Figure 4

Rocks with Interesting Textures

Pumice, shown in *Figure 5*, feels like sandpaper. It is often light enough to float in water. Pumice forms when the gases inside the lava effervesce (bubble off) as pressure is released, just the way bubbles form when a can of soda, under pressure, is opened. The lava cools into a rock that is mostly tiny holes, with only very thin walls of rock between the holes. Much of the material blown out at first around Mt. St. Helens was pumice.

Figure 5 Pumice sample from Mt. St. Helens.

Obsidian is a glassy rock that was cooled so quickly from lava that crystals did not form. A broken surface of obsidian is usually smooth and shiny, just like a broken piece of glass (which it is!).

Tuff is a rock composed of pyroclastic material deposited from explosive eruptions. Some tuffs consist of ash that was put into the atmosphere by the eruption and then fell back to the surface downwind of the volcano. These tuffs may be lightweight, soft, and fragile, if they have not been buried deeply by later deposits. Other tuffs form from pyroclastic flows. These tuffs are often welded tightly together by the heat of the pyroclastic flow.

Scoria is a crusty-looking rock filled with holes made by gas bubbles trying to escape from the lava as it solidified. It is usually red or black, depending on the degree of oxidation of iron. Scoria is heavier to lift than pumice, because it doesn't have as great an abundance of holes. If you get to visit a cinder cone that is being mined, look at the cinders in the center and compare them to the cinders on the flank. The central cinders will be more oxidized and red and the outer samples darker and more like the basalt associated with the cone.

Types of Volcanic Rocks (based on silica content)

Basalt is dark and fine-grained. It is a low-silica rock. It is the most common rock on Earth and makes up the floors of all the oceans. Basalt covers about 70 percent of the surface of Earth. Basalt underlies the sediments on the floors of all the deep oceans of the Earth. Because the deep oceans occupy about 60 percent of the Earth's surface, basalt is the most common rock near the Earth's surface. On the continents, however, sediments and sedimentary rocks are much more common than basalt near the Earth's surface.

Andesite is a rock that is intermediate in silica content. It is also intermediate in color, typically gray, green, or brown. Andesite is typical of volcanic rocks from around the "Ring of Fire" in the Pacific Ocean. The ash from Mt. St. Helens was andesitic.

Dacite is a silica-rich rock that is between a rhyolite and an andesite in composition, somewhat like the difference between blue and violet in the color spectrum. It is lighter in color but not quite as silica rich as a rhyolite. It, too, is fine-grained but may have some visible crystals.

Rhyolite is usually light in color and has a fine-grained background. It forms from viscous lava that is high in silica. One of the identifying aspects of rhyolite is that it has visible quartz crystals that are not often seen in other volcanic rocks. These quartz crystals form because of (and are evidence of) rhyolite's high silica content. Perhaps the most famous rhyolite is the rock found in Yellowstone National Park. It came from one of the most massive volcanic eruptions known in Earth history.

Check Your Understanding

1. How can you distinguish intrusive igneous rock from extrusive igneous rock?

2. What is the difference between the way intrusive and extrusive igneous rock is formed?

3. Which chemical elements tend to make minerals dark?

Understanding and Applying What You Have Learned

1. Examine the sketch of the volcano. Some rocks shown on the sketch are extrusive, some are intrusive, and others are not igneous. Label a copy of the sketch to point out the following:

   a) Two locations of intrusive igneous rocks.

   b) Two locations of extrusive igneous rocks.

2. Consider what you know about volcanic hazards.

   a) Why is it important to study rocks?

   b) What information can be learned by looking at the geologic evidence in your community?

Cross section of a composite volcano. Rocks that form from the material shown in this diagram include extrusive igneous (erupted out of the volcano), intrusive igneous (cooled and crystallized inside the volcano or elsewhere beneath the surface), and non-igneous (metamorphic or sedimentary rock).

Preparing for the Chapter Challenge

You have now completed a variety of studies about volcanoes and volcanic rocks. Think about what you have learned about the relationships between rock types and volcanoes. Use the information you have learned about the distribution of rocks in your region to write a paragraph about the potential for a volcano to affect your community. Design a creative way to include geologic evidence about past eruptions in your movie script.

Inquiring Further

1. Volcanoes as natural resources

Volcanoes provide energy to some communities. The Geothermal Education Office explains how heat associated with volcanic activity can be converted into electrical energy.

Volcanoes also provide materials that are used for a variety of applications. Visit the Volcano World web site and compile a list of metals and other materials mined from volcanoes.

Consult the AGI *EarthComm* web site for current addresses.

2. Simulating gases in igneous rock

How does all the gas in pumice form? Acquire some bread dough (from the store or make from a recipe). Note the characteristics of the dough. Cut an end off to examine the interior. Set the dough out for a while and watch what happens. If you bake it, look at the interior again. It is not exactly the same as a volcano because the bubble holes in lava come from gases evolving out of solution, but it will give you the idea of how the gases are formed.

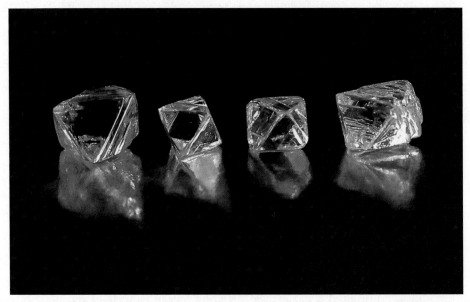

Volcanoes yield many materials that we use. Diamonds are just one example. Can you think of any others?

Activity 7   Monitoring Active Volcanoes

Goals

In this activity you will:

- Understand some of the changes that occur prior to volcanic eruptions.

- Describe volcanic monitoring systems.

- Design and build an instrument to monitor changes which occur prior to volcanic eruptions.

Think about It

Many volcanologists (geoscientists who study volcanoes) have been killed while observing active volcanoes, when the activity of the volcano increases unexpectedly.

- **How would you be able to tell that a volcano was about to erupt?**

What do you think? Record your ideas about this question in your *EarthComm* notebook. Be prepared to discuss your responses with your small group and the class.

935

Investigate

1. Read the report issued from the Montserrat Volcano Observatory. The report was released one day before an eruption.

2. Identify all the evidence that signaled an impending eruption.

   a) In your notebook, write down each kind of event that signals volcanic activity.

Montserrat Volcano Observatory Daily Report 3/25/99 to 3/26/99

A slight change in the nature and level of seismicity was observed during the period under review. There were nineteen (19) earthquakes, nineteen (19) rockfalls and three (3) regional events. An increase in small earthquakes that are thought to be associated with rock fracturing due to dome growth was noted. About one hundred and thirty-four (134) of these small events were recorded on the Gages and Chances Peak Seismographs. The southern and eastern EDM [electronic distance measurement] triangles were measured today. The very small changes in the elevation of the land measured were consistent with the recent trends. Visual observations were made early this morning from the helicopter in excellent viewing conditions and subsequently from Chances Peak. The dome continues to steam and vent gas from various locations. The focus of activity has shifted from the two previously active areas (near Farrell's and to the north of Castle Peak) to two new areas located on the eastern and western sectors of the dome. Rockfalls may continue to occur in the area north of Castle Peak. The rockfalls in this area have been the source of the recent ash clouds.

---

3. As a group, select one of the changes that occurs prior to an eruption. Discuss how you might design an instrument (or a model of an instrument) that would monitor the change. The instrument should be one that you could transport to an observation site. You should also be able to construct it from readily available materials. To learn more about volcano monitoring techniques, read the **Digging Deeper** section: "Volcano Monitoring."

   a) Sketch your design in your notebook.

   b) Label the parts of the instrument and list the materials from which it will be built.

4. Write a brief manual for the instrument. Be sure to specify the following:

   a) What the instrument does.

   b) Where on or near the volcano to place it.

   c) How to position it on the site.

   d) How to operate it.

   e) How to record observations.

5. Exchange your manual with another group and construct the instrument they have designed. If possible, take it to a test site and follow the procedures to take and record observations.

Reflecting on the Activity and the Challenge

Understanding how volcanoes are monitored can help you appreciate how scientists strive to keep people informed and safe from volcanic hazards. You looked at the aspects of the design of instruments that made them easy to understand, construct, and operate as well as the relative costs and savings associated with developing and maintaining monitoring systems. You can now share with your audience the general advantages and disadvantages of different monitoring systems. Perhaps you could have a scene in your movie showing a debate in which a town council must decide whether or not to purchase monitoring equipment.

Geo Words

seismology: the study of earthquakes and of the structure of the Earth.

Digging Deeper

VOLCANO MONITORING

The technology that geologists use is likely to be more sophisticated and expensive than the instruments you designed and constructed. However, they follow the same principles and observe the same changes. In emergencies, particularly in remote regions or in underdeveloped countries, simple monitoring techniques may provide the response time needed to save lives and property. Since the United States Geological Survey (USGS) cannot afford to monitor all volcanoes, on-site and local monitoring information can provide clues to the need for more sophisticated equipment.

Only 25% of the world's active volcanoes are monitored. Most potentially active volcanoes are not monitored due to a lack of funding. In response to this problem, the USGS developed a mobile volcano monitoring system. This system allows scientists to quickly install monitoring equipment and assess the potential hazard when a volcano becomes restless. The USGS sent this system to Mt. Pinatubo, Philippines in 1991 to help the Philippine Institute of Volcanology and Seismology monitor their volcano. (**Seismology** is the study of earthquakes and the structure of the Earth.) The lessons learned from monitoring volcanoes around the world help USGS scientists further prepare for eruptions in the United States. This system is not capable of stopping nor can it be designed to stop any natural occurrence. Its purpose is to provide forecasts to local populations and agencies for the health and welfare of their communities. Everyone must realize that the forces involved in a volcanic eruption cannot be stopped and must be allowed to take their course.

Figure 1 Many potentially active volcanoes around the world are not monitored due to a lack of funding.

Mt. Pinatubo erupted cataclysmically in June 1991. Fortunately for the residents of the island, monitoring prevented property losses of at least $250 million and saved an estimated 5000 lives. While 300 lives were lost and some $50 million were spent preparing for and monitoring the eruption, the estimated savings in human life far outweighed the costs associated with the monitoring program.

Volcano monitoring involves the recording and analyses of measurable phenomena such as ground movements, earthquakes, variations in gas compositions, and deviations in local electrical and magnetic fields that reflect pressure and stresses induced by the subterranean magma movements. To date, monitoring of earthquakes and ground deformation before, during, and following eruptions has provided the most reliable criteria in predicting volcanic activity, although other geochemical and geophysical techniques hold great promise.

Most of the commonly used monitoring methods were largely pioneered and developed by the Hawaiian Volcano Observatory (HVO). The HVO began monitoring in 1912. It has been operated continuously by the U.S. Geological Survey since 1948. Years of continuous observations of Kilauea and Mauna Loa, two of the world's most active volcanoes, have led to new instruments and measurement techniques. This work has increasingly been used in the study of other active volcanoes the world over. Moreover, early major advances in seismic research at HVO contributed significantly to subsequent systematic investigations of earthquake and related crustal processes, now conducted as part of the U.S. Geological Survey's Earthquake Hazards Reduction Program.

The volcanic plumbing and reservoir system beneath Kilauea can be pictured schematically as a balloon buried under thin layers of sand and plaster. When magma is fed into the reservoir (analogous to air filling a balloon), the internal pressure increases, and the sand-plaster surface layers are pushed upward and outward in order to accommodate the swelling or inflation. The net effects of such inflation include the steepening of slope of the volcano's surface; increases in horizontal and vertical distances between points on the

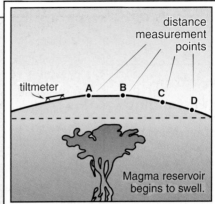

Stage 1: Inflation Begins

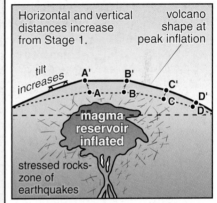

Stage 2: Inflation at Peak

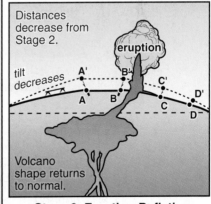

Stage 3: Eruption-Deflation

Figure 2 A schematic representation of changes in a volcano prior to and following an eruption.

939

Check Your Understanding

1. Can monitoring equipment be used to prevent a volcano from erupting? In your answer explain what monitoring equipment can do.

2. What does monitoring a volcano involve?

3. What external evidence indicates that a reservoir in a volcano is being "fed" with magma?

4. Is a tiltmeter a sensitive measuring device? Explain your answer.

surface; and, in places, the fracturing of rock layers stretched beyond the breaking point. Such rupturing of materials adjusting to magma-movement pressures results in earthquakes. A shrinking or rapidly draining reservoir to feed surface eruptions (analogous to deflating or popping the balloon) would produce the opposite effects: flattening of slopes, reduction in distances between surface points, and decrease in earthquake frequency.

Changes in slope in a real volcano can be measured precisely by various electronic mechanical "tiltmeters" or field tilt surveying techniques. Tiltmeters can detect the change in slope of a kilometer-long board raised by the thickness of a dime placed under one end. Similarly, an instrument that uses a laser beam can measure minute changes in horizontal distances. Tiny changes in vertical distances can be measured by making a series of precise leveling surveys. Such changes can be easily detected to a precision of only a few parts per million. The notion of one part per million can be compared to putting one drop of Kool-Aid® in 16 gallons of water. The frequency, location, and magnitude of earthquakes generated by magma movement can easily and accurately be determined by data obtained from a properly designed seismic network. The Hawaiian Volcano Observatory has recently expanded its seismic networks to 45 stations to continuously record the earthquake activity of Kilauea and Mauna Loa.

The monitoring of earthquakes and changes in volcano shape is not sufficient for predicting eruptions, however. The proper analysis of data requires a basic understanding of the prehistoric eruptive record and behavior of a volcano. Such glimpses into a volcano's prehistoric past are critically important because historic records extend over too short a time to permit the making of reliable predictions of future behavior.

Understanding and Applying What You Have Learned

1. Use the following questions to analyze your experiences designing, devising written plans, understanding the plans of other groups, and using monitoring instruments. In your notebook, record your responses.

   a) What aspect of your design would have worked well?

   b) What aspect of another group's design would have worked well?

   c) How useful do you think such devices would be in an emergency where expensive monitoring instruments were not readily available?

   d) How does elevation and slope of a volcano change prior to an eruption?

e) How could you measure changes in the elevation of a volcanic dome?

f) What changes in volcanic gases indicate an ensuing eruption?

g) How does seismic activity change prior to an eruption?

h) Did the class design at least one piece of equipment to measure change in slope? Elevation? Gases? Seismicity?

i) If you answered no to one of the questions in (h), think of two potential reasons why the instrument was not developed.

2. Describe three methods of monitoring an active volcano.

a) Which monitoring system is least reliable, and why?

b) Which monitoring systems are most dangerous to those who monitor an active volcano, and why?

c) Which monitoring systems would be most useful in your community, and why?

d) What aspects of volcano monitoring are most challenging for geologists? For communities? Explain your answers.

3. Why is weather monitored more closely than volcanic activity in the United States?

Preparing for the Chapter Challenge

Summarize the monitoring systems you would recommend for your community. Include a justification of their cost in relationship to the hazard information they might provide. Also, indicate the type of monitoring equipment you would recommend be used in the community in which you "stage" a volcanic eruption. Use this summary in your response to the **Chapter Challenge**.

Describe how the movie relates to what you have studied about volcanic hazards and volcano monitoring systems in this chapter.

Inquiring Further

1. Volcanic hazards of Mt. Rainier

View the movie *Perilous Beauty, The Hidden Dangers of Mount Rainier*, U.S. Department of the Interior, USGS. This movie is the result of in-depth research into the volcanic hazards of Mt. Rainier.

Earth Science at Work

ATMOSPHERE: *Air Traffic Controller*
Air traffic controllers are responsible for the safe passage of aircraft in their airspace.

BIOSPHERE: *Produce Manager*
Customers expect to find an ample supply of fresh produce throughout the year.

CRYOSPHERE: *Geohydrologist*
Some geologists are involved in the study of global water, its properties, circulation, and distribution. Mountain rivers and streams would be one area that they would investigate.

GEOSPHERE: *Commodities Supplier*
Some industries rely on an ample supply of sulfur in their manufacturing processes.

HYDROSPHERE: *Fisheries Worker*
In the fishing industry, income depends on the quality and the quantity of the catch.

How is each person's work related to the Earth system, and to Volcanoes?

Chapter 15

Plate Tectonics
...and Your Community

Getting Started

Plate tectonics is a scientific theory. A scientific theory is a well-tested concept that is supported by experimental or observational evidence. It explains a wide range of observations. The theory of plate tectonics explains the formation, movement, and changes of the outer, rigid layer of the Earth, called the lithosphere.

- How do you think the Earth's lithosphere formed?
- How do you know that the Earth's lithosphere moves?
- List some ways that the Earth's lithosphere changes.

What do you think? Write down your ideas about these questions in your notebook. Be prepared to discuss your ideas with your small group and the class.

Scenario

A middle-school science teacher in your community has been asked to teach a unit on plate tectonics to her class and has asked for your help. To her students, the fact that they are riding on Earth on a floating plate makes about as much sense as riding on a magic carpet. The students feel that because plate tectonics is "just a theory" that scientists have "made up," there is no need for them to learn about it, or try to understand it. Also, they believe it is totally irrelevant to their lives.

The teacher hopes that her students might be more likely to listen to local high school students. She would like you to convince her class of the significance of plate tectonics in their lives. She wants you to explain to her class the key science concepts behind the theory. She thinks that it would help her students if they could see how research, evidence, modeling, and technology support plate tectonics.

Can the EarthComm students meet this challenge and help a middle-school class to understand why the theory of plate tectonics is important to learn about and understand?

Chapter Challenge

Think about how you can use the theory of plate tectonics to help middle-school students understand scientific theories. You might want to prepare a PowerPoint™ presentation, a web page, or a three-panel poster display. Your project will need to address:

- Evidence for movement and changes in the geosphere over time
- The flow of matter and energy in the Earth
- The nature of the Earth's interior
- How plate tectonics accounts for the features and processes that geoscientists observe in the Earth
- How the theory of plate tectonics developed

Assessment Criteria

Think about what you have been asked to do. Scan ahead through the chapter activities to see how they might help you meet your challenge. Work with your classmates and your teachers to define the criteria for judging your work. Record all this information. Make sure that you understand the criteria as well as you can before you begin. Your teacher may provide you with a sample rubric to help you get started.

Activity 1

Taking a Ride on a Lithospheric Plate

Goals

In this activity you will:

- Determine the direction and rate of movement of positions within the plate on which your community is located, using data from the Global Positioning System and a computer model.

- Predict the position of your community in the near future, and "retrodict" its position in the recent past, by extrapolating from data already collected.

- Recognize that the rate and direction of plate motion is not necessarily constant.

- Describe several lines of evidence for plate motion.

Think about It

The motion of anything (you, your automobile, a lithospheric plate, or the Milky Way Galaxy!) has to be described in relation to something else.

- **How can you locate your position on the Earth's surface?**

- **How would you be able to determine whether your position on the Earth has moved?**

What do you think? Record your ideas about these questions in your *EarthComm* notebook. Include sketches as necessary. Be prepared to discuss your responses with your small group and the class.

Investigate

Part A: Data from the Global Positioning System

1. Data from Global Positioning System (GPS) satellites will help you find out if the position of your community has changed over time. The map shows measurements of movements at GPS recording stations in North America. Each station has a four-character symbol. Arrows show the rate and direction of motion of the Earth's surface at that station. Longer arrows indicate faster motion than shorter arrows. The motions shown are relative to the GPS frame of reference, which you can think of as "attached" to the Earth's axis of rotation.

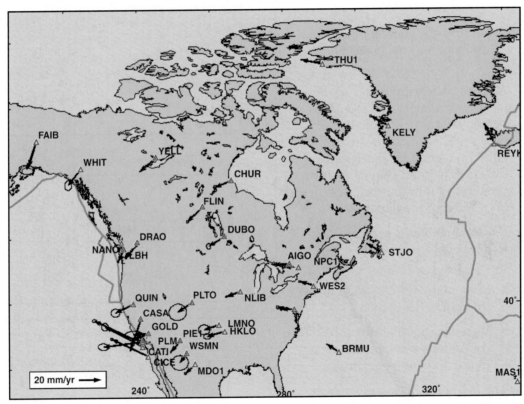

Measurements of movements at GPS recording stations in North America.

a) Find the WES2 station (in the northeastern United States). How do you know that the WES2 station has moved over time?

b) In what compass direction is the WES2 station moving? Be specific.

Coordinated Science for the 21st Century

c) The arrow in the lower left corner of the map is a scale. It shows the length of a "20 mm/yr" arrow. Is the WES2 station moving more than or less than 20 mm/yr? Explain.

d) Are all stations on the map moving at the same speed? Explain.

e) Are all stations on the map moving in the same direction? Explain.

f) What is the general or average direction of movement of North America?

2. A series of measurements of the location and elevation of a GPS station over time is called a GPS time series. The graph shows the GPS time series for the WES2 station. The solid sloping lines on the three graphs are the "best-fit" lines through the data points.

Use the map and the time series to answer the following questions:

a) How many years of data does the time series show?

b) Were measurements recorded continuously or only at certain times? Explain your answer.

3. The top graph shows movement of the station to the north or south. Northward movement is indicated by positive values, and southward movement is shown by negative values. Find the calculation above the top graph. How many millimeters per year did WES2 move? Convert

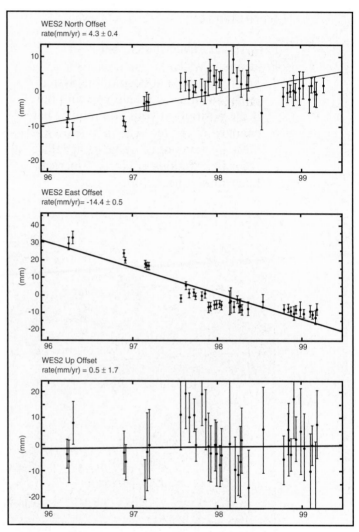

The location and elevation of GPS station WES2 over time. The vertical lines above and below each point are called "error bars". They show the uncertainty in the measurement. They tell you that the real value might lie anywhere within the error bar.

this value to centimeters per year. In which direction did it move?

a) Record the rate (in cm/yr) and the direction of motion in your notebook.

4. The middle graph shows movement to the east or west. Eastward movement is indicated by positive values, and westward movement is shown by negative values. Find the calculation above the middle graph. How many millimeters per year did WES2 move? Convert this value to centimeters per year. In which direction did it move?

   a) Record the rate (in cm/yr) and direction of motion in your notebook.

5. The bottom graph shows the movement up or down.

   a) Has the WES2 station always stayed at the same elevation? Explain.

6. Do the speed and direction of motion of WES2 shown in the graphs match the direction and length of the arrow shown on the map?

7. Look at the "best-fit" line in the top and middle graphs.

   a) Did the WES2 station move at a constant speed since 1996? Explain your answer.

   b) What additional data would you need to decide whether the differences between the measured data points and the best-fit straight line are due to the overall motion of the plate or are caused by processes in the local area around the WES2 station?

8. Obtain a GPS time series for a station nearest your community.

   a) Record the directions and rates of motion in cm/yr for the station nearest your community.

Part B. Data from a Computer Model

1. Computer models that use geologic data also provide information about the changes in position of your community over time. To use this model, you will need to know the latitude and longitude of your school in decimal format.

   Find your school (or another familiar place) on your local topographic map. Latitude and longitude are used to identify a position on the Earth's surface. Latitude is a measure of location in degrees, minutes, and seconds north or south of the equator. Therefore, it is found on the left or right side of the map. Longitude is a measure of location in degrees, minutes, and seconds east or west of the Prime Meridian, which passes through Greenwich, England. Therefore, it is found on the top or bottom of the map.

   a) Record the latitude and longitude of the position you chose in degrees, minutes, and seconds. (These "minutes" and "seconds" are *not* the same as the familiar minutes and seconds of time! They describe positions on a circular arc.)

   b) Convert the latitude and longitude values to a decimal format. An example for you to follow is provided on the following page.

Example:

> 42° (degrees) 40' (minutes) 30" (seconds) north latitude
> Each minute has 60".
> 30" divided by 60" equals 0.5'.
> This gives a latitude of 42° 40.5' north.
> Each degree has 60'.
> 40.5' divided by 60' equals 0.675°.
> The latitude in decimal format is 42.675° north.

2. Obtain a world outline map showing lithospheric plates, similar to the one shown.

   a) Place a dot on the map to represent the location of your community and label it with an abbreviation.

b) In your notebook, record the name of the plate your community lies within. Record the name of a plate next to your community.

3. Visit the Relative Plate Motion (RPM) Calculator web site. (See the *EarthComm* web site for more information.) The RPM calculator determines how fast your plate is moving relative to another plate that is assumed to be "fixed" (non-moving). At the web site, enter the following information:
   - The latitude and longitude of your community (decimal format)
   - The name of the plate on which your community is located
   - The name of the "fixed" reference plate adjacent to your plate. Use the African Plate as the reference plate

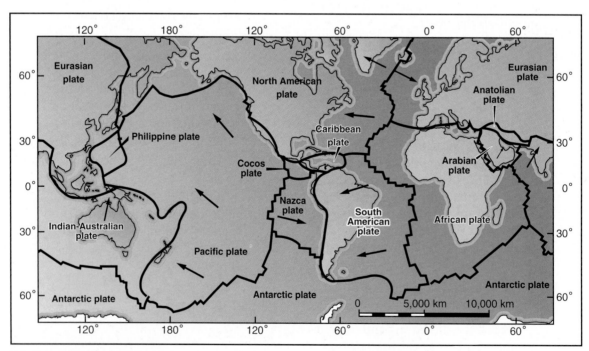

World map of major lithospheric plates. Arrows show the relative motions of the plates relative to the African Plate, which happens to be moving slowest relative to the Earth's axis of rotation.

Once you have entered the data, run the model. Print the results. (A sample printout for the location of station WES2 is provided for you.) Record the following information in your notebook:

a) The rate of movement of the plate on which your community is located (in centimeters per year).

b) Its direction of motion. (Note that direction is given in degrees, starting from 0°, clockwise from north. For example, 90° is directly east, 180° is directly south and 270° is directly west.)

c) In your own words, describe the motion of your plate over time.

d) How do the results from the computer model compare to those obtained from GPS data?

> NUVEL-1A Calculation Results
> Calculation results are as follows:
>
> Relatively fixed plate = Africa
> Relatively moving plate = North America
> Latitude of Euler pole = 78.8 degree
> Longitude of Euler pole = 38.3 degree
> Angular velocity = 0.24 degree/million years
> Latitude inputted = 42.364799 degree
> Longitude inputted = –71.293503 degree
> Velocity = 2.11 cm
> Direction = 283.39 degree

e) What data does GPS provide that the plate motion calculator does not?

Reflecting on the Activity and the Challenge

So far, you have learned that when you use GPS data collected over time, you can find the speed and direction of movement of your plate. You can also use data from a computer model to show that your community is moving as part of the movement of a large plate. You are riding on a piece of the Earth's lithosphere! Your maps show how your community, along with the plate on which it rides, has moved over time. You can also predict its future position. You have now gathered some evidence to help you explain to the middle school students that their community is in fact moving.

crust: the thin outermost layer of the Earth. Continental crust is relatively thick and mostly very old. Oceanic crust is relatively thin, and is always geologically very young.

mantle: the zone of the Earth below the crust and above the core. It is divided into the upper mantle and lower mantle with a transition zone between.

lithosphere: the outermost layer of the Earth, consisting of the Earth's crust and part of the upper mantle. The lithosphere behaves as a rigid layer, in contrast to the underlying asthenosphere.

asthenosphere: the part of the mantle beneath the lithosphere. The asthenosphere undergoes slow flow, rather than behaving as a rigid block, like the overlying lithosphere.

Digging Deeper

MEASURING THE MOTION OF LITHOSPHERIC PLATES
The Interior Structure of the Earth

Refer to *Figure 1* as you read this section. The thin, outermost layer of the Earth is called the **crust**. There are two kinds of crust: continental and oceanic. The continental crust forms the Earth's continents. It is generally 30–50 km thick, and most of it is very old. Some continental crust has been dated as old as four billion years! The geological structure of the continental crust is generally very complicated, as you will learn in a later chapter. In contrast, the oceanic crust is only 5–10 km thick, and it is young in terms of geologic time. All of the oceanic crust on the Earth is younger than about 200 million years.

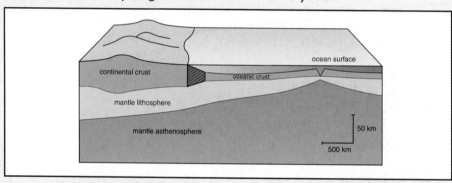

Figure 1 A schematic cross section through the outer part of the Earth. Note that the vertical and horizontal scales are very different. The diagram has a lot of "vertical exaggeration". If the diagram had been drawn without distortion, all of the layers would look much thinner. The boundary between continental crust and oceanic crust, shown by the shaded box, will be described later in the chapter.

Beneath the Earth's crust is the **mantle**. The rocks of the mantle are very different in composition from the crust, and the boundary between the crust and the mantle is sharp and well defined. The uppermost part of the mantle, which is cooler than below, moves as a rigid block, carrying the crust with it. The upper rigid part of the mantle, together with the crust, is called the **lithosphere**. The Earth's plates are composed of the lithosphere. At greater depths, the mantle is hot enough that it can flow very slowly, just like a very stiff liquid. That part of the mantle is called the **asthenosphere**.

In terms of composition and origin, the crust and the mantle are very different, but in terms of how they move, they behave in the same way. On the other hand, the lithosphere part of the mantle is the same in composition as the asthenosphere part of the mantle, but in terms of how they move, they behave very differently.

Measuring Plate Motions with GPS

The **Global Positioning System (GPS)** consists of 24 satellites that orbit the Earth at a height of 20,200 km. Receivers at stations on Earth (such as WES2 in Westport, Massachusetts) use the signals from satellites to calculate the location of the station. Geoscientists have set up a network of targets all over the world in order to monitor the movement of lithospheric plates. Steel spikes pounded into the ground (preferably embedded in solid rock) make up the targets, as shown in *Figure 2*. A high-precision GPS receiver is then mounted on a tripod and positioned directly above the target. The targets are revisited over a period of months or years. The receiver measures the distance to four or more GPS satellites and then uses stored data on satellite locations to compute the location of the target. Changes in horizontal and vertical positions can be detected within several millimeters.

Geo Words

Global Positioning System (GPS): a satellite-based system for accurate location of points on the Earth.

Figure 2 A GPS receiver mounted in rock is used to measure changes in the elevation of this volcano.

GPS data collected at stations all over the world confirm that the surface of the Earth is moving. However, GPS time series show data for only the last several years. GPS is a new technology, and a global network of GPS stations has existed for less than a decade. How do we know that the surface of the Earth has been moving for a longer period of time? The answer to this question comes from the study of rocks.

Sea-Floor Spreading

The computer model at the Plate Motion Calculator web site uses several sources of geologic data. One source comes from the study of the magnetism of rocks that make up the sea floor. All magnets and

953

materials that have magnetism have a north and south direction, or magnetic polarity. In the middle of the 20th century, geoscientists noted that they could group rocks by their magnetic polarity. Rocks with normal magnetic polarity match that of the Earth's magnetic field (the north end of the rock's "compass needle" points toward magnetic north.) The other group has magnetic minerals with reversed polarity(the north end of the rock's compass needle points south).

It was known that as lava cools to form **basalt** (an iron-rich volcanic rock that makes up the ocean floor), its iron minerals (such as magnetite) become magnetized and "lock in" the polarity of the Earth's magnetic field. Beginning in the 1950's, scientists began noting patterns in the magnetism of rocks on the ocean floor, as shown in *Figure 3*. The alternating belts of higher and lower than average magnetic field strength were of normal and reverse polarity, respectively.

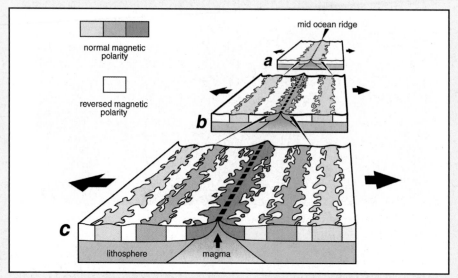

Figure 3 The formation of magnetic striping. New oceanic crust forms continuously at the crest of the mid-ocean ridge. It cools and becomes increasingly older as it moves away from the ridge crest with sea-floor spreading: a. the spreading ridge about 5 million years ago; b. about 2 to 3 million years ago; and c. present day.

In 1963, two scientists, F.J. Vine and D. H. Matthews, proposed the revolutionary theory of sea-floor spreading to explain this pattern. According to their theory, the matching patterns on either side of the **mid-ocean ridge** could be explained by new ocean crust forming at the ridge and spreading away from it. As ocean crust forms, it obtains the polarity of the Earth's magnetic field at that time. Over time, the strength of the Earth's magnetic field changes. When new ocean crust forms at the center of the spreading , it obtains a new kind of magnetic polarity. Over time, a series of magnetic "stripes" are formed.

Since the theory of sea-floor spreading was proposed, core samples of volcanic rock taken from the ocean floor have shown that the age of the rock increases from the crest of the ridge, just as the theory predicts. What's more, by measuring both the age and magnetic polarity of rocks on land, geologists have developed a time scale that shows when the magnetic field has reversed its polarity. Because the magnetic striping on the ocean floor records the reversals of the Earth's magnetic field, geoscientists can calculate the average rate of plate movement during a given time span. These rates range widely. The Arctic Ridge has the slowest rate (less than 2.5 cm/yr), and the East Pacific Ridge has the fastest rate (more than 15 cm/yr). The computer model in the plate motion calculator uses spreading rates from ocean ridges throughout the world to compute plate motion.

Geologic data is also used to find the direction of movement of the plate. Surveys of the depth of the ocean floor, mainly since the 1950s, revealed a great mountain range on the ocean floor that encircles the Earth, as shown in *Figure 4*. This mid-ocean ridge zig-zags between the continents, winding its way around the globe like the seams on a baseball. The mid-ocean ridge is

Geo Words

basalt: a kind of volcanic igneous rock, usually dark colored, with a high content of iron.

mid-ocean ridge: a chain of undersea ridges extending throughout all of the Earth's ocean basins, and formed by sea-floor spreading.

Figure 4 Map of the world ocean floor. The crest of the mid-ocean ridge system is shown as a broad light blue line throughout the ocean floor. The flanks of the mid-ocean ridges slope gradually down to the deeper part of the oceans, nearer the continents.

Check Your Understanding

1. What is the difference between the lithosphere and the asthenosphere?

2. What does the abbreviation GPS stand for?

3. From where does a GPS receiver get its signal?

4. Why are GPS data not enough to confirm that the Earth's surface has been moving for many years?

5. What has caused the "zebra pattern" in the rock of the ocean floor?

6. What is the significance of the patterns of offsets along mid-ocean ridges?

not straight; it is offset in many places (*Figure 5*). The offsets are perpendicular to the axis of the ridge. When combined with knowledge that the ocean floor is spreading apart at mid-ocean ridges, geologists realized that the offsets are parallel to the direction the plates are moving. By mapping the orientations of these offsets, and entering this data into the computer model, the plate motion calculator is able to generate the directions of plate motions. Comparisons between GPS measurements and results from geologic computer models show very good agreement, within 4%.

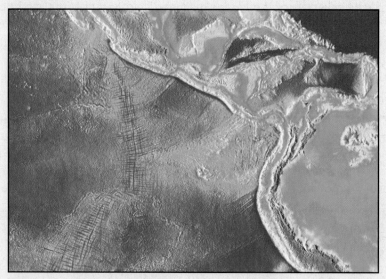

Figure 5 This map shows the network of fractures along the mid-ocean ridge in the eastern Pacific Ocean floor. Because the ocean floor spreads away from both sides of the ridge, the fractures indicate the direction of plate motion.

Understanding and Applying What You Have Learned

1. Describe the direction and the rate of movement for the plate on which you live.

2. Examine the scale of the USGS topographic map of your community. Given the rate of plate motion in your community, estimate the minimum number of years that it would take for a change in the location of your school to be detected on a topographic map.

3. How does GPS provide evidence that the surface of the Earth moves over time?

4. Describe at least one advantage of using GPS technology to gather evidence of plate motion.

5. How do studies of the magnetism of rocks on the sea floor provide evidence that the surface of the Earth moves over time?

6. What evidence examined in this activity suggests that the direction and rate of motion of plates is not constant?

7. Look at the world map in the **Investigate** section showing the relative motion of plates. This map shows how plates move relative to each other.

   a) Look at the names of the plates. On what basis does it appear that the plates were named?

   b) Write down the name of your plate and all the plates that border it. Describe the motion of your plate relative to all the plates that border it.

   c) How might the differences in motion of these plates affect the Earth's lithosphere?

Preparing for the Chapter Challenge

In your notebook, write a paragraph to convince the middle-school students that your community is moving as part of the movement of a much larger segment of the Earth's lithosphere. Describe the evidence used to make this determination.

Inquiring Further

1. **Technology used to detect plate motions**

   Explore how GPS allows plate movement to be measured. Excellent web sites that describe how GPS works can be found on the *EarthComm* web site.

2. **Investigating scales of motion**

   Plate motion is extremely slow. Make a list of other things you know about (or have heard about) that move or take place slowly. Possible examples include growth of fingernails, grass, tree height, tree-trunk diameter, and so on. Find out how fast they move. Compare the rate of these motions to the rate of movements of plates.

3. **Study animations of plate motions**

   Visit the *EarthComm* web site for the address of animated images of the motions of lithospheric plates. Describe how the motions shown in the animations match your analysis from this activity.

Activity 2 — Plate Boundaries and Plate Interactions

Goals

In this activity you will:

• Classify and label the types of movement at plate boundaries, using a world map that shows relative plate motion.

• Identify the distribution of plates by means of the world map of relative plate motions.

• Describe the present plate-tectonic setting of your community, and infer possible past plate-tectonic activity based on your knowledge.

Think about It

New plates are created at certain places on Earth, and existing plates are consumed at certain other places. The total surface area of the Earth stays the same, so the creation of new plates has to be exactly equal to the consumption of existing plates.

• **Where do you suppose you would have the most "interesting" ride on a plate? Would it be at the center, on a leading edge, on a trailing edge, or somewhere else on the plate?**

What do you think? Record your ideas about these questions in your *EarthComm* notebook. Include sketches as necessary. Be prepared to discuss your response with your small group and the class.

Investigate

Part A: Observing Plate Motions and Plate Interactions

1. Obtain the equipment shown in the following diagram.

2. Use the equipment to model a steady sea-floor spreading and subduction, as follows: One student holds the two rolled-up dowels in one place, loosely, so that they can turn but not shift their position. Another student holds the stapled piece of 2 x 4 lumber "continent" and pulls it away from the rolled-up dowels. A third student holds the dowel and piece of 2 x 4 lumber "subduction zone" at the other end loosely in place. A fourth student pulls the paper strip from under the piece of 2 x 4 lumber "subduction zone." Be sure to unroll the paper strips at the same rate, so that the numbers of the stripes stay matched up as they appear.

a) What does the area at the rolled paper strips on the dowels represent?

b) What does the section of paper between the dowels and the continental lithosphere (the piece of 2 x 4 lumber) represent?

c) What happens to the length of this section of paper as the dowels are unrolled?

d) As the dowels are unrolled, what happens to the width of the section of paper between the dowels and the subduction zone (the other piece of 2 x 4 lumber)?

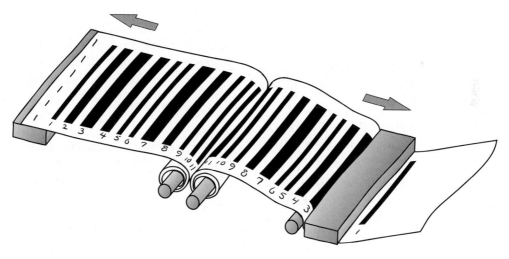

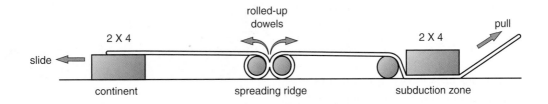

3. Use the equipment to model a collision of a spreading ridge and a subduction zone, as follows: Begin with the materials arranged in the same way at the end of **Step 1**. While two students pull the paper strips to unwind the two rolled-up dowels, the student holding those dowels slides them slowly toward the subduction zone. The student operating the subduction zone needs to make sure that the stripes appearing at the spreading ridge continue to have their numbers matched up.

a) What happens to the length of the strip of paper between the dowels and the "continent" side in this situation?

b) What happens to the length of the strip of paper between the dowels and the "subduction" zone?

c) At what "place" does the spreading ridge eventually arrive?

4. Think about the following questions, and write a brief answer to each in your notebook.

a) In the first part of the modeling (**Step 2**), how long will the ocean on the "subduction" side last?

b) In the second part of the modeling (**Step 3**), what do you think would happen in real life when the spreading ridge arrives at the subduction zone?

c) In the second part of the modeling (**Step 3**), how would the ocean on the "continent" side change after the spreading ridge arrives at the subduction zone?

d) In both cases, what do you think would happen in real life if the continent became blocked in its movement away from the spreading ridge by something happening on the other side of the continent?

Part B: How Transform Faults Are Formed

1. When the theory of plate tectonics was young, back in the 1970s, there was an experiment to see how divergent plate boundaries are formed. Scientists might describe this experiment as "beautiful" or "elegant." It was small enough to be done on a table top. You can run it as a "thought experiment." A diagram of the setup is shown below. Use a shallow square pan. Fill it with melted wax, and heat the bottom of the pan to keep the wax melted. Cool the surface of the wax with a fan, so that it crusts over to form "lithosphere." Install roller cylinders along opposite edges of the top of the pan. These two rollers can be rotated outward in opposite directions to pull the solid surface of the wax apart.

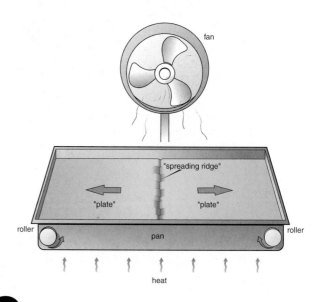

a) Predict what will happen to the crust of wax when the rollers are rotated. Assume that everything is adjusted right (composition of the wax; heating at the bottom; cooling at the top; speed of the rollers).

b) If a break were to occur in the surface wax, what do you think would happen to the liquid wax below?

2. In the actual experiment, the "lithosphere" broke along a crack across the middle of the pan. As the two "plates" on either side of the crack were formed, new "lithosphere" formed as the liquid wax welled up into the crack and solidified. The amazing thing about the experiment was that the original crack was formed with exactly the same pattern of ridge-crest segments and transform faults that can be seen in the real mid-ocean ridges!

a) Were your predictions correct? How do you explain any differences between what happened and your predictions?

b) What does the experiment suggest about the age of the transform faults compared to the mid-ocean ridges?

Part C: Plate Boundaries on World Maps

1. Look at the world map of Major Lithospheric Plates in the **Investigate** section of the previous activity, which shows the relative motion of the plates. Observe what it shows about how plates move relative to each other. Answer the following:

a) Name two plates that are moving toward each other (colliding/converging).

b) Name two plates that are moving apart (diverging).

c) Name two plates that are sliding past each other.

2. Use a blank world map to make a map that shows the three major types of plate boundaries.

a) On the map, color the boundary line that separates two converging plates. Do not outline both of the plates completely; highlight only the boundary between the two plates.

b) Using two other colors, highlight the divergent plate boundaries, (where plates are moving away from one another) and the plate boundaries, where plates slide past one another (called transform boundaries). Make a key that shows this color code.

Part D: The Plate-Tectonic Setting of Your Community

1. Describe the plate-tectonic setting of your community. Refer to your world map and the map *This Dynamic Planet* (United States Geological Survey) in your work.

a) How far is your community from the nearest plate boundary?

b) What type of plate boundary is it?

c) How might your community change its position relative to plate boundaries in the future?

Reflecting on the Activity and the Challenge

You have seen that plates can interact in three different ways: they can converge, diverge, or move parallel to each other. You have gained some experience in recognizing the three kinds of plate boundaries on world maps. You are better prepared for a later activity, on the different kinds of landforms that develop at the different kinds of plate boundaries, and how earthquakes and volcanoes are caused at or near plate boundaries. You are now in a better position to help middle school students understand one of the key concepts of plate tectonics: the material of the ocean floors is always much younger than the oceans themselves!

Geo Words

divergent plate boundary: a plate boundary where two plates move away from one another.

Digging Deeper

THE DYNAMICS OF PLATE BOUNDARIES

Types of Plate Boundaries

Plate boundaries are geologically interesting areas because they are where the action is! Geologists use three descriptive terms to classify the boundaries between plates: 1) divergent boundaries are where two plates move away from each other; 2) convergent boundaries are where two plates move toward each other; and 3) transform boundaries are where two plates slide parallel to each other.

Divergent Plate Boundaries

You have already learned some things about **divergent plate boundaries** in **Activity 1**, because mid-ocean ridges are divergent plate boundaries. The mid-ocean ridges are places where mantle asthenosphere rises slowly upward. As it rises, some of the rock melts to form magma. Why does melting happen there? To understand that, you need to know that the melting temperature of rock decreases as the pressure on the rock decreases. As the mantle rock rises, its temperature stays about the same, because cooling takes a long time. However, the pressure from the overlying rock is less, so some of the rock melts. The magma then rises up, because it is less dense than the rock. It forms volcanoes in the central valley of the mid-ocean ridge. Geoscientists in deep-diving submersibles can watch these undersea volcanoes! Because of the great water pressure in the deep ocean, and also the cooling effect of the water, these volcanoes behave differently from volcanoes on land. The lava oozes out of cracks in the rocks, like toothpaste out of a tube. Some of the magma stays below the sea floor and crystallizes into rock there. All of these new igneous rocks, at the sea floor and below, make new oceanic crust, which then moves away from the ridge crest. This would be a good time to go back to **Activity 1** and review the material on sea-floor spreading.

In the investigation you modeled how the "continent" moved farther and farther from the "spreading ridge." Look back again at the world map of lithospheric plates in **Activity 1**. In both the North Atlantic Ocean and the South Atlantic Ocean, there is no plate boundary along the coastlines on either side of the ocean. That tells you that the Atlantic Ocean is getting wider as time goes on. Why? Because new lithosphere is being created all the time at the mid-ocean ridge but is not being consumed at the edges of the continents. Does that make you wonder what would happen if you could make time run backward and watch the ocean shrink? At some time in the past, there was no Atlantic Ocean!

A new ocean begins when hot mantle material begins to move upward beneath a continent. Geoscientists are still not certain about why that happens. The lithosphere of the continent bulges upward and is stretched sideways. Eventually it breaks along a long crack, called a rift. See *Figure 1* for what a newly formed **rift valley** looks like. Magma rises up to feed volcanoes in the rift. As the rift widens, the ocean invades the rift. A new ocean basin has now been formed, and it gets wider as time goes on.

Geo Words

rift valley: a large, long valley on a continent, formed where the continent is pulled apart by forces produced when mantle material rises up beneath the continent.

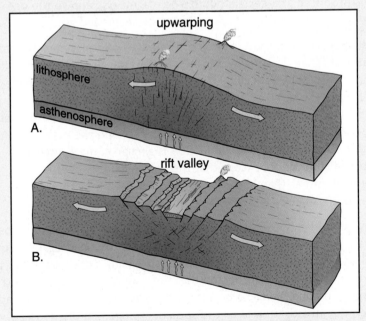

Figure 1 The formation of a rift valley on a newly rifted continent.

Geo Words

convergent plate boundary: a plate boundary where two plates move toward one another.

subduction: the movement of one plate downward into the mantle beneath the edge of the other plate at a convergent plate boundary. The downgoing plate always is oceanic lithosphere. The plate that stays at the surface can have either oceanic lithosphere or continental lithosphere.

Convergent Plate Boundaries

At a **convergent plate boundary**, two plates are moving toward each other. Your common sense tells you that one of them has to go under the other. (Would it surprise you to hear that common sense is important to a scientist, even though sometimes common sense can fool you?) There are three kinds of places where this happens. These are described below. In a later activity you will learn much more about the landforms that are produced at convergent plates boundaries, and how earthquakes and volcanoes are associated with convergent plate boundaries.

In some places, two oceanic lithospheric plates are converging. There are good examples along the western edge of the Pacific Ocean. Look back at the world map of lithospheric plates in **Activity 1**. The Pacific Plate and the Indo-Australian Plate are moving toward one another in the South Pacific, and the Pacific Plate and the Philippine Plate are moving toward one another in the Western Pacific. Look at the lower part of *Figure 2*. One plate stays at the surface, and the other plate dives down beneath it at some angle. This process is called **subduction**, and so these plate boundaries are called subduction zones. Where the downgoing plate first bends downward, a deep trench is formed on the ocean floor. These trenches are where the very deepest ocean depths are found. As the plate goes down into the mantle asthenosphere, magma is produced at a certain depth. The magma rises up to the ocean floor to form a chain of volcanic islands, called a volcanic island arc.

Other subduction zones are located at the edges of continents. Look at the upper part of *Figure 2*. The west coast of South America, where the Nazca Plate and the South American Plate are converging, is a good example. In places like this, the downgoing plate is always oceanic lithosphere and the plate that remains at the surface is always continental lithosphere. That's because the continental lithosphere is less dense than the oceanic lithosphere. Ocean–continent subduction is similar in many ways to ocean–ocean subduction, except that the volcanic arc is built at the edge of the continent, rather than in the ocean. The Andes mountain range in western South America is an example of a continental volcanic arc.

The third kind of convergent boundary is where two continental lithospheric plates have collided with each other. Think back to the investigation. You were asked what would happen when the "spreading ridge" arrives at the "subduction zone." If you suspected that the spreading ridge would go down the subduction zone, never to be seen again, you were right! If you had continued the investigation after the spreading ridge had disappeared, what would have happened? As the plate that was on the other side of the

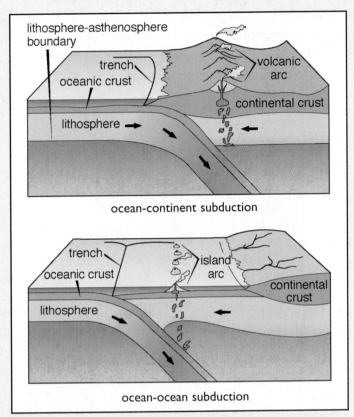

Geo Words

suture zone: the zone on the Earth's surface where two continents have collided and have been welded together to form a single continent.

ocean-continent subduction

ocean-ocean subduction

Figure 2 Cross sections through subduction zones. The upper cross section shows subduction of an oceanic lithospheric plate beneath a continental lithospheric plate (ocean–continent subduction). The lower cross section shows subduction of an oceanic lithospheric plate beneath another oceanic lithospheric plate (ocean–ocean subduction).

spreading ridge was consumed down the subduction zone, the continent on the other side of the ocean would have moved closer and closer to the subduction zone. This is how two continents can come together at a subduction zone. Remember that continental lithosphere is much less dense than the mantle, so continental lithosphere cannot be subducted. The subduction stops! The continent that was coming along toward the subduction zone keeps working its way under the other continent, for hundreds of kilometers, until finally the friction between the two continents is so great that plate movement stops. The zone where two continents have met and become welded into a single continent is called a **suture zone**. There is only one good example on today's Earth: the Indian Plate has collided with the Eurasian Plate and is still working its way under it. ➡

Geo Words

transform plate boundary: a plate boundary where two plates slide parallel to one another.

transform fault: a vertical surface of slippage between two lithospheric plates along an offset between two segments of a spreading ridge.

See *Figure 3*. There is a good reason why the Tibetan Plateau is the largest area of very high elevations in the world: the continental lithosphere is much thicker there because one continent has moved under another.

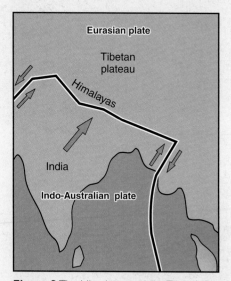

Figure 3 The Himalayas and the Tibetan Plateau were formed by the collision of the Indian Plate with the Eurasian Plate. The Indian Plate is being shoved horizontally underneath the Eurasian Plate, so the continental crust in the Himalayas and the Tibetan Plateau is much thicker than normal.

Transform Plate Boundaries

At **transform boundaries**, plates slide past one another. The surface along which the plates slide is called a **transform fault**. As you saw in **Activity 1** in the material about mid-ocean ridges, transform faults connect the offsets along mid-ocean spreading ridges. Most are short, but a few are very long. The most famous transform fault forms the boundary between the North American Plate and the Pacific Plate, in California. It is several hundred kilometers long. You can see from the map in *Figure 4* that the San Andreas Fault connects short segments of spreading ridges at its northern and southern ends.

The movement along the transform fault is limited to the distance between the two segments of ridge crest. *Figure 5* is a sketch map of a mid-ocean ridge, showing segments of the ridge crest offset by transform faults. Between Points 1 and 2, Plates A and B are sliding past each other.

The transform fault extends only between Points 1 and 2. To the left of Point 1, Plate A is in contact with itself along the east-west line, and to the right of Point 2, Plate B is in contact with itself along the east-west line. Since the plates are rigid, there is no slipping movement along the east-west lines, which are like "dead" transform faults!

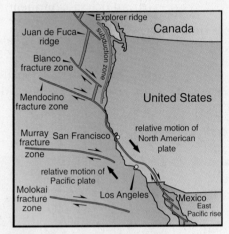

Figure 4 Plates showing a transform boundary.

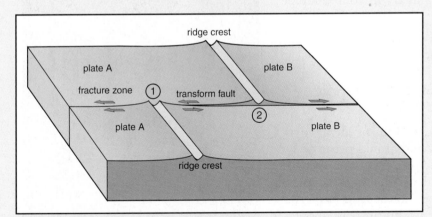

Figure 5 Sketch map of a mid-ocean ridge, showing segments of the ridge crest offset by transform faults.

Check Your Understanding

1. Name the three types of boundaries between lithospheric plates.

2. How and where are rift valleys formed?

3. How can ocean basins change in size?

4. Convergent plate boundaries can be in three different settings. What are they?

5. Describe subduction.

6. Why is it that transform faults can be used to figure out the directions of plate movements? Why can't subduction zones also be used for that?

7. What happens when two continents collide along a convergent plate boundary?

Coordinated Science for the 21st Century

Understanding and Applying What You Have Learned

1. Design an investigation, with the materials like the ones you used in this activity, to model the creation of a new ocean basin.

2. Identify on a world map of lithospheric plates one example of each of the following settings:

   a) an established divergent boundary

   b) a young divergent zone (continental rift zone)

   c) an ocean–ocean convergent boundary

   d) an ocean–continent convergent boundary

   e) a continent–continent collision zone

   f) the interior of a plate

   g) a transform plate boundary

Preparing for the Chapter Challenge

Reflect on your answers to the above. Write a short summary (one or two paragraphs) describing what you have learned so far about how plate movements and their interactions at plate boundaries can change the arrangement of continents and ocean basins on the Earth. Be prepared to include this summary in your **Chapter Challenge**.

Inquiring Further

1. **Evolution of the Biosphere at Mid-Ocean Ridges**

   Recently, new forms of life have been discovered at mid-ocean ridges. They thrive in the presence of superheated, mineral-rich water. This life does not depend upon the Sun for energy, but instead upon the energy and matter from Earth's interior. How has life evolved in such environments? For further information check the *EarthComm* web site.

Black smokers form at the mid-ocean ridges.

Activity 3

What Drives the Plates?

Goals

In this activity you will:

- Calculate the density of liquids and compare their densities with their position in a column of liquid.

- Observe the effects of temperature on the density of a material.

- Examine natural heat flow from within the Earth.

- Understand the results of uneven heating within the Earth.

- Understand the causes of the movement of lithospheric plates.

Think about It

Geoscientists are still uncertain about the most important forces that drive the plates.

- **What causes the movement of the lithospheric plates?**

What do you think? Record your ideas about this question in your *EarthComm* notebook. Be prepared to discuss your responses with your small group and the class.

Investigate

Part A: Effects of Density on the Position of Material

1. Obtain 30 mL each of water, pancake syrup, and vegetable oil. Suppose you were to carefully pour a small volume of each liquid into one graduated cylinder or clear tube.

   a) Predict what you think will happen. Sketch and explain your prediction.

2. One at a time, carefully pour 10 mL of each liquid into a cylinder or clear tube.

   a) Record your observations.

   b) Do your observations support your predictions?

   c) Does the order in which you pour the liquids make a difference in what you observe?

3. Develop a method to determine the density of each of the three liquids using a graduated cylinder, 10 mL of each liquid, and a balance scale. Density is mass per unit volume. Thus, the density of each liquid equals the mass of liquid (in grams) divided by the volume (10 mL).

   a) Write down your procedure for finding the density of each liquid.

   b) Make a data table to record your measurements and calculations for each liquid.

   c) After your teacher has approved your procedure, determine the density of each liquid.

⚠ Follow your teacher's safety advice about using a heat source. Hot corn syrup can cause burns. Clean up spills immediately.

4. Compare your calculations with your observations in **Step 2**.

   a) Describe how the densities you calculated explain what you observed.

   b) If layers of materials of different densities within the Earth behave like layers of liquids of different densities, what would you predict about the position of the rock layers of different densities in the Earth?

Part B: Effects of Temperature on Density of a Material

1. Pour about a 5 cm thick layer of corn syrup into a Pyrex® beaker or wide aluminum pan. Place the pan on a heat source.

2. Place three pieces of balsa wood on the syrup.

   a) Predict what you think will happen to the wood as the corn syrup is heated. Record your ideas in your notebook.

3. Observe the wood. Record any changes every 5 min for 20 to 30 min.

   a) Use diagrams to record the changes you observe.

   b) Do your observations support your predictions? What do you think caused the results you observed?

sandstone

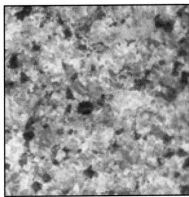
granite

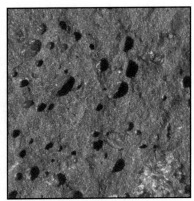
basalt

Part C: Density of Earth Materials

1. Collect samples of rock from your community and also obtain samples of granite, basalt, and sandstone.

2. If you can, predict qualitatively the density of the samples. Which sample appears to be least dense? Which appears to be most dense?

   a) Record your predictions in your notebook.

3. Develop a method to find the density of each rock sample using the sample, water, a graduated cylinder, and a balance scale. Density is mass per unit volume. Thus, the density of each rock equals the mass of rock (in grams) divided by the volume of rock (in cubic centimeters). Note that 1 mL = 1 cm^3.

   a) Write down your procedure for finding the density of each rock sample.

   b) Make a data table to record your measurements and calculations for each rock.

   c) After your teacher has approved your procedure, determine the density of each rock sample.

4. Compare your calculations with your predictions.

   a) How does the density of the rock from your community compare with the densities of granite, sandstone, and basalt?

Part D: Forces Causing Subduction of Lithospheric Plates

1. Partly fill a large, rectangular tub with warm water. Wait until any tiny air bubbles have disappeared. The water has to be perfectly clear.

2. Very slowly and carefully, put a few ounces of liquid dish detergent in the water and mix it slowly and carefully with a mixing spoon. If any soap bubbles or foam remain on the water surface, scrape them off with a damp sponge.

3. Cut a piece of the vinyl plastic to be about six inches wide and about twelve inches long. Trim a flat, clear-plastic ruler with the scissors to be the same width as the plastic sheet. (The ruler should sink in water.) Tape the ruler to one end of the plastic sheet.

4. Dip the ruler end of the plastic sheet into the water to a depth of about 1 cm. Immediately place the plastic sheet on the water surface. Do this by holding the ends up, and letting the sagging middle part of the sheet touch the water surface first, to avoid trapping air bubbles under the sheet. Observe what happens. Repeat this step as many times as you need to make careful observations.

a) Record your observations. Include a description of the motion of the plastic sheet in the water.

b) What is the force that makes the plastic behave as it did?

c) How does this demonstration show what happens in a subduction zone?

Reflecting on the Activity and the Challenge

You have seen evidence that liquids of varying densities will form layers in a container with the densest liquid on the bottom. You have also shown by modeling that a solid floating on a liquid seems to move away from a source of heat. You have seen evidence that different rocks are likely to have varying densities. Finally, you have made a direct observation of one of the main forces that cause subduction. These investigations will help you understand the Earth's interior and the flow of matter and energy in the Earth. You should now be able to explain why lithospheric plates can float and what might cause them to move.

 Keep work area clean and dry.

 Have paper towels ready for the wet plastic that is taken out of the tub.

Digging Deeper

THE EARTH'S INTERIOR STRUCTURE

Evidence for Earth's Layered Structure

Density (mass per unit volume) refers to how concentrated the mass (atoms and molecules) in an object or material is. Less dense material tends to rise upward and float on more dense material. Here are some everyday examples: a less dense solid floats in a more dense liquid; a more dense solid sinks to the bottom of a less dense liquid; a less dense liquid floats on a more dense liquid. Rocks in the Earth's crust (oceanic crust consists mainly of basalt; continental crust consists mainly of less dense rocks like granite) are less dense than the rocks of the underlying mantle. The crust "floats" on the more dense interior material.

Several kinds of evidence reveal that density varies within the Earth. Laboratory experiments in high-pressure apparatus show that rocks deep in the Earth are more dense than the same rocks when they are at the surface. The weight of the overlying rock puts pressure on rock below, making it more dense. The most dense material should be at the center of the Earth, where the pressure is greatest.

A second line of evidence comes from calculations of the average density of the Earth. You cannot put the Earth on a balance scale to find its mass, but the mass can be found indirectly using Newton's Law of Gravitation. According to that law, the gravitational force (F) between any two objects in the universe can be expressed this way:

$$F = \frac{g m_1 m_2}{d^2}$$

where m_1 and m_2 stand for the masses of two objects,

d stands for the distance between them, and

g stands for the gravitational constant (known from experiments).

Because the Earth exerts a certain force on a body (like you) with a certain mass m_1 on the Earth's surface, some 6400 km from its center, the known values can be substituted into the equation and the mass of the Earth (m_2) can be calculated. Dividing the mass of the Earth by its volume gives an average density of the Earth (in metric units) of 5.5 g/cm^3. The density of the rocks commonly found at the surface (granite, basalt, and sandstone) is much lower. The average density of surface rocks is 2.8 g/cm^3. The density of

Geo Words

density: the mass per unit volume of a material or substance.

973

Geo Words

core: the solid, innermost part of the Earth consisting mainly of iron.

the Earth's interior must be much greater than 2.8 g/cm³ for the entire Earth to average 5.5 g/cm³. This is partly due to the effect of compression, but also partly because the material in the Earth's core is mostly iron, which is much more dense than rocks, even when it is not under great pressure.

The speed of earthquake (seismic) waves within the Earth generated by earthquakes also provides convincing evidence about the properties of rock in the Earth. Scientists have learned that these waves travel faster the deeper they are in the Earth. It's known from laboratory experiments that earthquake (seismic) waves that travel at 4.8 km/s at the surface travel at 6.4 km/s at a depth of 1600 km. The reason why the speed of seismic waves increases downward in the mantle is complicated. In the laboratory, scientists use special equipment to measure the speeds of seismic waves in different rocks. They can determine how the speed of seismic waves changes with changes in temperature, pressure, and rock type.

Studies also show that change in density is not uniform with depth. Instead, there are distinct jumps or changes in density. By studying the changes in the speed of earthquake (seismic) waves as they pass through the Earth, scientists have concluded that the Earth's interior structure is layered. The thickness and the composition of the layers are shown in the table in *Figure 1*.

Layer	Thickness (km)	Composition	Temperature (°C)	Density (g/cm³)
Continental crust	30–60	Granitic silicate rock (>60% silica)	20–600	~2.7
Oceanic crust	5–8	Basaltic silicate rock (<50% silica)	20–1300	~3.0
Mantle	2800	Solid silicate	100–3000	~5
Outer core	2150	Liquid iron-nickel	3000–6500	~12
Inner core	1230	Solid iron-nickel	7000	~12

Figure 1 The composition of the layers of the Earth.

Using the evidence they have observed, geoscientists divide Earth into four main layers: the inner **core**, the outer core, the mantle, and the **crust**, as shown in *Figure 2*. The core is composed mostly of iron. It is so hot that the outer core is molten. The inner core is also hot, but under such great pressure that it remains solid. Most of the Earth's mass is in the mantle. The mantle is composed of iron, magnesium, and aluminum silicate minerals. At over 1000°C, the mantle is solid rock, but it can deform slowly in a plastic manner. The crust is much thinner than any of the other layers, and is composed of the least dense rocks.

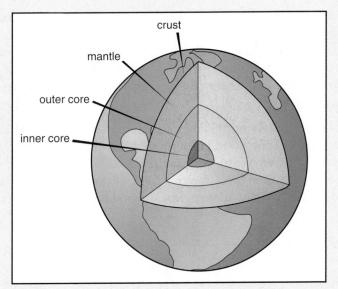

Figure 2 Schematic diagram showing the layered structure of the Earth's interior.

Labels: crust, mantle, outer core, inner core

Geo Words

thermal convection:
a pattern of movement in a fluid caused by heating from below and cooling from above. Thermal convection transfers heat energy from the bottom of the convection cell to the top.

The Flow of Matter and Energy within the Earth

The temperature of the Earth increases with depth. This can be observed directly in mines and in oil wells. Sources of the Earth's internal heat include the decay of radioactive elements, the original heat of Earth's formation, and heating by the impact of meteorites early in Earth's history. The Earth can be thought of as a massive heat engine. The transfer of heat from Earth's interior to its surface drives the movements of the Earth's crust and mantle.

Temperature affects the density of materials. Hot-air balloons show this effect well. When the air inside a balloon is heated it expands (increases in volume). The mass of the balloon stays the same, but the volume increases. When the ratio of mass to volume drops, the density drops. Therefore, heating makes the balloon less dense than the surrounding air. The hot-air balloon begins to rise. Similarly, as rocks in the interior of the Earth are heated enough, their density decreases. The less dense rock rises slowly over time, unless the rocks are too rigid to allow flow.

In the activity, you heated corn syrup and observed the movement of balsa wood. Why did the balsa wood move? The answer lies in the process of **thermal convection**. Heating lowers the density of the corn syrup at the bottom of the container. This causes it to rise. As the corn syrup approaches the upper surface, it flows to the side, making room for more corn ➡️ syrup rising from below. As it moves to the side, it cools. As it cools, it becomes more dense, and it sinks back to the bottom of the container. At the bottom of the container it is heated and rises again. This kind of density-driven circulation is called thermal convection, as shown in *Figure 3*. Thermal convection transfers heat energy from one place to another by the movement of material.

In 1929, the geologist Arthur Holmes elaborated on the idea that the mantle undergoes thermal convection. He suggested that this thermal convection is like a conveyor belt. He reasoned that rising mantle material can break a continent apart and then force the two parts of the broken continent in opposite directions. The continents would then be carried by the convection currents.

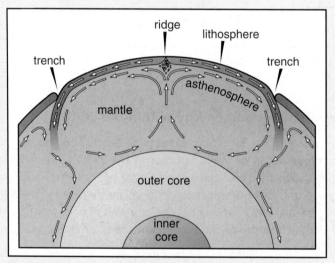

Figure 3 One possible pattern of thermal convection in the Earth's mantle. Convection cells like this might provide at least some of the driving force for the movement of lithospheric plates.

According to this hypothesis of mantle convection, material is heated at the core–mantle boundary. It rises upward, spreads out horizontally, cools, and sinks back into the interior. These extremely slow-moving convection cells might provide the driving force that moves the lithospheric plates (see *Figure 3*). Material rises to the surface at places where lithospheric plates spread apart from one another. Material sinks back into the Earth where plates converge. Although the idea was not widely appreciated during Holmes' time, mantle convection cells became instrumental in the development of the theory of plate tectonics.

Mantle convection can't be observed directly, the way you could have observed convection in the corn syrup if you had put some tiny marker grains in the syrup. Geoscientists are sure that the mantle is convecting, but they are still unsure of the patterns of convection. The patterns probably don't look much like what is shown in *Figure 3*! Geoscientists now think that the lithospheric plates themselves play a major part in driving the convection, rather than just being passive riders on top of the convection cells. Do you remember from **Activity I** that the mid-ocean ridges are broad, and they slope gradually down to the deep ocean nearer the continents? That means that the plates on either side of the ridge crest slope downward away from the ridge crest, and they tend to slide downhill under the pull of gravity! In this way, they help the convection cell to keep moving, instead of the other way around. Also, you probably know that most materials expand when they are heated and shrink when they are cooled. As the plates in the ocean cool, they become more dense than the deeper mantle because they have almost the same composition but they are not as hot. They sink into the mantle of their own accord, just as in **Part D** of the investigation. In that way they help to keep the convection cell moving.

Check Your Understanding

1. How can the density of the Earth be calculated?

2. How does the density of the Earth provide evidence that the interior of the Earth is denser than the surface?

3. Name three main layers of the Earth.

4. Why is the inner core of the Earth solid, even though it is hot?

5. How are convection currents set up?

6. What part of the Earth's interior layers are in motion due to density differences?

Understanding and Applying What You Have Learned

1. Look at the map of lithospheric plates near South America and the relative "horizontal" motion between these plates.

   a) At point A, the two plates are moving away from each other. What is happening between them?

   b) At point B, two plates are moving toward each other. What happens as they continue to push toward each other if they have:

      (i) Different densities?

      (ii) The same density?

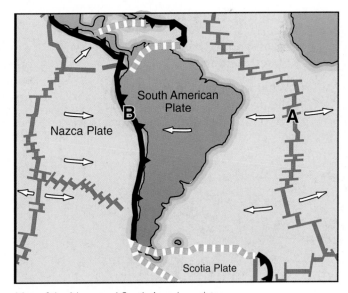

Map of the Nazca and South American plates.

2. Draw two pictures side by side. Make one the experiment with corn syrup and balsa wood. Make the other the Earth's interior structure. Show where heating and cooling occur, and use arrows to indicate the movement of material (the flow of matter and energy in both systems). Label the parts in each diagram and show how they correspond to each other.

3. What evidence is there at the Earth's surface for unequal heating somewhere within the Earth?

4. List some natural processes that occur when heat from the Earth's interior is transferred to the surface.

5. Use your understanding of density to calculate the missing values in the table below:

Object	Mass (g)	Volume (cm³)	Density (g/cm³)
Iron	41.8		7.6
Quartz	39.75	15.0	
Gold		8.0	19.3

Preparing for the Chapter Challenge

Reflect on your answers to the above. Write an essay that describes the Earth's interior structure and the flow of matter and energy within the Earth. Refer to the evidence you examined and the models you explored. Sketch and label a drawing or two to illustrate the main ideas. Be prepared to include this summary in your chapter report.

Inquiring Further

1. **Investigating Driving Forces for Plate Motions**

   What questions do you have about the driving forces behind plate tectonics? Develop a plan that would help you find an answer to one of your questions. Record your plan in your notebook. What further information might help you answer your questions?

Activity 4   Effects of Plate Tectonics

Goals

In this activity you will:

- Use maps to examine the distribution of earthquakes and volcanoes to the location of plate boundaries.

- Explain the location, nature, and cause of volcanic arcs in terms of plate tectonics.

- Explain the location, nature, and cause of hot spots.

- Explain how plate-tectonic processes have caused continents to grow through geologic time.

- Explain how plate-tectonic processes produce landforms.

- Explain how plate tectonics can affect the interior of a continent.

Think about It

Rocks high in the Himalayas, almost 8000 m (29,028 ft) above sea level, contain fossils of marine animals.

- **Why are most high mountain ranges located at or near plate boundaries?**

What do you think? Record your ideas about this question in your *EarthComm* notebook. Include sketches as necessary. Be prepared to discuss your response with your small group and the class.

Investigate

1. Compare the distribution of volcanoes and earthquakes shown on the following map with the map of crustal plates you created in **Activity 2.**

   a) Describe any differences between the distribution of volcanoes along plate boundaries and within plate interiors.

   b) Describe any differences between the distribution of earthquakes along plate boundaries and within plate interiors.

2. To model the rise of magma through the Earth, fill a tall, transparent jar almost up to the brim with honey. Put a very small volume of vegetable oil (about 5 mL, one teaspoon) on top of the honey. Screw on the lid. Try not to leave any air in the container. Turn the container upside down quickly and set it on a tabletop. Make your observations and answer the following questions:

   a) What does the honey represent?

   b) What does the vegetable oil represent?

   c) Describe and explain the behavior of the vegetable oil.

   d) What do you think are the similarities and differences between the behavior of this model and the rise of magma in the Earth?

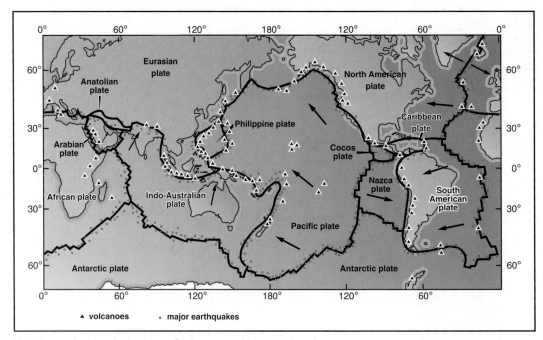

World map showing the location of volcanoes and large earthquakes.

⚠ Be sure the lid is on tight before the jar is turned over. Clean up spills immediately.

3. The diagram shows a cross section with two subduction zones, a spreading ridge, and a continental rift. Note the two zones where an oceanic plate is being subducted (plunged) under another plate. Volcanoes are common in a zone that is located a certain distance away from the trench where the subducted plate first bends downward.

In your group, develop as many hypotheses as you can think of for why the volcanoes occur, and why they are located where they are. Here are four important facts you might need to use or think about in developing your hypotheses:

(i) The downgoing plate is cooler than the mantle.

(ii) The composition of the downgoing plate is slightly different from the composition of the mantle. (Why?)

(iii) The oceanic crust on the downgoing plate contains water, which was added to the igneous rocks when they formed.

(iv) Friction generates heat.

a) Record your group's hypothesis.

b) In a class discussion, compare the hypotheses you developed in your group with those of other groups. Were their hypotheses different from yours and still seemed reasonable?

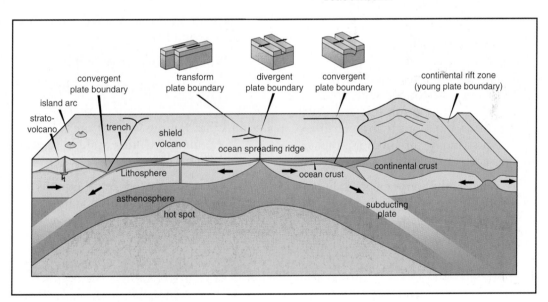

Cross-section with two subduction zones.

4. Look again at the diagram of the cross section with two subduction zones. Use the cross section to answer the following questions:

   a) Under what two types of plates is the oceanic lithosphere being subducted?

   b) What differences between oceanic volcanic arcs and continental volcanic island arcs can you see or infer from the cross section?

   c) On a copy of the map of volcanoes and earthquakes, circle two continental volcanic arcs and three oceanic volcanic arcs.

   d) In which part of the world are most volcanic arcs located? What does that suggest about the plate-tectonic setting of that part of the world?

5. To understand why volcanic arcs are called arcs, look at the Andes Mountains on a topographic map. The Andes are topped with volcanoes that are part of a volcanic arc. They appear to run along a straight line. Run a string or thread along their length on a globe.

   a) What is the shape of the line on the globe? Why are lines of volcanoes called arcs?

   b) Using the cross section of the subduction zones, explain why few volcanoes occur very far inland within a continent.

   c) If volcanic rock is found far inland within a continent, what is one possible reason why it is there?

6. One reason why volcanic rock might be located in the interior of a plate is illustrated on the cross section of the subduction zones where a hot spot is shown in the middle of one of the oceanic plates.

   a) Where does it appear that the hot spot originated?

   b) Is it related to subduction?

   c) Where does the hot spot begin to produce a pool of magma?

7. The map on the next page shows hot spots around the world:

   a) Where are most hot spots?

   b) Are they clustered or randomly located?

   c) What famous area of the continental United States sits over a hot spot?

   d) What sits atop another famous hot spot in the United States?

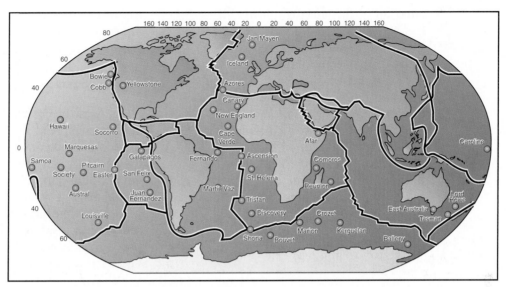

Map of hot spots around the world.

8. To model the growth in size at a subduction zone, set up pieces of lumber and a plastic sheet as shown in the diagram. With a table knife, spread a thin (about 2 mm) layer of cream cheese on the part of the plastic sheet that rests on the upper surface of the 2 x 6. Try to keep the layer as even in thickness as you can. Spread a layer of the cheese spread, of about the same thickness, over the layer of cream cheese.

Have one student hold the piece of 2 × 4 in place, pressing down on it gently. Another student very slowly pulls the loose end of the sheet of plastic out from under the 2 x 4. Observe what happens to the layer of cream cheese and cheese spread as "subduction" proceeds.

a) In your notebook, write a description of the process you observe. Be sure to note how the body of material at the subduction zone changes its shape, as well as its volume through time.

⚠ Keep work area clean. Have damp paper towels ready to wipe up spills and clean hands as required.

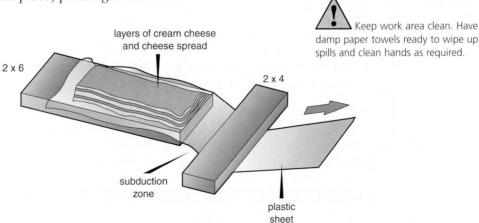

layers of cream cheese and cheese spread

2 x 6

2 x 4

subduction zone

plastic sheet

983

9. After all of the sheet of plastic has been consumed down the subduction zone, make a cut through the accumulated cream cheese and cheese spread, to see the internal structure of the new material at the subduction zone. The cut should be vertical, and parallel to the direction of subduction. Clear away all the material from one side of the cut and examine the face of the cut.

   a) In your notebook, draw a sketch of what the cut looks like. Show the topographic profile of the top of the material, and also any internal structures you observe.

   b) Is there a limit to the height of the "mountain range" that is formed at the subduction zone?

   c) Did any of the cream cheese and cheese spread go down the subduction zone (under the 2 × 4)? In a real subduction zone, what would happen to such material? (This is not a perfect model of what happens to oceanic sediment at a subduction zone, but it gives you the flavor of it. Geoscientists love to make puns.)

Reflecting on the Activity and the Challenge

In this activity you have observed features on the Earth's surface that are created as a result of crustal plate movements and activity within the geosphere. You have also seen that plate-tectonic processes can cause plates to grow in size. Finally, you have seen that plate movements can cause earthquakes that can be associated with specific plate boundaries. You will need to consider these surface features and disturbances created by plate motions in order to complete your **Chapter Challenge**.

Geo Words

plate tectonics: the field of study of plate motion.

Digging Deeper

"BUILDING" FEATURES ON EARTH'S SURFACE

Plate Tectonics

You can now see why the field of study of plate motion is called **plate tectonics**. Tectonics comes from the Greek word *tekton*, which means builder. Plate tectonics refers to the building of the features on Earth's surface due to deformation caused by plate movements.

You have learned that plate movements create mountain ranges, trenches, and rift valleys at or near plate boundaries. Also, there is a clear relationship between volcanoes and plate boundaries, and between earthquakes and plate boundaries. This is particularly evident around the rim of the Pacific Ocean, where the subduction of oceanic plates around much of the rim results in volcanic arcs and earthquakes.

Oceanic Trenches

The deepest valleys on Earth are in the ocean, where they can't be seen except from special deep-diving submersibles. Where an oceanic plate is subducted under another plate, it bends downward as it enters the subduction zone. The valley that is formed above the zone of bending is called a trench. Oceanic trenches are very deep. Many are deeper than 10,000 m, which is twice the average depth of the deep ocean. The word "valley" is a bit misleading, because trenches are wide, and their side slopes are not very steep. You can easily spot the locations of trenches in most world atlases, because they are shown with the darkest blue shading on maps of the world's oceans. Trenches are common in many places in the western Pacific, where there is ocean–ocean subduction. There is a long trench along the west coast of South America, where the Nazca Plate is being subducted under the continent.

Volcanoes at Plate Boundaries

You know already that volcanoes are common along mid-ocean ridges, where basalt magma rises up from the asthenosphere to form new oceanic crust. Volcanoes are also common along subduction zones, where they form volcanic arcs. At a depth of 200 to 300 kilometers, magma is produced above the subducted plate, and rises toward the surface because it is less dense than the surrounding rock. At first it was thought that the magma was produced as rock near the top of the downgoing plate and was heated by friction, but geoscientists are now convinced that the melting is for a different reason. When the oceanic crust is first produced, at the mid-ocean ridges, a lot of water is combined with certain minerals in the igneous rocks. As the pressure and temperature increase down the subduction zone, this water is driven off, and it rises upward from the plate. It's known that the melting temperature of the mantle rock above the plate is lowered when water is added to it. This causes some of the mantle rock to melt. This is a good way to explain why melting doesn't start until the plate has reached a certain depth down the subduction zone, and then stops at a slightly deeper depth. The "Ring of Fire" around the Pacific Ocean is caused by this melting at subduction zones all around the Pacific.

Hot Spots

Not all volcanoes are associated with mid-ocean ridges and subduction zones. Hot spots, which originate at the boundary between the mantle

and the outer core, are narrow plumes of unusually hot mantle material. These plumes rise up through the mantle and melt the rock at the base of the lithosphere, creating pools of magma. This magma then rises to the surface, resulting in hot spot volcanoes.

Some hot spots are located under continents. The hot spot producing the hot springs at Yellowstone National Park is an example. One theory suggests that the bulge created by a hot spot may initiate the rifting of a continent. It is thought that a hot spot lies below, and is responsible for, the Rift Valley of Africa. There is also evidence suggesting that the New Madrid fault, which runs down the Mississippi River Valley, may represent an aborted rift zone originally created by a series of hot spots. The largest series of earthquakes in the United States, outside of Alaska, occurred on the New Madrid fault in the early 19th century, ringing bells as far away as Philadelphia and causing the Mississippi River to run backwards for a short time! In this way, plate tectonics can even affect areas that are within the heart of a continent.

Mountains at Plate Boundaries

Most of the great mountain ranges of the world are located near convergent plate boundaries. When someone says "mountain ranges" to you, which of them do you think of? The Alps in Europe, the Himalayas in southern Asia, the Andes in South America, and the coastal mountain ranges in western North America are some examples. Mountain ranges like those are built in mainly two ways. You already know that the magma that is generated above the subducted plate rises up to form a chain of volcanoes. Much of the magma remains below the surface and cools to form large underground masses of igneous rock called batholiths. The combination of volcanoes at the surface and batholiths deep in the Earth adds a lot of new rock to the area above the subduction zone, and makes the elevation of the land much higher. Also, many subduction zones experience compression, when the two plates are pushed together by plate movements elsewhere. In places like that, great masses of rock are pushed together and stacked on top of one another in complicated structures, to form high mountains. This happens also where two continents collide with each other, as in the Himalayas.

Growth of Continents at Subduction Zones.

During the long travels of an oceanic plate from a mid-ocean ridge to a subduction zone, a hundred meters or more of oceanic sediment is deposited on the top of the plate. In **Part 6** of the investigation, you saw how a lot of this sediment is scraped off and added to the edge of the other plate. This material, which is deformed into very complicated structures, is turned into rock by heat and pressure. It becomes a solid part of the other plate. When material is added to the edge of a continent in this way, the continent grows larger at its edge. Continents also grow as the igneous rock of volcanoes and batholiths are added to the continent above the subduction zone, as described above. The growth of a continent along its edge in these ways is called **continental accretion**. This has been going on through geologic time, making the continents larger and larger.

Earthquakes and Plate Tectonics

As plates move past each other at plate boundaries, they don't always slide smoothly. In many places the rocks hold together for a long time and then slip suddenly. Earthquakes along mid-ocean ridges are common because of movement along the transform faults that connect segments of the ridge crest. Only where transform faults are on land or close to land, as in California, are these earthquakes likely to be hazardous. As you saw in **Activity 2**, the famous San Andreas Fault in California is a transform fault. Earthquakes at subduction zones and continent–continent collision zones are a bigger problem for human society, because these areas are so common on the Earth, especially around the rim of the Pacific and along a belt that stretches from the Mediterranean to southeast Asia. Earthquakes in subduction zones happen at depths that range from very shallow, near the trench, to as deep as hundreds of kilometers, along the subducted plate. Earthquakes in continent–continent collision zones happen over wide areas as one continent is pushed under the other.

Geo Words

continental accretion: the growth of a continent along its edges.

Check Your Understanding

1. Why is plate tectonics a suitable name for the study of plate motion. Explain.

2. What geographic features would you expect to see at plate boundaries.

3. How do geoscientists suggest that "hot spots" are related to plate tectonics.

4. In your own words explain the process of continental accretion.

Understanding and Applying What You Have Learned

1. Review your work with earthquakes and volcanoes one more time.

   a) Summarize where most earthquakes are located compared to plate boundaries.

   b) Summarize where most volcanoes are located compared to plate boundaries.

2. Although most earthquakes and volcanoes are associated with plate boundaries, they are not always located directly along the boundaries. Considering boundaries between oceanic and continental plates:

   a) Why are volcanoes usually found on the continental side of a plate boundary?

   b) Why are earthquakes usually found on the continental side of a plate boundary?

3. Many volcanoes and earthquakes are found far from modern plate boundaries. Write a paragraph giving one idea you think might explain how at least some of them have formed. Be sure to point out examples by describing their location.

4. Make a list of the various plate-tectonic settings where mountain ranges are likely to be produced. For each item on the list, draw a cross section that shows the mountain range and how it relates to the plate-tectonic setting. For each item, give an example from somewhere in the world.

Preparing for the Chapter Challenge

In this activity you have learned that plate tectonics has an influence on every part of the world. Surface features and movements caused by plate tectonics are not only at plate boundaries, but also at locations far from modern plate boundaries. Write a paragraph, based on what you now know of movements or features in your region that indicate plate-tectonic activity (whether it is nearby or far away). Carefully list your evidence so you can present it to the middle school class.

Inquiring Further

1. **Plate tectonics and the local climate**

   Distant mountain ranges and plateaus created by plate tectonics can affect air flow in many ways, affecting local climate and thus vegetation, soil, wildlife, and drainage patterns. Research and report on how your local region has been affected directly or indirectly by plate tectonics. You may even wish to include some of this research in your **Chapter Challenge.**

Activity 5

The Changing Geography of Your Community

Goals

In this activity you will:

- Use several present-day distributions of minerals, rock formations, and fossils to help figure out the distribution of continents.

- Construct a map showing the position of continents 250 million years ago by reversing the present direction of plate motion.

- Recognize a convergence of presently widely scattered minerals, rock formations, and fossils when all the continents were part of Pangea.

- Compare present average community motions with that of the past 250 million years, by calculating the average yearly rate of motion over the last 250 million years.

- Describe the context in which the hypothesis of continental drift was proposed and why it was subjected to criticism.

- Show that your community has moved through different ecological regions over time.

Think about It

The plates forming the Earth crust can be compared to pieces of newspaper torn from the same page.

- **How would you be able to decide if the pieces all came from the same page?**

- **How could you convince someone else that the pieces came from the same page?**

What do you think? Record your ideas about these questions in your *EarthComm* notebook. Include sketches as necessary. Be prepared to discuss your responses with your small group and the class.

Investigate

1. Begin your work individually. Obtain four copies of the diagram, showing the outline of the continents at sea level, as well as the boundary between the continental crust and the oceanic crust. The diagram also shows the location of rock formations, mountain ranges, and fossil plants and animals. Cut out the continents on the first sheet along the edges of the continental shelves, which in most places are close to the boundaries betwen the oceanic lithosphere and the continental lithosphere.

    a) Why cut the pieces at the boundaries between the continental and oceanic crust?

    b) In which ecological region is your community today: tropics, subtropics, mid-latitudes, subpolar, or polar?

    c) Coal deposits originated in the swamps of tropical forests. Are the coal deposits shown on the map in the tropics today?

    d) Where do you find mountains similar in structure to the Appalachian Mountains?

    e) Where do you find rock formations similar to those in South America?

    f) Glossopteris is an extinct seed fern that had leaves like ferns of today. It produced seeds too large to travel by air or float on water. Where are fossils of these ferns located today?

    g) Mesosaurus is an extinct freshwater reptile that thrived during the Triassic Period (245 to 208 million years ago). Where are fossils of this reptile found today?

2. Rearrange the cut pieces on a blank piece of paper as the continents now appear and tape them in place.

    a) Label the outlines "Present."

    b) Draw a border around the map.

    c) Sketch in and label the Equator and latitude lines at 30° and 60° north and south.

    d) Title the map "Present."

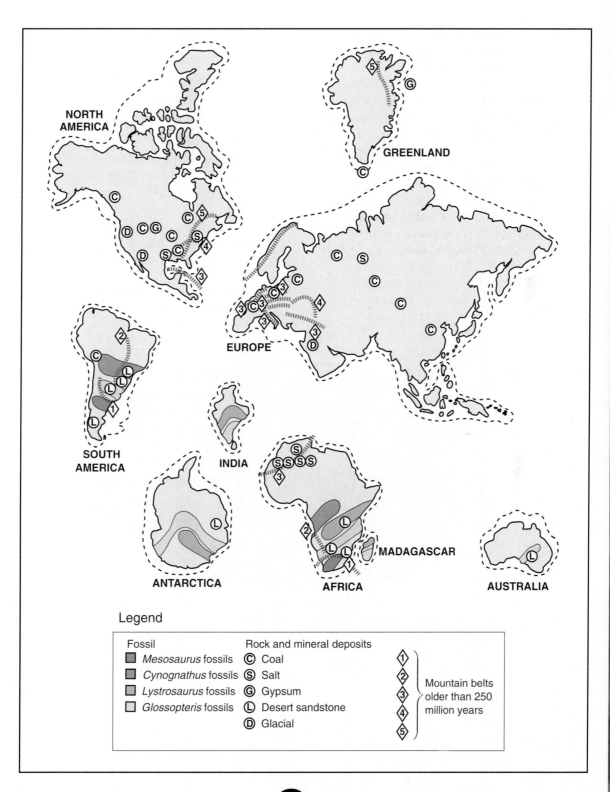

NORTH
AMERICA

GREENLAND

EUROPE

SOUTH
AMERICA

INDIA

ANTARCTICA

AFRICA

MADAGASCAR

AUSTRALIA

Legend

Fossil

- ▨ *Mesosaurus* fossils
- ▨ *Cynognathus* fossils
- ▨ *Lystrosaurus* fossils
- ▨ *Glossopteris* fossils

Rock and mineral deposits

- Ⓒ Coal
- Ⓢ Salt
- Ⓖ Gypsum
- Ⓛ Desert sandstone
- Ⓓ Glacial

◇1
◇2
◇3
◇4
◇5

Mountain belts
older than 250
million years

3. Cut out the continents from the second sheet. Try to arrange them on another piece of paper, as they would have appeared 250 million years ago, before the Atlantic Ocean and the Indian Ocean began to open. You can do this by using two methods: (a) moving each continent in the direction opposite of that shown by the arrows on the map of plate motions; (b) matching similar rock formations, mountain ranges, and fossils from continent to continent. Try to move each of the continents at the speeds given by the lengths of the arrows, until they all meet. Tape the continents together.

a) Draw a border around the map.

b) Sketch in and label the Equator and latitude lines at 30° and 60° north and south.

c) Title this map, "250 million years ago."

d) The following diagram shows the reconstruction of Pangea that is generally accepted by geoscientists. Your reconstruction is likely to be somewhat different, because the evidence you had is less detailed. Compare your map with the following map, and adjust the positions of the continents on your map as required.

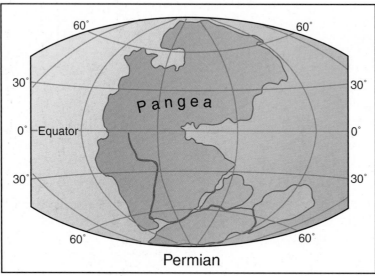

Generally accepted reconstruction of Pangea, in the Permian period of geologic time, about 250 million years ago.

4. Use your adjusted map, "250 million years ago" to answer the following questions:

   a) Which two continents fit together best?

   b) Why do you think the continents do not fit together exactly?

   c) From the map of Pangea, what can you say about the latitude and longitude of your community 250 million years ago?

   d) In what ecological region was your community 250 million years ago?

   e) Many coal deposits were created before 280 million years ago in the tropics. Where were they 250 million years ago? Does this make sense? Explain your answer.

   f) Do the Appalachian Mountains line up with other mountain ranges that they resemble?

   g) Do rock formations in South America line up with other formations that they resemble?

   h) How does Glossopteris appear to have migrated to its present fossil distribution, since its seeds could not be carried by the wind or float on water?

   (i) How does Mesosaurus appear to have migrated to its present fossil locations, since it could not swim in the salty ocean?

5. Cut out the continents from the third sheet, in the same way as before. Arrange them on a new piece of paper, as they might appear 250 million years in the future. You can do so by starting with the present distribution of the continents and then moving each in the direction and at the speed shown by the arrows on the map of plate motions. Remember that some plates will be subducted under others.

   a) What will fill the spaces between the continents in the future?

   b) What will happen to the Mediterranean Sea? What will be created in southern Europe?

   c) Where will the southern coast of California be in 250 million years?

   d) In what latitude and in which ecological region might your community lie in 250 million years?

   e) How might the change in ecological region affect your community?

   f) Why might your prediction regarding the future location of your community and continent be in error?

Reflecting on the Activity and the Challenge

You have seen that by moving the continents "in reverse," in directions opposite to their present movement, you can make the continents fit together fairly well as a single continent. You have also seen that features like rock formations, mountain ranges, and fossil plants and animals that are similar but are now separated by wide oceans are brought together when the continents are assembled into the single large continent of Pangea. You have gotten some idea about how far a particular place on a continent might have moved in the 250 million years since Pangea broke apart.

Digging Deeper

DEVELOPMENT OF THE PLATE TECTONICS THEORY

In this activity, you examined some of the evidence that supports the idea that the continents of the Earth have moved during geologic time. Two features of the Earth were the subject of intense study in the late 1800s—the discovery of similar fossils on continents that are now separated widely by oceans, and the origin of mountain ranges. Both played a part in the early stages of the development of theory of plate tectonics.

In the late 1800s, an Austrian geologist named Eduard Suess (1831–1914) tried to solve a basic geological question: how do mountain ranges form? He based his model of mountain formation on some of the same principles that you explored in this chapter. Suess stated that as the Earth cooled from a molten state, the more dense materials contracted and sank toward the center, and the least dense materials "floated" and cooled to form the crust. He then speculated that mountain ranges formed from the contraction and cooling of the Earth. He likened this to the way that an apple wrinkles and folds as it dries out and shrinks.

Suess went on to explain the origins of oceans, continents, and the similarities of fossils on different continents now separated by oceans. In his model, during the cooling process, parts of the Earth sank deeper than others, forming the ocean basins. Suess claimed that certain parts of the sea floor and continents could rise and sink as they adjusted to changes in the cooling earth. This led him to propose land bridges between continents. Suess coined the term Gondwanaland for a former continent made up of central and southern Africa, Madagascar, and peninsular India. These areas all contained similar fossils that were hundreds of millions of years old. According to Suess, the land bridges allowed various animals and plants to migrate and spread without crossing an ocean.

Although other geologists proposed different models to explain mountains, oceans, and fossils, all generally agreed that the Earth's crust moved up and down, but not very far sideways. Land bridges were often cited as allowing various kinds of organisms to move between continents now separated by oceans. According to Suess and others, the land bridges sank into the ocean long ago and no longer exist.

Not all geologists accepted the theory of a contracting Earth. In 1912, the German geoscientist Alfred Wegener (1880–1930) proposed the hypothesis of continental drift. He saw a variety of problems with the contraction theory. One difficulty was the severe compression of the Alps. The Alps are a young mountain range. Rock layers in the Alps are severely folded and stacked up on top of one another, indicating a great horizontal shortening of original distances, as shown in *Figure 1*. Wegener thought that contraction could not produce such great shortening of the Earth's crust. He also thought that contraction should produce uniform "wrinkles" in the Earth, not narrow zones of folding. The discovery of radioactive heat in Wegener's time also provided evidence against cooling. Heat from radioactive decay in the Earth would work against the cooling and contraction process.

Figure 1 Wegener used the severe compression of the Alps as evidence to support his hypothesis of continental drift.

According to Wegener, about 200 millions years ago, a huge supercontinent called **Pangea** (Greek for *all land*) broke into separate continents that moved apart. Wegener claimed that compression at the leading edge of the moving continent led to the formation of mountains. Wegener's hypothesis allowed him to explain the different ages of the different mountain belts. He claimed that the timing of the breakup was variable, with some parts of Pangea separating earlier than others. His evidence included the puzzle-like fit of the continents and the similarity of rocks, geologic structures, and fossils on opposite sides of the Atlantic Ocean. Wegener's hypothesis eliminated the need for (now sunken) land bridges that once connected widely separated continents.

But how did continents move? Wegener thought that the material beneath the Earth's lithosphere acts like a slow-moving fluid. If this is true for vertical movements, it should also be true for horizontal movements. To visualize Wegener's argument, think about a piece of candy taffy or "silly putty." At the right temperature, taffy that will shatter when struck with a hammer will deform by flowing rather than by breaking when a force is applied slowly and constantly. Although other geologists saw folded mountains as evidence of contraction, Wegener saw folded mountains as evidence of horizontal compression caused by movement of the continents. The presence of folded mountains convinced Wegener that forces within the Earth are powerful enough to move continents. A quote from Wegener summarizes his ideas about the way that all the geological evidence "fit together":

"It is just as if we were to refit the torn pieces of a newspaper by matching their edges and then check whether the lines of print run smoothly across. If they do, there is nothing left to conclude but that the pieces were in fact joined this way. If only one line was available to the test, we would still have found a high probablility for the accuracy of fit, but if we have *n* number of lines, this probability is raised to the *nth* power."

The reaction to Wegener's hypothesis was mixed. Some scientists accepted his arguments. Others argued that it would be impossible for continents to "plow through" the ocean floor (see *Figure 2*). Most geoscientists rejected Wegener's hypothesis. At an international meeting in 1926 devoted to the discussion of continental drift, only a handful of scientists were sympathetic to Wegener's ideas. One scientist raised 18 different arguments against Wegener's evidence! Although his evidence for drift was strong, the mechanism he proposed for the drift of the continents was inadequate.

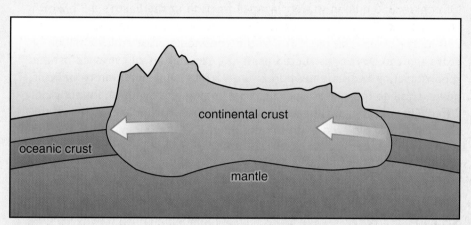

Figure 2 Wegener's proposal that continents plowed through oceanic crust was not accepted by many other geologists.

Convincing evidence began to emerge after World War II, as the sea floor was explored and mapped extensively. By the late 1960s, the theory of plate tectonics had been developed based on many types of evidence. Today, this evidence is considered so abundant and convincing that almost all geoscientists accept the theory. Much of the evidence that Wegener used to support his hypothesis supports plate tectonics. However, new evidence has emerged that provides a more plausible mechanism for the movement of the Earth's lithosphere.

Using evidence such as magnetic striping on the ocean floor (described in **Activity 1**), ages of ocean-floor basalts, outlines of continental plates, and the locations of similar fossils and rock types on widely spaced continents, geologists have reconstructed the record of the breakup of the supercontinent Pangea. Pangea started to break up about 200 million years ago, as continental rifts (divergent zones) began to open and oceanic crust began to form. As Pangea continued to be rifted apart, oceanic crust formed between the northern continents (called Laurasia) and the southern continents (called Gondwana). New ocean floor also was formed between Antarctica and Australia and between Africa and South America. India started to separate from Antarctica and travel northward.

The maps shown in *Figure 3* summarize what has been reconstructed as the breakup of Pangea, from about 250 million years ago to the present. As you can see, continents that are now connected were not always that way, and continents that are now widely separated once were part of the same land mass.

Plate Tectonics

Geo Words

supercontinent: a large continent consisting of all of the Earth's continental lithosphere. Supercontinents are assembled by plate-tectonic processes of subduction and continent–continent collision.

Of course, 250 million years is a small fraction of the Earth's 4.5 billion year history. There may be rocks in your community much older than 250 million years. The positions of the continents prior to 250 million years ago can be reconstructed using the same types of evidence used for reconstructing Pangea, shown in *Figure 4*. The task is much more difficult, however, since the oldest oceanic crust geologists have ever found is only 200 million years old. Thus, the evidence for the earlier geography of the Earth must be gathered from the continents. Old mountain belts such as the Appalachians of North America and the Urals (which separate Europe from Asia) help locate ancient collision zones between continents of the past. Rock types and fossils provide evidence for the locations of ancient seas, glaciers, mountains and ecological regions. Continents like Pangea, which consist of all of the Earth's continental lithosphere in one single piece, are called **supercontinents**. Geoscientists are fairly sure that there was at least one earlier supercontinent before Pangea, and maybe others as well. The cycle of assembly, breakup, and reassembly of supercontinents is called the Wilson Cycle.

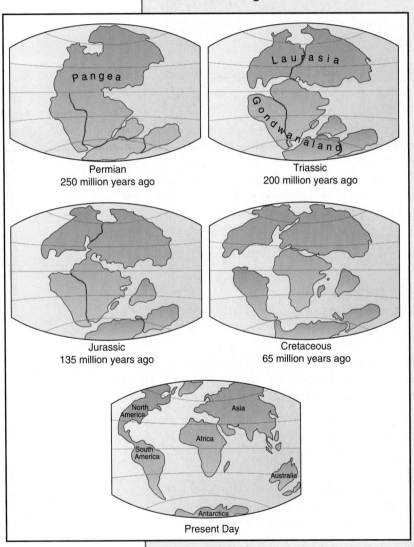

Figure 3 The breakup of Pangea.

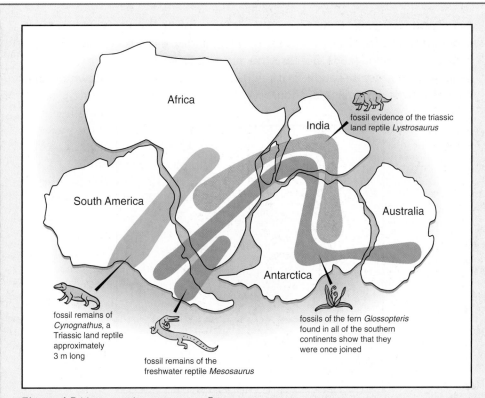

Figure 4 Evidence used to reconstruct Pangea.

Geo Words

paleomagnetism: the record of the past orientation and polarity of the Earth's magnetic field recorded in rocks containing the mineral magnetite.

Paleomagnetism

You learned in **Activity 1** that the mineral magnetite "locks" the Earth's magnetic field into its atomic structure as it cools. Geoscientsts collect rock samples containing magnetite and measure the past magnetism the rocks record (called **paleomagnetism**). They do this by putting the sample in a special room that is arranged so that the present Earth's magnetic field is canceled out. The Earth's magnetic field has the same pattern that would be observed if there were a giant bar magnet inside the Earth, lying along the Earth's axis of rotation. There isn't really a big magnet in the Earth; the magnetic field is thought to exist because of movements of liquid iron in the Earth's core. *Figure 5* shows how the lines of the Earth's magnetic field are arranged. The angle that the magnetic field lines make with the Earth's surface changes from the Equator to the poles. Near the Equator the lines are nearly horizontal, and near the poles they are nearly vertical. This means that the paleomagnetism of a rock sample can tell you the latitude of the sample when it formed, called the paleolatitude. Measurement of

1. How did Suess explain the formation of mountain ranges?

2. What evidence was found to contradict Suess's proposal that the Earth is cooling and shrinking?

3. What evidence did Wegener use to support his theory of the breakup of Pangea?

4. How did Wegener propose that the continents move horizontally?

5. How was fossil evidence used to reconstruct Pangea?

paleolatitudes is one of the things geoscientists use to reconstruct past supercontinents like Pangea. The big problem is that there is no way of measuring paleolongitude, because the magnetic field lines are always oriented about north–south! That's why no longitude lines are shown on the map of Pangea in the investigation.

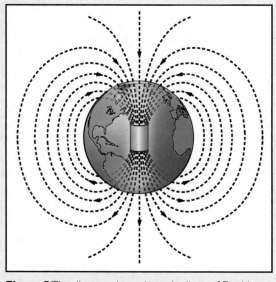

Figure 5 The diagram shows how the lines of Earth's magnetic field are arranged.

Understanding and Applying What You Have Learned

1. Geoscientists often try to figure out paleogeography (the geography of land and sea in the geologic past) using the clues given on your continent puzzle pieces. What additional evidence would you need to be more confident about your "250 Million Years Ago" map?

2. Paleoclimatology is also used to show how continents were connected in the past. What type of climate data might have been helpful to you in making your "250 Million Years Ago" map?

3. Why was the theory of continental drift questioned when it was first proposed by Alfred Wegener?

4. What discoveries helped scientists begin to accept the idea that parts of the Earth's lithosphere move? Why were the more modern clues not available in Wegener's time?

5. New scientific theories often take many years to be accepted by the scientific community. Explain why this is so, using the theory of plate tectonics as an example.

6. The theory of plate tectonics is now accepted by almost all geoscientists. The theory overcomes the objections many scientists had to the idea of continents moving around the globe. How does plate tectonics explain the seeming movement of continents through rigid oceanic crust?

7. Describe what has happened to the lithosphere under the Atlantic Ocean during the last 200 million years. What has happened to the lithosphere under the Pacific Ocean? How does this information support the theory of plate tectonics?

Preparing for the Chapter Challenge

In your notebook, write a brief essay describing how the arrangement of continents and oceans in the geologic past can be figured out, and what are the limitations to doing this.

Try to devise a way of animating the movements of all of the continents from the time of the breakup of Pangea to the present time, so that the middle-school students can watch the movements with their own eyes. Write up your ideas in your notebook. Compare your ideas with those of others in the class.

Inquiring Further

1. **History of science**

   The history of the development of the theory of plate tectonics is a fascinating one. A very important piece of evidence that supported plate tectonic theory was the discovery of paleomagnetism in ocean bottom basalts. How was this paleomagnetic evidence of sea-floor spreading discovered?

   Write down in your notebook at least one additional question you have about the geologic history of your community. How would you go about gathering information to answer these questions? Write your ideas in your notebook.

2. **Plate tectonics and the Earth system**

   Write an essay explaining how Earth systems would change if plate tectonics were to "stop." You might begin with something directly connected to plate tectonics, such as volcanism or mountain building. For example, *"If plate tectonics were to cease, then global volcanism..."*

Earth Science at Work

ATMOSPHERE: *Meteorologist*
In predicting weather patterns, meteorologists must take into account the geological features of the surrounding land.

BIOSPHERE: *Paleontologist*
Geoscientists study the fossils of plants and animals and their relationship to existing plants and animals in order to "reconstruct" the geologic past.

CRYOSPHERE: *Tour Guides*
Some of the most spectacular scenery can be found on and off the shores of Alaska. Tourists are interested in the geography of the areas they visit.

GEOSPHERE: *Civil Engineer*
Transportation networks, whether within a city, or within the country, are a vital lifeline in today's society. The public relies on safe roads and bridges.

HYDROSPHERE: *Harbor Master*
Harbor masters and marina operators are responsible for the vessels that are tied up at their docks. Millions of dollars worth of boats could be housed in a single marina.

How is each person's work related to the Earth system, and to Plate Tectonics?

EARTH SYSTEMS

The **atmosphere** is the gaseous envelope that surrounds the Earth and consists of a mixture of gases composed primarily of nitrogen, oxygen, carbon dioxide, argon, and water vapor.

The **biosphere** is the life zone of the Earth and includes all living organisms, including humans, and all organic matter that has not yet decomposed.

The **cryosphere** is the portion of the climatic system consisting of the world's ice masses and snow deposits. This includes ice sheets, ice shelves, ice caps and other glaciers, sea ice, seasonal snow cover, lake and river ice, and seasonally frozen ground and permafrost.

The **geosphere** is the solid Earth that includes the continental and oceanic crust as well as the various layers of the Earth's interior.

The **hydrosphere** includes the water of the Earth, including surface lakes, streams, oceans, underground water, and water in the atmosphere.

Glossary

abiotic: the nonliving components of an ecosystem

absolute zero: the temperature at which all vibrations of the atoms and molecules of matter cease; the lowest possible temperature

absorption spectrum: a continuous spectrum interrupted by absorption lines or a continuous spectrum having a number of discrete wavelengths missing or reduced in intensity

acceleration: the change in velocity per unit time

$$a = \frac{\Delta v}{\Delta t}$$

accretion: the process whereby dust and gas accumulated into larger bodies like stars and planets

accuracy: how close the measured value is to the standard or accepted value of that measurement

acid: a compound or solution with a concentration of hydrogen ions greater than the neutral value (corresponding to a pH value of less than 7)

acid: a substance that produces hydrogen ions in water, or is a proton donor

acid-base indicator: a dye that has a certain color in an acid solution and a different color in a base solution

adaptation: an inherited trait or set of traits that improve the chance of survival and reproduction of an organism

adaptive radiation: the diversification by natural selection, over evolutionary time, of a species or group of species into several different species that are typically adapted to different ecological niches

air resistance: a force by the air on a moving object; the force is dependent on the speed, volume, and mass of the object as well as on the properties of the air, like density

albedo: the reflective property of a non-luminous object. A perfect mirror would have an albedo of 100% while a black hole would have an albedo of 0%.

alloy: A substance that has metal characteristics and consists of two or more different elements

alternating current: an electric current that reverses in direction

ampere: the SI unit for electric current; one ampere (1 A) is the flow of one coulomb of charge every second

amplitude: the maximum displacement of a particle as a wave passes; the height of a wave crest; it is related to a wave's energy

anatomy: the (study of) internal structure of organisms

angle of incidence: the angle a ray of light makes with the normal to the surface at the point of incidence

angle of reflection: the angle a reflected ray makes with the normal to the surface at the point of reflection

anion: a negatively charged ion

anthropogenic: generated or produced by human activities

antinode: a point on a standing wave where the displacement of the medium is at its maximum

aphelion: the point in the Earth's orbit that is farthest from the Sun. Currently, the Earth reaches aphelion in early July.

aquifer: any body of sediment or rock that has sufficient size and sufficiently high porosity and permeability to provide an adequate supply of water from wells

ash: fine pyroclastic material (less than 2 mm in diameter)

asteroid: a small planetary body in orbit around the Sun, larger than a meteoroid (a particle in space, less than a few meters in diameter) but smaller than a planet. Many asteroids can be found in a belt between the orbits of Mars and Jupiter.

asthenosphere: the part of the mantle beneath the lithosphere. The asthenosphere undergoes slow flow, rather than behaving as a rigid block, like the overlying lithosphere.

astronomical unit: a unit of measurement equal to the average distance between the Sun and Earth, i.e., about 149,600,000 (1.496×10^8) km

atom: the smallest particle of an element that has all the element's properties; it consists of a nucleus surrounded by electrons

atom: the smallest representative part of an element

atomic mass unit (amu): a unit of mass defined as one-twelfth of the mass of a carbon-12 atom

atomic mass unit: a standard unit of atomic mass based on the mass of the carbon atom, which is assigned the value of 12

atomic mass: atomic mass is determined by the mass of the protons and neutrons of the atom

aurora: the bright emission of atoms and molecules near the Earth's poles caused by charged particles entering the upper atmosphere

autotroph: an organism that is capable of obtaining its energy (food) directly from the physical environment

Avogadro's number: The number of atoms contained in 12 grams of carbon-12. The number is 6.022×10^{23}.

axial precession: the wobble in the Earth's polar axis

background extinction: normal extinction of species that occurs as a result of changes in local environmental conditions

basalt: a kind of volcanic igneous rock, usually dark colored, with a high content of iron

base: a compound or solution with a concentration of hydrogen ions less than the neutral value (corresponding to a pH value of greater than 7)

base: a substance that releases hydroxide ions (OH^-) in water, or is a proton acceptor

bias: a purposeful or accidental distortion of observations, data, or calculations in a systematic or nonrandom manner

bilateral symmetry: a body plan that divides the body into symmetrical left and right halves

binary compound: a compound formed from the combining of two different elements

biodiversity: the diversity of different biologic species and/or the genetic variability among individuals within each species

biodiversity: the sum of all the different types of organisms living on Earth

biodiversity curve: a graph that shows changes in the diversity of organisms as a function of geologic time

biomass: the total mass of living matter in the form of one or more kinds of organisms present in a particular habitat

biome: a recognizable assemblage of plants and animals that characterizes a large geographic area of the Earth; a number of different biomes have been recognized, and the distribution of the biomes is controlled mainly by climate

biosphere: the area on Earth where living organisms can be found

biotic: the living components of an ecosystem

birthrate (natality): the rate at which reproduction increases the size of a population

body fossil: any remains or imprint of actual organic material from a creature or plant that has been preserved in the geologic record (like a bone)

bounce: the ability of an object to rebound to its original position when dropped from a given height

caldera: a large basin-shaped volcanic depression, more or less circular, the diameter of which is many times greater than that of the included vent or vents

carbon cycle: the global cycle of movement of carbon, in all of its forms, from one reservoir to another

carbon cycle: the process in which carbon is passed from one organism to another, then to the abiotic community, and finally back to the plants

carnivore: an animal that feeds exclusively on other animals

carrying capacity: the maximum population that can be sustained by a given supply of resources

catalyst: a substance that changes the speed of a chemical reaction without being permanently changed itself

cation: a positively charged ion

cement: new mineral material precipitated around the particles of sediment when it is buried below the Earth's surface

center of mass: the point at which all the mass of an object is considered to be concentrated for calculations concerning motion of the object

centripetal acceleration: the inward radial acceleration of an object moving at a constant speed in a circle

$$a = \frac{v^2}{R}$$

centripetal force: a force directed towards the center that causes an object to follow a circular path

$$F = \frac{mv^2}{R}$$

chemical change: a change that converts the chemical composition of a substance into different substance(s) with different chemical composition

chemical energy: energy stored in a chemical compound, which can be released during chemical reactions, including combustion

chemical formula: the combination of the symbols of the elements in a definite numerical proportion used to represent molecules, compounds, radicals, ions, etc.

chemical group: a family of elements in the periodic table that have similar electron configurations

chemical property: a characteristic that a substance undergoes in a chemical reaction that produces new substance(s)

chemical reaction: a process in which new substance(s) are formed from starting substance(s)

chemical test: a physical procedure or chemical reaction used to identify a substance

chromosome: threads of genetic material found in the nucleus of cells

chromosphere: a reddish layer in the Sun's atmosphere, the transition between the outermost layer of the Sun's atmosphere, or corona

climate: the general pattern of weather conditions for a region over a long period of time (at least 30 years)

climate proxy: any feature or set of data that has a predictable relationship to climatic factors and can therefore be used to indirectly measure those factors

climax community: the final, quite stable community reached during the stages of succession

closed system: a physical system on which no outside influences act; closed so that nothing gets in or out of the system and nothing from outside can influence the system's observable behavior or properties

closed system: a system in which material moves from place to place but is not gained or lost from the system

coal: a combustible rock that had its origin in the deposition and burial of plant material

colligative property: a property such as freezing-point depression or boiling-point elevation whose physical value depends on only the ratio of the particles of solute and solvent particles and not on their chemical identities

colloid: a mixture containing particles larger than the solute but small enough to remain suspended in the continuous phase of another component. This is also called a colloidal dispersion.

coma: a spherical cloud of material surrounding the head of a comet. This material is mostly gas that the Sun has caused to boil off the comet's icy nucleus. A cometary coma can extend up to a million miles from the nucleus.

combustion: the rapid reaction of a material with oxygen accompanied by rapid evolution of flame and heat

comet: a chunk of frozen gasses, ice, and rocky debris that orbits the Sun

community: all the populations of organisms occupying a given area

composite: a solid heterogeneous mixture of two or more substances that make use of the properties of each component

composite cone (stratovolcano): a volcano that is constructed of alternating layers of lava and pyroclastic deposits

compound: a material that consists of two or more elements united together in definite proportion

concave lens: a lens that causes parallel light rays to diverge; a lens that is thicker at its edges than in the center

concentration: a measure of the composition of a solution, often given in terms of moles of solute per liter of solution

condensation: the process of changing from a gas to a liquid

conduction: (of heat) the energy transfer from one material or particle to another when the materials or particles are in direct contact

conduction: a process of heat transfer by which the more vigorous vibrations of relatively hot matter are transferred to adjacent relatively cold matter, thus tending to even out the difference in temperature between the two regions of matter

conductivity: the property of transmitting heat and electricity

constellation: a grouping of stars in the night sky into a recognizable pattern. Most of the constellations get their name from the Latin translation of one of the ancient Greek star patterns that lies within it. In more recent times, more modern astronomers introduced a number of additional groups, and there are now 88 standard configurations recognized.

constructive interference: the result of superimposing different waves so that two or more waves overlap to produce a wave with a greater amplitude

consumer: a heterotrophic organism

continental accretion: the growth of a continent along its edges

contour interval: the vertical distance between the elevations represented by two successive contour lines on a topographic map

contour line: a line on a map that connects points of equal elevation of the land surface

convection: motion of a fluid in which the fluid moves in a pattern of closed circulation

convection: the heat transfer resulting from the movement of the heated substance, such as air or water currents

convection cell: a pattern of motion in a fluid in which the fluid moves in a pattern of a closed circulation

convergent plate boundary: a plate boundary where two plates move toward one another

converging lens: parallel beams of light passing through the lens are brought to a real point or focus (convex lens) (if the outside index of refraction is less than that of the lens material); also called a convex lens

convex lens: a lens that causes parallel light rays to converge (if the outside index of refraction is less than that of the lens material); a lens that is thinner at its edges than in the center

core: the solid, innermost part of the Earth, consisting mainly of iron

corona: the outermost atmosphere of a star (including the Sun), millions of kilometers in extent, and consisting of highly rarefied gas heated to temperatures of millions of degrees

correlation: a mutual relationship or connection

cosmologist: a scientist who studies the origin and dynamics of the universe

coulomb: the SI unit for electric charge; one coulomb (1 C) is equal to the charge of 6.25×10^{18} electrons

Coulomb's Law: the relationship among electrical force, charges, and the distance between the charges

$$F = k\frac{q_1 q_2}{d^2}$$

covalent bond: a bond formed when two atoms combine and share their paired electrons with each other

crest: the highest point of displacement of a wave

critical angle: the angle of incidence for which a light ray passing from one medium to another has an angle of refraction of 90° degrees

crude oil: (see **petroleum**)

crust: the thin outermost layer of the Earth. Continental crust is relatively thick and mostly very old. Oceanic crust is relatively thin, and is always geologically very young.

death rate (mortality rate): the rate at which death decreases the size of a population

decomposers: organisms that break down the remains or wastes of other organisms to obtain their nutrients

decomposition: a chemical reaction in which a single compound reacts to give two or more products

decomposition: the chemical process of separation of matter into simpler chemical compounds

denitrification: the conversion of nitrates and nitrites to nitrogen gas, which is released into the atmosphere

density: the mass per unit volume of a material or substance

density: the mass per unit volume of a material

destructive interference: the result of superimposing different waves so that two or more waves overlap to produce a wave with a decreased amplitude

diffraction: the ability of a wave to spread out as it emerges from an opening or moves beyond an obstruction

digital: a description of data that is stored or transmitted as a sequence of discrete symbols; usually this means binary data (1s and 0s) represented using electronic or electromagnetic signals

displacement: the difference in position between a final position and an initial position; it depends only on the endpoints, not on the path; displacement is a vector; it has magnitude and direction

divergent plate boundary: a plate boundary where two plates move away from one another.

dominant: used to describe the gene that determines the expression of a genetic trait; the trait shows up even when the gene is present as a single copy

Doppler Effect: change in frequency of a wave of light or sound due to the motion of the source or the receiver

double-displacement reaction: a chemical reaction in which two ionic compounds "exchange" cations to produce two new compounds

ductility: a property that describes how easy it is to pull a substance into a new permanent shape, such as, pulling into wires

earthquake: a sudden motion or shaking in the Earth caused by the abrupt release of slowly accumulated strain.

eccentricity: the ratio of the distance between the foci and the length of the major axis of an ellipse

ecosystem: a community and the physical environment that it occupies

ecosystem: a unit in ecology consisting of the environment with its living elements, plus the nonliving factors that exist in it and affect it

efficiency: the ratio of the useful energy obtained from a machine or device to the energy supplied to it during the same time period

elasticity: the property of a material to resist deformation and return to its normal size or shape after a force has been applied to it

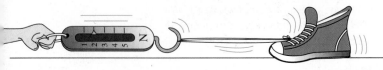

elastic rebound: the return of a bent elastic solid to its original shape after the deforming force is removed

electric charge: a fundamental property of matter; charge is either positive or negative

electric circuit: an electrical device that provides a conductive path for electrical current to move continuously

electric current: the flow of electric charges through a conductor; electric current is measured in amperes

electric energy: energy associated with the generation and transmission of electricity

electric field: the region of electric influence defined as the force per unit charge

electric power: power associated with the generation and transmission of electricity

electrical resistance: opposition of a material to the flow of electrical charge through it: it is measured in ohms (Ω); the ratio of the potential difference to the current

$$R = \frac{V}{I}$$

electrolysis: the conduction of electricity through a solution that contains ions or through a molten ionic compound that will induce chemical change

electromagnet: a device that uses an electric current to produce a concentrated magnetic field

electromagnetic radiation: the energy propagated through space by oscillating electric and magnetic fields. It travels at 3×10^8 m/s in a vacuum and includes (in order of increasing energy) radio, infrared, visible light (optical), ultraviolet, x-rays, and gamma rays.

electromagnetic radiation: the movement of energy, at the speed of light, in the form of electromagnetic waves

electromagnetic spectrum: the complete spectrum of electromagnetic radiation, such as radio waves, microwaves, infrared, visible, ultraviolet, x-rays, and gamma rays

electromagnetic waves: transverse waves that are composed of perpendicular electric and magnetic fields that travel at 3×10^8 m/s in a vacuum; examples of electromagnetic waves in increasing wavelength are gamma rays, x-rays, ultraviolet radiation, visible light, infrared radiation, microwaves, and radio waves

electron: a negatively charged particle with a charge of 1.6×10^{-19} coulombs and a mass of 9.1×10^{-31} kg

electron: a subatomic particle that occurs outside of the nucleus and has a charge of -1 and mass of 9.109×10^{-28}g

electron configuration: the distribution of electrons in an atom's energy levels

element: a substance in which all of the atoms have the same atomic number

elevation: the height of the land surface relative to sea level

emigration: the number of individuals of a species that move out of an existing population

emission spectrum: a spectrum containing bright lines or a set of discrete wavelengths produced by an element. Each element has its own unique emission spectrum.

emulsion: a colloid or colloidal dispersion of one liquid suspended in another

endothermic change or reaction: a change in which energy in the form of heat is absorbed from the surrounding environment resulting in an increase in the internal energy of the system

evaporation: the process of changing from a liquid to a gas

evolution: a gradual change in the characteristics of a population of organisms during successive generations, as a result of natural selection acting on genetic variation

exothermic change or reaction: a change in which energy in the form of heat is released from a system resulting in a decrease in the internal energy of the system

extinction: the permanent disappearance of a species from Earth

extrusive igneous rock: an igneous rock that has formed by eruption of lava onto the surface of the Earth

fault: a fracture or fracture zone in rock, along which the rock masses have moved relative to one another parallel to the fracture

feedback loops: the processes where the output of a system causes positive or negative changes to some measured component of the system

feedstocks: raw materials (for example, petroleum) that are supplied to a machine or processing plant that produces manufactured material (for example, plastics)

fertilizer: a material used to provide or replace soil nutrients

First Law of Thermodynamics: the law that energy can be converted from one form to another but be neither created nor destroyed

fission: the process of breaking apart nuclei into smaller nuclei and with the release of a large amount of energy

flame test: an experimental technique or process in identifying a metal from its characteristic flame color

focal length: the distance between the center of a lens and either focal point

focus: the place at which light rays converge or from which they appear to diverge after refraction or reflection; also called focal point

focus: the point of an earthquake within the Earth where rupture first occurs to cause an earthquake

food chain: a series of organisms through which food energy is passed in an ecosystem

food web: a complex relationship formed by interconnecting food chains in an ecosystem representing the transfer of energy through different levels

foraminifera: an order of single-celled organisms (protozoans) that live in marine (usually) and freshwater (rarely) environments. Forams typically have a shell of one or more chambers that is typically made of calcium carbonate.

force: a push or a pull that is able to accelerate an object; force is measured in newtons; force is a vector quantity

force: a push or pull exerted on a body of matter

fossil: any remains, trace, or imprint of a plant or animal that has been preserved in the Earth's crust since some past geologic or prehistoric time

fossil fuel: fuel derived from materials (mainly coal, petroleum, and natural gas) that were generated from fossil organic matter and stored deep in the Earth for geologically long times

fossiliferous rock: a rock containing fossils

frame of reference: a vantage point with respect to which position and motion may be described

free fall: a fall under the influence of only gravity

frequency: the number of waves per second or cycles per second or hertz (Hz)

frequency: the number of waves produced per unit time; the frequency is the reciprocal of the amount of time it takes for a single wavelength to pass a point

$$f = \frac{v}{\lambda}$$

friction: a force that acts to resist the relative motion or attempted motion of objects that are in contact with each other

friction: the force exerted by a body of matter when it slides past another body of matter

fusion: nuclei of lighter atoms combining to form nuclei with greater mass and release of a large amount of energy

galvanometer: an instrument used to detect, measure, and determine the direction of small electric currents

gas giant planets: the outer solar system planets: Jupiter, Saturn, Uranus, and Neptune, composed mostly of hydrogen, helium and methane, and having a density of less than 2 gm/cm^2

gene: a unit of instruction located on a chromosome that produces or influences a specific trait in the offspring

genotype: the genes of an individual

geothermal energy: energy derived from hot rocks and/or fluids beneath the Earth's surface

glacial period: an interval in time that is marked by one or more major advances of glacier ice. Note that the time interval is not necessarily of the same magnitude as the "Period" rank of the geologic time scale.

glacier: a large, long-lasting accumulation of snow and ice that develops on land and flows under its own weight

global climate: the mean climatic conditions over the surface of the Earth as determined by the averaging of a large number of observations spatially distributed throughout the entire region of the globe

Global Positioning System (GPS): a satellite-based system for accurate location of points on the Earth

gravimeter: an instrument for measuring variations in Earth's gravitational field

gravitational potential energy: the energy a body possesses as a result of its position in a gravitational field   GPE = mgh

gravity: the force of attraction between two bodies due to their masses

greenhouse gases: gases in the Earth's atmosphere that absorb certain wavelengths of the long-wavelength radiation emitted to outer space by the Earth's surface

greenhouse gases: gases responsible for the greenhouse effect. These gases include: water vapor (H_2O), carbon dioxide (CO_2); methane (CH_4); nitrous oxide (N_2O); chlorofluorocarbons (CF_xCl_x); and tropospheric ozone (O_3).

ground water: the part of the subsurface water that is in the zone of saturation, including underground streams

groundwater: water contained in pore spaces in sediments and rocks beneath the Earth's surface

growth rate: the rate at which the size of a population increases as a result of death rate, birthrate, immigration, and emigration

halogens: Group VIIA (17) on the periodic table consisting of fluorine, chlorine, bromine, iodine, and astatine

hazard: a natural event, like an earthquake, that has the potential to do damage or harm

heat: kinetic energy of atoms or molecules associated with the temperature of a body of material

heat capacity: the quantity of heat energy required to increase the temperature of a material or system; typically referenced as the amount of heat energy required to generate a 1°C rise in the temperature of 1 g of a given material that is at atmospheric pressure and 20°C

heat energy: a form of energy associated with the motion of atoms or molecules

heat transfer: the movement of heat from one region to another

herbivore: a heterotroph that feeds exclusively on plant materials

heredity: the passing of traits from parent to offspring

heterotroph: an organism that must obtain its energy from autotrophs or other heterotrophs

horsepower: a unit of power

hot spot: a fixed source of abundant rising magma that forms a volcanic center that has persisted for tens of millions of years

hydrocarbon: a molecular compound containing only hydrogen and carbon

hydroelectric power: electrical power derived from the flow of water on the Earth's surface

hypothesis: a statement that can be proved or disproved by experimental or observational evidence

igneous rock: rock or mineral that solidified from molten or partly molten material, i.e., from magma

immigration: the number of individuals of a species that move into an existing population

impulse: the product of force and the interval of time during which the force acts; impulse results in a change in momentum

$$Ft = \Delta(mv)$$

inclination: the angle between the orbital plane of the solar system and the actual orbit of an object around the Sun

index fossil: a fossil of an organism that was widespread but lived for only a short interval of geological time

index of refraction: a property of a medium that is related to the speed of light through it; it is calculated by dividing the speed of light in vacuum by the speed of light in the medium

inertia: the natural tendency of an object to remain at rest or to remain moving with constant speed in a straight line

inertial frame of reference: unaccelerated point of view in which Newton's Laws hold true

infiltration: the movement of water through pores or small openings into the soil and porous rock

infrared: electromagnetic radiation with wavelengths between about 0.7 to 1000 mm. Infrared waves are not visible to the human eye.

inorganic compound: a compound not based on molecular compounds of carbon

insolation: the direct or diffused shortwave solar radiation that is received in the Earth's atmosphere or at its surface

insolation: the rate at which a given area of land receives solar energy

interglacial period: the period of time during an ice age when glaciers retreated because of milder temperatures

intrusive igneous rock (plutonic igneous rock): igneous rock formed at considerable depth by the crystallization of magma

invasive species: a nonnative species whose introduction does or is likely to cause economic or environmental harm or harm to human health

inverse square relation: the relationship of a force to the inverse square of the distance from the mass (for gravitational forces) or the charge (for electrostatic forces)

inverse-square law: a scientific law that states that the amount of radiation passing through a specific area is inversely proportional to the square of the distance of that area from the energy source

ion: an atom with one or more electrons removed (or added), giving it a positive (or negative) charge

ion: an electrically charged atom or group of atoms that has acquired a net charge, either negative or positive

ionic bond: the attraction between oppositely charged ions.

ionic compound: a compound consisting of positive or negative ions

ionization energy: the energy required to free an electron from an atom

ionization energy: the energy required to remove an electron from a gaseous atom at ground state

ionosphere: the part of the Earth's atmosphere above about 50 km where the atoms are significantly ionized and affect the propagation of radio waves

ions: atoms that have an electric charge because one or more electrons (particles with a negative electric charge, which orbit around the nucleus of the atom) have been added to the atom or removed from the atom

isoseismal map: a map showing the lines connecting points on the Earth's surface at which earthquake intensity is the same

isotope: atoms of the same element but different atomic masses due to different number of neutrons

isotope: one of two or more kinds of atoms of a given chemical element that differ in mass because of different numbers of neutrons in the nucleus of the atoms

joule: the SI unit for work and all other forms of energy; one joule (1J) of work is done when a force of one newton moves an object one meter in the direction of the force

kinetic energy: the energy an object possesses because of its motion

$$KE = \frac{1}{2}mv^2$$

lahar: a wet mixture of water, mud, and volcanic rock fragments, with the consistency of wet concrete, that flows down the slopes of a volcano and its river valleys

lake-effect snow: the snow that is precipitated when an air mass which has gained moisture by moving over a relatively warm water body is cooled as it passes over relatively cold land.
This cooling triggers condensation of clouds and precipitation.

lapilli: pyroclastics in the general size range of 2 to 64 mm

latitude: a north-south measurement of position on the Earth. It is defined by the angle measured from the Earth's equatorial plane.

lava: molten rock that issues from a volcano or fissure

lava flow: an outpouring of molten lava from a vent or fissure; also, the solidified body of rock so formed

Law of Conservation of Matter: in chemical reactions, the quantity of matter does not change

Law of Definite Proportions: the composition of a pure substance is always the same or the elements of the compound always combine in the same proportion by mass

leeward: the downwind side of an elevated area like a mountain, opposite of windward

lepton: a group of elementary particles that are not affected by the nuclear force; electrons belong to this group

light-year: a unit of measurement equal to the distance light travels in one year, i.e., 9.46×10^{12} km

lithosphere: the outermost layer of the Earth, consisting of the Earth's crust and part of the upper mantle. The lithosphere behaves as a rigid layer, in contrast to the underlying asthenosphere.

lithospheric plate: a rigid, thin segment of the outermost layer of the Earth, consisting of the Earth's crust and part of the upper mantle. The plate can be assumed to move horizontally and adjoins other plates.

Little Ice Age: the time period from A.D. 1350 to 1850. During this period, global temperatures were at their coldest since the beginning of the Holocene.

loess: the deposits of wind-blown silt laid down over vast areas of the mid-latitudes during glacial and postglacial times

longitudinal pulse or wave: a pulse or wave in which the motion of the medium is parallel to the direction of the motion of the wave

luminosity: the total amount of energy radiated by an object every second

luster: the reflection of light from the surface of a material described by its quality and intensity

magma: naturally occurring molten rock material generated within the Earth. Magma also contains dissolved gases, and sometimes solid crystals.

magma: naturally occurring molten rock material, generated within the Earth, from which igneous rocks are derived through solidification and related processes

magnetic field: the region of magnetic influence around a magnetic pole or a moving charged particle

magnetometer: an instrument for measuring variations in Earth's magnetic field

malleability: the property of a material to be able to be hammered into various shapes without breaking

mantle: the zone of the Earth below the crust and above the core. It is divided into the upper mantle and lower mantle with a transition zone between

map projections: the process of systematically transforming positions on the Earth's spherical surface to a flat map while maintaining spatial relationships

mass extinction: a catastrophic, widespread perturbation where major groups of species become extinct in a relatively short time compared to normal background extinction

mechanical energy: the sum of the kinetic energy and the potential energy of a body of matter

Mercator projection: a map projection in which the Equator is represented by a straight line true to scale, the meridians by parallel straight lines perpendicular to the Equator and equally spaced according to their distance apart at the Equator, and the parallels by straight lines perpendicular to the meridians and the same length as the Equator. There is a great distortion of distances, areas, and shapes at the polar regions.

metal: classes of materials that exhibit the properties of conductivity, malleability, reactivity, and ductility. Metal elements readily lose electrons to form positive ions.

meteor: the luminous phenomenon seen when a meteoroid enters the atmosphere (commonly known as a shooting star)

meteorite: a part of a meteoroid that survives through the Earth's atmosphere

meteoroid: a small rock in space

mid-ocean ridge: a chain of undersea ridges extending throughout all of the Earth's ocean basins, and formed by sea-floor spreading

Milankovitch cycles: the cyclical changes in the geometric relationship between the Earth and the Sun that cause variations in solar radiation received at the Earth's surface

mineral: a naturally occurring inorganic, solid material that consists of atoms and/or molecules that are arranged in a regular pattern and has characteristic chemical composition, crystal form, and physical properties

model: a representation of a process, system, or object

mole: a collection of objects that contains Avogadro's number (6.022×10^{23})

molecular cloud: a large, cold cloud made up mostly of molecular hydrogen and helium, but with some other gases, too, like carbon monoxide. It is in these clouds that new stars are born.

molecular compound: two or more atoms bond together by sharing electrons

momentum: the product of the mass and the velocity of an object; momentum is a vector quantity

$$p = mv$$

morphology: the (study of the) features that comprise or describe the shape, form, and structure of an object or organism

motile: having the ability to move spontaneously

native-element mineral: a mineral consisting of only one element

natural gas: a gas, consisting mainly of methane, that is produced in sediments and sedimentary rocks during burial of organic matter

natural selection: the differences in survival and reproduction among members of a population

neap tide: the tides of decreased range occurring semimonthly near the times of the first and last quarter of the Moon

nebula: a general term used for any "fuzzy" patch on the sky, either light or dark; a cloud of interstellar gas and dust

neutral: having a concentration of hydrogen ions that corresponds to a value of pH of 7

neutralization: the process of an acid and base reacting to form water and salt

neutron: a subatomic particle that is part of the structure of the atomic nucleus; a neutron is electrically neutral with a mass of

$$1.675 \times 10^{-24}\,\text{g}$$

located in the nuclei of the atom

neutron star: the imploded core of a massive star produced by a supernova explosion

Newton's Laws of Motion:

Newton's First Law of Motion: an object at rest stays at rest and an object in motion stays in motion unless acted upon by an unbalanced, external force

Newton's Law of Gravitational Attraction: the relationship among gravitational force, masses, and the distance between the masses

$$F = \frac{Gm_1 m_2}{d^2}$$

Newton's Second Law of Motion: if a body is acted upon by an external force, it will accelerate in the direction of the unbalanced force with an acceleration proportional to the force and inversely proportional to the mass

$$F = ma$$

Newton's Third Law of Motion: forces come in pairs; the force of object A on object B is equal and opposite to the force of object B on object A

niche: the ecological role of a species; the set of resources it consumes and habitats it occupies in an ecosystem

nitrogen cycle: the movement of nitrogen through ecosystems, the soil, and the atmosphere

nitrogen fixation: the process by which certain organisms produce nitrogen compounds from the gaseous nitrogen in the atmosphere

noble gas: a family of elements (Group 18 or VIIIA) of the periodic table

node: a point on a standing wave where the medium is motionless

nonconsumptive water: water that is returned, in liquid form, to the natural environment after use

nonmetal: elements that do not exhibit the properties of conductivity, malleability, reactivity, and ductility. These elements tend to form negative ions. The oxides of the elements are acidic.

nonnative (exotic, alien, introduced, or non-indigenous) species: any species, including its seeds, eggs, spores, or other biological material capable of propagating that species, that is not native to that ecosystem

nonrenewable resource: an energy source that is powered by materials that exist in the Earth and are not replaced nearly as fast as they are consumed

normal: at right angles or perpendicular to

normal boiling point: the temperature at which the vapor pressure of the pure liquid equals 1 atm

normal freezing point: the characteristic temperature, at 1 atm, at which a material changes from a liquid state to its solid state

normal melting point: the characteristic temperature, at 1 atm, at which a material changes from a solid state to its liquid state

nuclear fission: a nuclear reaction in which a massive, unstable nucleus splits into two or more smaller nuclei with a release of a large amount of energy

nuclear fission: the process by which large atoms are split into two parts, with conversion of a small part of the matter into energy

nuclear (strong) force: a strong force that holds neutrons and protons together in the nucleus of an atom; the force operates only over very short distances

nuclear fusion: a nuclear process that releases energy when lightweight nuclei combine to form heavier nuclei

nuclear fusion: a nuclear reaction in which nuclei combine to form more massive nuclei with the release of a large amount of energy

nucleon: the building block of the nucleus of an atom; either a neutron or a proton

nucleus: (of an atom): the positively charged dense center of an atom containing neutrons and protons

nucleus: the very dense core of the atom that contains the neutrons and protons

obliquity: the tilt of the Earth's rotation axis as measured from the perpendicular to the plane of the Earth's orbit around the Sun. The angle of this tilt varies from 22.5° to 24.5° over a 41,000-year period. Current obliquity is 23.5°.

ohm: the SI unit of electrical resistance; the symbol for ohm is Ω

omnivores: a heterotroph that feeds on both plant materials and animals

open population: a natural population in which all four factors that affect population size (death rate, birthrate, immigration, and emigration) are functioning

open system: a physical system on which outside influences are able to act; open so that energy can be added and/or lost from the system

orbit: the path of the electron in its motion around the nucleus of Bohr's hydrogen atom

orbital: in the quantum mechanical model of an atom, it is the region surrounding the atomic nucleus in which the electron distribution is given by a wave function

orbital parameters: any one of a number of factors that describe the orientation and/or movement of an orbiting body or the shape of its orbital path

orbital plane: (also called the ecliptic or plane of the ecliptic). A plane formed by the path of the Earth around the Sun.

orbital precession: rotation about the Sun of the major axis of the Earth's elliptical orbit

organic (carbon-based) molecules: molecules with the chemical element carbon as a base

organic compound: a molecular compound of carbon

organism: an individual living thing

oxidation: the chemical process by which certain kinds of matter are combined with oxygen

oxidation: the process of a substance losing one or more electrons

oxidation number: a number assigned to an element in a compound designating the number of electrons the element has lost, gained, or shared in forming that compound

paleoclimate: the climatic conditions in the geological past reconstructed from a direct or indirect data source

paleoclimatologist: a scientist who studies the Earth's past climate

paleoclimatology: the scientific study of the Earth's climate during the past

paleomagnetism: the record of the past orientation and polarity of the Earth's magnetic field recorded in rocks containing the mineral magnetite

paleontologist: a scientist who studies the fossilized remains of animals and/or plants

Pangea: Earth's most recent supercontinent which was rifted apart about 200 million years ago

parallax: the apparent difference of position of an object as seen from two different places, or points of view

parsec: a unit used in astronomy to describe large distances. One parsec equals 3.26 light-years.

peak wavelength: the wavelength of light with the most electromagnetic energy emitted by any object

peat: a porous deposit of partly decomposed plant material at or not far below the Earth's surface

perihelion: the point in the Earth's orbit that is closest to the Sun. Currently, the Earth reaches perihelion in early January.

period: a horizontal row of elements in the periodic table

period: the time required to complete one cycle of a wave

periodic wave: a repetitive series of pulses; a wave train in which the particles of the medium undergo periodic motion (after a set amount of time the medium returns to its starting point and begins to repeat its motion)

permeability: the ease with which a fluid can be forced to flow through a porous material by a difference in fluid pressure from place to place in the material

petroleum: an oily, flammable liquid, consisting of a variety of organic compounds, that is produced in sediments and sedimentary rocks during burial of organic matter; also called crude oil

pH: a quantity used to represent the acidity of a solution based on the concentration of hydrogen ions (pH $= -\log[H+]$)

phase change: the conversion of a substance from one state to another state at a specific temperature and pressure. Example: solid to liquid, liquid to gas, or solid to gas (sublimation).

phenotype: the observable traits of an organism that result because of the interaction of genes and the environment

phosphorous cycle: the cycling of environmental phosphorous through a long-term cycle involving rocks on the Earth's crust, and through a shorter cycle involving living organisms

photosphere: the visible surface of the Sun, lying just above the uppermost layer of the Sun's interior, and just below the chromosphere

photosynthesis: the process by which plants use solar energy, together with carbon dioxide and nutrients, to synthesize plant tissues

photovoltaic energy: energy associated with the direct conversion of solar radiation to electricity

physical change: a change that involves changes in the state or form of a substance but does not cause any change in chemical composition

physical property: a property that can be measured without causing a change in the substance's chemical composition

phytoplankton: small photosynthetic organisms, mostly algae and bacteria, found inhabiting aquatic ecosystems

pioneer community: the first species to appear during succession

pitch: the quality of a sound dependent primarily on the frequency of the sound waves produced by its source

Planck's constant: a proportionality constant of the energy of a photon to its frequency, derived by Max Planck in 1900. His equation was: $E = hf$ and Planck's constant is $(h) = 6.626 \times 10^{-34}$ J $\cdot$ s

planetesimal: one of the small bodies (usually micrometers to kilometers in diameter) that formed from the solar nebula and eventually grew into protoplanets

plasma: a state of matter wherein all atoms are ionized; a mixture of free electrons and free atomic nuclei

plate tectonics: the field of study of plate motion

plate tectonics: the study of the movement and interaction of the Earth's lithospheric plates

polarized waves: disturbances where the medium vibrates in only one plane

pollen: a collective term for pollen grains, which are microspores containing the several-celled microgametophyte (male gametophyte) of seed plants

polyatomic ion: an ion that consists of 2 or more atoms that are covalently bonded and have either a positive or negative charge

polymer: a substance that is a macromolecule consisting of many similar small molecules (monomers) linked together in long chains

polymerization: a chemical reaction that converts small molecules (monomers) into large molecules (polymers)

population: a group of organisms of the same species occupying a given area

porosity: the ratio of pore space to total volume of a rock or sediment, multiplied by 100 to be expressed as a percentage

potential energy: energy that is dependent on the position of the object

potential energy: mechanical energy associated with position in a gravity field; matter farther away from the center of the Earth has higher potential energy

potential energy: stored energy of the material as a result of its position in an electric, magnetic, or gravitational field

power: that time rate at which work is done on a body or at which energy is produced or consumed

power: the time rate at which work is done and energy is transformed

$$P = \frac{W}{t}$$

precession: slow motion of the axis of the Earth around a cone, one cycle in about 26,000 years, due to gravitational tugs by the Sun, Moon, and major planets

precipitate: an insoluble solid formed in a liquid solution as a result of some chemical reactions

precipitation: water that falls to the Earth's surface from the atmosphere as liquid or solid material in the form of rain, snow, hail, or sleet

precision: the closeness of agreement of several measurements of the same quantity

primary succession: the occupation by plant life of an area previously not covered with vegetation

primary wave (P wave): a seismic wave that involves particle motion (compression and expansion) in the direction in which the wave is traveling. It is the fastest of the seismic waves.

probability: a measure of the likelihood of a given event occurring

producer: an organism that is capable of making its own food

product: the substance(s) produced in a chemical reaction

projectile: an object traveling through the air

proton: a positively charged subatomic particle contained in the nucleus of an atom. The mass of a proton is 1.673×10^{-24}g and it has a charge of +1

proton: a subatomic particle that is part of the structure of the atomic nucleus; a proton is positively charged

protoplanetary body: a clump of material, formed in the early stages of solar system formation, which was the forerunner of the planets we see today

pure material: an element or compound that has a defined composition and properties

pure substance: a substance that contains only one kind of particle

pyramid of energy: a pyramid developed on the basis of the energy at each trophic level

pyramid of living matter: a pyramid developed on the basis of the mass of dry living matter at each trophic level

pyroclastic flow: a high-density mixture of hot ash and rock fragments with hot gases formed by a volcanic explosion or aerial expulsion from a volcanic vent

radial symmetry: a body plan that is symmetrical about a center axis

radiation: (heat transfer): electromagnetic radiation strikes a material that can absorb it, causing the particles in the material to have more energy often resulting in a higher temperature

radioactive: a term applied to an atom that has an unstable nucleus and can spontaneously emit a particle and become the nucleus of another atom

radioactive: an atom that has an unstable nuclei and will emit alpha, positron, or beta particles in order to achieve more stable nuclei

radio telescope: an instrument used to observe longer wavelengths of radiation (radio waves), with large dishes to collect and concentrate the radiation onto antennae

rain shadow: the reduction of precipitation commonly found on the leeward side of a mountain

ray: the path followed by a very thin beam of light

reactants: the starting materials in a chemical reaction

reactivity: a property that describes how readily a material will react with other materials

real image: an image that will project on a screen or on the film of a camera; the rays of light actually pass through the image

recessive: used to describe the gene that is overruled by a dominant gene; the trait is masked

reduction: a process in which the substance under consideration gains electron(s)

reforestation: the replanting of trees on land where existing forest was previously cut for other uses, such as agriculture or pasture

refraction: the change in direction (bending) of a light beam as it passes obliquely from one medium to a different one

relativity: the study of the way in which observations from moving frames of reference affect your perceptions of the world

relief: the physical configuration of a part of the Earth's surface, with reference to variations of height and slope or to irregularities of the land surface

renewable energy source: an energy source that is powered by solar radiation at the present time rather than by fuels stored in the Earth

reservoir: 1. a place in the Earth system that holds water; 2. a large body of porous and permeable sedimentary rock that contains economically valuable petroleum and/or natural gas

resonance: a condition in which a vibration affecting an object has about the same period as the natural vibration period of the object

respiration: physical and chemical processes by which an organism supplies its cells and tissues with oxygen needed for metabolism

rift valley: a large, long valley on a continent, formed where the continent is pulled apart by forces produced when mantle material rises up beneath the continent

rift valley: the deep central cleft in the crest of the mid-oceanic ridge

risk: the potential impact of a natural hazard on people or property

rocks: naturally occurring aggregates of mineral grains

runoff: the part of the precipitation appearing in surface streams

saturated solution: the maximum amount of solute that can be dissolved at a given temperature and pressure

scalar: a quantity that has magnitude, but no direction

scrubbing: removal of sulfur dioxide, ash, and other harmful byproducts of the burning of fossil fuels as the combustion products pass upward through a stack or flue

seal: an impermeable layer or mass of sedimentary rock that forms the convex-upward top or roof of a petroleum reservoir

seamount: a peaked or flat-topped underwater mountain rising from the ocean floor

Second Law of Thermodynamics: the law that heat cannot be completely converted into a more useful form of energy

secondary recovery: the use of techniques to recover oil still trapped among sediment particles after years of production

secondary succession: the occupation by plant life of an area that was previously covered with vegetation and still has soil

secondary wave (S wave): a seismic wave produced by a shearing motion that involves vibration perpendicular to the direction in which the wave is traveling. It does not travel through liquids, like the outer core of the Earth. It arrives later than the P wave.

sedimentary basin: an area of the Earth's crust where sediments accumulate to great thicknesses

sedimentary rock: a rock resulting from the consolidation of accumulated sediments

sedimentary rock: a rock, usually layered, that results from the consolidation or lithification of sediment

sediments: loose particulate materials that are derived from breakdown of rocks or precipitation of solids in water

sediments: solid fragmental material that originates from weathering of rocks and is transported or deposited by air, water, or ice, or that accumulates by other natural agents, such as chemical precipitation from solution or secretion by organisms

seismic (earthquake) waves: a general term for all elastic waves in the Earth, produced by earthquakes or generated artificially by explosions

seismic wave: a general term for all elastic waves in the Earth, produced by earthquakes or generated artificially by explosions

seismogram: the record made by a seismometer

seismology: the study of earthquakes and of the structure of the Earth

seismometer: an instrument that measures seismic waves. It receives seismic impulses and converts them into a signal like an electronic voltage.

seral stages: the communities in between the pioneer and climax community during the stages of succession

sessile (non-motile): an organism that is permanently attached rather than free-moving

shear strength: the shear force needed to break a solid material

shield volcano: a broad, gently sloping volcanic cone of flat-dome shape, usually several tens or hundreds of square miles in extent

silica: material with composition SiO_2

single-displacement reaction: a reaction in which an element displaces or replaces another element in a compound

Snell's Law: describes the relationship between the index of refraction and the ratio of the sine of the angle of incidence and the sine of the angle of refraction

$$n = \frac{\sin \angle i}{\sin \angle R}$$

solar wind: a flow of hot charged particles leaving the Sun

solenoid: a coil of wire

solute: the substance that dissolves in a solvent to form a solution

solution: a homogeneous mixture of two or more substances

solvent: the substance in which a solute dissolves to form a solution

source rocks: sedimentary rocks, containing significant concentrations of organic matter, in which petroleum and natural gas are generated during burial of the deposits

Special Theory of Relativity: the theory of space and time

species: a group of organisms that can interbreed under natural conditions and produce fertile offspring

species: a group of organisms, either plant or animal, that can interbreed and produce fertile offspring

specific gravity: the ratio of the weight of a given volume of a substance to the weight of an equal amount of water

specific heat: the amount of energy required to raise the temperature of 1 kg of a material by 1°C

$$E = mc\Delta t$$

spectroscope: an instrument consisting of, at a minimum, a slit and grating (or prism) which produces a spectrum for visual observation

spectroscopy: the science that studies the way light interacts with matter

speed: the change in distance per unit time; speed is a scalar, it has no direction

spore: a typically unicellular reproductive structure capable of developing independently into an adult organism either directly if asexual or after union with another spore if sexual

spring potential energy: the internal energy of a spring due to its compression or stretch

spring tide: the tides of increased range occurring semimonthly near the times of full Moon and new Moon

stellar black hole: the leftover core of a massive single star after a supernova. Black holes exert such large gravitational pull that not even light can escape.

steppe: an extensive, treeless grassland found in semiarid mid-latitude regions. Steppes are typically considered to be drier than the prairie.

strength: the property of how well a material withstands the application of a force

subduction zone: a long, narrow belt in which one plate descends beneath another

subduction: the movement of one plate downward into the mantle beneath the edge of the other plate at a convergent plate boundary. The downgoing plate always is oceanic lithosphere. The plate that stays at the surface can have either oceanic lithosphere or continental lithosphere.

sublimation: the change of state of a solid material to a gas without going through the liquid state

succession: the slow and orderly replacement of community replacement, one following the other

supercontinent: a large continent consisting of all of the Earth's continental lithosphere. Supercontinents are assembled by plate-tectonic processes of subduction and continent–continent collision.

supernova: the death explosion of a massive star whose core has completely burned out. Supernova explosions can temporarily outshine a galaxy.

supersaturated solution: a solution containing more solute than a saturated solution and therefore not at equilibrium. This solution is not stable and cannot be maintained indefinitely.

surface area: changing the nature of the reactants into smaller particles increases the surface exposed to react. Successful reaction depends on collision and increasing the area of the reactant increases the chance of a successful collision taking place. Lighting a log is more difficult than lighting wood shavings. The shavings have a greater surface area and speed up the reaction.

surface wave: a seismic wave that travels along the surface of the Earth

suspension: heterogeneous mixture that contains fine solid or liquid particles in a fluid that will settle out spontaneously. By shaking the container they will again be dispersed throughout the fluid.

suture zone: the zone on the Earth's surface where two continents have collided and have been welded together to form a single continent

synthesis reaction: a chemical reaction in which two or more substances combine to form a compound

taxonomy: the theory and practice of classifying plants and animals

temperature: a measure of the average kinetic energy of the molecules of a material

temperature: a measure of the energy of vibrations of the atoms or molecules of a body of matter

tephra: a collective term for all the particles ejected from a volcano and transported through the air. It includes volcanic dust, ash, cinders, lapilli, scoria, pumice, bombs, and blocks.

terrestrial planets: any of the planets Mercury, Venus, Earth, or Mars, or a planet similar in size, composition, and density to the Earth. A planet that consists mainly of rocky material.

texture: the characteristics of the surface of a material, like how smooth, rough, or coarse it is

theory: a proven and generally accepted truth

thermal convection: a pattern of movement in a fluid caused by heating from below and cooling from above. Thermal convection transfers heat energy from the bottom of the convection cell to the top.

thermal insulator: a material that impedes or slows heat transfer

thermodynamics: a branch of physics that deals with the relationships and transformations of energy

thermodynamics: the study of energy transformations described by laws

thermohaline circulation: the vertical movement of seawater, generated by density differences that are caused by variations in temperature and salinity

titration: an analytical procedure in which the concentration of an unknown solution is added to a standard solution until a color change of some indicator indicates that equivalent quantities have reacted

topographic map: a map showing the topographic features of the land surface

trace fossil: a fossilized track, trail, burrow, tube, boring, tunnel, or other remnant resulting from the life activities of an animal

trace fossil: any evidence of the life activities of a plant or animal that lived in the geologic past (but not including the fossil organism itself)

trait: an aspect of an organism that can be described or measured

trajectory: the path followed by an object that is launched into the air

transform fault: a vertical surface of slippage between two lithospheric plates along an offset between two segments of a spreading ridge

transform plate boundary: a plate boundary where two plates slide parallel to one another

transpiration: the emission of water vapor from pores of plants as part of their life processes

transpiration: the process by which water absorbed by plants, usually through the roots, is evaporated into the atmosphere from the plant surface in the form of water vapor

transverse pulse or wave: a pulse or wave in which the motion of the medium is perpendicular to the motion of the wave

trophic level: the number of energy transfers an organism is from the original solar energy entering an ecosystem; the feeding level of one or more populations in a food web

trough: the lowest point on a wave

tsunami: a great sea wave produced by a submarine earthquake (or volcanic eruption or landslide)

turbine: a rotating machine or device that converts the mechanical energy of fluid flow into mechanical energy of rotation of a shaft

Tyndall Effect: the scattering of a light beam as it passes through a colloid

ultraviolet: electromagnetic radiation at wavelengths shorter than the violet end of visible light; with wavelengths ranging from 5 to 400 nm

unconfined aquifer: an aquifer that has a free connection upward to the surface

uniformity: the property of how consistent a material is throughout

urban heat-island effect: the observed condition that urban areas tend to be warmer than surrounding rural areas

valence electrons: the outermost electrons of an atom. These electrons are used for chemical bonding of atoms.

vaporization: the change of state from a liquid to a gas

vector: a quantity that has both magnitude and direction

velocity: speed in a given direction; displacement divided by the time interval; velocity is a vector quantity, it has magnitude and direction

virtual image: an image from which rays of reflected or refracted light appear to diverge, as from an image seen in a plane mirror; no light comes directly from or passes through the image

viscosity: the property of a fluid to offer internal resistance to flow.

visible spectrum: part of the electromagnetic spectrum that is detectable by human eyes. The wavelengths range from 350 to 780 nm. (A nanometer is a billionth of a meter.)

volcanic bomb: a blob of lava that was ejected while viscous and received a rounded shape (larger than 64 mm in diameter) while in flight

Volcanic Explosivity Index: the percentage of pyroclastics among the total products of a volcanic eruption

volt: the SI unit of electric potential; one volt (1 V) is equal to one joule per coulomb (J/C)

water (hydrologic) cycle: the cycle or network of pathways taken by water in all three of its forms (solid, liquid, and vapor) among the various places where is it temporarily stored on, below, and above the Earth's surface

water budget: an accounting of the sources of water supply and water demand, and of how the supply is divided among the various uses that make up the demand

water cycle (or hydrologic cycle): the constant circulation of water from the sea, through the atmosphere, to the land, and its eventual return to the atmosphere by way of transpiration and evaporation from the land and evaporation from the sea

water table: the surface between the saturated zone and the unsaturated zone (zone of aeration)

watt: a unit of power

wavelength: the distance between two identical points in consecutive cycles of a wave

wavelength: the distance measured from crest to crest of one complete wave or cycle.

weather: the condition of the Earth's atmosphere, specifically, its temperature, barometric pressure, wind velocity, humidity, clouds, and precipitation

weight: the vertical, downward force exerted on a mass as a result of gravity

$$F_g = mg$$

windward: the upwind side or side directly influenced to the direction that the wind blows from, opposite of leeward

work: the product of the displacement and the force in the direction of the displacement; work is a scalar quantity

$$W = F \cdot d$$

work: the product of the force exerted on a body and the distance the body moves in the direction of that force; work is equivalent to a change in the mechanical energy of the body

x-ray telescope: instrument used to detect stellar and interstellar x-ray emission. Because the Earth's atmosphere absorbs x-rays, x-ray telescopes are placed high above the Earth's surface.

Glosario

abiótico (abiotic): los componentes sin vida de un ecosistema

aceite crudo (crude oil): (ver petróleo)

aceleración (acceleration): el cambio en la velocidad por unidad de tiempo

$$a = \frac{\Delta v}{\Delta t}$$

aceleración centrípeta (centripetal acceleration): la aceleración radial hacia adentro de un objeto moviéndose a una velocidad constante en un círculo

ácido (acid): un compuesto o solución con una concentración de iones de hidrógeno mayor que su valor neutral (correspondiente al valor de pH menor que 7)

ácido (acid): una sustancia que produce iones de hidrógeno en agua, o es un donante de protones

acuífero (aquifer): cualquier cuerpo de sedimento o roca que tiene suficiente tamaño, nivel de porosidad y permeabilidad para permitir un suministro adecuado de agua de los pozos

acuífero sin límite/sin restricción (unconfined aquifer): un acuífero que tiene una conexión libre hacia arriba en la superficie

adaptación (adaptation): un rasgo o conjunto de rasgos heredados que mejoran las posibilidades de supervivencia y reproducción de un organismo

administración de agua (water budget): una contabilidad de los recursos del agua y su demanda, y cómo el suministro es dividido entre varios usos que hacen la demanda

afelio (aphelion): el punto en la órbita de la Tierra que está más alejado del sol. Actualmente, la Tierra alcanza el afelio a principios de julio.

agua que no se consume (nonconsumptive water): agua que es devuelta , en forma líquida a su medio ambiente después de usarse

agua subterránea (ground water): la parte de agua debajo de la superficie que está en la zona de saturación, incluyendo arroyos subterráneos

agua subterránea (groundwater): agua que se encuentra en los poros de las rocas en el sedimento debajo de la superficie de la Tierra

agujero negro estelar (stellar black hole): las sobras del centro de una estrella masiva y sola después de un supernova. Los agujeros negros ejercen una fuerza gravitacional tal fuerte que ni la luz se puede escapar.

aislante termal (thermal insulator): un material que obstruye o retrasa la transferencia de calor

albedo (albedo): la propiedad reflexiva de un objeto no luminoso. Un espejo perfecto tendría que tener un albedo de 100% mientras que un agujero negro tendra un albedo de 0%.

aleación (alloy): una sustancia que tiene características de metal y consiste de dos o más elementos diferentes

ampere (ampere): la unidad SI para una corriente eléctrica; un ampere (1 A) es el flujo de un columbio de carga cada segundo

amplitud (amplitude): el desplazamiento máximo de una partícula según pasa por una onda; la altura de la cresta de una onda; está relacionada con la energía de la onda

anatomía (anatomy): el estudio de la estructura interna de los organismos

ángulo crítico (critical angle): el ángulo de incidencia por el cual el rayo de luz pasando de un medio a otro tiene un ángulo de refracción de 90° grados

ángulo de incidencia (angle of incidence): el ángulo que se forma entre el rayo de luz y la normal de la superficie en el punto de incidencia

ángulo de reflexión (angle of reflection): el ángulo que hace un rayo reflejado con la normal de la superficie en el punto de reflexión

anión (anion): Un ión con carga negativa

año luz (light-year): una unidad de medida igual a la distancia que viaja la luz en un año ej. 9.46×10^{12} km.

antropopegénico (anthropopegenic): generado o producido por actividades humanas

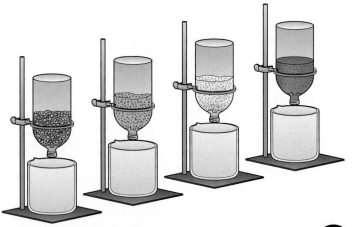

área de superficie (surface area): cambiar la naturaleza de los reactivos a pequeñas partículas aumenta la superficie expuesta a la reacción. Una reacción exitosa depende en el colisión y aumento del área de los reactantes aumentando así la oportunidad de que un colisión exitoso se lleve a cabo. Encender un tronco es más difícil que encender virutas de madera. Las virutas tienen un área de mayor superficie y aceleran la reacción.

astenósfera (asthenosphere): la parte del manto debajo de la litosfera. La astenósfera experimenta un movimiento lento, en vez de actuar como un bloque rígido, como es la capa encima de la litosfera.

asteroide (asteroid): un pequeño cuerpo planetario en la órbita alrededor del sol, más grande que un meteoride (una partícula en el espacio, menos de unos metros de diámetro) pero más pequeño que un planeta. Muchos asteroides se pueden encontrar en el cinturón entre las órbitas de Martes y Júpiter.

átomo (atom): la parte representativa más pequeña de un elemento.

átomo (atom): la partícula más pequeña de un elemento que tiene todas las propiedades del elemento; consiste de un núcleo rodeado de electrones

aumento (accretion): el proceso donde el polvo y el gas son acumulados en grandes cuerpos como las estrellas y los planetas

aurora (aurora): una emisión brillante de átomos y moléculas cerca de los polos de la Tierra causado por partículas cargadas entrando la atmósfera superior

autótrofo químico-sintético (autotroph): un organismo que puede sintetizar su propia energía (alimento) directamente del ambiente físico

basalto (basalt): tipo de piedra ígnea volcánica, generalmente de color oscuro, con un alto contenido de hierro

base (base): un compuesto o solución con una concentración de iones de hidrógeno menor que su valor neutral (correspondiente al valor pH mayor que 7)

base (base): una sustancia que libera iones de hidróxido (OH^-) en agua, o es un aceptor de protones.

bias (bias): una distorsión a propósito o accidental de observaciones, datos o cálculos en una manera sistemática o salteada

biodiversidad (biodiversity): la diversidad de diferentes especies biológicas y/o la variación genética entre los individuos dentro de una misma especie

biodiversidad (biodiversity): la suma de todos los tipos diferentes de organismos vivos en la Tierra

biomasa (biomass): la masa total de materia viviente en la forma de uno o más tipos de organismos presentes en un habitat particular

biome (biome): un ensamblaje reconocido de plantas y animales que caracterizan un área geográfica grande de la Tierra; un numero de diferentes biomes han sido reconocidos, y la distribución de los biomes es controlado mayormente por el clima

biosfera (biosphere): el área en la Tierra donde se pueden encontrar organismos vivos

biótico (biotic): los componentes vivos en un ecosistema

bomba volcánica (volcanic bomb): una masa de lava que fue expulsada cuando estaba viscosa y recibió una forma redonda (de más de 64 mm en diámetro) mientras estaba en vuelo

brillo/lustre (luster): el reflejo de la luz desde la superficie de un material descrito por su calidad e intensidad

bucles de reacción (feedback loops): los procesos donde los sistemas de salida causa cambios positivos o negativos a algún componente de medida del sistema

caballos de fuerza (horsepower): una unidad de potencia

cadena alimenticia (food chain): una serie de organismos por los cuales pasa energía alimenticia en un ecosistema

caída libre (free fall): una caída bajo la influencia de la gravedad solamente

caldera (caldera): una gran depresión volcánica en forma de cuenca, más o menos circular, la cual muchas veces tiene un diámetro más grande que cualquiera de las aberturas

calor (heat): energía cinética de átomos o moléculas asociadas con la temperatura del cuerpo de un material

calor específico (specific heat): la cantidad de energía requerida para elevar la temperatura de 1 kg. de un material por 1°C

$$E = mc\Delta t$$

cambio de fase (phase change): la conversión de una sustancia de un estado a otro a una temperatura y presión específica. Ejemplo: sólido a líquido, líquido a gaseoso, o sólido a gaseoso (sublimación).

cambio físico (physical change): un cambio que envuelve cambios en el estado o forma de una sustancia pero que no causa ningún cambio en su composición química

cambio químico (chemical change): un cambio que convierte la composición química de una sustancia a una(s) sustancia(s) diferente(s) con una composición química diferente

campo eléctrico (electric field): la región de influencia eléctrica definida como la fuerza por unidad de carga

campo magnético (magnetic field): la región de influencia magnética alrededor de un polo magnético o una partícula de carga en movimiento

cancelar (scrubbing): extirpación del dióxido sulfúrico, cenizas, y otros derivados dañinos de la quemada de combustible fósil como los productos de combustión pasando hacia arriba a través de una pila o cañón

capacidad de calor (heat capacity): la cantidad de energía de calor requerida para aumentar la temperatura de un material o sistema; típicamente referida como la cantidad de energía de calor requerida para generar un aumento a 1°C en la temperatura de 1 g en un material dado que está a una presión atmosférica y a 20°C

capacidad de carga (carrying capacity): la población máxima que puede ser sostenida por un suministro determinado de recursos

carbón (coal): una roca combustible que ha tenido sus orígenes en la disposición y entierro de material vegetal

carga eléctrica (electric charge): una propiedad fundamental de la materia; la carga es positiva o negativa

carnívoro (carnivore): un animal que se alimenta exclusivamente de otros animales

catalizador (catalyst): una sustancia que cambia la velocidad de una reacción química sin cambiarse a sí misma permanentemente

catión (cation): un ión con carga positiva

célula de convección (convection cell): un patrón de movimiento en un líquido en el cual el líquido se mueve en un patrón de circulación cerrada

cemento (cement): un material mineral nuevo precipitado alrededor de las partículas de sedimento cuando ha sido enterrado bajo la superficie de la Tierra

cenizas (ash): material fino piroclástos (de menos de 2 mm en diámetro)

centro de masa (center of mass): el punto en el cual toda la masa de un objeto es considerado concentrado por los cálculos que concierne al movimiento de ese objeto

cero absoluto (absolute zero): la temperatura en la cual todas las vibraciones de los átomos y las moléculas de la materia cesan; la más baja temperatura posible

ciclo (hidrológico) del agua (water–hydrologic–cycle): el ciclo o la red de senderos tomados por el agua en todas sus tres formas (sólida, líquida y gaseosa) entre los varios lugares donde es guardada temporalmente en, bajo o sobre la superficie de la Tierra

ciclo de agua (o ciclo hidrológico) (water cycle or hydrologic cycle): la circulación constante del agua de mar, a través de la atmósfera a la tierra y el regreso eventual a la atmósfera a través de la transpiración, evaporación de la tierra y evaporación del mar

ciclo de carbono (carbon cycle): el ciclo global del movimiento del carbono, en todas sus formas, de una reserva a otra

ciclo de carbono (carbon cycle): el proceso en el cual el carbono es pasado de un organismo a otro, de ahí a la comunidad abiótica y finalmente de vuelta a las plantas

ciclo de fósforo (phosphorous cycle): el ciclo del fósforo ambiental a través de un ciclo a largo plazo envolviendo rocas en la corteza de la Tierra y a través de un ciclo corto envolviendo organismos vivos

ciclo de nitrógeno (nitrogen cycle): el movimiento del nitrógeno a través de los ecosistemas, el suelo y la atmósfera

ciclos Milankovitch (Milankovitch cycles): los cambios cíclicos en la relación geométrica entre la Tierra y el sol que causa variaciones en la radiación solar recibida en la superficie de la Tierra

cima de la onda de longitud (peak wavelength): la longitud de onda de la luz con la mayor cantidad de energía electromagnética emitida por cualquier objeto

circuito eléctrico (electric circuit): un aparato eléctrico que provee un camino conductivo para las corrientes eléctricas más continuas

circulación termoalina (thermohaline circulation): el movimiento vertical del agua de mar, generada por las diferencias en la densidad que son causadas por variaciones en temperatura y salinidad

clima (climate): el patrón general de las condiciones del tiempo para una región sobre un largo período de tiempo (por lo menos 30 años)

clima global (global climate): las condiciones promedio sobre la superficie de la Tierra según es determinado por el promedio de los grandes números de las observaciones espaciales distribuidas a través de la región entera del globo

clima paleolítico (paleoclimate): las condiciones climáticas en el pasado geológico reconstruido de fuentes de datos directos e indirectos

climatología paleolítica (paleoclimatology): el estudio científico del clima de la Tierra durante el pasado

climatólogo paleolítico (paleoclimatologist): un científico que estudia el clima pasado de la Tierra

coloide (colloid): una mezcla que contiene partículas más grande que el soluto pero suficientemente pequeñas para mantenerse suspendidas en un fase continúa en otro componente. Esta mezcla también es conocida como una dispersión coloidal.

coma (coma): una nube esférica de material que rodea la cabeza de un cometa. Este material es mayormente gas que el sol ha ocasionado que hierva del núcleo helado del cometa. Una coma cometaria se puede extender a un millón de millas del núcleo.

combustión (combustion): la reacción rápida de un material con oxígeno acompañado por una evolución rápida de flama y calor

cometa (comet): un grupo de gases helados, hielo, y residuos rocosos que orbitan el sol

compuesto (composite): una mezcla heterogénea sólida de dos o más sustancias que hacen uso de las propiedades de cada componente

compuesto (compound): un material que consiste de dos o más elementos unidos en una proporción definida

compuesto binario (binary compound): un compuesto formado por la combinación de dos elementos diferentes

compuesto inorgánico (inorganic compound): un compuesto que no está basado en compuestos moleculares de carbono

compuesto iónico (ionic compound): un compuesto que consiste de iones positivos y negativos

compuesto molecular (molecular compound): dos o más átomos enlazados al compartir electrones

compuesto orgánico (organic compound): un compuesto molecular de carbono

comunidad (community): todas las poblaciones de organismos ocupando una área determinada

comunidad culminante (climax community): el final que una comunidad estable alcanza durante las etapas de sucesión

comunidad pionera (pioneer community): las primeras especies en aparecer durante la sucesión

concentración (concentration): la medida de la composición de una solución, algunas veces dada en términos de moles de un soluto por litro de solución

condensación (condensation): el proceso de cambio de gas a líquido

conducción (conduction): (de calor) la energía se transfiere de un material o partícula a otro cuando los materiales o partículas están en contacto directo

conducción (conduction): el proceso de transferencia de calor por el cual las más vigorosas vibraciones de materia relativamente caliente es transferida a materia adyacente relativamente fría, entonces tiende a nivelar la diferencia en la temperatura entre las dos regiones de materia

conductibilidad (conductivity): la propiedad de transmitir calor y electricidad

configuración de un electrón (electron configuration): la distribución de electrones en los niveles de energía de un átomo.

cono de escoria - estratovolcanes (composite cone - stratovolcano): un volcán que es construido de capas alternas de lava y depósitos piroclástos

constante de Planck (Planck's constant): una constante proporcional de energía de un fotón a su frecuencia, derivada por Max Planck en el 1900. Su ecuación fue $E = hf$ y la constante de Planck es $(h) = 6.626 \times 10^{-34}$ J · s

constelación (constellation): un grupo de estrellas en la noche en un patrón reconocible. La mayoría de las constelaciones obtienen su nombre de la traducción en latín de unos de los patrones de estrellas Griegas. En los más recientes tiempos, más astrónomos modernos introdujeron grupos adicionales, hay ahora 88 configuraciones estandares reconocidas.

consumidor (consumer): un organismo heterótrofo

convección (convection): la transferencia de calor como resultado del movimiento de la sustancia calentada, como el aire o las corrientes de agua

convección (convection): movimiento de un líquido en el cual el líquido se mueve en un patrón de circulación cerrada

convección termal (thermal convection): un patrón de movimiento en un flujo causado por calentamiento de abajo y enfriamiento de arriba. El calentamiento termal transfiere energía de calor del fondo de la célula de calor a la superficie.

cordilleras submarinas (mid-ocean ridge): una cadena de cordilleras debajo del agua que se extienden a través de todas las cuencas del océano, formadas y extendidas por el suelo marino

corona (corona): la parte de afuera de la atmósfera de una estrella (incluyendo el sol), millones de kilómetros en extensión, y consistiendo de gases enrarecidos calentados a millones de grados de temperatura

correlación (correlation): una relación o conexión mutua

corriente alterna (alternating current): una corriente eléctrica que se devuelve en la dirección

corriente eléctrica (electric current): el flujo de las cargas eléctricas a través de un conductor; la corriente eléctrica se mide en amperes

corteza (crust): la capa externa más delgada de la Tierra. La corteza continental es relativamente gruesa y mayormente muy vieja. La corteza oceánica es relativamente delgada, y siempre es geológicamente muy joven.

cosmólogo (cosmologist): un científico que estudia el origen y dinámicas del universo

cresta (crest): el punto más alto de desplazamiento una onda

cromosfera (chromosphere): una capa rojiza en la atmósfera del sol, la transición entre la capa más afuera de la atmósfera del sol, o corona

cromosoma (chromosome): hilo de material genético encontrado en el núcleo de las células

cuerpo fósil (body fossil): cualquier residuo o huella de material orgánico actual de una criatura o planta que ha sido preservado en un archivo geológico (como un hueso)

cuerpo protoplanetario (protoplanetary body): un bloque de material, formados en las etapas de formación tempranas del sistema solar, el cual fue el predecesor de los planetas que vemos hoy

culombio (coulomb): la unidad SI para una carga eléctrica; un culombio (1 C) es igual a la carga de 6.25×10^{18} de electrones

curva de biodiversidad (biodiversity curve): una gráfica que muestra cambios en la diversidad de organismos como función de tiempo geológico

densidad (density): a masa por unidad de volumen de un material o una sustancia

densidad (density): la masa por unidad de volumen de un material

descomposición (decomposition): el proceso quíco de la separación de la materia a compuestos químicos más simples

descomposición (decomposition): una reacción química en la cual un compuesto sencillo reacciona para dar dos o más productos

desnitrificación (denitrification): la conversión de nitratos y nitritos a gas nitrógeno, el cual es liberado en la atmósfera

desplazamiento (displacement): la diferencia en posición entre una posición final y una posición inicial; depende sólo de los puntos finales, no del camino; el desplazamiento es un vector, tiene magnitud y dirección

difracción (diffraction): la habilidad de una onda para extenderse según emerge de una apertura o se mueve más allá de una obstrucción

digital (digital): una descripción de datos que han sido guardados o transmitidos como una secuencia de símbolos discretos; normalmente esto significa datos binarios (1s y 0s) representados usando señales electrónicas o electromagnéticas

dominante (dominant): se usa para describir el gene que determina la expresión de un rasgo genético; el rasgo se muestra aún cuando el gene se presenta como una copia individual

ductibilidad (ductility): una propiedad que describe cuán fácil es convertir una sustancia a una forma nueva y permanente, como es, cambiarla a alambres

ecosistema (ecosystem): una unidad en ecología que consiste del medio ambiente y sus elementos vivos, más los factores no vivientes que existen en él y que lo afectan

ecosistema (ecosystem): una comunidad y el medio ambiente físico que ésta ocupa

Efecto Doppler (Doppler Effect): cambio en la frecuencia de una onda de luz o sonido debido al movimiento de la fuente o del recibidor

Efecto Tyndall (Tyndall Effect): la diseminación de un rayo de luz cuando este pasa a través de un coloide

efecto urbano de calor de isla (urban island-heat effect): la condición observada de que las áreas urbanas tienden a ser más calientes que las áreas rurales de los alrededores

eficiencia (efficiency): la proporción de energía útil obtenida por una maquina o aparato a la energía suministrada durante el mismo período de tiempo

elasticidad (elasticity): la propiedad de un material al resistir una deformación y regresar a su tamaño o forma normal después que se le ha aplicado fuerza

electricidad (power): la proporción de tiempo en el cual el trabajo es hecho en un cuerpo o al cual la energía es producida o consumida

electricidad (power): la proporción de tiempo en el cual el trabajo es hecho y la energía es transformada

$$P = \frac{W}{t}$$

electrólisis (electrolysis): la conducción de electricidad a través de una solución que contiene iones o a través de un compuesto iónico fundido que induzca cambio químico

electromagneto (electromagnet): un aparato que usa una corriente eléctrica para producir un campo magnético concentrado

electrón (electron): una partícula de carga negativa con una carga de 1.6×10^{-19} coulombs y una masa de 9.1×10^{-31} kg.

electrón (electron): una partícula subatómica que ocurre fuera del núcleo y que tiene una carga de -1 y una masa de 9.109×10^{-28}g

electrones de Valencia (Valence electrons): los electrones más externos de un átomo. Estos electrones son usados para enlazar a los átomos químicamente.

elemento (element): una sustancia en la cual todos los átomos tienen el mismo número atómico

elevación (elevation): la altura de la superficie de la tierra relativa al nivel del mar

embalse (reservoir): 1. un lugar en sistema de la Tierra que retiene agua; 2. un cuerpo grande de roca sedimentaria permeable y poroso que contiene petróleo económicamente valioso y/o gas natural

emigración (emigration): el número de individuos en una especie que se muda fuera de una población existente

emulsión (emulsion): un coloide o una dispersión coloidal de un líquido suspendido en otro

energía cinética (kinetic energy): la energía que posee un objeto por su movimiento

$$KE = \frac{1}{2}mv^2$$

energía de calor (heat energy): una forma de energía asociada con el movimiento de átomos o moléculas

energía eléctrica (electric energy): energía asociada con la generación y transmisión de electricidad

energía fotovoltaica (photovoltaic energy): energía asociada con la conversión directa de la radiación solar a la electricidad

energía geotermal (geothermal energy): energía derivada de rocas calientes y/o líquidos debajo de la superficie de la Tierra

energía ionizante (ionization energy): la energía requerida para liberar un electrón de un átomo

energía ionizante (ionization energy): la energía requerida para remover un electrón en un átomo gaseoso en un estado firme

energía mecánica (mechanical energy): la suma de la energía cinética y la energía potencial del cuerpo de un material

energía potencial (potential energy): energía almacenada de un material como resultado de su posición en un campo eléctrico, magnético o gravitacional

energía potencial (potential energy): energía mecánica asociada con la posición en el campo de gravedad; materia más lejana del centro de la Tierra que tiene la más alta energía potencial

energía potencial (potential energy): energía que es dependiente de la posición de un objeto

energía potencial de gravitación (gravitational potencial energy): la energía que un cuerpo posee como resultado de su posición en el campo gravitacional

$$GPE = mgh$$

energía potencial de un resorte (spring potential energy): la energía interna de un resorte debido a su estiramiento o compresión

energía química (chemical energy): energía guardada en un compuesto químico, la cual se puede liberar durante reacciones químicas, incluyendo la combustión

enlace covalente (covalent bond): un enlace formado cuando dos átomos se combinan y comparten pares de electrones entre ellos

enlace iónico (ionic bond): la atracción entre iones con cargas opuestas

escalar (scalar): una cantidad que tiene magnitud, pero no tiene dirección

escurrimiento superficial (runoff): la parte de la precipitación que aparece en chorros superficiales

especies (species): un grupo de organismos que pueden ser reproducidos entre sí bajo condiciones naturales y producir descendencia fértil

especies (species): un grupo de organismos, sea vegetal o animal, que puede ser injertado o cruzados y producir hijos fértiles

especies invasivas (invasive species): una especie que no es natural del área y que su introducción causa o puede causar daño económico, ambiental o a la salud humana

especies no naturales – exóticas, extranjeras, introducidas o no indígenas (nonnative - exotic, alien, introduced, or non-indigenous) species: cualquier especie, incluyendo sus semillas, huevos, esporas u otro material biológico capaz de propagar la especie y que no es natural de ese ecosistema

espectro de absorción (absorption spectrum): un espectro continuo sin interrupciones por líneas de absorción o un espectro con un número discreto de longitud de ondas faltantes o reducidas en intensidad

espectro de emisión (emission spectrum): un espectro conteniendo líneas brillantes o un conjunto de longitudes de ondas discretas producidas por un elemento. Cada elemento tiene su propio espectro de emisión único.

espectro electromagnético (electromagnetic spectrum): el espectro completo de radiación electromagnética, como son las ondas de radio, microondas, infrarrojas, visible, ultravioleta, rayos-x y rayos gama

espectro visible (visible spectrum): parte de un espectro electromagnético que es detectable por el ojo humano. Las ondas de longitud tienen un alcance de 350 a 780 nm. (Un nanómetro es una billonésima de un metro.)

espectroscopia (spectroscopy): la ciencia que estudia la manera en que la luz interacciona con la materia

espectroscopio (spectroscope): un instrumento consistiendo de, como mínimo, un prisma el cual produce un espectro para observación visual

espora (spore): estructura típica unicelular reproductiva capaz de desarrollarse independientemente en un organismo adulto tanto directamente si es asexual o después de la unión con otra espora si es sexual

estado de referencia (frame of reference): un punto de ventaja con respecto a cuál posición y movimiento puede ser descrito

estepa (steppe): una pradera extensa y sin árboles encontrada en regiones semiáridas en latitudes medias. Las estepas son regularmente consideradas más secas que las sabanas.

estrella de neutrón (neutron star): explosión supernova producida por la implosión del centro de una estrella masiva

etapas serales (seral stages): son las comunidades entre la comunidad pionera y la culminante durante las etapas de sucesión

evaporación (evaporation): el proceso de cambio de un líquido a un gas

evolución (evolution): un cambio gradual en las características de la población de organismos durante generaciones sucesivas, como resultado de la selección natural actuando en la variación genética

exactitud (accuracy): cuán cerca es el valor de la medida al estándar o al valor aceptado de esa medida

excentricidad (eccentricity): la proporción de la distancia entre el foci y la longitud del eje mayor de un elipse

extinción (extinction): la desaparición permanente de una especie en la Tierra

extinción de fondo (background extinction): extinción normal de especies que ocurre como resultado de los cambios en las condiciones locales del ambiente

extinción en masa (mass extinction): una perturbación catastrófica universal donde grupos mayores de especies es extinguen en un corto tiempo comparado con la historia de extinción normal

falla (fault): la rotura o la zona de rotura en la roca, por la cual las masas de roca se han movido relativamente de una a otra paralelamente a la rotura

falla transformante (transform fault): una superficie vertical de deslizamiento entre dos placas litosféricas a lo largo de una desviación entre dos segmentos de una cima extendida

fenotipo (phenotype): los rasgos observables de un organismo que son el resultado de una interacción de los genes y el medio ambiente

fertilizador (fertilizer): un material usado para proveer o reemplazar nutrientes en el suelo

fijación del nitrógeno (nitrogen fixation): el proceso por el cual algunos organismos producen compuestos de nitrógeno del nitrógeno gaseoso en la atmósfera

fisión (fission): el proceso de dividir el núcleo en núcleos más pequeños y con la liberación de grandes cantidades de energía

fisión nuclear (nuclear fission): el proceso en el cual átomos grandes se separan en dos partes, convirtiendo una pequeña parte de la materia en energía

fisión nuclear (nuclear fission): una reacción nuclear en la cual un núcleo masivo y desequilibrado se separa en dos o más núcleos pequeños con la liberación de una gran cantidad de energía

fitoplancton (phytoplankton): pequeños organismos fotosintéticos, mayormente alga y bacteria, encontrado habitando los ecosistemas acuáticos

flujo de piroclástos (pyroclastic flow): una mezcla de alta densidad de ceniza caliente y fragmentos de rocas con gases calientes formados por una explosión volcánica o una expulsión aérea desde una abertura volcánica

foco (focus): el lugar en el cual los rayos de luz convergen o donde éstos parecen divergir después de la refracción o reflejo; también llamado punto focal

foco (focus): el punto de un terremoto dentro de la Tierra donde primero ocurre la ruptura que causa el terremoto

foraminífera (foraminifera): un orden de organismos de una sola célula (protozoarios) que viven en ambientes marinos (usualmente) y de agua fresca (raramente). Los forams típicamente tienen un caparazón de uno o más compartimientos que son usualmente hechos de carbonato de calcio.

fórmula química (chemical formula): la combinación de los símbolos de los elementos en una proporción numérica definida usada para representar moléculas, compuestos, radicales, iones, etc.

fósil (fossil): cualquier resto, rasgo o característica de una planta o animal que ha sido preservado en la corteza de la Tierra desde algún pasado geológico o tiempo prehistórico

fósil índice (index fossil): un fósil de un organismo que se extendió pero vivió solamente por un corto intervalo de una era geológica

fotosfera (photosphere): la superficie visible del sol, encontrada justamente encima de la capa interior más alta del sol, y exactamente debajo de la cromosfera

fotosíntesis (photosynthesis): el proceso por el cual las plantas usan energía solar, junto con el dióxido de carbono y nutrientes, para sintetizar los tejidos de las plantas

frecuencia (frequency): el número de ondas por segundo o ciclos por segundo o hertz (Hz).

$$f = \frac{v}{\lambda}$$

frecuencia (frequency): el número de ondas producidas por unidad de tiempo; la frecuencia es el recíproco de la cantidad de tiempo que le toma a una sola longitud de onda pasar un punto

fricción (friction): la fuerza ejercida por un cuerpo de materia cuando pasaba por otro cuerpo de materia

fricción (friction): una fuerza que actúa resistiendo el movimiento relativo o movimiento atentado de objetos que están en contacto con cada uno

fuente de energía renovable (renewable energy source): una fuente de energía que es impulsada por radiación solar en el tiempo presente en vez de por combustibles almacenados en la Tierra

fuente de rocas (source rocks): rocas sedimentarias, que contienen concentración significativa de materia orgánica, en el cual el petróleo y el gas natural son generados durante el entierro de los depósitos

fuerza (force): un empujón o halón que puede acelerar un objeto; la fuerza se mide en newtons; la fuerza es un vector de cantidad

fuerza (strength): la propiedad de cuán bien un material resiste el empleo de una fuerza

fuerza centrípeta (centripetal force): la fuerza directa hacia el centro que causa que un objeto siga un camino circular

$$F = \frac{mv^2}{R}$$

fuerza de rompimiento (shear strength): la fuerza de rompimiento necesaria para romper un material sólido

fuerza hidroeléctrica (hydroelectric power): energía eléctrica derivada del flujo de agua en la superficie de la Tierra

fuerza nuclear - fuerte (nuclear–strong–force): una fuerza que mantiene neutrones y protones juntos en el núcleo de un átomo; la fuerza opera solamente sobre distancias bien cortas

fusión (fusion): núcleos de átomos ligeros combinados para formar núcleos con mayor masa y liberando una cantidad mayor de energía

fusión nuclear (nuclear fusion): un proceso nuclear que libera energía cuando núcleos de peso liviano se combinan para formar núcleos más pesados

fusión nuclear (nuclear fusion): una reacción nuclear en la cual núcleos se combinan para formar núcleos masivos con la liberación de una gran cantidad de energía

galvanómetro (galvanometer): un instrumento usado para detectar, medir, y determinar la dirección de corrientes eléctricas pequeñas

gas natural (natural gas): un gas, que consiste mayormente de metano, este es producido en sedimentos y rocas sedimentarias durante el entierro de una materia orgánica

gas noble (noble gas): una familia de elementos (Grupo 18 o VIIIA) de la tabla periódica

gasolina fósil (fossil fuel): combustible derivado de materiales (mayormente brea, petróleo, y gas natural) que fueron generados por material de fósil orgánico y guardado profundamente en la Tierra por largos tiempos geológicamente

gene (gene): una unidad de instrucción ubicada en un cromosoma que produce o influye un rasgo específico en la descendencia

genotipo (genotype): los genes de un individuo

glaciar (glacier): una gran y duradera acumulación de nieve y hielo que se desarrolla en el suelo y fluye bajo su propio peso

gravedad (gravity): la fuerza de atracción entre dos cuerpos debido a sus masas

gravedad específica (specific gravity): la proporción del peso de un volumen dado de una sustancia al peso de una cantidad igual de agua

gravímetro (gravimeter): un instrumento para medir las variaciones de gravitación en los campos de la Tierra

grupo químico (chemical group): una familia de elementos en la tabla periódica que tiene la configuración de electrones similares

guyote (seamount): una montaña debajo del agua que surge del suelo oceánico con una cima plana

halógenos (halogens): Grupo VIIA (17) en la tabla periódica consistiendo de flúor, cloro, bromo, yodo, y astatín

herencia (heredity): el paso de rasgos de los padres a los descendientes

hervíboro (herbivore): un heterótrofo que se alimenta exclusivamente de plantas

heterótrofo (heterotroph): un organismo que debe obtener su energía de autótrofos u otros heterótrofos

hidrocarbono (hydrocarbon): un compuesto molecular conteniendo solamente hidrógeno y carbóno

hipótesis (hypothesis): proposición que puede ser probada o refutada por evidencia experimental o de observación

hoya sedimentaria (sedimentary basin): un área de la corteza de la Tierra donde los sedimentos se acumulan para formar grandes capas

huella de fósil (trace fossil): cualquier evidencia de las actividades de vida de una planta o un animal que vivió en el pasado geológico (pero sin incluir el organismo fósil)

huella fósil (trace fossil): cualquier evidencia de las actividades de vida de plantas o animales que vivieron en el pasado geológico (pero que no incluye el mismo organismo fósil)

imagen real (real image): una imagen que se proyecta sobre una pantalla o en la película de una cámara; los rayos de luz pasan a través de la imagen

imagen virtual (virtual image): una imagen de la cual los rayos reflejados o la luz refractada aparece divergida, como es la imagen vista en un espejo plano, no hay luz que venga directamente o que pase a través de la imagen

impulso (impulse): el producto de fuerza y el intervalo de tiempo durante el cual la fuerza actúa; el impulso resulta en el cambio del momento

$$Ft = \Delta(mv)$$

inclinación (inclination): el ángulo entre el plano orbital del sistema solar y la órbita actual de un objeto alrededor del sol

incremento continental (continental accretion): el crecimiento de un continente a lo largo de sus orillas

indicador a base de ácido (acid-base indicator): una tinta que tiene cierto color en una solución de ácido y un color diferente en una solución de base

Índice de explosividad volcánica (Volcanic Explosivity Index): el porcentaje de piroclástos entre el total de productos de una erupción volcánica

índice de refracción (index of refraction): una propiedad de un medio que está relacionado con la velocidad de la luz a través de este; es calculado dividiendo la velocidad de la luz en el vacío por la velocidad de la luz en el medio

inercia (inertia): la tendencia natural de un objeto de mantenerse en reposo o mantenerse en movimiento con una velocidad constante en una línea recta

infiltración (infiltration): el movimiento del agua pasando por poros o pequeñas aberturas en el suelo o en roca porosa

infrarroja (infrared): radiación electromagnética con longitudes de ondas entre 0.7 a 1000 mm. Ondas infrarrojas no son visibles al ojo humano

inmigración (immigration): el número de individuos en una especie que se mueve adentro de una población ya existente

insolación (insolation): el ritmo en el cual un área de la tierra recibe energía solar

insolación (insolation): ondas cortas directas o difundidas de radiación solar que son recibidas en la atmósfera de la Tierra o en su superficie

interferencia constructiva (constructive interference): el resultado de diferentes ondas para que dos o más ondas se unan para producir una onda de mayor amplitud

interferencia destructiva (destructive interference): el resultado de diferentes ondas para que dos o más ondas se unan para producir una onda de menor amplitud

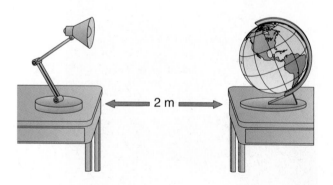

2 m

intervalo de curvas de nivel (contour interval): la distancia vertical entre las elevaciones representadas por dos líneas sucesivas de curvas de nivel en un mapa topográfico

invernadero de gases (greenhouse gases): gases en la atmósfera de la Tierra que absorben ciertas longitudes de ondas de las ondas de radiación de las largas ondas radiales emitidas al espacio por la superficie de la Tierra

invernadero de gases (greenhouse gases): los gases responsables por el efecto de invernadero. Estos gases incluyen: vapor de agua (H_2O), dióxido de carbono (CO_2); metano (CH_4); óxido nitro (N_2O); clorofluorocarbonos (CF_xCl_x); y ozono troposférico (O_3).

ión (ion): un átomo con uno o más electrones removidos (o añadidos), dando una carga positiva (o negativa)

ión (ion): un átomo o grupo de átomos cargados eléctricamente que han adquirido una carga neta, positiva o negativa

ión poliatómico (polyatomic ion): un ión que consiste de 2 ó más átomos que están enlazados covalentemente y tiene una carga positiva o negativa

iones (ions): átomos que tienen una carga eléctrica porque uno o más electrones (partículas con una carga eléctrica negativa, la cual órbita alrededor del núcleo de un átomo) han sido añadidas al átomo o removidas del átomo

ionosfera (ionosphere): la parte de la atmósfera de la Tierra sobre 50 Km. donde los átomos son significantemente ionizados y afecta la propagación de las ondas radiales

isótopo (isotope): uno de dos o más clases de átomos de un elemento químico dado que se diferencia en la masa debido a los números diferentes de neutrones en el núcleo de un átomo

isótopo (isotope): átomos del mismo elemento pero con diferente masa atómica debido a diferentes números de neutrones

julio (joule): la unidad SI para el trabajo y todas otras formas de energía; un julio (1J) de trabajo está hecho cuando la fuerza de un newton mueve un objeto un metro en la dirección de la fuerza

lahar (lahar): una mezcla húmeda de agua, barro, y fragmentos de roca volcánica, con la consistencia de concreto húmedo, que fluye por las pendientes de un volcán y sus valles

lapilli (lapilli): piroclástos con un tamaño general entre 2 a 64 mm

latitud (latitude): una medida de norte a sur de la posición de la Tierra. Es definido por el ángulo medido del plano ecuatorial de la Tierra.

lava (lava): roca fundida que sale de un volcán o una fisura

lente cóncavo (concave lens): un lente que causa que los rayos de luz paralelos se divergen; un lente que es más grueso en las orillas que en el centro

lente convergente (converging lens): un rayo de luz paralelo pasando a través del lente es conducido a un punto real o enfoque (lente cóncavo) (si la parte de afuera del índice de refracción es menor que la del lente del material); también llamado lente convexo

lente convexo (convex lens): un lente que causa que los rayos de luz paralelos convergen; (si la parte de afuera del índice de refracción es menor que la del lente del material) un lente más delgado en las orillas que en el centro

lepton (lepton): un grupo de partículas elementales que no son afectadas por la fuerza nuclear; los electrones pertenecen a este grupo

Ley de Coulomb (Coulomb's Law): la relación entre la fuerza eléctrica, carga y la distancia entre las cargas

$$F = k\frac{q_1 q_2}{d^2}$$

Ley de la Conservación de la Materia (Law of Conservation of Matter): en reacciones químicas, la cantidad de la materia no cambia

Ley de proporciones definida (Law of Definite Proportions): la composición de una sustancia pura es siempre la misma o los elementos del compuesto siempre se combinan en la misma proporción por masa

Ley de Snell (Snell's Law): describe la relación entre el índice de refracción y la proporción del seno de un ángulo de incidencia y el seno del ángulo de refracción

$$n = \frac{\sin \angle i}{\sin \angle R}$$

Ley del inverso cuadrado (inverse-square law): una ley científica que dice que la cantidad de radiación que pasa a través de un área en específico es proporcionalmente inversa al cuadrado de la distancia de esa área de los recursos de energía

Leyes de Movimiento de Newton (Newton's Law of Motion):

Primera Ley de Movimiento de Newton (Newton's First Law of Motion): un objeto en reposo se mantiene en reposo y un objeto en movimiento se mantiene en movimiento al menos que una fuerza extrema desequilibrada sea impulsado en él

Ley de Atracción Gravitacional de Newton (Newton's Law of Gravitational Attraction): la relación entre la fuerza de gravitación, masas y la distancia entre las masas

$$F = \frac{Gm_1 m_2}{d^2}$$

Segunda Ley en Movimiento de Newton (Newton's Second Law of Motion): si un cuerpo es accionado por una fuerza externa, este acelerará en la dirección de la fuerza desequilibrada con una fuerza proporcional a la fuerza y proporcionalmente inversa a la masa

$$F = ma$$

Tercera Ley de Movimiento de Newton (Newton's Third Law of Motion): las fuerzas vienen en pares; la fuerza del objeto A en el objeto B es igual y opuesta a la fuerza del objeto B en el objeto A

límite convergente (convergent plate boundary): el límite de placas donde dos placas se mueven una en dirección a la otra

límite de transformación (transform plate boundary): el límite de placas donde dos placas se deslizan paralelamente una hacia la otra

límite divergente (divergent plate boundary): el límite de placas donde dos placas se mueven en dirección contraria una de la otra

línea de contorno (contour line): una línea en el mapa que conecta puntos con una elevación igual a la de la superficie terrestre

litosfera (lithosphere): la capa más externa de la Tierra, que consiste de la corteza terrestre y parte del manto superior. La litosfera actúa como una capa rígida, en contraste con la astenósfera.

loess (loess): los depósitos de sedimentación que han sido volados por el viento a vastas áreas de las latitudes medias durante y después de las eras glaciales

longitud de onda (wavelength): la distancia entre dos puntos idénticos en ciclos consecutivos de una onda

longitud de onda (wavelength): la distancia medida de una cresta a otra en una onda o ciclo completo

longitud focal (focal length): la distancia desde el centro de un lente o espejo hasta el punto focal

luminosidad (luminosity): la cantidad de energía total radiada por un objeto cada segundo

magma (magma): material fundido de roca que ocurre naturalmente dentro de la Tierra por la cual rocas ígneas son derivadas a través de la solidificación y procesos relacionados

magma (magma): material fundido de roca que ocurre naturalmente dentro de la Tierra. El magma contiene gases disueltos y algunas veces cristales

magnetómetro (magnetometer): un instrumento para medir las variaciones de los campos magnéticos de la Tierra

maleabilidad (malleability): la propiedad de un material que puede ser martillado a varias formas sin romperse

manto (mantle): la zona de la Tierra debajo de la corteza y sobre el centro terrestre. Está dividida entre el manto superior y el manto inferior con una zona de transición entre ellas.

mapa isosísmico (isoseismal map): un mapa que muestra las líneas que conectan los puntos en la superficie de la Tierra en el cual la intensidad de un terremoto es igual

mapa topográfico (topographic map): un mapa mostrando las características topográficas de la superficie terrestre

marco inerte de referencia (inertial frame of reference): punto de vista sin aceleración en el cual las Leyes de Newton son ciertas

marea de primavera (spring tide): las mareas de un alcance que aumenta y que ocurre dos veces al mes cerca de cuando hay luna llena o luna nueva

marea muerta (neap tide): las mareas que disminuyen de alcance que ocurren dos veces al mes cerca de los tiempos del primer y último cuarto menguante

masa atómica (atomic mass): la masa atómica es determinada por la masa de protones y neutrones de un átomo

material puro (pure material): un elemento o compuesto que tiene una composición y propiedades definidas

mesa de agua (water table): la superficie entre la zona saturada y no saturada (zona de aereación)

metales (metal): tipos de materiales que exhiben las propiedades de conductividad, maleabilidad, reactividad y ductilidad. Los elementos de metales fácilmente pierden electrones para formar iones positivos.

meteorito (meteorite): la parte de un meteoride que sobrevive a través de la atmósfera de la Tierra

meteoro (meteor): el fenómeno luminoso visto cuando un meteoride entra en la atmósfera (comúnmente conocido como estrella fugaz)

meteoride (meteoroid): una roca pequeña en el espacio

mineral (mineral): material sólido que consiste de átomos y/o moléculas que han sido arreglados en un patrón regular y tiene características de composición química, de formación de cristal y propiedades físicas que ocurre naturalmente inorgánicamente

mineral del elemento-nativo (native-element mineral): un mineral que consiste de solamente un elemento

modelo (model): una representación de un proceso, sistema u objeto

mol (mole): una colección de objetos que contienen el número de Avogadro (6.022×10^{23})

moléculas orgánicas-basadas en carbono (organic-carbon based-molecules): moléculas con el elemento químico carbono como base

momento (momentum): el producto de la masa y la velocidad de un objeto; el momento es un vector de cantidad

$$p = mv$$

morfología (morphology): las (estudio de) características que constan o describen la figura, forma y estructura de un objeto u organismo

móvil (motile): tener la habilidad de moverse espontáneamente

movimiento de lava (lava flow): un derramamiento de lava fundida de una abertura o fisura; también, el cuerpo sólido de una roca formada de esta manera

nebulosa (nebula): un término general utilizado para cualquier parcho "borroso" en el cielo, ya sea claro u oscuro; una nube de gas interestelar y con polvo

neutral (neutral): teniendo una concentración de iones de hidrógeno que corresponden al valor de un pH de 7

neutralización (neutralization): el proceso de un ácido y una base reaccionando para formar agua y sal

neutrón (neutron): una partícula subatómica que es parte de la estructura de un núcleo atómico; un neutrón es eléctricamente neutral

neutrón (neutron): partícula neutral subatómica con una masa de 1.675×10^{-24} g situada en el núcleo del átomo.

nicho (niche): el rol ecológico de las especies; el conjunto de recursos que consume y los habitats ocupado en un ecosistema

nieve de efecto de lago (lake-effect snow): la nieve que es precipitada cuando una masa de aire, la cual ha ganado humedad por moverse sobre un cuerpo de agua relativamente templado, se enfría según pasa a tierra relativamente fría. Este enfriamiento desencadena la condensación de las nubes y la precipitación.

nivel trófico (trophic level): el número de energía que un organismo transfiere es de la energía solar original que está entrando en un ecosistema; el nivel de alimentación de una o más poblaciones en una red alimenticia

no metales (nonmetal): elementos que no exhiben las propiedades de conductividad, maleabilidad, reactividad, y ductilidad. Estos elementos tienden a formar iones negativos. El óxido de los elementos es acídico.

nodo (node): un punto de una onda estacionario donde su medio no tiene movimiento

normal (normal): en ángulos rectos o perpendicular

nube molecular (molecular cloud): una nube fría y grande hecha mayormente de hidrógeno molecular y helio, pero con otros gases también, como el monóxido de carbono. Es en estas nubes que las nuevas estrellas se forman.

núcleo - de un átomo (nucleus of an atom): el centro denso y cargado positivamente de un átomo que contiene neutrones y protones

núcleo (core): la parte sólida más interna de la Tierra, que consiste mayormente de hierro

nucleón (nucleon): el bloque edificador del núcleo de un átomo; lo mismo un neutrón o un protón

núcleos (nucleus): el centro muy denso de un átomo que contiene neutrones y protones

número de Avogadro (Avogadro's number): el número de átomos que hay en 12 gramos de carbono-12. El número es 6.022×10^{23}.

número de oxidación (oxidation number): un número asignado a un elemento en un compuesto designando el número de electrones que el elemento ha perdido, ganado o compartido al formar ese compuesto

oblicuidad (obliquity): la inclinación del eje de rotación de la Tierra como es medido de la perpendicular al plano de la órbita de la Tierra alrededor del sol. El ángulo de esta inclinación varia entre 22.5° a 24.5° sobre un período de 41,000 años. La oblicuidad actual es de 23.5°.

ohmios (ohm): la unidad de resistencia eléctrica SI, el símbolo para ohmio es Ω

omnívoro (omnivores): un heterótrofo que se alimenta tanto de plantas como de animales

onda de superficie (surface wave): una onda sísmica que viaja a lo largo de la superficie de la Tierra

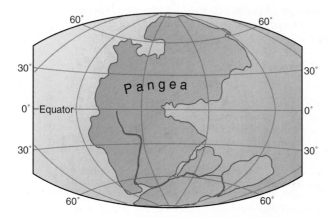

onda o pulso transversal (transverse pulse or wave): un pulso u onda en la cual el movimiento del medio es perpendicular al movimiento de la onda

onda periódica (periodic wave): una serie repetitiva de pulsaciones; una serie de ondas en la cuales las partículas de un medio atraviesan un movimiento periódico (después de un tiempo definido el medio vuelve a un punto de partida y comienza la repetición del movimiento)

onda primaria - onda P (primary wave - P wave): una onda sísmica que envuelve movimientos de partículas (compresión y expansión) en la dirección en la cual la onda viaja. Es la más rápida de las ondas sísmicas.

onda secundaria - onda S (secondary wave - S wave): una onda sísmica producida por un movimiento repetitivo que envuelve vibración perpendicular en la dirección en la cual la onda está viajando. Ésta no viaja a través de líquidos, como el núcleo exterior de la Tierra. Ésta llega más tarde que la onda P.

onda sísmica (seismic wave): un término general para todas las ondas elásticas de la Tierra, producidas por terremotos o generadas artificialmente por explosiones

ondas electromagnéticas (electromagnetic waves): ondas transversales que están compuestas de electricidad perpendicular y campos magnéticos que viajan a 3×10^8 m/s en un vacío; ejemplos de ondas electromagnéticas en longitud de onda son los rayos gama, rayos x, radiación ultravioleta, luz visible, radiación infrarroja, microondas y ondas de radio

ondas polarizadas (polarized waves): disturbios donde el medio vibra en un solo plano

ondas sísmicas (seismic "earthquake" waves): un término general para todas las ondas elásticas en la Tierra, producidas por terremotos o generadas artificialmente por explosiones

órbita (orbit): la trayectoria de un electrón en su movimiento alrededor del núcleo del átomo de hidrógeno de Bohr

orbital (orbital): en el modelo mecánico quanta de un átomo, es la región que rodea el núcleo atómico en el cual la distribución del electrón es dada por una función de onda

organismo (organism): un ser vivo individual

organismos de descomposición (decomposers): organismos que descomponen los restos o desperdicios de otros organismos para obtener sus nutrientes

oxidación (oxidation): el proceso de pérdida de uno o más electrones en una sustancia

oxidación (oxidation): el proceso químico por el cual ciertos tipos de materia son combinados con oxígeno

paleomagnetismo (paleomagnetism): el récord de la orientación pasada y la polaridad del campo magnético de la Tierra registrado en las rocas que contienen el mineral magnetita

paleontólogo (paleontologist): un científico que estudia los restos fosilizados de los animales y/o plantas

Pangea (Pangea): el supercontinente más reciente de la Tierra que fue separado alrededor de 200 millones de años atrás.

paralaje (parallax): el cambio aparente en la posición de un objeto que sucede cuando se observa desde distintos ángulos o puntos de vista

parámetros orbitales (orbital parameters): cualquier número de factores que describen la orientación y o el movimiento de un cuerpo orbitando o la forma de su sendero orbital

pársec (parsec): una unidad usada en astronomía para describir distancias largas. Un pársec es igual a 3.26 años luz.

peligro (hazard): un evento natural, como un terremoto, que tiene la posibilidad de hacer daño

Pequeña Era de Hielo (Little Ice Age): el período de tiempo desde 1350 A.D. a 1850. Durante este período, las temperaturas globales estaban bien frías hasta el comienzo del Holoceno.

perigeo (perihelion): el punto de la órbita terrestre que está más cerca del sol. Actualmente, la Tierra alcanza el perigeo a principios de enero.

período (period): el tiempo requerido para completar un ciclo de una onda

período (period): una línea horizontal de elementos en la tabla periódica

período glacial (glacial period): un intervalo en el tiempo marcado por uno o más avances del hielo glacial. Nota que el intervalo de tiempo no es necesariamente la misma magnitud como el "Período" marcado en la escala del tiempo geológico.

período interglacial (interglacial period): el período de tiempo durante una era de hielo cuando los glaciados se retiran por temperaturas apacibles

permeabilidad (permeability): la facilidad con la cual un fluido puede ser forzado a fluir a través de un material poroso por una diferencia en la presión del fluido de un lugar a otro en el material

peso (weight): la fuerza hacia abajo, vertical ejercida en una masa como resultado de la gravedad

petróleo (petroleum): líquido inflamable y aceitoso que consiste de una variedad de componentes orgánicos, que es producido en sedimentos y rocas sedimentarias durante el entierro de materia orgánica; también llamado crudo

pH (pH): la cantidad usada para representar la acidez en una solución basada en la concentración de iones de hidrogeno (pH = $-\log[H+]$)

pirámide de energía (pyramid of energy): una pirámide desarrollada basada en energía en cada nivel trófico

pirámide de material viviente (pyramid of living matter): una pirámide desarrollada basada en la masa de material viviente seca en cada nivel trófico

placas litosféricos (lithospheric plates): un segmento fino y rígido de la capa mas externa de la Tierra, que consiste de la corteza de la Tierra y la parte superior del manto. Esta placa se puede asumir que se ha movido horizontalmente y ha sido adjunto a otras placas.

placas tectónicas (plate tectonics): el campo de estudio del movimiento de placas

placas tectónicas (plate tectonics): el estudio del movimiento y la interacción de las placas litosferitas de la Tierra

planestesimal (planetesimal): uno de los cuerpos pequeños (regularmente de micrómetros a kilómetros en diámetro) que se forman de la nebulosa solar y eventualmente crecen en foto planetas

planetas de gases gigantes (gas giant planets): los planetas del sistema solar más extremos: Júpiter, Saturno, Urano, y Neptuno, compuesto mayormente de hidrógeno, helio y metano, y que tienen una densidad menor de 2 gm/cm^2

planetas terrestres (terrestrial planets): cualquiera de los planetas Mercurio, Venus, Tierra o Marte, o un planeta similar en tamaño, composición y densidad a la Tierra. Un planeta que consiste mayormente en material rocoso.

plano orbital (orbital plane): (también llamado eclíptico o plano del eclíptico). Un plano formado por el sendero de la Tierra alrededor del sol.

plasma (plasma): un estado de materia en donde todos los átomos están ionizados; una mezcla de electrones libres y núcleos atómicos libres

población (population): un grupo de organismos de la misma especie ocupando una área determinada

población abierta (open population): una población natural en la cual los cuatro factores que afectan el tamaño de la población (tasa de mortalidad, tasa de natalidad, inmigración y emigración) están en funcionamiento

poder del clima (climate proxy): cualquier conjunto de datos que tiene una relación predecible de los factores del clima y pueden ser entonces usados para indirectamente medir estos factores

poder eléctrico (electric power): energía asociada con la generación y transmisión de electricidad

polen (pollen): un término colectivo para granos de polen, los cuales son microscópicos conteniendo los micro gametofitos de muchas células (gametofito macho) de semillas de plantas

polimerización (polymerization): una reacción química que convierte moléculas pequeñas (monómeros) a moléculas grandes (polímeros)

polímero (polymer): una sustancia que es una macromolécula consistiendo de muchas moléculas pequeñas similares (monómeros) enlazadas en largas cadenas

porosidad (porosity): la proporción de espacio de poros por el volumen total de una roca o sedimento, multiplicado por 100 para ser expresado como un porcentaje

precipitación (precipitation): agua que cae a la superficie desde la atmósfera en forma de lluvia, nieve, granizo o aguanieve

precipitación (precipitation): agua que cae de la atmósfera sobre la superficie de la Tierra como material líquido o sólido en forma de lluvia, nieve, cellisca o granizo

precipitar (precipitate): un sólido insoluble formado en una solución líquida como resultado de algunas reacciones químicas

precisión (precision): lo cercano del acuerdo de muchas medidas de la misma cantidad

precisión axial (axial presesion): el tambaleo en el eje polar de la Tierra

presesión (precession): movimiento lento de los ejes de la Tierra alrededor de un cono, un ciclo en acerca de 26,000 años, debido a tirones gravitacionales del sol, la luna y planetas mayores

presesión orbital (orbital precession): rotación sobre el sol por el eje mayor de la órbita elíptica de la Tierra

Primera Ley de Termodinámica (First Law of Thermodynamics): la ley de energía se puede convertir de una forma a otra pero no puede ser creada ni destruida

primera sucesión (primary succession): la ocupación de una vida vegetal en un área que anteriormente no estaba cubierta por vegetación

probabilidad (probability): una medida de la probabilidad de un evento dado ocurriendo

producto (product): la sustancia producida en una reacción química

productor (producer): un organismo que sintetiza su propio alimento

propiedad coligativa (colligative property): una propiedad como la depresión del punto de congelación o la elevación del punto de ebullición en las cuales su valor físico depende solamente en la proporción de las partículas del soluto y el solvente y no en sus identidades químicas

propiedad física (physical property): una propiedad que puede ser medida sin causar un cambio en la composición química de la sustancia

propiedad química (chemical property): una característica que sufre una sustancia en una reacción química que produce una nueva sustancia

protón (proton): una partícula subatómica cargada positivamente contenida en el núcleo de un átomo. La masa de un protón es 1.673×10^{-24}g y tiene una carga de +1.

protón (proton): una partícula subatómica que es parte de la estructura de un núcleo atómico; un protón está cargado positivamente

proyección de Mercator (Mercator projection): una proyección de mapa en la cual el Ecuador es representado por una línea recta hecha a escala, los meridianos por líneas paralelas rectas perpendiculares al Ecuador y con espacios iguales de acuerdo a la distancia que las separa del este, y los paralelos por líneas rectas perpendiculares a los meridianos en la misma longitud al Ecuador. Hay una gran distorsión de las distancias, áreas y formas en las regiones polares.

proyecciones de mapa (map projections): el proceso sistemático de transformar posiciones en la superficie esférica de la Tierra a un mapa plano mientras se mantienen relaciones espaciales

proyectil (projectile): un objeto viajando a través del aire

prueba de llama (flame test): una técnica o proceso experimental para identificar a un metal por sus características de color en una llama

prueba química (chemical test): un procedimiento físico o reacción química usada para identificar una sustancia

pulsaciones u ondas longitudinales (longitudinal pulse or wave): una vibración u onda en la cual el movimiento del medio es paralelo a la dirección del movimiento de la onda

punto caliente (hot spot): una fuente fija de abundante magma que forma un centro volcánico y que ha existido por decenas de millones de años

punto normal de congelación (normal freezing point): la temperatura característica, de 1 atm, en la cual un material cambia de un estado líquido a uno sólido

punto normal de ebullición (normal boiling point): la temperatura en la cual la presión de vapor de un líquido puro iguala 1 atm

punto normal de fusión (normal melting point): la temperatura característica, de 1 atm, en la cual un material cambia de un estado sólido a uno líquido

radiación adaptiva (adaptive radiation): la diversificación por selección natural, a través de tiempo evolutivo, de especies o grupos de especies a muchas diferentes especies que se adaptan regularmente a diferentes nichos ecológicos

radiación electromagnética (electromagnetic radiation): el movimiento de energía, a velocidad luz, en la forma de ondas electromagnéticas

radiación electromagnética (electromagnetic radiation): la energía propagada a través del espacio oscilando electricidad y campos magnéticos. Viaja a 3×10^8 m/s en un vacio e incluye (en orden del ascenso de energía) radio, infrarroja, luz visible (óptica), ultravioleta, rayos x, y rayos gama.

radiación-transferencias de calor (radiation-heat transfer): radiación electromagnética que golpea el material que puede ser absorbida, causando que las partículas en el material tengan más energía a veces resultando en altas temperaturas

radiactivo (radioactive): un átomo que tiene un núcleo inestable y emitirá partículas alfa, positrón o beta para poder conseguir un núcleo más estable

radiactivo (radioactive): un término aplicado a un átomo que tiene un núcleo desequilibrado y que puede emitir espontáneamente una partícula y convertirse en el núcleo de otro átomo

radio telescopio (radio telescope): un instrumento usado para observar ondas de longitud de radiación más larga (ondas radiales), con grandes platos para coleccionar y concentrar la radiación en una antena

rapidez (speed): el cambio en distancia por unidad de tiempo; rapidez es un escalar, no tiene dirección

rasgo (trait): característica de un organismo que puede ser descrito o medido

rayo (ray): la trayectoria seguida por un rayo muy fino de luz

reacción de reemplazo simple (single-displacement reaction): una reacción en la cual un elemento desplaza o reemplaza a otro elemento en un compuesto

reacción de síntesis (synthesis reaction): una reacción química en la cual dos o más

reacción doble de desplazamiento (double-displacement reaction): una reacción química en la cual dos compuestos iónicos "intercambian" cationes para producir dos nuevos compuestos

reacción o cambio endotérmico (endothermic change or reaction): un cambio en el cual la energía en la forma de calor es absorbida por el ambiente circundante resultando en un aumento en la energía interna del sistema

reacción o cambio exotérmico (exothermic change or reaction): un cambio en el cual la energía en la forma de calor es liberada por el ambiente circundante resultando en una disminución en la energía interna del sistema

reacción química (chemical reaction): un proceso en el cual una nueva sustancia es formada por una primera sustancia

reactividad (reactivity): una propiedad que describe cuán fácilmente un material reaccionará con otro

reactivos/reactantes (reactants): los materiales que empiezan una reacción química

rebote (bounce): la habilidad de un objeto para regresar a su posición original cuando es tirado desde cierta altura.

rebote elástico (elastic rebound): el regreso de un sólido elástico doblado a su forma original después de remover la fuerza deformante

recesivo (recessive): se usa para describir el gene que es anulado por un gene dominante; el rasgo está encubierto

recuperación secundaria (secondary recovery): el uso de técnicas para recuperar el aceite todavía atrapado entre las partículas sedimentarias después de años de producción

recurso no renovable (nonrenewable resource): una fuente de energía que es impulsada por materiales que existen en la Tierra y que no se reemplazan tan rápidamente como son consumidas

red alimenticia (food web): una relación compleja formada por cadenas alimenticias interconectadas en un ecosistema representando el traslado de energía a través de diferentes niveles

reducción (reduction): un proceso en el cual la sustancia bajo consideración gana un electrón(es)

reforestación (reforestation): el volver a plantar árboles en la tierra donde existían bosques que fueron cortados anteriormente para otros usos, como el de agricultura o pasturaje

refracción (refraction): el cambio en la dirección de los rayos de luz que pasan indirectamente entre dos medios diferentes

relación del inverso cuadrado (inverse square relation): la relación de la fuerza al inverso cuadrado de la distancia de la masa (para fuerzas gravitacionales) o la carga (para fuerzas electroestáticas)

relatividad (relativity): el estudio de la manera en la cual observaciones de una cuadro de referencia en movimiento afecta la percepción del mundo

relieve (relief): la configuración física de una parte de la superficie de la Tierra, con referencia a las variaciones de altura y pendiente o las irregularidades en la superficie terrestre

resistencia de aire (air resistance): una fuerza de aire en un objeto en movimiento; la fuerza es dependiente de la velocidad, volumen, y masa del objeto al igual que de las propiedades del aire, como la densidad

resistencia eléctrica (electrical resistance): oposición de un material al flujo de carga eléctrica a través de la misma; es medida en ohmios (Ω); la proporción de la diferencia potencial de la corriente

$$R = \frac{V}{I}$$

resonancia (resonance): una condición en la cual una vibración afectando un objeto tiene el mismo período de tiempo a la vibración natural del objeto

respiración (respiration): procesos físicos y químicos por los cuales un organismo suple a sus células y tejidos con el oxígeno necesario para el metabolismo

riesgo (risk): el impacto potencial de un peligro natural en personas o en propiedad

roca fosilífera (fossiliferous rock): una roca que contiene fósiles

roca ígnea (igneous rock): roca o mineral que se ha solidificado de material fundido o parcialmente fundido, Ej. de magma

roca ígnea extrusiva - roca ígnea volcánica (extrusive igneous rock): una roca ígnea que se ha formado por la erupción de lava en la superficie de la Tierra

roca ígnea intrusiva - roca ígnea plutónica (intrusive igneous rock - plutonic igneous rock): roca ígnea formada a una profundidad considerable por la cristalización de magma

roca sedimentaria (sedimentary rock): una roca formada como resultado de la consolidación de sedimentos acumulados

roca sedimentaria (sedimentary rock): una roca, regularmente en capas, como resultado de la consolidación de sedimentos

rocas (rocks): agregados naturales de granos minerales

sección ventral (antinode): un punto en una onda parada donde el desplazamiento de la mediana esta al máximo

sedimentos (sediments): material sólido fragmentario que se origina por la desintegración de las rocas por acción atmosférica y es transportado o depositado por aire, agua, hielo o que se acumula por otros agentes naturales, como es la precipitación química de solución o secreción de organismos

sedimentos (sediments): partículas sueltas de materiales que son derivados de las rocas desechas o la precipitación de sólidos en agua

Segunda Ley de Termodinámicas (Second Law of Thermodynamics): la ley que establece que el calor no puede ser completamente convertido en una forma más útil de energía

selección natural (natural selection): la diferencia en supervivencia y reproducción entre los miembros de una población

sello (seal): una capa impermeable o masa de roca sedimentaria que forma la cima convexa ascendente o techo de una embalse de petróleo

seno (trough): el punto más bajo de la onda

sésil; sin movimiento (sessile; non-motile): un organismo que está permanentemente unido, sin movimiento libre

sílice (silica): material con composición SiO_2

simetría bilateral (bilateral symmetry): disposición de las partes de un organismo que se divide en dos mitades simétricas, derecha e izquierda

simetría radial (radial symmetry): disposición de las partes de un organismo que tiene simetría sobre el eje central

sismógrafo (seismometer): un instrumento que mide las ondas sísmicas. Recibe impulsos sísmicos y los convierte en una señal con un voltaje electrónico

sismograma (seismogram): el registro hecho por un sismógrafo

sismología (seismology): el estudio de terremotos y la estructura de la Tierra

sistema abierto (open system): un sistema físico en el cual influencias externas pueden actuar; abierto para que la energía pueda ser añadida, y o pérdida (liberada) del sistema

sistema cerrado (closed system): un sistema en el cual el material se mueve de lugar a lugar pero no se ha ganado o perdido del sistema

sistema cerrado (closed system): un sistema físico en el cual no actúan influencias externas; cerrado para que nada entre o salga del sistema y nada externo puede influenciar el comportamiento observable del sistema o sus propiedades

sistema de posición global (Global Positioning System GPS): un sistema satelital para la localización precisa de los puntos en la Tierra

solenoide (solenoid): rollo de alambre

solución (solution): una mezcla homogénea de dos o más sustancias

solución saturada (saturated solution): la cantidad máxima de un soluto que puede ser disuelta a cierta temperatura y presión

solución sobresaturada (supersaturated solution): una solución conteniendo más soluto que una solución saturada y que por lo tanto está desequilibrada. Esta solución no es estable y no se puede mantener indefinidamente.

soluto (solute): la sustancia que se disuelve en un solvente para formar una solución

solvente (solvent): la sustancia en la cual un soluto se disuelve para formar una solución

sombra de lluvia (rain shadow): la reducción de precipitación comúnmente encontrado en el lado de sotavento de una montaña

subducción (subduction): movimiento de una placa hacia abajo del manto, debajo del borde de la otra placa en el límite convergente de la placa. La placa de abajo es siempre la litosfera oceánica. La placa que permanece en la superficie puede tener tanto litosfera oceánica o litosfera continental.

sublimación (sublimation): el cambio en el estado de un material sólido a gaseoso sin pasar por el estado líquido

substancia pura (pure substance): una sustancia que solamente contiene un tipo de partícula

sucesión (succession): el lento y ordenado reemplazo de la comunidad de reemplazo, una siguiendo a la otra

sucesión secundaria (secondary succession): la ocupación de vida vegetal en una área que estaba cubierta anteriormente por vegetación y que todavía tiene tierra

suministro de materiales sin procesar (feedstocks): materiales crudos (por ejemplo, petróleo) que son suplidos a una máquina o planta procesadora que produce materiales manufacturados (por ejemplo, plásticos)

supercontinente (supercontinent): un continente grande consistiendo de toda la litosfera continental de la Tierra. Los supercontinentes son formados por procesos placa-tectónico de conducción subterránea y colisión de continente a continente.

supernova (supernova): la muerte por explosión de una estrella masiva en la cual su centro se ha quemado completamente. Explosiones de supernova pueden opacar temporalmente una galaxia.

suspensión (suspension): mezcla heterogénea que contiene partículas sólidas finas o líquidas en un líquido y que se asentará espontáneamente. Al agitar el envase volverán a dispersarse a través del líquido.

tasa de crecimiento (growth rate): la proporción a la cual el tamaño de la población aumenta como resultado de la tasa de mortalidad, la tasa de natalidad, la inmigración y la emigración

tasa de mortalidad (death rate – mortality rate): la proporción a la cual las muertes disminuyen el tamaño de la población

tasa de natalidad (birthrate – natality): la proporción a la cual la reproducción aumenta el tamaño de la población

taxonomía (taxonomy): la teoría y práctica de clasificar plantas y animales

tefra (tephra): un término colectivo para todas las partículas expulsadas de un volcán y transportadas a través del aire. Incluye polvo volcánico, cenizas, rescoldos, lapilli, escoria, piedra pómez, bombas volcánicas y bloques volcánicos.

telescopio de rayos x (x-ray telescope): instrumento utilizado para detectar emisiones de rayos x estelares e interestelares. Ya que la atmósfera de la Tierra absorbe los rayos x, los telescopios están localizados muy altos sobre la superficie de la Tierra.

temperatura (temperature): una medida de la energía cinética promedio de las moléculas de un material

temperatura (temperature): una medida de la energía de la vibración de los átomos y moléculas del cuerpo de una materia

teoría (theory): una verdad aprobada y generalmente aceptada

Teoría Especial de la Relatividad (Special Theory of Relativity): la teoría de espacio y tiempo

termodinámicas (thermodynamics): una rama de la física que trata con las relaciones y transformaciones de la energía

termodinámicas (thermodynamics): el estudio de la transformación de energía descrita por las leyes

terremoto (earthquake): un movimiento rápido o temblor en la Tierra causado por una liberación repentina de una tensión acumulada lentamente

textura (texture): las características en la superficie de un material, como cuán suave, áspera, o gruesa es

tiempo (weather): la condición de la atmósfera de la Tierra, específicamente, su temperatura, presión barométrica, velocidad del viento, humedad, nubosidad, y precipitación

titulaje (titration): un procedimiento analítico en el cual la concentración de una solución desconocida es añadida a una solución estándar hasta que el cambio del color de algún indicador indique que las cantidades equivalentes han reaccionado

tono (pitch): la calidad de un sonido depende principalmente en la frecuencia que las ondas de sonido producen por su fuente

trabajo (work): el producto de la fuerza ejercida en un cuerpo y la distancia que el cuerpo mueve en dirección hacia la fuerza; el trabajo es equivalente al cambio en la energía mecánica del cuerpo

trabajo (work): el producto del desplazamiento y la fuerza en la dirección del desplazamiento; trabajo es una cantidad de escala

transferencia de energía (heat transfer): el movimiento de calor de una región a otra

transpiración (transpiration): el proceso por el cual el agua es absorbida por plantas, regularmente a través de las raíces, es evaporada en la atmósfera de la superficie de la planta en la forma de vapor de agua

transpiración (transpiration): la emisión de vapor de agua de los poros de las plantas como parte de sus procesos de vida

trayectoria (trajectory): la senda seguida por un objeto que es lanzado al aire

tsunami (tsunami): Una gran ola producida por un terremoto submarino (o una erupción volcánica o deslizamiento de tierra)

turba (peat): un depósito poroso de un material vegetal parcialmente descompuesto en o no más bajo de la superficie de la Tierra

turbina (turbine): una máquina de rotación u otro mecanismo que convierte la energía mecánica de un líquido que fluye en energía mecánica de rotación de un eje

ultravioleta (ultraviolet): radiación electromagnética en longitudes de onda más cortas que el extremo violeta de la luz visible; con longitudes de onda alcanzando de 5 a 400 NM

unidad astronómica (astronomical unit): una unidad de medida igual a la distancia promedio entre el sol y la Tierra, Ej. cerca de 149, 600,000 (1.496×10^8) Km.

unidad de masa atómica (atomic mass unit - amu): una unidad de masa definida como una doceava parte de la masa de un átomo de carbono-12

unidad de masa atómica (atomic mass unit): una unidad estándar de masa atómica basada en la masa del átomo de carbono, la cual ha sido asignado el valor de 12

uniformidad (uniformity): la propiedad de consistencia a lo largo de un material

valle de hendiduras (rift valley): la profunda grieta central en la cima de la cordillera medio-oceánica

valle de hendiduras (rift valley): un valle largo y grande en un continente, formado donde el continente es separado por fuerzas producidas cuando los materiales del manto se levantan bajo el continente

vaporización (vaporization): el cambio de estado líquido a gaseoso

vatio (watt): una unidad de poder

vector (vector): una cantidad que tiene ambos magnitud y dirección

velocidad (velocity): la velocidad en una dirección dada; desplazamiento dividido por un intervalo de tiempo; la velocidad es un vector de cantidad, tiene magnitud y dirección

viento de barlovento (leeward): la parte que baja de un área elevada como una montaña, opuesto al barlovento

viento de sotavento (windward): la parte del viento hacia arriba o hacia el lado directamente influenciado a la dirección que los vientos soplan de, opuestos al viento de barlovento

viento solar (solar wind): un flujo de partículas calientes cargadas dejando el sol

viscosidad (viscosity): una propiedad de un líquido que ofrece resistencia interna a la fluidez

volcán escudo (shield volcano): un cono volcánico de cúpula plana, ancha y con una pendiente suave, generalmente con una extensión de varias decenas o cientos de millas cuadradas

voltio (volt): la unidad SI de electricidad potencial; un voltio (1 V) es igual que un julio por coulomb (J/C)

$$W = F \cdot d$$

zona de subducción (subduction zone): un cinturón largo y estrecho en el cual una placa desciende bajo otra

zona de sutura (suture zone): la zona en la superficie de la Tierra donde dos continentes han chocado y se han soldado para formar un sólo continente

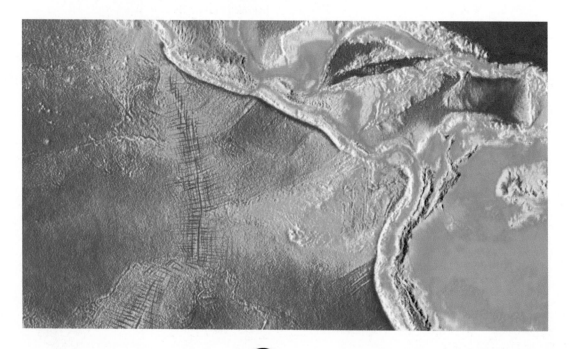

Index

Amplitude = 20 cm

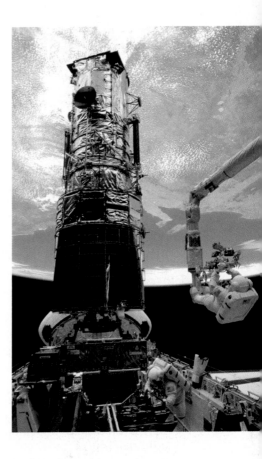

Coordinated Science for the 21st Century

Charts/Graphs/Tables